UNIT CONVERSION

This book employs SI units. However, other units are sometimes preferable. Some conversion factors are listed below.

Length
1 Angstrom (Å) = 10^{-8} meter
1 foot = 0.305 meter
1 light year = 9.46×10^{15} meters
1 parsec = 3.26 light years

Volume
1 liter = 1000 centimeter^3 = 10^{-3} meter^3
1 U.S. gallon = 0.83 imperial gallon = 3.78 liters

Time
1 hour = 3600 seconds
1 day = 8.64×10^4 seconds
1 hertz (hz) = 1 second^{-1}

Mass
1 atomic mass unit (amu) = 1.6605×10^{-27} kilogram

Force
1 pound (lb) = 4.45 newtons

Energy and Power
1 erg = 10^{-7} joule
1 kcal = 1 Cal = 1000 cal = 4.184×10^3 joules
1 electron volt (eV) = 1.602×10^{-19} joule
1 foot-pound = 1.36 joules
1 horsepower = 746 watts

Pressure
1 atmosphere = 1.013 bar = 1.013×10^5 newtons/meter^2 = 14.7 pounds/inch^2 = 760 torr
1 pascal = 1 newton/meter^2

Temperature
x°C = $(273.16 + x)$ K
x°F = $5(x - 32)/9$ °C
1 eV = $k_B \times 11{,}605$ K

NUMERICAL AND ANALYTICAL METHODS FOR SCIENTISTS AND ENGINEERS USING *MATHEMATICA*

NUMERICAL AND ANALYTICAL METHODS FOR SCIENTISTS AND ENGINEERS USING *MATHEMATICA*

DANIEL DUBIN

A JOHN WILEY & SONS, INC., PUBLICATION

Cover Image: Breaking wave, theory and experiment photograph by Rob Keith.

Published by John Wiley & Sons, Inc., Hoboken, New Jersey.
Published simultaneously in Canada.

Library of Congress Cataloging-in-Publication Data is available.

ISBN 0-471-26610-8

Printed in the United States of America

10 9 8 7 6 5 4 3 2 1

CONTENTS

PREFACE

TO THE STUDENT

Up to this point in your career you have been asked to use mathematics to solve rather elementary problems in the physical sciences. However, when you graduate and become a working scientist or engineer you will often be confronted with complex real-world problems. Understanding the material in this book is a first step toward developing the mathematical tools that you will need to solve such problems.

Much of the work detailed in the following chapters requires standard pencil-and-paper (i.e., analytical) methods. These methods include solution techniques for the partial differential equations of mathematical physics such as Poisson's equation, the wave equation, and Schrödinger's equation, Fourier series and transforms, and elementary probability theory and statistical methods. These methods are taught from the standpoint of a working scientist, not a mathematician. This means that in many cases, important theorems will be stated, not proved (although the ideas behind the proofs will usually be discussed). Physical intuition will be called upon more often than mathematical rigor.

Mastery of analytical techniques has always been and probably always will be of fundamental importance to a student's scientific education. However, of increasing importance in today's world are numerical methods. The numerical methods taught in this book will allow you to solve problems that cannot be solved analytically, and will also allow you to inspect the solutions to your problems using plots, animations, and even sounds, gaining intuition that is sometimes difficult to extract from dry algebra.

In an attempt to present these numerical methods in the most straightforward manner possible, this book employs the software package *Mathematica*. There are many other computational environments that we could have used instead—for example, software packages such as *Matlab* or *Maple* have similar graphical and numerical capabilities to *Mathematica*. Once the principles of one such package are learned, it is relatively easy to master the other packages. I chose *Mathematica* for this book because, in my opinion, it is the most flexible and sophisticated of such packages.

Another approach to learning numerical methods might be to write your own programs from scratch, using a language such as C or Fortran. This is an excellent way to learn the elements of numerical analysis, and eventually in your scientific careers you will probably be required to program in one or another of these languages. However, *Mathematica* provides us with a computational environment where it is much easier to quickly learn the *ideas* behind the various numerical methods, without the additional baggage of learning an operating system, mathematical and graphical libraries, or the complexities of the computer language itself.

An important feature of *Mathematica* is its ability to perform *analytical* calculations, such as the analytical solution of linear and nonlinear equations, integrals and derivatives, and Fourier transforms. You will find that these features can help to free you from the tedium of performing complicated algebra by hand, just as your calculator has freed you from having to do long division.

However, as with everything else in life, using *Mathematica* presents us with certain trade-offs. For instance, in part because it has been developed to provide a straightforward interface to the user, *Mathematica* is not suited for truly large-scale computations such as large molecular dynamics simulations with 1000 particles or more, or inversions of 100,000-by-100,000 matrices, for example. Such applications require a stripped-down precompiled code, running on a mainframe computer. Nevertheless, for the sort of introductory numerical problems covered in this book, the speed of *Mathematica* on a PC platform is more than sufficient. Once these numerical techniques have been learned using *Mathematica*, it should be relatively easy to transfer your new skills to a mainframe computing environment.

I should note here that this limitation does not affect the usefulness of *Mathematica* in the solution of the sort of small to intermediate-scale problems that working scientists often confront from day to day. In my own experience, hardly a day goes by when I do not fire up *Mathematica* to evaluate an integral or plot a function. For more than a decade now I have found this program to be truly useful, and I hope and expect that you will as well. (No, I am not receiving any kickbacks from Stephen Wolfram!)

There is another limitation to *Mathematica*. You will find that although *Mathematica* knows a lot of tricks, it is still a dumb program in the sense that it requires precise input from the user. A missing bracket or semicolon often will result in long paroxysms of error statements and less often will result in a dangerous lack of error messages and a subsequent incorrect answer. It is still true for this (or for any other software) package that garbage in = garbage out. Science fiction movies involving intelligent computers aside, this aphorism will probably hold for the foreseeable future. This means that, at least at first, you will spend a good fraction of your time cursing the computer screen. My advice is to get used to it—this is a process that you will go through over and over again as you use computers in your career. I guarantee that you will find it very satisfying when, after a long debugging session, you finally get the output you wanted. Eventually, with practice, you will become *Mathematica* masters.

I developed this book from course notes for two junior-level classes in mathematical methods that I have taught at UCSD for several years. The book is oriented toward students in the physical sciences and in engineering, at either the advanced undergraduate (junior or senior) or graduate level. It assumes an understanding of introductory calculus and ordinary differential equations. Chapters 1–8 also require a basic working knowledge of *Mathematica*. Chapter 9, included only in electronic form on the CD that accompanies this book, presents an introduction to the software's capabilities. I recommend that *Mathematica* novices read this chapter first, and do the exercises.

Some of the material in the book is rather advanced, and will be of more interest to graduate students or professionals. This material can obviously be skipped when the book is used in an undergraduate course. In order to reduce printing costs, four advanced topics appear only in the electronic chapters on the CD: Section 5.3 on wave action; Section 6.3 on numerically determined eigenmodes; Section 7.3 on the particle-in-cell method; and Section 8.3 on the Rosenbluth–Teller–Metropolis Monte Carlo method. These extra sections are highlighted in red in the electronic version.

Aside from these differences, the text and equations in the electronic and printed versions are, *in theory*, identical. However, I take sole responsibility for any inadvertent discrepancies, as the good people at Wiley were not involved in typesetting the electronic textbook.

The electronic version of this book has several features that are not available in printed textbooks:

1. ***Hyperlinks.*** There are hyperlinks in the text that can be used to view material from the web. Also, when the text refers to an equation, the equation number itself is a hyperlink that will take you to that equation. Furthermore, all items in the index and contents are linked to the corresponding material in the book, (For these features to work properly, all chapters must be located in the same directory on your computer.) You can return to the original reference using the `Go Back` command, located in the main menu under `Find`.
2. ***Mathematica Code.*** Certain portions of the book are *Mathematica* calculations that you can use to graph functions, solve differential equations, etc. These calculations can be modified at the reader's pleasure, and run *in situ*.
3. ***Animations and Interactive 3D Renderings.*** Some of the displayed figures are interactive three-dimensional renderings of curves or surfaces, which can be viewed from different angles using the mouse. An example is Fig. 1.13, the strange attractor for the Lorenz system. Also, some of the other figures are actually animations. Creating animations and interactive 3D plots is covered in Sections 9.6.7 and 9.6.6, respectively.
4. ***Searchable text.*** Using the commands in the `Find` menu, you can search through the text for words or phrases.

Equations or text may sometimes be typeset in a font that is too small to be read easily at the current magnification. You can increase (or decrease) the magnification of the notebook under the **Format** entry of the main menu (choose **Magnification**), or by choosing a magnification setting from the small window at the bottom left side of the notebook.

A number of individuals made important contributions to this project: Professor Tom O'Neil, who originally suggested that the electronic version should be written in *Mathematica* notebook format; Professor C. Fred Driscoll, who invented some of the problems on sound and hearing; Jo Ann Christina, who helped with the proofreading and indexing; and Dr. Jay Albert, who actually waded through the entire manuscript, found many errors and typos, and helped clear up fuzzy thinking in several places. Finally, to the many students who have passed through my computational physics classes here at UCSD: You have been subjected to two experiments—a *Mathematica*-based course that combines analytical and computational methods; and a book that allows the reader to interactively explore variations in the examples. Although you were beset by many vicissitudes (crashing computers, balky code, debugging sessions stretching into the wee hours) your interest, energy, and good humor were unflagging (for the most part!) and a constant source of inspiration. Thank you.

DANIEL DUBIN

La Jolla, California
March, 2003

CHAPTER 1

ORDINARY DIFFERENTIAL EQUATIONS IN THE PHYSICAL SCIENCES

1.1 INTRODUCTION

1.1.1 Definitions

Differential Equations, Unknown Functions, and Initial Conditions Three centuries ago, the great British mathematician, scientist, and curmudgeon Sir Isaac Newton and the German mathematician Gottfried von Liebniz independently introduced the world to calculus, and in so doing ushered in the modern scientific era. It has since been established in countless experiments that natural phenomena of all kinds can be described, often in exquisite detail, by the solutions to *differential equations*.

Differential equations involve derivatives of an *unknown function* or functions, whose form we try to determine through solution of the equations. For example, consider the motion (in one dimension) of a point particle of mass m under the action of a prescribed time-dependent force $F(t)$. The particle's velocity $v(t)$ satisfies Newton's second law

$$m\frac{dv}{dt}=F(t). \tag{1.1.1}$$

This is a differential equation for the unknown function $v(t)$.

Equation (1.1.1) is probably the simplest differential equation that one can write down. It can be solved by applying the *fundamental theorem of calculus*: for any function $f(t)$ whose derivative exists and is integrable on the interval $[a, b]$,

$$\int_a^b \frac{df}{dt}dt=f(b)-f(a). \tag{1.1.2}$$

Integrating both sides of Eq. (1.1.1) from an initial time $t=0$ to time t and using Eq. (1.1.2) yields

$$\int_0^t \frac{dv}{dt}\,dt = v(t) - v(0) = \frac{1}{m}\int_0^t F(t)\,dt. \tag{1.1.3}$$

Therefore, the solution of Eq. (1.1.1) for the velocity at time t is given by the integral over time of the force, a known function, and an *initial condition*, the velocity at time $t=0$. This initial condition can be thought of mathematically as a constant of integration that appears when the integral is applied to Eq. (1.1.1). Physically, the requirement that we need to know the initial velocity in order to find the velocity at later times is intuitively obvious. However, it also implies that the differential equation (1.1.1) by itself is not enough to completely determine a solution for $v(t)$; the initial velocity must also be provided. This is a general feature of differential equations:

> Extra conditions beyond the equation itself must be supplied in order to completely determine a solution of a differential equation.

If the initial condition is *not* known, so that $v(0)$ is an undetermined constant in Eq. (1.1.3), then we call Eq. (1.1.3) a *general solution* to the differential equation, because different choices of the undetermined constant allow the solution to satisfy different initial conditions.

As a second example of a differential equation, let's now assume that the force in Eq. (1.1.1) depends on the position $x(t)$ of the particle according to Hooke's law:

$$F(t) = -kx(t), \tag{1.1.4}$$

where k is a constant (the spring constant). Then, using the definition of velocity as the rate of change of position,

$$v = \frac{dx}{dt}. \tag{1.1.5}$$

Eq. (1.1.1) becomes a differential equation for the unknown function $x(t)$:

$$\frac{d^2x}{dt^2} = -\frac{k}{m}x(t). \tag{1.1.6}$$

This familiar differential equation, the *harmonic oscillator* equation, has a general solution in terms of the trigonometric functions $\sin x$ and $\cos x$, and *two* undetermined constants C_1 and C_2:

$$x(t) = C_1\cos(\omega_0 t) + C_2\sin(\omega_0 t), \tag{1.1.7}$$

where $\omega_0 = \sqrt{k/m}$ is the natural frequency of the oscillation. The two constants can be determined by two initial conditions, on the initial position and velocity:

$$x(0) = x_0, \qquad v(0) = v_0. \tag{1.1.8}$$

Since Eq. (1.1.7) implies that $x(0) = C_1$ and $x'(0) = v(0) = \omega_0 C_2$, the solution can be written directly in terms of the initial conditions as

$$x(t) = x_0 \cos(\omega_0 t) + \frac{v_0}{\omega_0} \sin(\omega_0 t). \tag{1.1.9}$$

We can easily verify that this solution satisfies the differential equation by substituting it into Eq. (1.1.6):

Cell 1.1

```
x[t_] = x0 Cos[ω0 t] + v0/ω0 Sin[ ω0 t];
Simplify[x"[t] == -ω0^2 x[t]]

True
```

We can also verify that the solution matches the initial conditions:

Cell 1.2

```
x[0]

x0
```

Cell 1.3

```
x'[0]

v0
```

Order of a Differential Equation The *order* of a differential equation is the order of the highest derivative of the unknown function that appears in the equation. Since only a first derivative of $v(t)$ appears in Eq. (1.1.1), the equation is a *first-order* differential equation for $v(t)$. On the other hand, Equation (1.1.6) is a *second-order* differential equation.

Note that the general solution (1.1.3) of the first-order equation (1.1.1) involved one undetermined constant, but for the second-order equation, *two* undetermined constants were required in Eq. (1.1.7). It's easy to see why this must be so—an Nth-order differential equation involves the Nth derivative of the unknown function. To determine this function one needs to integrate the equation N times, giving N constants of integration.

The number of undetermined constants that enter the general solution of an ordinary differential equation equals the order of the equation.

Partial Differential Equations This statement applies only to *ordinary* differential equations (ODEs), which are differential equations for which derivatives of the unknown function are taken with respect to only a single variable. However, this book will also consider *partial* differential equations (PDEs), which involve derivatives of the unknown functions with respect to *several* variables. One example of a PDE is Poisson's equation, relating the electrostatic potential $\phi(x, y, z)$ to the charge density $\rho(x, y, z)$ of a distribution of charges:

$$\nabla^2\phi(x, y, z) = -\frac{\rho(x, y, z)}{\epsilon_0}. \tag{1.1.10}$$

Here ϵ_0 is a constant (the dielectric permittivity of free space, given by $\epsilon_0 = 8.85\ldots \times 10^{-12}$ F/m), and ∇^2 is the *Laplacian operator*,

$$\nabla^2 = \frac{\partial^2}{\partial x^2} + \frac{\partial^2}{\partial y^2} + \frac{\partial^2}{\partial z^2}. \tag{1.1.11}$$

We will find that ∇^2 appears frequently in the equations of mathematical physics.

Like ODEs, PDEs must be supplemented with extra conditions in order to obtain a specific solution. However, the form of these conditions become more complex than for ODEs. In the case of Poisson's equation, *boundary conditions* must be specified over one or more surfaces that bound the volume within which the solution for $\phi(x, y, z)$ is determined.

A discussion of solutions to Poisson's equation and other PDEs of mathematical physics can be found in Chapter 3 and later chapters. For now we will confine ourselves to ODEs. Many of the techniques used to solve ODEs can also be applied to PDEs.

An ODE involves derivatives of the unknown function with respect to only a single variable. A PDE involves derivatives of the unknown function with respect to more than one variable.

Initial-Value and Boundary-Value Problems Even if we limit discussion to ODEs, there is still an important distinction to be made, between *initial-value problems* and *boundary-value problems*. In initial-value problems, the unknown function is required in some time domain $t > 0$ and all conditions to specify the solution are given at *one* end of this domain, at $t = 0$. Equations (1.1.3) and (1.1.9) are solutions of initial-value problems.

However, in boundary-value problems, conditions that specify the solution are given at different times or places. Examples of boundary-value problems in ODEs may be found in Sec. 1.5. (Problems involving PDEs are often boundary-value problems; Poisson's equation (1.1.10) is an example. In Chapter 3 we will find that some PDEs involving both time and space derivatives are solved as both boundary- *and* initial-value problems.)

For now, we will stick to a discussion of ODE initial-value problems.

In initial-value problems, all conditions to specify a solution are given at one point in time or space, and are termed *initial conditions*. In boundary-value problems, the conditions are given at several points in time or space, and are termed *boundary conditions*. For ODEs, the boundary conditions are usually given at two points, between which the solution to the ODE must be determined.

EXERCISES FOR SEC. 1.1

(1) Is Eq. (1.1.1) still a differential equation if the velocity $v(t)$ is given and the force $F(t)$ is the unknown function?

(2) Determine by substitution whether the following functions satisfy the given differential equation, and if so, state whether the functions are a general solution to the equation:

(a) $\dfrac{d^2x}{dt^2} = x(t),\ x(t) = C_1 \sinh t + C_2 e^{-t}$.

(b) $\left(\dfrac{dx}{dt}\right)^2 = x(t),\ x(t) = \frac{1}{4}(a^2 + t^2) - \dfrac{at}{2}$.

(c) $\dfrac{d^4x}{dt^4} - 3\dfrac{d^3x}{dt^3} - 7\dfrac{d^2x}{dt^2} + 15\dfrac{dx}{dt} + 18x = 12t^2,\quad x(t) = a e^{3t} t + b e^{-2t} + \dfrac{2t^2}{3} - \dfrac{10t}{9} + \dfrac{13}{9}$.

(3) Prove by substitution that the following functions are general solutions to the given differential equations, and find values for the undetermined constants in order to match the boundary or initial conditions. Plot the solutions:

(a) $\dfrac{dx}{dt} = 5x(t) - 3,\ x(0) = 1;\ x(t) = C e^{5t} + 3/5$.

(b) $\dfrac{d^2x}{dt^2} + 4\dfrac{dx}{dt} + 4x(t) = 0,\ x(0) = 0,\ x'(1) = -3;\ x(t) = C_1 e^{-2t} + C_2 t e^{-2t}$.

(c) $\dfrac{d^3x}{dt^3} + \dfrac{dx}{dt} = t,\quad x(0) = 0,\quad x'(0) = 1,\quad x''(\pi) = 0;\quad x(t) = t^2/2 + C_1 \sin t + C_2 \cos t + C_3$.

1.2 GRAPHICAL SOLUTION OF INITIAL-VALUE PROBLEMS

1.2.1 Direction Fields; Existence and Uniqueness of Solutions

In an initial-value problem, how do we know when the initial conditions specify a *unique* solution to an ODE? And how do we know that the solution will even exist? These fundamental questions are addressed by the following theorem:

Theorem 1.1 Consider a general initial-value problem involving an Nth-order ODE of the form

$$\frac{d^N x}{dt^N} = f\left(t, x, \frac{dx}{dt}, \frac{d^2 x}{dt^2}, \ldots, \frac{d^{N-1} x}{dt^{N-1}}\right) \tag{1.2.1}$$

for some function f. The ODE is supplemented by N initial conditions on x and its derivatives of order $N-1$ and lower:

$$x(0) = x_0, \qquad \frac{dx}{dt} = v_0, \qquad \frac{d^2 x}{dt^2} = a_0, \ldots, \qquad \frac{d^{N-1}}{dt^{N-1}} = u_0.$$

Then, if the derivative of f in each of its arguments is continuous over some domain encompassing this initial condition, the solution to this problem exists and is unique for some length of time around the initial time.

Now, we are not going to give the proof to this theorem. (See, for instance, Boyce and Diprima for an accessible discussion of the proof.) But trying to understand it qualitatively is useful. To do so, let's consider a simple example of Eq. (1.2.1): the first-order ODE

$$\frac{dv}{dt} = f(t, v). \tag{1.2.2}$$

This equation can be thought of as Newton's second law for motion in one dimension due to a force that depends on both velocity and time.

Let's consider a graphical depiction of Eq. (1.2.2) in the (t, v) plane. At every point (t, v), the function $f(t, v)$ specifies the slope dv/dt of the solution $v(t)$. An example of one such solution is given in Fig. 1.1. At each point along the curve, the slope dv/dt is determined through Eq. (1.2.2) by $f(t, v)$. This slope is, geometrically speaking, an infinitesimal vector that is tangent to the curve at each of its points. A schematic representation of three of these infinitesimal vectors is shown in the figure.

The components of these vectors are

$$(dt, dv) = dt\left(1, \frac{dv}{dt}\right) = dt(1, f(t, v)). \tag{1.2.3}$$

The vectors $dt(1, f(t, v))$ form a type of *vector field* (a set of vectors, each member of which is associated with a separate point in some spatial domain) called a *direction field*. This field specifies the *direction* of the solutions at all points in the

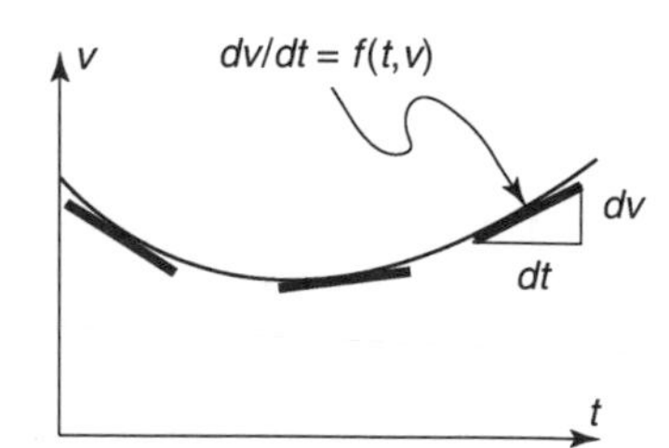

Fig. 1.1 A solution to $dv/dt = f(t, v)$.

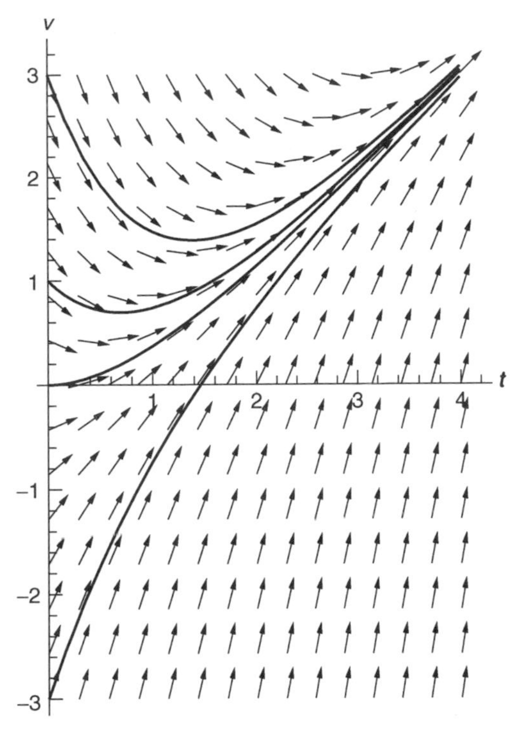

Fig. 1.2 Direction field for $dv/dt = t - v$, along with four solutions.

(t, v) plane: every solution to Eq. (1.2.2) for every initial condition must be a curve that runs tangent to the direction field. Individual vectors in the direction field are called *tangent vectors*.

By drawing these tangent vectors at a grid of points in the (t, v) plane (not infinitesimal vectors, of course; we will take dt to be finite so that we can see the vectors), we get an overall qualitative picture of solutions to the ODE. An example is shown in Figure 1.2. This direction field is drawn for the particular case of an acceleration given by

$$f(t, v) = t - v. \tag{1.2.4}$$

Along with the direction field, four solutions of Eq. (1.2.2) with different initial v's are shown. One can see that the direction field is tangent to each solution.

Figure 1.2 was created using a graphics function, available in *Mathematica*'s graphical add-on packages, that is made for plotting two-dimensional vector fields: **PlotVectorField**. The syntax for this function is given below:

```
PlotVectorField[{vx[x,y],vy[x,y]}, {x,xmin,xmax},{y,ymin,ymax},options].
```

The vector field in Fig. 1.2 was drawn with the following *Mathematica* commands:

Cell 1.4

```
<< Graphics`
```

Cell 1.5

```
f[t_, v_] = -v + t;
PlotVectorField[{1, f[t, v]}, {t, 0, 4}, {v, -3, 3},
Axes → True, ScaleFunction→ (1 &), AxesLabel→{"t", "v"}]
```

The option `ScaleFunction->(1&)` makes all the vectors the same length. The plot shows that you don't really need the four superimposed solutions in order to see the qualitative behavior of solutions for different initial conditions—you can trace them by eye just by following the arrows.

However, for completeness we give the general solution of Eqs. (1.2.2) and (1.2.4) below:

$$v(t) = Ce^{-t} + t - 1, \qquad (1.2.5)$$

which can be verified by substitution. In Fig. 1.2, the solutions traced out by the solid lines are for $C = [4, 2, 1 - 2]$. (These solutions were plotted with the `Plot` function and then superimposed on the vector field using the `Show` command.) One can see that for $t < \infty$, the different solutions never cross. Thus, specifying an initial condition leads to a *unique* solution of the differential equation. There are no places in the direction field where one sees convergence of two different solutions, except perhaps as $t \to \infty$. This is guaranteed by the differentiability of the function f in each of its arguments.

A simple example of what can happen when the function f is nondifferentiable at some point or points is given below. Consider the case

$$f(t, v) = v/t. \qquad (1.2.6)$$

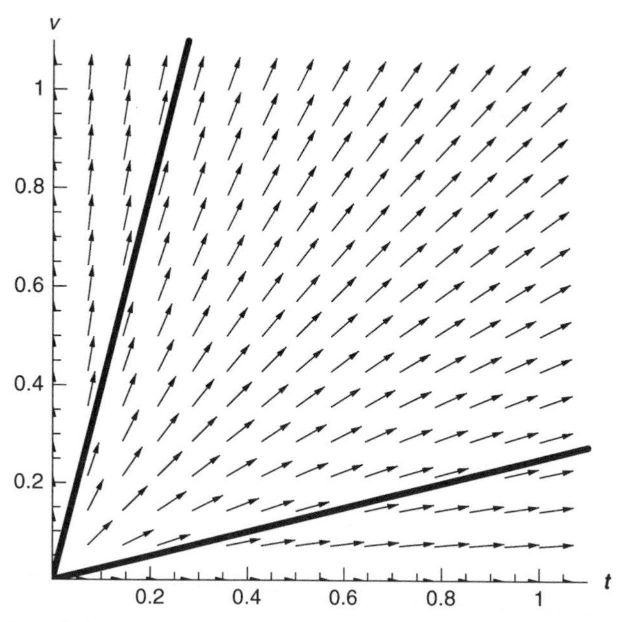

Fig. 1.3 Direction field for $dv/dt = v/t$, along with two solutions, both with initial condition $v(0) = 0$.

This function is not differentiable at $t = 0$. The general solution to Eqs. (1.2.2) and (1.2.6) is

$$v(t) = Ct, \tag{1.2.7}$$

as can be seen by direct substitution. This implies that all solutions to the ODE emanate from the point $v(0) = 0$. Therefore, the solution with initial condition $v(0) = 0$ is *not unique*. This can easily be seen in a plot of the direction field, Fig. 1.3. Furthermore, Eq. (1.2.7) shows that solutions with $v(0) \neq 0$ do not exist. When f is differentiable, this kind of singular behavior in the direction field cannot occur, and as a result the solution for a given initial condition exists and is unique.

1.2.2 Direction Fields for Second-Order ODEs: Phase-Space Portraits

Phase-Space We have seen that the direction field provides a global picture of all solutions to a first-order ODE. The direction field is also a useful visualization tool for higher-order ODEs, although the field becomes difficult to view in three or more dimensions. A nontrivial case that *can* be easily visualized is the direction field for second-order ODEs of the form

$$\frac{d^2x}{dt^2} = f\left(x, \frac{dx}{dt}\right). \tag{1.2.8}$$

Equation (1.2.8) is a special case of Eq. (1.2.1) for which the function f is *time-independent* and the ODE is *second-order*. Equations like this often appear in mechanics problems. One simple example is the harmonic oscillator with a frictional damping force added, so that the acceleration depends linearly on both oscillator position x and velocity $v = dx/dt$:

$$f(x, v) = -\omega_0^2 x - \gamma v, \tag{1.2.9}$$

where ω_0 is the oscillator frequency and γ is a frictional damping rate.

The direction field consists of a set of vectors tangent to the solution curves of this ODE in (t, x, v) space. Consider a given solution curve, as shown schematically in Fig. 1.4. In a time interval dt the solution changes by dx and dv in the x and v directions respectively. The tangent to this curve is the vector

$$(dt, dx, dv) = dt\left(1, \frac{dx}{dt}, \frac{dv}{dt}\right) = dt(1, v, f(x, v)). \tag{1.2.10}$$

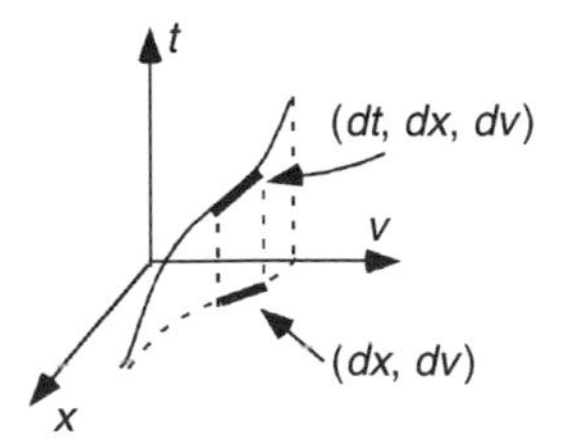

Fig. 1.4 A solution curve to Eq. (1.2.8), a tangent vector, and the projection onto the (x, v) plane.

Note that this tangent vector is independent of time. The direction field is the same in every time slice, so the trajectory of the particle can be understood by projecting solutions onto the (x, v) plane as shown in Fig. 1.4. The (x, v) plane is often referred to as *phase-space*, and the plot of a solution curve in the (x, v) plane is called a *phase-space portrait*.

Often, momentum $p = mv$ is used as a phase-space coordinate rather than v, so that the phase-space portrait is in the (x, p) plane rather than the (x, v) plane. This sometimes simplifies things (especially for motion in magnetic fields, where the relation between p and v is more complicated than just $p = mv$), but for now we will stick with plots in the (x, v) plane.

The projection of the direction field onto phase-space, created as usual with the **PlotVectorField** function, provides us with a global picture of the solution for all initial conditions (x_0, v_0). This projection is shown in Cell 1.6 for the case of a damped oscillator with acceleration given by Eq. (1.2.9), taking $\omega_0 = \gamma = 1$. One can see from this plot that all solutions spiral into the origin, which is expected, since the oscillator loses energy through frictional damping and eventually comes to rest.

Vectors in the direction field point toward the origin, in a manner reminiscent of the singularity in Fig. 1.3, even though $f(x, v)$ is differentiable. However, particles actually require an infinite amount of time to reach the origin, and if placed at the origin will not move from it (the origin is an *attracting fixed point*), so this field does not violate Theorem 1.1, and all initial conditions result in unique trajectories.

Cell 1.6

```
<<Graphics`;
f[x_, v_] = -x - v;
PlotVectorField[{v, f[x, v]}, {x, -1, 1}, {v, -1, 1},
  Axes → True, ScaleFunction→ (1&), AxesLabel→ {"x", "v"}];
```

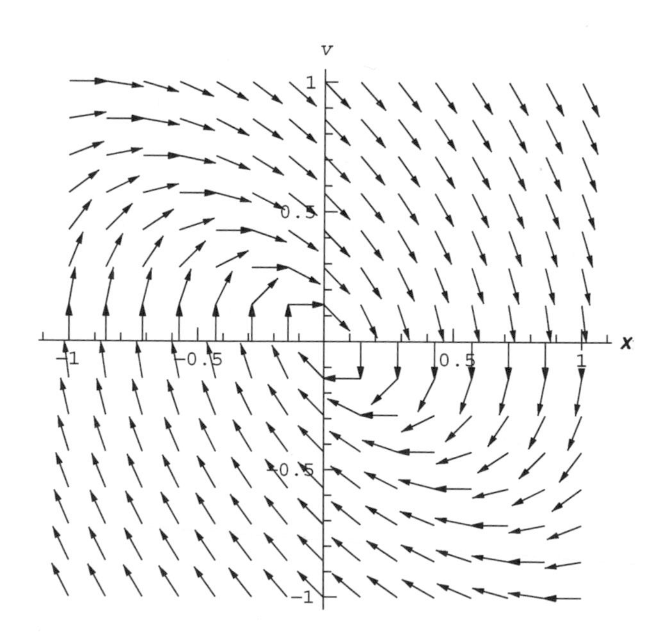

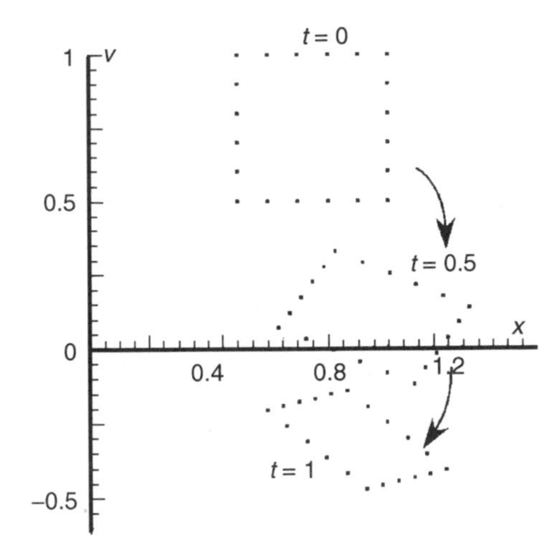

Fig. 1.5 Flow of a set of initial conditions for $f(x, v) = -x - v$.

Conservation of Phase-Space Area The solutions of the damped oscillator ODE do not conserve phase-space area. By this we mean the following: consider an area of phase-space, say a square, whose boundary is mapped out by a collection of initial conditions. As these points evolve in time according to the ODE, the square changes shape. The area of the square shrinks as all points are attracted toward the origin. (See Fig. 1.5.)

Dissipative systems—systems that lose energy—have the property that phase-space area shrinks over time. On the other hand, nondissipative systems, which conserve energy, can be shown to conserve phase-space area. Consider, for example, the direction field associated with motion in a potential $V(x)$. Newton's equation of motion is $m\,d^2x/dt^2 = -\partial V/\partial x$, or in terms of phase-space coordinates (x, v),

$$\begin{aligned} \frac{dx}{dt} &= v, \\ \frac{dv}{dt} &= -\frac{1}{m}\frac{\partial V}{\partial x}. \end{aligned} \tag{1.2.11}$$

According to Eq. (1.2.10), the projection of the direction field onto the (x, v) plane has components $(v, -(1/m)\partial V/\partial x)$. One can prove that this flow is area-conserving by showing that it is *divergence-free*. It is easiest at first to discuss such flows in the (x, y) plane, rather than the (x, v) plane. A flow in the (x, y) plane, described by a vector field $\mathbf{v}(x, y) = (v_x(x, y), v_y(x, y))$, is divergence-free if the flow satisfies

$$\nabla \cdot \mathbf{v}(x, y) = \left.\frac{\partial v_x}{\partial x}\right|_y + \left.\frac{\partial v_y}{\partial y}\right|_x = 0, \tag{1.2.12}$$

where we havc explicitly shown what is held fixed in the derivatives. The connection between this divergence and the area of the flow can be understood by

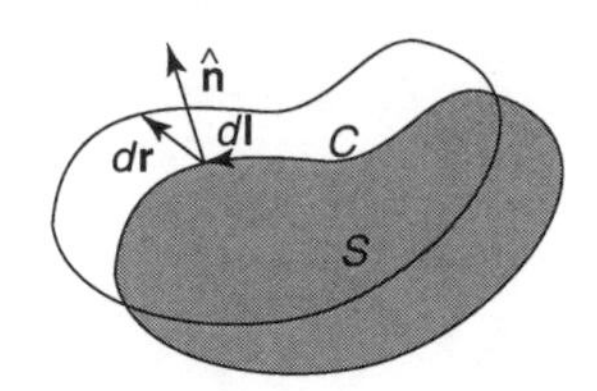

Fig. 1.6 Surface S moving by $d\mathbf{r}$ in time dt.

examining Fig. 1.6, which depicts an area S moving with the flow. The differential change dS in the area as the boundary C moves by $d\mathbf{r}$ is $dS = \oint_C d\mathbf{r}\cdot\hat{\mathbf{n}}\,dl$, where dl is a line element along C, and $\hat{\mathbf{n}}$ is the unit vector normal to the edge, pointing out from the surface. Dividing by dt, using $\mathbf{v} = d\mathbf{r}/dt$, and applying the *divergence theorem*, we obtain

$$\frac{dS}{dt} = \oint_C \mathbf{v}\cdot\hat{\mathbf{n}}\,dl = \int_S \nabla\cdot\mathbf{v}\,d^2\mathbf{r}. \tag{1.2.13}$$

Thus, the rate of change of the area dS/dt equals zero if $\nabla\cdot\mathbf{v} = 0$, proving that divergence-free flows are area-conserving.

Returning to the flow of the direction field in the (x, v) plane given by Eqs. (1.2.11), the x-component of the flow field is v, and the v-component is $-(1/m)\,\partial V/\partial x$. The divergence of this flow is, by analogy to Eq. (1.2.12),

$$\left.\frac{\partial v}{\partial x}\right|_v + \frac{\partial}{\partial v}\left(-\frac{1}{m}\frac{\partial V}{\partial x}\right)\bigg|_x = 0. \tag{1.2.14}$$

Therefore, the flow is area-conserving.

Why should we care whether a flow is area-conserving? Because the direction field for area-conserving flows looks very different than that for a non-area-conserving flow such as the damped harmonic oscillator. In area-conserving flows, there are no attracting fixed points toward which orbits fall; rather, the orbits tend to circulate indefinitely. This property is epitomized by the phase-space flow for the undamped harmonic oscillator, shown in Fig. 1.7.

Hamiltonian Systems Equations (1.2.11) are a specific example of a more general class of area-conserving flows called Hamiltonian flows. These flows have equations of motion of the form

$$\begin{aligned}\frac{dx}{dt} &= \frac{\partial H(x,p,t)}{\partial p},\\ \frac{dp}{dt} &= -\frac{\partial H(x,p,t)}{\partial x},\end{aligned} \tag{1.2.15}$$

where p is the *momentum associated with the variable* x. The function $H(x, p, t)$ is

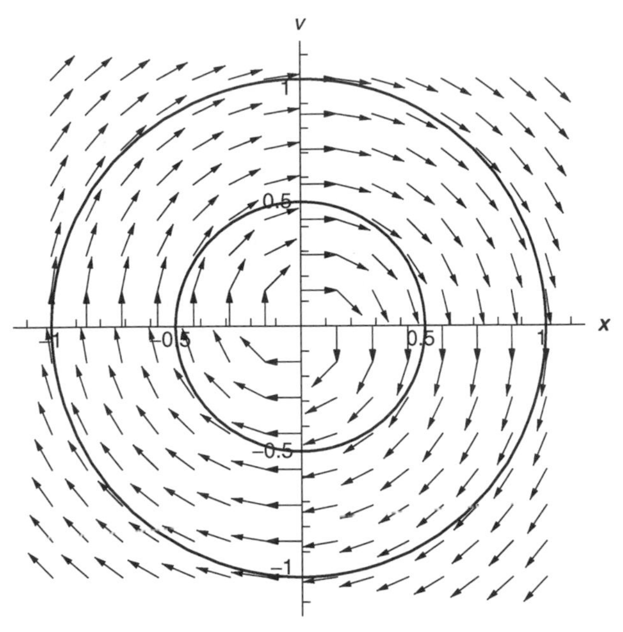

Fig. 1.7 Phase-space flow and constant-H curves for the undamped harmonic oscillator, $f(x,v) = -x$.

the *Hamiltonian* of the system. These flows are area-conserving, because their phase-space divergence is zero:

$$\frac{\partial}{\partial x}\frac{dx}{dt} + \frac{\partial}{\partial p}\frac{dp}{dt} = \frac{\partial^2 H(x,p,t)}{\partial x\,\partial p} - \frac{\partial^2 H(x,p,t)}{\partial x\,\partial p} = 0. \tag{1.2.16}$$

For Eqs. (1.2.11), the momentum is $p = mv$, and the Hamiltonian is the total energy of the system, given by the sum of kinetic and potential energies:

$$H = \frac{mv^2}{2} + V(x) = \frac{p^2}{2m} + V(x). \tag{1.2.17}$$

If the Hamiltonian is time-independent, it can easily be seen that the direction field is everywhere tangent to surfaces of constant H. Consider the change dH in the value of H as a particle follows along the flow for a time dt. This change is given by

$$dH = dx\frac{\partial H}{\partial x} + dp\frac{\partial H}{\partial p} = dt\left(\frac{\partial H}{\partial x}\frac{dx}{dt} + \frac{\partial H}{\partial p}\frac{dp}{dt}\right).$$

Using the equations of motion, we have

$$dH = dt\left[\frac{\partial H}{\partial x}\frac{\partial H}{\partial p} + \frac{\partial H}{\partial p}\left(-\frac{\partial H}{\partial x}\right)\right] = 0. \tag{1.2.18}$$

In other words energy is conserved, so that the flow is along constant-H surfaces. Some of these constant-H surfaces are shown in Fig. 1.7 for the harmonic oscillator. As usual, we plot the direction field in the (x, v) plane rather than in the (x, p) plane.

For a time-independent Hamiltonian $H(x, p)$, curves of constant H are nested curves in phase space, which describe the orbits. Even for very complicated Hamiltonian functions, these constant-H curves must be nested (think of contours of constant altitude on a topographic map). The resulting orbits must always remain on a given constant-H contour in a given region of phase space. Different regions of phase-space are isolated from one another by these contours. Such motion is said to be *integrable*.

However, this situation can change if the Hamiltonian depends explicitly on time so that energy is not conserved, or if phase-space has four or more dimensions [as, for example, can occur for two coupled oscillators, which have phase-space (x_1, v_1, x_2, v_2)]. Now energy surfaces no longer necessarily isolate different regions of phase-space. In these situations, it is possible for particles to explore large regions of phase space. The study of such systems is a burgeoning area of mathematical physics called *chaos theory*. A comprehensive examination of the properties of chaotic systems would take too far afield, but we will consider a few basic properties of chaotic systems in Sec. 1.4.

EXERCISES FOR SEC. 1.2

(1) Find, by hand, three valid solutions to $(d^2x/dt^2)^3 = tx(t)$, $x(0) = x'(0) = 0$. (Hint: Try solutions of the form at^n for some constants a and n.)

(2) Plot the direction field for the following differential equations in the given ranges, and discuss the qualitative behavior of solutions for initial conditions in the given ranges of y:

(a) $\dfrac{dv}{dt} = \sqrt{t}\, y^2$, $0 < t < 4$, $-2 < y < 2$.

(b) $\dfrac{dy}{dt} = \sin(t + y)$, $0 < t < 15$, $-8 < t < 8$.

(Hint: You can increase the resolution of the vector field using the `PlotPoints` option, as in `PlotPoints → 25`.)

(3) For a Hamiltonian $H(x, v, t)$ that depends explicitly on time, show that rate of change of energy dH/dt along a particular trajectory in phase space is given by

$$\frac{dH}{dt} = \left.\frac{\partial H}{\partial t}\right|_{x,v}. \tag{1.2.19}$$

(4) A simple pendulum follows the differential equation $\theta''(t) = -(g/l)\sin\theta(t)$, where θ is the angle the pendulum makes with the vertical, $g = 9.8\ \mathrm{m/s^2}$ is the acceleration of gravity, and l is the length of the pendulum. (See Fig. 1.8.) Plot the direction field for this equation projected into the phase space (θ, θ'), in the ranges $-\pi < \theta < \pi$ and $-4\ \mathrm{s^{-1}} < \theta' < 4\ \mathrm{s^{-1}}$, assuming a length l of 10 m.

(a) Discuss the qualitative features of the solutions. Do all phase-space trajectories circle the origin? If not, why not? What do these trajectories correspond to physically?

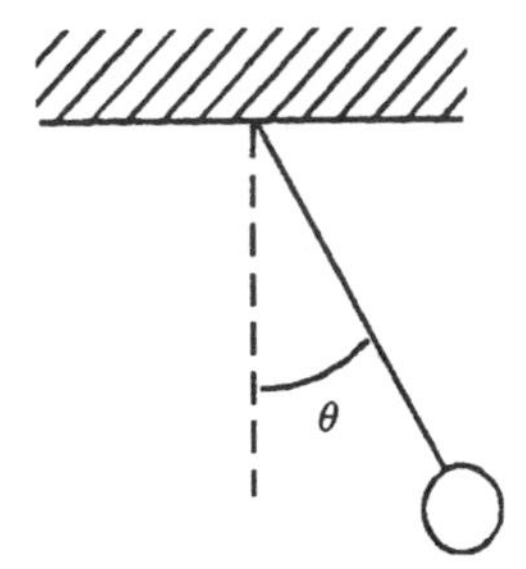

Fig. 1.8 Simple pendulum.

(b) Find the energy H for this motion in terms of θ and θ'. Plot several curves of constant H on top of your direction field, to verify that the field is tangent to them.

(5) Find an expression for the momentum p_θ associated with the variable θ, so that one can write the equations of motion for the pendulum in Hamiltonian form,

$$\frac{d\theta}{dt} = \frac{\partial H(\theta, p_\theta, t)}{\partial p_\theta},$$

$$\frac{dp_\theta}{dt} = -\frac{\partial H(\theta, p_\theta, t)}{\partial \theta}.$$

(6) The Van der Pol oscillator ODE models some properties of excitable systems, such as heart muscle or certain electronic circuits. The ODE is

$$x'' + (x^2 - 1)x' + x = 0. \tag{1.2.20}$$

The restoring force is a simple Hooke's law, but the "drag force" is more complicated, actually accelerating "particles" with $|x| < 1$. (Here, x could actually mean the oscillation amplitude of a chunk of muscle, or the current in a nonlinear electrical circuit.) At low amplitudes the oscillations build up, but at large amplitudes they decay.

(a) Draw the direction field projected into the phase space (x, x') for $-2 < x < 2$, $-2 < x' < 2$. Discuss the qualitative behavior of solutions that begin (i) near the origin, (ii) far from the origin.

(b) Does this system conserve phase-space area, where (x, x') is the phase-space?

(7) A particle orbiting around a stationary mass M (the sun, for example) follows the following differential equation for radius as a function of time, $r(t)$ where r is the distance measured from the stationary mass:

$$\frac{d^2r}{dt^2} = \frac{L^2}{r^3} - \frac{GM}{r^2}. \tag{1.2.21}$$

Here, G is the gravitational constant, and L is a constant of the motion—the specific angular momentum of the particle, determined by radius r and angular velocity $\dot{\theta}$ as

$$L = r^2\dot{\theta}. \tag{1.2.22}$$

(a) Assuming that L is nonzero, find a transformation of time and spatial scales, $\bar{r} = r/a$, $\bar{t} = t/t_0$, that puts this equation into the dimensionless form

$$\frac{d^2\bar{r}}{d\bar{t}^2} = \frac{1}{\bar{r}^3} - \frac{1}{\bar{r}^2}. \tag{1.2.23}$$

(b) Plot the projection of the direction field for this equation into the phase space $(\bar{r}, \bar{r}')$ in the range $0.1 < \bar{r} < 4$, $-0.7 < \bar{r}' < 0.7$.

(i) What is the physical significance of the point $(\bar{r}, \bar{r}') = (1, 0)$?

(ii) What happens to particles that start with large radial velocities at large radii, $\bar{r} \gg 1$?

(iii) What happens to particles with zero radial velocities at small radius, $\bar{r} \ll 1$? Explain this in physical terms.

(iv) For particles that start with velocities close to the point $(\bar{r}, \bar{r}') = (1, 0)$, the closed trajectories correspond to elliptical orbits, with the two points where $\bar{r}' = 0$ corresponding to distance of closest approach r_0 (perihelion) and farthest distance r_1 (aphelion) from the fixed mass. Therefore, one closed orbit in the $(\bar{r}, \bar{r}')$ plane corresponds to a whole set of actual orbits with different scale parameters a and t_0 but the same elliptical shape. How do the periods of this set of orbits scale with the size of the orbit? (This scaling is sometimes referred to as Kepler's third law.)

(c) Find the Hamiltonian $H(\bar{r}, \bar{r}')$ associated with the motion described above. Plot a few curves of constant H on top of the direction field, verifying that the field is everywhere tangent to the flow.

(8) Magnetic and electric fields are often visualized by drawing the *field lines* associated with these fields. These field lines are the trajectories through space that are everywhere tangent to the given field. Thus, they are analogous to the trajectories followed by particles as they propagate tangent to the direction field. Consider a field line that passes through the point $\mathbf{r} = \mathbf{r}_0$. We parametrize this field line by the displacement s measured along the field line from the point $\mathbf{r}_0$. Thus, the field line is given by a curve through space, $\mathbf{r} = \mathbf{r}(s)$, where $\mathbf{r}(0) = \mathbf{r}_0$. A displacement $d\mathbf{r}$ along the field line with magnitude ds is in the direction of the local field: $d\mathbf{r} = ds\,\mathbf{E}(\mathbf{r})/|\mathbf{E}(\mathbf{r})|$. Dividing by ds yields the following differential equation for the field line:

$$\frac{d}{ds}\mathbf{r}(s) = \frac{\mathbf{E}(\mathbf{r})}{|\mathbf{E}(\mathbf{r})|}. \tag{1.2.24}$$

Equation (1.2.24) is a set of coupled first-order ODEs for the components of $\mathbf{r}(s)$, with initial condition $\mathbf{r}(0) = \mathbf{r}_0$.

(a) Using **PlotVectorField**, plot the electric field $\mathbf{E}(x, y) = -\nabla\phi(x, y)$ that arises from the following electrostatic potential $\phi(x, y)$:

$$\phi(x, y) = x^2 - y^2.$$

(This field satisfies Laplace's equation, $\nabla^2\phi = 0$.) Make the plot in the ranges $-2 < x < 2$, $-2 < y < 2$.

(b) Show that for this potential, Eq. (1.2.24) implies that $dy/dx = -y/x$ along a field line. Solve this ODE analytically to obtain the general solution for $y(x)$, and plot the resulting field lines in the (x, y) plane for initial conditions $(x_0, y_0) = (m, n)$ where $m = -1, 0, 1$ and $n = -1, 0, 1$ (nine plots in all). Then superimpose these plots on the previous plot of the field. [Hint 1: Make a table of plots; then use a single **Show** command to superimpose them. Hint 2: $(dx/ds)/(dy/ds) = dx/dy$.]

1.3 ANALYTIC SOLUTION OF INITIAL-VALUE PROBLEMS VIA DSOLVE

1.3.1 DSolve

The solution to some (but not all) ODEs can be determined analytically. This section will discuss how to use *Mathematica's* analytic differential equation solver **DSolve** in order to find these analytic solutions.

Consider a simple differential equation with an analytic solution, such as the harmonic oscillator equation

$$\frac{d^2x}{dt^2} = -\omega_0^2 x. \tag{1.3.1}$$

DSolve can provide the general solution to this second-order ODE. The syntax is as follows:

DSolve[*ODE*, *unknown function*, *independent variable*].

The ODE is written as a *logical* expression, **x"(t) == -ω₀²x(t)**. *Note that in the ODE you must refer to* **x[t]**, *not merely* **x** *as we did in Eq.* (1.3.1). The unknown function is $x(t)$ in this example. Then we specify the independent variable t, and evaluate the cell:

Cell 1.7

```
DSolve[x"[t] == -ω0^2 x[t], x[t], t]
{{x[t] → C[2] Cos[tω0] + C[1] Sin[tω0]}}
```

The result is a list of solutions (in this case there is only one solution), written in terms of two undetermined constants, **C[1]** and **C[2]**. As we know, these constants are set by specifying initial conditions.

It is possible to obtain a unique solution to the ODE by specifying particular initial conditions in **DSolve**. Now the syntax is

DSolve[{*ODE*, *initial conditions*}, *unknown function*, *independent variable*].

Just as with the ODE, the initial conditions are specified by *logical* expressions, *not* assignments, for example, **x[0]==x0, v[0]==v0**:

Cell 1.8

```
DSolve[ {x"[t]== -ω0^2 x[t],
          x[0] == x0, x'[0] == v0}, x[t], t]
```

$$\{\{x[t] \to x0\ Cos[t\ \omega_0] + \frac{v0\ Sin[t\omega_0]}{\omega_0}\}\}$$

As expected, the result matches our previous solution, Eq. (1.1.9).

DSolve can also be used to provide solutions to systems of coupled ODEs. Now, one provides a list of ODEs in the first argument, along with a list of the unknown functions in the second argument. For instance, consider the following coupled ODEs, which describe a set of two coupled harmonic oscillators with positions **x1[t]** and **x2[t]**, and with given initial conditions:

Cell 1.9

```
DSolve[{x1"[t] == -x1[t] + 2 (x2[t] - x1[t]),
        x2"[t] == -x2[t] + 2 (x1[t] - x2[t]),
  x1[0] == 0, x1'[0] == 0, x2[0] == 1, x2'[0] == 0},
  {x1[t], x2[t]}, t]
```

$$\{\{x1[t] \to -\frac{1}{4}e^{-i\ t-i\sqrt{5}t}(e^{it} - e^{i\sqrt{5}t} - e^{2it+i\sqrt{5}t} + e^{it+2i\sqrt{5}t}),$$
$$x2[t] \to \frac{1}{4}e^{-i\ t-i\sqrt{5}t}\ (e^{it} + e^{i\sqrt{5}t} + e^{2it+i\sqrt{5}t} + e^{it-2\ i\sqrt{5}t})\}\}$$

Mathematica found the solution, although it is not in the simplest possible form. For example, **x1[t]** can be simplified by applying **FullSimplify**:

Cell 1.10

```
FullSimplify[x1[t]/. %[[1]]]
```

$$\frac{1}{2}\ (Cos[t] - Cos[\sqrt{5}\ t])$$

Mathematica knows how to solve a large number of quite complex ODEs analytically. For example, it can find the solution to a harmonic oscillator ODE where the square of natural frequency ω_0 is time-dependent, decreasing linearly with time: $\omega_0^2 = -t$. This ODE is called the *Airy equation*:

$$x''(t) = tx(t). \tag{1.3.2}$$

The general solution to this equation is

Cell 1.11

```
DSolve[x"[t] - tx[t] == 0, x[t], t]
```

{{x[t] → AiryAi[t] C[1] + AiryBi[t] C[2]}}

The two independent solutions to the ODE are special functions called *Airy* functions, $\mathrm{Ai}(x)$ and $\mathrm{Bi}(x)$. These are called special functions in order to distinguish them from the elementary functions such as $\sin x$ or $\log x$ that appear on your calculator. *Mathematica* refers to these functions as **AiryAi[x]** and

AiryBi[x]. These are only two of the huge number of special functions that *Mathematica* knows. Just as for the elementary functions, one can plot these special functions, as shown in Cell 1.12.

Cell 1.12

```
<<Graphics`;
Plot [{AiryAi[x], AiryBi[x]}, {x, -10, 3},
   PlotStyle→{Red, Green},
  PlotLabel→TableForm[{{StyleForm["Ai[x]",
  FontColor→RGBColor [1, 0, 0]], ", ", StyleForm["Bi[x]",
  FontColor→Green]}}, TableSpacing→0]];
```

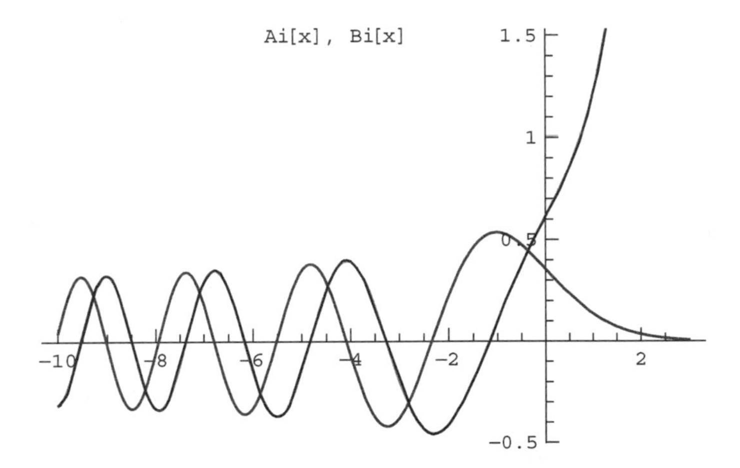

On the other hand, there are many seemingly straightforward ODEs that have no solution in terms of either special functions or elementary functions. Here is an example:

Cell 1.13

```
DSolve[x'[t] == t/(x[t] + t)^2, x[t], t]
```

$$\text{DSolve}\left[x'[t] == \frac{t}{(t+x[t])^2},\ x[t],\ t\right]$$

Mathematica could not find an analytic solution for this simple first-order ODE, although if we wished we could plot the direction field to find the qualitative form of the solutions. Of course, that doesn't mean that there is no analytic solution in terms of predefined functions—after all, *Mathematica* is not omniscient. However, as far as I know there really is no such solution to this equation.

You may wonder why a reasonably simple first-order ODE has no analytic solution, but a second-order ODE like the Airy equation does have an analytic solution. The reason in this instance is mainly historical, not mathematical. The solutions of the Airy equation are of physical interest, and were explored originally by the British mathematician George B. Airy. The equation is important in the

Table 1.1. **`DSolve`**

`DSolve[`*eqn*`,x[t], t]`	Solve a differential equation for `x[t]`
`DSolve[{`*eqn1*, *eqn2*,...`},{x1[t], x2[t],...},t]`	Solve coupled differential equations

study of wave propagation through inhomogeneous media, and in the quantum theory of tunneling, as we will see in Chapter 5. Many of the special functions that we will encounter in this course—Bessel functions, Mathieu functions, Legendre functions, etc.—have a similar history: they were originally studied because of their importance to some area of science or mathematics.

Our simple first-order ODE, above, has no analytic solution (as far as I know) simply because no one has ever felt the need to define one. Perhaps some day the need will arise, and the solutions will then be detailed and named.

However, there are many ODEs for which *no exact analytic solution can be written down*. These ODEs have *chaotic* solutions that are so complex that they cannot be predicted on the basis of analytic formulae. Over long times, the solutions cannot even be predicted numerically with any accuracy (as we will see in the next section).

The syntax for **`DSolve`** is summarized in Table 1.1.

EXERCISES FOR SEC. 1.3

(1) In the process of radioactive decay, an atom spontaneously changes form by emitting particles from the nucleus. The rate υ at which this decay happens is defined as the fraction of nuclei that decay per unit of time in a sample of material. Write down and solve a differential equation for the mass of radioactive material remaining at time t, $m(t)$, given an amount m_0 at $t = 0$. How long does it take for half the material to decay?

(2) A spaceship undergoing constant acceleration $g = 9.8\ \mathrm{m/s^2}$ (as felt by the passengers) will follow Newton's second law, with the addition (by Einstein) that the apparent mass of the ship as seen from a stationary observer will increase with velocity $\upsilon(t)$ in proportion to the factor $1/\sqrt{1-\upsilon^2/c^2}$. This implies that the velocity satisfies the following first order ODE:

$$\frac{d}{dt}\left(\frac{\upsilon}{\sqrt{1-\upsilon^2/c^2}}\right) = g.$$

(a) Find the general solution, using pencil and paper, for the position as a function of time.

(b) After 100 earth years of acceleration, starting from rest, how far has the ship gone in light-years (one light-year $= 9.45 \times 10^{15}$ m)?

(c) Thanks to relativistic time dilation, the amount of time τ that has passed onboard the ship is considerably shorter than 100 years, and is given by the solution to the differential equation $d\tau/dt = \sqrt{1-\upsilon(t)^2/c^2}$, $\tau(0) = 0$.

Solve this ODE, using **DSolve** and the solution for $v(t)$ from part (b), above, to find the amount of time that has gone by for passengers on the ship. (Note: The nearest star is only about 4.3 light years from earth.) What was the average speed of the ship over the course of the trip, in units of c, as far as the passengers are concerned?

(3) The charge $Q(t)$ on the capacitor in an *LRC* electrical circuit obeys a second-order differential equation,

$$LQ'' + RQ' + Q/C = V(t). \tag{1.3.3}$$

(a) Find the general solution to the equation, taking $V(t) = 0$.

(b) Plot this solution for the case $Q(0) = 10^{-5}$ coulomb, $Q'(0) = 0$, taking $R = 10^4$ ohms, $C = 10^{-5}$ farad, $L = 0.1$ henry. What is the frequency of the oscillation being plotted (in radians per second)? What is the rate of decay of the envelope (in inverse seconds)?

(4) A man throws a pebble straight up. Its height $y(t)$ satisfies the differential equation $y'' + \gamma y' = -g$, where g is the acceleration of gravity and γ is the damping rate due to frictional drag with the air.

(a) Find the general solution to this ODE.

(b) Find the solution $y(t)$ for the case where the initial speed is 6 m/s, $y(0) = 0$, and $\gamma = 0.2\ \text{s}^{-1}$. Plot this solution vs. time.

(c) Find the time when the pebble returns to the ground (this may require a numerical solution of an algebraic equation).

(5) Atomic hydrogen (H) recombines into molecular hydrogen (H_2) according to the simple chemical reaction $H + H \rightleftharpoons H_2$. The rate of the forward recombination reaction (number of reactions per unit time per unit volume) is $v_1 n_H^2$, where n_H is the number density (in atoms per cubic meter) of atomic hydrogen, and v_1 is a constant. The rate of the reverse reaction (spontaneous decomposition into atomic hydrogen) is $v_2 n_{H_2}$, where n_{H_2} is the number density of molecular hydrogen.

(a) Write down two coupled first-order ODEs for the densities of molecular and atomic hydrogen as a function of time.

(b) Solve these equations for general initial densities.

(c) Show that the solution to these equations satisfy $n_H + 2n_{H_2} = \text{const}$. Take the constant equal to n_0 (the total number density of hydrogen atoms in the system, counting those that are combined into molecules), and find the ratio of densities *in equilibrium*.

(d) Plot the densities as a function of time for the initial condition $n_{H_2} = 1$, $n_H = 0$, $v_1 = 3$, and $v_2 = 1$.

(6) A charged particle, of mass m and charge q, moves in uniform magnetic and electric fields $\mathbf{B} = (0, 0, B_0)$, $\mathbf{E} = (E_\perp, 0, E_z)$. The particle satisfies the nonrelativistic equations of motion,

$$m\frac{d\mathbf{v}}{dt} = q(\mathbf{E} + \mathbf{v} \times \mathbf{B}). \tag{1.3.4}$$

(a) Find, using **DSolve**, the general solution of these coupled first-order ODEs for the velocity $\mathbf{v}(t) = (v_x(t), v_y(t), v_z(t))$.

(b) Note that in general there is a net constant velocity perpendicular to **B**, on which there is superimposed circular motion. The constant velocity is called an $\mathbf{E} \times \mathbf{B}$ *drift*. The circular motion is called a *cyclotron orbit*. What is the frequency of the cyclotron orbit? What are the magnitude and direction of the $\mathbf{E} \times \mathbf{B}$ drift?

(c) Find $\mathbf{v}(t)$ for the case for an electron with $\mathbf{r}(0) = \mathbf{v}(0) = 0$, $E_z = 0$, $E_\perp = 5000$ V/m, and $B_0 = 0.005$ tesla (these are the proper units for the equation of motion as written above). Plot $v_x(t)$ and $v_y(t)$ vs. t for a time of 10^{-7} s.

(d) Use the results of part (b) to obtain $x(t)$ and $y(t)$. Plot x vs. y using a parametric plot to look at the trajectory of the electron. Plot the trajectory again in a frame moving at the $\mathbf{E} \times \mathbf{B}$ drift speed. The radius of the circle is called the *cyclotron radius*. What is the magnitude of the radius (in meters) for this example?

(7) The trajectory $r(\theta)$ of a particle orbiting a fixed mass M at the origin of the (r, θ) plane satisfies the following differential equation:

$$\frac{L}{r^2}\frac{d}{d\theta}\left(\frac{L}{r^2}\frac{dr}{d\theta}\right) - \frac{L^2}{r^3} = -\frac{GM}{r^2}, \tag{1.3.5}$$

where L is the specific angular momentum, as in Eq. (1.2.22).

(a) Introduce a scaled radius $\bar{r} = r/a$ to show that with proper choice of a this equation can be written in dimensionless form as

$$\frac{1}{\bar{r}^2}\frac{d}{d\theta}\left(\frac{1}{\bar{r}^2}\frac{d\bar{r}}{d\theta}\right) - \frac{1}{\bar{r}^3} = -\frac{1}{\bar{r}^2}.$$

(b) Find the general solution for the trajectory, and show that *Mathematica*'s expression is equivalent to the expression $1/r(\theta) = (GM/L^2)\,[e\cos(\theta - \theta_0) + 1]$, where e is the eccentricity and θ_0 is the angular position of perihelion. [See Eq. (1.2.21) and Exercise (3) of Sec. 9.6].

(8) Consider the electric field from a unit point dipole at the origin. This field is given by $\mathbf{E} = -\nabla\phi(\rho, z)$ in cylindrical coordinates (ρ, θ, z), where $\phi = z/(\rho^2 + z^2)^{3/2}$. In cylindrical coordinates the field lines equation, Eq. (1.2.24), has components

$$\begin{aligned} \frac{d\rho}{ds} &= \frac{E_\rho(\mathbf{r})}{|\mathbf{E}(\mathbf{r})|}, \\ \rho\frac{d\theta}{ds} &= \frac{E_\theta(\mathbf{r})}{|\mathbf{E}(\mathbf{r})|}, \\ \frac{dz}{ds} &= \frac{E_z(\mathbf{r})}{|\mathbf{E}(\mathbf{r})|}. \end{aligned} \tag{1.3.6}$$

(a) Use **DSolve** to determine the field lines for the dipole that pass through the following points: $(\rho_0, z_0) = (1, 0.5n)$, where $n = 1, 2, 3, 4$. Make a table of **ParametricPlot** graphics images of these field lines in the (ρ, z) plane for $-10 < s < 10$, and superimpose them all with a **Show** command to visualize the field lines from this dipole. (Hint: Create a vector function $\mathbf{r}(s, z_0) = \{\rho(s, z_0), z(s, z_0)\}$ using the **DSolve** solution, for initial condition $\rho(0) = \rho_0 = 1$, $z(0) = z_0$. Then plot that vector function for the given values of z_0.)

(b) A simple analytic form for the field lines from a dipole can be found in spherical coordinates (r, θ, ϕ). In these coordinates $\mathbf{r}(s) = (r(s), \theta(s), \phi(s))$ and Eq. (1.2.24) becomes

$$\frac{dr}{ds} = \frac{E_r(\mathbf{r})}{|\mathbf{E}(\mathbf{r})|},$$
$$r\frac{d\theta}{ds} = \frac{E_\theta(\mathbf{r})}{|\mathbf{E}(\mathbf{r})|}, \tag{1.3.7}$$
$$r \sin\theta \frac{d\phi}{ds} = \frac{E_\phi(\mathbf{r})}{|\mathbf{E}(\mathbf{r})|}.$$

Also, since $r = \sqrt{\rho^2 + z^2}$ and $z = r\cos\theta$, the dipole potential has the form $\phi(r, \theta) = (\cos\theta)/r$. An equation for the variation of r with θ along a field line can be obtained as follows:

$$\frac{1}{r}\frac{dr}{d\theta} = \frac{dr/ds}{r\,d\theta/ds} = \frac{E_r(r, \theta)}{E_\theta(r, \theta)}.$$

Solve this differential equation for $r(\theta)$ with initial condition $r(\theta_0) = r_0$ to show that the equation for the field lines of a point dipole in spherical coordinates is

$$r(\theta) = r_0 \frac{\sin^2\theta}{\sin^2\theta_0}. \tag{1.3.8}$$

Superimpose plots of $r(\theta)$ for $r_0 = 1, 2, 3, 4$ and $\theta_0 = \pi/2$.

1.4 NUMERICAL SOLUTION OF INITIAL-VALUE PROBLEMS

1.4.1 NDSolve

Mathematica can solve ODE initial-value problems numerically via the intrinsic function **NDSolve**. The syntax for **NDSolve** is almost identical to that for **DSolve**:

```
NDSolve[{ODE, initial conditions}, x[t],{t,tmin,tmax}]
```

Three things must be remembered when using **NDSolve**.

> (1) Initial conditions must always be specified.
> (2) No nonnumerical constants can appear in the list of ODEs or the initial conditions.
> (3) A finite interval of time must be provided, over which the solution for $x(t)$ is to be determined.

As an example, we will solve the problem from the previous section that had no analytic solution, for the specific initial condition x(0) = 1:

Cell 1.14

```
NDSolve[{x'[t] == t/(x[t] + t)^2, x[0] == 1}, x[t],
  {t, 0, 10}]

{{x[t] → InterpolatingFunction [{{0., 10.}}, <>][t]}}
```

The result is a list of possible substitutions for $x(t)$, just as when using **DSolve**. However, the function $x(t)$ is now determined numerically via an **InterpolatingFunction**. These **InterpolatingFunctions** are also used for interpolating lists of data (see Sec. 9.11). The reason why an **InterpolatingFunction** is used by **NDSolve** will become clear in the next section, but can be briefly stated as follows: When **NDSolve** numerically solves an ODE, it finds values for $x(t)$ only at specific values of t between *tmin* and *tmax*, and then uses an **InterpolatingFunction** to interpolate between these values of t.

As discussed in Sec. 9.11, the **InterpolatingFunction** can be evaluated at any point in its range of validity from *tmin* to *tmax*. For example, we can plot the solution by first extracting the function from the list of possible solutions,

Cell 1.15

```
x[t]/. %[[1]]

InterpolatingFunction[{{0., 10.}}, <>] [t]
```

and then plotting the result as shown in Cell 1.16.

Cell 1.16

```
Plot[%, {t,0,10}];
```

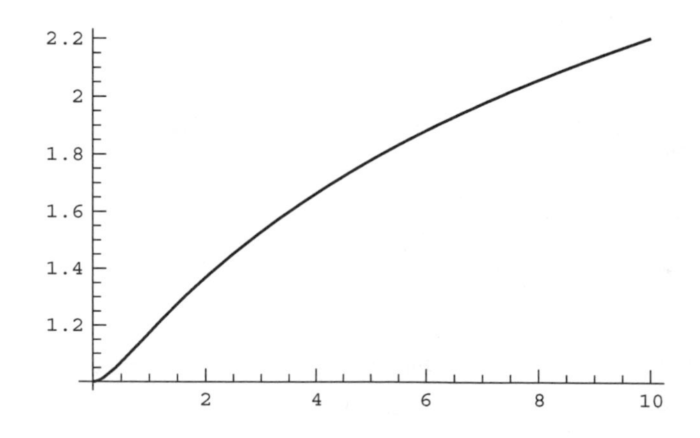

Now we come to an important question: how do we know that the answer provided by **NDSolve** is correct? The numerical solution clearly matches the initial condition, $x(0) = 1$. How do we tell if it also solves the ODE? One way to tell this is to plug the solution back into the ODE to see if the ODE is satisfied. We can do this just as we have done with previous analytic solutions, except that the answer will now evaluate to a numerical function of time, which must then be plotted to see how much it differs from zero (see Cell 1.17).

Cell 1.17

```
x[t_] = %%;
error[t_] = x'[t] - t/(x[t] + t)^2;
Plot[error[t], {t, 0, 10}];
```

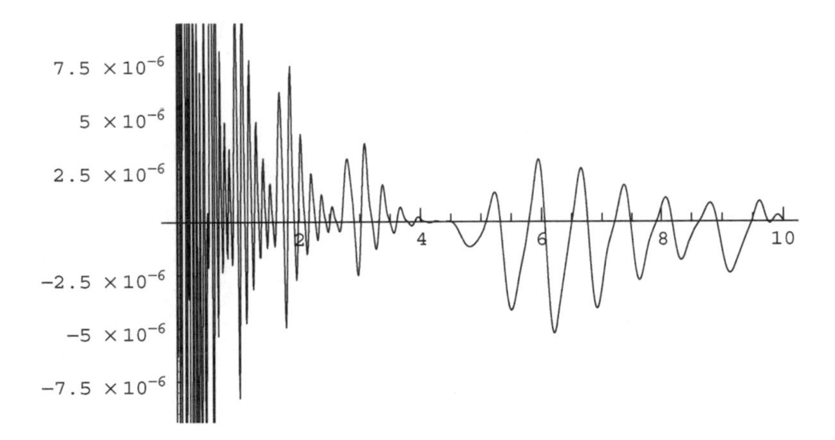

The plot shows that the error in the solution is small, but nonzero.

In order to further investigate the accuracy of **NDSolve**, we will solve a problem with an analytic solution: the harmonic oscillator with frequency $\omega_0 = 1$ and with initial condition $x(0) = 1$, $x'(0) = 0$. The exact solution is $x(t) = \cos t$. **NDSolve** provides a numerical solution that can be compared with the exact solution, in Cell 1.20.

Cell 1.18

```
Clear[x];

NDSolve[{x"[t] == -x[t], x[0] == 1, x'[0] == 0}, x[t],
  {t, 0, 30}]

{{x[t] → InterpolatingFunction [{{0., 30.}}, <>] [t]}}
```

Cell 1.19

```
x[t]/.%[[1]]

InterpolatingFunction[{{0., 30.}}, <>] [t]
```

Cell 1.20

```
Plot[% - Cos[t], {t, 0, 30}];
```

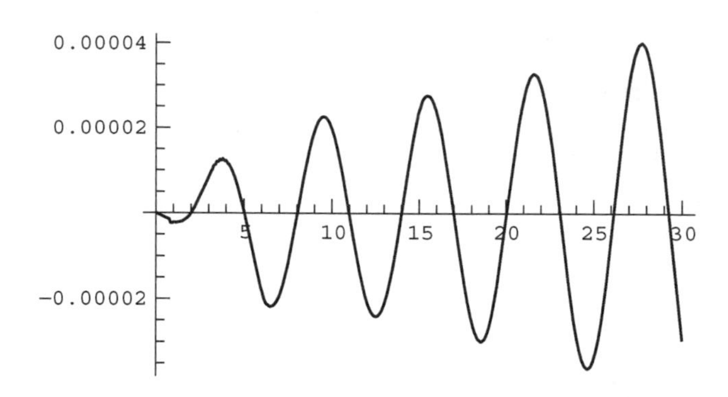

The difference between **NDSolve**'s solution and $\cos t$ is finite, and is growing with time. This is typical behavior for numerical solutions of initial-value problems: the errors tend to accumulate over time. If this level of error is too large, the error can be reduced by using two options for **NDSolve**: **AccuracyGoal** and **PrecisionGoal**. The default values of these options is **Automatic**, meaning that *Mathematica* decides what the accuracy of the solution will be. We can intercede, however, choosing our own number of significant figures for the accuracy. It is best to set both **AccuracyGoal** and **PrecisionGoal** to about the same number, and to have this number smaller than **$MachinePrecision** (otherwise the requested accuracy cannot be achieved, due to numerical roundoff error). Good values for my computer (with **$MachinePrecision** of 16) are **AccuracyGoal → 13, PrecisionGoal → 13**:

Cell 1.21

```
xsol[t_] = x[t]/. NDSolve[{x"[t] == -x[t], x[0] == 1,
      x'[0] == 0}, x[t], {t, 0, 30}, AccuracyGoal→13,
       PrecisionGoal→13] [[1]];
```

The results are shown in Cell 1.22. The error in the solution has now been considerably reduced. (Note that I have saved a little space by directly defining the solution of **NDSolve** to be the function **xsol[t]**, all in one line of code.)

Cell 1.22

```
Plot[xsol[t] - Cos [t], {t, 0, 30}];
```

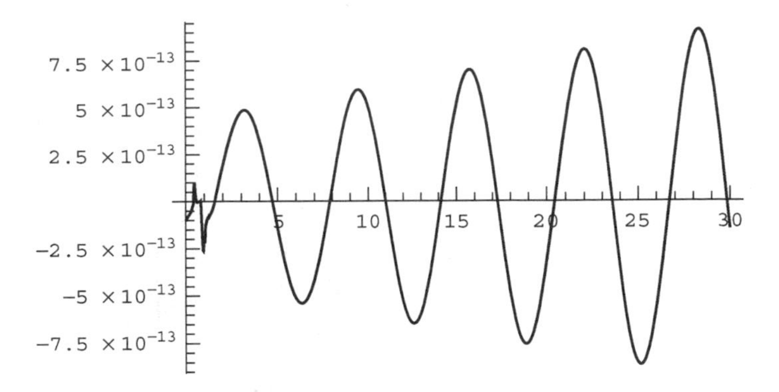

1.4.2 Error in Chaotic Systems

A Chaotic System: The Driven Pendulum The problem of error accumulation in numerical solutions of ODEs is radically worse when the solutions display chaotic behavior. Consider the following equation of motion for a pendulum of length l (see Fig. 1.8):

$$l\theta''(t) = -g \sin \theta - f \sin(\theta - \omega t). \tag{1.4.1}$$

The first term on the right-hand side is the usual acceleration due to gravity, and the second term is an added time-dependent force that can drive the pendulum into chaotic motion. This term can arise if one rotates the pivot of the pendulum in a small circle, at frequency ω. (Think of a noisemaker on New Year's Eve.)

We can numerically integrate this equation of motion using **NDSolve**. In Fig. 1.9 we show $\theta(t)$ for $0 < t < 200$, taking $l = g = f = 1$ and $\omega = 2$, and initial conditions $\theta(0) = -0.5, \theta'(0) = 0$. One can see that $\theta(t)$ increases with time in a rather complicated manner as the pendulum rotates about the pivot, and sometimes doesn't quite make it over the top. (Values of θ larger than 2π mean that the pendulum has undergone one or more rotations about the pivot.)

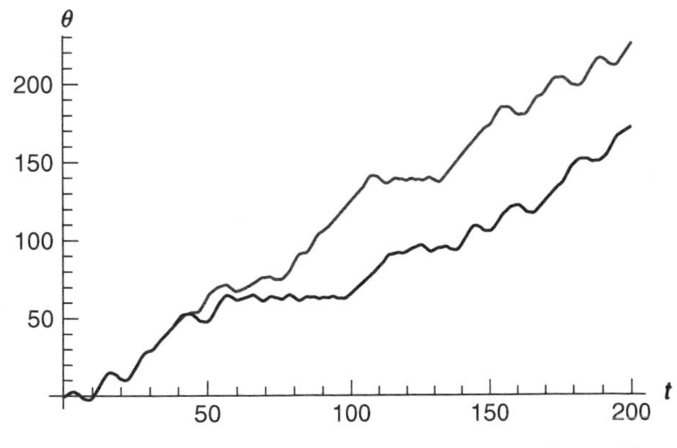

Fig. 1.9 Two trajectories starting from the same initial conditions. The upper trajectory is integrated with higher accuracy than the lower trajectory.

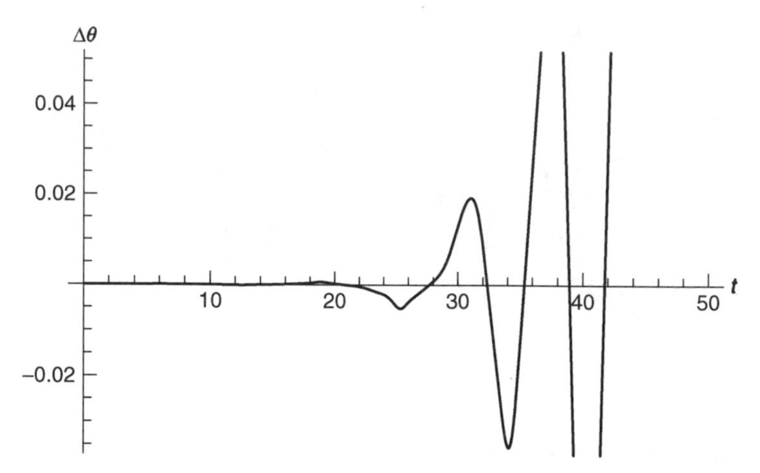

Fig. 1.10 Difference between the trajectories of Fig. 1.9.

If we repeat this trajectory, but increase the accuracy by setting **`AccuracyGoal->13`** and **`PrecisionGoal->13`**, the results for the two trajectories at late times bear little or no resemblance to one-another (Fig. 1.9).

The difference $\Delta\theta$ between the two $\theta(t)$ results is shown in Fig. 1.10. The error is exploding with time, as opposed to the relatively gentle increase observed in our previous examples. The explosive growth of accumulated error is a general feature of chaotic systems. In fact, one can show that if one compares a given trajectory with groups of nearby trajectories, on average the difference between these trajectories increases *exponentially* with time: $|\Delta\theta| \sim \exp(\lambda t)$. The rate of exponentiation of error, λ, is called the *Lyapunov exponent*.

This rapid error accumulation is the signature of a chaotic system. It makes it impossible to determine trajectories accurately over times long compared to $1/\lambda$. One can easily see why this is so: a fast computer working at double–double precision (32-digit accuracy) can integrate for times up to roughly $(\ln 10^{32})/\lambda \sim 70/\lambda$ before roundoff error in the 32nd digit causes order-unity deviation from the exact trajectory. To integrate accurately up to $7000/\lambda$, the accuracy would have to be increased by a factor of $e^{100} = 2.6 \times 10^{43}$!

For chaotic systems, small errors in computing the trajectory, or in the initial conditions, lead to exponentially large errors at later times, making the trajectory unpredictable.

Chaotic trajectories are not an isolated feature of only a few unusual dynamical systems. Rather, chaos is the norm. It has been shown that almost all dynamical systems are chaotic. Integrable systems such as the harmonic oscillator are the truly unusual cases, even though such cases are emphasized in elementary books on mechanics (because they are analytically tractable).

Since almost all systems are chaotic, and since chaotic systems are unpredictable, one might question the usefulness of Newton's formulation of dynamics, wherein a given initial condition, together with the force law, is supposed to provide all the information necessary to predict future behavior. For chaotic systems this predictive power is lost.

Fortunately, even chaotic systems have features that can be predicted and reproducibly observed. Although specific particle trajectories cannot be predicted over long times, average values based on many particle trajectories are reproducible. We will examine this statistical approach to complex dynamical systems in Chapter 8.

The Lyapunov Exponent One example of a reproducible average quantity for a chaotic system is the Lyapunov exponent itself. In order to define the Lyapunov exponent, note that since error accumulates on average as $\exp(\lambda t)$, the *logarithm* of the error should increase *linearly* with t, with a slope equal to λ.

Therefore, λ is defined as follows: Consider a given initial condition $\mathbf{z}_0 = (x_0, v_0)$. For this choice, the phase-space trajectory is $\mathbf{z}(t, \mathbf{z}_0) = (x(t, x_0, v_0), v(t, x_0, v_0))$. Now consider a small displacement $\mathbf{d}_0 = (\Delta x_0, \Delta v_0)$ to a nearby initial condition $\mathbf{z}_0 + \mathbf{d}_0$. The Lyapunov exponent is defined as

$$\lambda(\mathbf{z}_0) = \lim_{t \to \infty, |\mathbf{d}_0| \to 0} \left\langle \frac{1}{t} \ln \left(\frac{|\mathbf{z}(t, \mathbf{z}_0 + \mathbf{d}_0) - \mathbf{z}(t, \mathbf{z}_0)|}{|\mathbf{d}_0|} \right) \right\rangle, \tag{1.4.2}$$

where the $\langle\ \rangle$ stands for an average over many infinitesimal initial displacements $\mathbf{d}_0$ in different directions, and $|\mathbf{z}|$ corresponds to a vector magnitude in the phase space. [Units are unimportant in this vector magnitude: both position and momentum can be regarded as dimensionless, so that $\mathbf{z}$ can be thought of as a dimensionless vector for the purposes of Eq. (1.4.2).]

We can numerically evaluate the Lyapunov exponent by averaging over a number of orbits nearby to a given initial condition, all with small $|\mathbf{d}_0|$. Then by plotting the right-hand side of Eq. (1.4.2) as a function of time for $0 < t < 50$, we can observe that this function asymptotes to a constant value, equal to λ.

We will do this for our previous example of pendulum motion using the following *Mathematica* statements. In keeping with the notation of this subsection, we use the notation (x, v) for the pendulum phase space, rather than (θ, θ').

First, we create a test trajectory $\mathbf{z}(t, \mathbf{z}_0)$ using the initial conditions, $x(0) = -0.5, v(0) = 0$:

Cell 1.23

```
z =
  {x[t], v[t]}/. NDSolve[{x'[t] == v[t],
   v'[t] == -Sin [x[t]] - Sin [x[t] - 2t], x[0] == -0.5,
     v[0] == 0}, {x[t], v[t]}, {t, 0, 50}] [[1]];
```

This trajectory is the same as the lower trajectory shown in Fig. 1.9. Next, we create 40 nearby initial conditions by choosing values of $x(0)$ and $v(0)$ scattered randomly around the point $(-0.5, 0)$:

Cell 1.24

```
z0 = Table [{-0.5 + 10^-5 (2 Random[] - 1),
     10^-5 (2 Random[] -1) }, {m, 1, 40}];
```

Then, we integrate these initial conditions forward in time using **NDSolve**, and evaluate the vector displacement between the resulting trajectories and the test trajectory **z**:

Cell 1.25

```
Δz[t_] = Table[Sqrt[({x[t], v[t]} - z).({x[t], v[t]} - z)]/.
NDSolve[{x'[t] == v[t], v'[t] == -Sin[x[t]] - Sin[x[t] - 2t],
x[0] == z0 [[m, 1]], v[0] == z0[[m, 2]]}, {x[t], v[t]},
{t, 0, 50}][[1]], {m, 1, 40}];
```

Finally, we evaluate $\ln[\Delta z(t)/\Delta z(0)]/t$, averaged over the 40 trajectories, and plot the result in Cell 1.26.

Cell 1.26

```
λ[t_] = 1/40 Sum[Log[Δz[t][[n]]/Δz[0][[n]]], {n, 1, 40}]/t;
Plot[λ[t], {t, 0, 50}, PlotRange→{0, 0.5},
  AxesLabel→{"t", "λ(t)"}];
```

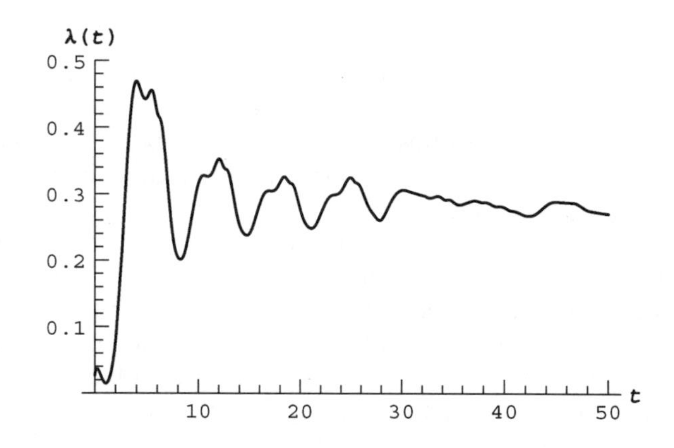

The Lyapunov exponent can be seen to asymptote to a fairly constant value of about 0.3 at large times. Fluctuations in the result can be reduced by keeping more trajectories in the average. Thus, over a time $t = 50$, nearby orbits diverge in phase space by a factor of $e^{0.3\times 50} = 3.\times 10^6$, on average. Over a time $t = 500$, initially nearby orbits diverge by the huge factor of 10^{65}. Even a tiny initial error gets blown up to a massive deviation over this length of time, leading to complete loss of predictability.

Note that the choice of $\mathbf{d}_0$ in the above numerical work is a rather tricky business. It must be larger than the intrinsic error of the numerical method; otherwise the effect of $\mathbf{d}_0$ on the trajectory is swamped by numerical error. But on the other hand, $\mathbf{d}_0$ must not be so large that orbits diverge by a large amount over the plotted time interval; otherwise we are not properly evaluating the difference between infinitesimally nearby trajectories; that is, we require $|\mathbf{d}_0|e^{\lambda t} \ll 1$. As a result of these trade-offs, this method for determining λ is not particularly accurate, but it will do for our purposes. More accurate (and complicated) methods exist for determining a Lyapunov exponent: see, for example, Lichtenberg and Lieberman (1992, p. 315).

1.4.3 Euler's Method

In this section, we will create our own ODE solver by first considering a simple (and crude) numerical method called *Euler's method*. Most ODE solvers are at their core merely more complex and accurate versions of Euler's method, so we will begin by examining this simplest of numerical ODE solvers. We will then consider more complicated numerical methods.

Euler's method applies to the simple first-order ODE discussed in Sec. 1.2:

$$\frac{dv}{dt} = f(t, v), \qquad v(0) = v_0. \tag{1.4.3}$$

Later, we will see how to modify Euler's method to apply to more general ODEs.

To apply Euler's method to Eq. (1.4.3), we first *discretize* the time variable, defining evenly spaced discrete *timesteps*

$$t_n = n\,\Delta t, \qquad n = 0, 1, 2, 3, \dots. \tag{1.4.4}$$

The quantity Δt is called *step size*. (Confusingly, some authors also refer to Δt as the timestep.) We will evaluate the solution $v(t)$ *only* at the discrete timesteps given in Eq. (1.4.4). Later, we will interpolate to find $v(t)$ at times between the timesteps.

Next, we integrate both sides of Eq. (1.4.3), and apply the fundamental theorem of calculus:

$$\int_{t_{n-1}}^{t_n} \frac{dv}{dt}\,dt = v(t_n) - v(t_{n-1}) = \int_{t_{n-1}}^{t_n} f(t, v(t))\,dt. \tag{1.4.5}$$

Note that we must take account of the time variation of $v(t)$ in the integral over f on the right-hand side.

So far, no approximation has been made. However, we will now approximate the integral over f, assuming that Δt is so small that f does not vary appreciably over the range of integration:

$$\int_{t_{n-1}}^{t_n} f(t, v(t))\,dt \approx \Delta t f(t_{n-1}, v(t_{n-1})) + O(\Delta t^2). \tag{1.4.6}$$

The error in this approximation scales as Δt^2 (see the exercises), and we use the notation $O(\Delta t^2)$ to indicate this fact. The same notation is, used in power series expansions, and indicates that if Δt is reduced by a factor of 2, the error in Eq. (1.4.6) is reduced by a factor of 4 (for small Δt).

Equation (1.4.6) is a very crude approximation to the integral, but it has the distinct advantage that, when used in Eq. (1.4.5), the result is a simple *recursion relation* for $v(t_n)$:

$$v(t_n) = v(t_{n-1}) + \Delta t f(t_{n-1}, v(t_{n-1})) + O(\Delta t^2). \tag{1.4.7}$$

Equation (1.4.7) is Euler's method. It is called a recursion relation because the value of v at the nth step is determined by quantities from the $n-1$st step, which were themselves determined in terms of variables at the $n-2$nd step, and so on back to the initial condition at $t=0$. Recursion relations like this are at the heart of most numerical ODE solvers. Their differences lie mainly in the degree of approximation to the integral in Eq. (1.4.6).

To see how Euler's method works, we will write a program that can be used to solve any ODE of the form of Eq. (1.4.3). In our code, we will employ the following simple notation for the velocity at timestep n: a function **v[n]**, defined for integer arguments. We will employ the same notation for the time t_n defining a function **t[n]=nΔt**. Then Euler's method can be written in *Mathematica* as

Cell 1.27

```
t[n_] := n Δt;
v[n_] := v[n-1] + Δt f[t[n-1], v[n-1]];
v[0] := v0
```

The first defines the time at step n, the second is Eq. (1.4.7), and the third is the initial condition. Note that delayed evaluation is used for all three lines, since we have not yet specified a step size Δt, an initial condition v_0, or the function f. However, even if these quantities are already specified, delayed evaluation *must be used* in the second line, since it is a recursion relation: **v[n]** is determined in terms of previous values of **v**, and can therefore only be evaluated for a given specific integer value of **n**.

Note that it is somewhat dangerous to write the code in the form given in Cell 1.27, because it is up to us to ask only for nonnegative integer values of n. If, for example, we ask for **v[0.5]**, the second line will evaluate this in terms of **v[-0.5]**, which is then evaluated in terms of **v[-1.5]**, etc., leading to an *infinite recursion*:

Cell 1.28

```
v[0.5]

$RecursionLimit :: reclim : Recursion depth of 256 exceeded.

$RecursionLimit :: reclim : Recursion depth of 256 exceeded.

$RecursionLimit :: reclim : Recursion depth of 256 exceeded.

General :: stop : Further output of
   $RecursionLimit :: reclim will be suppressed during this
   calculation.
```

In such errors, the kernel will often grind away fruitlessly for many minutes trying to evaluate the recursive tree, and the only way to stop the process is to quit the kernel. We can improve the code by adding conditions to the definition of **v[n]** that require **n** to be a positive integer:

Cell 1.29

```
Clear[v]
```

Cell 1.30

```
t[n_] := nΔt;
v[n_] := v[n-1] + Δt f[t[n-1], v[n-1]]/;
  n>0 && n∈Integers;
v[0] := v0
```

Here we have used the statement **n∈Integers**, which stands for the logical statement "n is an element of the integers," evaluating to a result of either **True** or **False**. The symbol ∈ stands for the intrinsic function **Element** and is available on the BasicInput palette. (Don't confuse ∈ with the Greek letter epsilon, ϵ.)

If we now ask for **v[0.5]**, there is no error because we have only defined **v[n]** for positive integer argument:

Cell 1.31

```
v[0.5]

v[0.5]
```

In principle, we could now run this code simply by asking for any value of **v[n]** for **n∈Integers** and **n>0**. *Mathematica* will then evaluate **v[n-1]** in terms of **v[n-2]**, and so on until it reaches **v[0]=v0**. The code stops here because the definition **v[0]=v0** takes precedence over the recursion relation.

However, there are a few pitfalls that should be avoided. First, it would not be a good idea to begin evaluating the code right now. We have not yet defined the function f, the step size Δt, or the initial condition v_0. Although *Mathematica* will return perfectly valid results if we ask for, say, **v[2]**, the result will be a complicated algebraic expression without much value. If we ask for **v[100]**, the result will be so long and complicated that we will probably have to abort the evaluation. Numerical methods are really made for solving *specific numerical instances* of the ODE in question.

Therefore, let us solve the following specific problem, which we encountered in Sec. 1.2.1:

$$f(t,v) = t - v, \qquad v(0) = 0. \tag{1.4.8}$$

The general solution was given in Eq. (1.2.5), and for $v(0) = 0$ is

$$v(t) = t + e^{-t} - 1. \tag{1.4.9}$$

Before we solve this problem using Euler's method, there is another pitfall that can be avoided by making a small change in the code. As it stands, the code will work, but it will be very slow, particularly if we ask for **v[n]** with **n≫1**. The reason is that every time we ask for **v[n]**, it evaluates the recursion relations all the way back to **v[0]**, even if it has previously evaluated the values of **v[n-1]**, **v[n-2]**, etc. This wastes time. It is better to make *Mathematica* remember values of the function **v[n]** that it has evaluated previously. This can be done as follows:

in the second line of the code, which specifies **v[n]**, write two equal signs:

Cell 1.32

```
v[n_] := ( v[n] = v[n-1] + Δt f[t[n-1], v[n-1]])/;
          n>0 && n∈Integers;
```

The second equal sign causes *Mathematica* to remember any value of **v[n]** that is evaluated by adding this value to the definition of **v**; then, if this value is asked for again, *Mathematica* uses this result rather than reevaluating the equation. Note that we have placed parentheses around part of the right-hand side of the equation. These parentheses must be included when the equation has conditions; otherwise the condition statement will not evaluate properly, because it will attach itself only to the second equality, not the first.

The modified Euler code is as follows:

Cell 1.33

```
t[n_] := n Δt;
v[n_] := (v[n] =
            v[n-1] + Δt f[t[n-1], v[n-1]])/;
          n >0 && n∈Integers;
v[0] := v0
```

Let's now evaluate a solution numerically, from $0 < t < 4$. To do so, first specify the step size, the function f, and the initial condition:

Cell 1.34

```
Δt = 0.2;
f[t_, v_] = t-v;
v0 = 0;
```

Next, make a list of data points **{t[n],v[n]}**, calling this result our numerical **solution**:

Cell 1.35

```
solution = Table[ {t[n], v[n]}, {n, 0, 4/Δt}]

{{0, 0}, {0.2, 0}, {0.4, 0.04}, {0.6, 0.112}, {0.8, 0.2096},
  {1., 0.32768}, {1.2, 0.462144}, {1.4, 0.609715},
  {1.6, 0.767772}, {1.8, 0.934218}, {2., 1.10737},
  {2.2, 1.2859}, {2.4, 1.46872}, {2.6, 1.65498},
  {2.8, 1.84398}, {3., 2.03518}, {3.2, 2.22815},
  {3.4, 2.42252}, {3.6, 2.61801}, {3.8, 2.81441},
  {4., 3.01153}}
```

Finally, plot these points with a **ListPlot**, and compare this Euler solution with the analytic solution of Eq. (1.4.9), by overlaying the two solutions in Cell 1.36. The Euler solution, shown by the dots, is quite close to the exact solution, shown by the solid line.

Cell 1.36

```
a =
  ListPlot[solution, PlotStyle→PointSize [0.015],
   DisplayFunction→Identity];
b = Plot[E^-t + t -1, {t, 0, 4},
   DisplayFunction→Identity];
Show[a, b, DisplayFunction→$DisplayFunction];
```

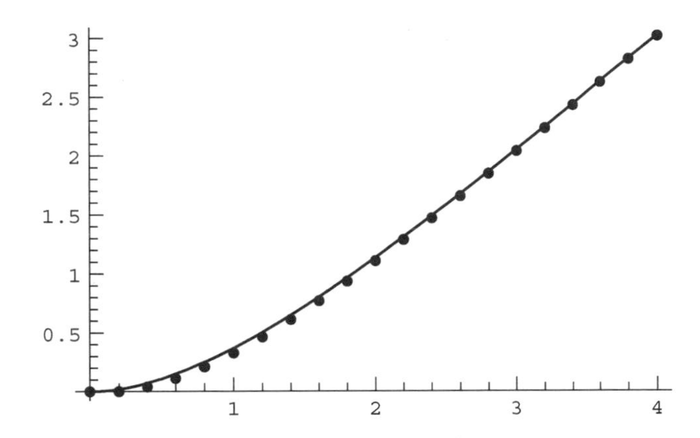

We have used an option in the **Plot** functions to turn off intermediate plots and thereby save space. The option **DisplayFunction→Identity** creates a plot, but does not display the result. After the plots are overlaid with the **Show** command, the display option was turned on again using **DisplayFunction→$DisplayFunction**.

If we wish to obtain the numerical solution at times between the timesteps, we can apply an interpolation to the data and define a numerical function **vEuler[t]**:

Cell 1.37

```
vEuler[t_] = Interpolation[solution][t]
InterpolatingFunction [{{0., 4.}}, <>][t]
```

One thing that we can do with this function is plot the difference between the numerical solution and the exact solution to see the error in the numerical method (see Cell 1.38).

Cell 1.38

```
vExact[t_] = E^-t + t - 1;
pl = Plot[vEuler[t] - vExact[t], {t, 0, 4}];
```

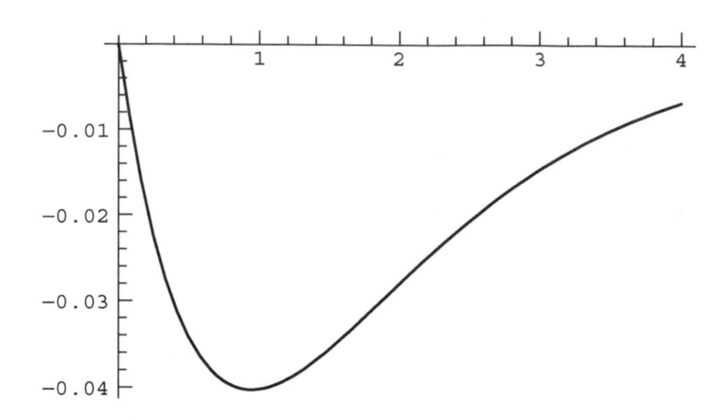

The error can be reduced by reducing the step size. To do this, we must go back and run the **Table** command again after setting the step size to a smaller value, and after applying the **Clear[v]** command. *We must* **Clear[v]** *before running the* **Table** *command again; otherwise the values of* **v[1],v[2],...,** *stored in the kernel's memory as a result of our using two equal signs in Eq.* (1.4.10), *will supersede the new evaluations.* After clearing **v**, we must then reevaluate the definition of **v** in Cell 2.33.

All of these reevaluations are starting to seem like work. There is a way to avoid having to reevaluate groups of cells over and over again. We can create a **Module**, which is a method of grouping a number of commands together to create a *Mathematica* function. Modules are the *Mathematica* version of C+ modules or Fortran subroutines, and have the following syntax:

Module[{*internal variables*}, *statements*] creates a module in *Mathematica*

The list of internal variables defines variables that are used only within the module. The definitions of these variables will not be remembered outside of the module.

Here is a version of the Euler solution that is written as a module, and assigned to a function **Eulersol[v0, time, Δt]**. This function finds the approximate solution **vEuler[t]** for $0 < t <$ time, with step size Δt. To use the module, all we need to do is specify the function $f(t, v)$ that enters the differential equation:

Cell 1.39

```
Eulersol[v0_, time_, Δt_] := Module[{t, v, solution},
  t[n_] := n Δt;
   v[n_]:= (v[n] =
            v[n-1] + Δt f[t[n-1], v[n-1]])/;
         n>0 && n∈Integers;
v[0] := v0;
solution = Table[{t[n], v[n]}, {n, 0, time/Δt}];
vEuler[t_] = Interpolation[solution][t];]
```

Note that we did not have to add a **Clear[v]** statement to the list of commands, because **v** is an internal variable that is not remembered outside the module, and is also not remembered from one application of the module to the next. Also, note that we don't really need the condition statements in the definition of **v[n]** anymore, since we only evaluate **v[n]** at positive integers, and the definition does not exist outside the **Module**.

Below we show a plot of the error vs. time as Δt is reduced to 0.1 and then to 0.05. The plot was made simply by running the **Eulersol** function at these two values of Δt and plotting the resulting error, then superimposing the results along with the original error plot at $\Delta t = 0.2$. As Δt decreases by factors of 2, the error can be seen to decrease roughly by factors of 2 as well. The error in the solution scales linearly with Δt: Error $\propto \Delta t$. In other words, the error is first order in Δt. (The same language is used in the discussion of power series expansions; see Sec. 9.9.2.) Euler's method is called a *first-order method.*

Cell 1.40

```
Eulersol[0, 4, 0.1];
p2 = Plot[vEuler[t]-vExact [t], {t, 0, 4},
     DisplayFunction→ Identity];
Eulersol[0, 4, 0.05];
p3 = Plot[vEuler[t]-vExact[t], {t, 0, 4},
     DisplayFunction→ Identity];
Show [p1, p2, p3, DisplayFunction→ $DisplayFunction,
  PlotLabel→ "Error for Δt = 0.2,0.1,0.05"];
```

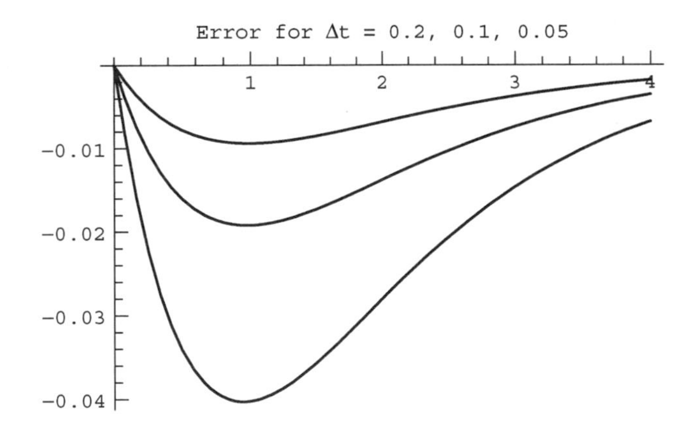

One can see why the error in this method is $O(\Delta t)$ from Eq. (1.4.7): the error in a single step is of order Δt^2. To integrate the solution over a fixed time interval T, N steps must be taken, with $N = T/\Delta t$ increasing as Δt decreases. The total error is the sum of all individual errors, and therefore scales as $N\Delta t^2 = T\Delta t$.

Euler's method is too crude to be of much practical use today. Clearly it would be a great improvement in efficiency if we could somehow modify Euler's method so that it is second-order, or even nth-order, with error scaling like Δt^n. Then, by reducing the step size by only a factor of 2, the error would be reduced by a factor of 2^n. In the next section we will see how to easily modify Euler's method to make it second order.

> The error of a numerical solution to an ODE is controlled by the step size Δt. Reducing the step size increases the accuracy of the solution, but also increases the number of steps required to find the solution over a fixed interval T. For a method with error that is of order n, the error in the solution, found over a fixed time interval T, scales like $(\Delta t)^n$.

1.4.4 The Predictor–Corrector Method of Order 2

The error in Euler's method arose from the crude approximation to the integral in Eq. (1.4.6). To improve the approximation, we need a more accurate value for this integral. Now, the required integral is just the area under the function $f(v(t), t)$, shown schematically in Fig. 1.11(a). Equation (1.4.6) approximates this area by the gray rectangle in Fig. 1.11(a), which is clearly a rather poor approximation to the area under the curve, if the function varies much over the step size Δt. A better approximation would be to use the average value of the function at the initial and final points in determining the area:

$$\int_{t_{n-1}}^{t_n} f(t, v(t))\, dt \approx \Delta t \frac{f(t_{n-1}, v(t_{n-1})) + f(t_n, v(t_n))}{2} + O(\Delta t^3). \quad (1.4.10)$$

This approximation would be exactly right if the shaded area above the curve in Fig. 1.11(b) equaled the unshaded area below the curve. If $f(v(t), t)$ were a linear function of t over this range, that would be true, and there would be no error. For Δt sufficiently small, $f(v(t), t)$ will be nearly linear in t if it is a smooth function of t, so for small Δt the error is small. In fact, one can easily show that the error in this approximation to the integral is of order Δt^3 (see the exercises at the end of this section), as opposed to the order-Δt^2 error made in a single step of the Euler's method [see Eq. (1.4.6)]. Therefore, this modification to Euler's method should improve the accuracy of the code to order Δt^3 in a single step.

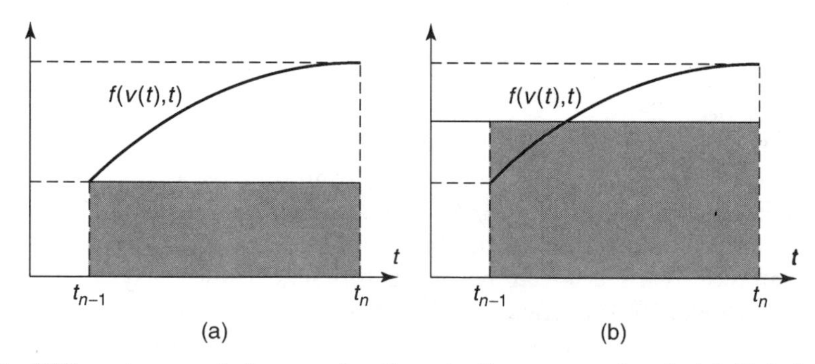

Fig. 1.11 Different numerical approximations to the area under f: (a) Euler's method, Eq. (1.4.6); (b) modified Euler's method, Eq. (1.4.10).

If we now use Eq. (1.4.10) in Eq. (1.4.5), we obtain the following result:

$$v(t_n) = v(t_{n-1}) + \Delta t \frac{f(t_{n-1}, v(t_{n-1})) + f(t_n, v(t_n))}{2} + O(\Delta t^3). \quad (1.4.11)$$

Since the error of the method is order Δt^3 in a single step, Eq. (1.4.11) is a distinct improvement over Euler's method, Eq. (1.4.7). However, there is a catch. Now $v(t_n)$ appears on the right-hand side of the recursion relation, so we can't use this equation as it stands to solve for $v(t_n)$. [We might try to solve this equation for $v(t_n)$, but for general f that is nontrivial. Such methods are called *implicit methods*, and will be discussed in Chapter 6.]

What we need is some way to replace $v(t_n)$ on the right-hand side: we need a *prediction* for the value of $v(t_n)$, which we will then use in Eq. (1.4.11) to get a better value. Fortunately, we have such a prediction available: Euler's method, Eq. (1.4.7), provides an approximation to $v(t_n)$, good to order Δt^2. This is sufficient for Eq. (1.4.11), since the $O(\Delta t^2)$ error in $v(t_n)$ is multiplied in Eq. (1.4.11) by another factor of Δt, making this error $O(\Delta t^3)$; but the right-hand side of Eq. (1.4.11) is already accurate only to $O(\Delta t^3)$.

The resulting recursion relation is called a *predictor–corrector method of order* 2. The method is second-order accurate, because over a fixed time interval T the number of steps taken is $T/\Delta t$ and the total error scales as $(T/\Delta t)\,\Delta t^3 = T\,\Delta t^2$. The method consists of the following two lines: an initial prediction for v at the nth step, which we assign to a variable v_1, and the improved correction step, given by Eq. (1.4.11), making use of the prediction:

$$\begin{aligned} v_1 &= v(t_{n-1}) + \Delta t f(t_{n-1}, v(t_{n-1})), \\ v(t_n) &= v(t_{n-1}) + \Delta t \frac{f(t_{n-1}, v(t_{n-1})) + f(t_n, v_1)}{2}. \end{aligned} \quad (1.4.12)$$

The following module, named **PCsol**, implements the predictor–corrector method in *Mathematica*:

Cell 1.41

```
PCsol[v0_, time_, Δt_] := Module[{t, v, f0, v1, solution},
  t[n_] = nΔt;
  v[0] = v0;
  f0 := f[t[n-1], v[n-1]];
  v1 := v[n-1] + Δt f0;
  v[n_] := v[n] = v[n-1] + Δt (f0 + f[t[n], v1])/2;
  solution = Table[{t[n], v[n]}, {n, 0, time/Δt}];
  vPC[t_] = Interpolation[solution][t];]
```

There is one extra trick that we have implemented in this module. We have assigned the value of f at the n-1st step to the variable **f0** (using delayed evaluation so that it is evaluated only when needed). The reason for doing so is that we used this value for f twice in Eq. (1.4.12). Rather than evaluating the function twice at the same point, we instead save its value in the variable **f0**. This does not save much time for simple functions, but can be a real time-saver if f is very complicated.

Also, note that we have paid a price in going to a second-order method. The code is more complicated, and we now need to evaluate the function at two points, instead of only one as we did in Euler's method.

But we have also gained something—accuracy. This relatively simple predictor–corrector method is much more accurate than Euler's method, as we can see by again evaluating the solution for three step sizes $\Delta t = \{0.2, 0.1, 0.05\}$, and plotting the error. We again choose our previous example:

Cell 1.42

```
f[t_, v_] = t - v;
vExact[t_] = E ^ -t + t - 1;
```

The resulting error is shown in Cell 1.43. Not only is the error much smaller than in Euler's method for the same step size, but the error also decreases much more rapidly as Δt is decreased. The maximum error goes from roughly 0.0029 to 0.0007 to 0.00017 as Δt goes from 0.2 to 0.1 to 0.05. In other words, the maximum error is reduced by roughly a factor of 4 every time Δt is reduced by a factor of 2. This is exactly what we expect for error that is $O(\Delta t^2)$.

Cell 1.43

```
PCsol[0, 4, 0.2];
p1 = Plot[vPC[t]-vExact [t], {t, 0, 4},
     DisplayFunction→ Identity];
PCsol[0, 4, 0.1];
p2 = Plot[vPC[t]-vExact [t], {t, 0, 4},
     DisplayFunction→ Identity];
PCsol[0, 4, 0.05];
p3 = Plot[vPC[t]-vExact [t], {t, 0, 4},
     DisplayFunction→ Identity];
Show[p1, p2, p3, DisplayFunction→ $DisplayFunction,
  PlotLabel→ "Error for Δt = 0.2,0.1,0.05"];
```

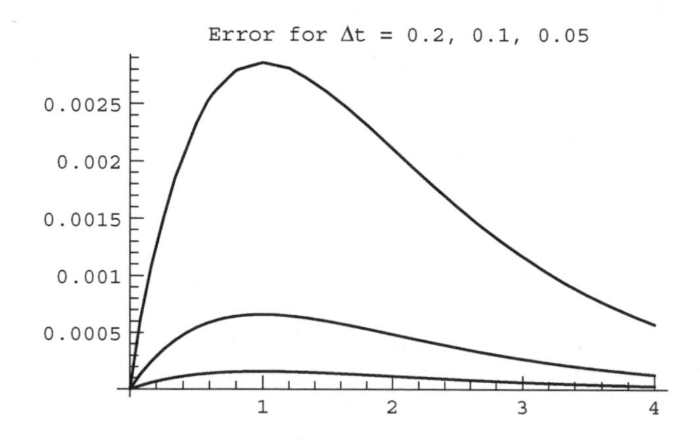

There are many higher-order methods that are even more accurate than this. Two of the more popular methods are the fourth-order Runge–Kutta method and the Bulirsch–Stoer method. These methods will not be discussed here, but the codes can be found in many other textbooks. See, for instance, Press et al. (1986).

Also, there are several second-order (and higher-order) methods that require only one force evaluation per timestep. These algorithms can be more efficient when the force evaluation is time-consuming. Three such methods are considered in the exercises: the leapfrog method and a centered-difference method for problems in Newtonian mechanics, and the Adams–Bashforth method for more general problems.

1.4.5 Euler's Method for Systems of ODEs

Consider the general second-order ODE

$$\frac{d^2x}{dt^2} = f\left(t, x, \frac{dx}{dt}\right), \qquad x(0) = x_0, \quad \frac{dx}{dt}(0) = v_0. \tag{1.4.13}$$

Since the ODE is second-order, Euler's method cannot be used to solve it numerically. However, we can modify the equation so that Euler's method can be used. By introducing a new variable $v(t) = dx/dt$, Eq. (1.4.13) can be written as the following system of first-order differential equations:

$$\frac{dx}{dt} = v(t), \quad \frac{dv}{dt} = f(t, x, v), \qquad x(0) = x_0, \quad v(0) = v_0. \tag{1.4.14}$$

Euler's method still does not apply, because it was written originally for a single first-order ODE. However, let us define a vector $\mathbf{z}(t) = \{x(t), v(t)\}$. Then Eqs. (1.4.14) can be written as a *vector ODE*:

$$\frac{d\mathbf{z}}{dt} = \mathbf{f}(t, \mathbf{z}), \qquad \mathbf{z}(0) = \mathbf{z}_0, \tag{1.4.15}$$

where $\mathbf{z}_0 = \{x_0, v_0\}$, and the vector function $\mathbf{f}(t, \mathbf{z})$ is defined as

$$\mathbf{f}(t, \mathbf{z}) = \{v(t), f(t, x, v)\}. \tag{1.4.16}$$

We can now apply Euler's method to this vector ODE, simply by reinterpreting the scalar quantities that appeared in Eq. (1.4.7) as vectors:

$$\mathbf{z}(t_n) = \mathbf{z}(t_{n-1}) + \Delta t\, \mathbf{f}(t_{n-1}, \mathbf{z}(t_{n-1})). \tag{1.4.17}$$

In fact, there is nothing about Eqs. (1.4.15) and (1.4.17) that limits them to two-dimensional vectors. An Nth-order ODE of the general form given by Eq. (1.2.1) can also be written in the form of Eq. (1.4.15) by defining a series of new variables

$$v(t) = \frac{dx}{dt}, \qquad a(t) = \frac{d^2x}{dt^2}, \dots, \qquad u(t) = \frac{d^{N-1}x}{dt^{N-1}}, \tag{1.4.18}$$

a vector

$$\mathbf{z}(t) = \{x(t), v(t), a(t), \ldots, u(t)\}, \tag{1.4.19}$$

and a force

$$\mathbf{f}(t, \mathbf{z}) = \{v, a, \ldots, u, f(t, x, v, a, \ldots, u)\}. \tag{1.4.20}$$

Thus, Euler's method in vector form, Eq. (1.4.17), can be applied to a general Nth-order ODE. Below, we provide the simple changes to the previous module **Eulersol** that allow it to work for a general ODE of order N:

Cell 1.44

```
Clear["Global`*"]
```

Cell 1.45

```
Eulersol[z0_, time_, Δt_] := Module[ {t, z, sol},
  t[n_] := n Δt;
  z[n_] := z[n] = z[n-1] + Δt f[t[n-1], z[n-1]];
  z[0] := z0;
  sol = Table[Table[{t[n], z[n][[m]]}, {n, 0, time/Δt}],
        {m, 1, Length[z0]}];
  zEuler = Table[Interpolation[sol[[m]]],
           {m, 1, Length[z0]}];]
```

Thanks to the ease with which *Mathematica* handles vector arithmetic, the module is nearly identical to the previous scalar version of the Euler method. In fact, except for renaming some variables, the first four lines are identical. Only the lines involving creation of the interpolating functions differ. This is because the solution list **sol** is created as a table of lists, each of which is a dataset of the form **{t[n], z**$_m$**[n]}**. Each element of **zEuler** is an interpolation of a component of **z**.

To use this module, we must first define a *force vector* $\mathbf{f}(t, z)$. Let's take the case of the 1D harmonic oscillator problem as an example. In this case $\mathbf{z} = (x, v)$ and $\mathbf{f} = (v, -x)$ (i.e. $dx/dt = v$, $dv/dt = -x$):

Cell 1.46

```
f[t_, z_] := {v, -x}/.{x→z[[1]], v→z[[2]]}
```

A delayed equality must be used in defining **f**; otherwise *Mathematica* will attempt to find the two elements of **z** when making the substitution, and this will lead to an error, since **z** has not been defined as a list yet.

Taking the initial condition $\mathbf{z}_0 = (1, 0)$ (i.e., $x_0 = 1, v_0 = 0$), in Cell 1.47 we run the Euler code and in Cell 1.48 plot the solution for $x(t)$, which is the first element of **zEuler**.

Cell 1.47

```
Eulersol[{1, 0}, 10, .02]
```

Cell 1.48

```
Plot[zEuler[[1]][t], {t, 0, 10}];
```

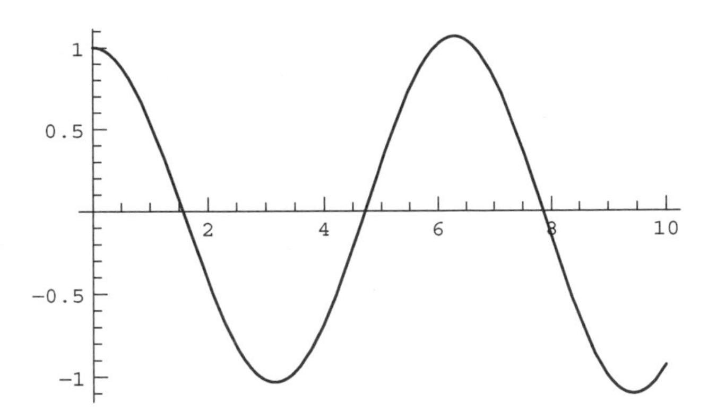

The code clearly works, but a keen eye can see that the expected cosine oscillations are actually growing slowly with time, even at the relatively small step size of 0.02. As already discussed, the Euler method is only first-order accurate. Nevertheless, the general methods discussed in this section also work for higher-order methods, such as the predictor–corrector code of the previous section. Examples may be found in the exercises at the end of the section.

1.4.6 The Numerical *N*-Body Problem: An Introduction to Molecular Dynamics

One way that systems of ODEs arise in the physical sciences is in the description of the motion of N interacting classical particles. Newton solved this problem for the case of two particles (the two-body problem) interacting via a central force. However, for three or more particles there is no general analytic solution, and numerical techniques are of great importance in understanding the motion.

In the numerical method known as *molecular dynamics*, the coupled equations of motion for the N particles are simply integrated forward in time using Newton's second law for each particle. There is nothing subtle about this—the numerical techniques learned in the previous sections are simply applied on a larger scale. The subtleties only arise when details such as error accumulation, code efficiency, and the like must be considered.

Below, we show how to use *Mathematica's* intrinsic function **NDSolve** to numerically solve the following N-body problem: For particles at positions $\mathbf{r}_1(t), \mathbf{r}_2(t), \dots, \mathbf{r}_N(t)$, the equations of motion are

$$m_i \frac{d^2 \mathbf{r}_i}{dt^2} = \sum_{\substack{j=1 \\ j \neq i}}^{N} \mathbf{F}_{ij}(\mathbf{r}_i - \mathbf{r}_j), \tag{1.4.21}$$

where $\mathbf{F}_{ij}$ is the force between particles i and j, and m_i is the mass of particle i.

To complete the problem, we must also specify initial conditions on position and velocity:

$$\mathbf{r}_i(0) = \mathbf{r}_{i0}, \quad \frac{d\mathbf{r}_i}{dt} = \mathbf{v}_{i0}, \qquad i = 1, \dots, n. \tag{1.4.22}$$

The following module **MDsol** solves this problem numerically:

Cell 1.49

```
Clear["Global`*"]
```

Cell 1.50

```
MDsol[z0_, time_] := Module[{},
  npart = Length[z0]/6;
  r[i_, t_] = {x[i][t], y[i][t], z[i][t]};
  v[i_, t_] = {vx[i][t], vy[i][t], vz[i][t]};
  Z[t_] = Flatten[Table[{r[i, t], v[i, t]}, {i, 1, npart}]];
  f[t_] = Flatten[Table[{v[i, t],
       (Sum[F[i, j, r[i, t]-r[j, t]], {j, 1, i-1}] +
           Sum[F[i, j, r[i, t]-r[j, t]],
               {j, i + 1, npart}]}/mass[[i]]},
     {i, 1, npart}]];
  ODEs = Flatten[Table[Z'[t][[n]] == f[t][[n]],
        {n, 1, 6 * npart}]];
  ics = Table[Z[0][[n]] == z0[[n]], {n, 1, 6*npart}];
  eqns = Join[ODEs, ics] ;
  NDSolve[eqns, Z[t], {t, 0, time}, MaxSteps→10^5]]
```

To understand what this module does, look at the last line. Here we see that **NDSolve** is used to integrate a list of equations called **eqns**, that the equations involve a vector of unknown functions **Z[t]**, and that the equations are integrated from **t=0** to **t=time**. The definition of **Z[t]** can be found a few lines higher in the module: it is a list of variables, **{r[i,t],v[i,t]}**. The ith particle position vector **r[i,t]** is defined in the second line as having components **{x[i][t],y[i][t],z[i][t]}**, and the velocity vector **v[i,t]** has components **{vx[i][t],vy[i][t],vz[i][t]}**. These functions use a notation we haven't seen before: the notation **x[i][t]** means the same thing as **x[i,t]**. The reason we use the former and not the latter notation is due to a vagary of **NDSolve**: **NDSolve** likes to work on functions of one variable; otherwise it gets confused and thinks it is solving a PDE. The notation **x[i][t]** fools **NDSolve** into thinking of **x** as a function of a single argument, the time **t**.

The **Flatten** function is used in the definition of **Z[t]** because **NDSolve** works only on a simple list of unknown functions, without sublists.

The list **eqns** can be seen to be a concatenation of a list of ODEs called **ODEs** and a list of initial conditions called **ics**. The initial conditions are given as an argument to the module, in terms of a list $\mathbf{z}_0$ of positions and velocities for each particle of the form

$$\mathbf{z}_0 = \texttt{Flatten[}\{\mathbf{r}_{10}, \mathbf{v}_{10}, \mathbf{r}_{20}, \mathbf{v}_{20}, \dots\}\texttt{]}.$$

The **Flatten** command is included explicitly here to ensure that $\mathbf{z}_0$ is a simple list, not an array.

The acceleration function $\mathbf{f}(t)$ is defined as

$$\mathbf{f}(t) = \left\{ \mathbf{v}_1, \frac{1}{m_1} \sum_{j=2}^{N} \mathbf{F}_{1j}(\mathbf{r}_1 - \mathbf{r}_j), \mathbf{v}_2, \frac{1}{m^2} \sum_{\substack{j=1 \\ j\neq 2}}^{N} \mathbf{F}_{2j}(\mathbf{r}_2 - \mathbf{r}_j), \ldots \right\}, \qquad (1.4.23)$$

and the result is flattened in order for each element to correspond to the proper element in **z**. The fact that the sum over individual forces must neglect the self-force term $j = i$ requires us to write the sum as two pieces, one from $j = 1$ to $i - 1$, and the other from $j = i + 1$ to N.

The value of N, called **npart** in the module (because **N** is a reserved function name), is determined in terms of the length of the initial condition vector **z0** in the first line of the code.

Finally, the module itself is given no internal variables (the internal-variable list is the null set { }), so that we can examine each variable if we wish.

In order to use this code, we must first define a force function $\mathbf{F}_{ij}(\mathbf{r})$. Let's consider the gravitational N-body problem, where the force obeys Newton's law of gravitation:

$$\mathbf{F}_{ij}(\mathbf{r}) = -Gm_i m_j \mathbf{r}/r^3, \qquad (1.4.24)$$

where $G = 6.67 \times 10^{-11}$ m^3/kg s^2. We can define this force using the command

Cell 1.51

```
F[i_, j_, r_] := -mass[[i]] mass[[j]] r/(r.r)^(3/2)
```

Here **mass** is a length-N list of the masses of all particles, and we have set the gravitational force constant $G = 1$ for simplicity.

Let's apply this molecular dynamics code to the simple problem of two gravitating bodies orbiting around one another. For initial conditions we will choose $\mathbf{r}_1 = \mathbf{v}_1 = \mathbf{0}$, and $\mathbf{r}_2 = (1, 0, 0), \mathbf{v}_2 = (0, 0.5, 0)$. Thus, the list of initial conditions is

Cell 1.52

```
z0 = Flatten[{{0, 0, 0, 0, 0, 0}, {1, 0, 0, 0, 0.5, 0}}]

{0, 0, 0, 0, 0, 0, 1, 0, 0, 0, 0.5, 0}
```

Also, we must not forget to assign masses to the two particles. Let's take one mass 3 times the other:

Cell 1.53

```
mass = {3, 1}

{3, 1}
```

Now we run the code for $0 < t < 4$:

Cell 1.54

```
S = MDsol[z0, 4]
```

```
{{x[1][t] → InterpolatingFunction[{{0., 4.}}, <>][t],
  y[1][t] → InterpolatingFunction[{{0., 4.}}, <>][t],
  z[1][t] → InterpolatingFunction[{{0., 4.}}, <>][t],
  vx[1][t] → InterpolatingFunction[{{0., 4.}}, <>][t],
  vy[1][t] → InterpolatingFunction[{{0., 4.}}, <>][t],
  vz[1][t] → InterpolatingFunction[{{0., 4.}}, <>][t],
  x[2][t] → InterpolatingFunction[{{0., 4.}}, <>][t],
  y[2][t] → InterpolatingFunction[{{0., 4.}}, <>][t],
  z[2][t] → InterpolatingFunction[{{0., 4.}}, <>][t],
  vx[2][t] → InterpolatingFunction[{{0., 4.}}, <>][t],
  vy[2][t] → InterpolatingFunction[{{0., 4.}}, <>][t],
  vz[2][t] → InterpolatingFunction[{{0., 4.}}, <>][t]}}
```

The result is a list of interpolating functions for each component of position and velocity of the two particles. We can do whatever we wish with these—perform more analysis, make plots, etc. One thing that is fun to do (but is difficult to show in a textbook) is to make an animation of the motion. An example is displayed in the electronic version of the book, plotting the (xy) positions of the two masses at a series of separate times. (The animation can be viewed by selecting the plot cells and choosing **Animate Selected Graphics** from the **Format** menu.) Only the command that creates the animation is given in the hard copy, in Cell 1.55.

Cell 1.55

```
Table[ListPlot[Table[{x[n][t], y[n][t]}/.%[[1]],
      {n, 1, npart}],
   PlotStyle → PointSize[0.015], AspectRatio → 1,
   PlotRange → {{- .1, 1.2}, {-.1, 1.2}}], {t, 0, 4, .1}];
```

Another thing one can do (that can be shown in a textbook!) is plot the orbits of the particles in the x-y plane. The parametric plots in Cell 1.56 do this, using the usual trick of turning off intermediate plot displays in order to save space. The mass-1 particle can be seen to move considerably farther in the x-direction than the mass-3 particle, as expected from conservation of momentum. Both particles drift in the y-direction, because the mass-1 particle had an initial y-velocity, which imparts momentum to the center of mass.

Cell 1.56

```
p1 = ParametericPlot[{x[1][t], y[1][t]}/.S[[1]],
     {t, 0, 4}, DisplayFunction → Identity];
p2 = ParametericPlot[{x[2][t], y[2][t]}/.S[[1]],  {t, 0, 4},
     DisplayFunction → Identity,
       PlotStyle → Dashing[{0.015, 0.015}]];
Show[p1, p2, DisplayFunction → $DisplayFunction,
PlotRange-> {{0, 1}, {0, 1}}, AspectRatio → 1];
```

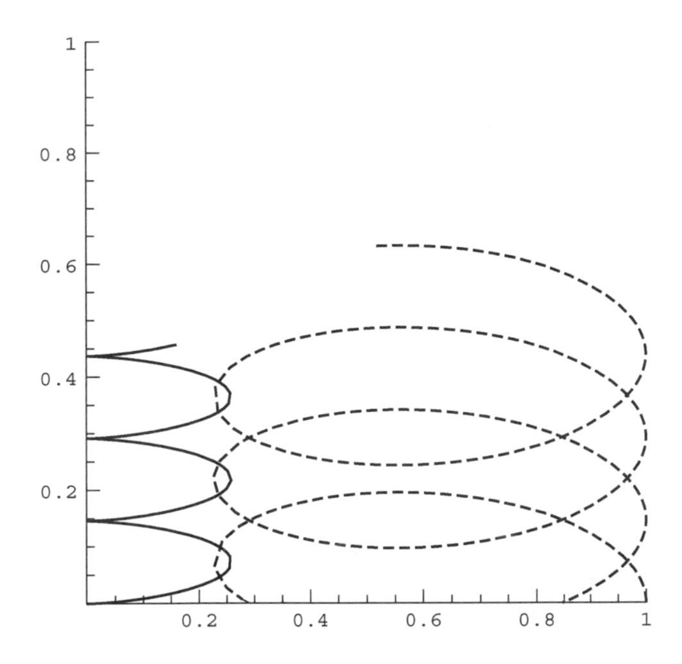

These orbits look a bit more complicated than one might have expected—aren't these two bodies simply supposed to perform elliptical orbits around the center of mass? The answer is yes, but the orbits look different depending on one's frame of reference. In a frame moving with the center of mass, the orbits do look like closed ellipses (see the exercises).

Of course, the orbits of two gravitating particles can be determined analytically, so there is really no need for molecular dynamics. However, for three or more bodies, no general analytic solution is available, and molecular dynamics is crucial for understanding the motion.

Take, for example, the solar system. Table 1.2 provides the positions and velocities of the major bodics in the solar system, with respect to the solar system center of mass, on January 1, 2001. We can use the information in this table as initial conditions to determine the subsequent motion of these planetary bodies.

Table 1.2. Positions and Velocities of Sun and Planets[a]

Body	Mass (kg)	x (m)	y (m)	z (m)	v_x (m/s)	v_y (m/s)	v_z (m/s)
Sun	1.9891×10^{30}	-7.0299×10^{8}	-7.5415×10^{8}	2.38988×10^{7}	14.1931	−6.9255	−0.31676
Mercury	3.302×10^{23}	2.60517×10^{10}	-6.1102×10^{10}	-7.3616×10^{9}	34796.	22185.2	−1379.78
Venus	4.8685×10^{24}	7.2129×10^{10}	7.9106×10^{10}	-3.0885×10^{9}	−25968.7	23441.6	1819.92
Earth	5.9736×10^{24}	-2.91204×10^{10}	1.43576×10^{11}	2.39614×10^{7}	−29699.8	−5883.3	0.050215
Mars	6.4185×10^{23}	-2.47064×10^{11}	-1.03161×10^{10}	5.8788×10^{9}	1862.73	−22150.6	−509.6
Jupiter	1.8986×10^{27}	2.67553×10^{11}	7.0482×10^{11}	-8.911×10^{9}	−12376.3	5259.2	255.192
Saturn	5.9846×10^{26}	6.999×10^{11}	1.16781×10^{12}	-4.817×10^{10}	−8792.6	4944.9	263.754
Uranus	1.0243×10^{26}	2.65363×10^{12}	-3.6396×10^{12}	1.37957×10^{10}	4356.6	3233.3	−166.986
Neptune	8.6832×10^{25}	2.2993×10^{12}	-1.90411×10^{12}	-3.6864×10^{10}	4293.6	4928.1	−37.32
Pluto	1.27×10^{22}	-1.31126×10^{12}	-4.2646×10^{12}	8.3563×10^{11}	5316.6	−2484.6	−1271.99

[a] 12 noon GMT, January 1, 2001, with respcct to the solar system center of mass. Data adapted from the Horizon system at JPL.

This data is also summarized in the following list, which makes it easy to use. Each element in the list corresponds to a column entry in the table:

Cell 1.57

```
sun = {1.9891*^30, -7.0299*^8,
    -7.5415*^8, 2.38988*^7, 14.1931, -6.9255, -0.31676};
mercury = {3.302*^23, 2.60517*^10, -6.1102*^10,
    -7.3616*^9, 34796., 22185.2, -1379.78};
venus = {4.8685*^24, 7.2129*^10, 7.9106*^10, -3.0885*^9,
    -25968.7, 23441.6, 1219.92};
earth = {5.9736*^24, -2.91204*^10, 1.43576*^11,
    2.39614*^7, -29699.8, -5883.3, 0.050215};
mars = {6.4185*^23, -2.47064*^11, -1.03161*^10,
    5.8788*^9, 1862.73, -22150.6, -509.6};
jupiter = {1.8986*^27, 2.67553*^11, 7.0482*^11,
    -8.911*^9, -12376.3, 5259.2, 255.192};
saturn = {5.9846*^26, 6.999*^11, 1.16781*^12,
    -4.817*^10, -8792.6, 4944.9, 263.754};
neptune = {8.6832*^25, 2.2993*^12, -1.90411*^12,
    -3.6864*^10, 4293.6, 4928.1, -37.32};
uranus = {1.0243*^26, 2.65363*^12, -3.6396*^12,
    1.37957*^10, 4356.6, 3233.3, -166.986};
pluto = {1.27*^22, -1.31126*^12, -4.2646*^12,
    8.3563*^11, 5316.6, -2484.6, -1271.99};
```

Cell 1.58

```
solarsys =
  {sun, mercury, venus, earth, mars, jupiter, saturn, uranus,
   neptune, pluto};
```

Let's use this data to try to answer the following important question: is the solar system stable? How do we know that planetary orbits do not have a nonzero Lyapunov exponent, so that they may eventually fly off their present courses, possibly colliding with one another or with the sun?

There has naturally been a considerable amount of very advanced work on this fundamental problem of celestial mechanics. Here, we will simply use our molecular dynamics algorithm to solve for the orbits of the planets, proving the system is stable over the next three hundred years. This is not very long compared to the age of the solar system, but it is about the best we can do using *Mathematica* unless we are willing to wait for long times for the code to complete. More advanced numerical integrators, run on mainframe computers, have evaluated the orbits over much longer time periods.

Because the inner planets are small and rotate rapidly about the sun, we will ignore Mercury, Venus, and Earth in order to speed up the numerical integration.

First we input the data for the sun and the outer planets into the mass list:

Cell 1.59

```
mass = Join[{sun[[1]]}, Table[solarsys[[n]][[1]],
       {n, 5, Length[solarsys]}]]
```

```
{1.9891×10^30, 6.4185×10^23, 1.8986×10^27,
 5.9846×10^26, 1.0243×10^26, 8.6832×10^25, 1.27×10^22}
```

Next, we create a list of initial conditions:

Cell 1.60

```
z0 = Flatten[Join[Table[sun[[j]], {j, 2, 7}],
    Table[Table[solarsys[[n]][[j]], {j, 2, 7}],
    {n, 5, Length[solarsys]}]]];
```

Finally, we define the force, this time keeping the correct magnitude for G:

Cell 1.61

```
G = 6.67 10 ^ -11;
F[i_, j_, r_] := -G mass[[i]] mass[[j]] r/(r.r) ^(3/2)
```

We now run the molecular dynamics code for the planet positions forward in time for 300 years:

Cell 1.62

```
solution1 = MDsol[z0, 300*365*24*3600];
```

This takes quite some time to run, even on a fast machine. In Cell 1.65 we plot the orbits in 3D with a parametric plot.

Cell 1.63

```
Table[{x[n][t], y[n][t], z[n][t]}/.solution[[1]], {n, 1, 7}];
```

Cell 1.64

```
orbits = Table[ParametricPlot3D[%[[n]],
   {t, 0, 3 10^2 365 24 3600},
   PlotPoints→5000, DisplayFunction→Identity], {n, 1, 7}];
```

Cell 1.65

```
Show[orbits, DisplayFunction→$DisplayFunction,
   PlotLabel→"Orbits of the outer planets for 300 years",
    PlotRange→All];
```

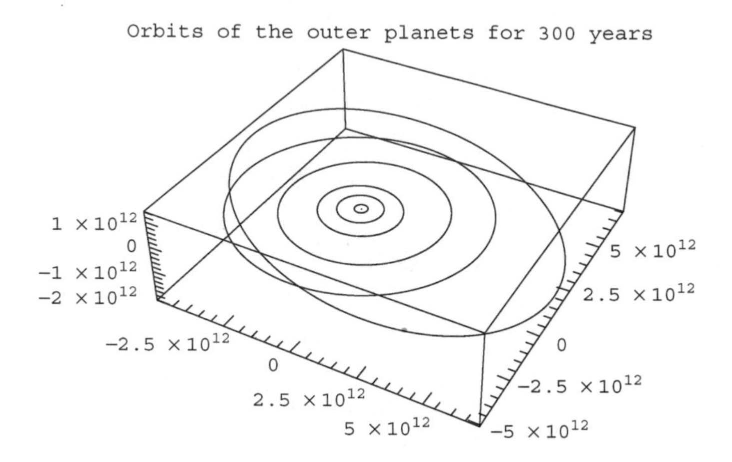

Evidently, the outer planets are stable, at least for the next 300 years! (If nothing else, this plot shows the huge scale of the outer planet's orbits compared to Mars, the innermost orbit in the plot. Earth's orbit would barely even show up as a spot at the center. Distances are in meters.)

EXERCISES FOR SEC. 1.4

(1) The drag force F on a blunt object moving through air is not linear in the velocity v except at very low speeds. A somewhat more realistic model for the drag in the regime where the wake is turbulent is $F = -cv^3$, where c is a constant proportional to the cross-sectional area of the object and the mass density of air. If we use this model for the drag force in the problem of a man throwing a pebble vertically [cf. Sec. 1.3, Exercise (4)], the equation for the height is now nonlinear:

$$\frac{d^2y}{dt^2} + \frac{c}{m}\left(\frac{dy}{dt}\right)^3 = -g.$$

(a) Solve this equation numerically using **NDSolve**, and plot the solution. Take $m = 1$ kg, $c = 1$ kg s/m^2, $y(0) = 0$, and $v(0) = 6$ m/s.

(b) Numerically determine the maximum height, and the time required for the rock to fall back to $y(0)$.

(2) Use **NDSolve** to find the trajectories $x(t)$ and $x'(t)$ for the Van der Pol oscillator, which satisfies Eq. (1.2.20), for initial conditions $(x, x') = (1, 1)$, $(0.1, 0.3)$, and $(3, 2)$ and $0 < t < 20$. Use a parametric plot to plot the trajectories in phase space (x, x'). Note how the trajectories converge onto a single curve, called a *limit cycle*. [See Sec. 1.2, Exercise (6), for the direction field associated with this oscillator.]

(3) Ming the Merciless drops Flash Gordon out the back of his spaceship (in a spacesuit, fortunately). The evil Ming has contrived to leave Flash initially motionless with respect to the earth, whose surface is 5,000 km below. Use Newton's $1/r^2$ force law and **NDSolve** to determine how long Flash has to be rescued before he makes a lovely display in the evening sky. (Hint: $M_{\text{earth}} = 5.98 \times 10^{24}$ kg. The radius of the earth is roughly 6,370 km, and the height of the atmosphere is about 100 km. The gravitational constant is $G = 6.67 \times 10^{-11}$ N m/kg^2. Remember that the $1/r^2$ force is measured with respect to the center of the earth.)

(4) Einstein's general theory of relativity generalizes Newton's theory of gravitation to encompass the situation where masses have large kinetic and/or potential energies (on the order of or larger than their rest masses). Even at low energies, the theory predicts a small correction to Newton's $1/r^2$ force law:

$$f(r) = -GM\left(\frac{1}{r^2} + \frac{3L^2}{c^2 r^4}\right),$$

where L is the specific angular momentum—see Eq. (1.2.22). This force per unit mass replaces that which appears on the right-hand side of the orbit equation (1.3.5).

(a) Use **NDSolve** to determine the new orbit $r(\theta)$ predicted by this equation, and plot it for $0 < \theta < 4\pi$, taking orbital parameters for the planet Mercury: $r(0) = 46.00 \times 10^6$ km (perihelion distance), $r'(0) = 0$, $L = 2.713 \times 10^{15}$ m^2/s. The mass of the sun is 1.9891×10^{30} kg.

(b) Show numerically that the orbit no longer closes, and that each successive perihelion precesses by an amount $\Delta\theta$. Find a numerical value for $\Delta\theta$. Be careful: the numerical integration must be performed very accurately. (The precession of Mercury's perihelion has been measured, and after successive refinements, removing extraneous effects, it was found to be in reasonable agreement with this result.)

(5) A cubic cavity has perfectly conducting walls of unit length, and supports electromagnetic standing waves. The magnetic field in the modes (assumed to be TE modes) is

$$B_{lmn}(x, y, z)$$
$$= B_0\left\{-\frac{l}{l^2+m^2}\sin(l\pi x)\cos(m\pi y)\cos(n\theta z),\right.$$
$$\left.-\frac{m}{l^2+m^2}\cos(l\pi x)\sin(m\pi y)\cos(n\pi z), \cos(l\pi x)\cos(m\pi y)\sin(n\pi z)\right\}.$$

For $(l, m, n) = (1, 1, 1)$ solve Eqs. (1.2.24) numerically for the field lines $\mathbf{r}(s, \mathbf{r}_0)$ for $-2 < s < 2$ and initial conditions $\mathbf{r}_0 = \{0.25i, 0.25j, 0.25k\}$, $i, j, k, = 1, 2, 3$. Use **ParametricPlot3D** to plot and superimpose the solutions. [The solution is shown in Fig. 1.12 for the mode with $(l, m, n) = (1, 2, 1)$.]

(6) Repeat the calculation of the Lyapunov exponent done in the text, but for an *integrable* system, the one-dimensional undamped harmonic oscillator with

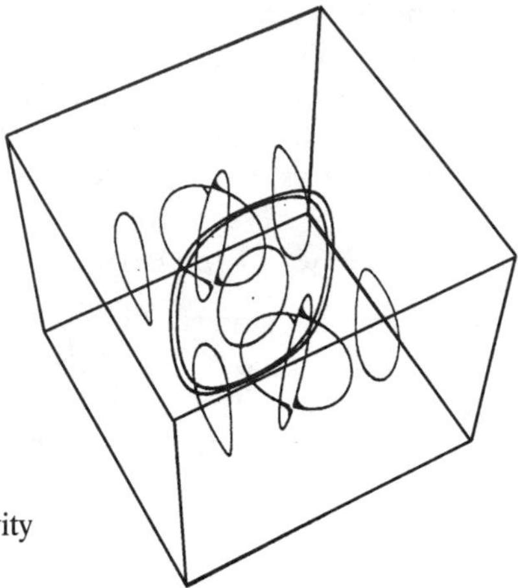

Fig. 1.12 Magnetic field lines in a TE(1, 2, 1) cavity mode.

dimensionless Hamiltonian $H = (v^2 + x^2)/2$, taking the initial condition $x = 0$, $v = 1$. What happens to $\lambda(t)$, the right-hand side of Eq. (1.4.2), at large times?

(7) Not all trajectories of a chaotic system have positive Lyapunov exponents. Certain regions of phase space can still be integrable, containing nested curves upon which the orbits lie. Take, for example, our chaotic system described by Eq. (1.4.1) with the same parameter values as discussed before ($V_0 = V_1 = k_0 = k_1 = m = 1$, $\omega = 2$), but a different initial condition, $x(0) = 3$, $v(0) = 3$. Repeat the evaluation of the Lyapunov exponent for this trajectory, again taking 40 adjacent trajectories with $|\mathbf{d}_0| < 10^{-5}$ and $0 < t < 50$.

(8) Hamiltonian systems are not the only systems that exhibit chaotic motion. Systems that have dissipation can also exhibit chaos. The fact that these systems no longer conserve phase-space volume implies that orbits can collapse onto weirdly shaped surfaces called *strange attractors*. Although motion becomes confined to this attracting surface, motion within the surface can be chaotic, exhibiting a positive Lyapunov exponent. The Lorenz system of ODEs is an example of a dissipative chaotic system with a strange attractor. This system models an unstable thermally convecting fluid, heated from below. The equations for the system are three coupled ODEs for functions $x(t)$, $y(t)$, and $z(t)$ (which are a normalized amplitude of convection, a temperature difference between ascending and descending fluid, and a distortion of the vertical temperature profile from a linear law, respectively):

$$\begin{aligned}
\frac{dx}{dt} &= \sigma(y - x), \\
\frac{dy}{dt} &= rx - y - xz, \qquad (1.4.25) \\
\frac{dz}{dt} &= xy - bz,
\end{aligned}$$

where σ, b, and r are constants. For sufficiently large r, this system exhibits chaos.

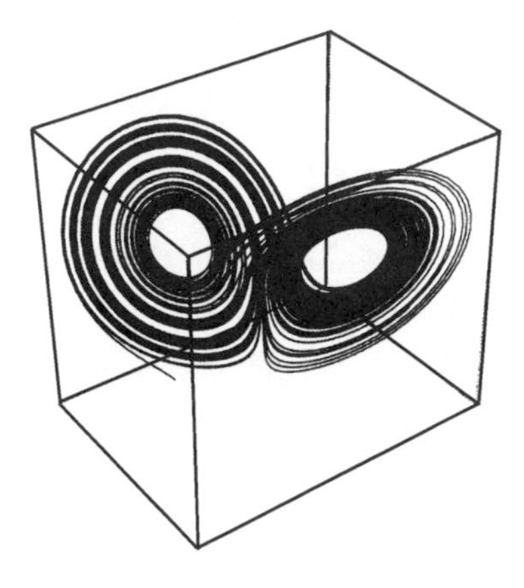

Fig. 1.13 Strange attractor for the Lorenz system.

(a) Taking characteristic values of $\sigma = 10$, $b = \frac{8}{3}$, and $r = 28$, integrate the Lorenz equations for $0 < t < 100$. Take as an initial condition $x = 1$, $y = 15$, $z = 10$. Use the function **ParametricPlot3D** to plot the $(x(t), y(t), z(t))$ orbit. This orbit will exhibit the strange attractor for this dissipative dynamical system. (Hints: To integrate for the required length of time you will need to increase the **MaxSteps** option in **NDSolve** to around 10,000 or so. Also, after plotting the strange attractor, it is fun to rotate it and view it at different angles. See the discussion of real-time 3D graphics in Chapter 9. You will need to increase the number of plot points used in the parametric plot, to **PlotPoints->5000** or more.)

(b) Repeat part (a) using higher accuracy, by taking **AccuracyGoal** and **PrecisionGoal** in **NDSolve** to their highest possible values for your computer system. Plot the displacement between the two trajectories as a function of time. Does this system exhibit the explosive growth in error characteristic of chaos?

(c) Calculate the Lyapunov exponent for this trajectory by plotting the right hand side of Eq. (1.4.2) for $0 < t < 15$. [Now $\mathbf{z} = (x, y, z)$ in Eq. (1.4.2).] Average over 20 nearby trajectories with $|\mathbf{d}_0| < 10^{-5}$. The solution of part (a) is shown in Fig. 1.13.

(9) Magnetic and electric field lines can also display chaotic behavior. For example, consider the following simple field:

$$\mathbf{B}_0(r, \theta, z) = 2r\hat{\mathbf{r}} \sin 2\theta + \hat{\boldsymbol{\theta}}\left(\frac{r^3}{4} + 2r\cos 2\theta\right) + \hat{\mathbf{z}}.$$

[One can easily show that this field satisfies $\nabla \cdot \mathbf{B} = 0$. It consists of a uniform solenoidal field superimposed on a quadrupole field created by external currents and the field from a current density $\mathbf{j}(r) \propto r^2 \hat{\mathbf{z}}$.] For this field is it useful to consider field line ODEs of the form

$$\frac{dr}{dz} = \frac{dr/ds}{dz/ds} = \frac{B_r(r, \theta, z)}{B_z(r, \theta, z)} \quad \text{and} \quad r\frac{d\theta}{dz} = \frac{r\,d\theta/ds}{dz/ds} = \frac{B_\theta(r, \theta, z)}{B_z(r, \theta, z)}.$$

[see Eqs. (1.3.6)].

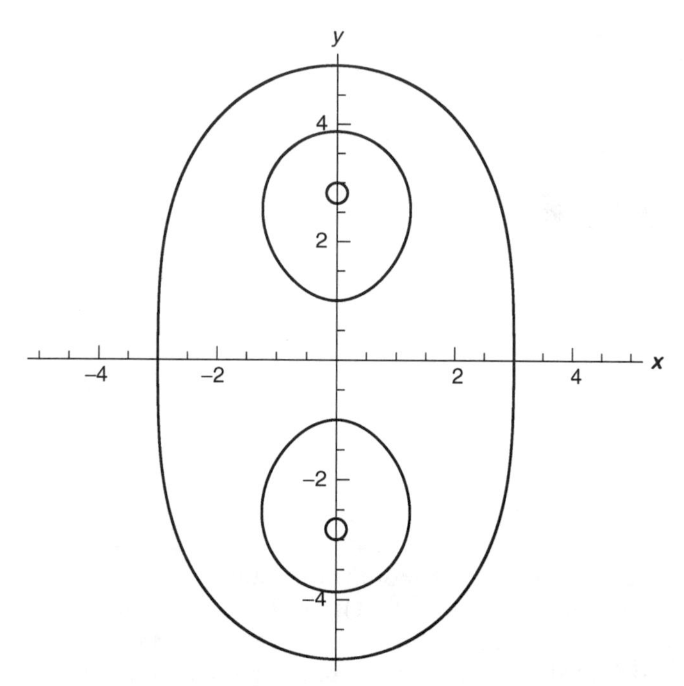

Fig. 1.14 Solution to Exercise (8)(b): Field lines for the magnetic field $\mathbf{B}_0$, projected into the (x, y) plane.

(a) Solve these coupled ODEs for $r(z)$ and $\theta(z)$ using **NDSolve**, and plot the resulting field line in (x, y, z) via **ParametricPlot3D** for $0 < z < 20$ and for initial conditions $\theta_0 = \pi/2$, and $r_0 = -5 + 2n$, $n = 0, 1, 2, 3, 4$. [Hint: Along the field line, $x(z) = r(z)\cos\theta(z)$, $y(z) = r(z)\sin\theta(z)$.]

(b) Although the result from part (a) appears to be very complicated, these field lines are not chaotic. One way to see this is to project the field lines into the (x, y) plane, since the field is independent of z. Do so, and show that the field lines created in part (a) fall on nested closed curves. (The solution is shown in Fig 1.14. Note the appearance of two *magnetic islands*, around which the field lines spiral.)

(c) A chaotic magnetic field can be created by adding another magnetic field to $\mathbf{B}_0$, writing $\mathbf{B}_s(r, \theta, z) = \mathbf{B}_0(r, \theta, z) + \epsilon[\hat{\mathbf{r}} r \sin(\theta - z) + \hat{\boldsymbol{\theta}}\, 2r \cos(\theta - z)]$. (This field also satisfies $\nabla \cdot \mathbf{B} = 0$.) For $\epsilon = \frac{1}{3}$ replot the field lines for this field in (x, y, z), using the same initial conditions as in part (a). The field lines now become a "tangle of spaghetti." Project them into the (x, y) plane. You will see that some of the field lines wrap around one magnetic island for a time, then get captured by the adjoining island. This complicated competition between the islands is responsible for the chaotic trajectory followed by the lines.

(d) One way to test visually whether some of the field lines are chaotic is to note that the magnetic field $\mathbf{B}_s(r, \theta, z)$ is periodic in z with period 2π. If $\mathbf{B}_s$ were not chaotic, it would create field lines that fell on closed surfaces,

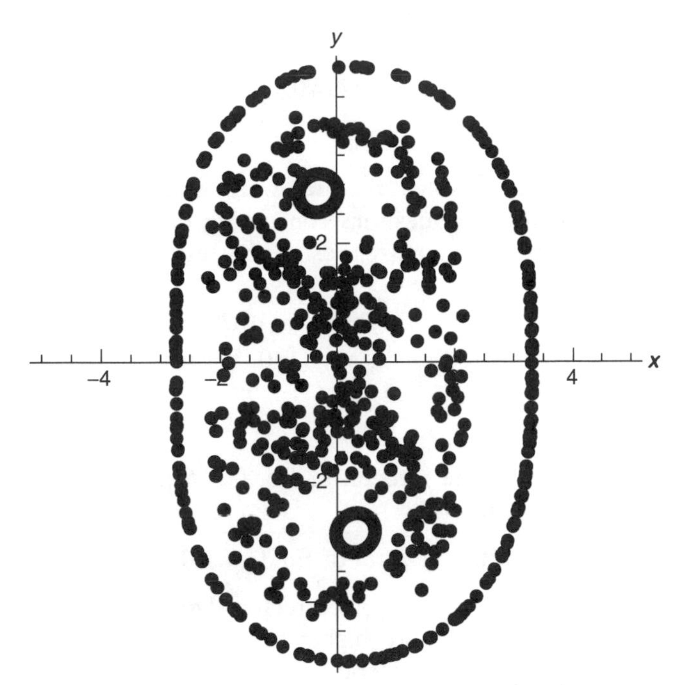

Fig. 1.15 Solution to Execicse (8)(d): Poincaré plot in the (x, y) plane for the magnetic field $\mathbf{B}_s$ for $\epsilon = \frac{1}{4}$.

and the surfaces would also have to be periodic in z. Therefore, if you plot values of $(r(z_n), \theta(z_n))$ for $z_n = 2\pi n$ in either the (r, θ) plane or the (x, y) plane, the resulting points must form closed curves for a non-chaotic field. This is called a *Poincaré plot*. However, for a chaotic field the lines are *not* on closed surfaces; rather, they fill space. A Poincaré plot will now show a chaotic collection of points filling a region in the (r, θ) plane [or the (x, y) plane]. For the same initial conditions as in part (a), use **NDSolve** to evaluate the field lines for the field $\mathbf{B}_s$ for $\epsilon = \frac{1}{3}$ and $0 \le z \le 400\pi$. (Increase **MaxSteps** to about 200,000.) Make a table of values of $(r(z_n), \theta(z_n))$ for $z_n = 2\pi n$. Use **ListPlot** to plot these values in the (x, y) plane. (The solution is shown in Fig. 1.15 for $\epsilon = \frac{1}{4}$.)

(10) Use Euler's method to

(a) solve the following ODEs with initial conditions over the given range of time, and for the given step size. Then,

(b) plot the solution;

(c) solve the ODE analytically, and plot the error in $x(t)$;

(d) in each case, using the results of (c), *predict* how small a step size is necessary for the error to be smaller than 10^{-4} over the course of the run.

(i) $\dfrac{dx}{dt} = \sin t - x,\ x(0) = x_0,\ 0 < t < 5,\ \Delta t = 0.1.$

(ii) $4x + \frac{dx}{dt} + \frac{d^2x}{dt^2} = 0$, $x(0) = 2$, $x'(0) = 1$, $0 < t < 10$, $\Delta t = 0.01$.

(iii) $x''' + 2x'' + x' + 2x = \cos t$, $x(0) = x'(0) = x''(0) = 0$,

$0 < t < 20$, $\Delta t = 0.02$.

(11) Use Euler's method to solve the following nonlinear ODE initial-value problems, and answer the questions concerning the solutions. By looking at solutions as you vary the step size, ensure that each answer is accurate to at least 10^{-4}.

(a) $x'(t) = t/(x+t)^2$, $x(0) = 3$. What is $x(10)$?

(b) $x''(t) = \cos t \cos x$, $x(0) = -1$, $x'(0) = 0$. What is $x(20)$?

(c) Eqs. (1.4.25), $\sigma = 10$, $b = \frac{8}{3}$, and $r = 28$; $x = 1$, $y = 1$, $z = 1$. What is $x(40)$?

(12) Prove that the error in the integral of Eq. (1.4.6) is $O(\Delta t^2)$. (Hint: Taylor-expand f about the initial time.)

(13) Prove that the error in the integral of Eq. (1.4.10) is $O(\Delta t^3)$.

(14) Modify our second-order predictor–corrector algorithm so that it can handle differential equations of order higher than one, or systems of coupled equations. Use the modified method to repeat

(a) Exercise (10)(ii),

(b) Exercise (10)(iii),

(c) Exercise (11)(b),

(d) Exercise (11)(c).

(15) *Centered-difference method.* The following discretization method can be used to solve a second-order differential equation of the form $d^2x/dt^2 = f(x,t)$, with initial condition $x(0) = x_0$, $x'(0) = v_0$. The method requires only one force evaluation per timestep. First, discretize time in the usual way, with $t_n = n\,\Delta t$. Approximate the second derivative as

$$\frac{d^2x}{dt^2}(t_n) = \frac{x(t_{n+1}) - 2x(t_n) + x(t_{n-1})}{\Delta t^2}. \qquad (1.4.26)$$

This approximation is referred to as a *centered-difference* form for the derivative, since the expression is symmetric about timestep t_n. (See the Appendix and Sec. 2.4.5.) The differential equation then becomes the recursion relation

$$x_{n+1} - 2x_n + x_{n-1} = \Delta t^2 f(x_n, t_n), \qquad n > 1. \qquad (1.4.27)$$

Note that in order to determine the first step, x_1, given the initial condition x_0, Eq. (1.4.27) requires x_{-1}, which is not defined. Therefore, we need a different equation to obtain x_1. Use

$$x_1 = x_0 + \Delta t\, v_0 + \frac{\Delta t^2}{2} f(x_0, t_0), \qquad (1.4.28)$$

which is the formula for position change due to constant acceleration.

(a) Write a module that implements this scheme.

(b) Use this module to solve the problem of a harmonic oscillator, with $f = -x$, taking $x_0 = 1$, and $v_0 = 0$ on the interval $0 < t < 20$, and taking $\Delta t = 0.1$. Plot the error in your solution, compared to the analytic solution $\cos t$.

(c) Repeat (b) with $\Delta t = 0.02$. By what factor has the error been reduced? What is the order of this method?

(16) *The leapfrog method.* The following discretization method can also be used to solve a second-order differential equation of the form $d^2x/dt^2 = f(x)$, with initial conditions $x(0) = x_0$, $x'(0) = v_0$. This method requires only one force evaluation per timestep. We first write this equation in terms of two first-order equations for position and velocity:

$$\frac{dx}{dt} = v(t), \qquad \frac{dv}{dt} = f(x).$$

We then discretize $x(t)$ on a grid $t_n = n\,\Delta t$. But we discretize the velocity $v(t)$ on the grid $t_{n+1/2} = (n + \frac{1}{2})\,\Delta t$. In each case we use centered-difference forms for the discretized first derivative:

$$\frac{v_{n+1/2} - v_{n-1/2}}{\Delta t} = f(x_n),$$

$$\frac{x_{n+1} - x_n}{\Delta t} = v_{n+1/2}.$$

In the first equation, the derivative of v is evaluated at timestep n using a centered-difference form for $\frac{dv}{dt}(t_n)$. (See the Appendix and Sec. 2.4.5.) In the second equation, the derivative of x is evaluated at timestep $n + \frac{1}{2}$, using the same centered-difference form. The method is started using a predictor–corrector step in order to obtain $v_{1/2}$ from x_0 and v_0:

$$x_{1/2} = x_0 + v_0\,\Delta t/2,$$

$$v_{1/2} = v_0 + \tfrac{1}{2}\left[f(x_0) + f(x_{1/2})\right]\Delta t/2.$$

(a) Write a module that implements this scheme.

(b) Use the module to solve the same problem as in Exercise (15)(b) and (c). What is the order of this method?

(17) *The Adams–Bashforth method.* Consider the following general first-order ODE (or system of ODEs): $dv/dt = f(v, t)$, with initial condition $v(0) = v_0$. We wish to obtain a second-order (or even higher-order) method for solving this problem, using only one force evaluation per step. First, we replace Eq. (1.4.11) by

$$v(t_n) = v(t_{n-1}) + \Delta t f\big(t_{n-1/2}, v(t_{n-1/2})\big) + O(\Delta t^3).$$

This formula is exact if f is a linear function of time, as is clear from Fig. 1.11.

(a) Show that the error in one step is in fact of order Δt^3.

(b) To obtain the force at the intermediate time $t_{n-1/2}$, we *extrapolate* from previous force evaluations. If at timestep t_n we call the force f_n, then by using a linear fit through f_{n-1} and f_{n-2} show that

$$f_{n-1/2} = \frac{3f_{n-1} - f_{n-2}}{2}.$$

(c) The code arising from this approach is called the second-order Adams–Bashforth method. It is

$$v(t_n) = v(t_{n-1}) + \Delta t \frac{3f_{n-1} - f_{n-2}}{2}. \quad (1.4.29)$$

This algorithm is an example of a *multistep method*, where force evaluations from previous steps are kept in memory and used to make the next step. One way to save previous force evaluations is to use the double-equal-sign trick: define a function `force[n_] := force[n] =` $f(t_n, v(t_n))$, and use this function in Eq. (1.4.29). Write a module for this algorithm. To take the first step, use the second-order predictor–corrector method. Use your module to solve the *coupled system*

$$\frac{dx}{dt} = -y - x, \qquad \frac{dy}{dt} = 2x - 3y + \sin t$$

for $0 < t < 10$, taking $\Delta t = 0.1$, with initial conditions $x(0) = 0$, $y(0) = 1$. Compare $y(t)$ with the exact solution found using `DSolve`, and verify that the method is second-order accurate by evaluating the error taking half the step size, then reducing by half again.

(18) For force laws that are derivable from a potential, such as the gravitational force, the equations (1.4.21) are Hamiltonian in form, with conserved energy

$$H = \sum_{i=1}^{N} \frac{m_i v_i^2}{2} + \sum_{i=1}^{N} \sum_{j=i+1}^{N} V_{ij}(\mathbf{r}_i - \mathbf{r}_j), \quad (1.4.30)$$

where V_{ij} is the potential energy of interaction between particles i and j. Evaluating the energy in molecular dynamics calculations provides a useful check of the accuracy of the numerics. This Hamiltonian also conserves total linear momentum,

$$\mathbf{P} = \sum_{i=1}^{N} m_i \mathbf{v}_i. \quad (1.4.31)$$

For central-force problems where the potential energy depends only on the *distance* between bodies (again, gravity is an example), the total angular momentum $\mathbf{L} = (L_x, L_y, L_z)$ is also conserved, and provides three more useful checks on the numerical accuracy of a code:

$$\mathbf{L} = \sum_{i=1}^{N} m_i \mathbf{r}_i \times \mathbf{v}_i. \quad (1.4.32)$$

(a) Run the example problem on two gravitating bodies (see Cell 1.56). Use the results to calculate and plot the energy, the total momentum, and the z-component of angular momentum as a function of time.

(b) Repeat (a), setting `AccuracyGoal` and `PrecisionGoal` to their highest possible values for your computer system.

(c) The center-of-mass velocity is defined as $\mathbf{V}_{\rm cm} = \mathbf{P}/\Sigma_{i=1}^{N} m_i$. Plot the orbits of the two planets as seen in a frame moving at the constant speed $\mathbf{V}_{\rm cm}$. What do they look like in this frame of reference?

(19) A classical model of a helium atom consists of a massive nucleus with mass roughly $4m_p$ (m_p being the proton mass) and charge $2e$, and two electrons with mass m_e and charges $-e$. In this model the electrons are equidistant from the nucleus, on opposite sides, and the electrons move in circular orbits. The charges interact via Coulomb's law, written below for charges q_i and q_j separated by displacement $\mathbf{r}$:

$$\mathbf{E}_{ij}(\mathbf{r}) = \frac{q_i q_j \mathbf{r}}{4\pi\epsilon_0 r^3}. \tag{1.4.33}$$

(a) Analytically solve for an equilibrium distance $d(v)$ of each electron from the nucleus, as a function of the orbital speed v. The orbital period of this motion is $T = 2\pi d/v$.

(b) We will numerically examine the stability of this equilibrium. Choose any value of the orbital speed v that you wish. Move the electrons a small distance, $0.05d(v)$, in a random direction from the equilibrium determined in part (a). Numerically evaluate the resulting motion for a time $5T$, and make a movie of the (x, y) motion, plotting every $0.1T$. Is this motion stable (i.e., do the electrons remain near the equilibrium orbit)?

(c) Repeat for a displacement from equilibrium of $0.3d(v)$.

(d) Repeat (a), (b), and (c) for lithium, nuclear charge $3e$, nuclear mass $7m_p$. The equilibrium now consists of three electrons arranged in an equilateral triangle around the nucleus.

(20) The great mass of the sun compared to that of the planets is essential to the long-term stability of the solar system. By integrating the solar system equations four times for 1000 years, keeping the initial positions of the outer planets the same in each case, but taking larger masses for the planets by factors of 10 in each consecutive run, determine roughly how massive the sun must be, as a multiple of that of Jupiter, in order to provide a stable equilibrium for outer-planet orbits over 10^3 years. Perform the integration only for the outer planets, from Jupiter on out. (Take care to check whether `NDSolve` is giving accurate results over this long time period.)

(21) An astronomer discovers that a minor asteroid has been kicked through a collision into an unusual orbit. The asteroid is initially located somewhere between Mars and Jupiter, and is heading at rather high speed into the inner solar system. As of January 1, 2001, at 12 noon, the asteroid has velocity $(11{,}060, -9817, -744)$ m/s and position $(-5.206 \times 10^{11}, 3.124 \times 10^{11}, 6.142 \times 10^{10})$ m (with respect to the solar system center of mass). Using the data of

Table 1.2 and the molecular dynamics code, determine which planet this asteroid is going to strike by making a movie of the (x, y) positions of all solar system bodies in Table 1.2 over a 2-year period, including the asteroid.

(22)(a) Modify the molecular dynamics code to allow for a drag force, so that equations of motion are of the form

$$m_i \frac{d^2 \mathbf{r}_i}{dt^2} = \sum_{\substack{j=1 \\ j \neq i}}^{N} \mathbf{F}_{ij}(\mathbf{r}_i - \mathbf{r}_j) - m_i \gamma \frac{d\mathbf{r}_i}{dt}.$$

(b) The Lenard-Jones potential is often used to model interatomic interactions classically. The form of the potential energy is

$$V(r) = \epsilon \left(\frac{1}{(r/a)^{12}} - \frac{2}{(r/a)^6} \right),$$

where a is a distance scale and ϵ is an energy scale. We are going to use molecular dynamics to determine the form of molecules that interact via the Leonard-Jones potential. To do so, start with N atoms distributed randomly in a cube with sides of length $N^{1/3}$. Take $a = m = \epsilon = 1$. Initially the atoms have zero velocities. Add some (small) damping to the motion of the atoms, and follow their motion until they fall into a minimum-energy state.

(i) Construct the energy function. Then, taking $\gamma = 0.05$, and for $N = 3, 4, 5, 6$,

(ii) follow the motion for $0 < t < 100$.

(iii) Evaluate and plot the energy vs. time. Does it appear that an energy minimum has been achieved? For each N-value, what is the minimum energy in units of ϵ?

(iv) Use `ParametricPlot3D` to plot the positions of the atoms. Can you describe the structure in words?

(v) If a minimum has not yet been achieved, repeat the process using the final state of the previous simulation as the new initial condition.

(vi) The larger the molecule, the more local minimum-energy states there are. For $N = 5$ and 6, repeat your simulation for five different random initial conditions to see if you can find any other minimum-energy states. How many did you find?

(23) Modify the molecular dynamics code to be able to handle applied magnetic and electric fields $\mathbf{B}(\mathbf{r}, t)$ and $\mathbf{E}(\mathbf{r}, t)$, and a damping force: The equations of motion are now

$$\frac{d^2 \mathbf{r}_i}{dt^2} = \sum_{\substack{j=1 \\ j \neq i}}^{N} \mathbf{F}_{ij}(\mathbf{r}_i - \mathbf{r}_j) + q_i[\mathbf{E}(\mathbf{r}_i, t) + \mathbf{v}_i \times \mathbf{B}(\mathbf{r}_i, t)] + m_i \gamma \hat{\mathbf{z}} \frac{dz_i}{dt}.$$

(The damping is allowed to work only on the z-motion.)

(a) Check that your new molecular dynamics code works by solving numerically for the motion of a single particle in a uniform electric and magnetic field $\mathbf{E} = (1, 0, 0), \mathbf{B} = (0, 0, 1)$, taking $m = q = 1$, $\gamma = 0$, and $\mathbf{r} = \mathbf{v} = 0$ at $t = 0$. Compare the numerical result from $0 < t < 10$ with the analytic result of Sec. 1.3, Exercise (6).

(b) A *Penning trap* is a trap for charged particles, which can hold the particles in vacuum, away from solid walls, using only static electric and magnetic fields. The trap works for particles that all have the same sign of charge. We will take all the charges to have charge 1 and mass 1, so that their interaction potential is $V(r) = 1/r$. The trap has the following applied electric and magnetic fields:

$$\mathbf{B} = (0, 0, 5), \qquad \mathbf{E} = (x/2, y/2, -z).$$

Consider four ions given random initial x, y, and z velocities in the ranges -0.1 to 0.1, and positions $(0, 0, \frac{1}{4})$, $(1, 0, 0)$, $(-1, 0, 0)$, $(0, 1, 0)$. Their z-motion is damped (using lasers) with rate $\gamma = 0.1$. (Only the z-motion is damped; otherwise the ions would be lost from the trap because of the applied torque from the damping.) Numerically integrate the motion of these ions until they settle into an equilibrium configuration. What is this configuration? [Hint: make (x, y) images of the ion positions. The equilibrium will actually rotate at a constant rate around the origin due to the $\mathbf{E} \times \mathbf{B}$ drift [Sec. 1.3, Exercise (6)].

(24) Consider a model of an elastic rod as a system of masses and springs. The equilibrium of such a system, is examined using an analytic model in Sec. 1.5, Exercise (1). Here we will examine the dynamics of this elastic rod, using the molecular dynamics method. We will consider a system of $M = 41$ masses. In the absence of gravity, the masses are arranged in equilibrium positions $\mathbf{R}_i$, which are the same as in the statics problem of Sec. 9.10, Exercise (5):

Cell 1.66

```
p = a {1/2, 0}; q = a {0, Sqrt[3]/2};
R[i_] = (i-1) p + Mod[i, 2] q;
```

As before nearest neighbor masses i and j interact via the isotropic force

Cell 1.67

```
F[i_, j_, r_Y] := -k[i, j] (r - a r/Sqrt[r.r])
```

The total force on mass i is given by interactions with its four nearest neighbors, assuming that the mass is not at the ends of the rod:

Cell 1.68

```
Ftot[i_] :=
  {0, -mg} + F[i, i-2, r[i, t] - r[i - 2, t]] +
F[i, i-1, r[i, t]-r[i-1, t]] +
  F[i, i + 1, r[i, t]-r[i + 1, t]] +
F[i, i + 2, r[i, t]-r[i + 2, t]]
```

where $\mathbf{r}_i(t)$ is the position of the ith mass. The end masses are fixed to the walls:

Cell 1.69

```
r[1, t_] = R[1];
r[2, t_] = R[2];
r[M, t_] = R[M];
r[M-1, t_] = R[M-1];
```

Modify the molecular dynamics code to handle this type of nearest neighbor force, and solve the following problem: $\mathbf{r}_i(t=0)=\mathbf{R}_i$, and $\mathbf{v}_j(t=0)=0$, for $m=1, g=0.5$, $k_{ij}=160$, over the time range $0<t<24$. [An analytic approach to a similar problem can be found in Sec. 4.2, Exercise (7).]

(25) Use the molecular dynamics code and the data in Table 1.2 to plot the $x-y$ velocities of the sun, $(v_x(t), v_y(t))$, as a parametric plot over the time range $0<t<30$ years. What is the magnitude of the maximum speed attained by the sun? This oscillatory motion has been discerned in distant stars through tiny oscillatory Doppler shifts in the starlight, providing indirect evidence for the existence of planetary systems beyond our own solar system [Marcy and Butler (1998)].

1.5 BOUNDARY-VALUE PROBLEMS

1.5.1 Introduction

In order to determine a unique solution to an initial-value problem for an Nth-order ODE, we have seen that N initial conditions must be specified. The N initial conditions are all given at the same point in time.

Boundary-value problems differ from initial-value problems in that the N conditions on the problem are provided at *more than one point*. Typically they are given at starting and finishing points—at, say, $t=0$ and $t=T$.

As a simple example of a boundary-value problem, consider trying to hit a moving object with an arrow. To further simplify the problem, let's assume that the arrow moves in a single dimension, vertically, under the influence only of gravity—no air drag or other forces will be kept. Then the position of the arrow, $y(t)$, satisfies

$$\frac{d^2y}{dt^2} = -g, \tag{1.5.1}$$

where $g=9.8\ \mathrm{m/s^2}$ is the acceleration of gravity. The arrow starts at $y=0$ at time $t=0$, and must be at $y=H$ at $t=T$ in order to hit an object that passes overhead at that instant. Therefore, the *boundary conditions* on the problem are

$$y(0)=0, \qquad y(T)=H. \tag{1.5.2}$$

In order to solve this problem, consider the general solution of Eq. (1.5.1), determined by integrating the ODE twice:

$$y(t) = C_1 + C_2 t - \frac{gt^2}{2}. \tag{1.5.3}$$

We determine the constants C_1 and C_2 using the boundary conditions (1.5.2). Since $y(0) = 0$, Eq. (1.5.3) evaluated at $t = 0$ implies that $C_1 = 0$. Since $y(T) = H$, we find that $C_2 = (H/T - gT/2)$, yielding the following solution for $y(t)$:

$$y(t) = \left(\frac{H}{T} - \frac{gT}{2}\right)t - \frac{gt^2}{2}. \tag{1.5.4}$$

Finding the solution of this boundary-value problem seems to be no different than finding the solution of an initial-value problem. However, there is a fundamental difference between these two types of problems: unlike solutions to initial-value problems that satisfy the conditions of Theorem 1.1,

> The solutions to boundary-value problems need not exist, and if they exist they need not be unique.

It is easy to find examples of boundary-value problems for which there is no solution. Consider the motion of a harmonic oscillator, whose position satisfies

$$\frac{d^2x}{dt^2} = -\omega_0^2 x. \tag{1.5.5}$$

Let's again try to hit a passing object with this oscillator (an arrow attached to a spring?). The object is assumed to pass through the point x_0 at a time $t = \pi/\omega_0$. Starting the oscillator from the origin at $t = 0$, the boundary conditions are

$$x(0) = 0, \qquad x(\pi/\omega_0) = x_0. \tag{1.5.6}$$

Using the first boundary condition in the general solution, given by Eq. (1.1.7), implies that $C_1 = 0$ and $x(t) = C_2 \sin(\omega_0 t)$. Now it seems like a simple task to find the value of C_2 using the second boundary condition, as we did in the previous example. Unfortunately, however, we are faced with a dilemma: at the requested time π/ω_0,

$$x(\pi/\omega_0) = C_2 \sin \pi = 0 \tag{1.5.7}$$

for *all* values of C_2, so it is impossible to satisfy the second boundary condition. Therefore, there is *no* solution to this boundary-value problem.

It is also easy to find boundary-value problems for which the solution is not unique. Consider again the previous harmonic oscillator problem, but this time take boundary conditions

$$x(0) = 0, \qquad x(T) = 0 \tag{1.5.8}$$

for some given time T: that is, we want the oscillator to pass back through the origin at time T. Now, for most values of T, there is only one solution to this problem: the trivial solution $x(t) = 0$, where the oscillator is stationary at the origin for all time. However, for special values of T there are other solutions. If $T = n\pi/\omega_0$ for some integer n, the solution

$$x(t) = C_2 \sin(\omega_0 t) \tag{1.5.9}$$

matches the boundary conditions *for any value of* C_2. Therefore, for these special values of T, the solution is not unique—in fact, there are an infinite number of solutions, corresponding to sine oscillations with arbitrary amplitude given by Eq. (1.5.9).

Another way to do this problem is to hold the time T fixed and instead allow the parameter ω_0 to vary. For most values of ω_0, the boundary conditions (1.5.8) are satisfied only for the trivial solution $x(t) = 0$. But at values of ω_0 given by

$$\omega_0 = n\pi/T, \tag{1.5.10}$$

the solution again is of the form of Eq. (1.5.9) with arbitrary amplitude C_2.

> The problem of determining the values of a parameter (such as ω_0) for which nontrivial (i.e., nonzero) solutions of a boundary-value problem exist is referred to as an *eigenvalue problem.*

These problems are called eigenvalue problems because as we will see, they are often equivalent to finding the eigenvalues of a matrix. (See Sec. 6.3.) Eigenvalue problems turn out to be very important in the solution of linear PDEs, so we will return to a discussion of their solution in later chapters.

1.5.2 Numerical Solution of Boundary-Value Problems: The Shooting Method

For the simple cases discussed above, general analytic solutions to the ODEs could be found, and the boundary-value problem could be solved analytically (when the solution existed). However, we have already seen that there are many ODEs for which no general analytic solution can be found. In these cases numerical methods must be employed. This section will consider one method that can be used to find a numerical solution to a boundary-value problem: the *shooting method.*

As an example of the shooting method, consider a general second-order ODE of the form

$$\frac{d^2y}{dt^2} = f\left(t, y, \frac{dy}{dt}\right) \tag{1.5.11}$$

and with boundary conditions

$$y(t_0) = y_0, \qquad y(t_1) = y_1. \tag{1.5.12}$$

We require the solution for $y(t)$ between the initial time t_0 and the final time t_1. An example of such a problem would be our previous archery problem, but with an acceleration determined by the function f.

In attempting to solve this problem numerically, we run into an immediate difficulty. All of the numerical methods that have been described so far in this book have dealt with initial value problems, where for *given* initial conditions, we take steps forward in time until the final time is reached. Here we don't know all of the required initial conditions, so we can't step forward in time. Although we are given $y(t_0)$, we don't know $y'(t_0)$. **NDSolve**, Euler's method, and the predictor–corrector method all require the initial position *and* velocity in order to integrate this problem forward in time.

The shooting method proposes the following solution to this difficulty: if you don't know all of the initial conditions, have a guess. Using the guess, integrate the solution forward and see if it matches the second boundary condition. If it misses, adjust the guess and try again, iterating until the guess gives a solution that does match the second boundary condition.

You can see immediately why this is called the shooting method: we are shooting an arrow, refining our guesses for the initial velocity until we make a hit at the required instant.

To do this problem in *Mathematica*, we will first define a function **Sol[v0]**, which solves the initial-value problem taking $y'(t_0) = v_0$:

Cell 1.70

```
Sol[v0_] :=
 NDSolve[{y"[t] == f[t, y[t], y'[t]], y[t0] == y0,
  y'[t0] == v0}, y, {t, t0, t1}]
```

The result of evaluating this cell is an **InterpolatingFunction** that gives a solution for the chosen initial velocity **v0**. In the second argument **NDSolve**, we have specified that the unknown function is y rather than $y(t)$, so that the output will be in terms of a pure function. We will see that this slightly simplifies the code for the shooting method.

To see an example, we must first define the function f and choose an initial time and position. Let us take for our example the problem of an arrow shot in the vertical direction, adding a drag force due to air on the motion of the arrow. The acceleration of the arrow is taken as

Cell 1.71

```
f[t_, y_, v_] := -g-γv;
g = 9.8; γ = 0.1;
```

We are working in units of meters and seconds. The added acceleration due to air drag is assumed to be a linear function of the velocity, $-\gamma v(t)$, and we have taken the drag coefficient γ to equal 0.1 s^{-1}. Also, to complete the problem we must choose initial and final times, and initial and final positions:

Cell 1.72

```
t0 = 0; y0 = 0;
t1 = 1; y1 =20;
```

Then for a given initial velocity of, say, 30 m/s, the arrow's position vs. time is

Cell 1.73

```
Sol[30]

{{y → InterpolatingFunction [{{0., 1.}}, <>]}}
```

In Cell 1.74 we plot this **InterpolatingFunction** to see how close it comes to the required solution. The plot shows that we have overshot the mark, reaching somewhat higher than 20 m at $t = 1$ s. We need to lower the initial velocity a bit. Of course, we could do this by hand, and by making several attempts eventually we would obtain the required boundary conditions. However, it is easier to automate this process.

Cell 1.74

```
Plot[y[t]/.%[[1]], {t, 0, t1}];
```

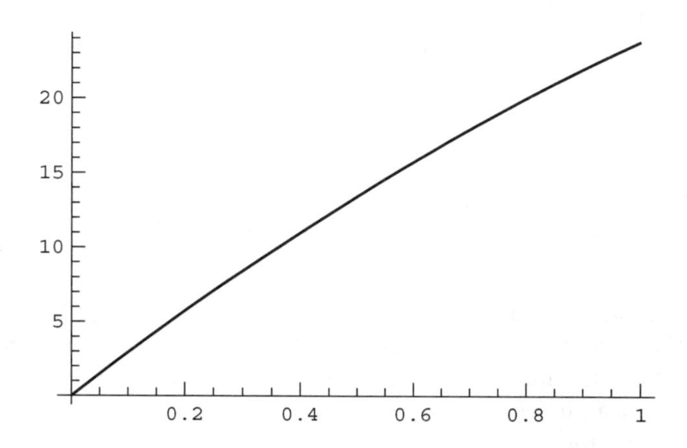

Let us define a function **yend[v0]**, the solution at the final time t_1:

Cell 1.75

```
yend[v0_] := y[t1]/.Sol[v0][[1]]
```

We can test whether this function is working by trying it out for the case of **v0** = 30 m/s:

Cell 1.76

```
yend[30]

23.8081
```

This value appears to agree with the trajectory plotted above.

Now we can apply the **FindRoot** function to solve the equation **yend[v0] == y1**. To do so, we will need to provide **FindRoot** with *two* initial guesses for **v0**, since the function **ysol[v0]** is not analytically differentiable. Since **v0 = 30** almost worked, we'll try two guesses near that:

Cell 1.77

```
FindRoot[yend[v0] == y1, {v0, 30, 29}]

{v0 → 25.9983}
```

Thus, a throw of about 26 m/s will hit the mark at 20-m height after one second. The trajectory is displayed in Cell 1.78 by evaluating **Sol** at this velocity and plotting the resulting function.

Cell 1.78

```
ysol = y/.Sol[v0/.%][[1]];
Plot[ysol[t], {t, 0, t1}];
```

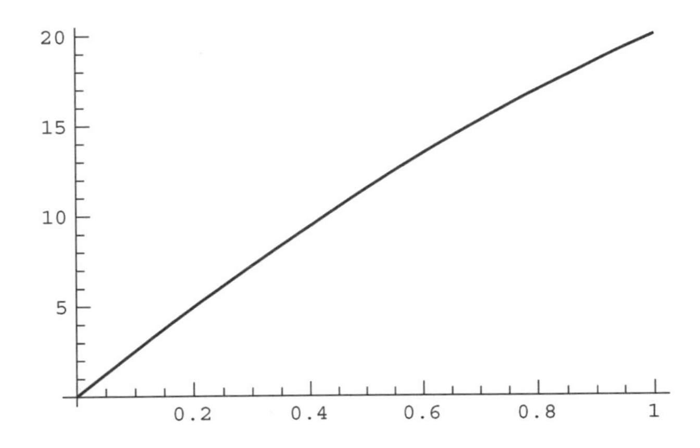

In this example of the shooting method, we found one solution to the boundary-value problem. How do we know that there are no other solutions? We don't. There could be other solutions that would be found if we made different choices for the initial velocity. (Actually, in this particular problem, one can show analytically that the above solution is unique, but for other problems this is not the case; see the exercises.)

This points out a major weakness in the shooting method:

> The shooting method only finds one solution at a time. To find a solution, reasonably accurate initial guesses must be made. Thus, it is possible to miss valid solutions to a boundary-value problem when using the shooting method.

EXERCISES FOR SEC. 1.5

(1) A thin rod of length L and mass ρ per unit length is clamped between two vertical walls at $x=0$ and $x=L$. In the absence of gravity, the rod would be horizontal, but in gravity the rod sags with a vertical displacement given by the function $y(x)$. According to the theory of elasticity, the shape of the rod satisfies the following boundary-value problem, assuming that the sag is small: $D(\partial^4/\partial x^4)y(x) = -\rho g$, where g is the acceleration of gravity and D depends on Young's modulus E and the cross-sectional area a of the rod according to $D = \alpha E a^2$, and where α is a dimensionless constant that depends on the shape of the cross section of the rod. The boundary conditions for a rod clamped at both ends are $y(0) = y'(0) = y(L) = y'(L) = 0$. Solve this problem analytically, and determine the shape $y(x)$ of the rod. How does the maximum sag scale with the length L of the rod, holding everything else fixed?

(2) A neutral plasma is a gas of freely moving charged particles, with equal amounts of positive and negative charge. If the plasma encounters a conductor to which a positive voltage is applied, negative plasma charges will be attracted to the conductor and positive charges will be repelled. As a result, an excess negative charge will surround a positively charged conductor. If the applied voltage is not too large, the plasma's net charge density $\rho(\mathbf{r})$ will satisfy a linear law:

$$\rho(\mathbf{r}) = -A\phi(\mathbf{r}), \tag{1.5.13}$$

where ϕ is the electrostatic potential in the plasma at position $\mathbf{r}$, and A is a positive constant that depends on the plasma density and temperature. The potential satisfies Poisson's equation (1.1.10), so Eq. (1.5.13) implies a linear PDE for the potential must be solved:

$$\nabla^2\phi = \frac{A}{\epsilon_0}\phi. \tag{1.5.14}$$

(a) Consider a plasma confined by conducting plates at $x=0$ and $x=L$. The plate at $x=0$ is biased to potential V, and the plate at $x=L$ is grounded. Solve analytically the 1D version of Eq. (1.5.14), $(\partial^2/\partial x^2)\phi=(A/\epsilon_0)\phi$, to obtain $\phi(x)$ between the plates.

(b) Repeat this solution numerically using the shooting method. Take $L=2$ and $A/\epsilon_0=1$. Plot the analytic and numerical results for $\phi(x)$.

(3) An artillery sergeant is asked to hit a fixed object at position $(x,y)=(d,0)$ with respect to his cannon. The muzzle velocity of the field piece is fixed at v_0, but the angle θ of the muzzle with respect to the horizontal can be varied.

(a) Solve this boundary-value problem analytically for θ. Show that there are two solutions for θ if the distance to the object is not too great, but that there is no solution if d exceeds a distance $d_{\max}$, and find $d_{\max}$. (Note that the time of impact is unimportant in this problem.)

(b) Create a module that will perform this solution using the shooting method, for given d. Use it to solve the problem where $v_0=1000$ m/s, $d=1$ km, and there is linear damping of the shell's velocity, with rate $\gamma=0.3$. Plot the (x,y) trajectory of the shell. By choosing a different initial guess, have the method converge to other solutions, if any; and plot all on the same graph of y vs. x.

(4) **(a)** A jet aircraft follows a straight trajectory given by $\mathbf{R}_{\text{jet}}(t)=(v_{\text{jet}}t+x_0, y_0, z_0)$, where $v_{\text{jet}}=250$ m/s, $x_0=-500$ m, $y_0=800$ m, and $z_0=5000$ m. An antiaircraft gun at $\mathbf{r}=0$ is trying to shoot the plane down. The muzzle velocity of the gun is $v_0=600$ m/s. If the gun fires a shell at $t=0$, where should it aim (i.e., what is the direction of the initial shell velocity)? Solve the problem *analytically* (using *Mathematica* to help with the algebra; the final equation needs to be solved numerically using `FindRoot`), keeping only the force of gravity on the shell. Plot the trajectory of the

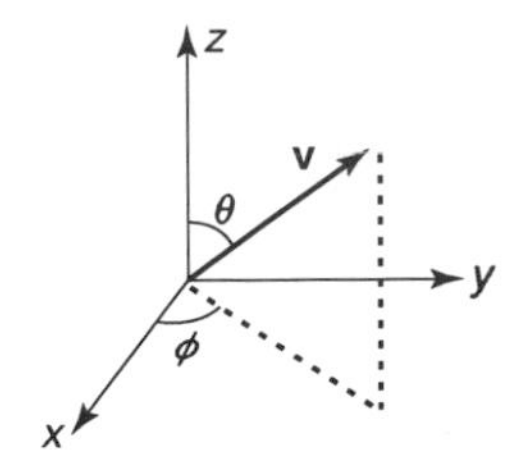

Fig. 1.16 Spherical polar angles (θ, ϕ) describing the direction of a vector **v** with respect to fixed axes.

shell and the plane using a three-dimensional parametric plot (`ParametricPlot3D`) up to the instant of impact. Is there more than one solution? [Hint: It is useful to introduce spherical polar angles (θ, ϕ) to describe the direction of the initial shell velocity: $\mathbf{v}_0 = v_0 (\sin\theta\cos\phi, \sin\theta\sin\phi, \cos\theta)$. See Fig. 1.16.

(b) Repeat the procedure using the shooting method, but now add a frictional deceleration of the form $-\gamma v$, where $\gamma = 0.075\ \mathrm{s}^{-1}$.

(5) James Bond, mass 85 kg, needs to jump off an overpass onto the bed of a passing truck 12 m below. He is attached to a bungee cord to break his fall, with a nonlinear spring force of $-1.1y^3$ newtons, where y is the displacement of Bond from the overpass measured in meters. A positive displacement corresponds to moving *down*. By eye he quickly calculates that the truck will be beneath him in 2.1 seconds. He immediately jumps.

(a) Use the shooting method to numerically determine what vertical velocity he must give himself, neglecting friction with the air, so that he lands on the truck at just the right instant. (A positive velocity corresponds to jumping down.) Plot Bond's trajectory $y(t)$.

(b) Can you find other, less appealing solutions for Bond's initial velocity that involve multiple oscillations at rather high speed?

(6) On January 1, 2001, at 12 noon GMT, a spacecraft is located 500 km above the earth's surface, on the night side, along the line directly connecting the earth to the sun. The computer controlling the spacecraft (a HAL9000, of course) has been asked to ensure that the ship will be at the future location of Jupiter exactly three years from this instant. (To be precise, the location is to be 100,000 km on the inboard side of Jupiter on a line toward the sun.) Use a shooting method and the information in Table 1.2 to determine the HAL9000's solution for the required initial velocity of the spacecraft, and plot the trajectory of the craft through the solar system. (Hint: To speed up the orbit integration, keep only the orbits of the earth, Mars, and Jupiter in the simulation. Use the molecular dynamics code developed in Sec. 2.4 to determine the orbits.) The solution is shown graphically in Fig. 1.17 as a plot of the orbits. (The plot can be viewed from different angles by dragging on it with the mouse.)

(7) The temperature of a thin rod of unit length satisfies $d^2T/dx^2 = T(x)^4$ (in suitably scaled units). [The T^4 term represents heat loss due to radiation, and the d^2T/dx^2 term arises from thermal conduction: see Chapter 3.] Find and plot $T(x)$ assuming the ends are held at fixed temperature: $T(0) = T(1) = 1$.

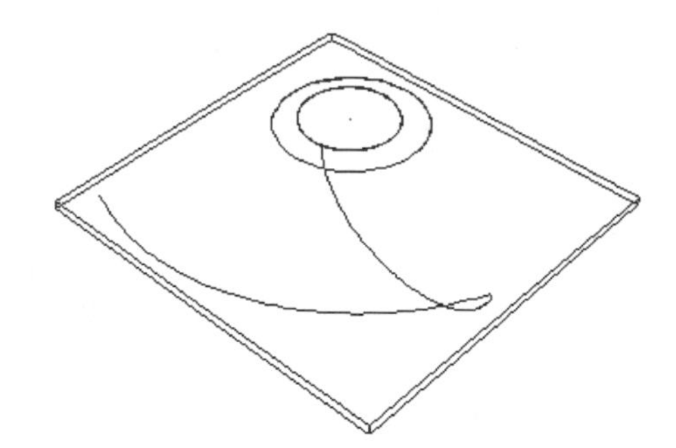

Fig. 1.17 Solution to Exercise (6).

1.6 LINEAR ODES

1.6.1 The Principle of Superposition

Linear ODEs of Order N A linear ODE is distinguished by the following property: the equation is *linear* in the unknown function; that is, only the first power of this function or its derivatives appear. An Nth-order linear ODE has the form

$$\frac{d^N x}{dt^N} + u_{N-1}(t)\frac{d^{N-1}x}{dt^{N-1}} + \cdots + u_1(t)\frac{dx}{dt} + u_0(t)x = f(t). \tag{1.6.1}$$

We have already seen several examples of linear differential equations, such as Eq. (1.1.1) (linear in v), or Eq. (1.1.6) (linear in x). Another example is the driven damped harmonic oscillator

$$\frac{d^2x}{dt^2} + \gamma\frac{dx}{dt} - \omega_0^2 x = f(t), \tag{1.6.2}$$

where γ and ω_0^2 are time-independent nonnegative constants. This equation describes damped harmonic motion with natural frequency ω_0 and damping rate γ, driven by an external force $mf(t)$ (where m is the oscillator's mass).

There are many, many other linear ODEs that have physical significance. Linear differential equations play a special role in the physical sciences, appearing in literally every discipline. As a consequence, the properties of these equations and their solutions have received considerable attention.

Linear Differential Operators An operator is simply a rule that transforms one function into another. An integral is an operator, and so is a square root. *Differential* operators are combinations of derivatives that act on a given function, transforming it into another function. For example, $\hat{L}f = e^{df/dt}$ defines a differential operator $\hat{L}$ (i.e., a rule) that takes the function $f(t)$ to the function $e^{df/dt}$.

Let us define a *linear differential operator* of Nth-order as

$$\hat{L} = \frac{d^N}{dt^N} + u_{N-1}(t)\frac{d^{N-1}}{dt^{N-1}} + \cdots + u_1(t)\frac{d}{dt} + u_0(t). \tag{1.6.3}$$

For the moment, think of this as merely a convenience, so that we can write Eq. (1.6.1) in the compact form $\hat{L}x = f$. Linear operators have the following two properties:

(1) For any two functions $f(t)$ and $g(t)$, $\hat{L}(f+g) = \hat{L}f + \hat{L}g$.
(2) For any function $f(t)$ and any constant C, $\hat{L}(Cf) = C\hat{L}f$.

It is easy to see that the operator in Eq. (1.6.3) satisfies these properties, and so it is a linear operator. It is also easy to see that the integral of a function is another linear operator (a linear *integral* operator). However, the operator defined by $\hat{L}f = e^{df/dt}$ does not satisfy either property. It is a *nonlinear* differential operator. For the most part, we will concentrate on the properties of linear operators in this book. Some examples of nonlinear operators with relevance to physics can be found in Chapter 7.

The Superposition Principle One important property of linear ODEs is called the *principle of superposition.* Consider the general solution of Eq. (1.6.1), *assuming that the forcing function vanishes*: $\hat{L}x = f(t) = 0$. In this case the equation is termed *homogeneous*.

Now, the general solution of the ODE involves N undetermined constants, as discussed in Sec. 1.1. Let us arbitrarily choose any two different sets of values for these constants, and thereby obtain two different possible solutions to the *homogeneous* equation, $x_1(t)$ and $x_2(t)$ (corresponding to different initial or boundary conditions). Then the principle of superposition states that the linear combination

$$C_1 x_1(t) + C_2 x_1(t) \tag{1.6.4}$$

is also a solution of the homogeneous ODE (corresponding to some other initial or boundary conditions). This follows directly from the linear nature of the differential equation, as we will now show.

By construction, the functions $x_1(t)$ and $x_2(t)$ have the property that $\hat{L}x_1 = \hat{L}x_2 = 0$. If we now substitute Eq. (1.6.4) into Eq. (1.6.1), we obtain

$$\hat{L}(C_1 x_1 + C_2 x_2) = \hat{L}(C_1 x_1) + \hat{L}(C_2 x_2) = C_1 \hat{L}x_1 + C_2 \hat{L}x_2 = 0, \tag{1.6.5}$$

verifying our contention that Eq. (1.6.4) satisfies the homogeneous ODE, and proving the principle of superposition.

The Principle of Superposition If $x_1(t)$ and $x_2(t)$ both satisfy the homogeneous linear ODE $\hat{L}x = 0$, then the linear combination $C_1 x_1(t) + C_2 x_2(t)$ also satisfies this ODE for any value of the constants C_1 and C_2.

1.6.2 The General Solution to the Homogeneous Equation

Introduction Let us return now to the discussion surrounding Eq. (1.6.3) regarding the general solution of the homogeneous equation $\hat{L}x = 0$. Rather than choosing only two sets of values for the N undetermined constants, let us choose

N different sets of values, so that we obtain N different functions, $x_1(t), x_2(t), \ldots, x_N(t)$. This means that *no one function can be obtained merely as a linear superposition of the others*. The functions are *linearly independent* of one another. Then it should be clear that the function obtained by superimposing these functions,

$$x(t) = C_1 x_1(t) + C_2 x_2(t) + \cdots + C_N x_N(t) \tag{1.6.6}$$

is a form of the general solution of the homogeneous ODE. Recall that the general solution has N undetermined constants that can be used to satisfy any particular initial condition. The fact that the functions $x_1(t), x_2(t), \ldots, x_N(t)$ are linearly independent means that Eq. (1.6.6) and its derivatives, evaluated at the initial time, *span* the space of possible initial conditions. By this I mean that any given initial condition can be met by appropriate choice of the constants. (Note the use of the term *span*, from linear algebra, denoting a set of vectors that can be made to sum to any other vector in a given vector space. As we have already mentioned, the connection between linear ODEs and linear algebra will be made clear in the next section.)

We have already seen an example of Eq. (1.6.6): Eq. (1.1.7) shows that the solution of the harmonic oscillator equation is a sum of the linearly independent solutions $\cos t$ and $\sin t$. Equation (1.6.6) shows that the general solution of a homogeneous linear ODE can always be written in this way, as a sum of N independent functions each of which satisfies the ODE.

> The general solution of a homogeneous linear ODE $\hat{L}x = 0$ can be written as a linear combination of N independent solutions to the ODE.

Let's consider possible analytic forms for these N independent solutions in the case that the functions $u_n(t)$ appearing in Eq. (1.6.1) are time-independent constants:

$$\hat{L}x = \frac{d^N x}{dt^N} + u_{N-1}\frac{d^{N-1}}{dt^{N-1}} + \cdots + u_1 \frac{dx}{dt} + u_0 x = f(t). \tag{1.6.7}$$

This important special case occurs, for example, in the driven damped harmonic oscillator, Eq. (1.6.2). We will guess the form $x(t) = e^{st}$ for some constant s. Using the fact that

$$\frac{d^n}{dt^n} e^{st} = s^n e^{st}, \tag{1.6.8}$$

the ODE $\hat{L}x = 0$ becomes a polynomial in s:

$$\left(s^N + u_{N-1}s^{N-1} + u_{N-2}s^{N-2} + \cdots + u_1 s + u_0\right) e^{st} = 0. \tag{1.6.9}$$

The bracket must be zero, so we are faced with finding the roots of this Nth-order polynomial in s. Although a general analytic solution cannot be found for $N > 4$, it is well known that there are always N roots (which may be complex), $\{s_1, s_2, \ldots, s_N\}$.

These N roots supply us with out N independent functions,

$$x_n(t) = e^{s_n t}, \qquad n = 1, 2, \ldots, N, \tag{1.6.10}$$

provided that none of the roots are the same. If two of the roots are the same, the roots are said to be *degenerate*. In this case only $N-1$ of the solutions have the form of Eq. (1.6.10). The Nth solution remains to be determined.

Let's assume that $s_1 = s_2$. Then consider any one of the constants in Eq. (1.6.9) to be a variable; take the constant u_0, for example, and replace it with a variable u. Now the roots all become functions of u, and in particular so do our two degenerate roots, $s_1 = s_1(u)$ and $s_2 = s_2(u)$. Furthermore, $s_1(u_0) = s_2(u_0)$, but in general, for $u \neq u_0$, $s_1(u) \neq s_2(u)$. Now let us write $u = u_0 + \epsilon$, and take the limit of the following superposition as ϵ vanishes:

$$\lim_{\epsilon \to 0} \frac{1}{\epsilon} \left(e^{s_1(\epsilon + u_0)t} - e^{s_2(\epsilon + u_0)t} \right). \tag{1.6.11}$$

According to the superposition principle, this sum is also a perfectly good solution to the equation. *Mathematica* can easily find the limit, obtaining a finite result:

Cell 1.79

```
s2[u0] = s1[u0] = s1;
Factor[Normal[Series[ϵ^-1 (E^(s1[u0 + ϵ] t)-E^(s2[u0 + ϵ] t)),
  {ϵ, 0, 0}]]]

e^s1 t t (s1'[u0]-s2'[u0])
```

The result, $te^{s_1 t}$ (neglecting the unimportant multiplicative constant), provides us with the new function necessary to complete the set of N independent solutions. The case of three or more degenerate roots, and the case where the multiplicative constant vanishes, can all be handled easily using similar methods to those detailed here, and will be left for the exercises.

Different Functional Forms for the General Solution Often it happens that the exponential form of the solutions in Eq. (1.6.10) is not the most convenient form. For example, for the undamped harmonic oscillator (1.1.6), the functions obtained via Eq. (1.6.10) are

$$x_1(t) = e^{i\omega_0 t}, \qquad x_2(t) = e^{-i\omega_0 t}. \tag{1.6.12}$$

For the damped harmonic oscillator (1.6.2), s satisfies a quadratic equation

$$s^2 + \gamma s + \omega_0^2 = 0, \tag{1.6.13}$$

which has solutions

$$\begin{aligned} s_1 &= -\frac{\gamma}{2} + i\sqrt{\omega_0^2 - \frac{\gamma^2}{4}}, \\ s_2 &= -\frac{\gamma}{2} - i\sqrt{\omega_0^2 - \frac{\gamma^2}{4}}. \end{aligned} \tag{1.6.14}$$

These solutions are complex (when $\omega_0 > \gamma/2$), and this can be an inconvenience in certain applications. Fortunately, the superposition principle says that we can replace the functions $x_1(t)$ and $x_2(t)$ with any linear combination of them. For example, the new functions

$$\bar{x}_1(t) = \tfrac{1}{2}[x_1(t) + x_2(t)]$$
$$\bar{x}_2(t) = \tfrac{1}{2i}[x_1(t) - x_2(t)] \tag{1.6.15}$$

form a useful set of independent solutions for damped or undamped oscillator problems, since standard trigonometric identities can be used show that these functions are real. For example, for the undamped harmonic oscillator, Eqs. (1.6.12) and (1.6.15) yield

$$\bar{x}_1(t) = \cos \omega_0 t,$$
$$\bar{x}_2(t) = \sin \omega_0 t, \tag{1.6.16}$$

which may be recognized as the usual real form for the independent solutions. We can then drop the overbars in Eq. (1.6.16) and treat these functions as our new independent solutions. Similarly, the real solutions to the damped harmonic oscillator equation are

$$\bar{x}_1(t) = e^{-\gamma t/2} \cos\left(\sqrt{\omega_0^2 - \frac{\gamma^2}{4}}\, t\right),$$
$$\bar{x}_2(t) = e^{-\gamma t/2} \sin\left(\sqrt{\omega_0^2 - \frac{\gamma^2}{4}}\, t\right). \tag{1.6.17}$$

These solutions are real, assuming that $\omega_0 > \gamma/2$. The solutions decay with time, and oscillate at a frequency less than ω_0 due to the drag force on the oscillator.

1.6.3 Linear Differential Operators and Linear Algebra

Consider the following homogeneous linear initial-value problem for the unknown function $x(t)$:

$$\hat{L}x = 0, \qquad x(t_0) = x_0, \quad x'(t_0) = v_0, \ldots, \tag{1.6.18}$$

where $\hat{L}$ is some linear differential operator. In this section we will show that the function $x(t)$ can be thought of as a vector, and the operator $\hat{L}$ can be thought of as a matrix that acts on this vector. We can then apply what we know about linear algebra to understand the behavior of solutions to linear ODEs.

To directly see the connection of Eq. (1.6.18) to linear algebra, consider trying to find a numerical solution to this ODE using Euler's method. We then discretize time, writing $t_n = t_0 + n\,\Delta t$. The function $x(t)$ is replaced by a set of values

$\{x(t_0), x(t_1), x(t_2), \ldots\}$, *which can be thought of as a vector* $\mathbf{x}$:

$$\mathbf{x} = \{x(t_0), x(t_1), x(t_2), \ldots\}.$$

Similarly, the ODE $\hat{L}x = 0$ becomes a series of linear equations for the components of $\mathbf{x}$, and therefore the operator $\hat{L}$ becomes a matrix $\mathbf{L}$ that acts on the vector $\mathbf{x}$. To see how this works in detail, consider the case of a simple first-order linear homogeneous ODE:

$$\hat{L}x = \frac{dx}{dt} + u_0(t)x = 0, \qquad x(t_0) = x_0. \tag{1.6.19}$$

Solving this ODE numerically via Euler's method, we replace Eq. (1.6.19) by

$$\begin{aligned} x(t_0) &= x_0, \\ x(t_1) - x(t_0) + \Delta t u_0(t_0) x(t_0) &= 0, \\ x(t_2) - x(t_1) + \Delta t u_0(t_1) x(t_1) &= 0, \\ &\vdots \end{aligned} \tag{1.6.20}$$

These linear equations can be replaced by the matrix equation

$$\begin{pmatrix} 1 & 0 & 0 & 0 & \cdots \\ -1 + u_0(t_0)\,\Delta t & 1 & 0 & 0 & \cdots \\ 0 & -1 + u_0(t_1)\,\Delta t & 1 & 0 & \cdots \\ 0 & 0 & -1 + u_0(t_2)\,\Delta t & 1 & \cdots \\ \vdots & \vdots & \vdots & \vdots & \ddots \end{pmatrix} \begin{pmatrix} x(t_0) \\ x(t_1) \\ x(t_2) \\ x(t_3) \\ \vdots \end{pmatrix} = \begin{pmatrix} x_0 \\ 0 \\ 0 \\ 0 \\ \vdots \end{pmatrix}. \tag{1.6.21}$$

The above matrix is a realization of the matrix $\mathbf{L}$ for this simple first-order ODE. All elements above the main diagonal are zero because the recursion relation determines the nth element of $\mathbf{x}$ in terms of earlier steps only. The right-hand side of Eq. (1.6.21) is a vector containing information about the initial condition. We will call this vector $\mathbf{x}_0 = \{x_0, 0, 0, \ldots\}$.

We can easily write the matrix $\mathbf{L}$ in terms of a special function called a Kronecker delta function, δ_{nm}. This function takes two integer arguments, n and m, and is defined as

$$\delta_{nm} = \begin{cases} 1, & n = m, \\ 0, & n \neq m. \end{cases} \tag{1.6.22}$$

The Kronecker delta function can be thought of as the (n, m) element of a matrix whose elements are all zero except along the diagonal $n = m$, where the elements arc cqual to one. This is the *unit matrix* unit, discussed in Sec. 9.5.2. In *Mathematica*, the function δ_{nm} is called `KroneckerDelta[n,m]`.

Using the Kronecker delta function, the components L_{mn} of **L** can be expressed as

$$L_{nm} = \delta_{nm} - \delta_{n-1,m}\left[1 - \Delta t u_0(t_m)\right]. \tag{1.6.23}$$

The matrix can then be created with a `Table` command:

Cell 1.80

```
L = Table[KroneckerDelta[n, m]-KroneckerDelta[n-1, m]
   (1-Δt u[m]), {n, 0, 3}, {m, 0, 3}];
MatrixForm[L]
```

$$\begin{pmatrix} 1 & 0 & 0 & 0 \\ -1+\Delta t\ u[0] & 1 & 0 & 0 \\ 0 & -1+\Delta t\ u[1] & 1 & 0 \\ 0 & 0 & -1+\Delta t\ u[2] & 1 \end{pmatrix}$$

Here we have only constructed four rows of the matrix, for ease of viewing.

Of course, the matrix and vectors of Eq. (1.6.21) are formally infinite-dimensional, but if we content ourselves with determining the solution only up to a finite time $t_f = M\Delta t + t_0$, we can make the matrices and vectors $M+1$-dimensional.

Note that Eq. (1.6.21) is only one of many different possible forms for the matrix equation. Recall that there are many different schemes for solving an ODE: the Euler method embodied by Eq. (1.6.21) is one, but the predictor–corrector method, for example, would lead to a different matrix **L** (see the exercises). This uncertainty shouldn't bother us, since the solution of the matrix equation always leads to an approximate solution of Eq. (1.6.19) that converges to the right solution as $\Delta t \to 0$, independent of the particular method used in obtaining the matrix equation.

We can write Eq. (1.6.21) in a more compact form using vector notation:

$$\mathbf{L}\cdot\mathbf{x} = \mathbf{x}_0. \tag{1.6.24}$$

This matrix equation is a discretized form of the ODE and initial condition, Eq. (1.6.19). It can be shown that *the more general ODE of Eq.* (1.6.18) *can also be put in this form*, although this takes more work. (Some examples can be found in the exercises.) A solution for **x** then follows simply by inverting the matrix:

$$\mathbf{x} = \mathbf{L}^{-1}\mathbf{x}_0. \tag{1.6.25}$$

Recall that it is not always possible to find the inverse of a matrix. However, according to Theorem 1.1, the solution to an initial-value problem always exists and is unique, at least for problems that satisfy the strictures of the theorem. For linear problems of this type, the matrix inverse can be taken, and the unique solution given by Eq. (1.6.25) can be found.

We can perform this matrix inversion numerically in *Mathematica*. But to do so, we must be more specific about the problem we are going to solve. Let's take the case $t_0 = 0$, $x_0 = 1$, and $u_0(t) = 1$, a constant damping rate. The equation we solve is then $dx/dt = -x$, $x(0) = 1$. Then the analytic solution to Eq. (1.6.19) is a simple exponential decay: $x(t) = \exp(-t)$.

To do the problem using matrix inversion, we choose a step size, say $\Delta t = 0.05$, and solve the problem only up to a finite time $t_f = 2$. This implies that the dimension M of the vector $\mathbf{x}_0$ is $2/0.05 + 1 = 41$ (the "+1" is necessary because $t = 0$ corresponds to the first element of $\mathbf{x}_0$), and the matrix $\mathbf{L}$ is 41 by 41. The following *Mathematica* statements set up the vector $\mathbf{x}_0$ and the matrix $\mathbf{L}$:

Cell 1.81

```
Δt = 0.05; u[n_] = 1; M = 40;

x0 = Table[0, {0, M}];
x0[[1]] = 1;

L = Table[KroneckerDelta[n, m]-KroneckerDelta[n-1, m]
   (1-Δt u[m]), {n, 0, M}, {m, 0, M}];
```

We then solve for $\mathbf{x}$ using Eq. (1.6.25), and create a data list **sol** consisting of times and positions $\{t_n, x_n\}$:

Cell 1.82

```
x = Inverse[L].x0;

sol = Table[{n Δt, x[[n + 1]]}, {n, 0, M}];
```

This solution can be plotted and compared with $\exp(-t)$ (see Cell 1.83), showing good agreement (which could be further improved by taking a smaller step size and increasing the dimension M of the system).

Cell 1.83

```
a = ListPlot[sol, PlotStyle→PointSize[0.012],
    DisplayFunction→Identity];
b = Plot[E^-t, {t, 0, 2}, DisplayFunction→Identity};
Show[a, b, DisplayFunction -> $Displayfunction,
  PlotLabel→"Matrix Inversion compared to E^-t",
    AxesLabel→{"t", " "}];
```

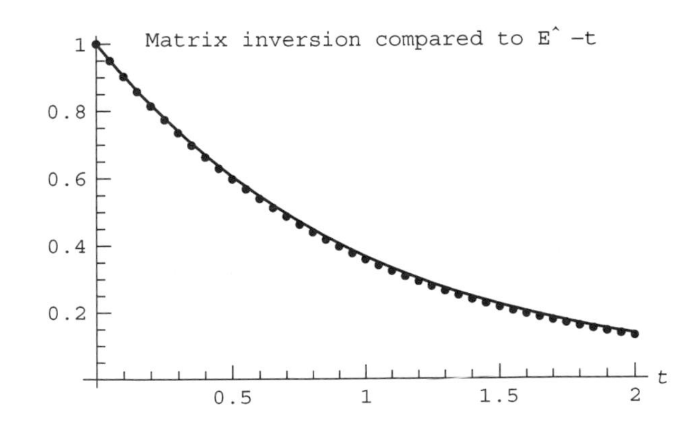

Note that the matrix-inverse method of solution, outlined above, is equivalent to the recursive solution of Eq. (1.6.20). In fact, performing the recursion in Eq. (1.6.20) can be thought of as just a way of performing the operations of taking the matrix inverse and applying the inverse to $\mathbf{x}_0$, so little is gained in any practical sense from using Eq. (1.6.25) rather than Eq. (1.6.20).

The matrix inverse method is really most useful in solving linear *boundary-value* problems, because matrix inversion solves the problem in a single step. This compares favorably with the shooting method for boundary-value problems (discussed in Sec. 1.5.2), which is an iterative process that requires several steps and an initial guess to find the solution.

Finally, we note the following: we have seen that a matrix $\mathbf{L}$ can be connected to any linear differential operator $\hat{L}$, and the inverse of the matrix, $\mathbf{L}^{-1}$, is useful in finding a solution to ODEs involving $\hat{L}$. Therefore, it may be useful to think about the *inverse of the operator itself*, which we might write as $\hat{L}^{-1}$. In fact, we will see in Chapter 2 that the inverse of a linear differential operator can be defined, that its discretized form *is* $\mathbf{L}^{-1}$, and that this operator inverse is connected to the idea of a *Green's function.*

1.6.4 Inhomogeneous Linear ODEs

Homogeneous and Particular Solutions In the preceding sections, we discussed solutions $x(t)$ to the homogeneous linear ODE for some linear differential operator $\hat{L}$. Let us now consider the case of an *inhomogeneous* linear ODE,

$$\hat{L}x = f. \tag{1.6.26}$$

We will examine the general solution of this problem, so that we do not have to specify boundary or initial conditions. Using the superposition principle, we write the general solution as a linear combination of two functions:

$$x(t) = x_h(t) + x_p(t), \tag{1.6.27}$$

where $x_h(t)$ is the general solution to the homogeneous problem, satisfying

$$\hat{L}x_h = 0, \tag{1.6.28}$$

and where $x_p(t)$ is any solution to the inhomogeneous problem, satisfying

$$\hat{L}x_p = f. \tag{1.6.29}$$

The function x_h is called the *homogeneous solution*, and the function x_p is called a *particular solution.*

Acting on Eq. (1.6.27) with $\hat{L}$, it is clear that $x(t)$ satisfies Eq. (1.6.26). It is also clear that Eq. (1.6.28) is the general solution of Eq. (1.6.26), since $x_h(t)$ contains all of the undetermined constants necessary to satisfy any given set of boundary or initial conditions.

We have already discussed how to find the homogeneous solution $x_h(t)$, in Sec. 1.6.2. The problem then comes down to finding a particular solution $x_p(t)$ to the inhomogeneous problem. This is actually rather nontrivial, and a complete and general answer will not be obtained until the end of Chapter 2. We will take the problem in steps of increasing difficulty.

Method of Undetermined Coefficients As a first step to finding a particular solution, we will consider the case where the ODE has constant coefficients; i.e., the functions $u_n(T)$ appearing in Eq. (1.6.1) are time-independent constants, so that the Nth-order ODE takes the form of Eq. (1.6.7). Also, we will assume that the forcing function $f(t)$ is of a very simple analytic form. With these assumptions, an analytic solution for $x_p(t)$ can be found simply by guessing a form for the solution. This is called the "method of undetermined coefficients" in elementary texts on ODEs.

Take, for example, the simple case of a linearly increasing force,

$$f(t) = a + bt. \tag{1.6.30}$$

For the response to this force, let's try the same form back again:

$$x_p(t) = A + Bt, \tag{1.6.31}$$

where A and B are undetermined coefficients. Acting on this guess with $\hat{L}$ yields

$$\hat{L}x_p = u_0(A + Bt) + u_1 B, \tag{1.6.32}$$

which is of the same form as f, provided that we now choose values for A and B correctly so as to satisfy Eq. (1.6.7):

$$u_0 A + u_1 B = a, \qquad u_0 B = b. \tag{1.6.33}$$

According to Eq. (1.6.31), one way that the system can respond to a linearly increasing force is for the amplitude to increase linearly as well: as you push harder on a spring, it stretches further. But this is only one possible solution; the spring could also oscillate. In fact, we know that the general solution to this problem is

$$x(t) = x_h(t) + A + Bt. \tag{1.6.34}$$

The oscillations are contained in the homogeneous solution $x_h(t)$, and their amplitude is set by the initial or boundary conditions.

Response to Sinusoidal Forcing There are many other analytically tractable forcing functions that we could consider. Of course, *Mathematica* could find such solutions for us, using `DSolve`. However, there is one more case that we will solve in detail by hand, because it will turn out to be of great importance to our future work: the case of an oscillating force of the form

$$f(t) = f_0 \cos \omega t. \tag{1.6.35}$$

A particular solution for this type of force can be found using the guess

$$x_p(t) = A\cos\omega t + B\sin\omega t, \tag{1.6.36}$$

where the constants A and B remain to be determined. In other words, the system responds to the oscillatory forcing with an oscillation *of the same frequency*. If we substitute this into Eq. (1.6.7) we obtain, after some work,

$$\begin{aligned}\hat{L}x_p = (\cos\omega t)\sum_{n=0}^{N/2}\left[Au_{2n}(-\omega^2)^n + B\omega u_{2n+1}(-\omega^2)^n\right]\\ +(\sin\omega t)\sum_{n=0}^{N/2}\left[Bu_{2n}(-\omega^2)^n - A\omega u_{2n+1}(-\omega^2)^n\right].\end{aligned} \tag{1.6.37}$$

This equation can be solved by choosing A and B so that the coefficient of $\sin\omega t$ vanishes and the coefficient of $\cos\omega t$ equals f_0.

A simpler alternative method of solution for this problem employs complex notation. We replace Eq. (1.6.35) by

$$f(t) = \operatorname{Re} f_0\, e^{-i\omega t}, \tag{1.6.38}$$

and we now try the form

$$x_p(t) = \operatorname{Re} C e^{-i\omega t}, \tag{1.6.39}$$

where $C = |C|e^{i\phi}$ is a complex number. The magnitude $|C|$ is the amplitude of the oscillation, and the argument ϕ is the phase of the oscillation with respect to the applied forcing. Using amplitude and phase notation, Eq. (1.6.39) can be written as

$$x_p(t) = |C|\cos(\omega t - \phi) = |C|\cos\phi\cos\omega t + |C|\sin\phi\sin\omega t. \tag{1.6.40}$$

By comparing Eq. (1.6.40) to Eq. (1.6.36) we can make the identifications $A = |C|\cos\phi$ and $B = |C|\sin\phi$.

The solution for $x_p(t)$ can again be found by substituting Eq. (1.6.39) into Eq. (1.6.7):

$$\hat{L}x_p = \operatorname{Re} C\hat{L}e^{-i\omega t} = \operatorname{Re} f_0\, e^{-i\omega t}, \tag{1.6.41}$$

where we have assumed that the coefficients u_n in Eq. (1.6.7) are real in order to take the operation Re through $\hat{L}$. Now, rather than solving only for the real part, we will solve the full complex equation. (If the full complex equation is satisfied, the real part will also be satisfied.) Also, we will use the fact that

$$\frac{d^n}{dt^n}e^{-i\omega t} = (-i\omega)^n e^{-i\omega t}, \tag{1.6.42}$$

so that Eq. (1.6.41) becomes

$$Ce^{-i\omega t}\sum_{n=0}^{N}(-i\omega)^n u_n = f_0\, e^{-i\omega t}. \tag{1.6.43}$$

Dividing through by the sum and using Eq. (1.6.39) allows us to write $x_p(t)$ in the following elegant form:

$$x_p(t) = \text{Re}\left(\frac{f_0}{\sum_{n=0}^{N} (-i\omega)^n u_n} e^{-i\omega t} \right). \tag{1.6.44}$$

In future chapters we will find that complex notation often simplifies algebraic expressions involving trigonometric functions.

Let us use Eq. (1.6.44) to explore the particular solution for the forced damped oscillator, Eq. (1.6.2). For the choice of u_n's corresponding to this ODE, Eq. (1.6.44) becomes

$$x_p(t) = \text{Re}\left(\frac{f_0}{-\omega^2 - i\omega\gamma + \omega_0^2} e^{-i\omega t} \right). \tag{1.6.45}$$

This particular solution oscillates at constant amplitude, and with the same frequency as the forcing. Since the homogeneous solutions decay with time [see Eq. (1.6.17)], Eq. (1.6.45) represents the form of the solution at times large compared to $1/\gamma$. At such large times, the oscillator has "forgotten" its initial conditions; every initial condition approaches Eq. (1.6.45). The convergence of different solutions can be seen directly in Fig. 1.18, which displays the time evolution of three different initial conditions. All three solutions converge to Eq. (1.6.45).

The loss of memory of initial conditions at long times is a general feature of linear driven damped systems. Nonlinear driven damped systems, such as the Van der Pol oscillator [Eq. (1.2.20)] with a driving term added, also display loss of memory of initial conditions; but initial conditions do not necessarily collapse onto a single trajectory as in Fig. 1.18. For instance, orbits can collapse onto a strange attractor, and subsequently wander chaotically across the surface of this attractor. A detailed analysis of the complex chaotic behavior of nonlinear driven damped systems is beyond the scope of this introductory text; see Ott (1993) for a discussion of this subject.

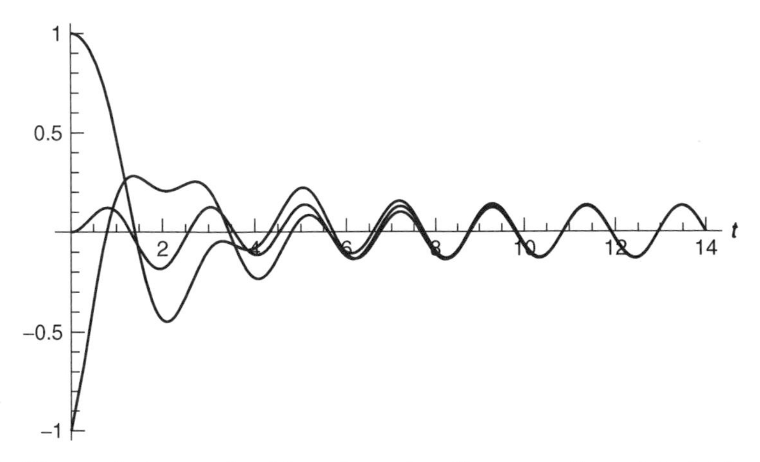

Fig. 1.18 Three solutions to the driven damped oscillator equation $x'' + x' + 2x = \cos 3t$.

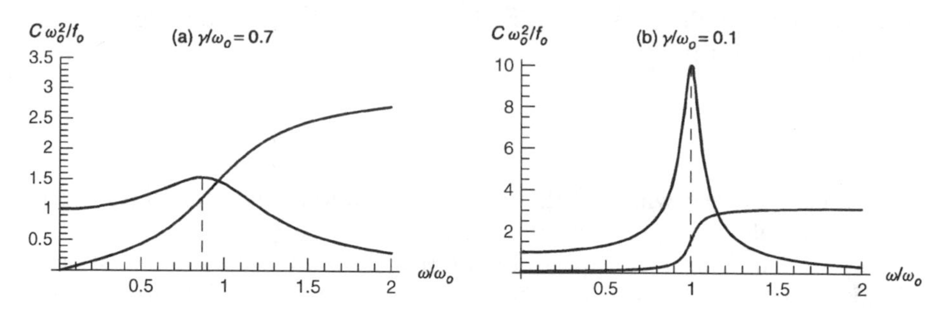

Fig. 1.19 Magnitude (thick line) and phase (thin line) of the amplitude of the particular solution to the driven damped oscillator: (a) heavy damping, $\gamma/\omega_0 = 0.7$; (b) light damping, $\gamma/\omega_0 = 0.1$. Also shown is the location of the peak amplitude, $\omega = \sqrt{\omega_0^2 - \gamma^2/2}$ (dashed line).

Resonance The particular solution to the driven damped oscillator equation has an amplitude that depends on the frequency ω at which the system is driven. According to Eq. (1.6.45), this complex amplitude is $C = f_0/(-\omega^2 - i\omega\gamma + \omega_0^2)$. Plots of the magnitude and phase of C/f_0 are shown in Fig. 1.19 as a function of ω. These plots show that a resonance occurs when the driving frequency ω is close to the natural frequency ω_0: the amplitude of oscillation has a maximum near ω_0, and the phase of the oscillation changes from a value near zero (the oscillation is in phase with the driving force) to one near π (the oscillation has the opposite sign to the driving force). The resonance becomes sharper as the damping γ becomes weaker. The frequency at which the amplitude is maximized can be easily shown to be equal to $\sqrt{\omega_0^2 - \gamma^2/2}$ (see the exercises). Note that this is *not* the same as the frequency of unforced oscillations in a damped oscillator, $\sqrt{\omega_0^2 - \gamma^2/4}$ [see Eq. (1.6.17)].

When the damping equals zero exactly, the undamped oscillator exhibits an *exact resonance* when driven at $\omega = \omega_0$. Here the amplitude C becomes infinite, and therefore the form of the solution, Eq. (1.6.45), is no longer valid.

One way to find the solution at an exact resonance is to use **DSolve**:

Cell 1.84

```
Clear[x];

Expand[
 FullSimplify[x[t] /. DSolve[x"[t] + ω0^2x[t] == f0 Cos[ω0 t],
  x[t], t][[1]]]]
```

$$C[2]\ \mathrm{Cos}[t\,\omega_0] + C[1]\ \mathrm{Sin}[t\,\omega_0] + \frac{\mathrm{Cos}[t\omega_0]\ f_0}{2\ \omega_0^2} + \frac{t\ \mathrm{Sin}[t\,\omega_0]\ f_0}{2\ \omega_0}$$

In addition to the usual cosine and sine terms, there is a new term proportional to $t \sin \omega_0 t$. This term is an oscillation that grows in amplitude over time. At an exact undamped resonance, the force is always in phase with the oscillation, adding energy to the motion in every cycle. Since this energy is not dissipated by damping, the oscillation increases without bound. Of course, in any real oscillator, some form of damping or nonlinearity will eventually come into play, stopping the growth of the oscillation.

In later chapters we will run across examples of other linear ODEs that exhibit exact resonance when driven at a natural frequency of the system. In each case, the response grows with time, and therefore must be treated as a special case for which Eq. (1.6.44) does not apply. The simplest approach is to apply **DSolve** or the method of undetermined coefficients for the case of resonance, and to use Eq. (1.6.44) otherwise.

Nevertheless, it is useful to understand mathematically how this resonant behavior arises. Consider an undamped oscillator driven at a frequency just off resonance, with forcing $f(t)=f_0\cos[(\omega_0-\epsilon)t]$. Then the particular solution is given by Eq. (1.6.45) with $\gamma=0$:

$$x_p(t)=\frac{f_0\cos[(\omega_0-\epsilon)t]}{2\omega_0\epsilon-\epsilon^2}. \tag{1.6.46}$$

This oscillation has very large amplitude, approaching infinity as $\epsilon\to 0$. However, consider a different particular solution, one that is chosen to be *zero* initially. Such a solution can be obtained by adding in a homogeneous solution to the oscillator equation. One choice for the homogeneous solution is simply $A\cos\omega_0 t$, with the appropriate choice of the constant A so that the solution is zero at $t=0$:

$$x_p(t)=\frac{f_0}{2\omega_0\epsilon-\epsilon^2}\{\cos[(\omega_0-\epsilon)t]-\cos\omega_0 t\}. \tag{1.6.47}$$

The two cosine functions are at nearly the same frequency, and therefore exhibit the phenomenon of beats, as shown in Cell 1.85 for the case $\epsilon=0.1$ and $\omega_0=1$. Oscillations grow for a time, then decay due to the interference between the two cosine functions. The smaller the frequency difference between the two cosine oscillations, the longer the beats become. (Try changing the frequency difference ϵ in Cell 1.85.) Finally, in the limit as the difference $\epsilon\to 0$, the length of time between beats goes to infinity, and the initial linear growth in amplitude of the oscillation continues indefinitely; the oscillation grows without bound. To see this from Eq. (1.6.47) mathematically, we can take a limit (see Cell 1.86).

Cell 1.85

```
ϵ = 0.1;
Plot [Cos[ (1 - ϵ) t] - Cos[t], {t, 0, 200}];
```

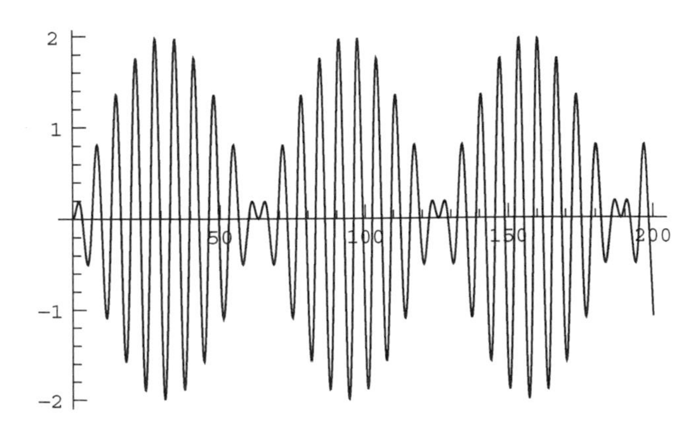

Cell 1.86

```
Limit[f0/(2ω0 ε - ε^2) (Cos[(ω0-ε) t]-Cos[ω0 t]), ε→0]
```

$$\frac{t\ \mathrm{Sin}[t\ \omega_0]\ f_0}{2\ \omega_0}$$

This limit reproduces the linear amplitude growth observed in the general solution of the undamped oscillator found previously using **DSolve**.

The examples we have seen so far in this section have all involved simple forcing functions. In the next chapter we will learn how to deal with general $f(t)$, and in the process develop an understanding of Fourier series and integrals.

EXERCISES FOR SEC. 1.6

(1) In the introduction to Sec. 1.6.1, we presented an operator $\hat{L}$ defined by $\hat{L}f = e^{df/dt}$, as an example of a nonlinear operator. On the other hand, consider the operator defined by $\hat{L}f = e^{d/dt}f$. Here,

$$e^{d/dt} = 1 + \frac{d}{dt} + \frac{1}{2!}\left(\frac{d}{dt}\right)^2 + \frac{1}{3!}\left(\frac{d}{dt}\right)^3 + \cdots$$

Is this operator linear or nonlinear? Find and plot the action of both operators on $f(t) = \sin t$, for $0 < t < 2\pi$. (Hint: The infinite series can be summed analytically.)

(2) Find, by hand, a complete set of independent solutions to the following linear homogeneous ODEs (you may check your results using *Mathematica*, of course):

(a) $x'''' + 2x''' + 3x'' + 2x' + 2x = 0$.

(b) $x'''' + 6x''' + 38x'' + 112x' + 104x = 0$.

(c) $x''' - 3x'' + 3x' - x = 0$.

(d) $x'' = 2(y - x) - x'$, $y'' = 2(x - y) - y'$.

(3) Use the matrix inversion technique to solve the following ODE numerically:

$$\hat{L}v = \frac{dv}{dt} + tv(t) = 0, \qquad v(0) = 1.$$

Use the Euler method form, Eq. (1.6.23), for the finite-differenced version of the operator $\hat{L}$ on the range $0 < t < 3$, with $\Delta t = 0.05$. Plot the solution on this time interval, and compare it with the exact solution found with *Mathematica*.

(4) A finite-difference method for second-order ODEs was discussed in Sec. 1.4; see Eqs. (1.4.27) and (1.4.28). Using this method, finite-difference *Airy's equation*

$$\frac{d^2x}{dt^2} = -tx(t)$$

with initial conditions $x(-1)=1$, $x'(-1)=0$. Write the ODE and initial conditions as a matrix equation of the form (1.6.24). Solve the ODE by matrix inversion, taking $\Delta t=0.1$, for $-1<t<5$, and plot the result along with the analytic result found from *Mathematica* using `DSolve`.

(5) **(a)** For the following general first-order linear ODE, find the matrix $\mathbf{L}$ that corresponds to the second-order predictor–corrector method, and write out the first four rows of $\mathbf{L}$:

$$\frac{dx}{dt}+u_0(t)x=0.$$

Use this matrix to solve the initial value problem where $u_0(t)=t/(1+t^2)$ and $x(0)=1$, for $0<t<5$, taking $\Delta t=0.1$. Plot $x(t)$ and, on the same plot, compare it with the exact solution found using `DSolve`.

(6) Add the following forcing functions to the right-hand sides of the problems listed in Exercise (2), and solve for a particular solution by hand using the method of undetermined coefficients (you can use *Mathematica* to help with the algebra, and to check your answers):

(a) $f(t)=\sin t$ to Exercise (2)(a) and (b).

(b) $f(t)=t^3$ to Exercise (2)(c).

(c) $x''=2(y-x)-x'$, $y''=2(x-y)+\cos 2\text{t}$.

(7) **(a)** Find the potential $\phi(x)$ between two parallel conducting plates located at $x=0$ and at $x=L$. The potential on the left plate is V_1, and that on the right plate is V_2. There is a charge density between the plates of the form $\rho=\rho_0\cos kx$. The potential satisfies $d^2\phi/dx^2=-\rho/\epsilon_0$.

(b) Discuss the behavior of the answer from part (a) for the case of constant charge density, $k=0$.

(8) Consider an *LRC* circuit driven by an oscillating voltage $V(t)=V_0\cos\omega t$. The charge on the capacitor satisfies Eq. (1.3.2). The homogeneous solution was found in Sect. 1.3, Exercise (3).

(a) Find a particular solution using the method of complex exponentials, $x_p(t)=\text{Re}(Ce^{-i\omega t})$.

(b) For $V_0=1$ volt, $R=2$ ohms, $C=100$ picofarads and $L=2\times10^{-3}$ henry, find the resonant frequency of the circuit. Plot the amplitude $|C|$ and phase ϕ of the particular solution vs. ω over a range of ω from zero to twice the resonant frequency.

(9) A damped linear oscillator has mass m and has a Hooke's-law force constant k and a linear damping force of the form $F_d=m\gamma v(t)$. The oscillator is driven by an external periodic force of the form $F_{\text{ext}}(t)=F_0\sin\omega t$.

(a) Find a particular solution in complex exponential form, $x_p(t)=\text{Re}(Ce^{-i\omega t})$

(b) The rate of work done by the external force on the mass is $dW_{\text{ext}}/dt=F_{\text{ext}}(t)v(t)$. Using the particular solution $v_p(t)$ from part (a), find a (time-independent) expression for $\langle dW_{\text{ext}}/dt\rangle$, the average rate of work done on the mass, averaged over an oscillation period $2\pi/\omega$. (Hint: Be careful!

When evaluating the work the real solution for $v_p(t)$ must be used. Use *Mathematica* to help do the required time integration over a period of the oscillation.)

(c) According to part (b), work is being done on the mass by the external force, but according to part (a) its amplitude of oscillation is not increasing. Where is the energy going?

(d) Work is done by the damping force F_d on the mass m at the rate $dW_d/dt = F_d(t)v(t)$. The work is negative, indicating that energy flows from the mass into the damper. What happens to this energy? Using the particular solution from part (a) for $v(t)$, show that $dW_d/dt + dW_{\text{ext}}/dt = 0$.

(10) Find the length of time between beats in the function $\cos[(\omega_0 - \epsilon)t] - \cos\omega_0 t$ (i.e., the time between maxima in the envelope of the oscillation). Show that this time goes to infinity as $\epsilon \to 0$. (Hint: Write this combination as the real part of complex exponential functions, and go on from there.)

(11) Show that the response of a damped oscillator to a harmonic driving force at frequency ω, Eq. (1.6.45), has a maximum amplitude of oscillation when $\omega = \sqrt{\omega_0^2 - \gamma^2/2}$.

REFERENCES

W. E. Boyce and R. C. DiPrima, *Elementary Differential Equations* (John Wiley and Sons, New York, 1969).

A. J. Lichtenberg and M. A. Lieberman, *Regular and Chaotic Dynamics* 2nd ed. (Springer-Verlag, New York, 1992).

G. W. Marcy and R. P. Butler, *Detection of extrasolar giant planets*, Ann. Rev. Astron. and Astroph. **36**, 57 (1998).

W. H. Press, S. A. Teukolsky, W. T. Vetterling, and B. P. Flannery, *Numerical Recipes* (Cambridge University Press, Cambridge, 1986).

E. Ott, *Chaos in Dynamical Systems* (Cambridge University Press, Cambridge, 1993).

CHAPTER 2

FOURIER SERIES AND TRANSFORMS

2.1 FOURIER REPRESENTATION OF PERIODIC FUNCTIONS

2.1.1 Introduction

A function $f(t)$ is *periodic with period* T when, for any value of t,

$$f(t) = f(t+T). \tag{2.1.1}$$

An example of a periodic function is shown in Fig. 2.1. We have already encountered simple examples of periodic functions: the functions $\sin t$, $\cos t$, and $\tan t$ are periodic with periods 2π, 2π, and π respectively.

Functions that have period T are also periodic over longer time intervals $2T, 3T, 4T, \ldots$. This follows directly from Eq. (2.1.1):

$$f(t) = f(t+T) = f(t+2T) = f(t+3T) = \cdots. \tag{2.1.2}$$

For example, $\sin t$ has period 2π, but also has period $4\pi, 6\pi, \ldots$. We can define the *fundamental period* of a periodic functions as the smallest period T for which Eq. (2.1.1) holds. So the fundamental period of $\sin t$ is 2π, and that of $\tan t$ is π. When we speak of the period of a function, we usually mean its fundamental period. We will also have occasion to discuss the *fundamental frequency* $\Delta\omega = 2\pi/T$ of the function.

Why should we care about periodic functions? They play an important role in our quest to determine the particular solution to an ODE due to arbitrary forcing. In Sec. 1.6, we found the response of an oscillator to a simple periodic sine or cosine forcing. However, this response will clearly be more complicated for periodic forcing of the type shown in Fig. 2.1. We can determine this response by first writing the periodic forcing function as a sum of simple sine and cosine functions. This superposition is called a *Fourier series*, after the French mathematician Jean Fourier, who first showed how such a series can be constructed. Once we

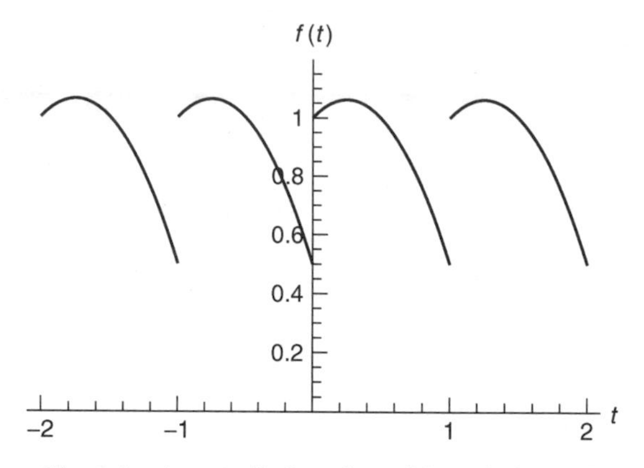

Fig. 2.1 A periodic function with period $T = 1$.

have this series for the forcing function, we can use the superposition principle to write the response of the oscillator as a sum of the individual responses to the individual cosine and sine terms in the series.

Later, we will find that Fourier series representation of periodic functions (and generalizations to the representation of nonperiodic functions) are also very useful in a number of other applications, such as the solution to certain common partial differential equations.

In order to expand a given periodic function $f(t)$ as a sum of sines and cosine functions, we must choose sines and cosines with the same periodicity as $f(t)$ itself. Since $f(t)$ has period T, we will therefore choose the functions $\sin 2\pi t/T, \sin 4\pi t/T, \sin 6\pi t/T \dots$ and $1, \cos 2\pi t/T, \cos 4\pi t/T, \cos 6\pi t/T, \dots$. These functions have fundamental periods T/n for integers $n = 0, 1, 2, 3, \dots,$ and therefore by Eq. (2.1.2) are also periodic with period T. Note that the constant function 1, with undefined period, is included. A few of these functions are shown in Cells 2.1 and 2.2.

Cell 2.1

```
<< Graphics`;
T = 1; Plot[{Sin[2Pi t/], Sin[4Pi t/T], Sin[6Pi t/T]},
 {t, 0, T}, PlotStyle→{Red, Blue, Purple}];
```

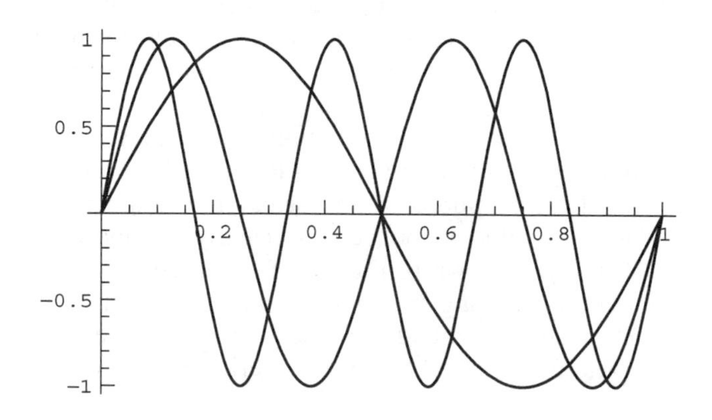

Cell 2.2

```
T = 1; Plot[{1, Cos[2 Pi t/T], Cos[4 Pi t/T]},
 {t, 0, T}, PlotStyle→{Red, Blue, Purple}];
```

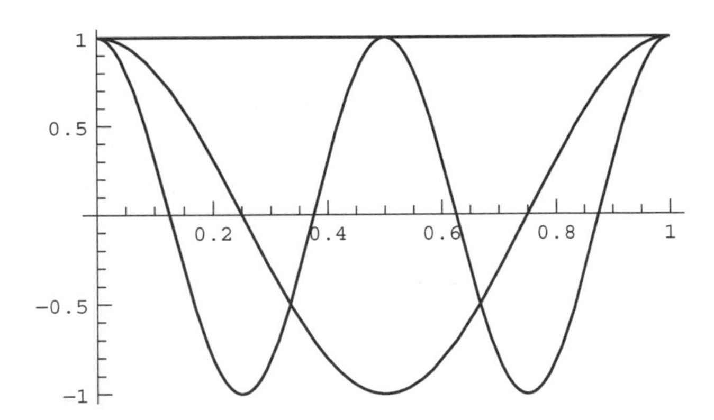

One can see that both $\cos 2\pi nt/T$ and $\sin 2\pi nt/T$ become more and more rapidly varying as n increases. The more rapidly varying functions will be useful in helping describe rapid variation in $f(t)$.

A general linear combination of these sines and cosines constitute a *Fourier series*, and has the form

$$a_0 + \sum_{n=1}^{\infty} \left(a_n \cos \frac{2\pi nt}{T} + b_n \sin \frac{2\pi nt}{T} \right),$$

where the constants a_n and b_n are called *Fourier coefficients*. The functions $\cos 2\pi nt/T$ and $\sin 2\pi nt/T$ are often referred to as *Fourier modes*.

It is easy to see that the above Fourier series has the correct period T. If we evaluate the series at time $t+T$, the nth cosine term is $\cos[2\pi n(t+T)/T] = \cos(2\pi nt/T + 2\pi n) = \cos 2\pi nt/T$, where in the last step we have used the fact that cosine functions have period 2π. Thus, the cosine series at time $t+T$ returns to the form it had at time t. A similar argument shows that the sine series evaluated at time $t+T$ also returns to its form at time t. Therefore, according to Eq. (2.1.1) the series is periodic with period T.

The fact that Fourier coefficients can be found that allow this series to equal a given periodic function $f(t)$ is a consequence of the following theorem:

Theorem 2.1 If a periodic function $f(t)$ is continuous, and its derivative is nowhere infinite and is *sectionally* continuous, then it is possible to construct a Fourier series that equals $f(t)$ for all t.

A sectionally continuous periodic function is one that is continuous in finite-size sections, with either no discontinuities at all or at most a finite number of discontinuities in one period of the function. Figure 2.1 is an example of a sectionally continuous periodic function. We will see later in this section what happens to the Fourier representation of $f(t)$ when $f(t)$ violates the restrictions placed on it by Theorem 2.1. For now, we assume that the function $f(t)$ satisfies the requirements of the theorem, in which case Fourier coefficients can be found such that

$$f(t) = a_0 + \sum_{n=1}^{\infty} \left(a_n \cos \frac{2\pi nt}{T} + b_n \sin \frac{2\pi nt}{T} \right) \qquad \text{for all } t. \tag{2.1.3}$$

2.1.2 Fourier Coefficients and Orthogonality Relations

We are now ready to find the Fourier coefficients a_n and b_n that enter into the Fourier series representation of a given periodic function $f(t)$. These coefficients can be found by using an important property of the sine and cosine functions that appear in Eq. (2.1.3): the property of *orthogonality*. Two real functions $g(t)$ and $h(t)$ are said to be *orthogonal on the interval* $[a, b]$ if they satisfy

$$\int_a^b g(t) h(t)\, dt = 0. \tag{2.1.4}$$

The sine and cosine Fourier modes in Eq. (2.1.3) have this property of orthogonality on the interval $[t_0, t_0 + T]$ for *any* choice of t_0. That is, for integers m and n the Fourier modes satisfy

$$\begin{aligned} &\int_{t_0}^{t_0+T} \sin \frac{2\pi nt}{T} \sin \frac{2\pi mt}{T}\, dt = \int_{t_0}^{t_0+T} \cos \frac{2\pi nt}{T} \cos \frac{2\pi mt}{T}\, dt = 0, \qquad m \neq n, \\ &\int_{t_0}^{t_0+T} \sin \frac{2\pi nt}{T} \cos \frac{2\pi mt}{T}\, dt = 0. \end{aligned} \tag{2.1.5}$$

In the first equations, the restriction $m \neq n$ was applied, because a real function cannot be orthogonal with itself: for any real function $g(t)$ that is nonzero on a finite range within $[a, b]$, $\int_a^b g^2(t)\, dt$ must be greater than zero. This follows simply because $g^2(t) \geq 0$, so there is a finite positive area under the $g^2(t)$ curve. For this reason, when $m = n$ in Eq. (2.1.5), the first and last integrals return a positive result:

$$\int_{t_0}^{t_0+T} \sin^2 \frac{2\pi nt}{T}\, dt = \int_{t_0}^{t_0+T} \cos^2 \frac{2\pi nt}{T}\, dt = \frac{T}{2}, \quad n > 0, \tag{2.1.6}$$

$$\int_{t_0}^{t_0+T} \cos^2 \frac{2\pi 0t}{T}\, dt = T. \tag{2.1.7}$$

The last equation follows because $\cos 0 = 1$. The analogous equation for the sine functions,

$$\int_{t_0}^{t_0+T} \sin^2 \frac{2\pi 0t}{T}\, dt = 0,$$

is not required, since $\sin 0 = 0$ is a trivial function that plays no role in our Fourier series. Equations (2.1.5) and (2.1.6) can be proven using *Mathematica*:

Cell 2.3

```
g = {Sin, Cos};
Table [Table[
  FullSimplify [Integrate[g[[i]] [2Pin t/T] g[[j]] [2Pim t/T],
   {t, t0, t0 + T}],
   n≠m && n∈Integers&&m∈Integers], {j, i, 2}], {i, 1, 2}]

{{0, 0}, {0}}
```

Cell 2.4

```
Table[FullSimplify[
  Integrate[g[[i]] [2Pin t/T] ^2, {t, t0, t0 + T}],
   n∈Integers], {i, 1, 2}]

{T/2, T/2}
```

Note that in the last two integrals, we did not specify that $n > 0$, yet *Mathematica* gave us results assuming that $n \neq 0$. This is a case where *Mathematica* has not been sufficiently careful. We also need to be careful: as we can see in Eqs. (2.1.5)–(2.1.7),

> The $n = 0$ Fourier cosine mode is a special case that must be dealt with separately from the other modes.

These orthogonality relations can be used to extract the Fourier coefficients from Eq. (2.1.3). For a given periodic function $f(t)$, we can determine the coefficient a_m by multiplying both sides of Eq. (2.1.3) by $\cos 2\pi mt/T$ and integrating over one period, from t_0 to $t_0 + T$ for some choice of t_0:

$$\int_{t_0}^{t_0+T} f(t) \cos\frac{2\pi mt}{T}\,dt = \sum_{n=0}^{\infty} a_n \int_{t_0}^{t_0+T} \cos\frac{2\pi nt}{T} \cos\frac{2\pi mt}{T}\,dt + \sum_{n=1}^{\infty} b_n \int_{t_0}^{t_0+T} \sin\frac{2\pi nt}{T} \cos\frac{2\pi mt}{T}\,dt. \tag{2.1.8}$$

The orthogonality of the sine and cosine Fourier modes, as shown by Eq. (2.1.5), implies that every term in the sum involving b_n vanishes. In the first sum, only the $n = m$ term provides a nonzero integral, equal to $T/2$ for $m \neq 0$ and T for $m = 0$ according to Eq. (2.1.6). Dividing through by these constants, we arrive at

$$\begin{aligned} a_0 &= \frac{1}{T} \int_{t_0}^{t_0+T} f(t)\,dt, \\ a_m &= \frac{2}{T} \int_{t_0}^{t_0+T} f(t) \cos\frac{2\pi mt}{T}\,dt, \qquad m > 0. \end{aligned} \tag{2.1.9}$$

Similarly, the b_n's are determined by multiplying both sides of Eq. (2.1.3) by $\sin 2\pi mt/T$ and integrating from t_0 to $t_0 + T$ for some choice of t_0. Now orthogonality causes all terms involving the a_n's to vanish, and only the term proportional to b_m survives. The result is

$$b_m = \frac{2}{T}\int_{t_0}^{t_0+T} f(t)\sin\frac{2\pi mt}{T}\,dt, \qquad m > 0. \tag{2.1.10}$$

2.1.3 Triangle Wave

Equations (2.1.3), (2.1.9), and (2.1.10) provide us with everything we need to determine a Fourier series for a given periodic function $f(t)$. Let's use these equations to construct Fourier series representations for some example functions. Our first example will be a triangle wave of period T. This function can be created from the following *Mathematica* commands, and is shown in Cell 2.5 for the case of $T = 1$:

Cell 2.5

```
f[t_] := 2t/T /; 0 ≤ t < T/2;

f[t_] := 2 - 2t/T /; T/2 ≤ t < T;

T = 1; Plot[f[t], {t, 0, T}];
```

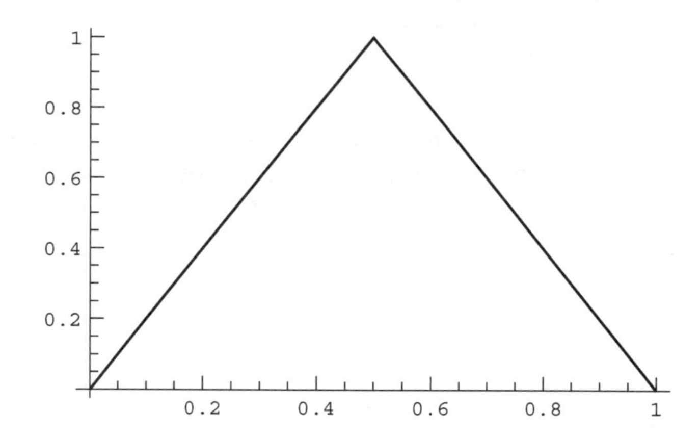

This is only one period of the wave. To create a periodic function, we need to define f for the rest of the real line. This can be done using Eq. (2.1.1), the definition of a periodic function, as a recursion relation for f:

Cell 2.6

```
f[t_] := f[t - T] /; t > T;
f[t_] := f[t + T] /; t < 0
```

Now we can plot the wave over several periods as shown in Cell 2.7.

Cell 2.7

```
Plot[f[t], {t, -3T, 3T}];
```

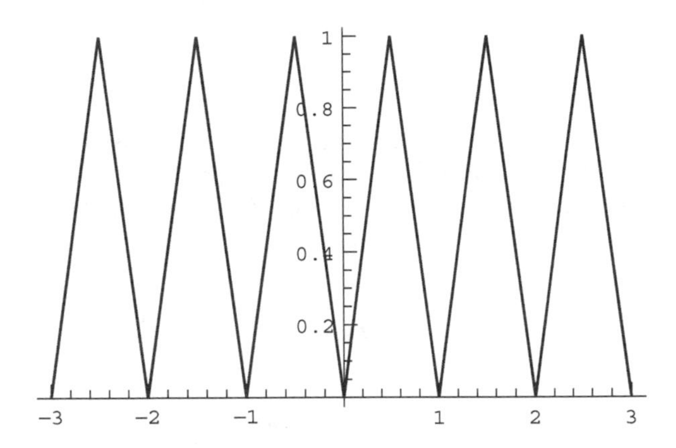

This function is continuous, and its derivative is sectionally continuous and not singular, so according to Theorem 2.1 the Fourier series representation of f should work. To test this conclusion, we first need to determine the Fourier coefficients. The a_n's are evaluated according to Eq. (2.1.9). We will perform this integral in *Mathematica* analytically, by choosing $t_0 = 0$ and breaking the integral over $f(t)$ into two pieces:

Cell 2.8

```
Clear[T];
a[n_] = FullSimplify[(2/T) Integrate[Cos[2Pin t/T] 2t/T,
  {t, 0, T/2}] +
   (2/T) Integrate[Cos[2Pin t/T] (2 - 2t/T), {t, T/2, T}],
    n ∈ Integers]
```

$$-\frac{2\ (-1)^n\ (-1+\ (-1)^n)}{n^2\pi^2}$$

Cell 2.9

```
a[0] = Simplify[(1/T) Integrate[2t/T, {t, 0, T/2}] +
    (1/T) Integrate[(2 - 2t/T), {t, T/2, T}]]
```

$$\frac{1}{2}$$

A list of a_n-values can now be constructed:

Cell 2.10

```
Table[a[n], {n, 0, 10}]
```

$$\left\{\frac{1}{2},\ -\frac{4}{\pi^2},\ 0,\ -\frac{4}{9\pi^2},\ 0,\ -\frac{4}{25\pi^2},\ 0,\ -\frac{4}{49\pi^2},\ 0,\ -\frac{4}{81\pi^2},\ 0\right\}$$

For future reference, we reproduce these results for our triangle wave below:

$$a_n = -\frac{4}{n^2\pi^2}, \qquad n \text{ odd}, \qquad\qquad (2.1.11)$$
$$a_0 = \tfrac{1}{2}.$$

Similarly, we can work out the b_n's by replacing the cosine functions in the above integrals with sine functions. However, we can save ourselves some work by noticing that $f(t)$ is an even function of t:

$$f(-t) = f(t). \qquad\qquad (2.1.12)$$

Since sine functions are odd in t, that is, $\sin(-\omega t) = -\sin \omega t$, the Fourier sum involving the sines is also an odd function of t, and therefore cannot enter into the representation of the even function f. This can be proven rigorously if we choose $t_0 = -T/2$ in Eq. (2.1.10). The integrand is an odd function multiplied by an even function, and is therefore odd. Integrating this odd function from $-T/2$ to $T/2$ must yield zero, so therefore $b_n = 0$.

> For an even function $f(t)$, the Fourier representation involves only Fourier cosine modes; for an odd function it involves only Fourier sine modes.

Thus, our triangle wave can be represented by a Fourier cosine series. We can construct this series in *Mathematica* provided that we keep only a finite number of terms; otherwise the evaluation of the series takes an infinitely long time. Let's keep only M terms in the series, and call the resulting function $f_{\text{approx}}(t, M)$:

Cell 2.11

```
fapprox[t_, M_] := Sum[a[n] Cos[2 Pi n t / T], {n, 0, M}]
```

For a given period T we can plot this function for increasing M and watch how the series converges to the triangle wave: see Cell 2.12. One can see that as M increases, the series approximation to f is converging quite nicely. This is to be expected: according to Eq. (2.1.11), the Fourier coefficients a_n fall off with increasing n like $1/n^2$, so coefficients with large n make a negligible contribution to the series.

Cell 2.12

```
T =1; Table[Plot[fapprox[t, M], {t, 0, 2}, PlotRange→
  {-.2, 1.2},
  PlotLabel→ "M ="<>ToString[M]], {M, 1, 11, 2}];
```

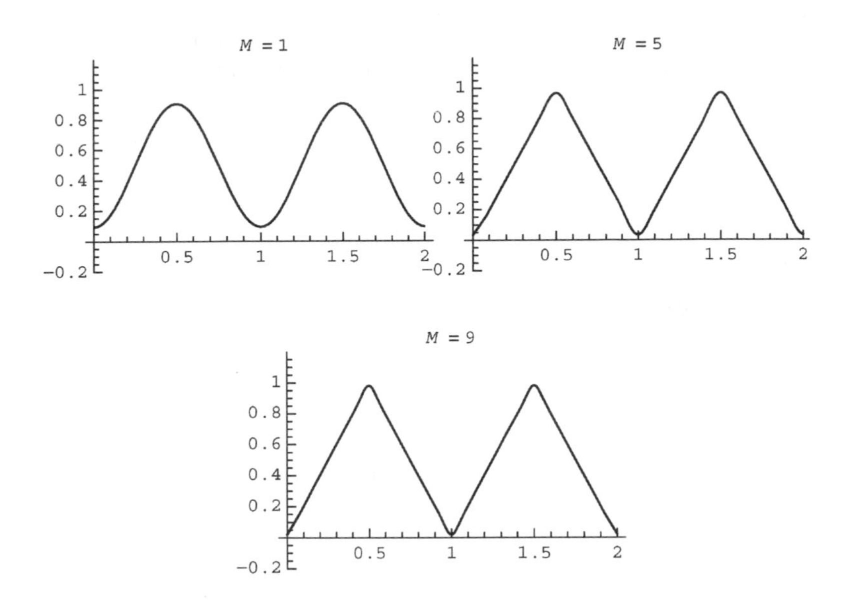

Although the series converges, convergence is more rapid in some places than others. The error in the series is greatest near the sharp points in the triangle wave. This should come as no surprise, since a sharp point introduces rapid variation that is difficult to reproduce by smoothly varying cosine Fourier modes. Functions with rapid variation must be described by rapid varying cosine and sine functions, which means that $n \gg 1$ terms must be kept in the Fourier series.

> Functions that vary smoothly can be well described by a finite Fourier series keeping a small number of terms. Functions with more rapid variation need more terms in the series.

Perhaps it is now starting to become clear as to why the restrictions on $f(t)$ are necessary in Theorem 2.1. If $f(t)$ has a discontinuity or its derivative is singular, it *cannot be represented properly by sine and cosine functions, because these functions do not have discontinuities or singularities*.

2.1.4 Square Wave

Our next example is a good illustration of what happens when a function violates the restrictions of Theorem 2.1. Consider a *square wave with period* T, defined by the following *Mathematica* commands:

Cell 2.13

```
Clear[f];
f[t_] := 1 /; 0 ≤ t < T/2;
f[t_] := -1 /; -T/2 ≤ t < 0
```

The definition of f is extended over the entire real line using the same recursive technique as for the triangle wave, as shown in Cell 2.14. Our square wave has been defined as an *odd function*, satisfying

$$f(-t) = -f(t), \tag{2.1.13}$$

Cell 2.14

```
f[t_] := f[t + T] /; t<-T/2;
f[t_] := f[t - T] /; t>T/2;

T = 1; Plot[f[t], {t, -3, 3}];
```

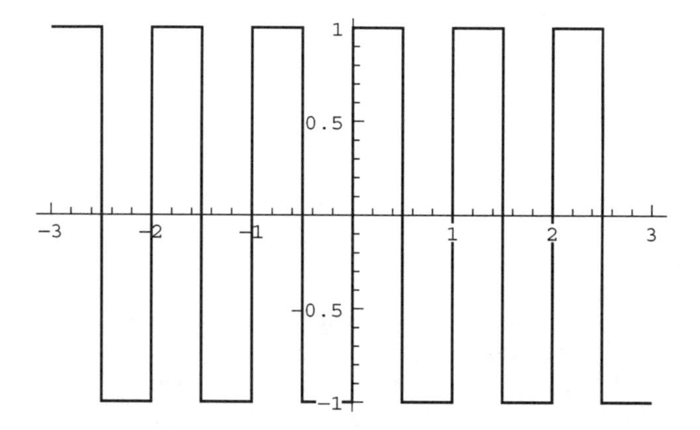

and therefore its Fourier representation will be as a sine series. The Fourier coefficients b_n follow from Eq. (2.1.10), and can be determined using *Mathematica* as follows:

Cell 2.15

```
b[n_] = FullSimplify[2/T (-Integrate[Sin[2Pin t/T],
    {t, -T/2, 0}] + Integrate[Sin[2Pin t/T],
     {t, 0, T/2}]), n∈Integers]

2 - 2 Cos[nπ]
-------------
     nπ
```

Thus, this Fourier series has the simple form

$$f_{\text{approx}}(t, M) = \frac{4}{\pi} \sum_{n=1(n\text{ odd})}^{M} \frac{1}{n} \sin \frac{2\pi nt}{T}. \tag{2.1.14}$$

The Fourier coefficients fall off rather slowly as n increases, like $1/n$. The coefficients for the triangle wave fell off more rapidly, as $1/n^2$ [see Eq. (2.1.11)]. This makes some sense, since the square wave is discontinuous and the triangle wave continuous, so the high-n terms in the square wave series have more weight. However, this is also a problem: because the high-n terms are so important, our finite approximation to the series will not converge the same way as for the triangle wave. Let's construct a finite series, $f_{\text{approx}}(t, M)$, and view its convergence with a table of plots as we did previously for the triangle wave. This is done in Cell 2.16. The series is clearly not converging as well as for the triangle wave. The discontinuity in the square wave is difficult to represent using a superposition of smoothly varying Fourier modes.

Cell 2.16

```
f_approx[t_, M_] := Sum[b[n] Sin[2Pi n t/T], {n, 1, M}];

T = 1; Table[Plot[f_approx[t, M], {t, -1, 1}, PlotRange→{-1.5,
  1.5}, PlotLabel→"M = " <>ToString[M]], {M, 4, 20, 4}];
```

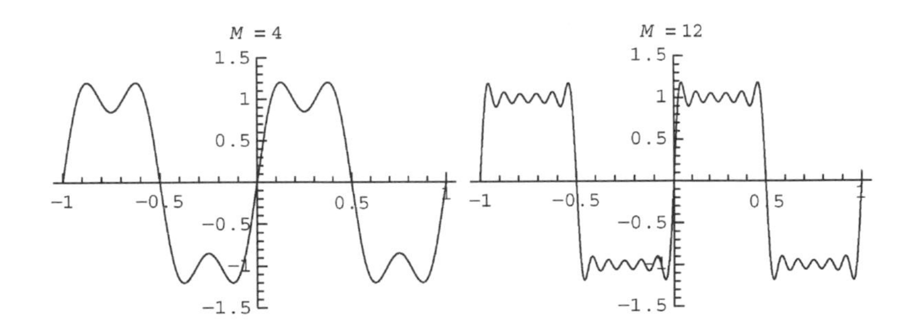

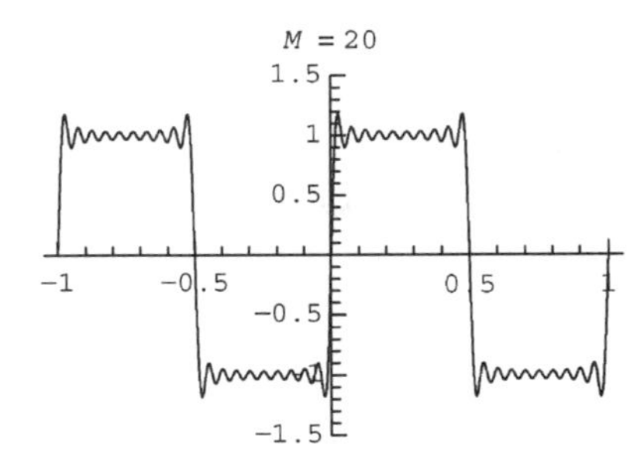

2.1.5 Uniform and Nonuniform Convergence

It is useful to consider the difference between the series approximation and the exact square wave as M increases. This difference is evaluated and plotted in Cell 2.17. The error has a maximum value of ± 1 at the discontinuity points $t = mT/2$, independent of M. This maximum error is easy to understand: the square wave takes on the values ± 1 at these points, but the Fourier series is zero there because at $t = mT/2$ the nth term in the series is proportional to $\sin(nm\pi) = 0$.

Cell 2.17

```
errorplot[M_] :=
  (a = f_approx[t, M]; Plot[a - f[t], {t, -0.5, 0.5},
  PlotRange→{-1, 1}, PlotPoints→100 M,
  PlotLabel→"Error, M = " <>ToString[M]]);

Table[errorplot[M], {M, 10, 50, 10}];
```

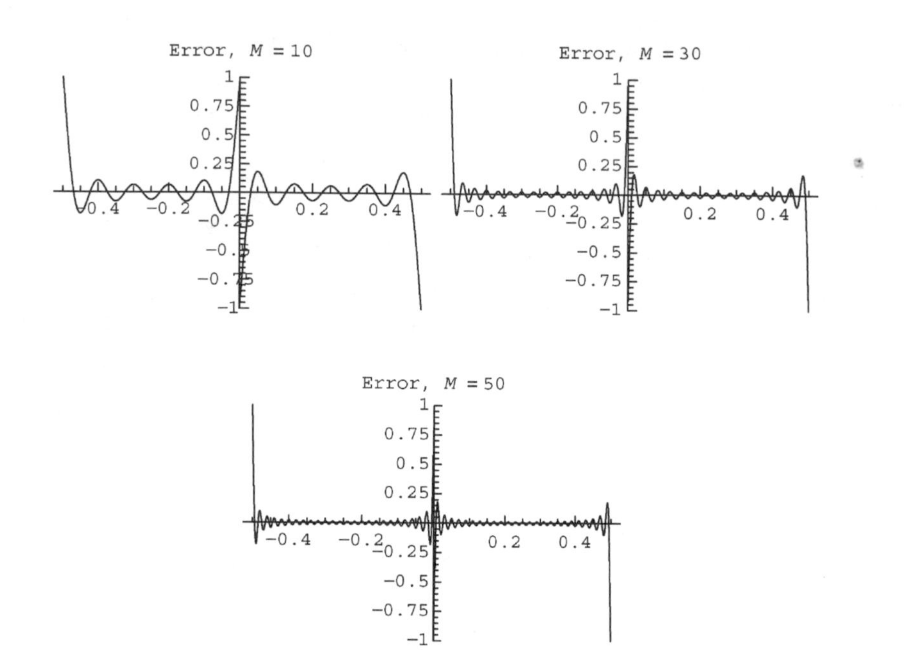

Furthermore, each successive peak in the error has a value that is independent of M: the first peak on the right side of the origin is at about 0.2, the next is at about 0.1, and so on, independent of M. In fact, in the next subsection we will show that the maximum size of error of the first peak is 0.1789..., that of the second is 0.0662..., independent of M. The constancy of these maximum errors as M increases is most easily observed by animating the above set of graphs. What you can see from the animation, however, is that while the height of the peaks is independent of M, the *width* of the peaks shrinks as M increases, and the peaks crowd in toward the origin. This strange behavior is called the *Gibbs phenomenon*.

As a result of the Gibbs phenomenon, for any finite value of M, no matter how large, there is always a small region around the discontinuity points where the magnitude of the error is independent of M. Although this region shrinks in size as M increases, the fact that the error is independent of M within this region distinguishes the behavior of this Fourier series from that of a function that satisfies the restrictions of Theorem 2.1, such as the triangle wave studied previously. There, the error in the series decreased *uniformly* as M increased. By this we mean that, as M increases, $|f_{\text{approx}}(t, M) - f(t)| \to 0$ for all t. This is called *uniform convergence of error*, and it is necessary in order for us to state that the left-hand and right-hand sides of Eq. (2.1.3) are strictly equal to one another for every t.

More precisely, as M increases, a uniformly convergent series satisfies

$$|f_{\text{approx}}(t, M) - f(t)| < \epsilon(M), \tag{2.1.15}$$

where $\epsilon(M)$ is some small number that is independent of t and that approaches zero as $M \to \infty$. Thus, the error in the series is bounded by $\epsilon(M)$, and this error goes to zero as M increases, independent of the particular value of t.

On the other hand, the behavior of the error in the series representation of the square wave is an example of *nonuniform convergence*. Here Eq. (2.1.15) is not satisfied for every value of t: we can find a small range of t-values around the discontinuities for which the error is not small, no matter how large we take M.

Functions that satisfy the restrictions of Theorem 2.1 *have Fourier series representations that converge uniformly*. But even for nonuniformly convergent series, the previous analysis of the square wave series show that the series can still provide a reasonably accurate representation of the function, provided that we stay away from the discontinuity points. Thus, Fourier series are often used to approximately describe functions that have discontinuities, and even singularities. The description is not exact for all t, but the error can be concentrated into small regions around the discontinuities and singularities by taking M large. This is often sufficient for many purposes in scientific applications, particularly in that there are no real discontinuities or singularities in nature; such discontinuities and singularities are always the result of an idealization, and therefore we usually need not be too concerned if the series representation of such functions does not quite describe the singular behavior.

2.1.6 Gibbs Phenomenon for the Square Wave

The fact that the width of the oscillations in the error decreases as M increases suggests that we attempt to understand the Gibbs phenomenon by applying a scale transformation to the time: let $\tau = Mt/T$. For constant τ the actual time t approaches zero as M increases. The hope is that in these scaled time units, the compression of the error toward the origin observed in the above animation will disappear, so that on this time scale the series will become independent of M as M increases.

In these scaled dimensionless time units, the series Eq. (2.1.14) takes the form

$$f_{\text{approx}}(\tau, M) = \frac{4}{\pi} \sum_{n=1(n\text{ odd})}^{M} \frac{1}{n} \sin \frac{2\pi n\tau}{M}. \tag{2.1.16}$$

There is still M-dependence in this function, so we will perform another scale transformation, defining $s = Mn$. Substituting this transformation into Eq. (2.1.16) yields

$$f_{\text{approx}}(\tau, M) = \frac{4}{\pi M} \sum_{s=1/M,\,3/M,\,5/M,\ldots}^{1} \frac{1}{s} \sin 2\pi s\tau. \tag{2.1.17}$$

The function $(\sin 2\pi s\tau)/s$ is independent of M and is well behaved as s varies on $[0, 1]$, taking the value $2\pi\tau$ at $s = 0$. Furthermore, the interval $\Delta s = 2/M$ between successive s-values decreases to zero as M increases, so we can replace the sum by an integral over s from 0 to 1:

$$\lim_{\Delta s \to 0} \left(\Delta s \sum_{s=\Delta s/2,\,3\Delta s/2,\,5\Delta s/2,\ldots}^{1} \frac{1}{s} \sin 2\pi s\tau \right) = \int_0^1 \frac{1}{s} \sin 2\pi s\tau \, ds. \tag{2.1.18}$$

Substituting this integral into Eq. (2.1.17) yields the following result for f_{approx}:

$$f_{\text{approx}}(\tau, M) = \frac{2}{\pi}\int_0^1 \frac{1}{s}\sin 2\pi s\tau\, ds. \tag{2.1.19}$$

As we hoped, f_{approx} is now independent of M when written in terms of the scaled time τ. It can be evaluated in terms of a special function called a *sine integral*:

Cell 2.18

```
fapprox[τ_] = Simplify[2/Pi Integrate[Sin[2Pi s τ]/s,
  {s, 0, 1}], Im[τ] == 0]
```

$$\frac{2\ \text{SinIntegral}[2\pi\tau]}{\pi}$$

A plot of this function vs. scaled time τ (Cell 2.19) reveals the characteristic oscillations of the Gibbs phenomenon that we observed previously. The largest error in the function occurs at the first extremum (see Cell 2.20).

Cell 2.19

```
Plot[fapprox[τ], {τ, -3, 3}];
```

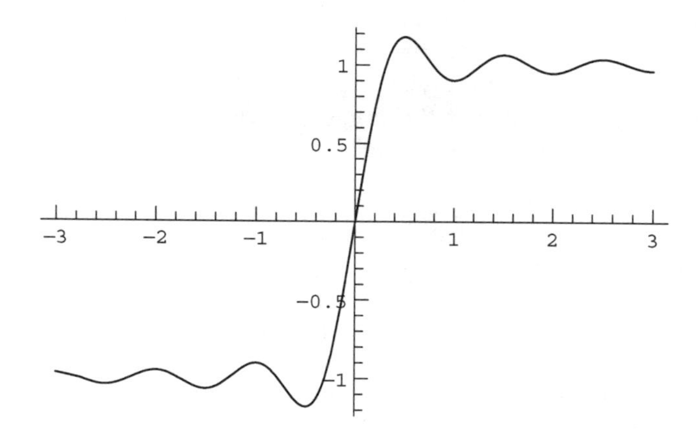

Cell 2.20

```
D[fapprox[τ], τ]
```

$$\frac{2\ \text{Sin}[2\ \pi\tau]}{\pi\tau}$$

This derivative vanishes at $\tau = n/2$, $n \neq 0$. The plot of f_{approx} shows that $\tau = \frac{1}{2}$ is the location of the first extremum. The maximum value of f_{approx} is therefore

Cell 2.21

```
fapprox[1/2]
```

$$\frac{2\ \text{SinIntegral}[\pi]}{\pi}$$

Cell 2.22

```
% // N
1.17898
```

Thus, the maximum overshoot of the oscillation above 1 is 0.17898.... The next maximum in f_{approx} occurs at $\tau = \frac{3}{2}$, with an overshoot above 1 of

Cell 2.23

```
fapprox[3/2] - 1 // N
0.0661865
```

If we return to regular time units and plot f_{approx} vs. t for different values of M, we can reproduce the manner in which the oscillations crowd toward the origin as M increases (see Cell 2.24). Near the time origin, the result looks identical to the behavior of the square wave Fourier series plotted in Cell 2.16, except that now f_{approx} is no longer periodic in t. The periodic nature of f_{approx} has been lost, because our integral approximation in Eq. (2.1.18) is correct only for t close to zero. [Equation (2.1.18) assumes τ remains finite as $M \to \infty$, so $t \to 0$.]

Cell 2.24

```
T = 1; Table [Plot[fapprox[Mt], {t, -1, 1}, PlotRange→
  {-1.5, 1.5},
  PlotLabel→"M = " <>ToString[M]], {M, 4, 20, 4}];
```

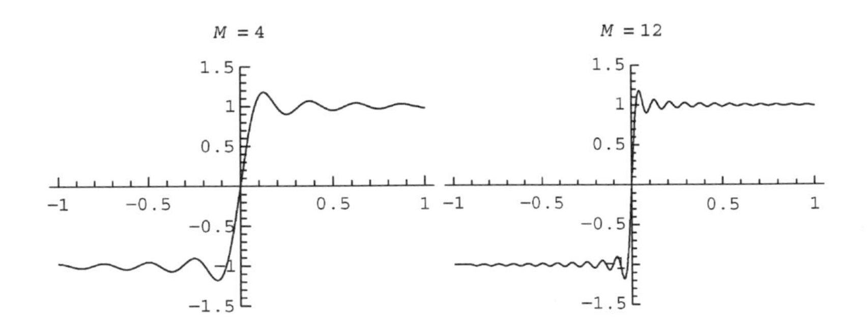

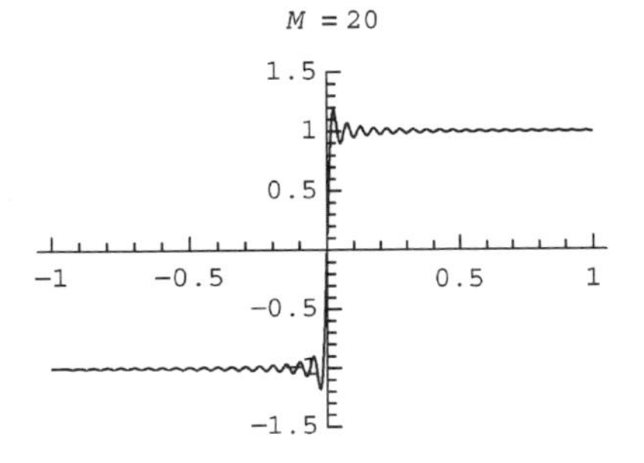

2.1.7 Exponential Notation for Fourier Series

In Sec. 1.6 we found that it could be useful to write a real periodic oscillation $a\cos\omega t + b\sin\omega t$ in the more compact complex notation, $\mathrm{Re}[C\exp(-i\omega t)]$, where C is a complex number. We can do the same thing for a Fourier series representation of a real periodic function of period T:

$$f(t) = a_0 + \sum_{n=1}^{\infty} a_n \cos(n\,\Delta\omega\, t) + \sum_{n=1}^{\infty} b_n \sin(n\,\Delta\omega\, t). \tag{2.1.20}$$

Here we have written the series in terms of the quantity $\Delta\omega = 2\pi/T$, which is the fundamental frequency of the periodic function (see Sec. 2.1.1).

In order to write this series in complex form, we will use the trigonometric identities

$$\cos x = \frac{e^{ix} + e^{-ix}}{2}, \qquad \sin x = \frac{e^{ix} - e^{-ix}}{2i}. \tag{2.1.21}$$

When these identities are employed in Eq. (2.1.20), and we combine the common terms involving $e^{in\,\Delta\omega\, t}$ and $e^{-in\,\Delta\omega\, t}$, we obtain

$$f(t) = a_0 + \sum_{n=1}^{\infty} \frac{a_n + ib_n}{2} e^{-in\,\Delta\omega\, t} + \sum_{n=1}^{\infty} \frac{a_n - ib_n}{2} e^{in\,\Delta\omega\, t}. \tag{2.1.22}$$

Note that, for real a_n and b_n, the second sum is the complex conjugate of the first sum. Using the fact that $z + z^* = 2\,\mathrm{Re}\, z$ for any complex number z, we see that Eq. (2.1.22) can be expressed as

$$f(t) = a_0 + \mathrm{Re}\left[\sum_{n=1}^{\infty} (a_n + ib_n)\, e^{-in\,\Delta\omega\, t}\right]. \tag{2.1.23}$$

If we now introduce complex Fourier coefficients C_n, defined as

$$\begin{aligned} C_0 &= a_0, \\ C_n &= a_n + ib_n, \qquad n > 0, \end{aligned} \tag{2.1.24}$$

we can write Eq. (2.1.22) in the following compact form:

$$f(t) = \mathrm{Re}\left[\sum_{n=0}^{\infty} C_n\, e^{-in\,\Delta\omega\, t}\right]. \tag{2.1.25}$$

Equation (2.1.25) is one form for an exponential Fourier series, valid for real functions $f(t)$. Another form that can also be useful follows from Eq. (2.1.22) by defining a different set of complex Fourier coefficients c_n:

$$c_0 = a_0,$$
$$c_n = \frac{a_n + ib_n}{2}, \qquad n > 0, \tag{2.1.26}$$
$$c_n = \frac{a_{-n} - ib_{-n}}{2}, \qquad n < 0.$$

The definition of these coefficients is extended to $n < 0$ for the following reason: this extension allows us to express the second sum in Eq. (2.1.22) as $\sum_{n=1}^{\infty} c_{-n}\, e^{in\,\Delta\omega\, t}$. Then by taking $n \to -n$ in this sum, and noting that this inversion changes the range of the sum to $-\infty$ to -1, we can combine the two sums and obtain

$$f(t) = \sum_{n=-\infty}^{\infty} c_n\, e^{-in\,\Delta\omega\, t}. \tag{2.1.27}$$

Equation (2.1.27) is a second form for the exponential Fourier series. It differs from the first form in that the real part is not taken, and instead the sum runs over both negative and positive n, from $-\infty$ to $+\infty$. Also, note that we did not assume that a_n and b_n are real, so Eq. (2.1.27) works for complex periodic functions $f(t)$ as well as for real periodic functions. For this reason, Eq. (2.1.27) is somewhat more general than Eq. (2.1.25), which applies only to real functions.

We are now left with the question of how to determine the complex Fourier coefficients. Of course, we could determine the real coefficients a_n and b_n and then use use either Eqs. (2.1.24) or Eqs. (2.1.26), but it would be better if we could determine the complex coefficients c_n (or C_n) directly without reference to the real coefficients. This can be done by using a new set of orthogonality relations, valid for complex exponential functions.

Before we can consider these orthogonality relations, we must first extend the notion of orthogonality, Eq. (2.1.4), to cover complex functions. Two complex functions $g(t)$ and $h(t)$ are said to be orthogonal on the interval $[a, b]$ if they satisfy

$$\int_a^b g(t)h(t)^*\, dt = 0. \tag{2.1.28}$$

The complex conjugation is added to the definition so that we can again say that a function cannot be orthogonal with itself: $\int_a^b g(t)g(t)^*\, dt = \int_a^b |g(t)|^2\, dt \geq 0$, with equality only for a function that equals zero across the interval. Of course, we could equally well have the complex conjugate of g rather than h in this definition, $\int_a^b g(t)^* h(t)\, dt = 0$.

The complex exponential Fourier modes, $e^{-in\,\Delta\omega\, t}$ (with $\Delta\omega = 2\pi/T$), satisfy the following orthogonality relations on the interval t_0 to $t_0 + T$, for any choice of t_0:

$$\int_{t_0}^{t_0+T} e^{-in\,\Delta\omega\, t}\left(e^{-im\,\Delta\omega\, t}\right)^*\, dt = 0, \qquad m \neq n. \tag{2.1.29}$$

This can easily be proven using a couple of lines of algebra:

$$\int_{t_0}^{t_0+T} e^{-in\Delta\omega t}(e^{-im\Delta\omega t})^* \, dt$$

$$= \int_{t_0}^{t_0+T} e^{-2\pi i(n-m)t/T} \, dt = \frac{T}{-2\pi i(n-m)} [e^{-2\pi i(n-m)(t_0+T)/T} - e^{-2\pi i(n-m)t_0/T}]$$

$$= \frac{T}{-2\pi i(n-m)} e^{-2\pi i(n-m)t_0/T}[e^{-2\pi i(n-m)} - 1] = 0. \tag{2.1.30}$$

The case where $m = n$ is even simpler:

$$\int_{t_0}^{t_0+T} e^{-in\Delta\omega t}(e^{-in\Delta\omega t})^* \, dt = \int_{t_0}^{t_0-T} e^{-i0} \, dt = T. \tag{2.1.31}$$

Equations (2.1.29) and (2.1.30) can now be used to determine the Fourier coefficients c_n for a given function $f(t)$. To do so, we multiply both sides of the equation by $(e^{-im\Delta\omega t})^*$, and integrate over one period:

$$\int_{t_0}^{t_0+T} (e^{-im\Delta\omega t})^* f(t) \, dt = \sum_{n=-\infty}^{\infty} c_n \int_{t_0}^{t_0+T} (e^{-im\Delta\omega t})^* e^{-in\Delta\omega t} \, dt. \tag{2.1.32}$$

Then, according to Eq. (2.1.29), all terms in the sum vanish, except for the $n = m$ term, which, after applying Eq. (2.1.31), equals $c_m T$. Thus, we find

$$c_m = \frac{1}{T} \int_{t_0}^{t_0+T} (e^{-im\Delta\omega t})^* f(t) \, dt. \tag{2.1.33}$$

Equations (2.1.33) and (2.1.27) allow us to write any periodic function $f(t)$ as an exponential Fourier series. Of course, the function must satisfy the requirements of Theorem 2.1 in order for the series to converge uniformly to f.

For real f we can also write the series in the form of Eq. (2.1.25) by using Eq. (2.1.33) along with the relations

$$\begin{aligned} C_0 &= c_0, \\ C_n &= 2c_n, \qquad n > 0, \end{aligned} \tag{2.1.34}$$

which follow from comparing Eqs. (2.1.24) and (2.1.26).

Two representations of an exponential Fourier series (the first is valid only for real $f(t)$):

(1) $f(t) = \text{Re}[\sum_{n=0}^{\infty} C_n e^{-in\Delta\omega t}]$,
(2) $f(t) = \sum_{n=-\infty}^{\infty} c_n e^{-in\Delta\omega t}$,

where

$C_0 = c_0$,
$C_n = 2c_n, \qquad n > 0$,

and

$c_n = \frac{1}{T}\int_{t_0}^{t_0+T} (e^{-in\Delta\omega t})^* f(t) \, dt \qquad$ for all n.

2.1.8 Response of a Damped Oscillator to Periodic Forcing

Armed with the knowledge we have gained in previous sections, we can now return to the question put forward at the beginning of the chapter: what is the response of a damped oscillator to general periodic forcing $f(t)$?

We will find a particular solution $x_p(t)$ to the oscillator equation (1.6.2) in the form of an exponential Fourier series:

$$x_p(t) = \sum_{n=-\infty}^{\infty} x_n e^{-in\,\Delta\omega\, t}, \tag{2.1.35}$$

where x_n is the complex Fourier coefficient of $x_p(t)$, $\Delta\omega = 2\pi/T$ is the fundamental frequency of the given periodic forcing function, and T is the fundamental period of the forcing. If we substitute Eq. (2.1.35) into Eq. (1.6.2), we obtain

$$\sum_{n=-\infty}^{\infty} x_n\left[-(n\,\Delta\omega)^2 - i\gamma n\,\Delta\omega + \omega_0^2\right] e^{-in\,\Delta\omega\, t} = f(t). \tag{2.1.36}$$

Finally, we can extract the Fourier coefficient x_n by multiplying both sides by $(e^{-in\Delta\omega t})^*$, integrating over a period of the force, and using the orthogonality relations Eq. (2.1.29) and (2.1.31). The result is

$$x_n = \frac{c_n}{-(n\,\Delta\omega)^2 - i\gamma n\,\Delta\omega + \omega_0^2}, \tag{2.1.37}$$

where c_n is the nth Fourier coefficient of the forcing function, given by Eq. (2.1.33).

This simple expression contains a considerable amount of physics. First, note that each Fourier mode in the force drives a single Fourier mode in the oscillator response. For the case of a strictly sinusoidal force, Eqs. (2.1.37) and (2.1.25) reduce to Eq. (1.6.45).

Second, the principle of superposition is implicitly coming into play: according to Eq. (2.1.35), the total oscillator response is a linear superposition of the responses from the individual Fourier modes. However, each mode is independently excited, and has no effect on other modes.

Third, note that for high-n Fourier modes, the response is roughly $x_n \sim -c_n/(n\,\Delta\omega)^2$, which approaches zero more rapidly with increasing n than do the forcing coefficients c_n. Basically, this is an effect due to inertia of the oscillator: a very high-frequency forcing causes almost no effect on an oscillator, because the oscillator's inertia doesn't allow it to respond before the force changes sign.

Fourth, for very low-frequency forcing, such that $(n\,\Delta\omega)^2 \ll \omega_0^2$ for all Fourier modes entering the force, the response is $x_n \sim c_n/\omega_0^2$. We can then re-sum the series according to Eq. (2.1.35) and (2.1.27), to find that the oscillator amplitude tracks the forcing as $x_p(t) \sim f(t)/\omega_0^2$. This makes sense intuitively: according to Hooke's law, when you slowly change the force on a spring, it responds by changing its length in proportion to the applied force.

Finally, note that some Fourier modes are excited to higher levels than other modes. For Fourier modes that satisfy $(n\,\Delta\omega)^2 \simeq \omega_0^2$, the denominator in Eq. (2.1.37) is close to zero if γ is small, and the system response exhibits the resonance phenomenon discussed in Sec. 1.6.4. These resonant modes are driven to large amplitude. For the case of an undamped oscillator ($\gamma = 0$) and exact resonance ($n\,\Delta\omega = \omega_0$ for some value of n), Eq. (2.1.37) does not apply. The resonant response is no longer described by a Fourier mode, but rather by a growing oscillation. The form of this oscillation can be found using the methods for exact resonance discussed in Sec. 1.6.4.

Resonance phenomena are of great importance in a number of systems, including the system to be discussed in the next section.

2.1.9 Fourier Analysis, Sound, and Hearing

The sound that a sinusoidal oscillation makes is a pure tone. *Mathematica* can play such sounds with the intrinsic function **Play**. For example, the sound of the pure note middle A is a sinusoid with frequency $\omega = 2\pi \times 440\ \mathrm{s}^{-1}$; the command and visible response are shown in Cell 2.25. **Play** assumes that the time t is given in seconds, so this command causes a pure middle-A tone to play for 1 second. The tone can be repeated by double-clicking on the upper corner of the inner cell box.

Cell 2.25

```
Play[Sin[2Pi 440t], {t, 0, 1}]
```

We can also play other sounds. For example (Cell 2.26), we can play the sound of a triangle wave, which has a distinctive buzzing quality. The visible response of **Play** is suppressed in order to save space. Here we have used the Fourier series for a triangle wave, with coefficients as listed in Eq. (2.1.11), keeping 30 coefficients, and neglecting the $n = 0$ term (since it merely produces a constant offset, of no importance to the sound). We have also added an option **PlayRange**, which is analogous to **PlotRange** for a plot, setting the range of amplitude levels to be included; it can be used to adjust the volume of the sound.

Cell 2.26

```
T = 1/440; a[n_] = -4/(n^2Pi^2);
f_approx[t_, 30] = Sum[a[n] Cos[2Pi n t/T], {n, 1, 30, 2}];
Play[f_approx[t, 30], {t, 0, 1}, PlayRange → {-0.6, 0.6)]
```

The harsh buzzing sound of the triangle wave compared to a pure sine wave is caused by the high harmonics of the fundamental middle-A tone that are kept in this series.

Let's now consider the following question: what happens to the sound of the triangle wave if we randomize the phases of the different Fourier modes with respect to one another? That is, let's replace $\cos(2\pi nt/T)$ with $\cos[2\pi(nt/T + \phi_n)]$, where ϕ_n is a random number between 0 and 2π. The resulting series can be written in terms of a sine and cosine series by using the trigonometric identity

$$\cos 2\pi(nt/T + \phi_n) = \cos \phi_n \cos 2\pi nt/T - \sin \phi_n \sin 2\pi nt/T.$$

If we plot the series, it certainly no longer looks like a triangle wave, although it remains periodic with period T, as shown in Cell 2.27. The waveform looks very different than a triangle wave, so there is no reason to expect that it would sound the same. However, if we play this waveform (using the same **PlayRange** as before, as shown in Cell 2.28), the sound is *indistinguishable* from that of the triangle wave. (Again, the visible output of **Play** is suppressed to save space.) This is surprising, given the difference between the shapes of these two waveforms. One can verify that this is not an accident. By reevaluating the random waveform one gets a different shape each time; but in each case the sound is identical. (Try it.)

Cell 2.27

```
T = 1/440; a[n_] = -4/(n^2Pi^2);
f_approx[t_, 30] = Sum[a[n] Cos[2Pi (n t/T + Random[])],
  {n, 1, 30, 2}];
Plot[f_approx[t, 30], {t, 0, 3T}];
```

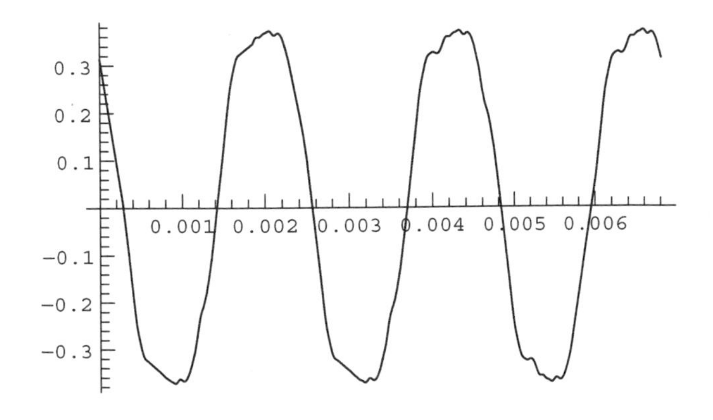

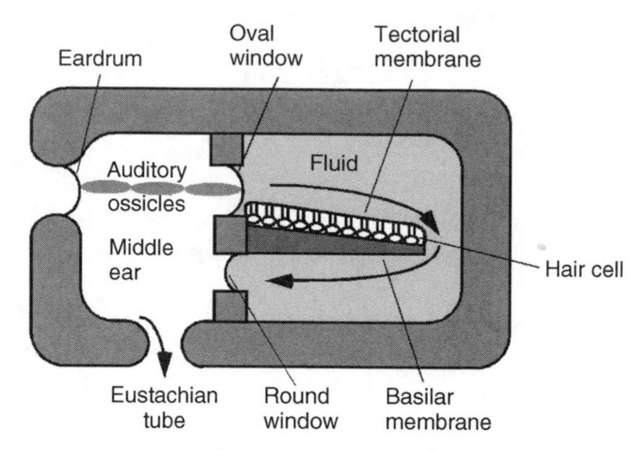

Fig. 2.2 Simplified diagram of the middle and inner ear (not to scale).

Cell 2.28

```
Play[f_approx[t, 30], {t, 0, 1}, PlayRange→{-0.6, 0.6)]
```

Why do different-shaped waveforms make the same sound? The reason has to do with how we perceive sound. Sound is perceived by the brain through the electrical signals sent by nerve cells that line the inner ear. Each nerve cell is attached to a hair that is sandwiched between two membranes—the basilar membrane and the tectorial membrane (see Fig. 2.2). As sound waves move through the fluid in the inner ear, these membranes, immersed in the fluid, move relative to one another in response. The basilar membrane is thicker and stiffer in some places than in others, so different parts of the membrane are resonant to different frequencies of sound. Therefore, a given sound frequency excites motion only in certain places on the membrane. (This correspondence between frequency and location is called a *tonotopic map*.) The motion excites the hairs at these locations, which in turn cause their respective neurons to fire at a rate that depends on the amplitude of the motion, sending signals to the brain that are interpreted as a sound of a given frequency and loudness.

If you think this system sounds complicated, you're right. After all, it is the product of millions of years of evolutionary trial and error. But one can think of it very roughly as just a system of oscillators having a range of resonant frequencies. Crudely speaking, the ear is doing a Fourier analysis of the incoming sound: the different frequency components of the sound resonantly excite different hairs, which through the tonotopic map are perceived as different frequencies. The *amplitude* of a hair's motion is translated into the amplitude of the sound at that frequency. The *phase* of the motion of the hairs relative to one another is apparently not important in what we perceive as the *quality* of the sound (as we have seen, the "sound" of the sound is unchanged by phase modulation of different frequency components).

However, as always seems to be the case in biological systems, things are really more complicated than this crude picture. The phase of the motion of the hairs is *not* completely ignored by the auditory system, at least for sounds with frequencies less than around 1–1.4 kHz. For this range, neurons are thought to be able to "phase lock" their firing to the phase of the sound—for instance, the neuron might fire only at the peak of the sine wave. (At higher frequencies, the neuron's firing rate apparently cannot keep up with the sound oscillation.) Experiments have shown that this phase information is used by the brain's auditory system to help locate the *source* of the sound, by comparing the phase in the right ear with that in the left. (See the exercises.)

Also, the auditory system is not passive. It has recently been shown that the resonant response of the hairs to a sound impulse is actually *amplified* by molecular motors in the membranes of the hair cells. This amplification allows the response of the system to be considerably more sharply peaked about the resonant frequency than would be the case for a purely passive system with the same damping rate. In fact, the molecular motors cause the hairs to vibrate continuously at a low level, and the sound this motion creates can be picked up by sensitive microphones *outside* the ear. The ear is not just a passive receiver: it also transmits (albeit at a level below our conscious perception).

In summary, two periodic waveforms will *look* different if their Fourier components have different relative *phases*, but they will still *sound* alike if their Fourier *amplitudes* are the same.

EXERCISES FOR SEC. 2.1

(1) Prove that for a periodic function $f(t)$ of period T, the following is true: $\int_0^T f(t)\,dt = \int_x^{T+x} f(t)\,dt$ for any x.

(2) **(a)** Do the following periodic functions meet the conditions of Theorem 2.1?

(i) $f(t) = |t|^3$ on $-\frac{1}{2} < t < \frac{1}{2}$; $f(t) = f(t+1)$.

(ii) $f(x) = 3x$ on $0 < x < 2$; $f(x) = f(x+2)$.

(iii) $f(t) = \exp(-t)$ on $0 < t < 3$; $f(t) = f(t+3)$.

(b) Find the Fourier series coefficients A_n and B_n for the periodic functions of part (a).

(c) Plot the resulting series for different numbers of coefficients M, $1 < M < 10$, and observe the convergence. Compare with the exact functions. Are the series converging?

(d) Plot the difference between the series and the actual functions as M increases, and determine using Eq. (2.1.15) whether the series are exhibiting uniform convergence.

(e) For the series of function (ii), evaluate the derivative of the series with respect to x, term by term. Compare it with the derivative of $3x$ on $0 < x < 2$ by plotting the result for $M = 10$, and $M = 50$. Does the derivative of the series give a good representation of $f'(x)$?

(3) **(a)** Theorem 2.1 provides sufficient conditions for convergence of a Fourier series. These conditions are not necessary, however. Functions that have singularities or singular derivatives can sometimes also have well-behaved convergent Fourier series. For example, use *Mathematica* to evaluate the

Fourier sine-cosine series of the periodic function

$$f(x) = \sqrt{x}\,(1-x) \quad \text{on } 0 \le x \le 1, \qquad f(x+1) = f(x),$$

and plot the result for $M = 4, 8, 12, 16, 20$. Does this series appear to be converging to $f(x)$ as M increases? (Hint: Don't be afraid of any special functions that *Mathematica* might spit out when evaluating Fourier coefficients. You don't need to know what they are (although you can look up their definitions in the *Mathematica* book if you want). Just use them in the series, stand back, and let *Mathematica* plot out the result.]

(b) At what value of x is the maximum error in the series occurring? Evaluate this maximum error for $M = 10, 30, 60, 90$. According to Eq. (2.1.15), is this series converging uniformly?

(4) Repeat Exercise (2)(b) and (c) using exponential Fourier series.

(5) A damped harmonic oscillator satisfies the equation $x'' + x' + 4x = f(t)$. The forcing function is given by $f(t) = t^2$, $-1 \le t \le 1$; $f(t+2) = f(t)$.

(a) Find a particular solution $x_p(t)$ to the forcing in terms of an exponential Fourier series.

(b) Find a homogeneous solution to add to your particular solution from part (a) so as to satisfy the ODE with initial conditions $x(0) = 1$, $x'(0) = 0$. Plot the solution for $0 < t < 20$.

(6) An undamped harmonic oscillator satisfies the equation $x'' + x = f(t)$. The forcing function is a square wave of period $\frac{1}{2}$: $f(t) = 1$, $0 < t < \frac{1}{4}$; $f(t) = 0$, $\frac{1}{4} < t < \frac{1}{2}$; $f(t + \frac{1}{2}) = f(t)$.

(a) Find a particular solution to this problem. (Be careful—there is an exact resonance.)

(b) Find a homogeneous solution to add to the particular solution so as to satisfy the ODE with initial conditions $x(0) = 0$, $x'(0) = 0$. Plot the solution for $0 < t < 20$.

(7) An *RC* circuit (a resistor and capacitor in series) is driven by a periodic sawtooth voltage, with period T, of the form $V(t) = V_0$ Mod$[t/T, 1]$. Plot $V(t)$ for $T = 0.002$, $V_0 = 1$ over a time range $0 \le t < 4T$.

(a) The charge $Q(t)$ on the capacitor satisfies the ODE $RQ' + Q/C = V(t)$. Find a particular solution for $Q(t)$ (in the form of an exponential Fourier series). Add a homogeneous solution to match the initial condition $Q(0) = 0$. Plot $Q(t)$ for $M = 5, 10, 20$ for the case $R = 500\ \Omega$, $C = 2\,\mu$F, $V_0 = 1$ V, $T = 0.002$ s. Compare the shape of $Q(t)$ for a few periods to the original voltage function.

(b) Play the sawtooth sound $V(t)$, and play the resulting $Q(t)$ as a sound. Play $Q(t)$ again, for $R = 5000\ \Omega$, all else the same. Would you characterize this circuit as one that filters out high frequencies or low frequencies?

(8) **(a)** Rewrite the exponential series of the previous problem for $Q(t)$ as a cos–sin series. For $M = 20$ compare with the exponential series by plotting both, showing they are identical.

(b) Randomize the phases of the Fourier modes in part (a) by replacing $\cos(n\pi\,\Delta\,\omega\,t)$ and $\sin(n\pi\,\Delta\,\omega\,t)$ with $\cos(n\pi\,\Delta\,\omega\,t+\phi_n)$ and $\sin[n\pi\,\Delta\,\omega\,t+\bar{\phi}_n]$, where ϕ_n and $\bar{\phi}_n$ are random phases in the range $(0,2\pi)$, different for each mode, and generated using `2πRandom[]`. Listen to the randomized series to see if you can tell the difference. Also plot the randomized series for a few periods to verify that the function looks completely different than $Q(t)$.

(9) Find a particular solution to the following set of coupled ODEs using exponential Fourier series:

$$x''(t) = -(x-2y) - x' + f_1(t),$$
$$y''(t) = -(y-x) + f_2(t),$$

where $f_1(t) = (\text{Mod}[t,1])^2$ and $f_2(t) = 2\,\text{Mod}[t,\sqrt{2}]$ are two sawtooth oscillations (with incommensurate periods). Plot the particular solution for $0 < t < 10$. Keep as many terms in the series as you feel are necessary to achieve good convergence.

(10) When a signal propagates from a source that is not directly in front of an observer, there is a time difference between when the signal arrives at the left and right ears. The human auditory system can use this time delay to help determine the direction from which the sound is coming. A phase difference between the left and right ears of even 1–2 degrees is detectable as a change in the apparent location of the sound source. This can be tested using `Play`. `Play` can take as its argument *two* sound waveforms, for left and right channels of a set of stereo headphones. For example, `Play[{Sin[440 2 π t], Sin[440 2 π t + ϕ(t)]},{t,0,10}]` plays a sine tone for 10 seconds in each ear, but with a phase advance $\phi(t)$ in the right ear.

Using a pair of stereo headphones with the above sound, see if you can determine an apparent location of a sound source. Try (a) $\phi(t) = 0.2t$, (b) $\phi(t) = -0.2t$; (c) $\phi(t) = 0$. Can you tell the difference? [See Hartmann (1999).] (Warning! The stereo effect in `Play` may not work on all platforms. Test it out by trying `Play[{0,Sin[440 2π t]},(t,0,1}]`: this should produce a tone only in the right ear. Repeat for the left ear. Also, make sure the volume is on a low setting, or else crosstalk between your ears may impede the directional effect.)

2.2 FOURIER REPRESENTATION OF FUNCTIONS DEFINED ON A FINITE INTERVAL

2.2.1 Periodic Extension of a Function

In the previous sections, Fourier series methods were applied to represent periodic functions. Here, we will apply Fourier methods to functions $f(t)$ defined only on an interval, $a \le t \le b$. Such functions often appear in boundary-value problems, where the solution of the differential equation is needed only between boundary points a and b.

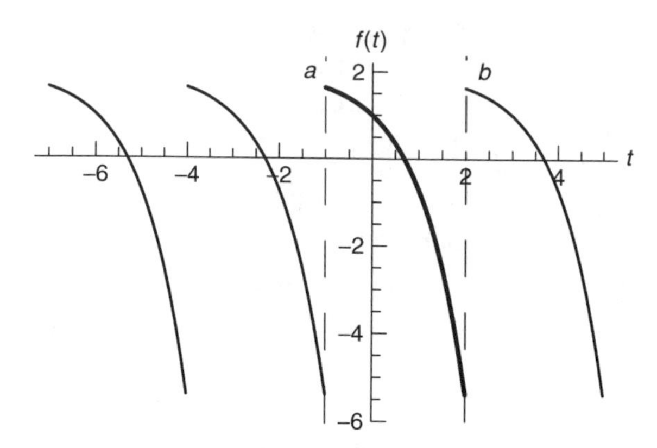

Fig. 2.3 Periodic extension of a function $f(t)$ defined on the interval $a \le t \le b$, with $a = -1$ and $b = 2$.

Functions defined only on an interval are not periodic, since they are not defined outside the interval in question and therefore do not satisfy Eq. (2.1.1) for all t. However, a Fourier representation can still be obtained by *replacing* $f(t)$ with a periodic function $f^{(p)}(t)$, defined on the entire real line $-\infty < t < \infty$.

There are several different choices for this periodic function. One choice requires $f^{(p)}(t)$ to equal $f(t)$ on $a \le t \le b$, and to have period $T = b - a$:

$$\begin{aligned} f^{(p)}(t) &= f(t), \qquad a < t < b, \\ f^{(p)}(t+T) &= f^{(p)}(t). \end{aligned} \tag{2.2.1}$$

The function $f^{(p)}(t)$ is called a *periodic extension* of $f(t)$. The type of periodic extension given by Eq. (2.2.1) is depicted in Fig. 2.3

Since $f^{(p)}$ is periodic with period T, it can be represented by a Fourier series,

$$f^{(p)}(t) = \sum_{n=-\infty}^{\infty} c_n \, e^{-in\,\Delta\omega\, t}, \tag{2.2.2}$$

where $\Delta\omega = 2\pi/T$. The Fourier coefficients c_n can be found using Eq. (2.1.33):

$$c_n = \frac{1}{T}\int_a^b f(t)\, e^{-in\,\Delta\omega\, t}\, dt. \tag{2.2.3}$$

The function could also be represented by a Fourier sine–cosine series,

$$f^{(p)}(t) = \sum_{n=0}^{\infty} a_n \cos(n\,\Delta\omega\, t) + \sum_{n=1}^{\infty} b_n \sin(n\,\Delta\omega\, t),$$

but usually the exponential form of the series is more convenient.

For example, say that $f(t) = t^2$ on $0 < t < 1$. The Fourier coefficients are then

Cell 2.29

```
c[n_] = Simplify[Integrate[t^2 Exp[I 2Pi n t], {t, 0, 1}],
  n ∈ Integers]
```

$$\frac{1 - i n \pi}{2 n^2 \pi^2}$$

Cell 2.30

```
c[0] = Integrate[t^2, {t, 0, 1}]
```

$$\frac{1}{3}$$

The M-term approximant to the Fourier series for f can then be constructed and plotted as in Cell 2.31. This $M = 50$ approximation to the complete exhibits the by now familiar Gibbs phenomenon, due to the discontinuity in the periodic extension of $f(t)$. For this reason, the series does not converge to $f(t)$ very rapidly; $f_n \sim -i/(2n\pi)$ for large n, which implies many terms in the series must be kept to achieve reasonable convergence.

Cell 2.31

```
f_approx[x_, M_] := Sum[c[n] Exp[-I 2Pi n t], {n, -M, M}];

func = f_approx[t, 50];

Plot[func, {t, -2, 2}, PlotRange→{-0.2, 1.2},
  PlotLabel→f^(p)(t) for f(t) = t^2 on 0<t<1"];
```

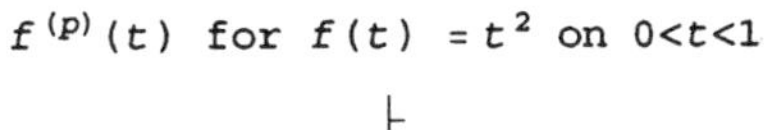

$f^{(p)}$ (t) for f (t) = t^2 on 0<t<1

$f^{(p)}$(t) for f(t) = t^2 on 0<t<1

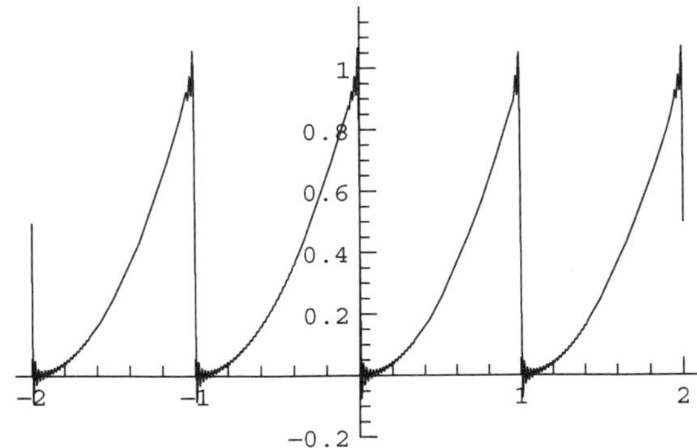

2.2.2 Even Periodic Extension

The problem of poor convergence can be avoided by using a different periodic extension of $f(t)$. Consider an *even* periodic extension of f, $f^{(e)}(t)$, with period $2T$

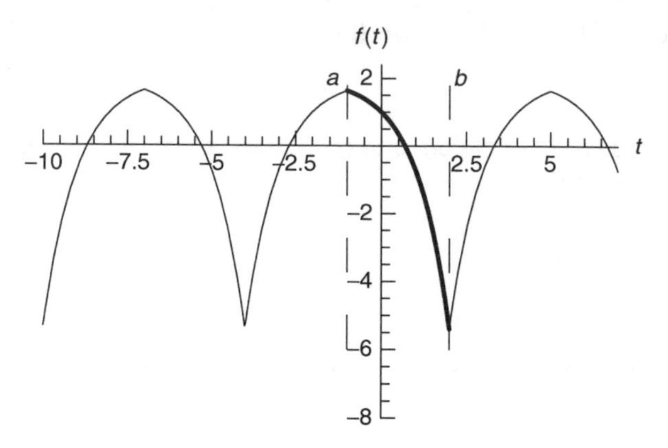

Fig. 2.4 Even periodic extension of a function $f(t)$ defined on the interval $-1<t<2$.

rather than T. This extension obeys

$$f^{(e)}(t) = \begin{cases} f(t), & a<t<a+T, \\ f(2a-t), & a-T<t<a, \end{cases} \tag{2.2.4}$$

$$f^{(e)}(t+2T) = f^{(e)}(t).$$

The even periodic extension of a function is depicted in Fig. 2.4. It is an even function of t about the point $t=a$, and for this reason no longer has a discontinuity. Therefore, we expect that the series for $f^{(e)}$ will converge more rapidly than that for $f^{(p)}$, and will no longer display the Gibbs phenomenon with nonuniform convergence.

Since the function is even around the point $t=a$, the series is of cosine form when time is evaluated with respect to an origin at a:

$$f^{(e)}(t) = \sum_{n=0}^{\infty} a_n \cos[n\,\Delta\omega\,(t-a)/2]. \tag{2.2.5}$$

Note that the period is now $2T$ rather than T, so the fundamental frequency is $\Delta\omega/2$.

In order to determine the Fourier coefficients for $f^{(e)}$, we must now integrate over an interval of $2T$. A good choice is the interval $a-T<t<a+T$, so that the Fourier coefficients have the form

$$a_n = \frac{1}{T}\int_{a-T}^{a+T} f^{(e)}(t)\cos[n\,\Delta\omega\,(t-a)/2]\,dt, \qquad n>0,$$
$$a_0 = \frac{1}{2T}\int_{a-T}^{a+T} f^{(e)}(t)\,dt. \tag{2.2.6}$$

Let's break the integrals up into two pieces, running from $a-T$ to a and from

a to $a+T=b$, and use Eq. (2.2.4):

$$a_n = \frac{1}{T}\int_{a-T}^{a} f(2a-t)\cos[n\,\Delta\omega\,(t-a)/2]\,dt$$
$$+\int_a^b f(t)\cos[n\,\Delta\omega\,(t-a)/2]\,dt, \qquad n>0, \qquad (2.2.7)$$
$$a_0 = \frac{1}{2T}\int_{a-T}^{a} f(2a-t)\,dt + \frac{1}{2T}\int_a^b f(t)\,dt.$$

Then, performing a change of variables in the first integral from t to $2a-t$, and using the fact that $\cos(-t)=\cos t$ for any t, we obtain

$$a_n = \frac{2}{T}\int_a^b f(t)\cos[n\,\Delta\omega\,(t-a)/2]\,dt, \qquad n>0,$$
$$a_0 = \frac{1}{T}\int_a^b f(t)\,dt. \qquad (2.2.8)$$

Equations (2.2.5) and (2.2.8) allow us to construct an even Fourier series of a function on the interval $[a,b]$, with no discontinuities. As an example, we will construct $f^{(e)}$ for the previous case of $f(t)=t^2$ on $[0,1]$:

Cell 2.32

```
T = 1;
a[n_] = 2/T Simplify[Integrate[t^2 Cos[2Pi n t/(2T)],
  {t, 0, T}], n ∈ Integers]
```

$$\frac{4\,(-1)^n}{n^2\pi^2}$$

Cell 2.33

```
a[0] = 1/T Integrate[t^2, {t, 0, 1}]
```

$$\frac{1}{3}$$

Terms in the series are now falling off like $1/n^2$ rather than $1/n$, so the series converges more rapidly than the previous series did. This can be seen directly from the plot in Cell 2.34. Note that we have only kept 10 terms in the series, but it still works quite well. There is no longer a Gibbs phenomenon; the series converges uniformly and rapidly to t^2 on the interval $[0,1]$.

Cell 2.34

```
f_approx[x_, M_] := Sum[a[n] Cos[2Pi n t/(2T)], {n, 0, M}];
func = f_approx[t, 10];
Plot[func, {t, -2, 2}, PlotRange → {-0.2, 1.2},
  PlotLabel → "f^(e) (t) for f(t)=t^2 on 0<t<1"];
```

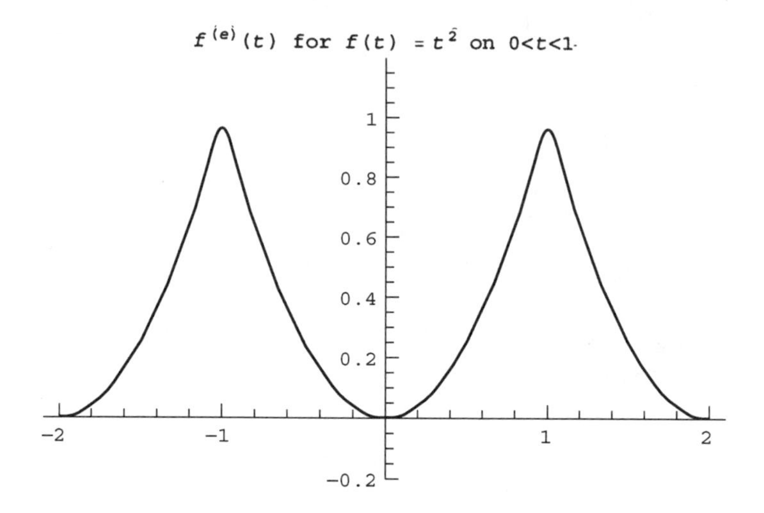

2.2.3 Odd Periodic Extension

It is also possible to define an odd periodic extension to a function $f(t)$, defined on the interval $[a, b]$. This extension, $f^{(o)}(t)$, is odd around the point $t = a$, and is defined by

$$f^{(o)}(t) = \begin{cases} f(t), & a < t < a + T, \\ -f(2a - t), & a - T < t < a, \end{cases}$$
$$f^{(o)}(t + 2T) = f^{(o)}(t). \tag{2.2.9}$$

This type of periodic extension is useful when one considers functions $f(t)$ for which $f(a) = f(b) = 0$. Although the periodic extension and the even periodic extension of such functions are both continuous at $t = a$ and b, the odd periodic extension also exhibits a continuous first derivative at the boundary points, as can be seen in Fig. 2.5. This makes the series converge even faster than the other types of periodic extension. However, if either $f(a)$ or $f(b)$ is unequal to zero, the convergence will be hampered by discontinuities in the odd periodic extension.

Like the even periodic extension, the odd periodic extension also has period $2T$. However, since it is odd about the point $t = a$, it can be written as a Fourier sine series with time measured with respect to an origin at $t = a$:

$$f^{(o)}(t) = \sum_{n=1}^{\infty} b_n \sin[n\,\Delta\omega\,(t - a)/2]. \tag{2.2.10}$$

The Fourier coefficients b_n can be determined by following an argument analogous to that which led to Eq. (2.2.8):

$$b_n = \frac{2}{T} \int_a^b f(t) \sin[n\,\Delta\omega\,(t - a)/2]\,dt. \tag{2.2.11}$$

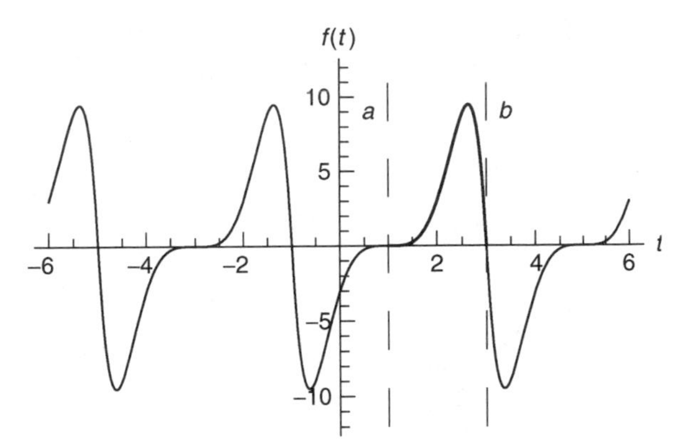

Fig. 2.5 Odd periodic extension of a function $f(t)$ defined on the interval $1 < t < 3$.

As an example, let's construct the odd periodic extension of the function $t(1 - t^2)$ on the interval [0, 1]. This function is zero at both $t = 0$ and $t = 1$, and so meets the conditions necessary for rapid convergence of the odd periodic extension.

The Fourier coefficients are given by

Cell 2.35

```
T = 1; b[n_] =
 2/T Simplify[Integrate[t (1 - t^2) Sin[2Pint/(2T)],
  {t, 0, T}], n∈Integers]
```

$$-\frac{12\ (-1)^n}{n^3 \pi^3}$$

These coefficients fall off as $1/n^3$, which is faster than either the coefficients of the even periodic extension,

Cell 2.36

```
T = 1; a[n_] =
 2/T Simplify[Integrate[t (1-t^2) Cos[2 Pi n t / (2T)],
  {t, 0, T}], n∈Integers]
```

$$\frac{2\ (6\ (-1 + (-1)^n) - (1 + 2\ (-1)^n)\ n^2 \pi^2)}{n^4 \pi^4}$$

or the regular periodic extension,

Cell 2.37

```
T = 1; c[n_] =
 1/T Simplify[Integrate[t (1 - t^2) Exp[I 2Pint/T],
  {t, 0, T}], n∈Integers]
```

$$-\frac{3\ (i + n\pi)}{4\ n^3\ \pi^3}$$

both of which can be seen to fall off as $1/n^2$ for large n.

A plot (Cell 2.38) of the resulting series for the odd periodic extension, keeping only five terms, illustrates the accuracy of the result when compared to the exact function.

Cell 2.38

```
f_approx[x_, M_] := Sum[b[n] Sin[2Pi n t/(2T)], {n, 1, M}];

func = f_approx[t, 5];
p1 = Plot[t (1 - t^2), {t, 0, 1},
  PlotStyle→{RGBColor[1, 0, 0], Thickness[0.012]},
  DisplayFunction→ Identity];
p2 = Plot[func, {t, -1, 2}, DisplayFunction→ Identity];
Show[p1, p2, DisplayFunction→ $DisplayFunction,
  PlotLabel→ "f^(o)(t) for f(t)=t(1- t^2) on 0<t<1"];
```

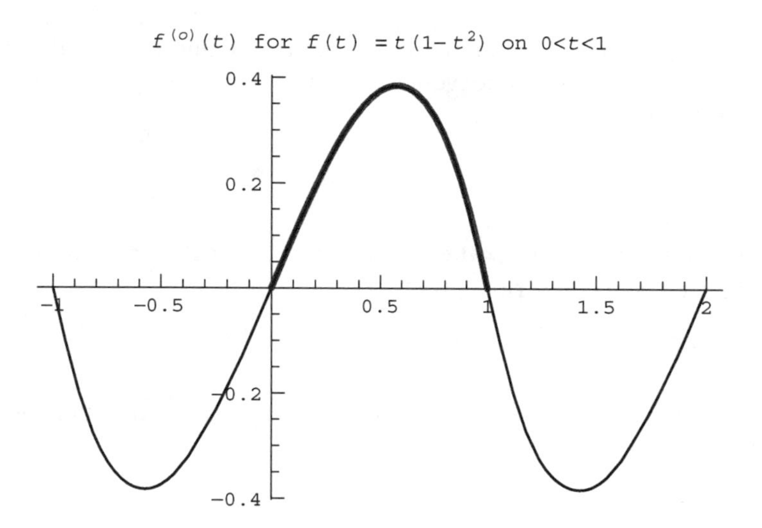

2.2.4 Solution of Boundary-Value Problems Using Fourier Series

In linear boundary-value problems, we are asked to solve for an unknown function $\phi(x)$ on a given interval $a < x < b$. (Here we consider a boundary-value problem in space rather than in time.) The general form of the linear ordinary differential equation is

$$\hat{L}\phi = \rho, \tag{2.2.12}$$

where $\rho(x)$ is a given function of x, and $\hat{L}$ is some linear differential operator. This equation is supplemented by boundary conditions on ϕ and/or its derivatives at a and b.

Fourier series methods are useful in finding a *particular* solution ϕ_p to this inhomogeneous ODE, provided that $\hat{L}$ is a differential operator with *constant coefficients*, of the form given in Eq. (2.6.7).

For example, consider the problem of determining the electrostatic potential between two parallel conducting plates at positions $x=a$ and $x=b$, between which there is some given distribution of charge, $\rho(x)$. The potential satisfies the Poisson equation (1.1.10), which in one spatial dimension takes the form

$$\frac{d^2\phi}{dx^2} = -\frac{\rho(x)}{\epsilon_0}, \qquad \phi(a)=V_1, \quad \phi(b)=V_2, \tag{2.2.13}$$

and where V_1 and V_2 are the voltages applied to the two plates.

This problem could be solved by the direct integration technique used to solve Eq. (1.1.1). Here, however, we will use Fourier techniques, since such techniques can also be applied to more complex problems that are not amenable to direct integration. We will run across many more problems of this type in future sections.

To solve this problem using Fourier techniques, we follow the procedure discussed in Sec. 1.6, and break ϕ into a homogeneous and a particular solution:

$$\phi = \phi_h + \phi_p, \tag{2.2.14}$$

where the particular solution ϕ_p is any solution that satisfies Eq. (2.2.13):

$$\frac{d^2\phi_p}{dx^2} = -\frac{\rho(x)}{\epsilon_0}. \tag{2.2.15}$$

After we have found a particular solution, we then solve for a homogeneous solution that satisfies the proper boundary conditions:

$$\frac{d^2\phi_h}{dx^2} = 0, \qquad \phi_h(a) = V_1 - \phi_p(a), \quad \phi_h(b) = V_2 - \phi_p(b). \tag{2.2.16}$$

In this way, the particular solution takes care of the inhomogeneity, and the homogeneous solution takes care of the boundary conditions. Adding Eqs. (2.2.15) and (2.2.16), one can easily see that the total potential, Eq. (2.2.15), satisfies both ODE and boundary conditions in Eq. (2.2.13).

For this simple ODE, the general homogeneous solution is $\phi_h(x) = C_1 + C_2 x$. To find the particular solution using Fourier methods, we replace ϕ_p by its periodic extension and employ an exponential Fourier series,

$$\phi_p(x) = \sum_{n=-\infty}^{\infty} \phi_n e^{i2\pi n x/L}, \tag{2.2.17}$$

where $L = b - a$ is the length of the interval.

Note the change in sign of the exponential compared to the previous section involving Fourier time series, Eq. (2.1.27). For spatial Fourier series, the above sign is conventional; for Fourier series in time, the opposite sign is used. Of course, either sign can be used as long as one is consistent throughout the calculation, but we will stick with convention and use different signs for time and space series. (Although these different conventions may seem arbitrary and confusing at first, there is a good reason for them, having to do with the form of traveling waves. See Sec. 5.1.1.)

The Fourier coefficients ϕ_n can then be found by substitution of Eq. (2.2.17) into the ODE and taking the derivative of the resulting series term by term:

$$\frac{d^2\phi_p}{dx^2} = \sum_{n=-\infty}^{\infty} \phi_n \left(\frac{i2\pi n}{L} \right)^2 e^{i2\pi nx/L} = -\frac{\rho(x)}{\epsilon_0}. \tag{2.2.18}$$

If we now multiply both sides of the equation by $(e^{i2\pi nx/l})^*$, integrate from a to b, and use the orthogonality relations (2.1.29), the result is

$$\phi_n = \left(\frac{L}{2\pi n} \right)^2 \frac{\rho_n}{\epsilon_0}, \qquad n \neq 0, \tag{2.2.19}$$

where ρ_n is the nth Fourier coefficient of $\rho(x)$, given by

$$\rho_n = \frac{1}{L} \int_a^b \rho(x)\, e^{-i2\pi nx/L}\, dx. \tag{2.2.20}$$

However, for the $n = 0$ term, the coefficient in front of ϕ_n in Eq. (2.2.18) vanishes, so a finite solution for ϕ_0 cannot be found. This is a case of an exact resonance, discussed in Sec. 1.6. For this special resonant term, the solution does not have the form of a Fourier mode. Rather, the right-hand side of Eq. (2.2.18) is a constant, and we must find a particular solution to the equation

$$\frac{d^2\phi_{p0}}{dx^2} = -\frac{\rho_0}{\epsilon_0}, \tag{2.2.21}$$

where ρ_0 is the $n = 0$ Fourier coefficient of $\rho(x)$. A particular solution, found by direct integration of Eq. (2.2.21), is

$$\phi_{p0}(x) = -\rho_0 x^2/2\epsilon_0, \tag{2.2.22}$$

which exhibits the secular growth typical of exact resonance. Thus, a particular solution to Eq. (2.2.13) is

$$\phi_p(x) = \sum_{\substack{n=-\infty \\ n\neq 0}}^{\infty} \phi_n\, e^{i2\pi nx/L} + \phi_{p0}(x), \tag{2.2.23}$$

with ϕ_n given by Eq. (2.2.19) and $\phi_{p0}(x)$ given by Eq. (2.2.22).

In order to complete the problem, we must add in the homogeneous solution $\phi_h(x) = C_1 + C_2 x$, with C_1 and C_2 chosen to satisfy the boundary conditions as described by Eq. (2.2.16). The result is

$$\phi(x) = V_1 - \phi_p(a) + \left[V_2 - \phi_p(b) - V_1 + \phi_p(a) \right] \frac{x-a}{L} + \phi_p(x). \tag{2.2.24}$$

Equation (2.2.24) is our solution to the boundary-value problem (2.2.13). Of course, this particular boundary-value problem is easy to solve using simple techniques such as direct integration. Applying Fourier methods to this problem is akin to using a jackhammer to crack an egg: it gets the job done, but the result is messier than necessary. However, in future sections we will find many situations for which the powerful machinery of Fourier series is essential to finding the solution.

EXERCISES FOR SEC. 2.2

(1) If $f(t)$ is continuous on $0 < t < T$, under what circumstances is (a) its periodic extension continuous? (b) its even periodic extension continuous? (c) its odd periodic extension continuous?

(2) Use the periodic extension for the following functions on the given intervals to determine an exponential Fourier series. In each case plot the resulting series, keeping $M = 20$ terms:

(a) $f(t) = e^t \sin 4t$, $0 \le t \le \pi/2$.

(b) $f(x) = x^4$, $-1 \le x \le 1$.

(c) $f(t) = t^2 - \frac{1}{2}$, $0 < t < 1$.

(3) For each case in Exercise (2), state which type of periodic extension (even or odd) will improve convergence the most. Evaluate the series for the chosen periodic extension with $M = 20$, and plot the result.

(4) The charge density between two grounded conducting plates ($\phi = 0$ at $x = -L$ and $x = L$) is given by $\rho(x) = Ax^2$. The electrostatic potential ϕ satisfies $d^2\phi/dx^2 = -\rho/\epsilon_0$.

(a) Find ϕ between the plates using an exponential Fourier series running from $-M$ to M. Taking $M = 10$, plot the shape of $\phi(x)$ taking $A/\epsilon_0 = 1$ and $L = 1$.

(b) Compare the approximate Fourier solution with the exact solution, found in any way you wish.

(5) Exercise (4) can also be solved using the trigonometric eigenfunctions of the operator d^2/dx^2 that satisfy $\psi = 0$ at $x = \pm L$. These eigenfunctions are $\sin[n\pi(x - L)/2L]$, $n = 1, 2, 3, \ldots$. Repeat Exercise (4) using these eigenfunctions.

(6) Exercise (4) can also be solved using the trigonometric eigenfunctions of the operator d^2/dx^2 that satisfy $\psi' = 0$ at $x = \pm L$. These eigenfunctions are $\cos[n\pi(x - L)/2L]$, $n = 0, 1, 2, 3 \ldots$. Repeat Exercise (4) using these eigenfunctions.

(7) **(a)** Find particular solutions to the following boundary-value problems using Fourier methods.

(b) Plot the particular solutions.

(c) Find homogeneous solutions to match the boundary conditions and solve the full problem. Plot the full solution.

(i) $\dfrac{d^2\phi}{dx^2} - \phi = x \sin 2x$, $\phi(-\pi) = 0$, $\phi(\pi) = 0$.

(ii) $\dfrac{d^2\phi}{dx^2} + 4\phi = \dfrac{1}{x}$, $\phi'(\pi) = 2$, $\phi(2\pi) = 0$. (Caution: there may be an exact resonance.)

(iii) $\dfrac{d^2\phi}{dx^2} + 8\dfrac{d\phi}{dx} + \phi = xe^{-x}$, $\phi'(0) = 0$, $\phi'(2) = 0$.

(iv) $\dfrac{d^2\phi}{dx^2} + 2\dfrac{d\phi}{dx} = \dfrac{x}{1+x^2}$, $\phi(0) = 0$, $\phi'(1) = 2$.

(v) $\dfrac{d^3\phi}{dx^3} - 2\dfrac{d^2\phi}{dx^2} + \dfrac{d\phi}{dx} - 2\phi = x^2 \cos x$, $\phi(0) = 0$, $\phi'(0) = 2$, $\phi(3) = 0$.

2.3 FOURIER TRANSFORMS

2.3.1 Fourier Representation of Functions on the Real Line

In Section 2.2, we learned how to create a Fourier series representation of a general function $f(t)$, defined on the interval $a \le t \le b$. In this section, we will extend this representation to general functions defined on the entire real line, $-\infty < t < \infty$.

As one might expect, this can be accomplished by taking a limit (carefully) of the previous series expressions as $a \to -\infty$ and $b \to \infty$. In this limit the period $T = b - a$ of the function's periodic extension approaches infinity, and the fundamental frequency $\Delta\omega = 2\pi/T$ approaches zero.

In the limit as $\Delta\omega \to 0$, let us consider the expression for an exponential Fourier series of $f(t)$, Eq. (2.1.27):

$$f(t) = \lim_{\Delta\omega\to 0} \alpha\,\Delta\omega \sum_{n=-\infty}^{\infty} \frac{c_n}{\alpha\,\Delta\omega} e^{-in\,\Delta\omega\, t}. \tag{2.3.1}$$

Here we have multiplied and divided the right-hand side by $\alpha\,\Delta\omega$, where α is a constant that we will choose in due course. The reason for doing so is that we can then convert the sum into an integral. Recall that for a function $g(\omega)$, the integral of g can be expressed as a Riemann sum,

$$\int_{-\infty}^{\infty} g(\omega)\, d\omega = \lim_{\Delta\omega\to 0} \Delta\omega \sum_{n=-\infty}^{\infty} g(n\,\Delta\omega). \tag{2.3.2}$$

Applying this result to Eq. (2.3.1) yields

$$f(t) = \alpha \int_{-\infty}^{\infty} \tilde{f}(\omega)\, e^{-i\omega t}\, d\omega, \tag{2.3.3}$$

where the function $\tilde{f}(\omega)$ is defined by $\tilde{f}(n\,\Delta\omega) \equiv c_n/(\alpha\,\Delta\omega)$. An expression for this function can be obtained using our previous result for the Fourier coefficient c_n, Eq. (2.1.33), and taking the limits $a \to -\infty$ and $b \to \infty$:

$$\tilde{f}(n\,\Delta\omega) = \lim_{\substack{a\to-\infty \\ b\to\infty}} \frac{1}{\alpha\,\Delta\omega\, T} \int_a^b f(t)\, e^{in\,\Delta\omega\, t}\, dt. \tag{2.3.4}$$

Substituting $\Delta\omega = 2\pi/T$ yields

$$\tilde{f}(\omega) = \frac{1}{2\pi\alpha}\int_{-\infty}^{\infty} f(t)\, e^{i\omega t}\, dt. \tag{2.3.5}$$

Equation (2.3.5) is called a *Fourier transform* of the function $f(t)$. Equation (2.3.3) is called an *inverse Fourier transform*, because it transforms the function $\tilde{f}(\omega)$ back to $f(t)$.

The Fourier transform transforms a function of time, $f(t)$, to a function of frequency, $\tilde{f}(\omega)$. This complex function of frequency, often called the *frequency spectrum* of $f(t)$, provides the complex amplitude of each Fourier mode making up $f(t)$. Since $f(t)$ is not periodic, $\tilde{f}(\omega)$ is nonzero for a continuous range of frequencies, as opposed to the discrete values of ω that enter a Fourier series.

Equations (2.3.3) and (2.3.5) are valid for any choice of the constant $\alpha \neq 0$. Different textbooks often choose different values for α. In the physical sciences, $\alpha = 1/(2\pi)$ is almost a universal convention, and is the choice we will adopt in this book.

Before we go on to evaluate some examples of Fourier transforms, we should mention one other convention in the physical sciences. Recall that for spatial Fourier series we reversed the sign in the exponentials [see Eq. (2.2.17)]. The same is done for spatial Fourier transforms. Also, it is conventional to replace the frequency argument ω of the Fourier transform function with the *wavenumber* k, with units of 1/length. These differing conventions for time and space transforms may seem confusing at first, but we will see in Chapter 5 when we discuss wave propagation that there are good reasons for adopting them. The table below provides a summary of our Fourier transform conventions.

To obtain a Fourier transform of a given function, we must evaluate the integral given in Table 2.1. For many functions, this integration can be performed analytically. For example, consider the Fourier transform of

$$f(t) = \frac{1}{1+s^2t^2}. \tag{2.3.6}$$

Using *Mathematica* to perform the integration,

Table 2.1. Fourier Transform Conventions

Time:	
$\tilde{f}(\omega) = \int_{-\infty}^{\infty} f(t)\, e^{i\omega t}\, dt$	Fourier transform
$f(t) = \int_{-\infty}^{\infty} \tilde{f}(\omega)\, e^{-i\omega t} \frac{d\omega}{2\pi}$	Inverse Fourier transform
Space:	
$\tilde{f}(k) = \int_{-\infty}^{\infty} f(x)\, e^{-ikx}\, dx$	Fourier transform
$f(x) = \int_{-\infty}^{\infty} \tilde{f}(k)\, e^{ikx} \frac{dk}{2\pi}$	Inverse Fourier transform

Cell 2.39

```
Integrate[Exp[I ω t] 1/(1 + s^2t^2), {t, -Infinity,
  Infinity}]
```

$$\text{If}[\text{Im}[\omega] == 0 \;\&\&\; \text{Arg}[s^2] \neq \pi,\; \frac{e^{\frac{-\omega \text{sign}[\omega]}{\sqrt{s2}}}\pi}{\sqrt{s^2}},\; \int_{-\infty}^{\infty} \frac{e^{it\omega}}{1 + s^2\, t^2}\, dt]$$

and noting that both inequalities in the output cell are satisfied, we obtain

$$\tilde{f}(\omega) = \pi\, e^{-|\omega/s|}/|s|.$$

By taking the inverse transform of $\tilde{f}(\omega)$, we should return to $f(t)$ as given by Eq. (2.3.6). However, because of the absolute value in $\tilde{f}(\omega)$, it is best to break the integral up into two pieces, $0 < \omega < \infty$ and $-\infty < \omega < 0$:

Cell 2.40

```
Simplify[Integrate[Exp[-I ω t] π Exp[-ω/s]/s,
  {ω, 0, Infinity}]/(2Pi) +
  Integrate[Exp[-I ω t] π Exp[ω / s] / s,
  {ω, -Infinity, 0}]/( Pi), s > 0 && Im[t] == 0]
```

$$\frac{1}{1 + s^2\, t^2}$$

As expected, the inverse transformation returns us to Eq. (2.3.6).

Mathematica has two intrinsic functions, **FourierTransform** and **InverseFourierTransform**. These two functions perform the integration needed for the transform and the inverse transform. However, the conventions adopted by these functions differ from those listed in Table 2.1: the value of α chosen in Eqs. (2.3.3) and (2.3.5) is $1/\sqrt{2\pi}$, and the transform functions use the time convention, not the space convention. To obtain our result for a time transform, the following syntax must be employed:

```
Sqrt[2Pi] FourierTransform[f[t],t,ω].
```

For example,

Cell 2.41

```
Simplify[Sqrt[2Pi] FourierTransform[1/(1 + s^2t^2), t, ω]]
```

$$\frac{e^{-\frac{\omega\, \text{Sign}[\omega]}{\sqrt{s2}}}\pi}{\sqrt{s^2}}$$

For the inverse transform, we must divide *Mathematica*'s function by $\sqrt{2\pi}$. The notation is

```
InverseFourierTransform[f[ω],ω,t]/Sqrt[2Pi].
```

For our example problem, we obtain the correct result by applying this function:

Cell 2.42

```
InverseFourierTransform[πExp[-Abs[ω]/s]/
  s, ω, t]/Sqrt[2Pi]
```

$$\frac{1}{1 + s^2\ t^2}$$

For spatial Fourier transforms, we must reverse the sign of the transform variable to match the sign convention for spatial transforms used in Table 2.1. The following table summarizes the proper usage of the intrinsic *Mathematica* functions so as to match our conventions.

Time:	
	$\sqrt{2\pi}$ `FourierTransform[f[t],t,ω]`
	`InverseFourierTransform[f[ω],ω,t]/`$\sqrt{2\pi}$
Space:	
	$\sqrt{2\pi}$ `FourierTransform[f[x],x,-k]`
	`InverseFourierTransform[f[k],k,-x]/`$\sqrt{2\pi}$

In following sections we will deal with time Fourier transforms unless otherwise indicated.

Fourier transforms have many important applications. One is in signal processing. For example, a digital bit may look like a square pulse, as shown in Cell 2.43.

Cell 2.43

```
f[t_] = UnitStep[t - 1] UnitStep [2 - t];
Plot[f[t], {t, 0, 3}, PlotStyle→Thickness[0.008],
  AxesLabel→{"t", "f(t)"}];
```

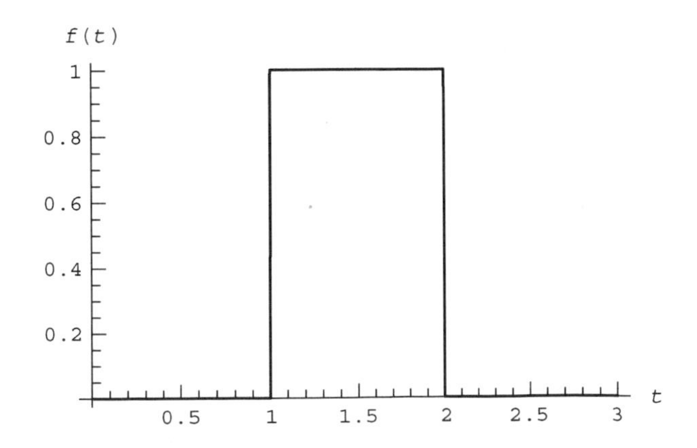

This signal has the following Fourier transform:

Cell 2.44

```
f̃[ω_] = Integrate[Exp[I ω t], {t, 1, 2}]
```

$$\frac{ie^{i\omega}}{\omega}-\frac{ie^{2i\omega}}{\omega}$$

This Fourier transform has both real and imaginary parts, as shown in Cell 2.45.

Cell 2.45

```
Plot[Re[f̃[ω]], {ω, -50, 50}, AxesLabel→{"ω", "Re[f̃(ω)]"},
  PlotRange →All];
Plot[Im[f̃[ω]], {ω, -50 50}, AxesLabel→{"ω", "Im[f̃(ω)]"},
  PlotRange→All];
```

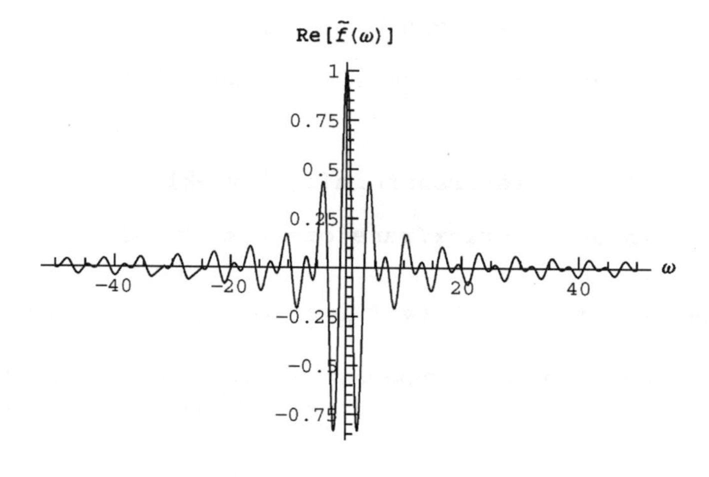

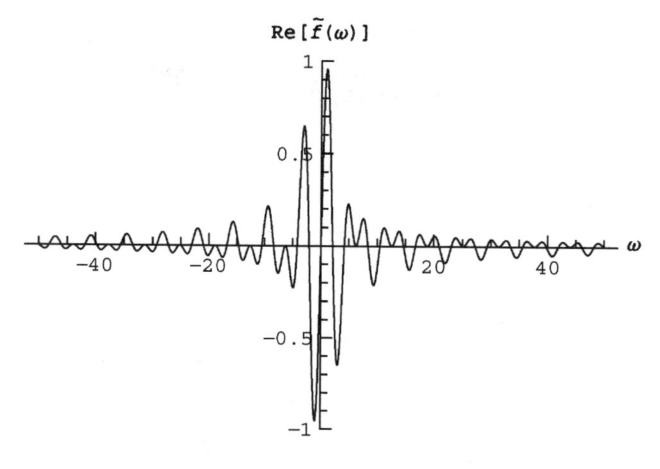

The real part of the transform $\tilde{f}(\omega)$ is an even function of ω, and the imaginary part an odd function. This follows from the fact that our function $f(t)$ was real. In general, for real $f(t)$,

$$\tilde{f}(-\omega) = \tilde{f}(\omega)^*. \tag{2.3.7}$$

Also note that the Fourier transform has nonnegligible high-frequency components. This is expected, because the function $f(t)$ has sharp jumps that require high-frequency Fourier modes.

However, the medium carrying the signal is often such that only frequencies within some range $\Delta\omega$ can propagate. This range is called the *bandwidth* of the

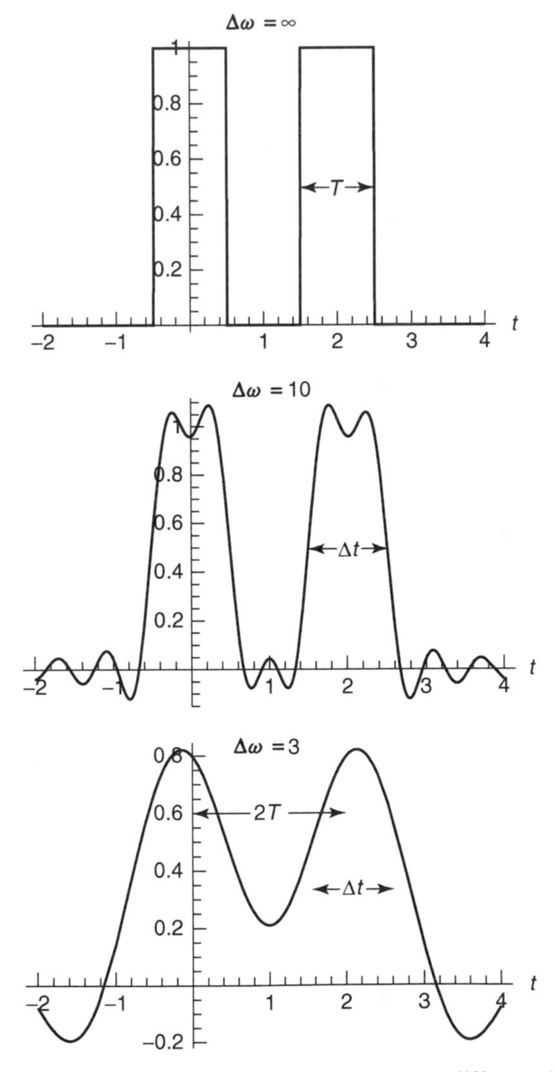

Fig. 2.6 The digital signal consisting of bits 101, for three different bandwidths $\Delta\omega$. As $\Delta\omega$ decreases, the width Δt of the pulses increases, until they begin to overlap and it is no longer possible to distinguish the bits.

medium. If, in our example, frequencies beyond 10 (in our dimensionless units) cannot propagate, then these components are cut out of the spectrum and the inverse transform of this signal is

Cell 2.46

```
f1[t_] = Integrate[Exp[-I ω t] f̃[ω], {ω, -10, 10}]/(2Pi);
```

This signal looks degraded and broadened due to the loss of the high-frequency components, as shown in Cell 2.47. If these pulses become so broad that they begin to overlap with neighboring pulses in the signal, then the signal will be garbled. For example, in order to distinguish a 0-bit traveling between two 1-bits, the length in time of each bit, T, must be larger than roughly half the width Δt of the degraded bits: $2T \gtrsim \Delta t$ (see Fig. 2.6).

Cell 2.47

```
Plot[f1[t], {t, 0, 3}, AxesLabel→{"t", " "},
  PlotLabel→"Signal degraded by finite bandwidth"];
```

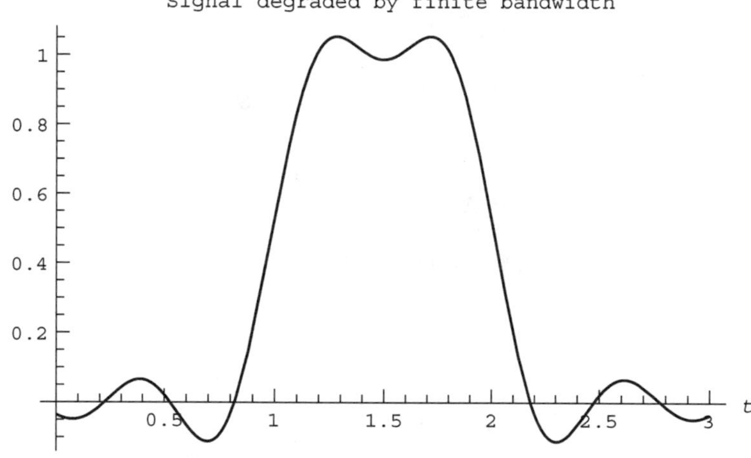

Also, Fig. 2.6 indicates that there is a connection between the degraded pulse width Δt and the bandwidth $\Delta\omega$: as $\Delta\omega$ decreases, Δt increases. In fact, in Sec. 2.3.3 we will show that $\Delta t \propto 1/\Delta\omega$: See Eq. (2.3.24). This implies that the minimum time between distinguishable pulses, $T_{\min}$, is proportional to $1/\Delta\omega$. The maximum rate $\upsilon_{\max}$ at which pulses can be sent is $\upsilon_{\max} = 1/T_{\min}$, so we find that

$$\upsilon_{\max} \propto \Delta\omega. \tag{2.3.8}$$

This important result shows that the maximum number of bits per second that can be sent through a medium is proportional to the bandwidth of the medium. For example, a telephone line has a bandwidth $\Delta\omega$ of roughly 4000 Hz, which limits the rate at which digital signals can be sent, as any modem user knows. However, an optical fiber has a bandwidth on the order of the frequency of light, around 10^{15}

Hz, which is why optical fibers can transmit much more information than phone lines. (Actually, the bandwidth quoted above for optical fiber is a theoretical bandwidth, applicable only to short fibers; in long fibers dispersion begins to limit the effective bandwidth. Dispersion will be discussed in Chapter 5. Also, the bandwidth of the receiver and the transmitter must be taken into account.)

2.3.2 Fourier Sine and Cosine Transforms

Sometimes one must Fourier-transform a function $f(t)$ that is defined only on a portion of the real line, $a \le t \le \infty$. For such functions, one can extend the definition of f to the range $-\infty < t < a$ in any way that one wishes, and then employ the usual transformation of Table 2.1 over the entire real line.

One simple choice is $f(t) = 0$ for $t < a$. In this case the Fourier transform is

$$\tilde{f}(\omega) = \int_a^\infty f(t)\, e^{i\omega t}\, dt, \tag{2.3.9}$$

and the inverse transformation remains unchanged.

For example, we may use Eq. (2.3.9) to take the Fourier transform of the function $f(t) = \exp(-t)$, $t > 0$. The integration in Eq. (2.3.9) can then be done by hand:

$$\tilde{f}(\omega) = \int_0^\infty e^{i\omega t - t}\, dt = \frac{1}{1 - i\omega}.$$

The inverse transformation should then return us to the original function $f(t)$. *Mathematica*'s intrinsic function can perform this task:

Cell 2.48

```
InverseFourierTransform[1/(1 - I ω), ω, t]/Sqrt [2Pi]
```

```
e^-t UnitStep[t]
```

The function **UnitStep**, also called the Heaviside step function, has been encountered previously, and is defined by Eq. (9.8.1). Since this function is zero for $t < 0$ and unity for $t > 0$, the inverse transform has reproduced $f(t)$, including the extension to $t < 0$.

It is sometimes useful to create an even or odd extension of the function, rather than setting the function equal to zero for $t < a$. In this case, the exponential transform is replaced by a *Fourier sine or cosine transform*.

For an odd extension of the function, we require that $f(a - t) = -f(a + t)$ for any t. The formulae for the Fourier sine transform then follow from a limiting procedure analogous to that done for the exponential transform, but instead applied to the Fourier sine series discussed in Sec. 2.2.3. Now one takes $b \to \infty$ but leaves a fixed. It is then an exercise to show that the result for the Fourier sine transform and the inverse sine transform is

$$\tilde{f}(\omega) = \int_a^{\infty} f(t) \sin \omega(t-a)\, dt,$$

$$f(t) = \int_{-\infty}^{\infty} \tilde{f}(\omega) \sin \omega(t-a)\, d\omega/\pi. \tag{2.3.10}$$

On the other hand, if we wish to use an even function, such that $f(a-t) = f(a+t)$ for any t, we can employ a cosine transform of the form

$$\tilde{f}(\omega) = \int_a^{\infty} f(t) \cos \omega(t-a)\, dt,$$

$$f(t) = \int_{-\infty}^{\infty} \tilde{f}(\omega) \cos \omega(t-a)\, d\omega/\pi. \tag{2.3.11}$$

The definitions for spatial sine and cosine transforms are identical, except for the convention of replacing ω by k and t by x.

As an example of a sine transform, we can again take $f(t) = \exp(-t)$ for $t > 0$. The sine transform is then given by

Cell 2.49

```
Simplify[Integrate[Exp[-t] Sin[ω t], {t, 0, Infinity}],
  Im[ ω] == 0]
```

$$\frac{\omega}{1 + \omega^2}$$

The inverse sine transform is

Cell 2.50

```
Simplify[Integrate[% Sin[ω t], {ω, -Infinity, Infinity}]/Pi,
  Im[t] == 0]
```

$$e^{-t\ \mathrm{Sign}[t]}\ \mathrm{Sign}[t]$$

which returns us to $f(t)$, but with an odd extension into the range $t < 0$. For an even extension of the same function, the cosine transform is

Cell 2.51

```
Simplify[Integrate[Exp[-t] Cos[ω t], {t, 0, Infinity]},
  Im[ω ] == 0]
```

$$\frac{1}{1 + \omega^2}$$

and the inverse cosine transform returns the correct even function:

Cell 2.52

```
Simplify[Integrate[% Cos[ω t], {ω, -Infinity, Infinity}]/Pi,
  Im[t] == 0]
```

```
e^-t sign[t]
```

2.3.3 Some Properties of Fourier Transforms

Fourier Transforms as Linear Integral Operators When one takes the Fourier transform of a function $f(t)$, the result is a new function $\tilde{f}(\omega)$. This is reminiscent of the manner in which a linear differential operator $\hat{L}$ transforms a function f to a different function $\hat{L}f$ by taking derivatives of f. In fact, a Fourier transform can also be thought of as a linear operator $\hat{F}$, defined by its operation on a given function f:

$$\hat{F}f = \int_{-\infty}^{\infty} e^{i\omega t} f(t)\, dt. \tag{2.3.12}$$

The result of the operation of $\hat{F}$ on a function f is a new function $\tilde{f}$, that is,

$$\hat{F}f = \tilde{f}. \tag{2.3.13}$$

The operator $\hat{F} = \int_{-\infty}^{\infty} e^{i\omega t}\, dt$ is a linear operator, since it satisfies $\hat{F}(Cf + g) = C\hat{F}f + \hat{F}g$ for any functions f and g and any constant C. However, this linear operator is an integral operator rather than a differential operator.

The inverse Fourier transform can also be thought of as an operator, $\hat{F}^{-1}$. This operator is defined by its action on a function $\tilde{f}(\omega)$, producing a function $f(t)$ according to

$$f = \hat{F}^{-1}\tilde{f} = \int_{-\infty}^{\infty} e^{-i\omega t}\tilde{f}(\omega)\, d\omega/2\pi. \tag{2.3.14}$$

The inverse transform has the property required of any inverse: for any function f,

$$\hat{F}^{-1}\hat{F}f = \hat{F}\hat{F}^{-1}f = f. \tag{2.3.15}$$

This follows directly from the definition of the inverse Fourier transform.

Fourier Transforms of Derivatives and Integrals The Fourier transform of the derivative of a function is related to the Fourier transform of the function itself. Consider

$$\hat{F}\frac{df}{dt} = \int_{-\infty}^{\infty} e^{i\omega t}\frac{df}{dt}\, dt. \tag{2.3.16}$$

An integration by parts, together with the assumption that $f(\pm\infty) = 0$ (required for the convergence of the Fourier integral), implies

$$\hat{F}\frac{df}{dt} = -\int_{-\infty}^{\infty} f(t)\frac{d}{dt}e^{i\omega t}\,dt = -i\omega\int_{-\infty}^{\infty} f(t)\,e^{i\omega t}\,dt = -i\omega\tilde{f}(\omega), \quad (2.3.17)$$

where $\tilde{f}=\hat{F}f$ is the Fourier transform of f.

We can immediately generalize Eq. (2.3.17) to the transform of derivatives of any order:

$$\hat{F}\frac{d^n f}{dt^n} = (-i\omega)^n\tilde{f}(\omega). \quad (2.3.18)$$

This simple result is of great importance in the analysis of particular solutions to linear differential equations with constant coefficients, as we will see in Sec. 2.3.6.

Also, it follows from Eq. (2.3.17) that the Fourier transform of the indefinite integral of a function is given by

$$\hat{F}\int^t f(t)\,dt = \frac{\tilde{f}(\omega)}{-i\omega}. \quad (2.3.19)$$

Convolution Theorem The *convolution* $h(t)$ of two functions, $f(t)$ and $g(t)$, is defined by the following integral:

$$h(t) = \int_{-\infty}^{\infty} f(t_1)g(t-t_1)\,dt_1 = \int_{-\infty}^{\infty} f(t-t_2)g(t_2)\,dt_2. \quad (2.3.20)$$

Either integral is a valid form for the convolution. The second form follows from a change of the integration variable from t_1 to $t_2 = t - t_1$.

Convolutions often appear in the physical sciences, as when we deal with Green's functions [see Eq. (2.3.73)]. The convolution theorem is a simple relation between the Fourier transforms of $h(t)$, $f(t)$, and $g(t)$:

$$\tilde{h}(\omega) = \tilde{f}(\omega)\tilde{g}(\omega). \quad (2.3.21)$$

To prove this result, we take the Fourier transform of $h(t)$:

$$\tilde{h}(\omega) = \int_{-\infty}^{\infty} dt\,e^{i\omega t}\int_{-\infty}^{\infty} f(t_1)g(t-t_1)\,dt_1.$$

Changing the integration variable in the t-integral to $t_2 = t - t_1$ yields

$$\tilde{h}(\omega) = \int_{-\infty}^{\infty} dt_2\int_{-\infty}^{\infty} e^{i\omega(t_2+t_1)}f(t_1)g(t_2)\,dt_1.$$

In this change of variables from t to t_2, t_1 is held fixed, so $dt = dt_2$ and the range of integration still runs from $-\infty$ to $+\infty$. We can now break the exponential into a product of exponentials, $e^{i\omega(t_2+t_1)} = e^{i\omega t_2}\,e^{i\omega t_1}$, and break the two integrals up into a product of Fourier transforms:

$$\hat{h}(\omega) = \int_{-\infty}^{\infty} dt_2\, e^{i\omega t_2} g(t_2) \int_{-\infty}^{\infty} e^{i\omega t_1} f(t_1)\, dt_1 = \tilde{g}(\omega)\tilde{f}(\omega),$$

proving the theorem.

The Uncertainty Principle of Fourier Analysis Consider a dimensionless function $f(\tau)$ that approaches zero when $|\tau| \gtrsim 1$. An example of such a function is $\exp(-|\tau|)$; another is $1/(1+\tau^2)$. A third example is the set of data bits plotted in Fig. 2.6. The Fourier transform of this function, $\tilde{f}(\omega)$, will typically approach zero for large $|\omega|$, because only frequencies up to some value are necessary to describe the function. Let us define the width of this transform function as Δ, that is, $\tilde{f}(\omega) \to 0$ for $|\omega| \gtrsim \Delta$. (Here Δ is a dimensionless number, on the order of unity for the three examples given above.)

Now consider a scale transformation of τ to a new time $t = \Delta t\, \tau$. When written in terms of the new time, the function $f(\tau)$ becomes a new function $g(t)$, defined by $g(t) = f(t/\Delta t)$. This function approaches zero for times $|t| > \Delta t$. Therefore, Δt is a measure of the width in time of the function $g(t)$.

An example of the function $g(t)$ is shown in Fig. 2.7 for different choices of Δt, taking $f(\tau) = \exp(-\tau^2)$. As Δt increases, the width of g increases. One can see that varying Δt defines a class of functions, all of the same shape, but with different widths.

Now consider the Fourier transform of $g(t)$:

$$\tilde{g}(\omega) = \int_{-\infty}^{\infty} dt\, e^{i\omega t} f(t/\Delta t). \tag{2.3.22}$$

We will now relate the width $\Delta\omega$ of the Fourier transform $\tilde{g}$ to the width Δt of g. This relation follows from a simple change of the integration variable in Eq. (2.3.22) back to $\tau = t/\Delta t$:

$$\tilde{g}(\omega) = \Delta t \int_{-\infty}^{\infty} d\tau\, e^{i\omega \Delta t \tau} f(\tau) = \Delta t \tilde{f}(\omega \Delta t). \tag{2.3.23}$$

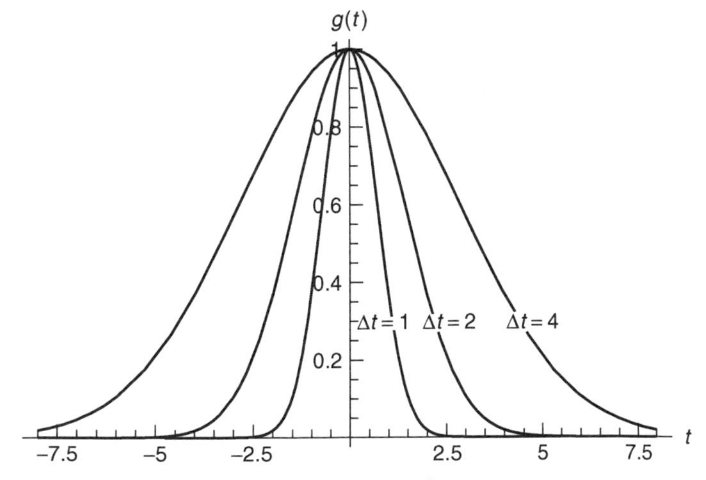

Fig. 2.7 The function $g(t) = e^{-(t/\Delta t)^2}$ for three choices of Δt.

Now, since the width of $\tilde{f}$ is Δ, Eq. (2.3.23) shows that the width $\Delta\omega$ of $\tilde{g}$ is $\Delta\omega = \Delta/\Delta t$, or in other words,

$$\Delta\omega\,\Delta t = \Delta, \tag{2.3.24}$$

where the constant Δ is a dimensionless number. This constant differs for different functions. However, if one defines the width of a function in a particular way, as the *rms width* [see Exercise (13)], then one can show that $\Delta \geq 1/2$, with equality only for *Gaussian* functions of the form $f(t) = f_0\, e^{-at^2}$.

Equation (2.3.24), along with the condition $\Delta \geq \frac{1}{2}$, is the uncertainty principle of Fourier analysis. It is called an uncertainty principle because it is the mathematical principle at the heart of Heisenberg's uncertainty principle in quantum mechanics. It shows that as a function becomes wider in time, its Fourier transform becomes narrower. This is sensible, because wider functions vary more slowly, and so require fewer Fourier modes to describe their variation. Alternatively, we see that very narrow, sharply peaked functions of time require a broad spectrum of Fourier modes in order to describe their variation.

As an example of the uncertainty principle, consider the Fourier transform of the function $g(t) = \exp[-(t/\Delta t)^2]$. This function is plotted in Fig. 2.7. The Fourier transform is

Cell 2.53

```
FourierTransform[Exp[-(t / Δt)^2], t, ω] Sqrt [2Pi]
```

$$e^{-\frac{1}{4}\Delta t2\ \omega 2}\sqrt{\pi}\sqrt{\Delta t^2}$$

This function is plotted in Fig. 2.8 for the same values of Δt as those in Fig. 2.7. One can see that as Δt increases, the transform function narrows.

An important application of the uncertainty principle is related to the data bit function of width Δt, plotted in Cell 2.43. We saw there that when a finite

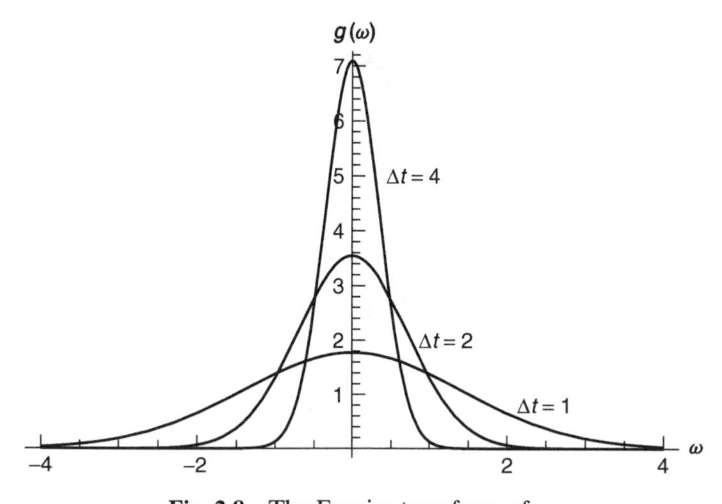

Fig. 2.8 The Fourier transform of g.

bandwidth $\Delta\omega$ cuts off the Fourier spectrum of a signal pulse, the width of the pulse in time, Δt, grows larger (see Fig. 2.6). We now see that this is a consequence of the uncertainty principle. This principle says that as the bandwidth $\Delta\omega$ of a medium decreases, the signal pulses must become broader in time according to Eq. (2.3.24), and hence the distance in time between pulses must be increased in order for the pulses to be distinguishable.

In turn, this implies that the maximum rate v_{max} at which signals can be propagated is proportional to the bandwidth of the medium: see Eq. (2.3.8).

2.3.4 The Dirac δ-Function

Introduction The function $g(t) = f(t/\Delta t)$ plotted in Fig. 2.7 increases in width as Δt increases. The area under the function also clearly increases with increasing Δt. One might expect that the area under the Fourier transform of $g(t)$ should *decrease* as the transform becomes narrower: see Fig. 2.8. However, we will now show that the area under $\tilde{g}(\omega)$ is actually *independent* of Δt.

This surprising result follows from the following property of the inverse transform:

$$g(t=0) = \int_{-\infty}^{\infty} \tilde{g}(\omega)\, e^{-i\omega 0}\, d\omega/2\pi = \int_{-\infty}^{\infty} \tilde{g}(\omega)\, d\omega/2\pi. \tag{2.3.25}$$

The area under $\tilde{g}(\omega)$ equals $2\pi g(0) = 2\pi f(0)$, independent of Δt.

Why is this important? Consider the limit as $\Delta t \to \infty$. In this limit, the width of $\tilde{g}(\omega)$ vanishes, but the area under the function remains constant, equaling $2\pi f(0)$. One can see from Fig. 2.8 that this can happen because the height of $\tilde{g}(\omega)$ approaches infinity as the width vanishes.

This strange curve is called a Dirac δ-function $\delta(\omega)$. To be precise,

$$\delta(\omega) = \lim_{\Delta t \to \infty} \frac{\tilde{g}(\omega)}{2\pi g(0)}. \tag{2.3.26}$$

This function is normalized so as to have unit area under the curve. However, since its width is vanishingly small, the function also has the properties that

$$\delta(\omega) = \begin{cases} 0, & \omega \neq 0, \\ \infty, & \omega = 0. \end{cases} \tag{2.3.27}$$

Therefore, the area integral need not involve the entire real line, because the δ-function is zero everywhere except at the origin. The integral over $\delta(\omega)$ equals unity for *any* range of integration that includes the origin, no matter how small:

$$\lim_{\epsilon \to 0} \int_{-\epsilon}^{\epsilon} \delta(\omega)\, d\omega = 1. \tag{2.3.28}$$

Dirac δ-functions often appear in the physical sciences. These functions have many useful properties, which are detailed in the following sections.

Integral of a δ-Function Equations (2.3.27) and (2.3.28) lead to the following useful result: for any function $h(t)$ that is continuous at $t = 0$, the following integral can be evaluated analytically:

$$\int_{-a}^{b} h(t)\,\delta(t)\,dt = \lim_{\epsilon\to 0}\int_{-\epsilon}^{\epsilon} h(t)\,\delta(t)\,dt = h(0)\lim_{\epsilon\to 0}\int_{-\epsilon}^{\epsilon}\delta(t)\,dt = h(0). \quad (2.3.29)$$

In the first step, we used the fact that $\delta(t)$ equals zero everywhere except at $t = 0$ in order to shrink the range of integration down to an infinitesimal range that includes the origin. Next, we used the assumption that $h(t)$ is continuous at $t = 0$, and finally, we employed Eq. (2.3.28).

δ-Function of More Complicated Arguments We will often have occasion to consider integrals over Dirac δ-function of the form

$$\int_{a}^{b} g(t)\,\delta(f(t))\,dt, \quad (2.3.30)$$

where $f(t)$ equals zero at one or more values of t in the interval $a < t < b$. Take, for example, the case $f(t) = ct$ for some constant c, with $a < 0 < b$. The integration in Eq. (2.3.30) can then be performed by making a change of variables: let $u = ct$. Then Eq. (2.3.30) becomes $(1/c)\int_{ca}^{cb} g(u/c)\delta(u)\,dt$.

Now, if $c > 0$, the result according to Eq. (2.3.29) is $g(0)/c$, but if $c < 0$, the result is $-g(0)/c$, because the range of integration is from a positive quantity ca to a negative quantity, cb. Therefore, we find

$$\int_{a}^{b} g(t)\,\delta(ct)\,dt = \frac{g(0)}{|c|}, \qquad \text{assuming} \quad a < 0 < b. \quad (2.3.31)$$

Equation (2.3.31) can be used to determine more general integrals. Let's assume that $f(t)$ passes through zero at M points in the range $a < t < b$, and let us label these points $t = t_n$, $n = 1, 2, 3, \ldots, M$. The integral can then be broken up into contributions from each one of these zeros:

$$\int_{a}^{b} g(t)\,\delta(f(t))\,dt = \sum_{n=1}^{M}\int_{t_n-\epsilon}^{t_n+\epsilon} g(t)\,\delta(f(t))\,dt. \quad (2.3.32)$$

Other regions of integration do not contribute, because the δ-function is only nonzero within the regions kept. Focusing on one of the zeros, $t = t_n$, we note that only values of t near t_n are needed, and so we make a change of variables from t to $\Delta t = t - t_n$:

$$\int_{t_n-\epsilon}^{t_n+\epsilon} g(t)\,\delta(f(t))\,dt = \int_{-\epsilon}^{\epsilon} g(t_n + \Delta t)\,\delta(f(t_n + \Delta t))\,d\Delta t.$$

Taylor-expanding $f(t)$ for small Δt, noting that $f(t_n) = 0$, and assuming that $f'(t_n) \neq 0$, we have

$$\int_{t_n-\epsilon}^{t_n+\epsilon} g(t)\,\delta(f(t))\,dt = \int_{-\epsilon}^{\epsilon} g(t_n + \Delta t)\,\delta(f'(t_n)\,\Delta t)\,d\Delta t = \frac{g(t_n)}{|f'(t_n)|},$$

where we used Eq. (2.3.31) in the last step. Therefore Eq. (2.3.32) becomes

$$\int_a^b g(t)\,\delta(f(t))\,dt = \sum_{n=1}^{M} \frac{g(t_n)}{|f'(t_n)|}. \tag{2.3.33}$$

Generalized Fourier Integrals The previous considerations lead us to a startling observation. Consider the Fourier transform of $\delta(t)$ itself:

$$\int_{-\infty}^{\infty} \delta(t)\, e^{i\omega t}\,dt = e^{i\omega 0} = 1, \tag{2.3.34}$$

where we have used Eq. (2.3.29). This result, that the Fourier transform of $\delta(t)$ equals one, is expected on one level: after all, since $\delta(t)$ is infinitely narrow, the uncertainty principle implies its transform must be infinitely broad.

However, if we write down the inverse Fourier transform of unity, which should return us to $\delta(t)$, we arrive at the strange result

$$\delta(t) = \int_{-\infty}^{\infty} e^{-i\omega t}\,d\omega/2\pi. \tag{2.3.35}$$

This integral is not convergent; the integrand does not decay to zero at large ω. But, somehow, it equals a δ-function!

We can try to understand this strange result by writing the inverse transform as

$$\delta(t) = \lim_{\Delta\omega\to\infty} \int_{-\Delta\omega}^{\Delta\omega} 1\, e^{-i\omega t}\,d\omega/2\pi.$$

This integral can be evaluated analytically. The result is

$$\delta(t) = \lim_{\Delta\omega\to\infty} \frac{\sin(\Delta\omega\, t)}{\pi t}. \tag{2.3.36}$$

For a fixed value of t, and as $\Delta\omega$ increases, the function $\sin(\Delta\omega\, t)/\pi t$ simply oscillates between the values $\pm 1/\pi t$. Since this oscillation continues indefinitely as $\Delta\omega \to \infty$, the limit is not well defined. How can this limit equal a δ-function?

The limit equals a δ-function in the following *average* sense. Consider an integral over this function multiplied by any continuous function $f(t)$:

$$\lim_{\Delta\omega\to\infty} \int_{-a}^{b} f(t) \frac{\sin(\Delta\omega\, t)}{\pi t}\,dt. \tag{2.3.37}$$

If we now make the change of variables to $\tau = \Delta\omega\, t$, the integral becomes

$$\lim_{\Delta\omega\to\infty} \int_{-\Delta\omega\, a}^{\Delta\omega\, b} f(\tau/\Delta\omega) \frac{\sin\tau}{\pi\tau}\,d\tau. \tag{2.3.38}$$

However, the function $(\sin\tau)/\pi\tau$ is peaked at the origin, with an area under the curve equal to unity:

$$\int_{-\infty}^{\infty} \frac{\sin\tau}{\pi\tau}\,dt = 1. \tag{2.3.39}$$

Since this integral is convergent, we can replace the limits of integration in Eq. (2.3.38) by $\pm\infty$. Furthermore, in the limit that $\Delta\omega \to \infty$, we can replace $f(\tau/\Delta\omega) \to f(0)$. Therefore, using Eq. (2.3.39) we find that

$$\lim_{\Delta\omega\to\infty} \int_{-a}^{b} \frac{\sin(\Delta\omega\, t)}{\pi t} f(t)\, dt = f(0), \tag{2.3.40}$$

Thus, the function $\lim_{\Delta\omega\to\infty}[\sin(\Delta\omega\, t)/\pi t$ has the most important property of a δ-function: it satisfies Eq. (2.3.29). If we take $f(t) = 1$, we can immediately see that the function also satisfies Eq. (2.3.28). On the other hand, it does not satisfy Eq. (2.3.27): it is not equal to zero for $t \neq 0$; rather it is undefined, oscillating rapidly between $\pm 1/\pi t$. Actually, "rapidly" is an understatement. In the limit as $\Delta\omega \to \infty$, the oscillations in the function become *infinitely rapid*. Fortunately, the nature of this variation allows us to call this function a δ-function. When evaluating an integral with respect to t over this function, the oscillations *average to zero* unless the origin is included in the range of integration. This can be seen in Cell 2.54, which displays the behavior of this function over a range of t as $\Delta\omega$ increases.

Cell 2.54

```
Table[Plot[Sin[Δω t] / (Pi t), {t, 1, 2}, PlotRange →
  {-1, 1}], {Δω, 10, 100, 10}];
```

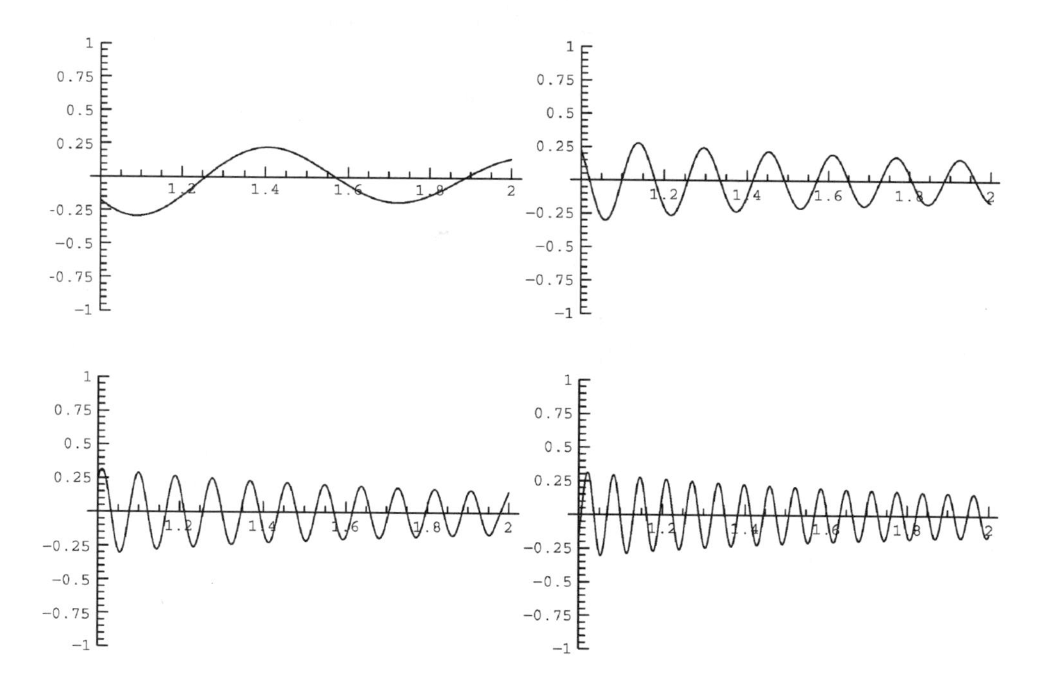

This sequence of plots shows that, if the origin is not included in the plots, the amplitude of the oscillations in $[\sin(\Delta\omega\, t]/\pi t$ does not change as $\Delta\omega$ increases; only the frequency increases until in the limit the function is zero *on average*. (Try changing the limits of the plot, and the range of $\Delta\omega$, to test this.)

However, if the origin is included, the amplitude of the peak at $t=0$ *increases* as $\Delta\omega$ becomes larger, since by l'Hospital's rule, $\lim_{t\to 0}[\sin(\Delta\omega t)]/\pi t = \Delta\omega/\pi$. This is what allows the area under the function to remain unity in the limit as $\Delta\omega$ becomes large.

Thus, Eq. (2.3.35) is a δ-function in an *average* sense: integrals over this function have the correct property given by Eq. (2.3.29). However, the function itself contains infinitely rapid oscillations.

Fourier integrals such as Eq. (2.3.35) are called *generalized* Fourier integrals: they do not converge in the usual sense to the term. Rather, the resulting functions contain infinitely rapid oscillations. We neglect these oscillations because all we use in applications are *integrals* over these functions, which average out the oscillations.

Derivatives of a δ-Function Several other generalized Fourier integrals are also of use. Consider, for example, the *derivative* of a δ-function, $\delta'(t) = d\delta(t)/dt$. According to Eq. (2.3.35), this derivative can be written as a generalized Fourier integral,

$$\delta'(t) = \frac{d}{dt}\int_{-\infty}^{\infty} e^{-i\omega t}\,d\omega/2\pi = \int_{-\infty}^{\infty}(-i\omega)\,e^{-i\omega t}\,d\omega/2\pi. \tag{2.3.41}$$

The integrand in Eq. (2.3.41) exhibits even worse convergence properties than Eq. (2.3.35). The resulting function has infinitely rapid oscillations of *infinite magnitude*. Nevertheless, integrals over this function are well behaved. We therefore may say, with a straight face, that the Fourier transform of a $\delta'(t)$ is $-i\omega$, and compute the inverse transform of this function. In fact, *Mathematica*'s **FourierTransform** function knows all about generalized Fourier integrals. For instance,

Cell 2.55

```
InverseFourierTransform[1, ω, t] / Sqrt [2Pi]
```

```
DiracDelta[t]
```

The intrinsic function **DiracDelta[t]** is the Dirac δ-function. Also,

Cell 2.56

```
InverseFourierTransform[-I ω, ω, t] / Sqrt[2 Pi]
```

```
DiracDelta'[t]
```

Of course, we can't plot these functions, because they are singular, but we know what they look like. The δ-function has a single positive spike at the origin. Since $\delta'(t)$ is the slope of $\delta(t)$, it has a positive spike just to the left of zero, and a negative spike just to the right.

The derivative of a δ-function has a useful property: for any function $f(t)$ that is differentiable at $t=0$, the following integral that includes the origin can be evaluated analytically via an integration by parts:

$$\int_{-a}^{b} f(t)\,\delta'(t)\,dt = -\int_{-a}^{b} f'(t)\,\delta(t)\,dt = -f'(0). \tag{2.3.42}$$

Similarly, the nth derivative of a δ-function has the property that

$$\int_{-a}^{b} f(t) \frac{d^n \delta(t)}{dt^n} dt = (-1)^n \frac{d^n f}{dt^n}(0). \tag{2.3.43}$$

The Fourier integral representation of $d^n\delta(t)/dt^n$ becomes progressively less convergent as n increases:

$$\frac{d^n \delta(t)}{dt^n} = \int_{-\infty}^{\infty} (-i\omega)^n e^{-i\omega t} d\omega/2\pi. \tag{2.3.44}$$

These generalized Fourier integrals allow us to do things that we couldn't do before. For instance, we can now compute the value of a nonconvergent integral, such as

$$\int_{-\infty}^{\infty} \frac{t^2}{1+t^2} \cos \omega t \, dt.$$

Normally, we would throw up our hands and declare that the integral does not exist. This is technically correct so far as it goes, but we still can compute its value as a generalized Fourier integral:

$$\begin{aligned}\int_{-\infty}^{\infty} \frac{t^2}{1+t^2} \cos \omega t \, dt &= \operatorname{Re} \int_{-\infty}^{\infty} \frac{1+t^2-1}{1+t^2} e^{i\omega t} \, dt \\ &= \operatorname{Re} \int_{-\infty}^{\infty} e^{i\omega t} \, dt - \operatorname{Re} \int_{-\infty}^{\infty} \frac{1}{1+t^2} e^{i\omega t} \, dt.\end{aligned}$$

The first integral is proportional to a δ-function, while the second integral is convergent, equaling $-\pi e^{-|\omega|}$. Thus, we obtain

$$\int_{-\infty}^{\infty} \frac{t^2}{1+t^2} \cos \omega t \, dt = 2\pi\delta(\omega) + \pi e^{-|\omega|}.$$

However, we must always remember that the equality is correct only in the average sense discussed above; the right-hand side neglects infinitely rapid oscillations.

Heaviside Step Function Before we move on to other topics, there is one more generalized Fourier integral of interest. Consider the *integral* of a δ-function,

$$h(t) = \int_{-\infty}^{t} \delta(t_1) \, dt_1. \tag{2.3.45}$$

This function equals zero if $t < 0$, but for $t > 0$ the range of integration includes the origin, and so, according to Eq. (2.3.28), $h(t) = 1$. Therefore, $h(t)$ is nothing other than the Heaviside step function **`UnitStep[t]`**, encountered previously in

Cell 2.48. For convenience, we reproduce the definition of $h(t)$ below:

$$h(t) = \begin{cases} 0, & t < 0, \\ 1, & t > 0. \end{cases} \tag{2.3.46}$$

We can find the Fourier integral of $h(t)$ by directly applying the transform:

$$\tilde{h}(\omega) = \int_{-\infty}^{\infty} e^{i\omega t} h(t)\, dt = \int_{0}^{\infty} e^{i\omega t}\, dt.$$

Now, this is not a convergent integral; rather, it is a generalized Fourier integral, and it provides an instructive example of some of the methods used to evaluate such integrals.

Breaking the integrand into real and imaginary parts via $e^{i\omega t} = \cos \omega t + i \sin \omega t$, we can write the result as

$$\tilde{h}(\omega) = \int_{0}^{\infty} \cos \omega t\, dt + i \int_{0}^{\infty} \sin \omega t\, dt.$$

In the first integral, note that $\cos \omega t$ is an even function of t, so we can double the range of integration to $-\infty < t < \infty$, and divide by $\frac{1}{2}$. Expressing the second integral as a limit, we can write

$$\tilde{h}(\omega) = \frac{1}{2} \operatorname{Re} \int_{-\infty}^{\infty} e^{i\omega t}\, dt + i \lim_{\Delta t \to \infty} \int_{0}^{\Delta t} \sin \omega t\, dt.$$

The first integral yields a δ-function via Eq. (2.3.34), and the second integral can be evaluated analytically:

$$\tilde{h}(\omega) = \pi \delta(\omega) - \frac{1}{i\omega} - i \lim_{\Delta t \to \infty} \frac{\cos(\omega \Delta t)}{\omega}. \tag{2.3.47}$$

As usual, we neglect the infinite oscillations. *Mathematica* can also deliver the same result:

Cell 2.57

```
Expand[FourierTransform[UnitStep[t], t, ω] Sqrt[2Pi]]
```

$$\frac{i}{\omega} + \pi\, \text{DiracDelta}[\omega]$$

Connection of Fourier Transforms to Fourier Series Since Fourier transforms can be used to represent any function as an integral over Fourier modes, they can be used to represent periodic functions as a special case. It is a useful exercise to see how this representation connects back to Fourier series.

As a first step, we will consider the Fourier series for a simple periodic function, the periodic δ-function of period T, $\delta^{(P)}(t)$. This function is a periodic extension of a Dirac δ-function, and can be written as

$$\delta^{(P)}(t) = \sum_{m=-\infty}^{\infty} \delta(t - mT). \tag{2.3.48}$$

Since this is a periodic function, it has a Fourier series representation of the form

$$\delta^{(P)}(t) = \sum_{n=-\infty}^{\infty} c_n \, e^{-i2\pi nt/T}.$$

The Fourier coefficients are given by an integral over one period of $\delta^{(P)}(t)$, which contains a single δ-function:

$$c_n = \int_{-T/2}^{T/2} \delta(t) \, e^{i2\pi nt/T} \, dt = 1.$$

Thus, the periodic δ-function of period T has a Fourier series of the form

$$\delta^{(P)}(t) = \sum_{m=-\infty}^{\infty} \delta(t - mT) = \sum_{n=-\infty}^{\infty} e^{-i2\pi nt/T}. \tag{2.3.49}$$

It is easy to see why Eq. (2.3.49) is a periodic δ-function. If $t = mT$ for any integer m, then $e^{-i2\pi nt/T} = e^{-i2\pi nm} = 1$ for all n, and the series sums to ∞. At these instants of time, each Fourier mode is *in phase*. However, it $t \neq mT$, the sum over $e^{-i2\pi nt/T}$ evaluates to a series of complex numbers, each with magnitude of unity, but with different phases. Adding together these complex numbers, there is *destructive interference* between the modes, causing the sum to equal zero. Thus, we get a function that is infinite for $t = mT$, and zero for $t \neq mT$.

However, the easiest way to see that this creates a periodic δ-function is to examine the series as a function of time using *Mathematica*. The following evaluation creates a periodic δ-function by keeping 300 terms in the Fourier series of Eq. (2.3.49). We choose a period of $\frac{1}{5}$, and note that the sum over n can be written in terms of cosines, since the sine functions are odd in n and cancel:

Cell 2.58

```
1 + 2 Sum[ Cos[2 Pi n 5 t], {n, 1, 300}];
```

We could plot this function, but it is more fun to `Play` it (Cell 2.59). It is necessary to keep several hundred terms in the series, because our ears can pick up frequencies of up to several thousand hertz. Since the fundamental frequency is 5 Hz, keeping 300 terms in the series keeps frequencies up to 1500 Hz. It would be even better to keep more terms; the sound of the "pops" then becomes higher pitched. However, keeping more terms makes the evaluation of the series rather slow.

Cell 2.59

```
Play [%, {t, -1.5, .5}, PlayRange → All[;
```

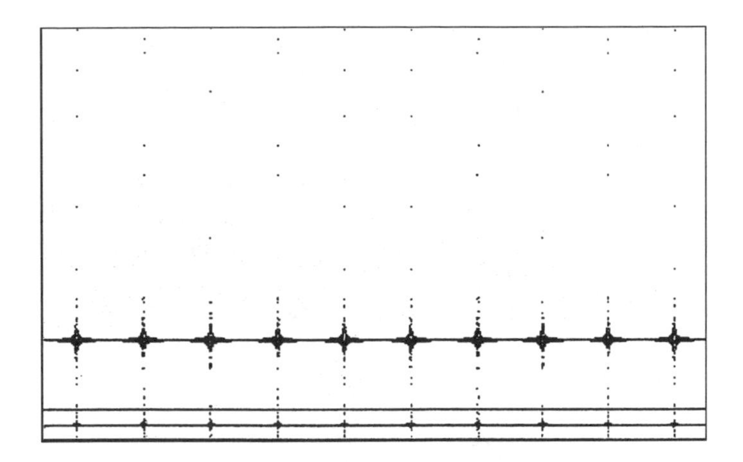

The periodic δ-function is useful in understanding the connection between Fourier transforms and Fourier series. Consider an arbitrary periodic function $f(t)$ with period T. This function has a Fourier transform $\tilde{f}(\omega)$, given by

$$\tilde{f}(\omega) = \int_{-\infty}^{\infty} f(t)\, e^{i\omega t}\, dt.$$

Using the periodic nature of $f(t)$, we can break the Fourier integral into a sum over separate periods:

$$\tilde{f}(\omega) = \sum_{n=-\infty}^{\infty} \int_{t_0-nT}^{t_0-nT+T} f(t)\, e^{i\omega t}\, dt,$$

where t_0 is an arbitrary time. Now we may change variables in the integral to $t_1 = t + nT$, and use Eq. (2.1.1) to obtain

$$\tilde{f}(\omega) = \sum_{n=-\infty}^{\infty} e^{i\omega nT} \int_{t_0}^{t_0+T} f(t_1)\, e^{i\omega t_1}\, dt_1. \tag{2.3.50}$$

However, Eq. (2.3.49) implies that $\sum_{n=-\infty}^{\infty} e^{i\omega nT} = \delta^{(P)}(\omega T^2/2\pi)$. Then Eq. (2.3.50) becomes

$$\begin{aligned}\tilde{f}(\omega) &= \sum_{m=-\infty}^{\infty} \delta\left(\frac{\omega T^2}{2\pi} - mT\right) \int_{t_0}^{t_0+T} f(t_1)\, e^{i\omega t_1}\, dt_1 \\ &= \sum_{m=-\infty}^{\infty} 2\pi\delta\left(\omega - \frac{2\pi m}{T}\right) \frac{1}{T} \int_{t_0}^{t_0+T} f(t_1)\, e^{i\omega t_1}\, dt_1, \end{aligned} \tag{2.3.51}$$

where we have used Eqs. (2.3.48) and (2.3.33). Furthermore, note that the integral in Eq. (2.3.51) is the expression for the mth Fourier coefficient, c_m, Eq. (2.1.33).

Therefore, we can write

$$\tilde{f}(\omega) = \sum_{m=-\infty}^{\infty} 2\pi\delta\left(\omega - \frac{2\pi m}{T}\right)c_m. \tag{2.3.52}$$

This equation connects the Fourier integral of a periodic function $f(t)$ to the function's exponential Fourier coefficients c_m. We see that in frequency space a periodic function consists of a sum of δ-functions at all harmonics of the fundamental frequency $\Delta\omega = 2\pi/T$.

Finally, applying the inverse transform to Eq. (2.3.52), the integral over each δ-function in the sum can be evaluated, and we return to our previous expression for $f(t)$ as a Fourier series:

$$f(t) = \sum_{m=-\infty}^{\infty} c_m e^{-im\,\Delta\omega\, t}.$$

2.3.5 Fast Fourier Transforms

Discrete Fourier Transforms In this section we consider methods for performing numerical Fourier transforms. The idea is that one is given a set of data $\{f_n\}$ measured at N evenly spaced discrete times $t_n = n\,\Delta t$, $n = 0,1,2,\dots,N-1$. From this data, one wishes to determine a numerical frequency spectrum.

This sort of problem arises in experimental physics as well as in numerical simulation methods, and in many other fields of science, including economics, engineering, signal processing, and acoustics.

One way to attack this problem is simply to discretize the integrals in the Fourier transform and the inverse transform. If we consider a discretized version of a Fourier transform, Eq. (2.3.12), with the time variable replaced by closely spaced discrete timesteps $t_n = n\,\Delta t$, and the frequency replaced by closely spaced discrete frequencies $\omega_m = m\,\Delta\omega$, one can immediately see that the operation of Fourier transformation, $\hat{F}f$, is equivalent to the dot product of a vector with a matrix:

$$\tilde{f}(\omega) = \hat{F}f(t) \to \tilde{f}_m = \sum_{n=-\infty}^{\infty} F_{mn} f_n, \tag{2.3.53}$$

where $f_n = f(n\,\Delta t)$ and $\tilde{f}_m = \tilde{f}(m\,\Delta\omega)$, and the matrix F_{mn} has components

$$F_{mn} = \Delta t\, e^{imn\,\Delta t\,\Delta\omega}. \tag{2.3.54}$$

This matrix is a discretized form of the Fourier transform operator $\hat{F}$. Equation (2.3.54) shows directly that there is an analogy between the linear integral operator $\hat{F}$ acting on functions and a matrix $\mathbf{F}$ acting on vectors. Previously, we saw that there was a similar analogy between linear *differential* operators and matrices.

When the inverse transform operator $\hat{F}^{-1}$ is discretized, it also becomes a matrix, with components

$$(\mathbf{F}^{-1})_{mn} = \Delta\omega\, e^{-imn\,\Delta t\,\Delta\omega}/2\pi. \tag{2.3.55}$$

This matrix can be applied to discretized functions of frequency $\tilde{\mathbf{f}}$ in order to reconstruct the corresponding discretized function of time, according to the matrix equation

$$\mathbf{F}^{-1} \cdot \tilde{\mathbf{f}} = \mathbf{f}. \tag{2.3.56}$$

So, in order to take a numerical Fourier transform of a data set $\{f_n\}$, it appears that all we need do is apply the matrix $\mathbf{F}$ to the vector $\mathbf{f}$ with components f_n, according to Eq. (2.3.53). To turn the resulting discretized spectral function $\tilde{\mathbf{f}}$ back into the time data, all we need do is apply the matrix $\mathbf{F}^{-1}$, defined by Eq. (2.3.55).

This is fine so far as it goes, but there are several problems hiding in this procedure. First, the matrices $\mathbf{F}$ and $\mathbf{F}^{-1}$ are formally infinite-dimensional. We can try to get around this problem by cutting off the Fourier integrals. Since the data runs from $0 \le t \le (N-1)\,\Delta t$, we can cut off the time integral in the Fourier transform beyond this range of times. For the frequency integral in the inverse transform, we can also impose a frequency cutoff, keeping only frequencies in the range $-M\,\Delta\,\omega \le \omega \le M\,\Delta\,\omega$ for some large integer value of M. The hope is that the frequency spectrum goes to zero for sufficiently large ω, so that this cutoff does not neglect anything important.

There is a second problem: although Δt is determined by the dataset, what should our choice for $\Delta\,\omega$ be? Apparently we can choose anything we want, so long as $\Delta\,\omega$ is "small." This is technically correct—if $\Delta\,\omega$ is small compared to the scale of variation of the frequency spectrum, then the discretized integral in the Fourier transform is well represented by the Riemann sum given by Eq. (2.3.56). However, we also must keep enough terms so that $M\,\Delta\,\omega$ is a large frequency—large enough to encompass the bulk of the frequency spectrum.

For very smooth time data with only low-frequency spectral components, this prescription works (see the exercises). However, for real data, which may have high-frequency spectral components and rapid variation in the spectrum, the above method becomes impractical, because we must take M very large. Also, the matrices $\mathbf{F}^{-1}$ and $\mathbf{F}$ are only approximately the inverses of one another, because of the errors introduced by discretizing the Fourier and inverse Fourier integrals.

One way to improve matters is to recognize that the time data, extending only over a finite time range $0 \le t \le (N-1)\,\Delta t$, can be replaced by a periodic extension with period $T = N\Delta t$. We take this time as the period [rather than, say, $(N-1)\,\Delta t$] so that the first data point beyond this interval, at time $N\Delta t$, has value f_0. This way, the data can be seen to repeat with period $T = N\Delta t$ (see Fig. 2.9).

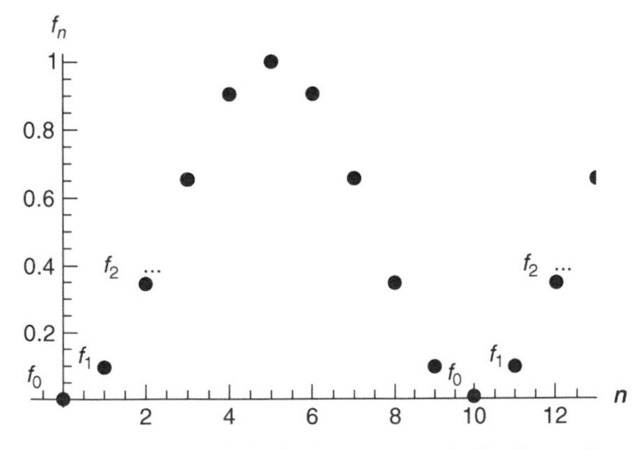

Fig. 2.9 A dataset with 10 elements, periodically replicated.

Since the periodic extension has period T, the data can be represented by a Fourier series rather than a transform, with a fundamental frequency $\Delta\omega = 2\pi/T$. This is the smallest frequency that can be represented by this data, and so is a good choice for our frequency discretization. Hearkening back to our equation for Fourier series, Eq. (2.2.2), we would like to write

$$f^{(p)}(t) = \sum_{m=-\infty}^{\infty} \tilde{f}_m \, e^{-im\,\Delta\omega\, t}, \tag{2.3.57}$$

where $f^{(p)}(t)$ is a periodic function of time that represents the discrete data (we will see what this function of continuous time looks like in a moment), and the $\tilde{f}_m$'s are the Fourier coefficients in the series. As in all Fourier series, the sum runs over all integers, and of course this is a problem in numerical applications. However, we will see momentarily that this problem can be solved in a natural way.

The Fourier components $\tilde{f}_m$ are determined by the integral:

$$\tilde{f}_m = \frac{1}{N\Delta t} \int_0^{N\Delta t} f(t)\, e^{im\,\Delta\omega\, t}\, dt \tag{2.3.58}$$

[see Eq. (2.2.3)]. However, since the time data is discrete, we replace this integral by the Riemann sum just as in Eq. (2.3.53):

$$\tilde{f}_m = \frac{1}{N\Delta t} \sum_{n=0}^{N-1} \Delta t f_n \, e^{im\,\Delta\omega\, n\,\Delta t} = \frac{1}{N} \sum_{n=0}^{N-1} f_n \, e^{i2\pi mn/N}, \tag{2.3.59}$$

where in the second step we have used $\Delta\omega = 2\pi/T$.

These Fourier components have an important property: they are themselves periodic, satisfying $\tilde{f}_{m+N} = \tilde{f}_m$. The proof is simple:

$$\tilde{f}_{m+N} = \frac{1}{N} \sum_{n=0}^{N-1} f_n \, e^{i2\pi(m+N)n/N} = \frac{1}{N} \sum_{n=0}^{N-1} f_n \, e^{i2\pi mn/N + i2\pi n} = \tilde{f}_m. \tag{2.3.60}$$

Now, since $\tilde{f}_m$ repeats periodically, we can rewrite the infinite sum over these $\tilde{f}_m$'s in Eq. (2.3.57) as sums over repeating intervals of size N:

$$f^{(p)}(t) = \sum_{p=-\infty}^{\infty} \sum_{m=Np}^{N-1+Np} \tilde{f}_m \, e^{-im\,\Delta\omega\, t} = \sum_{p=-\infty}^{\infty} e^{-iNp\,\Delta\omega\, t} \sum_{m=0}^{N-1} \tilde{f}_m \, e^{-im\,\Delta\omega\, t}.$$

However, the sum over p can be written as a periodic δ-function using Eq. (2.3.49):

$$f^{(p)}(t) = \Delta t \sum_{n=-\infty}^{\infty} \delta(t - n\,\Delta t) \sum_{m=0}^{N-1} \tilde{f}_m \, e^{-im\,\Delta\omega\, t}. \tag{2.3.61}$$

This equation represents our discrete data as a sum of δ-functions at the discrete times $n\,\Delta t$, periodically extended to the entire real line. This is not a bad way to think about the data, since the δ-functions provide a natural way of modeling the discrete data in continuous time. Also, we implicitly used this representation when we wrote the Riemann sum in Eq. (2.3.59). That is, if we define a function $f(t)$ for use in Eq. (2.3.58) according to

$$f(t) = \Delta t \sum_{n=0}^{N-1} f_n \delta(t - n\,\Delta t), \tag{2.3.62}$$

we directly obtain Eq. (2.3.59). The function $f^{(p)}(t)$ in Eq. (2.3.61) is merely the periodic extension of $f(t)$.

Furthermore, comparing Eq. (2.3.61) to (2.3.62) we see that the time data f_n can be written directly in terms of the Fourier coefficients $\tilde{f}_m$:

$$f_n = \sum_{m=0}^{N-1} \tilde{f}_m e^{-im\,\Delta\omega\, n\,\Delta t} = \sum_{m=0}^{N-1} \tilde{f}_m e^{-i2\pi mn/N}. \tag{2.3.63}$$

Equations (2.3.59) and (2.3.63) are called a discrete Fourier transform and a discrete inverse Fourier transform respectively. These two equations provide a method for taking a set of discrete time data f_n at times $n\,\Delta t$, $0 \le n \le N-1$, and obtaining a frequency spectrum $\tilde{f}_m$ at frequencies $m\,\Delta\omega$, $0 \le m \le N-1$, where $\Delta\omega = 2\pi/(N\Delta t)$ is the fundamental frequency of the periodic extension of the data.

Discrete Fourier transform of time data $\{f_n\}$:

$$\tilde{f}_m = \frac{1}{N} \sum_{n=0}^{N-1} f_n e^{i2\pi mn/N}.$$

Discrete inverse transform of frequency data $\{\tilde{f}_m\}$:

$$f_n = \sum_{m=0}^{N-1} \tilde{f}_m e^{-i2\pi mn/N}.$$

Equations (2.3.59) and (2.3.63) can be written as matrix equations, $\tilde{f} = \mathbf{F}\cdot f$ and $f = \mathbf{F}^{-1}\cdot\tilde{f}$. The N-by-N matrices $\mathbf{F}$ and $\mathbf{F}^{-1}$ have components

$$\begin{aligned} \mathbf{F}_{mn} &= \frac{1}{N} e^{i2\pi mn/N}, \\ (\mathbf{F}^{-1})_{mn} &= e^{-i2\pi mn/N}. \end{aligned} \tag{2.3.64}$$

These matrices are similar to the discretized forms for the Fourier transform operators, Eqs. (2.3.54) and (2.3.55). However, according to Eqs. (2.3.59) and (2.3.63), the matrices in Eq. (2.3.64) are exact inverses of one another, unlike the finite-dimensional versions of the discretized Fourier and inverse Fourier transforms, Eqs. (2.3.54) and (2.3.55). Therefore, these matrices are much more useful than the discretized Fourier transform operators.

After constructing these matrices, we can use them to take the discretized Fourier transform of a set of data. The matrices themselves are easy to create using a **Table** command. For instance, here are 100-by-100 versions:

Cell 2.60

```
nn = 100; F = Table[Exp[I 2. Pi m n/nn]/nn,
  {m, 0, nn - 1}, {n, 0, nn - 1}];
F1 =  Table[Exp[I 2. Pi m n/nn], {m, 0, nn - 1},
  {n, 0, nn - 1}];
```

Note the use of approximate numerical mathematics, rather than exact mathematics, in creating the matrices. When dealing with such large matrices, exact mathematical operations take too long.

We can use these matrices to Fourier analyze data. Let's create some artificial time data:

Cell 2.61

```
f = Table[N[Sin[40 Pi n/100] + (Random[] - 1/2)],
  {n, 100}];
```

This data is a single sine wave, $\sin(40\pi t)$, sampled at time $t_n = n/100$, with some noise added, as shown in Cell 2.62.

Cell 2.62

```
ListPlot[f, PlotJoined→True]
```

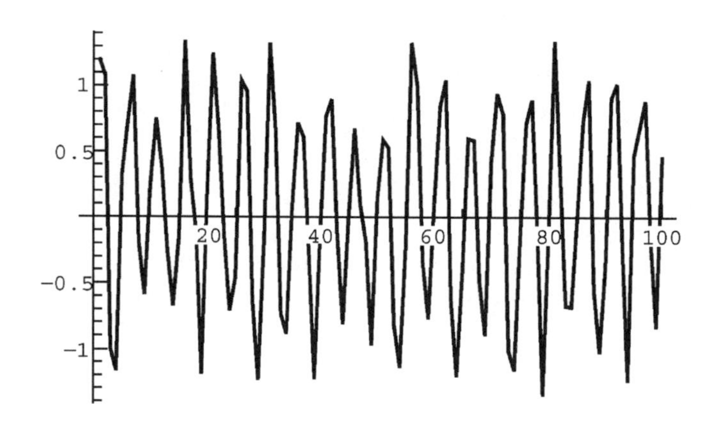

The frequency spectrum for this data is obtained by applying the Fourier matrix **F** to it:

Cell 2.63

```
f̃ = F.f;
```

Again, we will use a **ListPlot** to look at the spectrum. However, since $\tilde{f}$ is complex, we will plot the real and imaginary parts of the spectrum separately, in Cells 2.64 and 2.65. Evidently, the two peaks in the real and imaginary parts of the spectrum correspond to the two Fourier components of $\sin 40\pi t = e^{40\pi it} - e^{-40\pi it})/2i$. These components have frequencies of $\pm 40\pi$. Since the frequency is discretized in units of $\Delta\omega = 2\pi/(N\Delta t) = 2\pi$, the frequency 40π corresponds to the 21st element of $\tilde{\text{f}}$ (sine $\omega = 0$ corresponds to the first element). This agrees with the plots, which show a peak at the 21st element.

Cell 2.64

```
ListPlot[Re[f̃], PlotJoined→True, PlotRange→All];
```

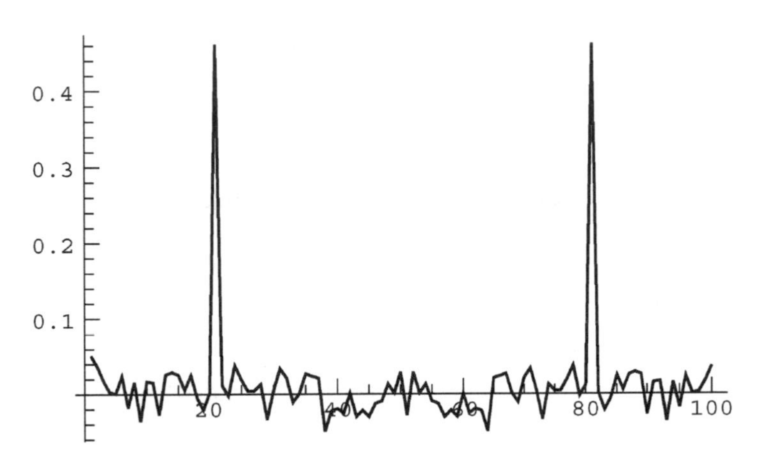

Cell 2.65

```
ListPlot[Im[f̃], PlotJoined→True, PlotRange→All];
```

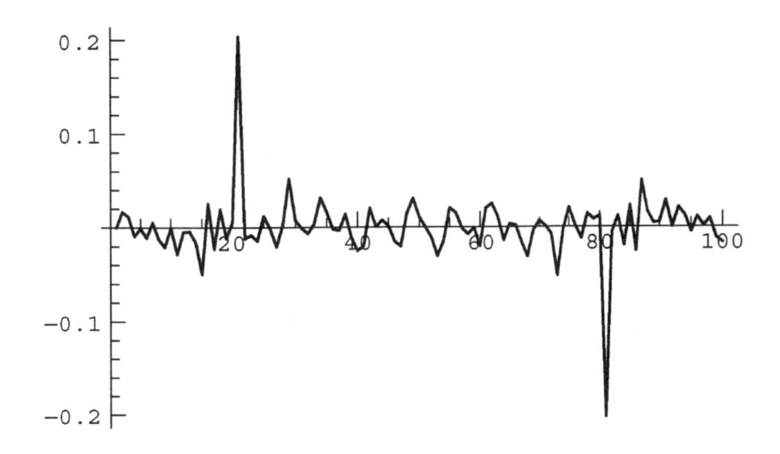

What about the other peak, which is supposed to occur at -40π? There are no negative frequencies in our spectrum. Rather, frequencies run from 0 to $(N-1)\Delta\omega$

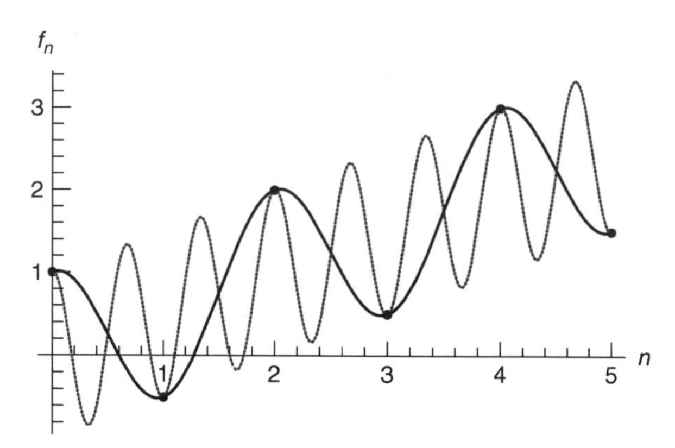

Fig. 2.10 The fastest sinusoidal oscillation that can be unambiguously identified in a set of data has the Nyquist frequency $\omega_{\max}$ and period $2\,\Delta t$ (black curve). Higher-frequency oscillations (dashed curve) can be translated back to lower frequency.

in units of $\Delta\omega = 2\pi$. Recall, however, that $\tilde{\mathbf{f}}$ is periodic with period N [see Eq. (2.3.60)]. In particular, it repeats over the negative frequency range. So we can think of the second peak in the plots, at element 80 of $\tilde{\mathbf{f}}$, as actually being at a negative frequency. Since $\tilde{f}_{100} = \tilde{f}_0$, then $\tilde{f}_{80} = \tilde{f}_{-20}$, corresponding to the spectral component with frequency $-20\,\Delta\omega = -40\pi$.

Thus, the upper half of the frequency spectrum should be thought of as corresponding to negative frequencies. This implies that as we count upwards through the elements of our frequency spectrum, starting with the first element, the frequency increases like $m\,\Delta\omega$ *until we get the center of the spectrum.* At this point the frequency jumps to negative values and approaches zero from the negative side. Therefore, the maximum frequency *magntiude* kept in our spectrum is

$$\omega_{\max} = N\Delta\,\omega/2 = \pi/\Delta t. \tag{2.3.65}$$

The frequency $\omega_{\max}$ is called the *Nyquist frequency* for the data. The physical reason why this is the maximum frequency in a set of data can be understood from Fig. 2.10, which displays some time data. The highest-frequency sinusoidal wave that we can interpolate through these data points has a *half* period equal to Δt: one needs at least two points to determine a sine wave, one at a minimum and one at a neighboring maximum. Thus, the minimum full wavelength defined by the data is $2\,\Delta t$, and the maximum frequency is given by the Nyquist frequency $\omega_{\max} = 2\pi/(2\,\Delta t)$, Eq. (2.3.65).

More rapid oscillations, with frequency $\omega_{\max} + pN\Delta\,\omega = (2p+1)\,\omega_{\max}$, for any integer p can also be made to go through these points as well—see the dashed curve in Fig. 2.10, which corresponds to $p = 1$. However, since these higher-frequency oscillations can always be referred back to the frequency $\omega_{\max}$, we use $\omega_{\max}$ to describe them—there is nothing in the data to distinguish them from an oscillation at $\omega_{\max}$. The identification of higher frequencies with a lower frequency is called *aliasing*. This identification arises from the discreteness of the data—there are no data points in between the given data points that can determine the locations of the peaks and troughs in the dashed curve.

There is something else that one can see in the real and imaginary parts of the plotted spectrum. Because the input data $\mathbf{f}$ was real, the spectral components $\tilde{\mathbf{f}}$ have the property that $\tilde{f}_{-m} = \tilde{f}_m^*$. This is analogous to Eq. (2.3.7) for continuous Fourier transforms, and follows directly from Eq. (2.3.59) for any real set of time data f_n. We can also see this symmetry in the previous plots if one remembers that the periodicity of $\tilde{f}_m$ implies that $\tilde{f}_{-m} = \tilde{f}_{N-m}$. Thus, for real data, the discrete Fourier transform has the property that

$$\tilde{f}_{-m} = \tilde{f}_{N-m} = \tilde{f}_m^*. \tag{2.3.66}$$

Fast Fourier Transforms Although the discrete Fourier transforms given by Eqs. (2.3.59) and (2.3.63) work, they are not very useful in practice. The reason is that they are too slow. To evaluate the spectrum for a data set of N elements, a sum over the data set must be done for each frequency in the spectrum. Since there are N elements in the sum and N frequencies in the spectrum, determining every frequency component of a dataset requires of order N^2 operations. For $N = 100$ this is not a problem (as we saw above), but for $N = 10^6$ it is out of the question. Datasets of this size are routinely created in all sorts of applications.

Fortunately, a method was developed that allows one to perform the discrete transform and inverse transform with far fewer than N^2 operations. The method is called the method of *fast Fourier-transforms* (FFT for short). Invented by several individuals working independently as far back as the 1940s, it later became well known through the work of Cooley and Tukey at IBM Research Center in the mid-1960s.

The method relies on symmetries of the discrete Fourier transform that allow one to divide the problem into a hierarchy of smaller problems, each of which can be done quickly (the *divide-and-conquer* approach). We will not examine the nuts and bolts of the procedure in detail. But it is worthwhile to briefly discuss the idea behind the method.

Given a discrete Fourier transform of a data set with N elements, N *assumed even*, one can break this transform up into two discrete transforms with $N/2$ elements each. One transform is formed from the even-numbered data points, the other from the odd-numbered data points:

$$\begin{aligned}
\tilde{f}_m &= \sum_{n=0}^{N-1} f_n \, e^{i2\pi mn/N} \\
&= \sum_{n=0}^{N/2-1} f_{2n} \, e^{i2\pi m2n/N} + \sum_{n=0}^{N/2-1} f_{2n+1} \, e^{i2\pi m(2n+1)/N} \\
&= \sum_{n=0}^{N/2-1} f_{2n} \, e^{i2\pi mn/(N/2)} + e^{i2\pi m/N} \sum_{n=0}^{N/2-1} f_{2n+1} \, e^{i2\pi m/(N/2)} \\
&= \tilde{f}_m^{(e)} + e^{i2\pi m/N} \tilde{f}_m^{(o)}.
\end{aligned} \tag{2.3.67}$$

In the last line, $\tilde{f}_m^{(e)}$ denotes the discrete Fourier transform over the $N/2$ even elements of f_n, and $\tilde{f}_m^{(o)}$ denotes the discrete Fourier transform over $N/2$ odd elements of the data.

At this point it appears that we have gained nothing. The two separate transforms each require of order $N/2$ operations, and there are still N separate frequencies $\tilde{f}_m$ to be calculated, for a total of N^2 operations again. However, according to Eq. (2.3.60), both $\tilde{f}_m^{(e)}$ and $\tilde{f}_m^{(o)}$ are periodic with period $N/2$. Therefore, we really only need to calculate the components $m = 1, 2, \dots, N/2$ for each transform.

In this single step, we have saved ourselves a factor of 2. Now, we apply this procedure again, assuming that $N/2$ is an even number, saving another factor of 2, and so on, repeating until there is only *one* element in each series. This assumes that $N = 2^P$ for some integer P. In fact, $N = 2^P$ are the only values allowed in the simplest implementations of the FFT method. (If $N \neq 2^P$ for some integer P, one typically pads the data with zeros until it reaches a length of 2^P for some P.)

It appears from the above argument that we have taken the original N^2 steps and reduced their number by a factor of $2^P = N$, resulting in a code that scales linearly with N. However, a more careful accounting shows that the resulting recursive algorithm actually scales as $N \log_2 N$, because the number of steps in the recursion equals $P = \log_2 N$. Nevertheless, this is a huge saving over the original N^2 operations required for the discrete Fourier transform.

Writing an efficient FFT code based on the above ideas is not entirely straightforward, and will not be pursued here. [An implementation of the code can be found in Press et al. (1986); see the references at the end of Chapter 1.] Fortunately, *Mathematica* has done the job for us, with the intrinsic functions **Fourier** and **InverseFourier**. **Fourier** acts on a list of data $\mathbf{f} = \{f_n\}$ to produce a frequency spectrum $\tilde{\mathbf{f}} = \{\tilde{f}_m\}$. The syntax is **Fourier[f]** and **InverseFourier[f̃]**. However, just as with continuous Fourier transforms, many different conventions exist for the definitions of the discrete transform. The default convention used by **Fourier** and **InverseFourier** differs from that used in Eqs. (2.3.59) and (2.3.63). For a dataset of length N our convention corresponds to

$$\tilde{\mathbf{f}} = \mathbf{Fourier[f]}/\sqrt{\mathbf{N}},$$
$$\mathbf{f} = \sqrt{\mathbf{N}}\,\mathbf{InverseFourier[\tilde{f}]}. \tag{2.3.68}$$

Of course, any convention can be used, provided that one is consistent in its application. We will stick to the convention of Eq. (2.3.68), since it corresponds to our previous discussion.

The length of the data sets taken as arguments in **Fourier** and **InverseFourier** need not be 2^P. For example, in Cell 2.66 we apply them to the original data of 100 elements created in the previous section. Comparing these plots with those generated in Cells 2.64 and 2.65 using the discrete Fourier transform method, one can see that they are identical.

Cell 2.66

```
nn = 100; f̃ = Fourier[f]/Sqrt[nn];
ListPlot[Re[f̃], PlotJoined→True, PlotRange→All];
ListPlot[Im[f̃], PlotJoined→True, PlotRange→All];
```

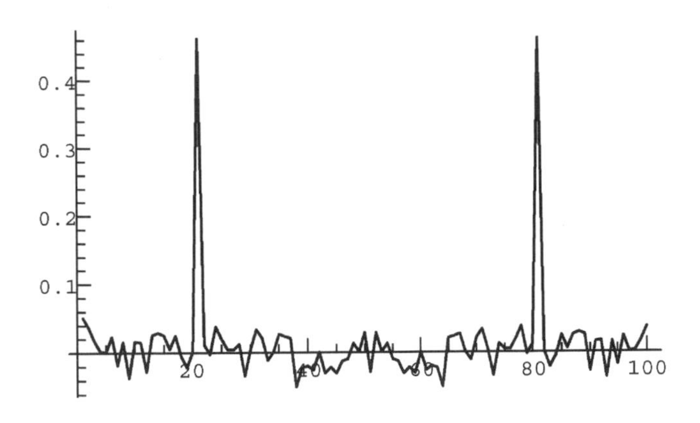

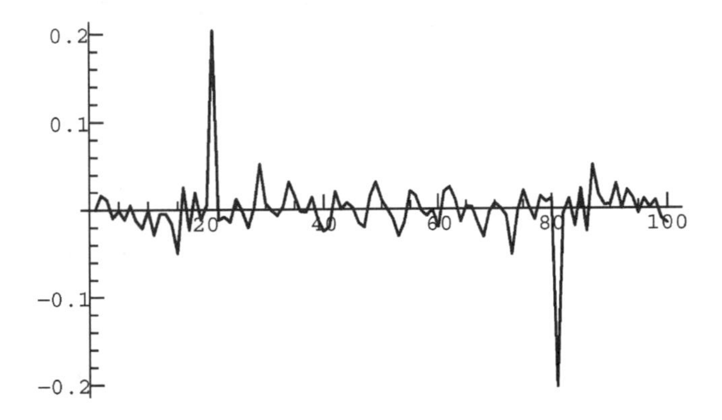

Let's apply Fourier analysis to a real data signal. *Mathematica* has the ability to read various types of audio files, such as AIFF format, μ-law encoding, or Microsoft WAV format. These sound files can be read using the intrinsic function **`Import`**. To read the following sound file of my voice, named **`ah.AIFF`**, first determine the current working directory on the hard disk with the command

Cell 2.67

```
Directory[]
```

```
/Users/dubin
```

Then either copy the file into this directory from the cd on which this book came, or if that is not possible, set the working directory to another hard disk location using the **`SetDirectory`** command. Once the file is in the current working directory, the file can be read:

Cell 2.68

```
snd = Import["ah.AIFF", "AIFF"]
```

```
- Sound -
```

As with any other sound in *Mathematica*, this sound can be played using the **Show** command (Cell 2.69).

Cell 2.69

```
Show[snd];
```

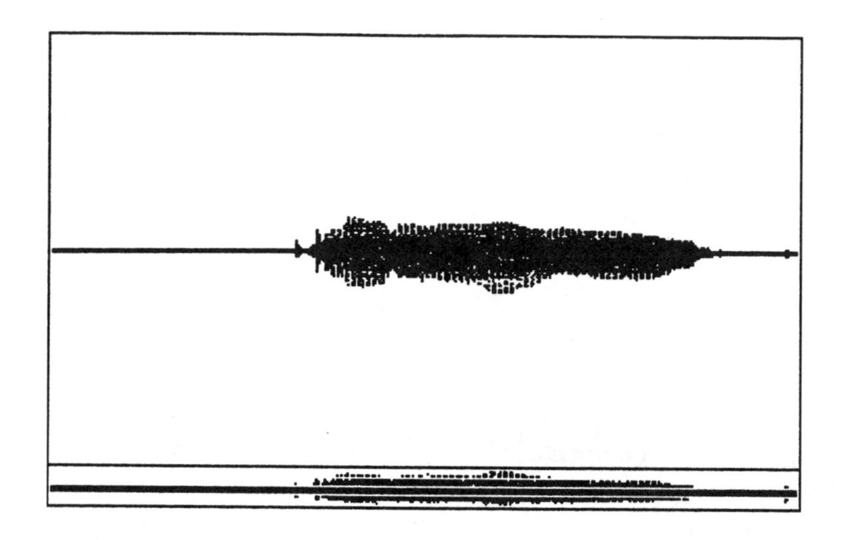

Let's take a look at what is contained in **snd** by looking at the internal form of the data:

Cell 2.70

```
Shallow[InputForm[snd]]
```

```
Sound[SampledSoundList[<<2>>]]
```

This data contains a **SampledSoundList**, which is a set of sound data in the form $\{\{f_0, f_1, f_2, \ldots\}, \textit{samplerate}\}$. The data $\{f_0, f_1, f_2, \ldots\}$ provide a list of sound levels, which are played consecutively at a rate given by *samplerate*. (We applied the command **Shallow** so that we didn't have to print out the full data list, just the higher levels of the data structure.) We can extract the sample rate using

Cell 2.71

```
samplerate = InputForm[snd][[1]][[1]][[2]]
```

```
22050.
```

which means the sample rate is 22,050 hertz, a common value used in digitized recordings. The time between samples, Δt, is just the reciprocal of the sample rate:

Cell 2.72

```
Δt = 1/samplerate
```

```
0.0000453515
```

The data itself can be extracted using

Cell 2.73

```
f = InputForm[snd][[1]][[1]][[1]];
```

The length of this data list is

Cell 2.74

```
nn = Length[f]

36864
```

Let's look at some of this data over a region where something is happening.

Cell 2.75

```
Table[f[[n]], {n, 25000, 25000 + 40}]

{0.03125, 0.015625, 0.0078125, 0, 0, 0, -0.0078125,
 -0.0078125, -0.0078125, 0, 0, 0, 0.0078125, 0.015625,
 0.0234375, 0.03125, 0.03125, 0.0390625, 0.03125,
 0.0234375, 0.0234375, 0.015625, 0.015625, 0.0234375,
 0.0234375, 0.0234375, 0.0234375, 0.0234375, 0.0234375,
 0.0234375, 0.0234375, 0.03125, 0.0390625, 0.046875,
 0.046875, 0.046875, 0.0390625, 0.03125, 0.015625,
 0.0078125, 0}
```

The numbers giving the sound levels are discretized in units of 0.0078125. This is because the sound has been digitized, so that amplitude levels are given by discrete levels rather than by continuous real numbers. Note that $1/0.0078125 = 128 = 2^7$. This corresponds to 8-bit digitization: the other bit is the sign of the data, $\pm$. (In base two, integers running from 0 to 127 require seven base-two digits, or bits, for their representation.) Because there are only 128 different possible sound levels in this data file, the sound is not very high quality, as you can tell from the playback above. There is quite a bit of high-frequency hiss, due at least in part to the rather large steps between amplitude levels that create high frequencies in the sound.

We can see this in the Fourier transform of the data:

Cell 2.76

```
f̃ = Fourier[f] / Sqrt[nn];
```

A plot of this Fourier transform (Cell 2.77) shows considerable structure at low frequencies, along with a low-level high-frequency background. It is easier to comprehend this data if we plot it in terms of actual frequencies rather than just the order of the elements of $\tilde{\mathbf{f}}$, so we will replot the data. The separation $\Delta\omega$ between adjacent Fourier modes is $2\pi/(N\Delta t)$, in radians per second. In hertz, the separation is $\Delta f = 1/(N\Delta t)$ (see Cell 2.78).

Cell 2.77

```
ListPlot[Abs[f̃], PlotRange → All];
```

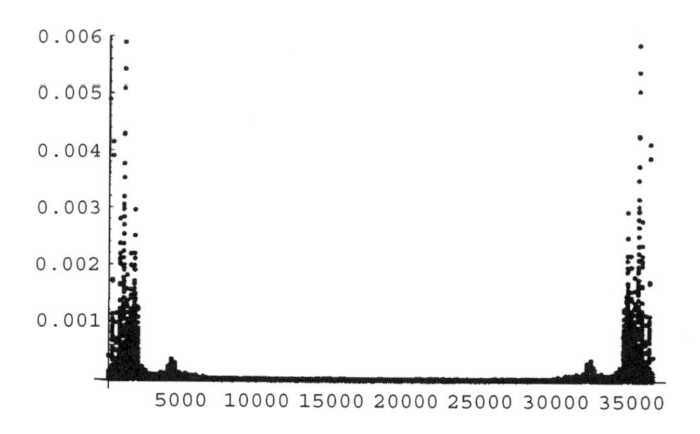

Cell 2.78

```
Δf = 1/(nn Δt)

General : : spell1 : Possible spelling error:
   new symbol name "Δf" is similar to existing symbol "Δt".

0.598145
```

We therefore create a data list, $\{n\,\Delta f, \tilde{f}_n\}$, $n = 0, 1, 2, \ldots, nn - 1$:

Cell 2.79

```
fdata = Table[{n Δf, f̃[[n + 1]]}, {n, 0, nn - 1}];
```

and in Cell 2.80 we plot this list over a range of low frequencies, up to 3000 hertz. A series of peaks are evident in the spectrum, which represent harmonics that are important to the "ahhh" sound. There is also a broad spectrum of low-level noise in the data, evident up to quite large frequencies. We can reduce this noise by applying a high-frequency filter to the frequency data. We will simply multiply all the data by an exponential factor that reduces the amplitude of the high frequencies (Cell 2.81).

Cell 2.80

```
ListPlot[Abs[fdata], PlotRange → {{0, 3000}, All},
AxesLabel → {"freq. (hz)", "|f̃|"}];
```

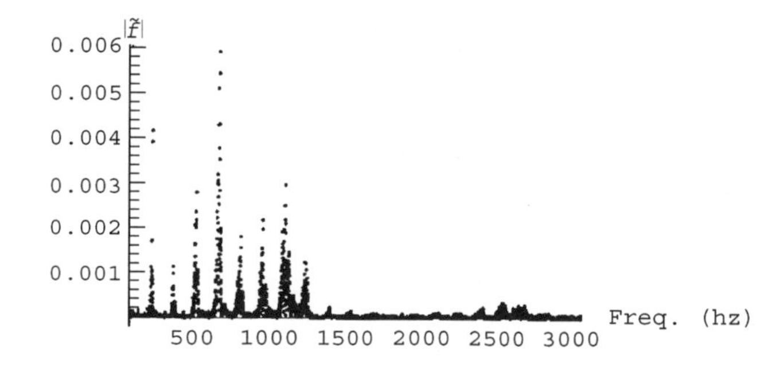

Cell 2.81

```
lowerhalf = Table[f̃[[n]] Exp[-n/600.], {n, 1, nn/2}];
```

We only run this filter over half of the data because the upper half of the spectrum corresponds to negative frequencies. To add in the negative-frequency half of the spectrum, it is easiest to simply use Eq. (2.3.66), which is just a statement that the upper half of the spectrum is the complex conjugate of the lower half, written in reverse order:

Cell 2.82

```
upperhalf = Reverse[Conjugate[lowerhalf]];
```

However, the lower half includes the zero-frequency point, while the upper half excludes this point. [Frequencies run from 0 to $(N-1)\,\Delta\,\omega$.] This zero-frequency point is the last point in the upper half, at position $\mathbf{n_{last}}$ = **Length[upperhalf]**:

Cell 2.83

```
nlast = Length[upperhalf]; upperhalf = Delete[upperhalf, nlast];
```

We now join these two lists together to get our new spectrum:

Cell 2.84

```
f̃new = Join[lowerhalf, upperhalf];
```

Since the length N of the original data was even, the new data has one less point, because in the above cut-and-paste process we have neglected the center point in the spectrum at the Nyquist frequency. This makes a negligible difference to the sound.

Cell 2.85

```
Length[f̃new]

36863
```

After inverse Fourier-transforming, we can play this resulting sound data using the **ListPlay** command, choosing to play at the original sample rate of 22050 hertz (Cell 2.86). Apart from the change in volume level (which can be adjusted using the **PlayRange** option) the main difference one can hear is that some high frequencies have been removed from this sound sample, as we expected. The audio filtering that we have applied here using an FFT is a crude example of the sort of operations that are employed in modern signal-processing applications. Some other examples of the use of FFTs may be found in the exercises, and in Secs. 6.2.2 and 7.3.

Cell 2.86

```
ffiltered = InverseFourier[f̃new];

ListPlay[ffiltered, SampleRate → samplerate];
```

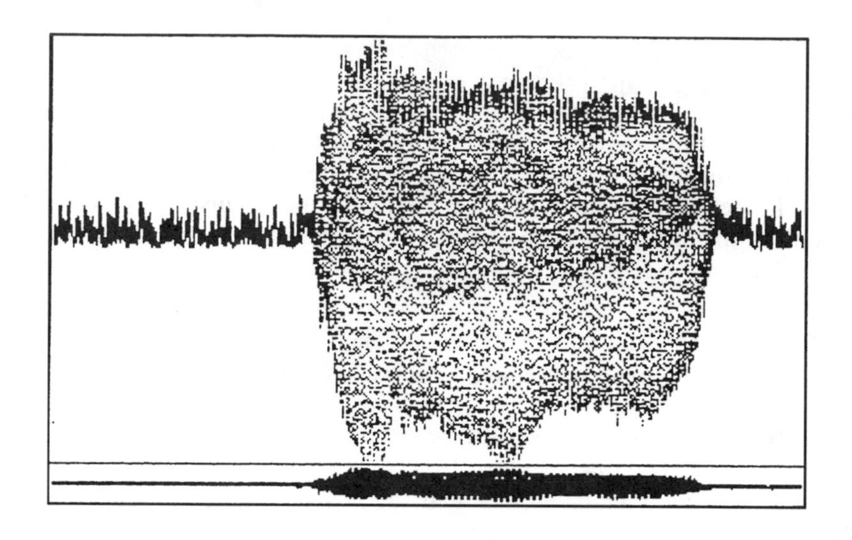

2.3.6 Response of a Damped Oscillator to General Forcing. Green's Function for the Oscillator

We are now ready to use Fourier transforms to solve a physics problem: the response of a damped harmonic oscillator to a general forcing function $f(t)$. Previously, only simple analytic forcing functions (Sec. 1.6) or periodic functions (Sec. 2.1) were considered. Using Fourier transforms, we can now deal with any form of forcing.

In order to obtain a particular solution $x_p(t)$ to the forced damped oscillator equation, Eq. (1.6.2), we act on the right and left-hand sides of the equation with a Fourier transform $\hat{F}$:

$$\hat{F}\left(x'' + \gamma x' + \omega_0^2 x\right) = \hat{F}f. \tag{2.3.69}$$

Defining the Fourier transform of the forcing function as $\tilde{f} = \hat{F}f$, and that of the particular solution as $\tilde{x}_p = \hat{F}x_p$, Eq. (2.3.69) is transformed from an ODE to a simple algebraic equation:

$$\left(-\omega^2 - i\omega\gamma + \omega_0^2\right)\tilde{x}_p(\omega) = \tilde{f}(\omega). \tag{2.3.70}$$

Here, we have applied Eq. (2.3.18) to the derivatives of x_p. We may then divide by the bracket, *assuming that the bracket is not zero*, to obtain

$$\tilde{x}_p(\omega) = \frac{\tilde{f}(\omega)}{-\omega^2 - i\omega\gamma + \omega_0^2}. \tag{2.3.71}$$

Equation (2.3.71) provides the amplitude $\tilde{x}_p(\omega)$ of all Fourier coefficients in the oscillator's response to the forcing. This equation shows that each Fourier mode in the response is excited only by its corresponding mode in the forcing. This is very similar in form to the response to periodic forcing, Eq. (2.1.37), except that now the frequency varies continuously rather than in discrete steps.

In order to determine the response in the time domain, an inverse transformation must be applied to Eq. (2.3.71):

$$x_p(t) = \int_{-\infty}^{\infty} \frac{e^{-i\omega t}\tilde{f}(\omega)}{-\omega^2 - i\omega\gamma + \omega_0^2} \frac{d\omega}{2\pi}. \tag{2.3.72}$$

This is as far as we can go in general. Without knowing the form of the forcing function, we cannot evaluate the frequency integral required in Eq. (2.3.72).

However, we can convert the frequency integral to a time integral by using the convolution theorem. Equation (2.3.71) can be seen to be a product of Fourier transforms; one transform is $\tilde{f}(\omega)$, and the other is $1/(-\omega^2 - i\omega\gamma + \omega_0^2) \equiv \tilde{g}(\omega)$. Since $\tilde{x}_p(\omega) = \tilde{f}(\omega)\tilde{g}(\omega)$, the convolution theorem implies that

$$x_p(t) = \int_{-\infty}^{\infty} f(t_0) g(t - t_0)\, dt_0, \tag{2.3.73}$$

where $g(t) = \hat{F}^{-1}\tilde{g}(\omega)$ is the inverse transform of $\tilde{g}$.

Although the time integral in Eq. (2.3.73) is not necessarily easier to evaluate than the frequency integral in Eq. (2.3.72), Eq. (2.3.73) has some conceptual and practical advantages. From a practical point of view, Eq. (2.3.73) deals only with $f(t)$, whereas Eq. (2.3.72) involves the Fourier transform of $f(t)$, which must be calculated separately. Thus, Eq. (2.3.73) saves us some work. Conceptually, the original differential equation is written in the time domain, and so is Eq. (2.3.73); so now there is no need to consider the frequency domain at all.

However, we do need to calculate the function $g(t - t_0)$. But we need do so only once, after which we can apply Eq. (2.3.73) to determine the response to any forcing function.

The function $g(t - t_0)$ is called the *Green's function* for the oscillator. We will soon see that Green's functions play a very important role in determining the particular solution to both ODE and PDEs.

The Green's function has a simple physical interpretation. If we take the forcing function in Eq. (2.3.73) to be a Dirac δ-function, $f(t_0) = \delta(t_0)$, then Eq. (2.3.73) yields

$$x_p(t) = g(t).$$

In other words, the Green's function $g(t)$ is a response of the oscillator to a δ-function force at time $t = 0$. The total impulse (i.e. momentum change) imparted by the force is proportional to $\int_{-\infty}^{\infty} \delta(t_0)\, dt_0 = 1$, so this force causes a finite change in the velocity of the oscillator. To understand this physically, think of a tuning fork. At $t = 0$, we tap the tuning fork with an instantaneous force that causes it to oscillate. The Green's function is this response. For this reason, Green's functions are often referred to as *response functions*.

Of course, the tuning fork could already be oscillating when it is tapped, and that would correspond to a different particular solution to the problem. In general, then, there are many different possible Green's functions, each corresponding to different initial (or boundary) conditions. We will see that when Fourier transforms are used to determine the Green's function for the damped oscillator, this method picks out the particular solution where the oscillator is at rest before the force is applied.

Before we proceed, it is enlightening to step back for a moment and contemplate Eq. (2.3.73). This equation is nothing more than another application of the superposition principle. We are decomposing the forcing function $f(t)$ into a sum of δ-function forces, each occurring at separate times $t = t_0$. Each force produces its own response $g(t - t_0)$, which is superimposed on the other responses to produce the total response $x_p(t)$.

Previously, we decomposed the function $f(t)$ into individual Fourier modes. Equations (2.3.72) and (2.3.73) show that the response to the force can be thought of either as a linear superposition of sinusoidal oscillations in response to each separate Fourier mode in the force, or as a superposition of the responses to a series of separate δ-function forces. Fourier decomposition into modes, and decomposition into δ-function responses, are both useful ways to think about the response of a system to forcing. Both rely on the principle of superposition.

We can determine the Green's function for the damped oscillator analytically by applying the inverse Fourier transform to the resonance function $\tilde{g}$:

$$g(t) = \int_{-\infty}^{\infty} \frac{e^{-i\omega t}}{-\omega^2 - i\omega\gamma + \omega_0^2} \frac{d\omega}{2\pi}. \tag{2.3.74}$$

To evaluate this integral, we must first simplify the integrand, noting that the denominator $-\omega^2 - i\omega\gamma + \omega_0^2$ can be written as $(i\omega_1 + s_1)(i\omega + s_2)$, where s_1 and s_2 are the roots of the homogeneous polynomial equation discussed in Sec. (2.6.2), and given by Eqs. (1.6.14). Then we may write Eq. (2.3.74) as

$$g(t) = \frac{1}{s_2 - s_1} \int_{-\infty}^{\infty} e^{-i\omega t}\left(\frac{1}{i\omega + s_1} - \frac{1}{i\omega + s_2}\right)\frac{d\omega}{2\pi}, \tag{2.3.75}$$

where we have separated the resonance function into its two components. It is best to perform each integral in Eq. (2.3.75) separately, and then combine the results. Integrals of this sort can be evaluated using **InverseFourierTransform**:

Cell 2.87

```
FullSimplify[
 InverseFourierTransform[1/(I ω-a), ω, t,
  Assumptions -> Re[a] ≥ 0]/Sqrt [2 Pi]]
```

$$-\frac{1}{2}e^{-at}\ (1 + \text{Sign}[t])$$

Noting that $[1 + \text{Sign}(t)]/2 = h(t)$, the Heaviside step function, and taking $a = -s_1$ in the above integral, we then have

$$\hat{F}^{-1}\frac{1}{i\omega + s_1} = -e^{ts_1}h(t), \qquad \text{Re}\, s_1 \le 0, \tag{2.3.76}$$

with a similar result for the inverse transform involving s_2. Since Eq. (1.6.14) implies that the real parts of s_1 and s_2 are identical, equaling the nonpositive

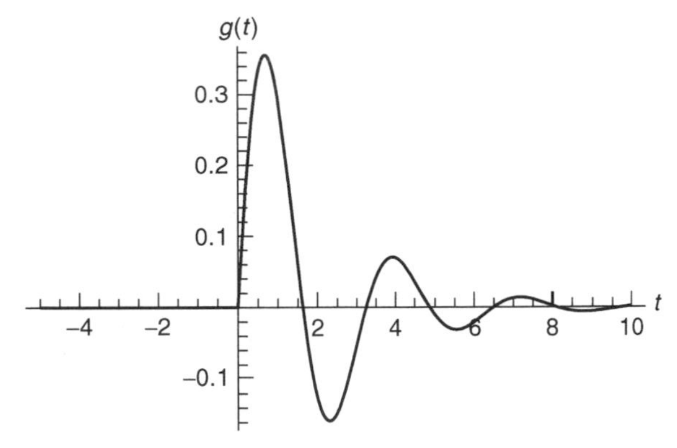

Fig. 2.11 Green's function for the linear oscillator: response to a δ-function force. Here $\omega_0 = 2$ and $\gamma = 1$.

quantity $-\gamma/2$, we can apply Eq. (2.3.76) to Eq. (2.3.75), yielding

$$g(t) = h(t)\frac{\mathbf{e}^{s_1 t} - e^{s_2 t}}{s_1 - s_2} = h(t)\frac{e^{-\gamma t/2}}{\sqrt{\omega_0^2 - \gamma^2/4}}\sin\left(\sqrt{\omega_0^2 - \frac{\gamma^2}{4}}\,t\right), \quad (2.3.77)$$

where we have used Eq. (1.6.14). A plot of this Green's function for particular choices of γ and ω_0 is shown in Fig. 2.11. This Green's function displays just the sort of behavior one would expect from an oscillator excited by a δ-function impulse. For $t < 0$, nothing is happening. Suddenly, at $t = 0$, the oscillator begins to display decaying oscillations. Note that for $t > 0$ this oscillation is simply a homogeneous solution to the ODE, as given by Eq. (1.6.17). This is expected, since for $t > 0$ the forcing has vanished, and the oscillator's motion decays freely according to the homogeneous ODE.

One can see that the oscillator is stationary just before $t = 0$ (referred to as $t = 0^-$), but begins moving with a finite velocity directly after the forcing, at $t = 0^+$. What determines the initial velocity of the oscillator?

Since the oscillator is responding to a δ-function force, the Green's function satisfies the differential equation

$$g'' + \gamma g' + \omega_0^2 g = \delta(t). \quad (2.3.78)$$

We can determine the initial velocity by integrating this equation from $t = 0^-$ to $t = 0^+$:

$$\int_{0^-}^{0^+} \left(g'' + \gamma g' + \omega_0^2 g\right) dt = \int_{0^-}^{0^+} \delta(t)\, dt = 1. \quad (2.3.79)$$

Applying the fundamental theorem of calculus to the derivatives on the left-hand side, we have

$$g'(0^+) - g'(0^-) + \gamma[g(0^+) - g(0^-)] + \omega_0^2 \int_{0^-}^{0^+} g\, dt = 1. \quad (2.3.80)$$

Since $g(t)$ is continuous and $g'(0^-) = 0$, the only term that is nonzero on the left-hand side is $g'(0^+)$, yielding the result for the initial slope of the Green's function,

$$g'(0^+) = 1. \tag{2.3.81}$$

In fact, if we take the limit of the derivative of Eq. (2.3.77) as $t \to 0^+$, we can obtain Eq. (2.3.81) directly from the Green's function itself.

Recalling our description of the Green's function as the response of a tuning fork to an impulse, we can now listen to the sound of this Green's function. In Cell 2.88 we take the frequency to be a high C (2093 Hz), with a damping rate of $\gamma = 4$ Hz.

Cell 2.88

```
Play[UnitStep[t] Exp[-2 t] Sin[2 Pi 2093 t], {t, -1, 4},
  PlayRange→{-1, 1}];
```

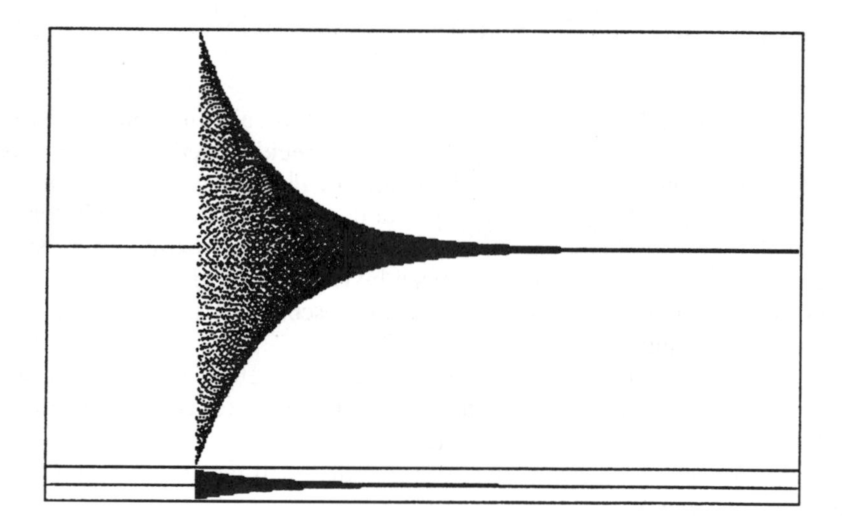

The Green's function technique can also be employed to solve for the particular solution of the general Nth-order linear ODE with constant coefficients, Eq. (1.6.7). Acting with a Fourier transform on this equation yields

$$\tilde{x}_p(\omega)(-i\omega - s_1)(-i\omega - s_2) \cdots (-i\omega - s_N) = \tilde{f}(\omega), \tag{2.3.82}$$

where $s_1, \ldots s_N$ are the roots of the characteristic polynomial for the homogeneous ODE, described in Sec. 1.6.2. Solving for $\tilde{x}_p(\omega)$, taking the inverse transform, and using the convolution theorem, we are again led to Eq. (2.3.73). Now, however, the

Green's function is given by the following inverse transform:

$$g(t) = \hat{F}^{-1} \frac{1}{(-i\omega - s_1)(-i\omega - s_2)\cdots(-i\omega - s_N)}. \tag{2.3.83}$$

This inverse transformation can be performed by separating the resonance function in Eq. (2.3.83) into individual resonances as in Eq. (2.3.75), assuming no degeneracies occur (i.e., $s_i \neq s_j$ for all $i \neq j$):

$$g(t) = -\sum_{i=1}^{N} \hat{F}^{-1} \frac{1}{i\omega - s_i} \frac{1}{\prod_{j=1, j\neq i}^{N} (s_i - s_j)}. \tag{2.3.84}$$

If we further assume that none of the system's modes are unstable, so that $\operatorname{Re} s_i < 0$ for all i, we can apply Eq. (2.3.76) to obtain the Green's function

$$g(t) = h(t) \sum_{i=1}^{N} \frac{e^{s_i t}}{\prod_{j=1, j\neq i}^{N} (s_i - s_j)}. \tag{2.3.85}$$

The case of a system exhibiting unstable oscillations, or the case of degeneracy, can also be easily handled using similar techniques, and is left to the exercises.

We have finally solved the problem, posed back in Sec. 1.6, of determining the response of an oscillator (or, more generally, a linear ODE of Nth-order with constant coefficients) to a forcing function $f(t)$ with arbitrary time dependence. Equation (2.3.85) shows that the response to a δ-function force at time $t = 0$ is a sum of the solution $e^{s_i t}$ to the homogeneous equation. This response, the Green's function for the system, can be employed to determine the response to an arbitrary force by applying Eq. (2.3.73).

As a simple example, say the forcing $f(t)$ grows linearly with time, starting at $t = 0$: $f(t) = th(t)$. Then Eq. (2.3.73) implies that a particular solution to this forcing is given by

$$x_p(t) = \int_{-\infty}^{\infty} t_0 h(t_0) g(t - t_0)\, dt_0. \tag{2.3.86}$$

Substituting Eq. (2.3.85) into Eq. (2.3.86), we find that a series of integrals of the form $\int_{-\infty}^{\infty} t_0 h(t_0) h(t - t_0)\, e^{s_i(t-t_0)}\, dt_0$ must be performed. The step functions in the integrand limit the range of integration to $0 < t_0 < t$, so each integral results in

$$\int_0^t t_0\, e^{s_i(t-t_0)}\, dt_0 = \frac{e^{s_i t} - (1 + s_i t)}{s_i^2}.$$

Finally, the total response is the sum of these individual terms:

$$x_p(t) = \sum_{i=1}^{N} \frac{e^{s_i t} - (1 + s_i t)}{s_i^2 \prod_{j=1, j\neq i}^{N} (s_i - s_j)}, \qquad t > 0. \tag{2.3.87}$$

We see that part of the response increases linearly with time, tracking the increasing applied force as one might expect. However, another part of the response is proportional to a sum of decaying homogeneous solution, $e^{s_i t}$. The particular solution given by Eq. (2.3.87) is the one for which the system is at rest before the forcing begins.

EXERCISES FOR SEC. 2.3

(1) Find the Fourier transform for the following functions. Use time transform conventions for functions of t, and space transform conventions for functions of x. Except where indicated, do the required integral by hand. (You may check your results using *Mathematica*.)

(a) $f(t)=h(t)\,e^{-at}\,\mathrm{Sin}\,\omega_0\,t$, $a>0$, where $h(t)$ is the Heaviside step function

(b) $f(t)=t$ for $-a<t<a$, and zero otherwise

(c) $f(t)=\cos\omega_0 t$ for $-a<t<a$, and zero otherwise

(d) $f(x)=x/(1+x^2)$. (You may use *Mathematica* to help with the required integral.)

(e) $f(x)=e^{-x^2}$. (You may use *Mathematica* to help with the required integral.)

(2) Verify by hand that inverse transformation of the functions $\tilde{f}(\omega)$ found in Exercise (1) returns the listed functions. [You may use *Mathematica* to help check integrals, and you may also use Eq. (2.3.76) without proving it by hand.]

(3) Plot the Fourier transform $|\tilde{f}(\omega)|$ arising from Exercise (1)(c), taking $\omega_0=3$ and $a=10, 20, 30$. Comment on the result. What will $\tilde{f}(\omega)$ converge to in the limit as $a\to\infty$?

(4) Starting with a Fourier sine series, Eq. (2.2.10), prove the relations for a Fourier transform, Eqs. (2.3.10).

(5) Repeat Exercise (1)(a) using a cosine transform on $0<x<\infty$.

(6) Repeat Exercise (1)(d) using a sine transform on $0<x<\infty$. Use *Mathematica* to help with the integral.

(7) Let $\omega/(iv+\omega)$ $(v>0)$ be the Fourier transform of $f'(t)$. Find $f(t)$ by hand.

(8) Find the value of the integral $\int_{-\infty}^{\infty} e^{-v|t_0|}\cos\omega_0(t-t_0)\,dt_0$ using the convolution theorem. Use paper and pencil methods to do all required transforms and inverse transforms.

(9) Show that the following functions approach Dirac δ-functions and/or their derivatives, and find the coefficients of these δ-functions:

(a) $f(t)=\lim_{\Delta t\to 0} t\{h(t-\Delta t)-h(t-\Delta t)\}/\Delta t^3$.

(b) $f(t)=\lim_{\Delta t\to 0}\exp[-(9-12t-2t^2+4t^3+t^4)/\Delta t^2]/|\Delta t|$.

(c) $f(t)=\lim_{\Delta\omega\to\infty}[\Delta\omega\, t\cos(\Delta\omega\, t)-\sin(\Delta\omega\, t)]/t^2$.

(Hint: In each case, consider the integral $\int_{t_0-\epsilon}^{t_0+\epsilon} f(t)g(t)\,dt$ for some function $g(t)$. Also, it is useful to plot the functions to see what they look like.)

(10) Evaluate the following integrals by hand:

(a) $\int_{-10}^{0} t^5\,\delta(t^3+8)\,dt$.

(b) $\int_{-\infty}^{\infty} \delta(\cos t)/t^2\,dt$.

(c) $\int_{3}^{\infty} \delta(\sin t^2)/t^2\,dt$.

(11) Evaluate the following generalized Fourier and inverse Fourier integrals by hand:

(a) $\int_0^\infty \cos\omega t\,d\omega/\pi$.

(b) $\int_{-\infty}^{\infty}[t/(ia+t)e^{i\omega t}\,dt$ (a real, $a>0$).

(c) $\int_{-\infty}^{\infty} -i\omega\,e^{-i\omega t}\,d\omega/2\pi$.

(d) $\int_{-\infty}^{\infty} th(t)\,e^{i\omega t}\,dt$, where $h(t)$ is the Heaviside step function.

(e) $\int_0^\infty \tanh\omega T\sin\omega t\,d\omega/\pi(T>0)$.

(12) Prove *Parseval's theorem*: for any function $f(t)$ with a Fourier transform $\tilde{f}(\omega)$,

$$\int_{-\infty}^{\infty}|f(t)|^2\,dt=\int_{-\infty}^{\infty}|\tilde{f}(\omega)|^2\,d\omega/2\pi. \tag{2.3.88}$$

(13) The uncertainty principle of Fourier analysis, Eq. (2.3.24), can be made more precise by defining the width of a real function $f(t)$ as follows. Let the square of the width of the function be $\Delta t^2=\int t^2f^2(t)\,dt/\int f^2(t)\,dt$. (We square the function because it may be negative in some regions.) Thus, Δt is the root mean square (rms) width of f^2.

Now define the rms width $\Delta\omega$ of the Fourier transform function, $\tilde{f}(\omega)$, in the same manner: $\Delta\omega^2=\int\omega^2(|\tilde{f}(\omega)|)^2\,d\omega/\int(|\tilde{f}(\omega)|)^2\,d\omega$. (We take the absolute value of $\tilde{f}$ because it may be complex.)

(a) Show that $\Delta\omega^2=\int(df/dt)^2\,dt/\int f^2(t)\,dt$. [Hint: Use Eq. (2.3.88).]

(b) Consider the function $u(t)=tf(t)+\lambda\,df/dt$. Show that $\int u^2(t)\,dt=(\Delta t^2+\lambda^2\Delta\omega^2-\lambda)\int f^2(t)\,dt$. [Hint: $f\,df/dt=(1/2)\,df^2/dt$.]

(c) Using the fact that $\int u^2(t)\,dt\ge 0$ for all real λ, prove that

$$\Delta\omega\Delta t\ge 1/2. \tag{2.3.89}$$

[Hint: The quadratic function of λ in part (b) cannot have distinct real roots.] Equation (2.3.89) is an improved version of the uncertainty principle, and is directly analogous to Heisenberg's uncertainty principle $\Delta\mathbf{E}\Delta t\ge\hbar/2$, where E is the energy.

(d) Show that equality in Eq. (2.3.89) is achieved only for Gaussian functions, $f(t)=f_0\exp(-at^2)$, $a>0$. Thus, Gaussians exhibit the "minimum uncertainty" in that the product $\Delta\omega\,\Delta t$ is minimized. [Hint: Show that $\Delta\omega\,\Delta t=1/2$ only if $u(t)=0$, and use this equation to solve for $f(t)$.]

(14) One can perform discrete Fourier transformations and inverse transformations on smooth data simply by discretizing the Fourier and inverse Fourier integrals.

(a) Take a numerical Fourier transform of the smooth function e^{-t^2} in the range $-3\le t\le 3$ by using Eqs. (2.3.53) and (2.3.54), taking $\Delta t=0.1$ and $\Delta\omega=0.4$, with ω in the range $-6\le\omega\le 6$. Compare the result with the

Table 2.2. Data for Exercise (14)

Cell 2.89

```
f = {3.42637, 2.26963, -1.70619, -2.65432, 0.24655, 1.40931,
0.470959, -0.162041, -0.336245, -0.225337, -0.112631, 0.447789,
-0.667762, -1.21989, 0.269703, 2.32636, 2.01974, -2.38678,
-3.75246, 0.298622, 3.86088, 2.27861, -1.77577, -2.46912,
0.134509, 1.02331, 0.715012, -0.339313, 0.0948633, -0.0859965,
0.0371488, 0.347241, -0.353479, -1.47499, -0.15022, 2.68935,
1.88084, -2.08172, -3.83105, 0.0629925, 3.87223, 2.13169,
-1.64515, -2.42553, -0.288646,1.4674, 0.315207, -0.480925,
-0.216251, 0.144092, -0.00670936, 0.382902, -0.495702, -1.38424,
0.256142, 2.22556, 2.02433, -2.33588, -3.60477, -0.163791,
3.55462, 2.17247, -1.94027, -2.41668, -0.0176065, 1.05511,
0.489467, -0.515668, -0.122057, -0.112292, -0.0326432, 0.489771,
-0.690393, -1.27071, 0.274066, 2.29677, 1.97186, -2.3131,
-3.99321, -0.228793, 3.95866, 1.84941, -1.95499, -2.2549,
0.104038, 1.29127, 0.769865, -0.362732, -0.271452, -0.0638439,
0.0734938, 0.0774499, -0.333983, -1.56588, -0.193863, 2.37758,
1.92296, -2.12179, -3.87906, -0.21919, 3.96223, 2.01793,
-2.05241, -2.7803, -0.296432, 1.18286, 0.687172, -0.449909,
-0.193565, 0.191591, 0.310403, 0.437337, -0.706701, -1.35889,
-0.0630913, 2.54978, 1.79384, -2.21964, -3.88036, -0.127792, 3.882,
2.32878, -1.56785, -2.6985, 0.219771, 1.32518, 0.669142, -0.44272,
0.123107, -0.15768, 0.375066, -0.0682963, -0.467915, -1.3636,
-0.235336, 2.28427, 1.80534, -1.83133, -3.58337, 0.0344805,
3.42263, 2.21493, -1.86957, -2.62763, -0.159368, 1.50048,
0.48287, -0.453638, -0.172236, -0.124694};
```

exact Fourier transform, found analytically, by plotting both on the same graph vs. frequency. (This requires you to determine the frequency associated with the position of a given element in the transformed data.)

(b) Take the inverse transform of the data found in part (a) using Eqs. (2.3.55) and (2.3.56). Does the result return to the original function? Plot the difference between the original function and this result.

(c) Repeat (a) and (b), but take the range of ω to be $-3 \leq \omega \leq 3$.

(15) Using a discrete Fourier transform that you create yourself (not a fast Fourier transform), analyze the noisy data in Table 2.2 and determine the main frequencies present, in hertz. The time between samples is $\Delta t = 0.0005$ s.

(16) A Fourier series or transform is a way to decompose a function $f(t)$ in terms of complex orthogonal basis functions $e^{-i\omega t}$. The method relies on the orthogonality of these basis functions. A discrete Fourier transform can be thought of as a way to decompose vectors **f** of dimension N in terms of N complex orthogonal basis vectors. The mth basis vector is $e^{(m)} = \{1, e^{-2\pi mi/N}, e^{-4\pi mi/N}, \ldots, e^{-2\pi mi(N-1)/N}\}$.

(a) Show directly that these basis vectors are orthogonal with respect to the following inner product defined for two vectors **f** and **g**: $(\mathbf{f}, \mathbf{g}) = \mathbf{f}^* \cdot \mathbf{g}$. [Hint: Use the following sum: $\Sigma_{n=0}^{N-1} x^N = (1 - x^N)/(1 - x)$.]

(b) Show that the decomposition of a vector **f** in terms of these basis vectors, $\mathbf{f} = \sum_{m=0}^{N-1} \tilde{f}_m \, \mathbf{e}^{(m)}$, leads directly to Eq. (2.3.63).

(c) Use the orthogonality of these basis vectors to determine the coefficients $\tilde{f}_m$, and compare the result with Eq. (2.3.59).

(17) A data file on the disk accompanying this book, entitled `aliendata.nb`, contains a (simulated) recording made by a (simulated) scientist listening for alien communications from nearby stars. Read in this data file using the command `<<aliendata.txt`. (A `Directory` and/or `SetDirectory` command may also be necessary.) Within this file is a data list of the form `f={ ··· }`. Reading in the file defines the data list `f`. Use `ListPlay` to play the data as a sound. Take the sample rate to be 22,050 hertz. As you can hear, the data is very noisy. By taking a Fourier transform, find a way to remove this noise by applying an appropriate filter function to the Fourier-transformed data. What is the aliens' message? Are they peaceful or warlike? [Caution: This data file is rather large. When manipulating it, always end your statements with a semicolon to stop any output of the file.]

(18) **(a)** A damped harmonic oscillator is driven by a force of the form $f(t) = h(t)t^2 \exp(-t)$, where $h(t)$ is a Heaviside step function. The oscillator satisfies the equation

$$x'' + 2x' + 4x = f(t).$$

Use pencil-and-paper methods involving Fourier transforms and inverse transforms to find the response of the oscillator, $x(t)$, assuming that $x(0) = 0$ and $x'(0) = 1$. Plot the solution for $x(t)$.

(b) Repeat the analysis of part (a) for a force of the form $f(t) = h(t)t \sin 4t$. Instead of Fourier methods, this time use the Green's function for this equation. Plot the solution for $0 < t < 10$.

(19) Use Fourier transform techniques to find and plot the current in amperes as a function of time in the circuit of Fig. 2.12 when the switch is closed at time $t = 0$.

(20) **(a)** Find the Green's function for the following ODE:

$$\hat{L}x = x'''' + 2x''' + 6x'' + 5x' + 2x.$$

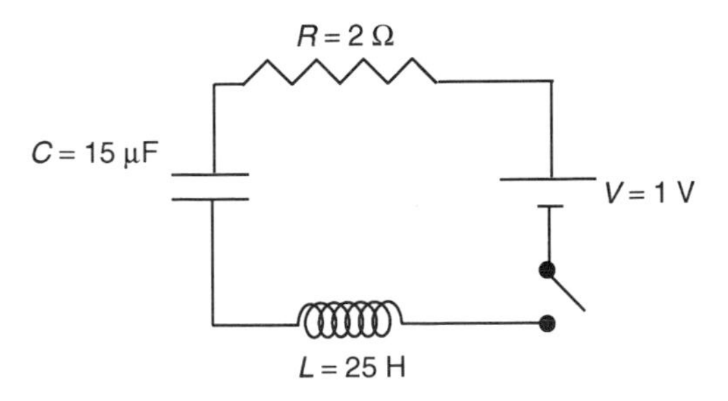

Fig. 2.12

(b) Use the Green's function found from part (a) to determine the solution $x(t)$ to $\hat{L}x = h(t)te^{-t}\cos t$, $x(0) = x'(0) = x''(0) = x'''(0) = 0$. Plot the solution.

(21) **(a)** The FFT can be used to solve numerically for particular solutions to differential equations with constant coefficients. This method is quite useful and important for solving certain types of PDEs numerically, such as Poisson's equation under *periodic boundary conditions* (see Chapters 6 and 7), but can also be used to solve ODEs. For example, consider the following ODE:

$$x'' + x' + x = h(t)te^{-t}\sin 3t.$$

Find a particular solution on the interval $0 \le t \le 4$ using an FFT. To do so, first discretize time as $t = n\,\Delta t$, $n = 0, 1, 2, \ldots, N-1$, with $N = 41$, taking $\Delta t = 0.1$. Make a table of the forcing function at these times, and take its FFT. Then, use what you know about Fourier transforms to determine a discretized form for the transform of x, $\tilde{x}$. Take the inverse FFT of this data and plot the resulting particular solution vs. time on $0 \le t \le 4$.

(b) This particular solution is periodic, with period $T = 4.1$, since the FFT is equivalent to a Fourier series solution with this periodicity. This solution is correct only in the first period, from $0 < t < T$; the analytic particular solution is not periodic, but should match the FFT solution in $0 < t < T$. To prove this, note that this particular solution satisfies *periodic boundary conditions* $x(0) = x(T)$, $x'(0) = x'(T)$. Solve the ODE analytically for these boundary conditions using **`DSolve`**, and plot the result on top of the FFT result from part (a) on $0 \le t \le T$.

(22) **(a)** The periodic δ-function $\delta^{(P)}(t)$ with period T given by Eq. (2.3.48) has Fourier components of equal amplitude over all frequencies, playing *continuously*. Do you think that the hairs in your ear responsible for hearing frequencies at around, say, 500 Hz are being excited continuously by the 500-Hz frequencies in the δ-function, or only when there is a chirp? Explain your reasoning in several sentences, with diagrams if necessary. (Hint: Hairs respond to a *range of frequencies* around their response peak, not just a single frequency. What is the effect on the amplitude of the hair's motion of adding together these different frequency components in the forcing? Think about constructive and destructive interference.)

(b) Find a particular solution for the response of a damped oscillator (a model of one of the hairs) to a periodic δ-function using Green's functions. The oscillator satisfies $x'' + \gamma x' + \omega_0^2 x = \delta^{(P)}(t)$. Plot the result over a time of $5T$ for $\omega_0 T = 60$ and for (i) $\gamma T = 0.01$, (ii) $\gamma T = 0.1$, (iii) $\gamma T = 1$, (iv) $\gamma T = 10$. Roughly how large must γT be for each response to a δ-function kick to be easily distinguishable from others?

(c) Create and play a periodic δ-function with different values of T, from $\frac{1}{10}$ to $\frac{1}{200}$ s. Determine the smallest value of T for which you can distinguish the chirps. Use the result of part (b) along with this result to crudely estimate a value of γ for the human auditory system.

(23) **(a)** Play the following periodic δ-function for which the phases have been randomized:

Cell 2.90

```
T = 0.2; 1 + 2 Sum[ Cos[2 Pi n t/T +
  2 Pi Random[]], {n, 1, 300}];
```

Does this still sound like a periodic δ-function? Previously, we found that randomizing the phases of the Fourier components in a waveform made no difference to the sound a waveform makes. Can you explain why there is a difference now? (Hint: Think about destructive interference.)

(b) The modes in a Fourier series or transform have amplitudes that are constant in time, but somehow the series or transform is able to produce sounds whose loudness (amplitude) varies in time, as in a periodic δ-function. Given the results of part (a), explain in a few words how this is accomplished.

(c) Reduce the period T of $\delta^{(P)}(t)$ to a value below that found in part (c) of the previous exercise. Does randomizing the phases make as much difference to the sound as for part (a)? Why or why not?

(24) Three coupled oscillators are initially at rest on a surface (at $t = -\infty$). Their equilibrium positions are $x_{10} = 2$, $x_{20} = 1$, $x_{30} = 0$. The oscillators satisfy the equations

$$x_1'' = -2(x_1 - x_2 - 1) - x_1',$$

$$x_2'' = -2(x_2 - x_1 + 1) - x_2' - 2(x_2 - x_3 - 1),$$

$$x_3'' = -2(x_3 - x_2 + 1) - x_3' + f(t).$$

The third oscillator in the chain is given a bump, $f(t) = 6\,e^{-4|t|}$. Use Fourier transform methods to solve for the motion of the oscillators. Plot the motion of all three oscillators as an animation by plotting their positions $x_i(t)$ as a set of `ListPlot`s at times from $-1 < t < 9$ in units of 0.2.

2.4 GREEN'S FUNCTIONS

2.4.1 Introduction

Consider a dynamical system described by a general Nth-order inhomogeneous linear ODE, such as Eq. (1.6.1). In operator notation this ODE can be written as

$$\hat{L}x(t) = f(t). \tag{2.4.1}$$

The Green's function can be used to find a particular solution to this ODE. The

Green's function is a function of two arguments, $g = g(t, t_0)$. This function satisfies

$$\hat{L}g(t, t_0) = \delta(t - t_0). \tag{2.4.2}$$

According to this equation, $g(t, t_0)$ is the response of the system to a δ-function forcing at time $t = t_0$.

The Green's function can be used to obtain a particular solution to Eq. (2.4.1.) by means of the following integral:

$$x_p(t) = \int_{-\infty}^{\infty} g(t, t_0) f(t_0)\, dt_0. \tag{2.4.3}$$

We have already seen a similar equation for the particular solution in terms of a Green's function for an ODE with constant coefficients, Eq. (2.3.73). There, the Green's function was written as $g(t - t_0)$ rather than $g(t, t_0)$, because when the ODE has constant coefficients the origin of time can be displaced to any arbitrary value, so only the time *difference* between t and t_0 is important in determining the response at time t to an impulse at time t_0.

We can easily show that Eq. (2.4.3) satisfies the ODE, Eq. (2.4.1). Acting with $\hat{L}$ on Eq. (2.4.3), we obtain

$$\hat{L}x_p(t) = \int_{-\infty}^{\infty} \hat{L}g(t, t_0) f(t_0)\, dt_0 = \int_{-\infty}^{\infty} \delta(t - t_0) f(t_0)\, dt_0 = f(t),$$

where we have used Eq. (2.4.2).

We have not yet specified initial or boundary conditions that go along with Eq. (2.4.2) for defining the Green's function. In initial-value problems, it is convenient to choose the initial condition that $g = 0$ for $t < t_0$, so that the system is at rest when the impulse is applied.

For boundary-value problems, other choices are made in order to satisfy boundary conditions, as we will see in Sec. 2.4.4.

The time integral in Eq. (2.4.3) runs all the way from $-\infty$ to ∞, so it seems that we need to know everything about both the past and future of the forcing function to determine x_p at the present time. However, the choice $g = 0$ for $t < t_0$ implies that the integral in Eq. (2.4.3) really runs only from $-\infty$ to t, so only past times are necessary. Also, in typical problems there is usually an initial time t_i (possibly far in the past) before which the forcing can be taken equal to zero (i.e. the beginning of the experiment). For these choices the integral really runs only from t_i to t:

$$x_p(t) = \int_{t_i}^{t} g(t, t_0) f(t_0)\, dt_0.$$

We can see that this particular solution will be zero for $t < t_i$, since $g(t, t_0) = 0$ for $t < t_0$. Thus, this particular solution is the one for which the system is at rest before the forcing begins.

We will now consider how to construct a solution to Eq. (2.4.2) for the Green's function. There are several methods for doing so. We have already seen one method using Fourier transforms in Sec. 2.3.6, applicable to ODEs with constant coefficients. Here, we consider a general analytic method for any linear ODE, where the Green's function is written as a sum of homogeneous solutions to the ODE. We also discuss numerical methods for determining the Green's function. In Chapter 4, we will consider another analytic method, applicable only to boundary-value problems, in which we decompose g in terms of operator eigenmodes, resulting in the *bilinear equation* for the Green's function.

2.4.2 Constructing the Green's Function from Homogeneous Solutions

Second-Order ODEs In this subsection we will construct the Green's function for a linear second-order ODE using homogeneous solutions. We will then discuss a generalization of the solution that is applicable to higher-order ODEs.

For a general linear second-order ODE of the form of Eq. (1.6.1), the Green's function satisfies

$$\hat{L}g(t,t_0) \equiv \frac{\partial^2}{\partial t^2}g(t,t_0) + u_1(t)\frac{\partial}{\partial t}g(t,t_0) + u_0(t)g(t,t_0) = \delta(t-t_0). \quad (2.4.4)$$

Assuming that we are solving an initial-value problem, we will take the initial condition that $g(t,t_0)=0$ for $t \le t_0$.

When $t > t_0$, the Green's function satisfies the homogeneous ODE $\hat{L}g(t,t_0)=0$, and therefore g can be written as a sum of the two independent homogeneous solutions to the ODE (see Sec. 2.6):

$$g(t,t_0) = C_1 x_1(t) + C_2 x_2(t), \qquad t > t_0. \quad (2.4.5)$$

We are then left with the task of determining the constants C_1 and C_2. One equation for these constants can be found by applying the initial condition that the system is at rest before the impulse is applied, so that $g=0$ at $t=t_0$. Applying this condition to Eq. (2.4.5) yields

$$g(t_0,t_0) = 0 = C_1 x_1(t_0) + C_2 x_2(t_0). \quad (2.4.6)$$

To obtain one more equation for the constants C_1 and C_2, we integrate Eq. (2.4.4) from a time just before t_0, t_0^-, to a time just after t_0, t_0^+:

$$\int_{t_0^-}^{t_0^+}\left(\frac{\partial^2}{\partial t^2}g(t,t_0) + u_1(t)\frac{\partial}{\partial t}g(t,t_0) + u_0(t)g(t,t_0)\right)dt = \int_{t_0^-}^{t_0^+}\delta(t-t_0)\,dt = 1.$$

Assuming that $u_0(t)$ and $u_1(t)$ are both continuous at $t=t_0$, we can replace them by their values at $t=t_0$ and take them outside the integral. We can then apply the fundamental theorem of calculus to the integral of the derivatives, yielding

$$\left.\frac{\partial}{\partial t}g(t,t_0)\right|_{t=t_0^+} - \left.\frac{\partial}{\partial t}g(t,t_0)\right|_{t=t_0^-}$$
$$+ u_1(t_0)\left[g(t_0^+,t_0) - g(t_0^-,t_0)\right] + u_0(t_0)\int_{t_0^-}^{t_0^+} g(t,t_0)\,dt = 1.$$

Since $g=0$ at $t \le t_0$, all terms on the left-hand side vanish except for the first term, yielding

$$\frac{\partial}{\partial t} g(t,t_0)\bigg|_{t=t_0^+} = 1.$$

Substituting for g using Eq. (2.4.5) yields the second equation for C_1 and C_2,

$$C_1 x_1'(t_0) + C_2 x_2'(t_0) = 1. \tag{2.4.7}$$

We may now solve for C_1 and C_2 using Eqs. (2.4.6) and (2.4.7). The result is

$$g(t,t_0) = \frac{x_2(t_0)x_1(t) - x_1(t_0)x_2(t)}{W(t_0)}, \qquad t > t_0, \tag{2.4.8}$$

where the function $W(t)$ is the *Wronskian*, defined as $W(t) \equiv x_1'(t)x_2(t) - x_2'(t)x_1(t)$.

If the Wronskian $W(t_0)$ is zero for some value of t_0, then according to Eq. (2.4.8) g is undefined. However, one can show (although we will not do so here) that the Wronskian is always nonzero if the homogeneous solutions x_1 and x_2 are linearly independent. A proof of this statement can be found in many elementary books on ODEs, such as Boyce and DiPrima (1969) (see the references at the end of Chapter 1).

For completeness, we also note our initial condition:

$$g(t,t_0) = 0, \qquad t < t_0. \tag{2.4.9}$$

Equations (2.4.8) and (2.4.9) are a solution for the Green's function for a general second-order ODE. This solution requires that we already know the independent homogeneous solutions to the ODE, $x_1(t)$ and $x_2(t)$. These solutions can often be found analytically, using **DSolve** for example, but we could also use numerical solutions to the homogeneous ODE, with two different sets of initial conditions so as to obtain independent numerical approximations to $x_1(t)$ and $x_2(t)$. Examples of both analytic and numerical methods for finding homogeneous solutions can be found in the exercises at the end of Sec. 1.6.

Equations (2.4.8) and (2.4.9) can now be used in Eq. (2.4.3) to obtain a particular solution to Eq. (2.4.1). Since $g(t,t_0)=0$ for $t_0 > t$, we obtain

$$\begin{aligned} x_p(t) &= \int_{-\infty}^{t} g(t,t_0) f(t_0)\, dt_0 \\ &= x_1(t) \int_{-\infty}^{t} \frac{x_2(t_0) f(t_0)}{W(t_0)}\, dt_0 - x_2(t) \int_{-\infty}^{t} \frac{x_1(t_0) f(t_0)}{W(t_0)}\, dt_0. \end{aligned} \tag{2.4.10}$$

Equation (2.4.10) is a form for the particular solution to the ODE that can also be found in elementary textbooks, based on the method of variation of parameters. Here we have used Green's-function techniques to obtain the same result. Note that one can add to Eq. (2.4.10) any homogeneous solution to the ODE in order to obtain other particular solutions. Equation (2.4.10) is the particular solution for which the system is at rest before the forcing begins.

Example As an example of the Green's function technique applied to a second-order ODE with time-varying coefficients, let us construct the Green's function for the operator

$$\hat{L}x = x'' - nx'/t, \qquad n \neq -1.$$

One can verify by substitution that this operator has two independent homogeneous solutions

$$x_1(t) = 1,$$
$$x_2(t) = t^{n+1}/(n+1).$$

Then the Wronskian is

$$W(t) = x_1'(t)x_2(t) - x_2'(t)x_1(t) = 0 - 1t^n = -t^n,$$

and Eq. (2.4.8) for the Green's function becomes

$$g(t,t_0) = \frac{1}{n+1}\left(t\left(\frac{t}{t_0}\right)^n - t_0\right), \qquad t > t_0.$$

We can use this Green's function to obtain a particular solution to $\hat{L}x(t) = f(t)$. Let's take the case $f(t) = t^{\alpha}h(t)$, where $h(t)$ is the Heaviside step function. Then the particular solution given by Eq. (2.4.3) is

$$x_p(t) = \int_{-\infty}^{t} g(t,t_0)t_0^{\alpha}h(t_0)\,dt_0 = \int_0^t \frac{1}{n+1}\left[t\left(\frac{t}{t_0}\right)^n - t_0\right]t_0^{\alpha}\,dt_0,$$

where on the right-hand side we have assumed that $t > 0$ in order to set $h(t_0) = 1$. If $t < 0$, then $h(t_0) = 0$ for the entire range of integration, and $x_p(t) = 0$. This is expected, since the Green's function has built in the initial condition that the system is at rest before forcing begins.

For this simple forcing function the integral can be performed analytically, yielding

$$x_p(t) = \frac{(1+n)t^{2+\alpha}}{(2+\alpha)(1+\alpha-n)}, \qquad t > 0.$$

Nth-Order ODEs The Green's function for a general linear ODE of order N can also be determined in terms of homogeneous solutions. The Green's function satisfies

$$\frac{d^N x}{dt^N} + u_{N-1}(t)\frac{d^{N-1}}{dt^{N-1}} + \cdots + u_1(t)\frac{dx}{dt} + u_0(t)x = \delta(t-t_0). \quad (2.4.11)$$

For $t < t_0$ we again apply Eq. (2.4.9) as our initial condition. For $t > t_0$ the Green's function is a sum of the N independent homogeneous solutions,

$$g(t,t_0) = \sum_{n=1}^{N} C_n x_n(t), \qquad t > t_0. \quad (2.4.12)$$

We now require N equations for the N coefficients C_n. One such equation is obtained by integrating the ODE across the δ-function from t_0^- to t_0^+. As with the case of the second-order ODE, only the highest derivative survives this operation, with the result

$$\left.\frac{d^{N-1}}{dt^{N-1}} g(t,t_0)\right|_{t=t_0^+} = 1. \tag{2.4.13}$$

Only this derivative exhibits a jump in response to the δ-function. All lower derivatives are continuous, and therefore equal zero as a result of Eq. (2.4.9):

$$\left.\frac{d^n}{dt^n} g(t,t_0)\right|_{t=t_0^+} = 0, \qquad n = 1,2,\dots,N-2. \tag{2.4.14}$$

Also, the Green's function itself is continuous,

$$g(t_0,t_0) = 0. \tag{2.4.15}$$

When Eq. (2.4.12) is used in Eqs. (2.4.13)–(2.4.15), the result is N equations for the N unknowns C_n. Solving these coupled linear equations allows us to determine the Green's function in terms of the homogeneous solutions to the ODE. For the case of an ODE with constant coefficients, the result returns us to Eq. (2.3.85). Knowledge of the Green's function, in turn, allows us to determine any particular solution. We leave specific examples of this procedure to the exercises.

2.4.3 Discretized Green's Function I: Initial-Value Problems by Matrix Inversion

Equation (2.4.3) can be thought of as an equation involving a linear integral operator $\hat{L}^{-1}$. This operator is defined by its action on any function $f(t)$:

$$\hat{L}^{-1} f \equiv \int_{-\infty}^{\infty} g(t,t_0) f(t_0)\, dt_0. \tag{2.4.16}$$

We have already encountered other linear integral operators: the Fourier transform and its inverse are both integral operators, (see Sec. 2.3.3). The operator $\hat{L}^{-1}$ appears in the particular solution given in Eq. (2.4.3), which may be written as

$$x_p = \hat{L}^{-1} f. \tag{2.4.17}$$

We call this operator $\hat{L}^{-1}$ because it is the *inverse* of the differential operator $\hat{L}$. We can see this by operating on Eq. (2.4.16) with $\hat{L}$, and using Eq. (2.4.3):

$$\hat{L}\hat{L}^{-1} f = \int_{-\infty}^{\infty} \hat{L} g(t,t_0) f(t_0)\, dt_0 = \int_{-\infty}^{\infty} \delta(t-t_0) f(t_0)\, dt_0 = f(t).$$

This shows that $\hat{L}^{-1}$ has the correct property for an inverse: $\hat{L}\hat{L}^{-1}$ returns without change any function to which it is applied.

We can see that $\hat{L}^{-1}$ is the inverse of $\hat{L}$ in another way. The ODE that x_p satisfies is $\hat{L}x_p = f$. If we apply $\hat{L}^{-1}$ to both sides, we obtain $\hat{L}^{-1}\hat{L}x_p = \hat{L}^{-1} f = x_p$, where in the last step we used Eq. (2.4.17). This shows that $\hat{L}^{-1}\hat{L}$ also returns without change any function to which it is applied. Since

$$\hat{L}\hat{L}^{-1}x_p = \hat{L}^{-1}\hat{L}x_p = x_p \tag{2.4.18}$$

for a function $x_p(t)$, this again proves that $\hat{L}^{-1}$ is the inverse of $\hat{L}$.

Note, however, that there seems to be something wrong with Eq. (2.4.18). There are functions $x_h(t)$ for which $\hat{L}x_h = 0$: the general homogeneous solutions to the ODE. For such functions, $\hat{L}^{-1}\hat{L}x_h = 0 \neq x_h$. In fact, one might think that since $\hat{L}x_h = 0$, $\hat{L}^{-1}$ cannot even exist, since matrices that have a finite null space do not have an inverse (and we already know that by discretizing time $\hat{L}$ can be thought of as a matrix).

The resolution to this paradox follows by considering the discretized form of the Green's function. By discretizing the ODE, we can write it as a matrix equation, and solve it via matrix inversion. In this approach, the operator $\hat{L}$ becomes a matrix $\mathbf{L}$, and the operator $\hat{L}^{-1}$ is simply the inverse of this matrix $\mathbf{L}^{-1}$.

We examined this procedure for homogeneous initial-value problems in Sec. 1.6.3. Adding the inhomogeneous term is really a very simple extension of the previous discussion. As an example, we will solve the first-order ODE

$$\hat{L}x = \frac{dx}{dt} + u_0(t)x = f(t), \qquad x(0) = x_0. \tag{2.4.19}$$

A discretized version of the ODE can be found using Euler's method, Eq. (1.4.7). Defining $t_n = n\,\Delta t$, Euler's method implies

$$\begin{gathered} x(0) = x_0, \\ \frac{x(t_n) - x(t_{n-1})}{\Delta t} - u_0(t_{n-1})x(t_{n-1}) = f(t_{n-1}), \qquad n > 0. \end{gathered} \tag{2.4.20}$$

These linear equations can be written as a matrix equation,

$$\mathbf{L}\cdot\mathbf{x} = \mathbf{f}, \tag{2.4.21}$$

where the vector $\mathbf{x} = \{x(0), x(t_1), x(t_2), \dots\}$, and the vector $\mathbf{f} = [x_0, f(t_0), f(t_1), \dots\}$ contains the force *and* initial condition. The matrix $\mathbf{L}$ is a discretized version of the operator $\hat{L}$:

$$\mathbf{L} = \begin{pmatrix} 1 & 0 & 0 & 0 & \cdots \\ -\frac{1}{\Delta t} + u_0(t_0) & \frac{1}{\Delta t} & 0 & 0 & \cdots \\ 0 & -\frac{1}{\Delta t} + u_0(t_1) & \frac{1}{\Delta t} & 0 & \cdots \\ 0 & 0 & -\frac{1}{\Delta t} + u_0(t_2) & \frac{1}{\Delta t} & \cdots \\ \vdots & \vdots & \vdots & \vdots & \ddots \end{pmatrix}. \tag{2.4.22}$$

A similar matrix was introduced for the homogeneous equation, in Sec. 1.6.3 [see Eq. (1.6.21)]. The only difference here is that we have multiplied by $1/\Delta t$

everywhere but in the first row, so that Eq. (2.4.21) has the same form as Eq. (2.4.19).

We can solve Eq. (2.4.20) for the unknown vector **x** by using matrix inversion:

$$\mathbf{x} = \mathbf{L}^{-1}\mathbf{f}. \tag{2.4.23}$$

Note the resemblance of this equation to Eq. (2.4.17). The matrix inverse appearing in Eq. (2.4.23) is a discretized version of the inverse operator $\hat{L}^{-1}$.

From a theoretical perspective, it is important to note that both the forcing and the initial condition are included in Eqs. (2.4.21) and (2.4.23). The initial condition $x(0) = x_0$ is built into the forcing function. If we wish to solve the *homogeneous* equation for a given initial condition, all that we need do is set all elements of **f** but the first equal to zero. In this case Eq. (2.4.21) returns to the form of Eq. (1.6.21). We can see now that *there is no difference between finding a solution to the homogeneous equation and finding a solution to the inhomogeneous equation.* After all, the matrix inversion technique is really just Euler's method, and Euler's method is the same whether or not the ODE is homogeneous.

We have seen how the forcing contains the inhomogeneous initial conditions for the discretized form of the ODE; but is this also possible for the actual ODE itself? The answer is yes. For instance, to define the initial condition $x(0) = x_0$ in the previous first order ODE, we can use a δ-function:

$$\frac{dx}{dt} + u_0(t)x = f(t) + x_0\delta(t). \tag{2.4.24}$$

If we integrate from 0^- to 0^+, we obtain $x(0^+) - x(0^-) = x_0$. Now let us *assume that* $x(t) = 0$ for $t < 0$. [This is the homogeneous initial condition associated with $\hat{L}^{-1}$: see Eq. (2.4.16), and recall that $g = 0$ for $t < t_0$.] We then arrive at the proper initial condition, $x(0^+) = x_0$. Furthermore, for $t > 0$ the ODE returns to its previous form, Eq. (2.4.19). By solving Eq. (2.4.24) with a homogeneous initial condition, we obtain the same answer as is found from Eq. (2.4.19).

Therefore, we can think of $\hat{L}^{-1}$ as the inverse of $\hat{L}$ *provided that* the ODE is solved using *homogeneous* initial conditions, $x(t) = 0$ for $t < 0$. The equation $\hat{L}x = f$ specifies both the ODE and the initial conditions at time $t = 0^+$. The inhomogeneous initial conditions are contained in the forcing function f, as in Eq. (2.4.24).

With homogeneous initial conditions attached to the ODE, it is no longer true that nontrivial solutions $x_h(t)$ exist for which $\hat{L}x_h = 0$; the only solution to this equation is $x_h = 0$. Thus, the null space of $\hat{L}$ is the empty set, and the operator does have an inverse. This resolves the apparent contradiction discussed in relation to Eq. (2.4.18).

Furthermore, application of $\hat{L}^{-1}$ via Eq. (2.4.16) no longer provides only a particular solution; it provides the full solution to the problem, *including the initial condition.* We can now see that the previous distinction between particular and homogeneous solutions to the ODE is artificial: both can be obtained from the Green's function via Eq. (2.4.16). For the above example of a first-order ODE, Eq. (2.4.24) implies that the general homogeneous solution for this first-order ODE is the Green's function itself: $x_h(t) = x_0 G(t, 0)$. This should not be surprising, given that the Green's function can be constructed from homogeneous solutions, as we saw in the previous section.

Let us now discuss Eq. (2.4.23) as a numerical method for obtaining the solution to the ODE. Now specific examples of the functions $f(t)$ and $u_0(t)$ must be chosen, and the solution can be found only over a finite time interval. Let us choose $u_0(t) = t$, $f(t) = 1$, and find the particular solution for $0 < t < 3$, taking a step size $\Delta t = 0.05$ with an initial condition $x_0 = 0$. Then, as discussed in Sec. 1.6.3, the operator can be constructed using Kronecker δ-functions using the following *Mathematica* commands:

Cell 2.91

```
Clear[u, Δt]

Δt = 0.05; u[n_] = Δt n; M = 60;
L = Table[KroneckerDelta[n, m] / Δt -
   KroneckerDelta[n, m + 1] (1 / Δt - u[m]), {n, 0, M},
   {m, 0, M}];
L[[1, 1]] = 1;
```

The force initial-condition vector is

Cell 2.92

```
f = Table [1, {i, 0, M}];
f[[1]] = 0;
```

Then we can solve for **x** via

Cell 2.93

```
x = Inverse[L].f;
```

and we can plot the solution by creating a table of (t, x) data values and using a **ListPlot**, as shown in Cell 2.94. The exact solution to this equation with the initial condition $x[0] = 0$ is in terms of a special function called an error function.

Cell 2.94

```
Table [{n Δt, x[[n + 1]]}, {n, 0, M}];
sol = ListPlot[%];
```

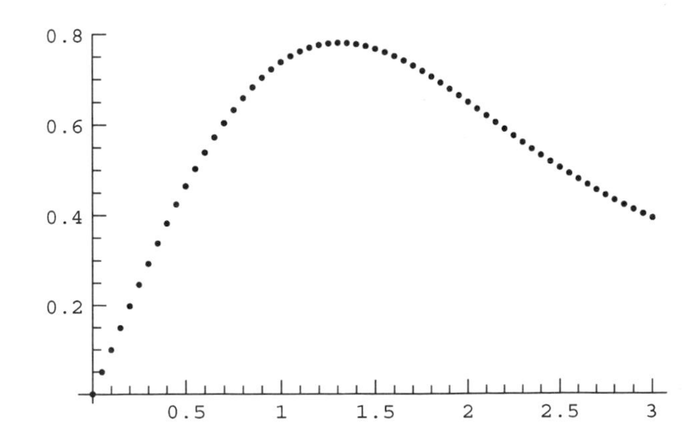

Cell 2.95

```
Clear[x];

x[t_] = x[t]/. DSolve[{x'[t] + t x[t] == 1, x[0] == 0},
  x[t], t][[1]]
```

$$e^{-t^2/2}\sqrt{\frac{\pi}{2}}\,\text{Erfi}\left[\frac{t}{\sqrt{2}}\right]$$

In Cell 2.96 we plot this solution and compare it with the numerics. The match is reasonably good, and could be improved by taking a smaller step size, or by using a higher-order method to discretize the ODE.

Cell 2.96

```
Plot[x[t], {t, 0, 3}, DisplayFunction → Identity];
Show[%, sol, DisplayFunction → $DisplayFunction];
```

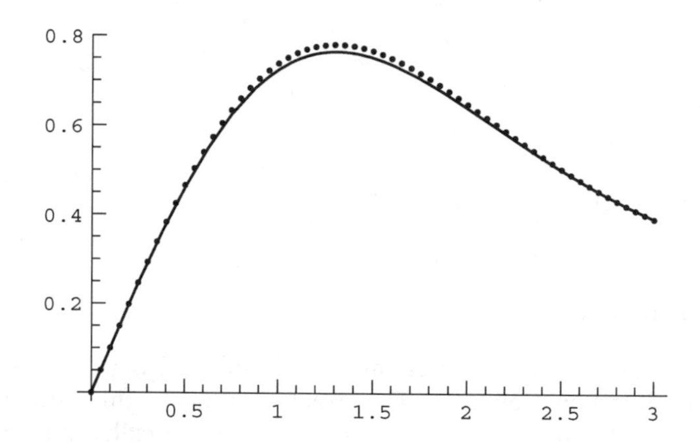

Higher-order inhomogeneous ODEs can also be written in the matrix form of Eq. (2.4.21) and solved by matrix inversion. Examples can be found in the exercises, and in Sec. 2.4.5, where boundary-value problems will be considered.

2.4.4 Green's Function for Boundary-Value Problems

Green's functions are often used to find particular solutions to inhomogeneous linear boundary-value problems. For example, Green's functions play a very important role in solutions to electrostatics problems. Such problems typically involve solution of Poisson's equation (1.1.10) for the potential $\phi(\mathbf{r})$ due to a given charge distribution $\rho(\mathbf{r})$, in the presence of conductors that determine the boundary conditions for the potential.

Poisson's equation is a PDE, so its complete solution via Green's function techniques will be left to Chapters 3 and 4. However, it is enlightening to use a Green's function for Poisson's equation in the case where there is only variation in one direction. If we call this direction x, then Poisson's equation is the following ODE:

$$\hat{L}\phi(x) = \rho(x), \tag{2.4.25}$$

where $\hat{L}$ is a second-order linear differential operator in x whose form depends on the geometry of the system. For example, for Cartesian coordinates, $\hat{L} = \partial^2/\partial x^2$, whereas for spherical coordinates with $x = r$, we have $\hat{L} = \partial^2/\partial x^2 + (2/x)\,\partial/\partial x$.

Assuming that there are conductors at $x = a$ and $x = b$ (assuming $a < b$) with fixed potentials V_a and V_b respectively, the boundary conditions are

$$\phi(a) = V_a, \qquad \phi(b) = V_b. \tag{2.4.26}$$

When solving Eqs. (2.4.25) and (2.4.26) analytically, as usual we break the solution into a homogeneous and a particular solution: $\phi = \phi_h + \phi_p$. The homogeneous solution ϕ_h satisfies the boundary conditions without the source term:

$$\hat{L}\phi_h(x) = 0, \qquad \phi(a) = V_a, \quad \phi(b) = V_b, \tag{2.4.27}$$

and the particular solution satisfies the inhomogeneous ODE,

$$\hat{L}\phi_p(x) = \rho(x), \qquad \phi_p(a) = \phi_p(b) = 0. \tag{2.4.28}$$

The particular boundary conditions chosen here are termed *homogeneous* boundary conditions. Such boundary conditions have the property that, for the forcing function $\rho = 0$, a solution to Eq. (2.4.28) is $\phi_p = 0$. We will have much more to say about homogeneous boundary conditions in Chapters 3 and 4.

The Green's function $g(x, x_0)$ is used to determine the particular solution to Eq. (2.4.28). The Green's function is the solution to

$$\hat{L}g(x, x_0) = \delta(x - x_0), \qquad g(a, x_0) = g(b, x_0) = 0. \tag{2.4.29}$$

Then Eq. (2.4.3) implies that the particular solution is

$$\phi_p(x) = \int_a^b g(x, x_0)\,\rho(x_0)\,dx_0. \tag{2.4.30}$$

The boundary conditions $\phi_p(a) = \phi_p(b) = 0$ are satisfied because $g(a, x_0) = g(b, x_0) = 0$ for all values of x_0.

We can construct the Green's function for this problem by using solutions to the homogeneous ODE in a manner analogous to the method used for initial-value problems, discussed previously. The only change is that the boundary conditions on the Green's function are now different.

The second-order ODE has two independent homogeneous solutions $\phi_1(x)$ and $\phi_2(x)$. We construct the Green's function by choosing two different linear combinations of these two solutions that match the boundary conditions in different regions. For $x < x_0$, we choose the linear combination

$$\overline{\phi}_1(x) = \phi_1(x) - \frac{\phi_1(a)}{\phi_2(a)}\phi_2(x), \tag{2.4.31}$$

and for $x > x_0$ we choose the combination

$$\overline{\phi}_2(x) = \phi_1(x) - \frac{\phi_1(b)}{\phi_2(b)}\phi_2(x). \tag{2.4.32}$$

These combinations are chosen so that $\overline{\phi}_1(a) = \overline{\phi}_2(b) = 0$. Then the Green's function is

$$g(x, x_0) = \begin{cases} C_1\overline{\phi}_1(x), & x < x_0, \\ C_2\overline{\phi}_2(x), & x > x_0, \end{cases} \tag{2.4.33}$$

where the constants C_1 and C_2 must still be determined. This Green's function satisfies the boundary conditions given in Eq. (2.4.29), and also satisfies the ODE for $x \neq x_0$, since $\hat{L}g(x, x_0) = 0$ for $x \neq x_0$.

To complete the solution, we need values for the constants C_1 and C_2. These are determined by the condition that the Green's function is continuous at $x = x_0$, so that

$$C_1\overline{\phi}_1(x_0) = C_2\overline{\phi}_2(x_0). \tag{2.4.34}$$

Also, a second equation is provided by the usual jump condition on the first derivative of g, obtained by integration of Eq. (2.4.29) from x_0^- to x_0^+,

$$\frac{\partial}{\partial x}g(x, x_0)\bigg|_{x=x_0^+} - \frac{\partial}{\partial x}g(x, x_0)\bigg|_{x=x_0^-} = 1.$$

Substitution of Eq. (2.4.33) then yields

$$C_2\overline{\phi}_2'(x_0) - C_1\overline{\phi}_1'(x_0) = 1. \tag{2.4.35}$$

Equations (2.4.34) and (2.4.35) can be solved for C_1 and C_2. The solution is

$$g(x, x_0) = \begin{cases} -\dfrac{\overline{\phi}_2(x_0)\overline{\phi}_1(x)}{W(x_0)}, & x < x_0, \\[2ex] -\dfrac{\overline{\phi}_1(x_0)\overline{\phi}_2(x)}{W(x_0)}, & x > x_0, \end{cases} \tag{2.4.36}$$

where the Wronskian $W(x) = \overline{\phi}_1'(x)\overline{\phi}_2(x) - \overline{\phi}_2'(x)\overline{\phi}_1(x)$ again makes an appearance [see Eq. (2.4.8)]. Equation (2.4.36) can be used in Eq. (2.4.30) to determine the particular solution for any given charge density $\rho(x)$. An alternate description of g in terms of eigenmodes can be found in Chapter 4; see Eq. (4.3.16).

2.4.5 Discretized Green's Functions II: Boundary-Value Problems by Matrix Inversion

Theory for Second-Order ODEs Equation (2.4.30) can be thought of as an operator equation,

$$\phi_p = \hat{L}^{-1}\rho, \tag{2.4.37}$$

involving the linear integral operator $\hat{L}^{-1}$ defined by

$$\hat{L}^{-1}\rho = \int_a^b g(x, x_0)\,\rho(x_0)\,dx_0. \tag{2.4.38}$$

This subsection discusses the matrix inverse method that follows from discretization of Eq. (2.4.37). This method, discussed previously for initial-value problems, is really most useful for determining the solution to boundary-value problems. We will apply this method to a general second-order ODE of the form

$$\hat{L}\phi = \frac{d^2}{dx^2}\phi + u_1(x)\frac{d}{dx}\phi + u_0(x)\phi = \rho(x) \tag{2.4.39}$$

with boundary conditions

$$\phi(a) = V_a, \qquad \phi(b) = V_b. \tag{2.4.40}$$

We solve this problem numerically on a grid of positions $x = x_n = a + n\,\Delta x$ specified by the step size $\Delta x = (b-a)/M$. Then Eq. (2.4.39) becomes a matrix equation that can be solved directly by matrix inversion.

First, we need to discretize the differential equation. To do so, we use a *centered-difference* scheme. For the first derivative, this involves the approximation

$$\frac{d\phi}{dx}(x_n) \simeq \frac{\phi(x_n + \Delta x) - \phi(x_n - \Delta x)}{2\,\Delta x} = \frac{\phi(x_{n+1}) - \phi(x_{n-1})}{2\,\Delta x}. \tag{2.4.41}$$

In the limit that Δx approaches zero, this approximation clearly approaches the slope of ϕ at $x = x_n$. Other schemes could also be used (Table 2.3), such as the

Table 2.3. Different Forms for Finite-Differenced First Derivatives

Approximation	Name
$\frac{d\phi}{dx}(x_n) \simeq \frac{\phi(x_{n+1}) - \phi(x_n)}{\Delta x}$	Forward difference
$\frac{d\phi}{dx}(x_n) \simeq \frac{\phi(x_n) - \phi(x_{n-1})}{\Delta x}$	Backward difference
$\frac{d\phi}{dx}(x_n) \simeq \frac{\phi(x_{n+1}) - \phi(x_{n-1})}{2\,\Delta x}$	Centered difference

scheme used in Euler's method. This is called a forward-difference scheme:

$$\frac{d\phi}{dx}(x_n) \simeq \frac{\phi(x_n + \Delta x) - \phi(x_n)}{\Delta x} = \frac{\phi(x_{n+1}) - \phi(x_n)}{\Delta x}. \qquad (2.4.42)$$

One might also choose the backward-difference scheme,

$$\frac{d\phi}{dx}(x_n) \simeq \frac{\phi(x_n) - \phi(x_n - \Delta x)}{\Delta x} = \frac{\phi(x_n) - \phi(x_{n-1})}{\Delta x}, \qquad (2.4.43)$$

but centered-differencing is the most accurate of the three methods (see the exercises). For the second derivative, we again use a centered difference:

$$\frac{d^2\phi}{dx^2}(x_n) \simeq \frac{\dfrac{d\phi}{dx}(x_{n+1/2}) - \dfrac{d\phi}{dx}(x_{n-1/2})}{\Delta x} = \frac{\phi(x_{n+1}) - 2\phi(x_n) + \phi(x_{n-1})}{\Delta x^2}. \qquad (2.4.44)$$

Higher-order derivatives can also be differenced in this fashion. Several of these derivatives are given in Table 2.4. These difference forms are derived in the Appendix. However, the first and second derivatives are all that we require here.

Using these centered-difference forms for the first and second derivatives, Eq. (2.4.39) becomes a series of coupled linear equations for $\phi(x_n)$:

$$\frac{\phi(x_{n+1}) - 2\phi(x_n) + \phi(x_{n-1})}{\Delta x^2} + u_1(x_n)\frac{\phi(x_{n+1}) - \phi(x_{n-1})}{2\,\Delta x} + u_0(x_n)\phi(x_n) = \rho(x_n), \qquad 0 < n < M. \qquad (2.4.45)$$

Table 2.4. Centered-Difference Forms for Some Derivatives[a]

$$\frac{d\phi}{dx}(x_n) \simeq \frac{\phi(x_{n+1}) - \phi(x_{n-1})}{2\,\Delta x}$$

$$\frac{d^2\phi}{dx^2}(x_n) \simeq \frac{\phi(x_{n+1}) - 2\phi(x_n) + \phi(x_{n-1})}{\Delta x^2}$$

$$\frac{d^3\phi}{dx^3}(x_n) \simeq \frac{\phi(x_{n+2}) - 2\phi(x_{n+1}) + 2\phi(x_{n-1}) - \phi(x_{n-2})}{2\,\Delta x^3}$$

$$\frac{d^4\phi}{dx^4}(x_n) \simeq \frac{\phi(x_{n+2}) - 4\phi(x_{n+1}) + 6\phi(x_n) - 4\phi(x_{n-1}) + \phi(x_{n-2})}{\Delta x^4}$$

$$\frac{d^5\phi}{dx^5}(x_n) \simeq \frac{\phi(x_{n+3}) - 4\phi(x_{n+2}) + 5\phi(x_{n+1}) - 5\phi(x_{n-1}) + 4\rho(x_{n-2}) - \phi(x_{n-3})}{2\,\Delta x^4}$$

$$\frac{d^6\phi}{dx^6}(x_n) \simeq \frac{\phi(x_{n+3}) - 6\phi(x_{n+2}) + 15\phi(x_{n+1}) - 20\phi(x_n) + 15\phi(x_{n-1}) - 6\phi(x_{n-2}) + \phi(x_{n-3})}{\Delta x^4}$$

[a]All forms shown are accurate to order Δx^2.

As indicated, this equation is correct only at interior points. At the end points $n = 0$ and $n = M$ we have the boundary conditions

$$\phi(x_0) = V_a, \qquad \phi(x_M) = V_b. \tag{2.4.46}$$

Equations (2.4.45) and (2.4.46) provide $M + 1$ equations in the $M + 1$ unknowns $\phi(x_n)$, $n = 0, 1, 2, \ldots, M$. We can write these equations as a matrix equation, and solve this equation directly by matrix inversion. When written as a matrix equation, Eqs. (2.4.45) and (2.4.46) take the form

$$\mathbf{L} \cdot \boldsymbol{\phi} = \boldsymbol{\rho}, \tag{2.4.47}$$

where

$$\begin{aligned} \boldsymbol{\phi} &= \{\phi(x_0), \phi(x_1), \ldots, \phi(x_M)\}, \\ \boldsymbol{\rho} &= \{V_a, \rho(x_1), \rho(x_2), \ldots, \rho(x_{M-1}), V_b\}, \end{aligned} \tag{2.4.48}$$

and the $M + 1$-by-$M + 1$ matrix $\mathbf{L}$ is determined in terms of Kronecker δ-functions for the interior points as follows:

$$L_{nm} = \frac{\delta_{n+1,m} - 2\,\delta_{nm} + \delta_{n-1,m}}{\Delta x^2} + u_1(x_n)\frac{\delta_{n+1,m} - \delta_{n-1,m}}{2\,\Delta x} + \delta_{nm} u_0(x_n). \tag{2.4.49}$$

A 4-by-4 version of the resulting matrix is displayed below:

Cell 2.97

```
M = 3; L = Table[KroneckerDelta[n, m] (u0[n] - 2 / Δx^2) +
   KroneckerDelta[n + 1, m] (1 / Δx^2 + u1[n] / (2 Δx)) +
   KroneckerDelta[n - 1, m] (1 / Δx^2 - u1[n] / (2 Δx)),
   {n, 0, M}, {m, 0, M}];
MatrixForm[
 L]
```

$$\begin{pmatrix} -\frac{2}{\Delta x^2} + u_0[0] & \frac{1}{\Delta x^2} + \frac{u_1[0]}{2\,\Delta x} & 0 & 0 \\ \frac{1}{\Delta x^2} - \frac{u_1[1]}{2\,\Delta x} & -\frac{2}{\Delta x^2} + u_0[1] & \frac{1}{\Delta x^2} + \frac{u_1[1]}{2\,\Delta x} & 0 \\ 0 & \frac{1}{\Delta x^2} - \frac{u_1[2]}{2\,\Delta x} & -\frac{2}{\Delta x^2} + u_0[2] & \frac{1}{\Delta x^2} + \frac{u_1[2]}{2\,\Delta x} \\ 0 & 0 & \frac{1}{\Delta x^2} - \frac{u_1[3]}{2\,\Delta x} & -\frac{2}{\Delta x^2} + u_0[3] \end{pmatrix}.$$

While the middle rows are correct, the first and last rows must be changed to provide the right boundary conditions:

Cell 2.98

```
L[[1, 1]] = 1; L[[1, 2]] = 0;
L[[M + 1, M + 1]] = 1; L[[M + 1, M]] = 0;
```

Now the matrix takes the form

Cell 2.99

```
MatrixForm[L]
```

$$\begin{pmatrix} 1 & 0 & 0 & 0 \\ \frac{1}{\Delta x^2} - \frac{u_1[1]}{2\,\Delta x} & -\frac{2}{\Delta x^2} + u_0[1] & \frac{1}{\Delta x^2} + \frac{u_1[1]}{2\,\Delta x} & 0 \\ 0 & \frac{1}{\Delta x^2} - \frac{u_1[2]}{2\,\Delta x} & -\frac{2}{\Delta x^2} + u_0[2] & \frac{1}{\Delta x^2} + \frac{u_1[2]}{2\,\Delta x} \\ 0 & 0 & 0 & 1 \end{pmatrix}$$

When this matrix is used in Eq. (2.4.47) together with Eq. (2.4.48), the first and last equations now provide the correct boundary conditions, $\phi(x_0 = V_a,\ \phi(x_M) = V_b$.

These boundary conditions are contained in the discretized forcing function ρ. One can also write the boundary conditions directly into the forcing function of the undiscretized ODE using δ-functions, in analogy to Eq. (2.4.24):

$$\frac{d^2}{dx^2}\phi + u_1(x)\frac{d}{dx}\phi + u_0(x)\phi = \rho(x) + V_a\delta'(x-a) - V_b\delta'(x-b), \quad (2.4.50)$$

with homogeneous boundary conditions $\phi = 0$ for $x < a$ and $x > b$. The proof that this is equivalent to Eqs. (2.4.39) and (2.4.40) is left as a an exercise). The full solution, satisfying the nonzero boundary conditions, is then determined by applying the Green's function to the new forcing function using Eq. (2.4.30). Again, we see that there is no real difference between homogeneous solutions to the ODE and inhomogeneous solutions: both can be written in terms of the Green's function. In fact, a general homogeneous solution can be determined directly in terms of the Green's function by applying Eqs. (2.4.37) and (2.3.42) to the forcing function in Eq. (2.4.50), taking $\rho = 0$. The result is

$$\phi_h(x) = -V_a\frac{d}{dx_0}G(x,x_0)\bigg|_{x_0=a} + V_b\frac{d}{dx_0}G(x,x_0)\bigg|_{x_0=b}. \quad (2.4.51)$$

One can see now why we have spent so much time discussing the particular solutions to ODEs as opposed to the homogeneous solutions: there is really no difference between them. Both types of solutions can be determined using the Green's function method. *Boundary or initial conditions are just another type of forcing, concentrated at the edges of the domain.* In later chapters we will often use this idea, or variations of it, to determine homogeneous solutions to boundary- and initial-value problems.

Now that we have the matrix **L**, we can solve the boundary-value problem (2.4.47) in a single step by matrix inversion:

$$\boldsymbol{\phi} = \mathbf{L}^{-1}\cdot\boldsymbol{\rho}. \quad (2.4.52)$$

Equation (2.4.52) is a very elegant numerical solution to the general linear boundary-value problem. It is somewhat analogous to finding the Green's function and applying Eq. (2.4.37) to $\rho(x)$, obtaining the particular solution that equals zero on the boundaries. However, in the Green's function method one must then add in a homogeneous solution to include the nonzero boundary conditions; but this is not necessary in Eq. (2.4.52).

A closer analytic analogy to Eq. (2.4.52) is the application of Eq. (2.4.37) to the forcing function of Eq. (2.4.50), which also has the nonzero boundary conditions built into the forcing function.

The matrix-inversion method has several distinct advantages compared to the shooting method discussed in Sec. 1.5, where initial guesses had to be made (which might be poor) and where the ODE had to be solved many times in order to converge to the correct boundary conditions. However, there are also several drawbacks to the matrix-inversion technique for boundary-value problems. First, the method works only for linear boundary-value problems, whereas the shooting method works for any boundary-value problem. Second, the matrix form of the operator $\hat{L}$ can be quite complicated, particularly for high-order ODEs. Third, matrix inversion is a computationally time-consuming operation, although very fast codes that can be run on mainframes do exist. Using *Mathematica*, the matrix inverse of a general 700-by-700 matrix takes roughly half a minute on a reasonably fast PC (as of summer 2001). This is fast enough for many problems. Furthermore, there are specialized methods of taking the inverse that rely on the fact that only terms near the diagonal of **L** are nonzero (the matrix is *sparse*). These methods are already built into *Mathematica*, and will not be discussed in detail here.

There is another important point to make concerning the matrix-inverse solution provided by Eq. (2.4.52). We know that for reasonable *initial-value* problems ("reasonable" in the sense of Theorem 1.1) a solution always exists and is unique. On the other hand, we know from Chapter 1 that for *boundary-value* problems the solution need not exist, and need not be unique. However, we have now seen that the solutions to both initial-value and boundary-value problems consist of inverting a matrix **L** and acting on a vector (called $\boldsymbol{\rho}$ for our boundary-value problem, and **f** for the initial-value problem of Sec. 2.4.3). There seems to be no obvious difference between the solutions of initial- and boundary-value problems when they are written in the form of a matrix equation. Why is one case solvable and the other not (necessarily)? The answer lies in the form of the matrix **L** for each case. We will see in the exercises that for boundary-value problems the changes that we have to make to the first and last few rows of **L**, required in order to satisfy the boundary conditions, can lead (for certain parameter choices) to the matrix having a zero determinant. In this event the matrix inverse does not exist (see Sec. 1.5.2), and the solution, if any, cannot be written in the form of Eq. (2.4.52). For examples of how this can happen, see Exercises (10)(e) and (f).

Example As an example of the matrix-inversion method, let's solve for the potential between two conducting spheres, with radii $a = 1$ m and $b = 2$ m respectively. The inner sphere is at $V_a = 5000$ V, and the outer sphere is grounded, $V_b = 0$. Between the spheres is a uniform charge density of $\rho_0 = 10^{-6}$ C/m^3. The potential satisfies the 1D Poisson equation

$$\frac{\partial^2}{\partial r^2}\phi + \frac{2}{r}\frac{\partial}{\partial r}\phi = -\frac{\rho_0}{\epsilon_0},$$

so we take $u_1(r) = 2/r$ and $u_0(r) = 0$ in Eq. (2.4.45).

The following *Mathematica* code will solve this problem, taking $M = 40$ grid points. First, we define the grid r_n and set up the functions $u_0(r)$ and $u_1(r)$ on the grid:

Cell 2.100

```
a = 1; b = 2; M = 40;
Δx = (b - a)/M;
r[n_] = a + n Δx;
u0[n_] = 0;
u1[n_] = 2/r[n];
```

We then reconstruct the matrix **L**, now using the full 41-by-41 form:

Cell 2.101

```
L = Table[KroneckerDelta[n, m] (u0[n] - 2 / Δx^2) +
  KroneckerDelta[n + 1, m] (1/Δx^2 + u1[n]/(2 Δx))+
  KroneckerDelta[n - 1, m] (1/Δx^2 - u1[n]/(2 Δx)),
    {n, 0, M}, {m, 0, M}];

L[[1, 1]] = 1; L[[1, 2]] = 0;
L[[M + 1, M + 1]] =1; L[[M + 1, M]] = 0;
```

Next, we set up the vector **ρ**:

Cell 2.102

```
ϵ0 = 8.85 10^-12; Va = 5000; Vb = 0; ρ0 = 10^-6;
ρ = Table [-ρ0/ϵ0, {n, 0, M}];
ρ[[1]] = Va;
ρ[[M + 1]] = Vb;
```

Finally, the electrostatic potential is determined by matrix inversion:

Cell 2.103

```
φ = Inverse[L].ρ;
```

It may be easily verified (using `DSolve`, for example) that the exact solution for the potential is given by

$$\phi(r) = -5000 + \frac{10000}{r} + \frac{7\rho_0}{6\epsilon_0} - \frac{\rho_0}{r\epsilon_0} - \frac{r^2\rho_0}{6\epsilon_0}$$

(in SI units). We compare the exact result with the numerical solution in Cell 2.104. This numerical method matches the analytic solution to the boundary-value problem quite nicely. For one-dimensional linear boundary-value problems, matrix inversion is an excellent numerical method of solution.

Cell 2.104

```
sol = Table[{r[n], φ[[n + 1]]}, {n, 0, M}];
pl = ListPlot[sol, DisplayFunction → Identity];
```

$$\phi\text{exact}[r\_] = -5000 + \frac{10000}{r} + \frac{7\rho 0}{6\epsilon_0} - \frac{\rho 0}{r\epsilon_0} - \frac{r^2\rho 0}{6\epsilon_0};$$

```
p2 = Plot[φexact[r], {r, 1, 2}, DisplayFunction → Identity];
Show[p1, p2, DisplayFunction → $DisplayFunction;
  PlotLabel → "electrostatic potential between two charged
    spheres",
  AxesLabel → {"r", " "}];
```

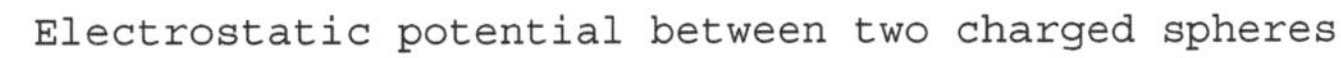

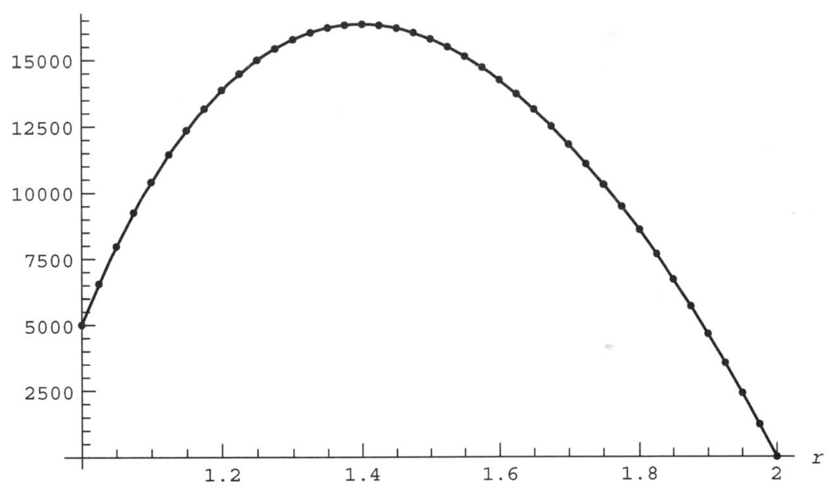

EXERCISES FOR SEC. 2.4

(1) **(a)** Find the Green's function for the following potential problem in terms of homogeneous solutions:

$$\frac{d^2}{dx^2}\phi = \rho(x), \qquad \phi(0) = 0, \quad \phi(b) = 0.$$

(b) Use the Green's function to find a particular solution for the case where $\rho(x) = x^3$.

(2) Use the method of Sec. 2.4.2 to solve the following ODEs for the Green's function, for the given homogeneous boundary or initial conditions:

(a) $G' + 3tG(t, t_0) = \delta(t - t_0)$, $G = 0$ for $t < t_0$. Plot $G(t, 0)$.

(b) $G'' + 4G(t, t_0) = \delta(t - t_0)$, $G = 0$ for $t < t_0$. Plot $G(t - t_0)$.

(c) $G'' + 2G' + G(t, t_0) = \delta(t - t_0)$, $G = 0$ for $t < t_0$. Plot $G(t - t_0)$.

(d) $G'' + tG(t,t_0) = \delta(t-t_0)$, $G=0$ for $t<t_0$. (Hint: The solution will be in terms of Airy functions.) Plot $G(t,0)$.

(e) $G''' + 2G'' - G' - 2G = \delta(t-t_0)$, $G=0$ for $t<t_0$. Plot $G(t-t_0)$.

(f) $G'' + \omega_0^2 G(x,x_0) = \delta(x-x_0)$, $G=0$ for $x=a$ and $x=b$ ($\omega_0>0$, $\omega_0 \neq n\pi/(b-a)$ for any integer n).

(g) $G'' - \kappa_0^2 G(x,x_0) = \delta(x-x_0)$, $G=0$ for $x=a$ and $x=b$ ($\kappa_0>0$).

(h) $G'' + nG'(x,x_0)/x = \delta(x-x_0)$, $G=0$ for $x=a>0$ and $x=b$, $n\neq 1$.

(i) $G''' + G'' + G' + G = \delta(x-x_0)$, $G=G'=0$ at $x=0$, $G=0$ at $x=1$. Plot $G(x-x_0)$.

(3) Use the Green's functions found from Exercise (2) to help determine solutions to the following problems. Plot the solution in each case.

(a) $x' + 3tx(t) = \sin t$, $x=0$ at $t=0$.

(b) $x'' + 4x(t) = \cos 2t$, $x(0)=0=x'(0)$.

(c) $x'' + 2x' + x(t) = t^2$, $x(0)=1$, $x'(0)=0$.

(d) $x'' + tx(t) = t$, $x(0)=1$, $x'(0)=0$.

(e) $x''' + 2x'' - x' - 2x = e^{-t}$, $x(0)=1$, $x'(0)=x''(0)=0$.

(f) $\phi'' + \phi(x) = x^3$, $\phi(0)=1$, $\phi(1)=0$.

(g) $\phi'' - \phi(x) = \sin x$, $\phi'(0)=0=\phi(1)$.

(h) $\phi'' + 3\phi'/x = x$, $\phi(1)=0$, $\phi'(2)=1$.

(i) $\phi''' + \phi'' + \phi' + \phi = xe^{-x}$, $\phi=\phi'=0$ at $x=0$, $\phi - 2\phi' = 1$ at $x=1$.

(4) Discretize the operator in Exercise (3)(a) using Euler's method, and solve the problem by matrix inversion for $0<t<8$. Take $\Delta t = 0.1$, and compare with the exact solution by plotting both in the same plot.

(5) **(a)** By writing the ODE in Exercise (3)(b) as a vector ODE in the unknown vector function $\mathbf{z}(t) = \{x(t), x'(t)\}$, discretize the ODE using the vector Euler's method, and solve by matrix inversion for $0<t<5$. Take $\Delta t = 0.05$. Plot the result along with the exact solution. (See Sec. 1.4.5.)

(b) Repeat for Exercise (3)(c).

(c) Repeat for Exercise (3)(d).

(d) Repeat for Exercise (3)(e), now taking the vector function as $\mathbf{z}(t) = \{x(t), x'(t), x''(t)\}$.

(6) **(a)** for the following general second-order ODE initial-value problem, find a way of including the initial conditions in the forcing function on the right-hand side, in analogy to Eq. (2.4.24), and state the proper homogeneous initial condition for the new ODE:

$$\frac{d^2x}{dt^2} + u_1(t)\frac{dx}{dt} + u_0(t)x(t) = f(t), \qquad x(t_0) = x_0, \qquad x'(t_0) = v_0.$$

(b) Use the result of part (a) along with Eq. (2.4.3), write a general homogeneous solution $x_h(t)$ to this problem in terms of the Green's function.

(c) Use the Green's function found in Exercise (2)(c) and the result from part (b) to solve the following problem:

$$\frac{d^2x}{dt^2} + 2\frac{dx}{dt} + x(t) = e^{-t}, \qquad x(0) = 1, \quad x'(0) = 2.$$

(7) In this problem we will show that the centered-difference scheme for taking a numerical derivative is more accurate than either forward- or backward-differencing. To do so, we take a smooth function $\phi(x)$ whose derivative we will determine numerically at a grid point $x = x_n$. Then the value of $\phi(x)$ at $x = x_{n+1}$ can be determined by Taylor expansion:

$$\phi_{n+1} = \phi_n + \Delta x\, \phi_n' + \Delta x^2 \frac{\phi_n''}{2} + \Delta x^3 \frac{\phi_n'''}{6} + \cdots,$$

with a similar result for ϕ_{n-1}.

(a) Use these Taylor expansions in the forward- and backward-difference methods, Eqs. (2.4.42) and (2.4.43), to show that the error in these methods is of order Δx.

(b) Use these Taylor expansions in the centered-difference derivative, Eq. (2.4.41), to show that the error in this method is of order Δx^2.

(c) Repeat for the centered-difference second derivative, Eq. (2.4.44), to show that its error is also of order Δx^2.

(8) **(a)** Prove Eq. (2.4.50).

(b) Prove Eq. (2.4.51).

(c) Test Eq. (2.4.51) directly for the potential problem given in Exercise (1) of this section: by applying the Green's function found in Exercise (1)(a) to Eq. (2.4.51), show that the correct homogeneous solution to this potential problem is recovered.

(9) **(a)** Find a way to include the boundary conditions in the forcing function for the following second-order boundary-value problem:

$$\frac{d^2}{dx^2}\phi + u_1(x)\frac{d}{dx}\phi + u_0(x)\phi = \rho(x), \qquad \phi'(a) = V_a, \quad \phi'(b) = V_b.$$

What homogeneous boundary conditions must be attached to the new ODE?

(b) Using the result of part (a) along with Eq. (2.4.3), write a general homogeneous solution $x_h(t)$ to this problem in terms of the Green's function.

(c) Using the results of part (a) and (b), find the appropriate Green's function in terms of homogeneous solutions, and solve the following boundary-value problem:

$$\frac{d^2}{dx^2}\phi + 4\frac{d}{dx}\phi + \phi = x, \qquad \phi'(0) = 0, \quad \phi'(1) = 1.$$

(10) Using the centered-difference discretization techniques discussed in Sec. 2.4.5, solve the following problems using matrix inversion, and compare each to the exact solution. In each case take $\Delta x = 0.05$:

- **(a)** Exercise (3)(f).
- **(b)** Exercise (3)(g). [Hint: To learn about setting the derivative equal to zero at the boundary, see Sec. 6.2.1, in the sub-subsection on von Neumann and mixed boundary conditions.]
- **(c)** Problem (3)(h) [see the hint for Exercise (10)(b)].
- **(d)** Problem (3)(i) [see the hint for Exercise (10)(b)].
- **(e)** $\phi''(x) = 1,\ \phi'(0) = 2;\ \phi'(1) = 0$. [Hint: First solve this problem analytically by direct integration to show that a solution *does not exist*. Then solve the problem by finite differencing. Also, see the hint for Exercise (10)(b).]
- **(f)** $\phi''(x) + \phi(x) = 0,\ \phi(0) = 0 = \phi(\pi)$. [Hint: Recall that there are an infinite number of solutions; see Eq. (1.5.9). Take $\Delta x = 0.05\ \pi$.]

REFERENCES

J. W. Bradbury and S. L. Vehrencamp, *Principles of Animal Communication* (Sinauer Associates, Sunderland, MA, 1998). An introductory reference on the auditory system.

W. F. Hartmann, *How we localize sound*, Phys. Today, November 1999, p. 24.

CHAPTER 3

INTRODUCTION TO LINEAR PARTIAL DIFFERENTIAL EQUATIONS

In this chapter we derive analytic solutions to some of the common linear partial differential equations (PDEs) of mathematical physics. We first examine PDEs in one spatial dimension, focusing on the wave equation and the heat equation. We then solve a PDE in more than one spatial dimension: Laplace's equation.

For the simple cases discussed in this chapter, we will find that solutions can be obtained in terms of Fourier series, using the *method of separation of variables*.

3.1 SEPARATION OF VARIABLES AND FOURIER SERIES METHODS IN SOLUTIONS OF THE WAVE AND HEAT EQUATIONS

3.1.1 Derivation of the Wave Equation

Introduction The first partial differential equation that we will discuss is the wave equation in one dimension. This equation describes (among other things) the transverse vibrations of a string under tension, such as a guitar or violin string. In the next two sub-subsections we will derive the wave equation for a string, from first principles.

String Equilibrium Consider a string, stretched tight in the x-direction between posts at $x=0$ and $x=L$. (See Fig. 3.1.) The tension in the string at point x is $T(x)$. Tension is a force, and so has units of newtons. The tension $T(x)$ at point x is defined as the force pulling on the piece of string to the left of point x as it is acted on by the string to the right of this point (see Fig. 3.1). According to Newton's third law, the force acting on the string to the right of point x as it is pulled by the string on the left is $-T(x)$: the forces of the two sections acting on one another are equal and opposite. Furthermore, tension always acts in the direction along the string. We will define positive tension forces as those acting to the right—in the positive x-direction.

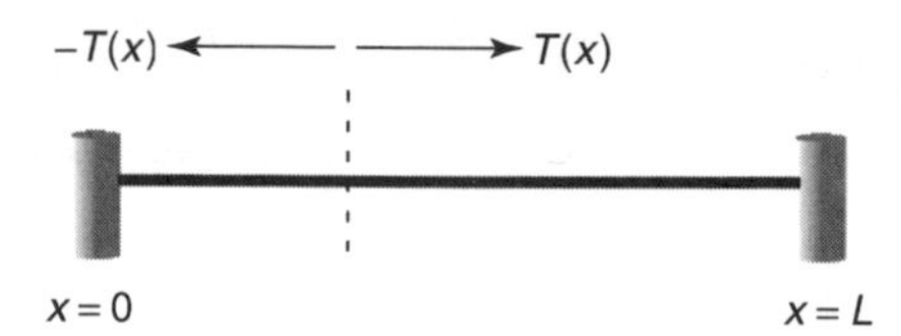

Fig. 3.1 Equilibrium of a string.

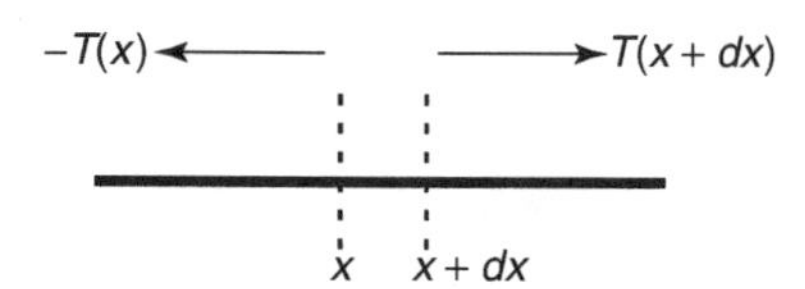

Fig. 3.2 Forces acting on a mass element in equilibrium.

The string in Fig. 3.1 is in equilibrium. In equilibrium, in the absence of gravity or other such external forces, the tension in the string is uniform, and the string is perfectly straight. This can be understood from the following argument. Consider an element of the string with mass dm and length dx, running from x to $x+dx$. If the string is straight, but the tension in the string is a function of position, then the tension pulling the element to the right is $T(x+dx)$ and that pulling the element to the left is $-T(x)$ (see Fig. 3.2). However, in equilibrium, the total force on dm must vanish, so $T(x+dx)-T(x)=0$, and therefore T must be independent of position in equilibrium. Also, since we have achieved force balance with a straight string, we have shown that a straight string is in equilibrium.

However, if an extra force dF, such as gravity, acts on the element dm in the x-direction, then T is not constant in equilibrium. Equilibrium force balance then yields $T(x+dx)+dF-T(x)=0$. Taylor expansion of this expression implies

$$\frac{dT}{dx} = -\frac{dF}{dx}. \tag{3.1.1}$$

For instance, if gravity points in the $-x$ direction, then $dF=-dm\,g$. Equation (3.1.1) then implies that $dT/dx=\rho g$, where

$$\rho(x) = dm/dx \tag{3.1.2}$$

is the mass per unit length. In this example the tension increases with increasing x, because more of the string weight must be supported by the remaining string as one moves up the string against gravity, in the $+x$ direction.

Since all forces have been assumed to act along the x-direction, the string remains straight. However, if time-independent forces, such as a force of gravity in the y-direction, act transverse to the string it will no longer remain straight in equilibrium, but will sag under the action of the transverse force. (This can be observed "experimentally" in spring–mass simulations of an elastic string, in Sec. 9.10.) In what follows, we neglect the effect of such forces on the equilibrium of the string, and assume that the equilibrium is a straight horizontal string along the x-axis, following the equation $y=0$. We will examine the effect of a gravitational force in the y-direction in Sec. 3.1.2.

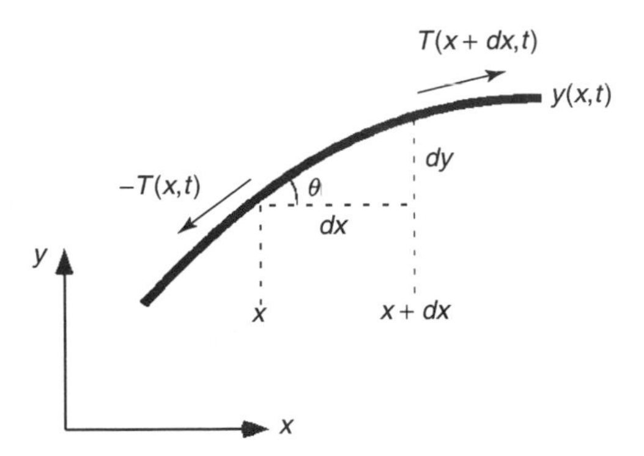

Fig. 3.3 Forces acting on a mass element in motion (displacement greatly exaggerated for ease of viewing).

String Dynamics: The Wave Equation Now let us consider a string that has been perturbed away from equilibrium by a small transverse displacement in the y-direction. The shape of the string is now a curve that changes with time, given by the function $y(x,t)$. (See Fig. 3.3.) Our job is to obtain an equation of motion for $y(x,t)$. To simplify this task, we will assume that the perturbed string is nearly straight: $|\partial y/\partial x| \ll 1$.

The equation of motion for $y(x,t)$ can be found by applying Newton's second law to the motion of an element $\boldsymbol{dm}$ of the string, located between x and $x + \boldsymbol{dx}$. The total force $\boldsymbol{dF}_y$ in the y-direction on this element is determined by the y-component of the tension $T_y(x,t)$ acting on each end of the string (assuming that no forces other than tension act in the y-direction):

$$\boldsymbol{dF}_y = T_y(x + \boldsymbol{dx}, t) - T_y(x,t) = \boldsymbol{dx}\frac{\partial}{\partial x}T_y(x,t). \tag{3.1.3}$$

By Newton's second law, this force determines the acceleration of the mass element in the y-direction:

$$\boldsymbol{dm}\frac{\partial^2}{\partial t^2}y(x,t) = \boldsymbol{dF}_y = \boldsymbol{dx}\frac{\partial}{\partial x}T_y(x,t). \tag{3.1.4}$$

Because tension forces act along the direction of the string, the y-component of the tension is related to the displacement of the string from equilibrium. According to Fig. 3.3, $T_y(x,t) = T(x,t)\sin\theta$, where $T(x,t)$ is the magnitude of the tension in the string at position x and time t, and θ is the angle of the element $\boldsymbol{dm}$ with respect to the horizontal. However, since the displacement from equilibrium is assumed small, θ must also be small, and therefore $\sin\theta \approx \theta$, which implies that

$$T_y(x,t) \approx T(x,t)\theta. \tag{3.1.5}$$

Furthermore, according to Fig. 3.3, θ is related to the displacement of the string through $\tan\theta = \partial y/\partial x$. For small angles $\tan\theta \approx \theta$, so this implies $\theta \approx \partial y/\partial x$. Combining this result with Eq. (3.1.5), we obtain $T_y(x,t) \approx T(x,t)\,\partial y/\partial x$. However, as we are interested only in small-amplitude transverse displacements of the string, we can replace $T(x,t)$ by the equilibrium tension $T(x)$. Therefore, we obtain

$$T_y(x,t) \approx T(x)\frac{\partial y}{\partial x}. \tag{3.1.6}$$

Applying this result to Eq. (4.1.4), dividing by dm, and using Eq. (3.1.2) yields

$$\frac{\partial^2}{\partial t^2} y(x,t) = \frac{1}{\rho(x)} \frac{\partial}{\partial x}\left(T(x) \frac{\partial}{\partial x} y(x,t) \right). \tag{3.1.7}$$

Equation (3.1.7) is the wave equation for a string with equilibrium tension $T(x)$ and mass per unit length $\rho(x)$. This equation describes the evolution of small transverse displacements $y(x,t)$ away from equilibrium. The equation is a linear partial differential equation in the unknown function $y(x,t)$, and is second-order in time and second-order in space. Since the equation is second-order in space, two boundary conditions are required: the ends of the string are fixed by the posts at each end, so $y=0$ at $x=0$ and at $x=L$.

Since the wave equation is second-order in time, two initial conditions are also needed. The position and transverse velocity of each element of the string must be specified initially in order to determine its subsequent motion. In other words, we require knowledge of $y(x,t=0)$ and $\frac{\partial}{\partial t}y(x,t=0)$ for all x in the range $0<x<L$. Thus, the solution of the wave equation is specified by boundary conditions

$$y(0,t) = y(L,t) = 0 \tag{3.1.8}$$

and by initial conditions

$$\begin{aligned} y(x,0) &= y_0(x), \\ \frac{\partial y}{\partial t}(x,0) &= v_0(x), \end{aligned} \tag{3.1.9}$$

for some initial transverse displacement $y_0(x)$ and initial transverse velocity $v_0(x)$.

The wave equation can be simplified in the case that the string tension is uniform $[T(x)=T]$ and the mass density is also uniform $[\rho(x)=\rho]$. Then Eq. (3.1.7) becomes

$$\frac{\partial^2}{\partial t^2} y(x,t) = c^2 \frac{\partial^2}{\partial x^2} y(x,t). \tag{3.1.10}$$

The constant c is

$$c = \sqrt{T/\rho}, \tag{3.1.11}$$

and has units of a velocity. In fact we will see that this quantity is the speed at which transverse disturbances propagate along the string.

The magnitude of c depends on the mass density and thickness of the string as well as the tension to which it is subjected. For instance, the high E-string on a steel string guitar is typically made of steel with a mass density $\mathcal{M}$ of roughly $\mathcal{M}=7.5\ \text{g/cm}^3$. The radius of this string is $r=0.15$ mm, giving a mass per unit length of

$$\rho = \pi r^2 \mathcal{M} = 0.0053\ \text{g/cm} = 6.3\times 10^{-4}\ \text{kg/m}. \tag{3.1.12}$$

According to Eq. (3.1.11), a tension of $T=500$ N yields a speed of $c=970$ m/s.

Although we have derived the wave equation for transverse waves on a string, the same PDE also applies to many other types of wave disturbances traveling in one spatial dimension. For example, the equation applies to the small-amplitude waves on the surface of shallow water (neglecting surface tension), with $y(x,t)$ now the height of the water surface [see Exercise (4) at the end of this section]. The equation also applies to pressure waves propagating through a medium such as air, with $y(x,t)$ now the pressure or density in the medium. The equation applies to propagation of electromagnetic waves such as visible light or radio waves, with $y(x,t)$ now identified with the electric or magnetic field in the wave. The wave equation itself is the same in each case, even though the physical processes for each type of wave are very different. Of course, in each instance, the speed of propagation c differs. Obviously, for sound waves c is the speed of sound, but for light waves c is the speed of light.

3.1.2 Solution of the Wave Equation Using Separation of Variables

Solution of the Wave Equation for a Uniform String and Fixed Ends

Separation of Variables for a Uniform String. We will now solve the wave equation for a *uniform string*, Eq. (3.1.10), with boundary conditions that the ends are fixed: $y=0$ at $x=0$ and $x=L$, and initial conditions given by Eq. (3.1.9). To solve this problem, we will apply the *method of separation of variables*. In this method, we look for a solution to the PDE of the form

$$y(x,t)=f(t)\psi(x), \tag{3.1.13}$$

where $f(t)$ and $\psi(x)$ are some functions that need to be determined in order to satisfy the PDE and match the boundary and initial conditions. In fact, we will find that there are many possible solutions of this form, each of which satisfy the PDE and the boundary conditions. We will create a superposition of these solutions in order to match the initial conditions.

If we substitute Eq. (3.1.13) into the wave equation, Eq. (3.1.10), and then divide by $f(t)\psi(x)$, we obtain

$$\frac{1}{f(t)}\frac{\partial^2 f}{\partial t^2}=c^2\frac{1}{\psi(x)}\frac{\partial^2\psi}{\partial x^2}. \tag{3.1.14}$$

This PDE can be separated into two ordinary differential equations by means of the following argument: the right-hand side of the equation is a function only of position x. Let us call this function $h(x)$. Then Eq. (3.1.14) implies that $[1/f(t)]\,\partial^2 f/\partial t^2=h(x)$. However, the left-hand side of this equation is independent of x. Therefore, $h(x)$ *must be a constant*. Let us call this constant $-\omega^2$, in anticipation of the fact that it is a negative quantity. Then Eq. (3.1.14) becomes two ODEs:

$$\frac{1}{f(t)}\frac{\partial^2 f}{\partial t^2}=-\omega^2, \tag{3.1.15}$$

$$c^2\frac{1}{\psi(x)}\frac{\partial^2\psi}{\partial x^2}=-\omega^2. \tag{3.1.16}$$

These two ODEs must be solved subject to the boundary and initial conditions. First, we consider the boundary conditions. The conditions, $y(0,t) = y(L,t) = 0$ imply that $\psi(0) = \psi(L) = 0$. These are *homogeneous boundary conditions,* introduced previously in relation to Eq. (2.4.28). With such boundary conditions, Eq. (3.1.16) can be recognized as a special kind of boundary-value problem: an *eigenvalue problem.*

The Eigenvalue Problem. An eigenvalue problem such as Eq. (3.1.16) is a linear boundary-value problem with homogeneous boundary conditions. Also, the differential equation must depend on a parameter that can be varied. In this case the parameter is ω. The homogeneous boundary conditions $\psi(0) = \psi(L) = 0$ imply that there is always a trivial solution to the problem, $\psi(x) = 0$. However, in eigenvalue problems, for special values of the parameter, there also exist *nontrivial* solutions for ψ, called *eigenmodes* or *eigenfunctions*. The special values of the parameter ω are called *eigenfrequencies* (since ω has units of a frequency), but more generally they are called *eigenvalues*.

To find the nontrivial solutions for ψ, we match the general solution of the differential equation to the boundary conditions. The general solution of Eq. (3.1.16) is

$$\psi(x) = C\cos\frac{\omega x}{c} + D\sin\frac{\omega x}{c}, \tag{3.1.17}$$

where C and D are constants. First, the condition $\psi(0) = 0$, when used in Eq. (3.1.17), implies that $C = 0$. Next, the condition $\psi(L) = 0$ implies that $D\sin(\omega L/c) = 0$. This equation can be satisfied in two ways. First, we could take $D = 0$, but this would then imply that $\psi = 0$, which is the trivial and uninteresting solution. The second possibility is that $\sin(\omega L/c) = 0$. This implies that $\omega L/c = n\pi$, n an integer.

Thus, we find that the nontrivial solutions of Eq. (3.1.16) with boundary conditions $\psi(0) = \psi(L) = 0$ are

$$\psi(x) = D\sin\frac{n\pi x}{L}, \qquad n = 1,2,3,\ldots \tag{3.1.18}$$

and also,

$$\omega = \omega_n = n\pi c/L. \tag{3.1.19}$$

These are the eigenfunctions and eigenfrequencies for this problem. We do not require values of n less than zero, because the corresponding eigenmodes are just opposite in sign to those with $n > 0$.

Recall from Chapter 1 that the solution to a boundary-value problem need not be unique. In eigenvalue problems, we have an example of this indeterminacy. When $\omega \neq \omega_n$, there is only one solution, $\psi = 0$, but when $\omega = \omega_n$, the constant D can take on any value, including zero, so there are many solutions. Fortunately, the specific value of this constant is not important in constructing the solutions to the wave equation, as we will now see.

The Solution for y(x, t). The time dependence of the solution is described by Eq. (3.1.15). The general solution of this equation is

$$f(t) = A\cos\omega t + B\sin\omega t,$$

where A and B are constants. Using Eq. (3.1.19) for the frequency, and Eq.

(3.1.18) for the eigenmodes, the solution for $y(x,t)=f(t)\psi(x)$ is

$$y(x,t) = (A\cos\omega_n t + B\sin\omega_n t)\sin\frac{n\pi x}{L},$$

where we have absorbed the constant D into A and B. However, this is not the full solution to the problem. There are an infinite number of solutions for the eigenmodes and eigenfrequencies, so we will create a *superposition* of these solutions, writing

$$y(x,t) = \sum_{n=1}^{\infty} (A_n\cos\omega_n t + B_n\sin\omega_n t)\sin\frac{n\pi x}{L}. \tag{3.1.20}$$

Equation (3.1.20) is the general solution to the wave equation for a uniform string with fixed ends. This equation describes a wealth of physics, so it is worthwhile to pause and study its implications.

Eigenmodes for a Uniform String. The eigenfunctions $\sin(n\pi x/L)$ are the normal modes of oscillation of the string. If only a single normal mode is excited by the initial condition, then the string executes a sinusoidal oscillation in time, and this oscillation persists forever. If several different eigenfunctions are excited by the initial conditions, each mode evolves in time independently from the others, with its own fixed frequency. Examples of single-mode oscillations are shown in Cell 3.1 for the first four normal modes, taking $c=L=1$. This figure displays three key features of the normal modes:

Cell 3.1

```
L = 1; c = 1;
ω[n_] = n Pi c/L;
plt[n_, t_] :=
  Plot[Cos[ω[n] t] Sin[n Pi x], {x, 0, L},
  DisplayFunction→Identity, PlotRange→{-1, 1},
  PlotLabel→"n = " <>ToString[n]];
Table[Show[GraphicsArray[Table[
    {{plt[1, t], plt[2, t]}, {plt[3, t], plt[4, t]}}]],
   DisplayFunction→$DisplayFunction], {t, 0, 2, .05}];
```

(1) Each single-mode oscillation forms a standing wave on the string, with a set of stationary nodes. At these nodes the amplitude of the oscillation is zero for all time.

(2) The number of nodes equals $n-1$ (excluding the end points). Consequently, modes with large n exhibit rapid spatial variation.

(3) One can plainly see (in the animation accompanying the electronic version of the text) that the modes with higher n oscillate more rapidly in time. For instance, the $n=4$ standing wave completes four oscillations for every single cycle completed by the $n=1$ mode. This follows from the expression for the frequencies of the modes, Eq. (3.1.19).

Equation (3.1.19) also shows that as the length of the string is reduced, the mode frequencies increase. This simple property of the wave equation has many applications. For instance, it is the principle behind the operation of stringed musical instruments. In order to increase the frequency of the sound, the musician reduces the effective length of the string, by placing some object (such as his or her finger) against the string, allowing only part of the string to vibrate when plucked.

When the string is plucked, the lowest, *fundamental* frequency ω_1 usually predominates in the response, and is primarily responsible for the pitch of the sound. However, the higher harmonics are also produced at multiples of ω_1, and the superposition of these harmonics are, in part, responsible for the characteristic sound of the instrument. (The manner in which these string vibrations couple to sound waves in the surrounding air is also of great importance to the sound produced. This coupling is a primary consideration in the design of the instrument.)

However, musicians often play tricks to alter the sound a string makes. For instance, musicians can create a high-frequency sound on an open string by placing their finger lightly at the location of the first node of the $n=2$ harmonic, in the middle of the string. This allows the $n=2$ mode to vibrate when the string is plucked, but suppresses the fundamental mode, creating a sound one octave above the fundamental (i.e., at twice the frequency).

Also, the frequency of the vibration increases as the propagation speed c increases. Thus, for very thin, high-tension strings such as the high E-string on a guitar, c is large and the fundamental frequency ω_1 of the string is correspondingly high. Thicker, more massive strings at lower tension have lower fundamental frequencies. By varying the tension in the string, a musician can change the frequency and tune his or her instrument.

Matching the Initial Conditions. Our final task is to determine the values of A_n and B_n in the general solution. These constants are found by matching the general solution to the initial conditions, Eqs. (3.1.9). At $t=0$, Eqs. (3.1.20) and (3.1.9) imply

$$y(x,0)=\sum_{n=1}^{\infty} A_n \sin\frac{n\pi x}{L}=y_0(x). \tag{3.1.21}$$

This is a Fourier sine series, and we can therefore determine the Fourier coefficients A_n using Eq. (3.2.11):

$$A_n=\frac{2}{L}\int_0^L y_0(x)\sin\frac{n\pi x}{L}\,dx. \tag{3.1.22}$$

Similarly, the constants B_n are determined by the second initial condition, $\partial y/\partial t|_{t=0} = v_0(x)$. Using Eq. (3.1.20) to evaluate $\partial y/\partial t|_{t=0}$, we find

$$\sum_{n=1}^{\infty} B_n \omega_n \sin\frac{n\pi x}{L} = v_0(x). \tag{3.1.23}$$

This equation is also a Fourier sine series, so B_n is given by

$$B_n \omega_n = \frac{2}{L}\int_0^L v_0(x)\sin\frac{n\pi x}{L}\,dx. \tag{3.1.24}$$

Equations (3.1.19), (3.1.20), (3.1.22), and (3.1.24) are the solution to the wave equation on a uniform string with fixed ends, for given initial conditions. We see that the solution matches the boundary conditions that $y(0,t) = y(L,t) = 0$ because each Fourier mode $\sin(n\pi x/L)$ satisfies these conditions. The solution matches the initial conditions because the Fourier coefficients A_n and B_n are chosen specifically to do so, via Eqs. (3.1.22) and (3.1.24).

Examples

Example 1: Plucked String We can use our general solution to determine the evolution of any given initial condition. For instance, consider the initial condition

$$y_0(x) = \begin{cases} ax, & 0 < x < L/2, \\ a(L-x), & L/2 < x < L, \end{cases}$$

$$v_0(x) = 0.$$

This initial condition, plotted in Cell 4.3, is formed by pulling sideways on the center of the string, and then letting it go. To find the subsequent motion, all we need do is determine the constant A_n and B_n in the general solution. Equation (3.1.24) implies that $B_n = 0$, and Eq. (3.1.20) implies that

$$A_n = \frac{2}{L}\int_0^L y_0(x)\sin\frac{n\pi x}{L}\,dx = \frac{2}{L}\left\{\int_0^{L/2} ax\sin\frac{n\pi x}{L}\,dx + \int_{L/2}^L a(L-x)\sin\frac{n\pi x}{L}\,dx\right\}.$$

These integrals can be evaluated analytically using *Mathematica*, and the result is as follows:

Cell 3.2

```
A[n_] = Simplify[2/L (Integrate[ a x Sin[n Pi x/L],
  {x, 0, L/2}] + Integrate[ a (L - x) Sin[n Pi x/L],
  {x, L/2, L}]), n∈Integers]
```

$$\frac{4\ \mathrm{a}\ \mathrm{L}\ \mathrm{Sin}\left[\frac{\mathrm{n}\pi}{2}\right]}{\mathrm{n}^2\pi^2}$$

Thus, this perturbation evolves according to

$$y(x,t) = \sum_{n=1}^{\infty} A_n \cos\frac{n\pi ct}{L}\sin\frac{n\pi x}{L}, \tag{3.1.25}$$

with A_n given in Cell 3.2. This sum is a Fourier sine series in space and a cosine series in time. The time dependence implies that $y(x,t)$ is periodic in time with fundamental period $2\pi/\omega_1 = 2L/c$. By cutting off the sum at some large but finite value M, we can observe the evolution of the string. We do so in taking $L = c = a = 1$.

Cell 3.3

```
M = 30; L = c = a = 1;
y[x_, t_] = Sum[A[n] Cos[n Pi c t/L] Sin[n Pi x/L],
  {n, 1, M}];
Table[Plot[y[x, t], {x, 0, L}, PlotRange→{{0,1}, {-1, 1}},
   PlotLabel→"t = "<>ToString[t], AxesLabel→{"x", "y"}],
   {t, 0, 1.9, .1}];
```

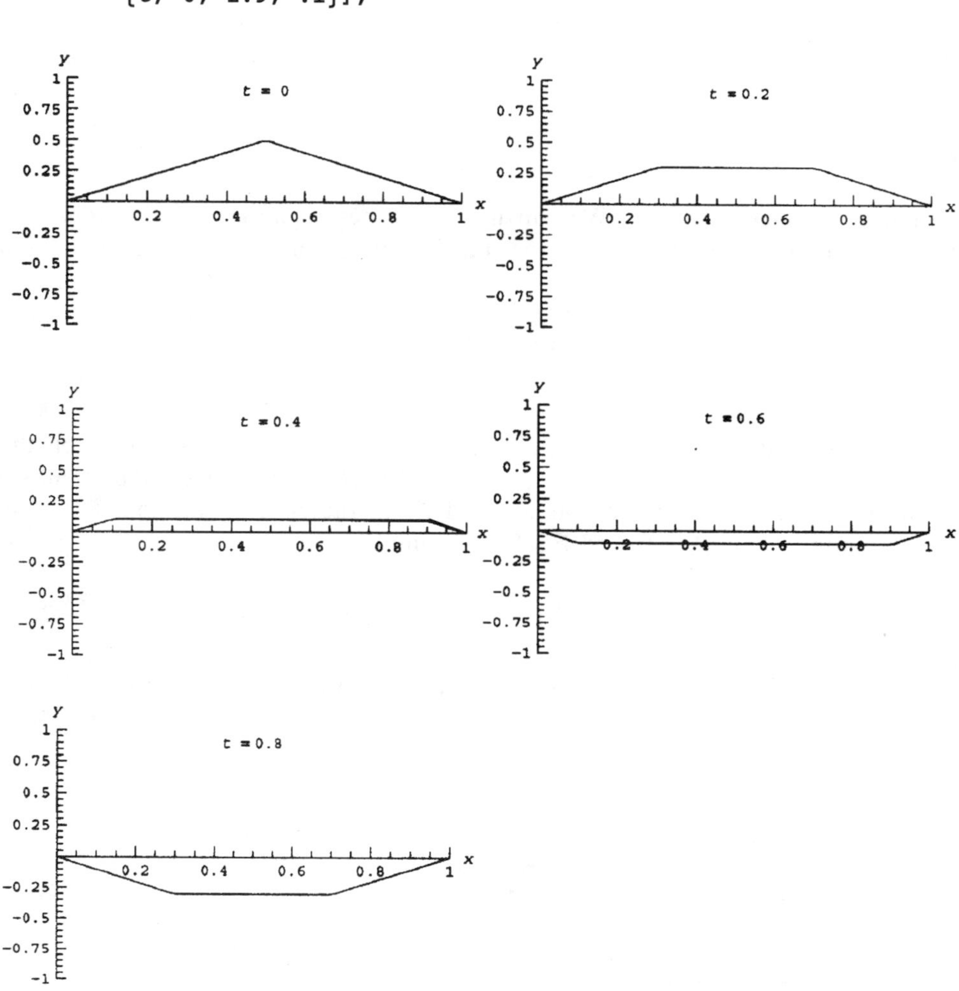

This evolution may seem strange. One might have expected that the entire triangle would simply change its amplitude in time, oscillating back and forth like the $n = 1$ normal mode shown in Cell 3.1. Instead, a flat area in the middle grows with time, and then decreases in size until the initial condition is re-formed. This is how an ideal undamped string actually behaves, and it is only because it is difficult to actually see the rapid evolution of a plucked string that this behavior seems unusual to us.

To better understand this motion, it is useful to look at the time evolution of the center of the string, superimposed on the evolution of a second point at $x = L/4$, as shown in Cell 3.4. The center of the string oscillates back and forth in a triangle wave (thin line). The point at $x = L/4$ only starts to move after a certain period of time (thick line), which is required by the fact that waves have a finite propagation speed on the string. Recall that our initial condition is formed by pulling on the center of the string, holding it off axis against the string tension, and then releasing the string. Once the center has been released, it takes some time for this information to propagate to points far from the center. We will soon show that the speed of propagation of this signal is c. Thus, the point at $L/4$ does not learn that the point at $L/2$ has been released until time $t = L/(4c)$ has elapsed. Only after this time does the point at $L/4$ begin to move.

Cell 3.4

```
In[9] := <<Graphics`;
Plot[{y[L/4, t], y[L/2, t]}, {t, 0, 2 c/L},
  PlotStyle→{{Blue, Thickness[0.01]}, Red}, AxesLabel→
   {"t", TableForm[{{StyleForm["y[L/4, t]", FontColor→Blue,
   FontWeight -> "Bold"], ", ", StyleForm["y[L/2, t]",
   FontColor→Red]}}, TableSpacing→0]}];
```

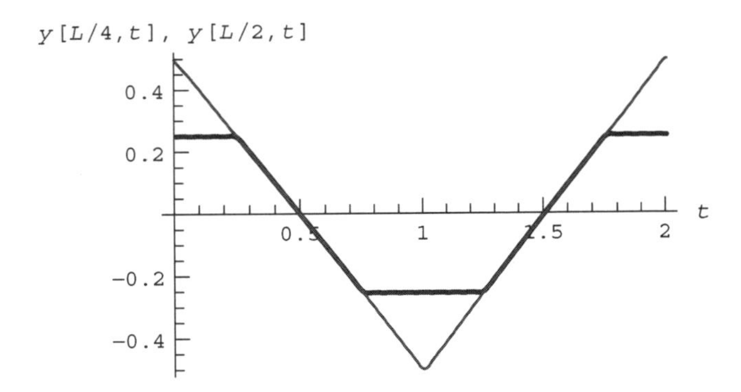

Example 2: Traveling Disturbance The finite propagation speed of traveling disturbances can be easily seen directly by considering a different initial condition, which is localized in the middle of the string:

$$y_0(x) = e^{-50(x/L-1/2)^2},$$

$$v_0(x) = 0.$$

The resulting Fourier series solution will still have the form of Eq. (3.1.25), but the coefficients A_n will be different. Now, it is best to simply solve for these coefficients numerically:

Cell 3.5

```
L = 1;
A[n_] := 2/L ( NIntegrate[ e^(-50 (x/L-1/2)^2) Sin[n Pi x/L],
  {x, 0, L}])
```

In Cell 3.6, we show the string motion resulting from this initial condition. (In order to reduce the computation time we have used the fact that only odd Fourier modes are present in the solution.)

Cell 3.6

```
M = 19; c = 1;
y[x_, t_] = Sum[A[n] Cos[n Pi c t/L] Sin[n Pi x/L],
  {n, 1, M, 2}];
Table[Plot[y[x, t], {x, 0, L}, PlotRange→{{0, 1}, {-1, 1}},
   PlotLabel→"t="<>ToString[t], AxesLabel→{"x", "y"}],
   {t, 0, 1.95, .05}];
```

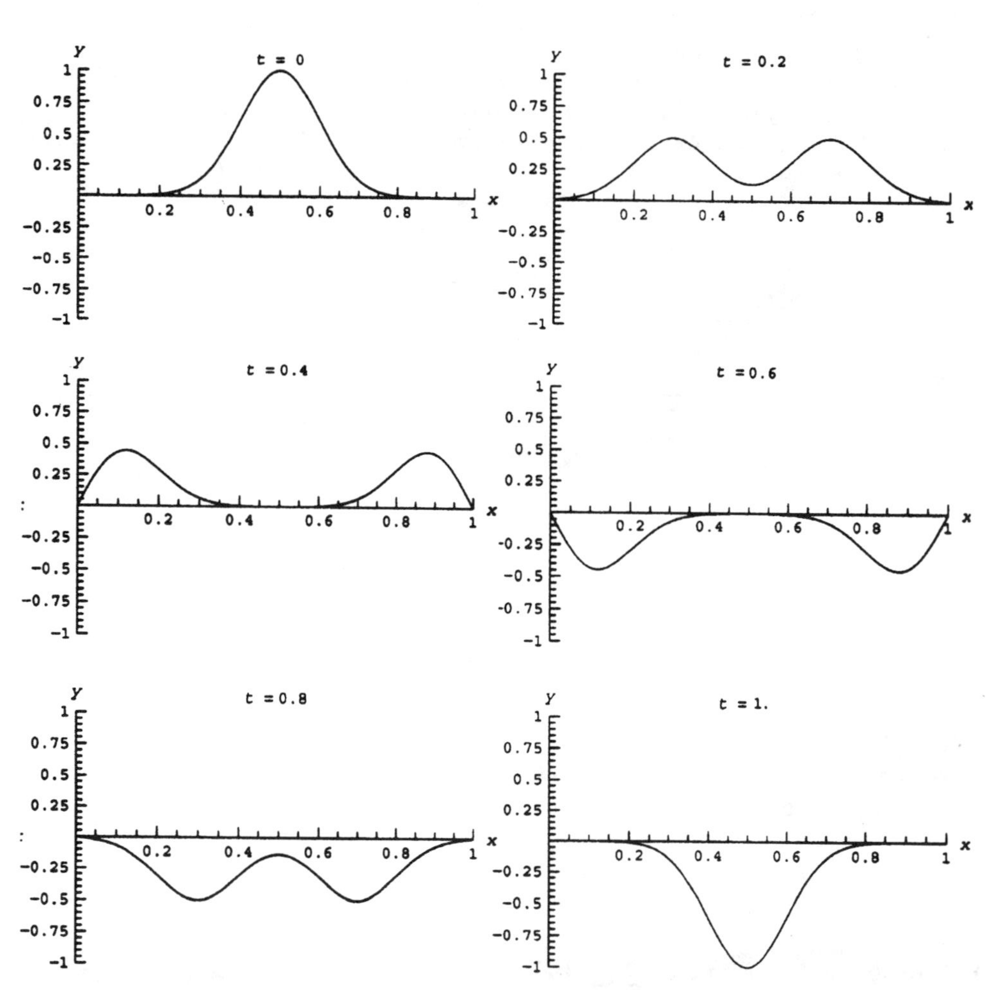

One can see that this initial condition breaks into two equal pulses, propagating in opposite directions on the string. This is expected from the left–right symmetry of the system: there is no reason why propagation of the pulse in one direction should be favored over the other direction. Also, the pulses do not change their shape until they reach the ends of the string, where they reflect and propagate back toward the center again, with opposite sign at time $t = 1$. This means that each pulse, having covered a distance $L = 1$ in time $t = 1$, has traveled with speed $c = 1$.

In fact, it is easy to see that any function of space and time of the form $f(x - ct)$ or $f(x + ct)$ satisfies the wave equation for a uniform string, Eq. (3.1.10). This is because, for such functions, the chain rule implies that $\partial f/\partial t = \pm c\, \partial f/\partial x$. Therefore, $\partial^2 f/\partial t^2 = c^2\, \partial^2 f/\partial x^2$, so $f(x \pm ct)$ satisfies the wave equation for any function f.

In Chapter 5 we will prove that the general solution to the wave equation for a uniform infinite string can be written as a superposition of two such functions, traveling in opposite directions:

$$y(x,t) = f(x - ct) + g(x + ct). \tag{3.1.26}$$

This form of the solution is called *d'Alembert's solution*, after its discoverer.

Disturbances of the form $f(x \pm ct)$ travel with speed c without changing shape. For example, if $f(x)$ has a maximum at $x = 0$, then this maximum point moves in time according to $x = \pm ct$. Every other point in the solution moves at the same speed, so the pulse does not change shape as it propagates. This is what we observed in Cell 3.6, up to the time that the pulses encountered the string ends.

Static Sources and Inhomogeneous Boundary Conditions

Dirichlet and von Neumann Boundary Conditions and Static Transverse Forces. In the previous wave equation examples, the boundary conditions $y(0,t) = y(L,t) = 0$ were not of the most general type that can be handled using separation of variables. The ends of the string need not be fixed at the same height. The boundary conditions are then

$$y(0,t) = y_1 \quad \text{and} \quad y(L,t) = y_2.$$

Boundary conditions of this sort, where the value of the unknown function is specified at the end points, are referred to as *Dirichlet* boundary conditions, or boundary conditions of the first kind. We will now consider the solution of the wave equation for these Dirichlet boundary conditions, *assuming that* the boundary conditions are fixed in time; so that y_1 and y_2 are constants. Time-dependent boundary conditions cannot be handled using the separation-of-variables method discussed here, and will be left to Chapter 4.

However, other types of *static* boundary conditions *can* be treated with separation-of-variables methods. For instance, the derivative $\partial y/\partial x$, rather than x itself, can be specified at the ends of the string. This type of boundary condition is called a *von Neumann* boundary condition. Such boundary conditions do not often occur for problems involving waves on a string, because the string tension is usually created by fixing the string ends to posts. Therefore, in this section we will limit discussion to Dirichlet boundary conditions. On the other hand, von Neumann boundary conditions can occur in other physical applications of the wave equation (see the exercises), and they also occur in applications involving other PDEs (see Sec. 3.1.3).

As a second generalization of the wave equation, we note that for any earth-bound string the force of gravity acts in the vertical direction, causing a horizontal string to sag under its own weight. This effect of gravity has been neglected so far, but can be incorporated into the wave equation as a source term (provided that the sag is small). To allow for gravity or any other transverse external force, it is necessary to refer back to Fig. 3.3. An extra force of magnitude dF now acts on the mass element in the y-direction. (For gravity, $dF = -dm\,g$.) This force must be added to the force acting to accelerate the element, on the right-hand side of Equation (3.1.4), which now reads

$$dm \frac{\partial^2}{\partial t^2} y(x,t) = dx \frac{\partial}{\partial x} T_y(x,t) + dF. \tag{3.1.27}$$

Dividing through by dm, substituting for the string tension via Eq. (3.1.6), and using Eq. (3.1.2), we obtain an inhomogeneous wave equation with a source term:

$$\frac{\partial^2}{\partial t^2} y(x,t) = \frac{1}{\rho(x)} \frac{\partial}{\partial x}\left(T(x) \frac{\partial}{\partial x} y(x,t) \right) + S(x), \tag{3.1.28}$$

where the source term $S(x) = dF/dm$ is the acceleration caused by the external transverse force. We *assume* that this external source is time-independent; time-dependent sources are treated in Chapter 4.

The Equilibrium Solution. As a first step to obtaining the full solution to this problem, we will first consider a time-independent solution of Eq. (3.1.28), subject to the boundary conditions $y(0,t) = y_1$ and $y(L,t) = y_2$. This is the equilibrium solution for the shape of the string $y = y_{\text{eq}}(x)$. This function satisfies the time-independent wave equation:

$$\frac{1}{\rho(x)} \frac{\partial}{\partial x}\left(T(x) \frac{\partial}{\partial x} y_{\text{eq}}(x,t) \right) = S(x), \qquad y_{\text{eq}}(0) = y_1, \quad y_{\text{eq}}(L) = y_2. \tag{3.1.29}$$

The general solution to this ODE can be obtained by direct integration:

$$y_{\text{eq}}(x) = -\int_0^x dx' \left(\frac{1}{T(x')} \int_0^{x'} S(x'')\rho(x'')\,dx'' \right) + C_1 + C_2 \int_0^x dx' \frac{1}{T(x')}. \tag{3.1.30}$$

The integration constants C_1 and C_2 are determined by the boundary conditions:

$$C_1 = y_1, \qquad C_2 = \frac{y_2 - y_1 - \int_0^L dx' \left(\frac{1}{T(x')} \int_0^{x'} S(x'')\rho(x'')\,dx'' \right)}{\int_0^L dx' \frac{1}{T(x')}}. \tag{3.1.31}$$

For instance, let us suppose that we are dealing with a gravitational force, $S(x) = -g$, that the string is uniform so that $\rho(x) = \rho =$ constant, and that $T(x)$ is also approximately uniform, $T(x) \approx T =$ constant. (The latter will be true if the tension in the string is large, so that the sag in the string is small.) Then Eq. (3.1.30) becomes

$$y_{\text{eq}}(x) = -\frac{g\rho}{2T}x(L-x) + \frac{(y_2 - y_1)x}{L} + y_1. \tag{3.1.32}$$

This parabolic sag in the string is the small-amplitude limit of the well-known *catenary* curve for a hanging cable. For the case of zero gravity, the string merely forms a straight line between the end points. Equation (3.1.32) is valid provided that the maximum displacement of the string due to gravity, $|y_{\max}| = g\rho L^2/8T$, is small compared to L. This requires that the tension satisfy the inequality $T \gg g\rho L/8$. For low string tension where this inequality is not satisfied, the sag is large and nonlinear terms in the wave equation must be kept. [A discussion of the catenary curve can be found in nearly any book on engineering mathematics. See, for instance, Zill and Cullen (2000).]

The Deviation from Equilibrium. Having dealt with static source terms and inhomogeneous boundary conditions, we now allow for time dependence in the full solution by writing $y(x,t)$ as a sum of the equilibrium solution, $y_{\text{eq}}(x)$, and a deviation from equilibrium, $\Delta y(x,t)$:

$$y(x,t) = y_{\text{eq}}(x) + \Delta y(x,t). \tag{3.1.33}$$

A PDE for Δy is obtained by substituting Eq. (3.1.33) into Eq. (3.1.28). Since $y_{\text{eq}}(x)$ already satisfies Eq. (3.1.28), Δy satisfies the *homogeneous* wave equation with *homogeneous* boundary conditions,

$$\frac{\partial^2}{\partial t^2}\Delta y(x,t) = \frac{1}{\rho(x)}\frac{\partial}{\partial x}\left(T(x)\frac{\partial}{\partial x}\Delta y(x,t)\right), \tag{3.1.34}$$

$$\Delta y(0,t) = \Delta y(L,t) = 0. \tag{3.1.35}$$

Initial conditions for Δy are obtained by substituting Eq. (3.1.33) into Eq. (3.1.9):

$$\begin{aligned} \Delta y(x,0) &= y_0(x) - y_{\text{eq}}(x), \\ \frac{\partial \Delta y}{\partial t}(x,0) &= v_0(x). \end{aligned} \tag{3.1.36}$$

For the case of a uniform string, we know how to solve for $\Delta y(x,t)$ by using separation of variables and a Fourier series. We will deal with a nonuniform string in Chapter 4.

This analysis shows that the static applied force and the inhomogeneous boundary conditions have no effect on the string normal modes, which are still given by Eqs. (3.1.18) and (3.1.19) for a uniform string. This is because of the linearity of the wave equation. Linearity allows the application of the superposition principle to the solution, so that we can separate the effect of the static force and inhomogeneous boundary conditions from the time-dependent response to an initial condition.

Summary In this subsection we derived the wave equation, and learned how to apply the method of separation of variables to solve the equation, for the case of a *uniform string* with a *static source* $S(x)$ and *time-independent Dirichlet* boundary conditions, $y(0,t)=y_1$, $y(L,t)=y_2$.

First, we determined the equilibrium shape of the string, $y_{eq}(x)$, as the time-independent solution to the PDE. Then, we found that the deviation from equilibrium $\Delta y(x,t)$ satisfies the *homogeneous* wave equation (i.e., no sources) with *homogeneous* boundary conditions $y(0,t)=y(L,t)=0$.

Homogeneity of the equation and the boundary conditions allowed us to apply separation of variables to the problem. The solution for Δy could be written as a Fourier sine series in x with time-dependent Fourier coefficients. Each Fourier mode was found to be a normal mode of oscillation—an eigenmode that matched the homogeneous boundary conditions and that satisfied an associated eigenvalue problem.

Other linear PDEs with time-independent sources and boundary conditions can also be solved using the method of separation of variables. In the next section we consider one such PDE: the heat equation.

3.1.3 Derivation of the Heat Equation

Heat Flux In a material for which the temperature T (measured in kelvins) is a function of position, the second law of thermodynamics requires that heat will flow from hot to cold regions so as to equalize the temperature. The flow of heat energy is described by an *energy flux* $\mathbf{\Gamma} = (\Gamma_x, \Gamma_y, \Gamma_z)$, with units of watts per square meter. This flux is a vector, with the direction giving the direction of the heat flow. In a time Δt, the amount of heat energy ΔE flowing through a surface of area A, oriented transverse to the direction of $\mathbf{\Gamma}$, is $\Delta E = A\Gamma\Delta t$.

Consider a piece of material in the form of a slab of thickness L and of cross-sectional area A, with a temperature $T(x)$ that varies only in the direction across the slab, the x-direction (see Fig. 3.4). It is an experimental fact that this temperature gradient results in a heat flux in the x-direction that is proportional to the temperature gradient,

$$\Gamma_x = -\kappa \frac{\partial T}{\partial x}, \tag{3.1.37}$$

where κ, the constant of proportionality, is the *thermal conductivity* of the material. This constant is an intrinsic property of the material in question. For example, pure water at room temperature and atmospheric pressure has $\kappa = 0.59$ W/(m K), but copper conducts heat much more rapidly, having a thermal conductivity of $\kappa = 400$ W/(m K). A temperature gradient of 1 K/m in water leads to an energy flux of 0.59 W/m^2 in the direction opposite to the temperature gradient, but in copper the flux is 400 W/m^2. Since the energy flows down the gradient, it acts to equalize the temperature.

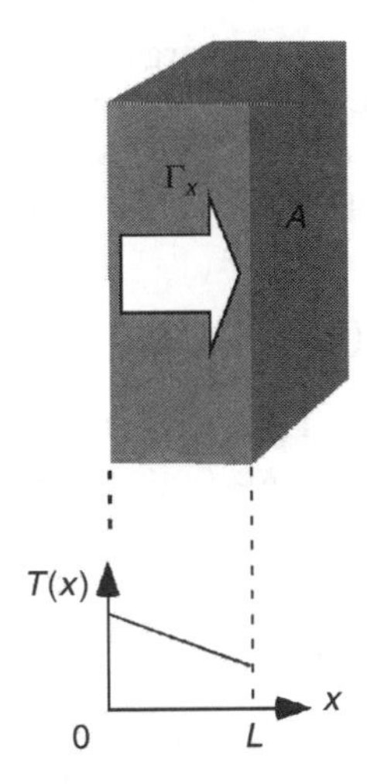

Fig. 3.4 Heat flux Γ_x in a slab of material of area A and thickness L, caused by a temperature $T(x)$ that varies in the x-direction.

Note that although thermal conductivity varies from one material to the next, it must be nonnegative; otherwise heat energy would flow up the temperature gradient from cold to hot regions, in contradiction to the second law of thermodynamics.

Energy Conservation When $T = T(x)$, Eq. (3.1.37) implies that the energy flux Γ_x in the x-direction is generally also a function of position. This, in turn, implies that the energy content of the material changes with time as energy builds up in some locations and is lost to other locations. We will now examine this process in detail.

Consider the volume element $\Delta V = A\,\Delta x$ shown in Fig. 3.5. This element, located at position x, has a time-varying thermal energy content $\epsilon(x,t)\,\Delta V$, where ϵ is the *energy density* of the material (energy per unit volume). The heat flux leaving this element from the right side has magnitude $\Gamma_x(x + \Delta x, t)$, but the flux entering the element on the left side has magnitude $\Gamma_x(x,t)$. The difference in these fluxes results in energy gained or lost to the element, at a rate $A[\Gamma_x(x,t) - \Gamma_x(x + \Delta x, t)]$. In addition, extra sources of heat such as chemical or nuclear

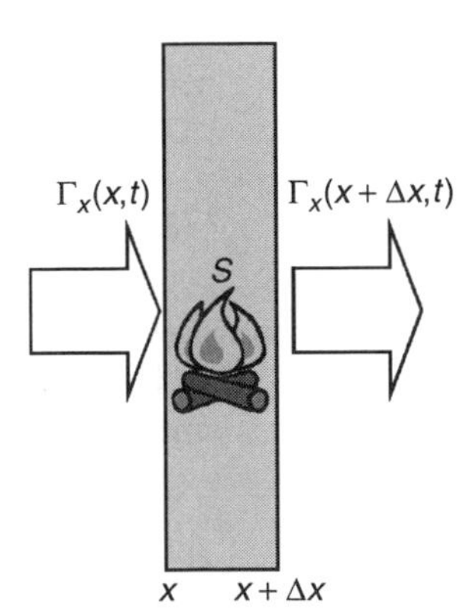

Fig. 3.5 Energy change of an element of material of width Δx due to a source S and due to heat flux Γ_x.

reactions can be occurring within the slab, adding or removing heat energy from the volume element at the rate $\Delta V S(x,t)$, where $S(x,t)$ represents a given source of heat energy per unit volume, with units of watts per cubic meter (i.e., it is the *power density* of the source). In a time dt, the total energy change of the element, $d(\epsilon \Delta V)$, is the sum of these two terms multiplied by dt:

$$d(\epsilon \Delta V) = dt\, A[\Gamma_x(x,t) - \Gamma_x(x+\Delta x,t)] + dt\, \Delta V S(x,t).$$

Taylor expansion of the heat flux implies that $dt\, A[\Gamma_x(x,t) - \Gamma_x(x+\Delta x,t)] = -dt\, A\, \Delta x\, \partial \Gamma_x/\partial x = -dt\, \Delta V \partial \Gamma_x/\partial x$. Dividing by the differentials, we obtain the *continuity equation* for the energy density,

$$\frac{\partial \epsilon}{\partial t} = -\frac{\partial \Gamma_x}{\partial x} + S(x,t). \tag{3.1.38}$$

The Heat and Diffusion Equations. Dirichlet, von Neumann, and Mixed Boundary Conditions When combined with Eq. (3.1.37), Eq. (3.1.38) yields the following partial differential equation:

$$\frac{\partial \epsilon}{\partial t} = \frac{\partial}{\partial x}\left(x \frac{\partial T}{\partial x}\right) + S(x,t). \tag{3.1.39}$$

This equation can be used to obtain a temperature evolution equation, because we can connect the energy density to the temperature via the laws of thermodynamics. A change in the thermal energy density of the material causes a change in the temperature according to the relation

$$d\epsilon = C\, dT, \tag{3.1.40}$$

where C is the *specific heat* of the material. This constant, with units of joules per cubic meter per kelvin, is another intrinsic property of the material. Like the thermal conductivity, the specific heat must be nonnegative. For water at room temperature and at constant atmospheric pressure, the specific heat is $C = 4.2 \times 10^6$ J/(m^3 K), meaning that 4.2 MJ of heat energy must be added to a cubic meter of water in order to raise its temperature by 1 K. [A typical hot tub contains a few cubic meters of water, so one can see that tens of megajoules of energy are required to heat it by several degrees. Fortunately, 1 MJ of energy costs only a few pennies (as of 2002).] The specific heat of copper is not very different from that of water, $C = 3.5 \times 10^6$ J/(m^3 K).

When we combine Eqs. (3.1.39) and (3.1.40), we arrive at the following partial differential equation for the temperature $T(x,t)$:

$$C(x)\frac{\partial T}{\partial t} = \frac{\partial}{\partial x}\left(\kappa(x)\frac{\partial T}{\partial x}\right) + S(x,t). \tag{3.1.41}$$

Equation (3.1.41) is the heat equation in one spatial dimension. In writing it, we have allowed for spatial variation in the material properties κ and C. This can occur in layered materials, for example. The equation simplifies when κ and C are independent of position:

$$\frac{\partial T}{\partial t} = \chi \frac{\partial^2 T}{\partial x^2} + \frac{S(x,t)}{C}, \tag{3.1.42}$$

where χ is the *thermal diffusivity* of the material, defined as

$$\chi = \kappa / C. \tag{3.1.43}$$

The thermal diffusivity has units of m^2/s. For water, $\chi = 1.4 \times 10^{-7}$ m^2/s, and for copper, $\chi = 1.1 \times 10^{-4}$ m^2/s. Equation (3.1.42) is sometimes referred to as the *diffusion equation.*

The heat equation is a first-order PDE in time and so requires a single initial condition, specifying the initial temperature in the slab as a function of position:

$$T(x,0) = T_0(x). \tag{3.1.44}$$

The equation is second-order in space, and so requires two boundary conditions to specify the solution. The boundary conditions one employs depend on the circumstances. For example, in some experiments one might fix the temperature of the slab faces to be given functions of time, by putting the faces in good thermal contact with heat reservoirs at given temperatures $T_1(t)$ and $T_2(t)$:

$$\begin{aligned} T(0,t) &= T_1(t), \\ T(L,t) &= T_2(t). \end{aligned} \tag{3.1.45}$$

These Dirichlet boundary conditions are of the same type as those encountered previously for the wave equation.

On the other hand, one might also insulate the faces of the slab, so that no heat flux can enter or leave the faces: $\Gamma_x(0) = \Gamma_x(L) = 0$. More generally, one might specify the heat flux entering or leaving the faces to be some function of time. Then according to Eq. (3.1.37), the temperature gradient at the faces is specified:

$$\begin{aligned} \frac{\partial T}{\partial x}(0,t) &= -\frac{\Gamma_{x1}(t)}{\kappa}, \\ \frac{\partial T}{\partial x}(L,t) &= -\frac{\Gamma_{x2}(t)}{\kappa}, \end{aligned} \tag{3.1.46}$$

where $\Gamma_{x1}(t)$ and $\Gamma_{x2}(t)$ are given functions of time, equaling zero if the faces are insulated. Boundary conditions where the gradient of the unknown function is specified are called *von Neumann* boundary conditions, or boundary conditions of the second kind.

There can also be circumstances where the flux of heat lost or gained from the slab faces is proportional to the temperature of the faces: for example, the hotter the face, the faster the heat loss. This leads to *mixed* boundary conditions at the faces:

$$\begin{aligned} \kappa\frac{\partial T}{\partial x}(0,t) &= a[T(0,t) - T_1(t)], \\ \kappa\frac{\partial T}{\partial x}(L,t) &= -b[T(L,t) - T_2(t)], \end{aligned} \tag{3.1.47}$$

where a and b are given (nonnegative) constants. The functions $T_1(t)$ and $T_2(t)$ are

Table 3.1. Possible Boundary Conditions for the Heat Equation[a]

Dirichlet	$T(0,t) = T_1(t)$
von Neumann	$\kappa \frac{\partial T}{\partial x}(0,t) = -\Gamma_{x1}(t)$
Mixed	$\kappa \frac{\partial T}{\partial x}(0,t) = a[T(0,t) - T_1(t)]$

[a]At one end, $x = 0$.

the temperatures of the surroundings: only if the face is hotter than the surroundings is there a flux of heat out of the face. If a and b are very large, representing good thermal contact with the surroundings, then the temperature of the faces is pinned to that of the surroundings and the boundary conditions are of Dirichlet form, Eq. (3.1.45). On the other hand, if a and b are very small, then the faces are insulating and obey homogeneous von Neumann conditions, $\partial T/\partial x = 0$ at $x = 0$ and $x = L$.

Finally, there can be situations where one face has a different type of boundary condition than the other; for instance, one side of a slab might be insulated, while on the other side the temperature might be fixed.

These possibilities are summarized in Table 3.1.

3.1.4 Solution of the Heat Equation Using Separation of Variables

Introduction The heat equation for a uniform medium, Eq. (3.1.41), can often be solved using the method of separation of variables. For this method to work, we require that an equilibrium solution $T_{eq}(x)$ for the temperature exist. A necessary (but not sufficient) requirement for equilibrium is time-independent boundary conditions and a time-independent source function.

If an equilibrium solution can be found, then this solution can be subtracted out, and the deviation from equilibrium, $\Delta T(x,t)$, can then be analyzed using separation of variables, just as for the wave equation. However, if an equilibrium solution does not exist, then other more general solution methods must be applied. Such methods will be discussed in Sec. 4.2.

One might have expected that time-independent boundary conditions and a time-independent source would necessarily imply that an equilibrium solution for the temperature exists. However, this is not always the case, as we will now show.

Static Boundary Conditions and a Static Source

The Equilibrium Solution. We consider time-independent boundary conditions, of either the Dirichlet, the von Neumann, or the mixed form, and a static source function, $S = S(x)$. We will look for an equilibrium solution $T_{eq}(x)$ that satisfies these boundary conditions, as well as the time-independent heat equation,

$$0 = \frac{\partial}{\partial x}\left(\kappa(x)\frac{\partial T_{eq}}{\partial x}\right) + S(x,t). \tag{3.1.48}$$

This equation has a general solution that can be found by direct integration:

$$T_{\text{eq}}(x) = -\int_0^x dx' \left(\frac{1}{\kappa(x')} \int_0^{x'} S(x'')\, dx'' \right) + C_1 + C_2 \int_0^x dx' \frac{1}{\kappa(x')}. \tag{3.1.49}$$

The constants C_1 and C_2 are chosen to match the boundary conditions. For example, for Dirichlet boundary conditions at each end, $T(0,t) = T_1$ and $T(L,t) = T_2$, the solution mirrors the equilibrium solution to the wave equation:

$$C_1 = T_1, \qquad C_2 = \frac{T_2 - T_1 + \int_0^L dx' \dfrac{1}{\kappa(x')} \int_0^{x'} S(x'')\, dx''}{\int_0^L dx' \dfrac{1}{\kappa(x')}}.$$

Similarly, one can also find a unique equilibrium solution for mixed boundary conditions at each end [Eq. (3.1.47) with T_1 and T_2 constants], although we will not write the solution here. In fact, one can always find a unique equilibrium solution for every possible combination of static boundary conditions at each end, *except one*: von Neumann conditions at each end.

Constraint Condition on the Existence of an Equilibrium for von Neumann Boundary Conditions. If we specify von Neumann boundary conditions, then according to Eq. (3.1.49) the derivative of the equilibrium temperature at each end must satisfy the following two equations:

$$\begin{aligned} \kappa(0) \frac{\partial T_{\text{eq}}}{\partial x}(0,t) &= -\Gamma_{x1} = C_2, \\ \kappa(L) \frac{\partial T_{\text{eq}}}{\partial x}(L,t) &= -\Gamma_{x2} = C_2 - \int_0^L S(x'')\, dx''. \end{aligned} \tag{3.1.50}$$

However, these are two equations in only one unknown, C_2. [The constant C_1 disappeared when the derivative of Eq. (3.1.49) was taken.] Therefore, a solution cannot necessarily be found to these equations. This should not be completely surprising. After all, Eq. (3.1.48) is being solved as a boundary-value problem, and we know that a solution to boundary-value problems need not exist.

Subtracting Eqs. (3.1.50) from one another implies that the equations have a solution for C_2 only if the external heat fluxes Γ_{x1} and Γ_{x2} are related to the heat source $S(x)$ through

$$\Gamma_{x1} - \Gamma_{x2} + \int_0^L S(x'')\, dx'' = 0. \tag{3.1.51}$$

If this equation is not satisfied, there is no equilibrium solution. The equation follows from the requirement that, in equilibrium, the overall energy content of the material must not change with time. In a slab with cross-sectional area A, the total energy content is $E = A\int_0^L \epsilon\, dx$, where ϵ is the energy density. Setting the time derivative of this expression equal to zero yields

$$\frac{dE}{dt} = 0 = A\int_0^L \frac{\partial \epsilon}{\partial t}(x,t)\,dx = A\int_0^L \left(-\frac{\partial \Gamma_x}{\partial x} + S(x,t)\right) dx,$$

where the second equality follows from energy conservation, Eq. (3.1.38). Performing the integral over the heat flux using the fundamental theorem of calculus then leads to Eq. (3.1.51).

Equation (3.1.51) is easy to understand intuitively: Take the case of an insulated slab, with $\Gamma_{x1} = \Gamma_{x2} = 0$. Then for a temperature equilibrium to exist, there can be no net heat energy injected into the slab by the source: $\int_0^L S(x'')\,dx'' = 0$; otherwise the temperature must rise or fall as the overall energy content in the slab varies with time. Similarly, if $\Gamma_{x1} - \Gamma_{x2} < 0$, then we are removing heat through the faces of the slab; and in equilibrium this energy must be replaced by the heat source $S(x)$.

For Dirichlet or mixed boundary conditions, where the external heat fluxes Γ_{x1} and Γ_{x2} are not directly specified, the fluxes through the slab faces are free to vary until the equilibrium condition of Eq. (3.1.51) is achieved. However, for von Neumann boundary conditions these fluxes are specified directly, and if they do not satisfy Eq. (3.1.51), the energy content of the slab will increase with time [if the left-hand side of Eq. (3.1.51) is greater than zero] or decrease with time [if it is less than zero], and no equilibrium will exist. Conditions such as this cannot be treated using separation of variables. A solution to this problem can be found in Chapter 4.

Separation of Variables for the Deviation from Equilibrium. Let's assume that the static boundary conditions and the source are such that Eq. (3.1.51) is satisfied, so that an equilibrium solution to the heat equation exists. We can then determine the evolution of $T(x,t)$ from a general initial condition, $T(x,0) = T_0(x)$, by following the prescription laid out for the wave equation.

We first subtract out the equilibrium and follow the deviation from equilibrium, $\Delta T(x,t)$, where

$$T(x,t) = T_{\text{eq}}(x) + \Delta T(x,t). \tag{3.1.52}$$

Substitution of this expression into the heat equation (3.1.41) and application of Eq. (3.1.48) implies that $\Delta T(x,t)$ satisfies the homogeneous heat equation

$$C(x)\frac{\partial \Delta T}{\partial t} = \frac{\partial}{\partial x}\left(\kappa(x)\frac{\partial \Delta T}{\partial x}\right) \tag{3.1.53}$$

with *homogeneous boundary conditions*, and initial condition

$$\Delta T(x,0) = T_0(x) - T_{\text{eq}}(x). \tag{3.1.54}$$

The boundary conditions are of the same type as the original equation, but are homogeneous. Recall that homogeneous boundary conditions are such that a trivial solution $\Delta T = 0$ exists. For instance, if the original equation had von

Neumann conditions at one end and Dirichlet conditions at the other [$T(0,t) = T_1$, $\partial T/\partial x|_{x=L} = -\Gamma_{x2}/\kappa$], then ΔT would satisfy homogeneous conditions of the same type, $\Delta T(0,t) = 0$, $\partial \Delta T/\partial x|_{x=L} = 0$.

The boundary conditions for the temperature deviation ΔT are of the same type as the original boundary conditions for T, but are homogenous. (This was the point of subtracting out the equilibrium solution.) Separation of variables only works if we can write down a PDE for ΔT that is accompanied by homogeneous boundary conditions.

The solution of Eq. (3.1.53) for ΔT can again be obtained using the method of separation of variables. Here for simplicity we will only consider the case where κ and C are constants, so that Eq. (3.1.53) becomes the diffusion equation,

$$\frac{\partial \Delta T}{\partial t} = \chi \frac{\partial^2 \Delta T}{\partial x^2}. \tag{3.1.55}$$

We look for a solution of the form $\Delta T(x,t) = f(t)\psi(x)$. Substituting this expression into Eq. (3.1.55) and dividing by $f(t)\psi(x)$ yields

$$\frac{1}{f(t)} \frac{\partial f}{\partial t} = \frac{\chi}{\psi(x)} \frac{\partial^2 \psi}{\partial x^2}, \tag{3.1.56}$$

which can be separated into two ODEs. The left-hand side is independent of x, and the right-hand side is independent of t. Therefore, Eq. (3.1.56) can only be satisfied if each side equals a constant, λ:

$$\frac{1}{f(t)} \frac{\partial f}{\partial t} = \lambda, \tag{3.1.57}$$

$$\frac{\chi}{\psi(x)} \frac{\partial^2 \psi}{\partial x^2} = \lambda. \tag{3.1.58}$$

The Eigenvalue Problem. The separation constant λ and the functions $\psi(x)$ are determined by the homogeneous boundary conditions. Let us assume Dirichlet boundary conditions, $\psi(0) = \psi(L) = 0$. With these boundary conditions, Eq. (3.1.58) may be recognized as an eigenvalue problem; in fact, it is the identical eigenvalue problem encountered previously for the wave equation! The solution for the eigenmodes is, as before,

$$\psi(x) = D \sin \frac{n\pi x}{L} \tag{3.1.59}$$

and

$$\lambda = \lambda_n = -\chi (n\pi/L)^2, \qquad n = 1,2,3,\ldots. \tag{3.1.60}$$

In addition, the solution of Eq. (3.1.57),

$$f(t) = A e^{\lambda t},$$

provides the time-dependent amplitude for each eigenmode. By forming a linear superposition of these solutions, we obtain the general solution to the heat equation for the temperature perturbation away from equilibrium:

$$\Delta T(x,t) = \sum_{n=1}^{\infty} A_n e^{\lambda_n t} \sin\frac{n\pi x}{L}. \tag{3.1.61}$$

As in the wave equation solution, we have absorbed the constant D into the Fourier coefficients A_n. These coefficients are found by matching the initial condition, Eq. (3.1.54):

$$\Delta T(x,0) = \sum_{n=1}^{\infty} A_n \sin\frac{n\pi x}{L} = T_0(x) - T_{eq}(x).$$

This is a Fourier sine series, so the A_n's are determined as

$$A_n = \frac{2}{L}\int_0^L \left[T_0(x) - T_{eq}(x)\right] \sin\frac{n\pi x}{L}\, dx. \tag{3.1.62}$$

Equations (3.1.61) and (3.1.62) are the solution for the deviation from equilibrium for the case of Dirichlet boundary conditions in a uniform slab. The solution is again made up of a sum of eigenmodes with the same spatial form as those for the wave equation, shown in Cell 3.1. But now the amplitudes of the modes decay with time with rate $|\lambda_n|$, rather than oscillating. The result is that $\Delta T \to 0$. Thus, the full solution for T, Eq. (3.1.48), approaches the equilibrium solution $T_{eq}(x)$ in the long-time limit. The time evolution of several of these eigenmodes is displayed in Cell 3.7, in the electronic version of the textbook. In the hardcopy, only the commands that create the animation are given.

Cell 3.7

```
L = 1; χ = 1;
λ[n_] = -χ (n Pi/L)^2;
p[n_, t_] :=
  Plot[Exp[λ[n] t] Sin[n Pi x], {x, 0, L},
   DisplayFunction→Identity, PlotRange→{-1, 1},
   PlotLabel→"n = " <>ToString[n]];
Table[Show[GraphicArray[Table [{{p[1, t], p[2, t]},
  {p[3, t], p[4, t]}}]],
  DisplayFunction→$DisplayFunction], {t, 0, 0.25, .0125}];
```

All of the modes approach zero amplitude as time progresses, because the boundary conditions on the eigenmodes dictate that the temperature equals zero at the slab faces. Thus, in equilibrium the temperature deviation ΔT is zero throughout the slab. The higher-order modes equilibrate more rapidly, because they have larger gradients and therefore larger heat fluxes according to Eq. (3.1.37).

Example In this example, we will assume a point heat source of the form $S(x) = \delta(x - L/2)$. We consider a slab of material of unit width, $L = 1$, with thermal diffusivity $\chi = 1$, and $\kappa = 1$ as well. Initially, the slab has a temperature distribution $T(x,0) = T_0(x) = x^3$, and the boundary conditions are $T(0,t) = 0, T(1,t) = 1$.

Then the equilibrium temperature distribution is given by Eq. (3.1.49) with C_1 and C_2 chosen to match the boundary conditions, $T_{eq}(0) = 0$ and $T_{eq}(L) = 1$. After performing the required integrals, we obtain $C_1 = 0, C_2 = \frac{3}{2}$, and

$$T_{eq}(x) = \frac{3x}{2} - \left(x - \frac{1}{2}\right) h\left(x - \frac{1}{2}\right), \tag{3.1.63}$$

where h is a Heaviside step function. This equilibrium temperature distribution is displayed in Cell 3.8.

Cell 3.8

```
Teq[x_] = 3 x/2 - (x - 1/2) UnitStep[x - 1/2];
Plot[Teq[x], {x, 0, 1}, AxesLabel→{"x", "Teq[x]"}];
```

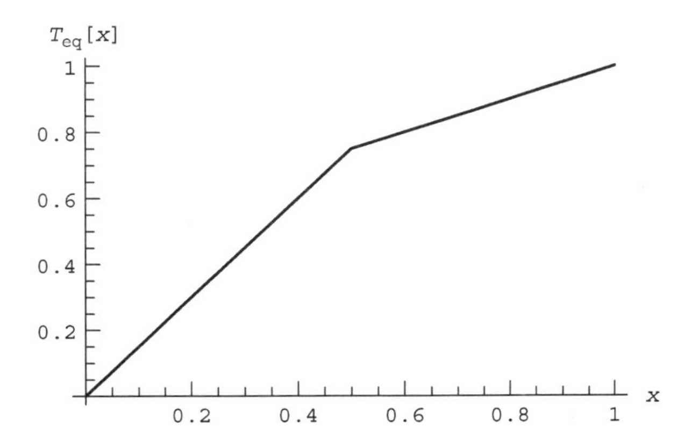

The solution for the deviation from equilibrium, $\Delta T(x,t)$, is given by Eq. (3.1.61). The constants A_n are determined by the initial condition, $T_0(x) = x^3$, via Eqs. (3.1.62) and (3.1.63). The required integrals are performed using *Mathematica* and the behavior of $T(x,t)$ is plotted in Cell 3.9 keeping 10 Fourier modes in the solution. The temperature rapidly approaches the equilibrium temperature $T_{eq}(x)$ as the point heat source raises the internal temperature of the slab.

Cell 3.9

```
(*parameters*)
χ = L = 1; M = 20;
 (* define the initial condition *)
T0[x_] = x³;
(*determine the constants Aₙ*)
A[n_] := 2/L Integrate[Sin[n π x/L] (T0[x] - Teq[x]),
  {x, 0, L}];
(*Fourier series for ΔT *)
ΔT[x_, t_] = Sum[A[n] Exp[- (n Pi/L)^2 χ t] Sin[n Pi x/L],
  {n, 1, M}];
```

```
(*The full solution*)
T[x_, t_] = Teq[x] + ΔT[x, t];
(*Plot the result*)
Table[Plot[T[x, t], {x, 0, L}, PlotRange→{{0, L}, {0, 1}},
    PlotLabel →"t = "<>ToString[t], AxesLabel→{"x", "T"},
    PlotStyle→Thickness[0.01]], {t, 0, 0.4, 0.4/20}];
```

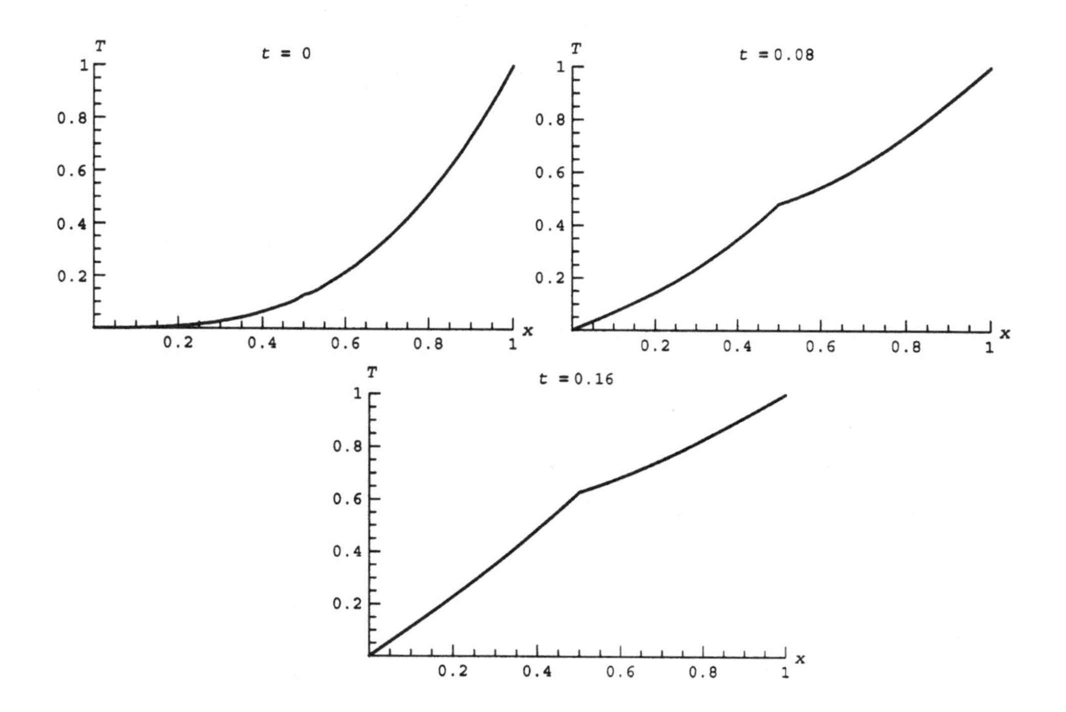

[Note the text within the brackets (* *), which is ignored by *Mathematica*. Comments such as these can be useful for documenting code.]

Observe that the rate at which the equilibrium is approached is mainly determined by the lowest eigenmode. This is because the time dependence of the eigenmode amplitudes is given by the factor $e^{-\chi(n\pi/L)^2 t}$. This factor decays rapidly for the modes with $n \gg 1$, so at large times the solution is approximately determined by the lowest mode:

$$T(x,t) \simeq T_{\text{eq}}(x) + A_1 e^{-\chi(\pi/L)^2 t} \sin\frac{\pi x}{L}. \tag{3.1.64}$$

This equation implies that, for this problem, the maximum temperature deviation from equilibrium occurs at the center of the slab ($x = L/2$), and has the time dependence $A_1 e^{-\chi(\pi/L)^2 t}$, where A_1 is determined from the initial condition by Eq. (3.1.62). Thus, at long times, the rate at which the slab temperature approaches equilibrium is $\chi(\pi/L)^2$. The larger the conductivity, the faster equilibrium is achieved. But the thicker the slab, the longer it takes for the heat to diffuse from the interior to the faces.

Homogeneous von Neumann Boundary Conditions

Separation-of-Variables Solution. Let us now consider the case of a uniform slab insulated on both sides, and with no source. These conditions clearly satisfy Eq. (3.1.51), so an equilibrium exists—in this case the trivial equilibrium $T_{\text{eq}} = \text{constant}$. The temperature within the slab now evolves according to

$$\frac{\partial T}{\partial t} = \chi \frac{\partial^2 T}{\partial x^2}, \tag{3.1.65}$$

with homogeneous von Neumann boundary conditions

$$\frac{\partial T}{\partial x}(0,t) = \frac{\partial T}{\partial x}(L,t) = 0 \tag{3.1.66}$$

and initial condition

$$T(x,0) = T_0(x). \tag{3.1.67}$$

The solution of Eq. (3.1.65) can again be obtained using the method of separation of variables. We now have no need to take out an equilibrium solution, since there is no source and the boundary conditions are already homogeneous. We look for a solution of the form $T(x,t) = f(t)\psi(x)$. Substituting this expression into Eq. (3.1.65), and dividing by $f(t)\psi(x)$, we have

$$\frac{1}{f(t)} \frac{\partial f}{\partial t} = \frac{\chi}{\psi(x)} \frac{\partial^2 \psi}{\partial x^2},$$

which can be separated into two ODEs, Eqs. (3.1.57) and (3.1.58), just as before. We repeat these equations below:

$$\frac{1}{f(t)} \frac{\partial f}{\partial t} = \lambda, \tag{3.1.68}$$

$$\frac{\chi}{\psi(x)} \frac{\partial^2 \psi}{\partial x^2} = \lambda. \tag{3.1.69}$$

The Eigenvalue Problem. The separation constant λ and the functions $\psi(x)$ are determined by the homogeneous von Neumann boundary conditions that $\psi'(0) = \psi'(L) = 0$. These boundary conditions yield another eigenvalue problem: for all but a special set of λ-values, the only solution to Eq. (3.1.69) that satisfies these boundary conditions is the trivial solution $\psi = 0$.

To find the nontrivial solutions, we apply the boundary conditions to the general solution of Eq. (3.1.69), $\psi(x) = C \cos kx + D \sin kx$, with $k = \sqrt{-\lambda/\chi}$. To match the condition that $\psi'(0) = 0$, we require that $D = 0$, and to match the condition that $\psi'(L) = 0$, we find that either $C = 0$ (the trivial solution) or $k \sin kx = 0$. This equation can be satisfied with the choices $k = n\pi/L$, $n = 0, 1, 2, \ldots$, so we find that

$$\psi(x) = C \cos \frac{n\pi x}{L} \tag{3.1.70}$$

and

$$\lambda = \lambda_n = -\chi(n\pi/L)^2, \qquad n = 0, 1, 2, 3, \ldots. \tag{3.1.71}$$

Forming a linear superposition of these solutions, we now obtain

$$T(x,t) = \sum_{n=0}^{\infty} A_n e^{\lambda_n t} \cos\frac{n\pi x}{L}, \tag{3.1.72}$$

where, as before, we absorb the constant C into the Fourier coefficients A_n. Note that we now keep the $n = 0$ term in the sum, since $\cos(0\pi x) = 1$ is a perfectly good eigenmode.

Just as before, the coefficients A_n are found by matching the initial condition, Eq. (3.1.67):

$$T(x,0) = \sum_{n=0}^{\infty} A_n \cos\frac{n\pi x}{L} = T_0(x).$$

This is a Fourier *cosine* series, so the A_n's are determined as

$$\begin{aligned} A_n &= \frac{2}{L}\int_0^L T_0(x)\cos\frac{n\pi x}{L}\,dx, \qquad n > 0, \\ A_0 &= \frac{1}{L}\int_0^L T_0(x)\,dx. \end{aligned} \tag{3.1.73}$$

The solution is again made up of a sum of eigenmodes. Although the eigenmodes differ from the previous case, they still form a set of orthogonal Fourier modes, which can be used to match to the initial conditions. Note that the $n = 0$ eigenmode has a zero eigenvalue, $\lambda_0 = 0$, so this mode does not decay with time. However, all of the higher-order modes do decay away. This has a simple physical interpretation: over time the initial temperature distribution simply becomes uniform within the slab, because the temperature equilibrates but heat cannot be conducted to the surroundings, due to the insulating boundary conditions.

Example Let us assume that an insulated slab of unit width has an initial temperature equal $T_0(x) = x^2/16 + x^3 - 65x^4/32 + x^5$. This somewhat complicated initial condition is displayed in the plot in Cell 3.13. It is chosen so as to have a peak in the center, and zero derivatives at each end point.

The Fourier coefficients A_n are determined according to Eqs. (3.1.73). The integrals are performed below, using *Mathematica*:

Cell 3.10

```
L = 1;

         x^2         65x^4
T0[x_] = --- + x^3 - ----- + x^5;
         16           32
A[n_] = (2/L) Integrate[T0[x] Cos[n Pi x], {x, 0, L}];
A[0] = (1/L) Integrate[T0[x], {x, 0, L}];
```

The forms for A_0 and A_n are

Cell 3.11

```
A[0]
```

$$\frac{1}{32}$$

Cell 3.12

```
Simplify[A[n], n ∈ Integers]
```

$$-\frac{3\ (-160\ (-1+(-1)^{n})+(-8+23\ (-1)^{n})\ n^{2}\pi^{2})}{2\ n^{6}\ \pi^{6}}$$

This solution is shown in Cell 3.13, keeping $M = 20$ terms in the sum over eigenmodes, and again taking $\chi = 1$.

Cell 3.13

```
(*parameters*)
χ = 1;
M = 20;
(* solution *)
T[x_, t_] = Sum[A[n] e^-(n Pi/L)^2χt Cos[n Pi x], {n, 0, M}];
(*Plot the result*)
Table[Plot[T[x, t], {x, 0, L},
   PlotRange→{{0, L}, {0, .07}},
   PlotLabel→"t = "<>ToString[t],
   AxesLabel→{"x", "T"}], {t, 0, 0.2, 0.2/20}];
```

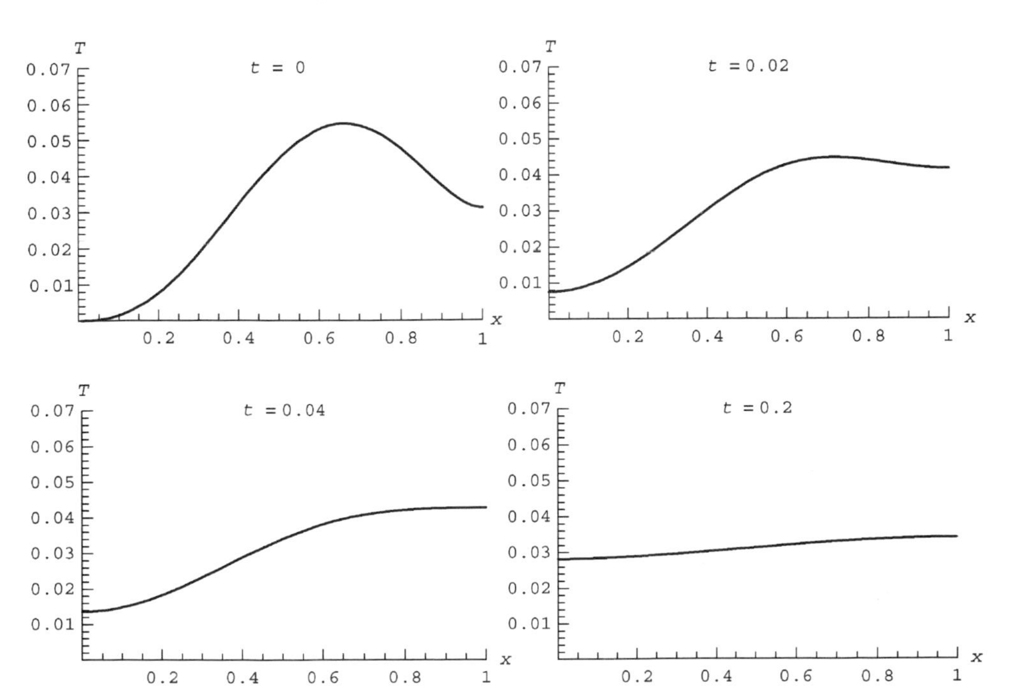

After a rapid period of relaxation, the $n > 0$ modes in the initial condition decay away and the solution settles down to a uniform temperature distribution. Again, in the long-time limit, the higher-order eigenmodes die away and the solution is approximately

$$y(x,t) \simeq A_0 + A_1 e^{-\chi(\pi/L)^2 t} \cos\frac{\pi x}{L}.$$

Now the maximum deviation from equilibrium occurs at the edges, as the temperature gradients within the insulated slab equilibrate.

Homogeneous Mixed Boundary Conditions Let us now turn to the case of no source and homogeneous mixed boundary conditions,

$$\kappa\frac{\partial T}{\partial x}(0,t) = aT(0,t),$$

$$\kappa\frac{\partial T}{\partial x}(L,t) = -bT(L,t).$$

If we again apply separation of variables to the heat equation, writing $T(x,t) = f(t)\psi(x)$, we are again led to Eqs. (3.1.68) and (3.1.69) for the functions f and ψ. Equation (3.1.68) still has the general solution $f(t) = Ae^{\lambda t}$, and Eq. (3.1.69) is still an eigenvalue problem, with the general solution $\psi(x) = A\cos kx + B\sin kx$, where $k = \sqrt{-\lambda/\chi}$. But now the boundary conditions on ψ are rather complicated:

$$\kappa\frac{\partial\psi}{\partial x}(0) = a\psi(0),$$

$$\kappa\frac{\partial\psi}{\partial x}(L) = -b\psi(L).$$

When these boundary conditions are used in the general solution, we obtain two coupled homogeneous equations for A and B:

$$\begin{aligned} \kappa kB &= aA, \\ \kappa k(-A\sin kL + B\cos kL) &= -b(A\cos kL + B\sin kL). \end{aligned} \tag{3.1.74}$$

A nontrivial solution to these coupled equations exists only for values of k that satisfy

$$(ab - \kappa^2 k^2)\sin kL = -(a+b)\kappa k\cos kL. \tag{3.1.75}$$

Again, we have an eigenvalue problem for the wavenumbers k. Unfortunately, this equation cannot be solved analytically for k. However, numerical solutions can be found for specific values of a and b using numerical techniques developed in Chapter 9. For example, if we take $a = b = 2\kappa/L$, then Eq. (3.1.75) can be written as $(4 - s^2)\sin s + 4s\cos s = 0$, where $s \equiv kL$.

The first four solutions for s can be found graphically as shown in Cell 3.14.

Cell 3.14

```
Plot[Sin[s] (-s^2 + 4) + 4 s Cos[s], {s, 0, 10},
  AxesLabel→{"s", ""}];
```

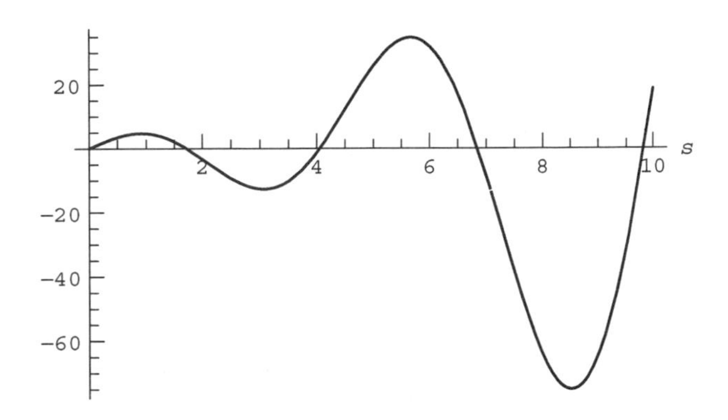

The plot shows that there are solution at $s \simeq 1.7$, 4, 7, and 9.5. The solution at $s = 0$ is trivial, and those at $s < 0$ provide modes that are merely opposite in sign to those with $s > 0$. The corresponding λ-values can then be picked out using the following table of **FindRoot** commands:

Cell 3.15

```
λ = -Table[s/. FindRoot[Sin[s] (-s^2 + 4) + 4 s Cos[s],
    {s, 1.8 + 2.5 n}] [[1]], {n, 0, 3}]^2 χ/L^2
```

$$\left\{-\frac{2.9607\,\chi}{L^2}, -\frac{16.4634\,\chi}{L^2}, -\frac{46.9394\,\chi}{L^2}, -\frac{96.5574\,\chi}{L^2}\right\}$$

In Cell 3.16, we show the form of the corresponding eigenmodes, taking $A = 1$ and with B given in terms of A by Eq. (3.1.74). The time dependence $e^{\lambda_n t}$ of the modes is also displayed in the electronic version of the text.

Cell 3.16

```
L = 1; χ = 1; a = b = 2 κ/L;
k = √(-λ / χ);
A = 1;
B = a A/ (κ k );
ψ[n_, x_] := A Cos[√(-λ[[n]] / χ) x] + B[[n]] Sin[√(-λ[[n]] / χ) x];

p[n_, t_] := Plot[Exp[λ[[n]] t] ψ[n, x], {x, 0, L},
    DisplayFunction→Identity,
    PlotRange→{-2, 2}, PlotLabel→"n = " <>ToString[n]];
Table[Show[GraphicsArray[Table[{{p[1, t], p[2, t]},
    {p[3, t], p[4, t]}}]],
    DisplayFunction→$DisplayFunction], {t, 0, 0.2, .01}];
```

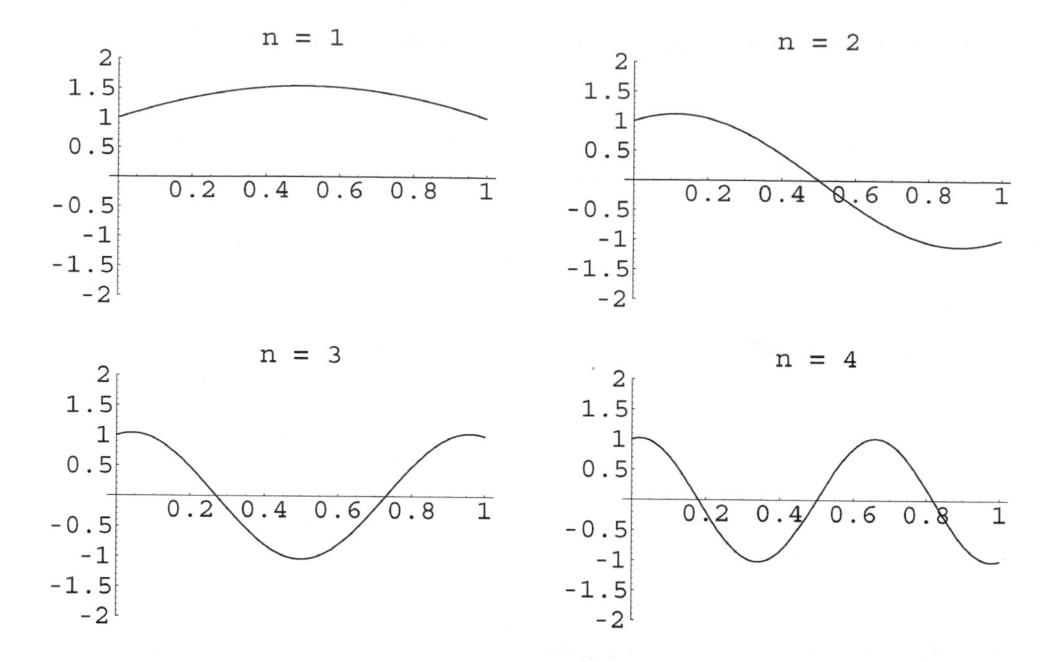

Because of the symmetry of the boundary conditions, the modes are either symmetric or antisymmetric about the center of the slab. These modes are a cross between insulating and fixed-temperature conditions: there is a small heat flux out of the slab faces that depends linearly on the temperature of the faces. In the long-time limit, the temperature throughout the slab is zero in the absence of a heat source.

We can still form a general superposition of these different solutions in order to satisfy the initial conditions:

$$T(x,t) = \sum_{n} A_n \, e^{\lambda_n t} \psi_n(x), \tag{3.1.76}$$

where $\psi_n(x)$ is the nth eigenmode, with corresponding eigenvalue λ_n. However, the eigenmodes are now complicated linear combinations of trigonometric functions, with wavenumbers that are no longer evenly spaced multiples of π/L. This is *not* a trigonometric Fourier series. How do we go about finding the constants A_n in this case?

Surprisingly, the eigenmodes are still orthogonal with respect to one another:

$$\int_0^L \psi_m(x)\,\psi_n(x)\,dx = 0 \qquad \text{if} \quad n \neq m.$$

One can check this numerically:

Cell 3.17

```
MatrixForm[Table[NIntegrate[ψ[n, x] ψ[m, x], {x, 0, L}],
  {n, 1, 4}, {m, 1, 4}]]
```

$$\begin{pmatrix} 1.85103 & -5.78329\times10^{-9} & 1.48091\times10^{-9} & -9.09982\times10^{-10} \\ -5.78329\times10^{-9} & 0.742963 & 4.253\times10^{-10} & 8.84096\times10^{-11} \\ 1.48091\times10^{-9} & 4.253\times10^{-10} & 0.585216 & -4.03937\times10^{-10} \\ -9.09982\times10^{-10} & 8.84096\times10^{-11} & -4.03937\times10^{-10} & 0.541426 \end{pmatrix}$$

The off-diagonal elements in this table of integrals are nearly zero, indicating orthogonality of the eigenmodes within the precision of the numerical integration.

Since the modes are orthogonal, they can still be used to determine the A_n's by matching to the initial conditions. At $t=0$, Eq. (3.1.76) implies

$$T(x,0)=\sum_n A_n\psi_n(x)=T_0(x).$$

Multiplying both sides by $\psi_m(x)$, and integrating over x from 0 to L, we obtain

$$\sum_n A_n\int_0^L \psi_m(x)\psi_n(x)\,dx=\int_0^L \psi_m(x)T_0(x)\,dx.$$

Orthogonality implies that each term in the sum vanishes except for the $n=m$ term, so we find that only the $n=m$ term survives, leaving a single equation for A_m: $A_m\int_0^L\psi_m^2(x)\,dx=\int_0^L\psi_m(x)T_0(x)\,dx$, which yields

$$A_m=\frac{\int_0^L \psi_m(x)T_0(x)\,dx}{\int_0^L \psi_m^2(x)\,dx}. \tag{3.1.77}$$

Equation (3.1.77) provides the coefficients A_m for every value of m. When used in Eq. (3.1.76), it yields the required solution $T(x,t)$ for any given initial temperature $T_0(x)$.

Summary In this subsection we found solutions to the heat equation using the method of separation of variables, the same approach as we employed when solving the wave equation. Just as for the wave equation, the approach only worked if we could find an equilibrium solution to the problem that took care of source terms and inhomogeneous boundary conditions. Then the deviation from equilibrium was expanded as a series consisting of orthogonal eigenmodes with time-dependent amplitudes.

This was all completely analogous to the wave equation solution. However, we allowed for more general boundary conditions such as can often occur in heat equation problems. Along with the Dirichlet conditions familiar from the wave equation, we also considered von Neumann and mixed conditions. These new boundary conditions brought with them several surprises.

First, we found that with static von Neumann boundary conditions, an equilibrium solution for the temperature is not possible unless the temperature gradients at the edge of the system satisfy an energy conservation condition, Eq. (3.1.51).

Second, we found that in the case of mixed boundary conditions the eigenmodes used in our series solution for ΔT were not simple trigonometric Fourier modes. Surprisingly, however, the modes still formed an orthogonal set, which could be used to find a solution just as in a Fourier series expansion. We will discover the reason for this amazing "coincidence" in Chapter 4.

EXERCISES FOR SEC. 3.1

(1) A mass M hangs at the lower end of a vertical string, in equilibrium under the force of gravity g. The string has constant mass density ρ per unit length. The mass is at $z=0$, and the string is attached to the ceiling at $z=L$. find the tension $T(z)$ in the string as a function of height z.

(2) A thin rod of height L is balanced vertically on its end at $z=0$. The rod has a nonuniform cross section. It is cylindrical, but has a radius that varies with height z as $r=z/10$. (That is, the rod is a cone, balanced on its tip.) The mass density of the material making up the cone is $\mathcal{M}=1000\ \mathrm{kg/m^3}$. Find the tension force in the rod vs. z (more aptly called compression force in this instance) due to gravity g.

(3) The following futuristic concept has been proposed for attaining orbit around a planetary body such as the earth: a very massive satellite, placed in geosynchronous orbit above the equator, lowers a thick rope (called a *tether*) all the way down to the ground. The mass of the satellite is assumed here for simplicity to be much larger than that of the tether. Astronauts, equipment, etc., simply ride an elevator up the tether until they are in space. Due to the huge tension forces in the tether, only fantastically strong materials can be used in the design, such as futuristic materials made of carbon nanotubes. Here, we will calculate the tension forces in the tether. [See also the article in the July–August 1997 issue of *American Scientist* on properties and uses of carbon nanotubes. The cover of this issue is reproduced in Fig. 3.6 (it depicts an open-ended tether design). Also, you may want to check out Sir Arthur Clarke's science fiction novel, *The Fountains of Paradise* (1979).]

(a) Assuming that the mass density ρ per unit length of the tether is constant, show that the tension $T(r)$ as a function of distance r from the center of the earth is

$$T(r) = \rho W(r) + T_0,$$

where $W(r) = GM_e/r + \omega^2 r^2/2$ is the potential energy per unit tether mass, including potential energy associated with centrifugal force, M_e is the mass of the earth, T_0 is an integration constant that depends on the load tension at the base of the tether, and $\omega = 2\pi/(24\ \text{hours})$. Evaluate and plot this tension versus r, assuming that $\rho = 50\ \mathrm{kg/m}$, that the tether is carrying no load, and that it is not attached to the earth at its base.

Fig. 3.6 Artist's depiction of a tether made of carbon nanotubes. Artist: D. M. Miller.

What is the maximum value of the tension (in newtons), and where does it occur? (Hint 1: There are two competing forces at play on a mass element dm in the tether: the centrifugal force due to the earth's rotation, $dm\,\omega^2 r\hat{\mathbf{r}}$, and the gravitational attraction of the tether to the earth $-dm\,GM_e\hat{\mathbf{r}}/r^2$. Neglect the attraction of the tether to itself and to the massive satellite. Hint 2: For the massive satellite, recall that for a point mass m the radius R_g of geosynchronous orbit is given by the solution to the force balance equation $GM_e m/R_g^2 = m\,\omega^2 R_g$, where $\omega = 2\pi/(24\text{ hours})$. Here, neglect the effect of the tether mass on the satellite orbit. Hint 3: Assuming that the tether is attached only to the satellite, not the earth, the tension force applied to the tether by the satellite must balance the total integrated centrifugal and gravitational forces on the tether.)

(b) The previous design is not optimized: the tension in the tether varies considerably with altitude, but the force F required to break the tether does not because the tether has uniform cross section. It is better to design a tether where the ratio of the breaking force to the tension is constant: $F/T = S$, where S is the safety factor, taken to be around 2 or 3 in many engineering designs. Now, the breaking force F of the tether is proportional to its cross-sectional area A, according to the equation $F = \tau A$, where τ is the tensile strength of the material (in newtons per square meter). Also, the density per unit length, ρ, is given by $\rho = \mathcal{M}A$, where $\mathscr{M}$ is the mass density (in kilograms per cubic meter). Find the cross-sectional area $A(r)$ such that the safety factor S is constant with altitude. For a boundary condition, assume that there is a loading tension T_l applied at the base of the tether, at $r = R_e$ (R_e being the radius of the earth). Show that

$$A(r) = \frac{ST_l}{\tau} e^{(\mathscr{M}S/\tau)[W(r) - W(R_e)]}.$$

Plot the radius of the tether, $\sqrt{A(r)/\pi}$ (in meters), assuming that it is made out of the strongest and lightest commercially available material (as of the year 2000), Spectra (a high-molecular-weight form of polyethylene), with a mass density of $\mathscr{M} = 970$ kg/m^3 and a tensile strength of $\tau = 3.5 \times 10^9$ N/m^2. Take as the safety factor $S = 2$. Take $T_l = 10{,}000$ N. What is the total mass of this tether?

(c) Redo the calculation and plot of part (b) for a tether made of carbon nanotubes. Take the same parameters as before, but a tensile strength 40 times greater than that of Spectra.

(4) *Waves in shallow water:* Waves on the surface of water with long wavelength compared to the water depth are also described by the wave equation. In this problem you will derive the equation from first principles using the following method, analogous to that for a wave on a string. During the wave motion, a mass of fluid in an element of unit width into the paper, equilibrium height h and length Δx moves a distance η and assumes a new height $h + z(x, t)$ and a length $(1 + \partial\eta/\partial x)\,\Delta x$, but remains of unit width. (See Fig. 3.7.)

(a) Using the incompressibility of the fluid (the volume of the element remains fixed during its motion) show that, to lowest approximation,

$$z = -h\frac{\partial \eta}{\partial x}.$$

(b) Neglecting surface tension, the net force in the x-direction on the element face of height $h + z$ arises from the difference in hydrostatic

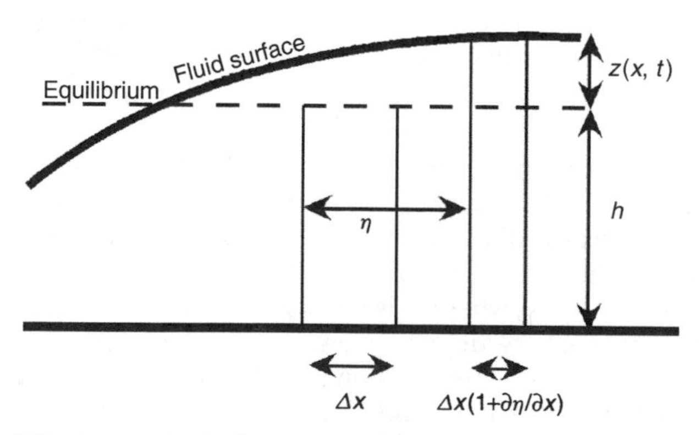

Fig. 3.7 A wave in shallow water, with motion greatly exaggerated.

pressure on each face of the element due to elements on the right and left of different heights. The hydrostatic pressure is a function of distance y from the bottom: $p = \mathcal{M} g[h + z(x,t) - y]$, where $\mathcal{M}$ is the mass density of the water and g is the acceleration of gravity. Show that the net force in the x-direction, ΔF_x, is given by

$$\Delta F_x = -\mathcal{M} gh \frac{\partial z}{\partial x} \Delta x.$$

(c) Using the results from part (a) and (b), and Newton's second law for the mass element, show that shallow water waves satisfy the wave equation $\partial^2 \eta / \partial t^2 = c^2 \, \partial^2 \eta / \partial x^2$, or alternatively,

$$\frac{\partial^2 z}{\partial t^2} = c^2 \frac{\partial^2 z}{\partial x^2}, \tag{3.1.78}$$

where the wave speed c is given by

$$c = \sqrt{gh}\,. \tag{3.1.79}$$

(d) A typical ocean depth is on the order of 4 km. Tidal waves have wavelengths that are considerably larger than this, and are therefore well described by the shallow-water wave equation. Calculate the wave speed (in kilometers per hour) for a tidal wave.

(5) In this problem we consider shallow-water waves, sloshing in the x-direction in a long channel of width L in the x-direction. Boundary conditions on the waves are that the horizontal fluid displacement in the x-direction, η (defined in the previous problem), equals zero at the channel boundaries (at $x = 0$ and $x = L$).

(a) Find the eigenmodes and eigenfrequencies for $\eta(x,t)$.

(b) Plot the wave height z vs. x for the first three sloshing modes.

(6) **(a)** The high E-string of a steel string guitar is about $L = 0.7$ m long from the fret to the post. It has a mass per unit length of $\rho = 5.3 \times 10^{-4}$ kg/m. Find the tension T required to properly tune the string, given that a high E has a frequency $f = 1318.51$ hertz.

(b) Assuming that the A-string is under the same tension as the E-string, is made of the same material, and is about the same length, what is the ratio of the thickness of this string to that of the E-string? An A-tone has a frequency of 440 hertz.

(7) Using an eigenmode expansion, find the solution for the motion of a string that is governed by the following wave equation:

$$\frac{\partial^2}{\partial t^2} y(x,t) = \frac{\partial^2}{\partial x^2} y(x,t),$$

with boundary conditions $y(0,t) = 0 = y(\pi,t)$, and initial conditions

(a) $y(x,0) = x(\pi - x)$, $\frac{\partial}{\partial t}y(x,0) = 0$.

(b) $y(x,0) = 0$, $\frac{\partial}{\partial t}y(x,0) = x^2 \sin x$.

(8) Transverse oscillations on a uniform string under uniform tension T and with mass density ρ need not be only in the y-direction—the oscillations can occur in any direction in a plane transverse to the equilibrium string. Call this plane the x-y plane, and the line of the equilibrium string the z-axis. Then, neglecting gravity effects, a displacement vector $\mathbf{r}(z,t) = (x(z,t), y(z,t))$ of a mass element away from equilibrium satisfies the *vector* wave equation,

$$\frac{\partial^2}{\partial t^2}\mathbf{r}(z,t) = c^2 \frac{\partial^2}{\partial z^2}\mathbf{r}(z,t). \tag{3.1.80}$$

(a) Find the spatial eigenmodes of this wave equation for boundary conditions $\mathbf{r} = \mathbf{0}$ at $z = 0$ and $z = L$. Show that there are two independent plane polarizations for the eigenmodes: an eigenmode involving motion only in the y-direction, and one only in the x-direction.

(b) Using the eigenmodes of part (a), write down a solution $\mathbf{r}(z,t)$ for a *rotating* mode with no nodes except at the ends (think of the motion of a skipping rope). (Hint: the x-motion is $\pi/2$ out of phase with the y motion.) Over one period, make an animation of the motion using `ParametricPlot3D`. This is an example of *circular* polarization.

(9) A rope that is $2L_1 = 12$ meters long is attached to posts at $x = \pm L_2$ that are at the same level, but are only $2L_2 = 10$ meters apart. Find and plot the equilibrium shape of the rope. (Hint: the element of length is

$$ds = \sqrt{dx^2 + dy^2} = dx\sqrt{1 + (dy/dx)^2} \approx dx\left[1 + \tfrac{1}{2}(dy/dx)^2\right],$$

assuming small perturbations away from a straight rope. Use the linear wave equation to determine the equilibrium solution for $y(x)$. Answer to lowest order in $(L_1 - L_2)/L_1$: $y(x) = -(g\rho/2T)(L_2^2 - x^2)$, where $T = \rho g L_2[L_2/6(L_1 - L_2)]^{1/2}$ is the rope tension.)

(10) A heavy string of length L and mass density ρ per unit length is spliced to a light string with equal length L and mass density $\rho/4$ per unit length. The combined string is fixed to posts and placed under tension T. The posts are both at the same height, so the string would be straight and horizontal if there were no gravity. Find and plot the shape of the string in the presence of gravity (assuming small displacement from horizontal). Take $L = 1$ m, $\rho = 0.5$ kg/m, and $T = 25$ N. Plot the shape.

(11) A mass of $m = 5$ kg is attached to the center of a rope that has a length of $2L_1 = 10$ meters. The rope is attached to posts at $x = \pm L_2$ that are at the same level but are only $2L_2 = 7$ meters apart. The mass of the rope is $M = 5$ kg. Find the shape of the rope and the tension applied to the posts. [Use the linear wave equation to determine the equilibrium solution for $y(x)$, assuming that $|y| \ll L_1$.] Plot the shape. [Answer to lowest order in $(L_1 - L_2)/L_1$: $y(x) = -(g/2T)(L_2 - x)(m + M + \rho x)$, $x > 0$, where $T = g[L_2(3m^2 + 3mM + M^2)/24(L_1 - L_2)]^{1/2}$ is the rope tension and $\rho = M/2L_1$ is the mass density.]

(12) **(a)** In the presence of gravity, the vector wave equation for a rope, Eq. (3.1.80), is modified to read

$$\frac{\partial^2}{\partial t^2}\mathbf{r}(z,t) = c^2 \frac{\partial^2}{\partial z^2}\mathbf{r}(z,t) - g\hat{\mathbf{y}}.$$

In equilibrium, the rope hangs with a parabolic shape given by Eq. (3.1.32). It is possible to oscillate the hanging rope back and forth like a pendulum. (Think of a footbridge swaying back and forth.) Find the frequency of this swaying motion, assuming that the length of the rope is L (Hint: Apply the principle of superposition to take care of the source term, and determine the eigenmodes of the system.)

(b) If one *assumes* that the rope oscillates like a rigid pendulum, show that the frequency of small oscillations is $\sqrt{10T/\rho L^2}$. (Recall that a rigid pendulum has frequency $\sqrt{Mg/I}$, where M is the mass and I is the moment of inertia about the pivot.) Why does this answer differ from the result of part (a)? Which answer is right?

(13) A quantum particle of mass m, moving in one dimension in a potential $V(x)$, is described by Schrödinger's equation,

$$i\hbar \frac{\partial}{\partial t}\Psi = \hat{H}\Psi, \tag{3.1.81}$$

where $\Psi(x,t)$ is the particle's wave function, the Hamiltonian operator $\hat{H}$ is given by

$$\hat{H} = -\frac{\hbar^2}{2m}\frac{\partial^2}{\partial x^2} + V(x), \tag{3.1.82}$$

and $\hbar = 1.055 \times 10^{-34}$ N m s is Planck's constant divided by 2π. A quantum particle moves in a box of width L. Boundary conditions on the wave function are therefore $\psi = 0$ at $x = 0$ and $x = L$. Use separation of variables to solve the following initial-value problem: $\Psi(0,x) = x^3(L - x)$. Animate the result for the probability density $|\Psi|^2$ over a time $0 < t < 20\hbar/mL^2$.

(14) In a slab with uniform conductivity κ, find the equilibrium temperature distribution $T_{\text{eq}}(x)$ under the listed conditions. Show directly that in each case Eq. (3.1.51) is satisfied.

(a) $S(x) = S_0$, $T(0) = T_1$, $T(L) = T_2$.

(b) $S(x) = x$, $\frac{\partial T}{\partial x}(0) = 0$, $T(L) = 0$.

(c) $S(x) = \alpha\delta(x - L/3)$, $\frac{\partial T}{\partial x}(0) = T(0)$, $\frac{\partial T}{\partial x}(L) = 0$.

(15) The ceiling of a Scottish castle consists of $\frac{1}{4}$-meter-thick material with thermal conductivity $\kappa = 0.5$ W/(m K). Most of the heat is lost through the ceiling. The exterior temperature is, on a average, 0°C, and the interior is kept at a chilly 15°C. The surface area of the ceiling is 2000 m^2. The cost per kilowatt-hour of heating power is 3 pence. How much does it cost, in pounds, to heat the castle for one month (30 days)? (Note: One British pound equals 100 Pence.)

(16) Solve the following heat equation problem on $0 < x < 1$, with given boundary and initial conditions, and plot the solution for $0 < t < 2$ by making a table of plots at a sequence of 40 times:

$$\frac{\partial T(x,t)}{\partial t} = \frac{\partial^2 T(x,t)}{\partial x^2} + S(x).$$

Boundary conditions, initial conditions, and source:

(a) $T(0,t) = 0$, $T(1,t) = 3$; initial condition $T(x,0) = 0$, $S(x) = 0$.

(b) Insulated at $x = 0$, $T(1,t) = 1$; initial condition $T(x,0) = x$, $S(x) = 1$.

(c) Insulated on both faces, $T(x,0) = 0$, $S(x) = \sin 2\pi x$.

(17) **(a)** A slab of fat, thickness 2 cm, thermal diffusivity $\chi = 10^{-6}$ m^2/s, and initially at temperature $T = 5$°C, is dropped into a pot of hot water at $T = 60$°C. Find $T(x,t)$ within the slab. Find the time t_0 needed for the center of the fat to reach a temperature of 55°C. Animate the solution for $T(x,t)$ up to this time.

(b) Over very long times, the behavior of $T(x,t)$ is dominated by the eigenmode with the lowest decay rate. Keeping only this mode in the evolution, use this approximate solution for T to rederive t_0 *analytically*, and compare the result with that found using the exact answer from part (a).

(18) A cold steak, initially at uniform temperature $T(x,0) = 8$°C, and of thickness $L = 3$ cm, is placed on a griddle (at $x = 0$) at temperature $T = 250$°C. The steak will be cooked medium rare when its minimum internal temperature reaches $T = 65$°C. How long does this take? Animate the solution for $T(x,t)$ up to this time. (The boundary condition on the upper face of the meat at $x = L$ can be taken to be approximately insulating. The thermal diffusivity of meat is about 3×10^{-7} m^2/s.)

(19) A sheet of copper has thickness L, thermal diffusivity χ, and specific heat C. It is heated uniformly with a constant power density $S = j^2\rho$ due to a current density j running through the sheet, where ρ is the resistivity of the copper. The faces of the sheet, each with area A and at temperature T_0 (to be determined), radiate into free space with a heat flux given by the Stefan–Boltzmann law for blackbody radiation: $\Gamma = (1-r)\sigma T_0^4$, where $\sigma = 5.67 \times 10^{-8}$ W/m^2 K^4) is the Stefan–Boltzmann constant, and r is the reflectivity of the material, equal to zero for a blackbody and 1 for a perfect reflector.

(a) Find the equilibrium temperature $T_{eq}(x)$ in the sheet. What is the maximum temperature T_{max} as a function of the current density j?

(b) If the temperature varies by an amount $\Delta(T(x,t)$ from equilibrium, the radiated power out of the face at $x=0$ changes by an amount

$$\Delta\Gamma = (1-r)\left\{\left[T_0 + \Delta T(0,t)\right]^4 - T_0^4\right\} \simeq 4(1-r)\sigma T_0^3\,\Delta T(0,t)$$

(assuming that $T_0 \gg \Delta T$), and similarly for the face at $x=L$. Thus, the boundary condition for small temperature deviations is mixed. Find the first three eigenmodes, and their rate of decay. Take $L=1$ cm, $r=0.8$, and $S=10^3$ W/m^3.

(20) Damped waves on a string of length π in gravity satisfy the wave equation

$$\frac{\partial^2 y}{\partial t^2} + \frac{1}{4}\frac{\partial y}{\partial t} = \frac{\partial^2 y}{\partial x^2} - 1, \qquad y(-\pi/2,t) = y(\pi/2,t) = 0.$$

For initial conditions $y(x,0) = \pi/2 - |x|$, $\dot{y}(x,0) = 0$, plot $y(x,t)$ for $0 < t < 20$.

3.2 LAPLACE'S EQUATION IN SOME SEPARABLE GEOMETRIES

In a region of space that is charge-free, the electrostatic potential $\phi(\mathbf{r})$ satisfies Poisson's equation without sources:

$$\nabla^2\phi(\mathbf{r}) = 0, \tag{3.2.1}$$

where $\mathbf{r} = (x, y, z)$ is the position vector, and the Laplacian operator ∇^2 is defined by

$$\nabla^2 = \frac{d^2}{dx^2} + \frac{d^2}{dy^2} + \frac{d^2}{dz^2}. \tag{3.2.2}$$

This PDE is called *Laplace's equation*. To solve for $\phi(\mathbf{r})$ within a specified volume V we require boundary conditions to be given on the surface S of the volume (see Fig. 3.8). We will consider boundary conditions that fall into three categories:

Dirichlet, where $\phi|_S = \phi_0(\mathbf{r})$ for some potential $\phi_0(\mathbf{r})$ applied to the surface S;

von Neumann, where the directional derivative of ϕ normal to the surface is determined: $\hat{\mathbf{n}}\cdot\nabla\phi|_S = E_0(\mathbf{r})$, where $\hat{\mathbf{n}}$ is a unit vector perpendicular to the surface S, or

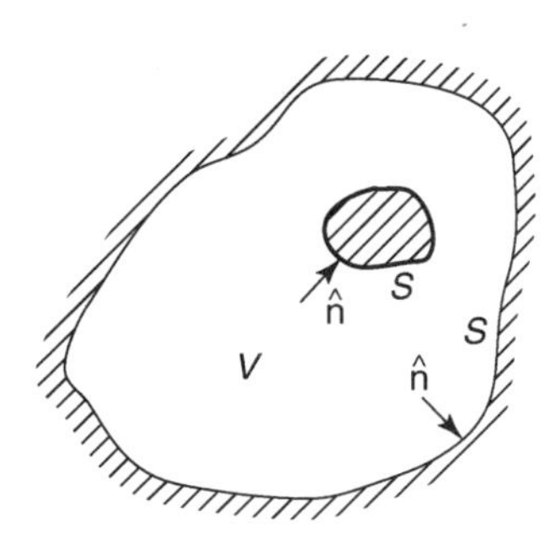

Fig. 3.8 Region V (unshaded) for solution of Poisson's equation. The surface of this region is S.

mixed, where $(f\phi + g\hat{\mathbf{n}}\cdot\nabla\phi)|_S = u_0(\mathbf{r})$ for some function $u_0(\mathbf{r})$ and (nonnegative) functions $f(\mathbf{r})$ and $g(\mathbf{r})$. The functions f and g cannot both vanish at the same point on S.

These three boundary conditions are straightforward generalizations of the conditions of the same name, considered previously for one-dimensional PDEs. Physically, Dirichlet conditions can occur in the case where the surface S is a set of one or more conductors that are held at fixed potentials. von Neumann conditions are less common in electrostatics problems, applying to the case where the normal component of the electric field is determined at the surface. Mixed conditions rarely occur in electrostatic problems, but can sometimes be found in applications of Poisson's equation to other areas of physics, such as thermal physics [see Eq. (3.1.47), for example].

3.2.1 Existence and Uniqueness of the Solution

For the above boundary conditions, can one always find a solution to Laplace's equation? And is this solution unique?

We will answer the second question first. If a solution exists, it is unique for Dirichlet or mixed boundary conditions. For von Neumann boundary conditions, the solution is unique only up to an additive constant. This can be proven through the following argument.

Say that two solutions exist with the same boundary conditions. Call these solutions ϕ_1 and ϕ_2. We will prove that these solutions must in fact be the same (up to an additive constant for von Neumann boundary conditions).

The difference between the solutions, $\Phi = \phi_1 - \phi_2$, also satisfies the Laplace equation $\nabla^2\Phi = 0$, and has *homogeneous* boundary conditions. To find the solution to $\nabla^2\Phi = 0$ with homogeneous boundary conditions, multiply this equation by Φ and integrate over the volume of the domain. Then apply Green's first identity:

$$0 = \int_V \Phi\,\nabla^2\Phi\,d^3r = \int_S \Phi\,\nabla\Phi\cdot\hat{\mathbf{n}}\,d^2r - \int_V \nabla\Phi\cdot\nabla\Phi\,d^3r.$$

Now, for Dirichlet or von Neumann boundary conditions, either $\Phi = 0$ or $\nabla\Phi\cdot\hat{\mathbf{n}} = 0$ on S, so the surface integral vanishes, and we are left with $\int_V |\nabla\Phi|^2\,d^3r = 0$. Furthermore, since $|\nabla\Phi|^2$ is always nonnegative, the only way this integral can be zero is if $\nabla\Phi = 0$ throughout the domain. Therefore, $\Phi = \text{constant}$ is the only solution. This implies that $\Phi = 0$ for Dirichlet boundary conditions, because $\Phi = 0$ on S; but for von Neumann conditions, $\Phi = \text{constant}$ satisfies the boundary conditions.

For mixed boundary conditions, Φ satisfies $(f\Phi + g\hat{\mathbf{n}}\cdot\nabla\Phi)|_S = 0$. Assuming that $f(\mathbf{r})$ is nonzero over some part of the surface S (call this portion S_1), and $g(\mathbf{r})$ is nonzero over remainder of the surface (call it S_1^c), Green's first identity becomes

$$0 = -\int_{S_1} \frac{g}{f}(\nabla\Phi\cdot\hat{\mathbf{n}})^2\,d^2r - \int_{S_1^c} \frac{f}{g}\Phi^2\,d^2r - \int_V \nabla\Phi\cdot\nabla\Phi\,d^3r.$$

Each integral is nonnegative (since both f and g are nonnegative by assumption), and therefore, by the same argument as before, the only possible solution is $\Phi = 0$.

Therefore, ϕ_1 and ϕ_2 are equal for Dirichlet or mixed boundary conditions, and for von Neumann conditions they differ at most by a constant. We have shown that the solution to the Laplace equation is unique for Dirichlet and mixed conditions, and unique up to an additive constant for von Neumann conditions.

One can also show that for Dirichlet and mixed boundary conditions, a solution can always be found. (We will later prove this by construction of the solution.) For von Neumann boundary conditions, however, a solution for the potential $\phi(\mathbf{r})$ only exists provided that the boundary conditions satisfy the following integral constraint:

$$\int_S \nabla\phi\cdot\hat{\mathbf{n}}\, d^2r = 0. \tag{3.2.3}$$

This constraint on the normal derivative of the potential at the domain surface follows from Laplace's equation through an application of the divergence theorem: $0 = \int_V \nabla^2\phi\, d^3r = \int_S \nabla\phi\cdot\hat{\mathbf{n}}\, d^2r$.

Students with some training in electrostatics will recognize Eq. (3.2.3) as a special case of Gauss's law, which states that the integral of the normal component to the electric field over a closed surface must equal the charge enclosed in the surface (which in this case is zero).

If Eq. (3.2.3) is not satisfied, then there is no solution. This only constrains von Neumann boundary conditions, since only von Neumann conditions directly specify the normal derivative of ϕ. For Dirichlet and mixed conditions, which do not directly specify $\nabla\phi\cdot\hat{\mathbf{n}}$, one can always find a solution that satisfies Eq. (3.2.3).

We now consider several geometries where the solution to Laplace's equation can be found analytically, using the method of separation of variables.

3.2.2 Rectangular Geometry

General Solution In rectangular coordinates (x, y), a solution to Laplace's equation for $\phi(x, y)$ can be found using separation of variables. Following the by now familiar argument, we write

$$\phi(x, y) = X(x)Y(y) \tag{3.2.4}$$

for some functions X and Y. If we substitute Eq. (3.2.4) into the Laplace equation and divide the result by ϕ, we obtain

$$\frac{1}{X(x)}\frac{d^2X}{dx^2} + \frac{1}{Y(y)}\frac{d^2Y}{dy^2} = 0.$$

As usual, this equation can be separated into two ODEs, one for X and the other for Y:

$$\begin{aligned} \frac{1}{X(x)}\frac{d^2X}{dx^2} &= \lambda, \\ \frac{1}{Y(y)}\frac{d^2Y}{dy^2} &= -\lambda, \end{aligned} \tag{3.2.5}$$

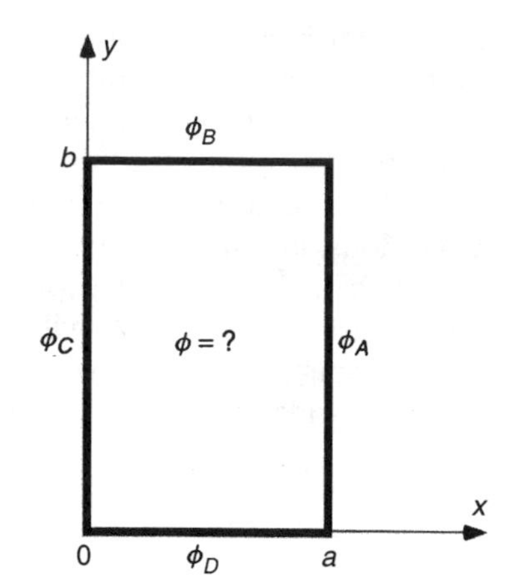

Fig. 3.9

where λ is the separation constant. The general solution to each ODE can be found easily:

$$X(x) = C_1 e^{\sqrt{\lambda}x} + C_2 e^{-\sqrt{\lambda}x},$$
$$Y(y) = C_3 e^{i\sqrt{\lambda}y} + C_4 e^{-i\sqrt{\lambda}y}. \tag{3.2.6}$$

Example 1: Dirichlet Boundary Conditions The general solution given by Eq. (3.2.6) is useful for boundary conditions specified on a rectangle, as shown in Fig. 3.9. The potential on the sides of the rectangle is specified by the four functions $\phi_A(y)$, $\phi_B(x)$, $\phi_C(y)$, and $\phi_D(x)$.

We first consider the special case where only $\phi_A(y)$ is nonzero. Then the homogeneous boundary conditions on the bottom and the top imply that $Y(0) = Y(b) = 0$. We again confront an eigenvalue problem for $Y(y)$. The condition $Y(0) = 0$ implies that $C_4 = -C_3$ in Eq. (3.2.6), so that $Y(y) = 2iC_3 \sin\sqrt{\lambda}\, y$. The condition that $Y(L) = 0$ then implies that either $C_3 = 0$ (trivial) or $\sqrt{\lambda}\, b = n\pi$. Thus, we find that

$$Y(y) = D \sin\frac{n\pi y}{b}, \tag{3.2.7}$$

where

$$\lambda = \lambda_n = (n\pi/b)^2, \qquad n = 1, 2, 3, \dots . \tag{3.2.8}$$

Since there are many solutions, we can superimpose them in order to match boundary conditions on the other two sides:

$$\phi(x, y) = \sum_n \left(C_{1n} e^{n\pi x/b} + C_{2n} e^{-n\pi x/b}\right) \sin\frac{n\pi y}{b}, \tag{3.2.9}$$

where, as usual, we have absorbed the constant D into the constants C_{1n} and C_{2n}.

Equation (3.2.9) has an oscillatory form in the y-direction, and an exponential form in the x-direction. This is because Laplace's equation implies that $\partial^2\phi/\partial x^2 = -\partial^2\phi/\partial y^2$. Therefore, a solution that oscillates sinusoidally in y, satisfying $\partial^2\phi/\partial y^2 = -\lambda_n\phi$, must be exponential in x. By the same token, one can also construct a solution that consists of sinusoids in x and exponentials in y. That solution does not match the given boundary conditions in this problem, but is required for other boundary conditions such as those for which $\phi_A = \phi_C = 0$.

To find the constants C_{1n} and C_{2n}, we now satisfy the remaining two boundary conditions. The potential is zero along the boundary specified by $x = 0$, which requires that we take $C_{2n} = -C_{1n}$ in Eq. (3.2.9). We then obtain

$$\phi(x, y) = \sum_n A_n \sinh\frac{n\pi x}{b} \sin\frac{n\pi y}{b}, \tag{3.2.10}$$

where $A_n = 2C_{1n}$. The constants A_n in Eq. (3.2.10) are determined by the boundary condition that $\phi(a, y) = \phi_A(y)$:

$$\phi_A(y) = \sum_{n=1}^{\infty} A_n \sinh\frac{n\pi a}{b} \sin\frac{n\pi y}{b}. \tag{3.2.11}$$

Equation (3.2.11) is a Fourier sine series for the function $\phi_A(y)$ defined on $0 < y < b$. The Fourier coefficient $A_n \sinh(n\pi a/b)$ is then determined by Eq. (3.2.11),

$$A_n \sinh\frac{n\pi a}{b} = \frac{2}{b}\int_0^b \phi_A(y) \sin\frac{n\pi y}{b}\,dy. \tag{3.2.12}$$

Equations (3.2.10) and (3.2.12) provide the solution for the potential within a rectangular enclosure for which the potential is zero on three sides, and equals $\phi_A(y)$ on the fourth side.

For instance, if $\phi_A(y) = \sin(n\pi y/b)$, then the solution for $\phi(x, y)$ consists of only a single term,

$$\phi(x, y) = \frac{\sinh(n\pi x/b)}{\sinh(n\pi a/b)} \sin\frac{n\pi y}{b}.$$

This solution is plotted in Cell 3.18 for the case $n = 3$ and $a = b = 1$. The reader is invited to vary the value of n in this solution, as well as the shape of the box.

Cell 3.18

```
a = b = 1; n = 3;

         Sinh[n Pi x/b]
φ[x_, y_] = -------------- Sin [n Pi y/b];
         Sinh[n Pi a/b]

Plot3D[φ[x, y], {x, 0, a}, {y, 0, b},
  PlotLabel → "Potential in a box",
  AxesLabel → {"x", "y", ""}, PlotRange → All, PlotPoints → 30];
```

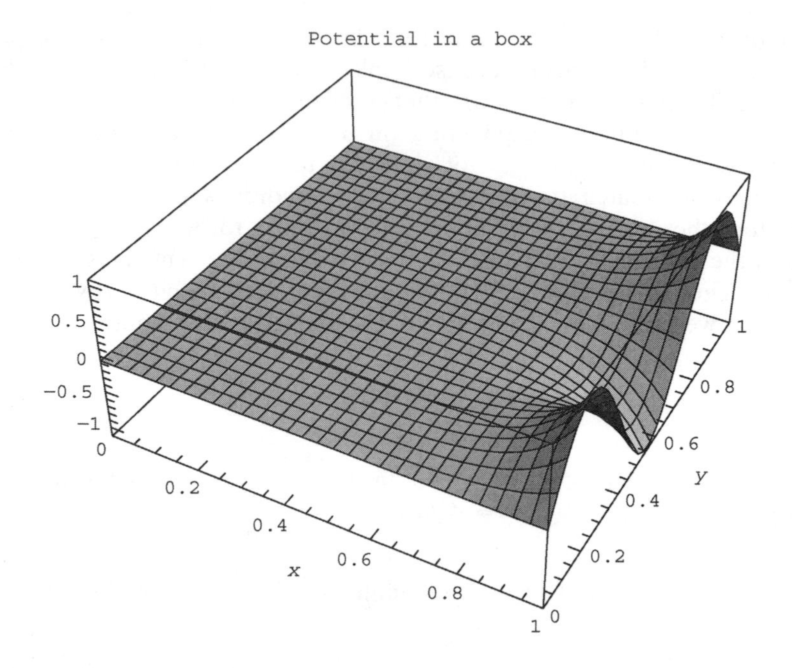

We can use this solution method to determine the general solution for the case of arbitrary nonzero potentials specified on all four sides at once by using the following argument. We can repeat our analysis, taking the case where $\phi_A = \phi_C = \phi_D = 0$ and only $\phi_B(x)$ is nonzero, next evaluating the case where $\phi_A = \phi_B = \phi_D = 0$ and $\phi_C(y)$ is nonzero, and finally taking the case where $\phi_A = \phi_B = \phi_C = 0$ and $\phi_D(x)$ is nonzero. We can then superimpose the results from these calculations to obtain the potential $\phi(x, y)$ for any combination of potentials specified on the rectangular boundary.

Example 2: von Neumann Boundary Conditions; the Current Density in a Conducting Wire Let's now study an example with von Neumann boundary conditions over part of the surface. Consider a current-carrying wire made of electrically conductive material, such as copper (see Fig. 3.10). The length of the wire is b, and

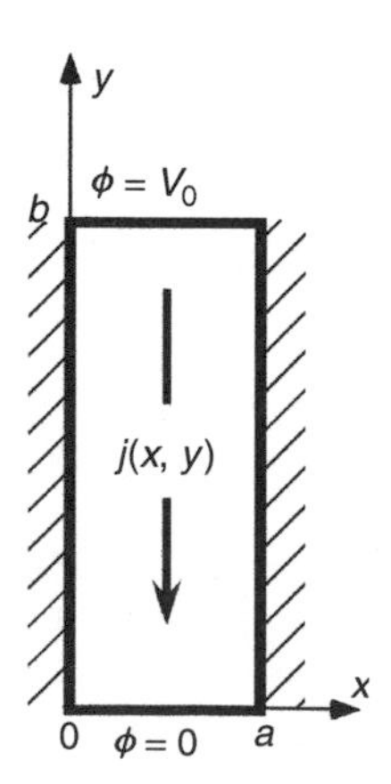

Fig. 3.10 Current density in a wire.

its width is a. (For simplicity, we neglect any z-dependence.) The left and right sides of the wire are insulated (coated with rubber, say). To the top of the wire, a voltage V_0 is applied, causing a current I (in amperes) to run through to the bottom, where it is extracted (see the figure). The question is to determine the distribution of current and the electric field inside the wire.

These two quantities are related: current flows because there is an electric field $\mathbf{E} = -\nabla\phi$ inside the conductor. The *current density* $\mathbf{j}$ inside the rod (in amperes per square meter) satisfies Ohm's law, $\mathbf{j} = \sigma\mathbf{E}$, where σ is the electrical conductivity of the material. We wish to determine $\mathbf{j}(x,y)$ and $\mathbf{E}(x,y)$ [or equivalently, $\phi(x,y)$].

The boundary conditions on the top and bottom faces of the wire are set by the applied potentials:

$$\phi(x,0) = 0,$$

$$\phi(x,b) = V_0.$$

However, the boundary conditions on the two insulated sides are determined by the fact that no current flows through these sides, so the current runs parallel to these faces. Therefore, according to Ohm's law, $j_x = \sigma E_x = -\sigma\,\partial\phi/\partial x = 0$ along these faces, so we have von Neumann conditions on these faces:

$$\frac{\partial\phi}{\partial x}(0,y) = \frac{\partial\phi}{\partial x}(a,y) = 0. \tag{3.2.13}$$

Furthermore, the potential in the interior of the conductor must satisfy Laplace's equation. This is because there are no sources or sinks of current in the interior of the rod, so $\nabla\cdot\mathbf{j} = 0$ [recall the discussion surrounding Eq. (1.2.12)]. Then Ohm's law implies

$$\nabla\cdot\sigma\mathbf{E} = -\sigma\nabla^2\phi = 0.$$

We can now solve this Laplace's equation for the potential, using separation of variables. Given the boundary conditions of Eq. (3.2.13), we expect from our standard separation-of-variables argument that $\phi(x,y)$ will consist of a sum of terms of the form $Y(y)X(x)$, with $X(x)$ being an eigenmode of the von Neumann form, $X(x) = \cos(n\pi x/a)$, $n = 0, 1, 2, \dots$. Substitution of this form into $\nabla^2\phi = 0$ then implies that $Y(y)$ satisfies

$$\frac{d^2Y}{dy^2} - \left(\frac{n\pi}{a}\right)^2 Y = 0.$$

For $n > 0$ the solution that is zero at $y = 0$ is $Y(y) = A_n \sinh(n\pi y/a)$. However, we must be careful: the $n = 0$ term must also be kept. For $n = 0$ the solution that is zero at $y = 0$ is $Y(y) = A_0 y$. Therefore, the potential has the form

$$\phi(x,y) = \sum_{n=1}^{\infty} A_n \sinh\frac{n\pi y}{a}\cos\frac{n\pi x}{a} + A_0 y. \tag{3.2.14}$$

This solution matches the boundary conditions on the bottom and sides. The constants A_n are then determined by satisfying the boundary condition on the top face, $V_0 = \Sigma_{n=1}^{\infty} A_n \sinh(n\pi b/a)\cos(n\pi x/a) + A_0 b$. This is simply a Fourier cosine series in x. The Fourier coefficients are $A_0 = V_0/b$ and

$$A_n = \frac{V_0}{\sinh(n\pi a/b)} \int_0^b \cos\frac{n\pi y}{b}\,dy, \qquad n > 0.$$

However, the integral over the cosine equals zero, so the only nonzero Fourier coefficient is $A_0 = V_0/b$. Therefore, the solution for the potential interior to the wire is simply $\phi(x, y) = V_0 y/b$. The electric field in the wire is uniform, of magnitude V_0/b. The current density runs vertically and uniformly throughout the conductor, with magnitude $j = \sigma V_0/b$.

3.2.3 2D Cylindrical Geometry

Separation of Variables The general solution to Laplace's equation can also be found in cylindrical geometry (r, θ, z). The cylindrical radius r and the angle θ are defined by the coordinate transformation $x = r\cos\theta$ and $y = r\sin\theta$. At first, for simplicity, we consider the case where the potential is independent of z, so that $\phi = \phi(r, \theta)$. In these 2D cylindrical coordinates, the method of separation of variables can again be used to solve Laplace's equation. We assume that the potential takes the form

$$\phi(r, \theta) = R(r)\Theta(\theta). \tag{3.2.15}$$

In cylindrical coordinates ∇^2 is given by

$$\nabla^2 = \frac{1}{r}\frac{\partial}{\partial r}\left(r\frac{\partial}{\partial r}\right) + \frac{1}{r^2}\frac{\partial^2}{\partial\theta^2} + \frac{\partial^2}{\partial z^2}. \tag{3.2.16}$$

Applying ∇^2 to Eq. (3.2.15) and dividing by $R(r)\Theta(\theta)$ yields

$$\frac{1}{rR(r)}\frac{\partial}{\partial r}\left(r\frac{\partial R}{\partial r}\right) + \frac{1}{r^2\Theta(\theta)}\frac{\partial^2\Theta}{\partial\theta^2} = 0. \tag{3.2.17}$$

Following the standard procedure, this equation can be separated into two equations for the r and θ dependence. The expression $[1/\Theta(\theta)]\,\partial^2\Theta/\partial\theta^2$ must equal some function of θ, which we call $f(\theta)$. Equation (3.2.17) can then be written as

$$\frac{1}{rR(r)}\frac{\partial}{\partial r}\left(r\frac{\partial R}{\partial r}\right) + \frac{f(\theta)}{r^2} = 0. \tag{3.2.18}$$

However, as the rest of the equation is independent of θ, it can only be satisfied if $f(\theta)$ is a constant, $f(\theta) = -m^2$ (where we have anticipated that the constant will be nonpositive). Then the equation for $\Theta(\theta)$ is

$$\frac{\partial^2\Theta}{\partial\theta^2} = -m^2\Theta(\theta). \tag{3.2.19}$$

The boundary conditions on this equation arise from the fact that the variable θ is periodic: the angles $\theta + 2\pi$ and θ are equivalent. This implies that we must require *periodic boundary conditions,* $\Theta(\theta + 2\pi) = \Theta(\theta)$. With these boundary conditions, we again have an eigenvalue problem, because these boundary conditions allow only the trivial solution $\Theta(\theta) = 0$, except for special values of m. These values may be found by examining the general solution to Eq. (3.2.19),

$$\Theta(\theta) = A e^{im\theta} + B e^{-im\theta}. \tag{3.2.20}$$

For nonzero A and B, the periodic boundary conditions imply that $e^{\pm im\theta} = e^{\pm im(\theta + 2\pi)}$, which requires that $e^{\pm i2\pi m} = 1$. This can only be satisfied if m is an integer. Therefore, we obtain

$$\Theta(\theta) = e^{im\theta}, \qquad m \in \text{Integers} \tag{3.2.21}$$

[Allowing m to run over both positive and negative integers accommodates both independent solutions in Eq. (3.2.20).]

Turning to the radial equation, we find that for $f(\theta) = -m^2$, Eq. (4.2.18) becomes

$$\frac{1}{r}\frac{\partial}{\partial r}\left(r\frac{\partial R}{\partial r}\right) - \frac{m^2}{r^2}R(r) = 0. \tag{3.2.22}$$

This equation has the following general solution:

$$R(r) = \begin{cases} A_m r^{|m|} + B_m / r^{|m|} & \text{if} \quad m \neq 0, \\ A_0 + B_0 \ln r & \text{if} \quad m = 0. \end{cases} \tag{3.2.23}$$

The general solution to the Laplace equation in cylindrical coordinates is a sum of these independent solutions:

$$\phi(r, \theta) = A_0 + B_0 \ln r + \sum_{\substack{m=-\infty \\ m \neq 0}}^{\infty} \left(A_m r^{|m|} + \frac{B_m}{r^{|m|}} \right) e^{im\theta}. \tag{3.2.24}$$

Example The general solution to Laplace's equation in cylindrical coordinates is useful when boundary conditions are provided on the surface of a cylinder or cylinders. For instance, say the potential is specified on a cylinder of radius a:

$$\phi(a, \theta) = V_a(\theta). \tag{3.2.25}$$

We require a solution to Laplace's equation in $0 \le r \le a$. This solution is given by Eq. (3.2.24); we only need to match the boundary conditions to determine the constants A_n and B_n. First, the fact that the potential must be finite at $r = 0$ implies that $B_n = 0$ for all n. Next, Eqs. (3.2.24) and (3.2.25) imply that

$$\sum_{m=-\infty}^{\infty} A_m a^{|m|} e^{im\theta} = V_a(\theta).$$

This is an exponential Fourier series, so the Fourier coefficients can be determined using Eq. (3.2.3):

$$A_m a^{|m|} = \frac{1}{2\pi} \int_0^{2\pi} V_a(\theta)\, e^{-im\theta}\, d\theta. \tag{3.2.26}$$

The solution for the potential is

$$\phi(r,\theta) = \sum_{m=-\infty}^{\infty} A_m r^{|m|} e^{im\theta}.$$

For instance, if $V_a(\theta) = V_0 \sin\theta$, then only the $m = \pm 1$ terms in the sum contribute to the solution, and the rest are zero. This is because $\sin\theta$ contains only $m = \pm 1$ Fourier components. The solution is then clearly of the form $\phi(r,\theta) = Ar \sin\theta$, and the constant A can be determined by matching the boundary condition that $V(a,\theta) = V_0 \sin\theta$. The result is

$$\phi(r,\theta) = V_0 \frac{r}{a} \sin\theta.$$

3.2.4 Spherical Geometry

Separation of Variables We next consider the solution to Laplace's equation written in spherical coordinates (r, θ, ϕ). These coordinates are defined by the transformation $x = r\sin\theta\cos\phi$, $y = r\sin\theta\sin\phi$, and $z = r\cos\theta$ (see Fig. 3.11). In these coordinates Laplace's equation becomes

$$\nabla^2 \Psi = \frac{1}{r^2}\frac{\partial}{\partial r}\left(r^2 \frac{\partial \Psi}{\partial r}\right) + \frac{1}{r^2 \sin\theta}\frac{\partial}{\partial \theta}\left(\sin\theta \frac{\partial \Psi}{\partial \theta}\right) + \frac{1}{r^2 \sin^2\theta}\frac{\partial^2 \Psi}{\partial \phi^2}. \tag{3.2.27}$$

(We use the symbol Ψ for potential in this section so as not to confuse it with the azimuthal angle ϕ.) We again employ the method of separation of variables, writing

$$\Psi(r,\theta,\phi) = R(r)\Theta(\theta)\Phi(\phi).$$

Then the equation $(\nabla^2\Psi)/\Psi = 0$ is

$$\frac{1}{r^2 R}\frac{\partial}{\partial r}\left(r^2 \frac{\partial R}{\partial r}\right) + \frac{1}{r^2 \sin\theta\, \Theta}\frac{\partial}{\partial \theta}\left(\sin\theta \frac{\partial \Theta}{\partial \theta}\right) + \frac{1}{r^2 \sin^2\theta\, \phi}\frac{\partial^2 \Phi}{\partial \phi^2} = 0. \tag{3.2.28}$$

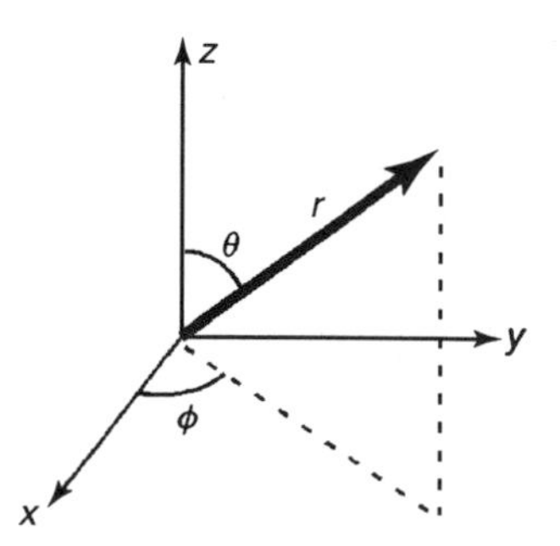

Fig. 3.11 Spherical coordinates (r, θ, ϕ).

The separation-of-variables analysis proceeds just as before. One finds that Eq. (3.2.28) separates into three ODEs for R, Θ, and Φ:

$$\frac{\partial^2 \Phi}{\partial \phi^2} = -m^2 \Phi(\phi), \tag{3.2.29}$$

$$\frac{1}{\sin\theta}\frac{\partial}{\partial\theta}\left(\sin\theta\frac{\partial\Theta}{\partial\theta}\right) - \frac{m^2}{\sin^2\theta}\Theta(\theta) = \lambda\Theta(\theta), \tag{3.2.30}$$

$$\frac{1}{r^2}\frac{\partial}{\partial r}\left(r^2\frac{\partial R}{\partial r}\right) + \frac{\lambda}{r^2}R(r) = 0, \tag{3.2.31}$$

where we have introduced the separation constants m and λ. Just as in cylindrical coordinates, periodicity of the angle ϕ implies that m must be an integer, so that the eigenvalue problem posed by Eq. (3.2.29) has the solution

$$\Phi(\phi) = e^{im\phi}, \qquad m \in \text{Integers}. \tag{3.2.32}$$

Eigenmodes in θ: Associated Legendre Functions Turning to Eq. (3.2.30), we make a coordinate transformation to the variable $x = \cos\theta$. Then by the chain rule,

$$\frac{\partial\Theta}{\partial\theta} = \frac{\partial x}{\partial\theta}\frac{\partial\Theta}{\partial x} = -\sin\theta\frac{\partial\Theta}{\partial x} = -\sqrt{1-x^2}\,\frac{\partial\Theta}{\partial x}.$$

When written in terms of x, Eq. (3.2.30) becomes

$$\frac{\partial}{\partial x}\left((1-x^2)\frac{\partial\Theta}{\partial x}\right) - \frac{m^2}{1-x^2}\Theta(x) = \lambda\Theta(x). \tag{3.2.33}$$

This ODE has regular singular points at $x = \pm 1$. Its general solution is in terms of special functions called *hypergeometric functions*. In general, these functions are singular at the end points because of the regular singular points there. However, for special values of λ the solution is finite at both ends. Again, we have an eigenvalue problem. In this case, the eigenvalues are

$$\lambda = -l(l+1) \qquad \text{for } l \text{ a positive integer taking on the values } l \geq |m|. \tag{3.2.34}$$

The corresponding eigenmodes are special cases of the hypergeometric functions called *associated Legendre functions*,

$$\Theta(\theta) = P_l^m(x), \qquad x = \cos\theta.$$

For $m = 0$ the functions $P_l^m(x)$ are simple polynomials called Legendre polynomials, but for $m \neq 0$ they have the form of a polynomial of order $l - |m|$ multiplied by the factor $(1-x^2)^{|m|/2}$. Some of these functions are listed in Table 3.2. The table shows, among other things, that the functional form of $P_l^m(x)$ differs only by a constant from $P_l^{-m}(x)$.

We can see using *Mathematica* that the associated Legendre functions $P_l^m(x)$ satisfy Eq. (3.2.33). In *Mathematica* these functions are called

Table 3.2. Associated Legendre Functions

	$P_l^m(x)$				
l	$m=2$	-1	0	1	2
0			1		
1		$\frac{1}{2}\sqrt{1-x^2}$	x	$-\sqrt{1-x^2}$	
2	$\frac{1}{8}(1-x^2)$	$\frac{1}{2}x\sqrt{1-x^2}$	$-\frac{1}{2}+\frac{3}{2}x^2$	$-3x\sqrt{1-x^2}$	$3(1-x^2)$

`LegendreP[l,m,x]`. The following cell tests whether each Legendre function satisfies Eq. (3.2.33), up to $l=5$:

Cell 3.19

```
Θ[x_] = LegendreP[l, m, x];
Table[
 Table[Simplify[D[(1-x^2) D[LegendreP[l, m, x], x], x],
  -m^2 Θ [x]/(1-x^2) ==
    -l (l + 1) Θ [x]], {m, -l, l}], {l, 0, 5}]

{{True}, {True, True, True}, {True, True, True, True, True},
 {True, True, True, True, True, True, True},
 {True, True, True, True, True, True, True, True, True},
 {True, True, True, True, True, True, True, True, True, True,
  True}}
```

Turning to the radial dependence of the potential, we require the solution of Eq. (3.2.31), with $\lambda = -l(l+1)$:

Cell 3.20

```
FullSimplify[R[r]/. DSolve[
    1/r^2 D[ r^2 D[R[r], r], r] -l (l + 1) R[r]/r^2 ==
     0, R[r], r][[1]], l > 0]

r^(-1-l) C[1] + r^l C[2]
```

Thus,

$$R(r) = A/r^{l+1} + Br^l,$$

where A and B are constants.

Since l and m can take on different values, we have actually found an infinite number of independent solutions to Laplace's equation. We can sum them together to obtain the general solution in spherical coordinates:

$$\Psi(r,\theta,\phi) = \sum_{l=0}^{\infty} \sum_{m=-l}^{l} \left(\frac{A_{lm}}{r^{l+1}} + B_{lm} r^l \right) e^{im\phi} P_l^m(\cos\theta). \tag{3.2.35}$$

Finally, we are left with the task of determining the constants A_{lm} and B_{lm} in terms of the boundary conditions. Say, for example, we know the potential on the surface of a sphere of radius a to be $\Psi(a,\theta,\phi) = V(\theta,\phi)$. If we require the solution within the sphere, we must then set $A_{lm}=0$ in order to keep the solution finite at the origin. At $r=a$ we have

$$\sum_{l=0}^{\infty}\sum_{m=-l}^{l} B_{lm} a^l e^{im\phi} P_l^m(\cos\theta) = V(\theta,\phi). \tag{3.2.36}$$

It would be useful if the associated Legendre functions formed an orthogonal set, so that we could determine the Fourier coefficients by extracting a single term from the sum. Amazingly, the associated Legendre function do satisfy an orthogonality relation:

$$\int_0^{\pi} P_l^m(\cos\theta) P_{\bar{l}}^m(\cos\theta)\sin\theta\, d\theta = \frac{2}{2l+1}\frac{(l+m)!}{(l-m)!}\delta_{l,\bar{l}}. \tag{3.2.37}$$

Thus, we can determine B_{lm} by multiplying both sides of Eq. (3.2.36) by $e^{-im\phi}P_l^m(\cos\theta)$, and then integrating over the surface of the sphere (i.e., applying $\int_0^{\pi}\sin\theta\, d\theta\int_0^{2\pi} d\phi$). This causes all terms in the sum to vanish except one, providing us with an equation for B_{lm}:

$$2\pi\frac{2}{2l+1}\frac{(l+m)!}{(l-m)!}B_{lm}a^l = \int_0^{\pi}\sin\theta\, d\theta\int_0^{2\pi} d\phi\, e^{-im\phi}P_l^m(\cos\theta)V(\theta,\phi), \tag{3.2.38}$$

where on the left-hand side we have used Eq. (3.2.37).

Again, we observe the surprising fact that the nontrigonometric eigenmodes $P_l^m(\cos\theta)$ form an orthogonal set on the interval $0<\theta<\pi$, just like trigonometric Fourier modes, but with respect to the integral given in Eq. (3.2.37). The reasons for this will be discussed in Chapter 4.

Spherical Harmonics It is often convenient to combine the associated Legendre functions with the Fourier modes $e^{im\phi}$. It is conventional to normalize the resulting functions of θ and ϕ, creating an orthonormal set called the *spherical harmonics* $Y_{l,m}(\theta,\phi)$:

$$Y_{l,m}(\theta,\phi) = \sqrt{\frac{2l+1}{4\pi}\frac{(l-m)!}{(l+m)!}}\, P_l^m(\cos\theta)\, e^{im\phi}. \tag{3.2.39}$$

Mathematica has already defined these functions, with the intrinsic function `SphericalHarmonicY[l,m,θ,ϕ]`. The spherical harmonics obey the following orthonormality condition with respect to an integral over the surface of a unit sphere:

$$\int_0^{2\pi} d\phi\int_0^{\pi} d\theta\,\sin\theta\, Y_{\bar{l},\bar{m}}^{*} Y_{l,m} = \delta_{l\bar{l}}\,\delta_{m\bar{m}}. \tag{3.2.40}$$

We have already seen in Eq. (3.2.38) that spherical harmonics are useful in decomposing functions defined on the surface of a sphere, $f(\theta,\phi)$, such as the potential on a spherical conductor. The spherical harmonics also enter in many other problems with spherical symmetry. For instance, they describe some of the normal modes of oscillation of a rubber ball, and this example provides a useful method for visualizing the spherical harmonics.

The plots in Cell 3.21 show the distortion of the ball for the $m = 0$ modes, and for $l = 0, \ldots, 5$. These modes are cylindrically symmetric because they have $m = 0$. In each case the dashed line is the unperturbed spherical surface of the ball. We then add to this surface a distortion proportional to the given spherical harmonic.

Cell 3.21

```
m = 0;
Do[r][θ_, ϕ_] = 1 + 0.2 Re[SphericalHarmonicY[l, m, θ, ϕ]];
  a = ParametricPlot[{Sin[θ], Cos[θ]}, {θ, 0, 2 Pi},
   DisplayFunction→ Identity,
   PlotStyle→Dashing[{0.03, 0.03}]];
  b = ParametricPlot[r[θ, 0] {Sin[θ] Cos[0], Cos[θ]},
   {θ, 0, Pi}, DisplayFunction→ Identity];
  c = ParametricPlot[r[θ, Pi] {Sin[θ] Cos[Pi], Cos[θ]},
   {θ, 0, Pi}, DisplayFunction→ Identity];
  d[l] = Show[a, b, c, PlotRange→{{-1.2, 1.2},
   {-1.2, 1.2}},
   AspectRatio→1, Frame→True,
   FrameLabel→ "m=0"<>ToString[m]<>",l=" <>ToString[l],
   RotateLabel→False, AxesLabel→{"x", "z"}], {l, 0, 5}];
Show[GraphicArray[{{d[0], d[1]}, {d[2], d[3]}, {d[4], d[5]}}],
 DisplayFunction→ $DisplayFunction,
 PlotLabel→ "m=<>ToString[m]<> Spherical Harmonics"];
```

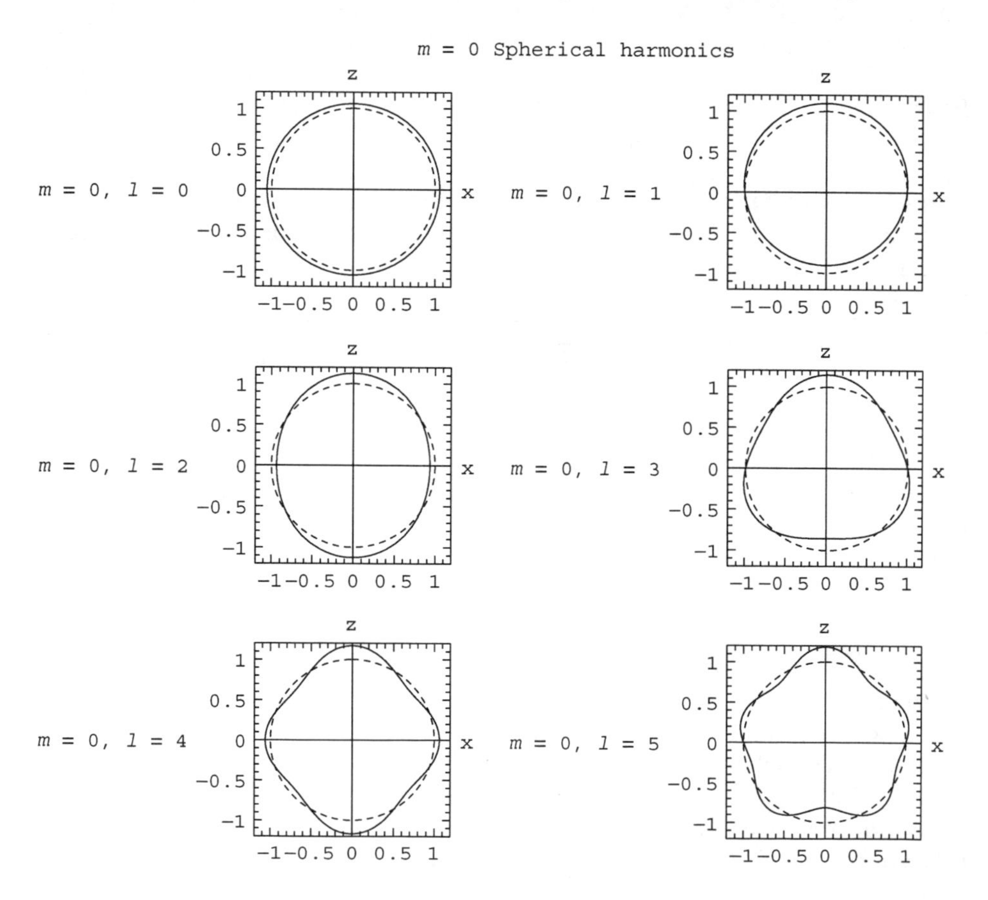

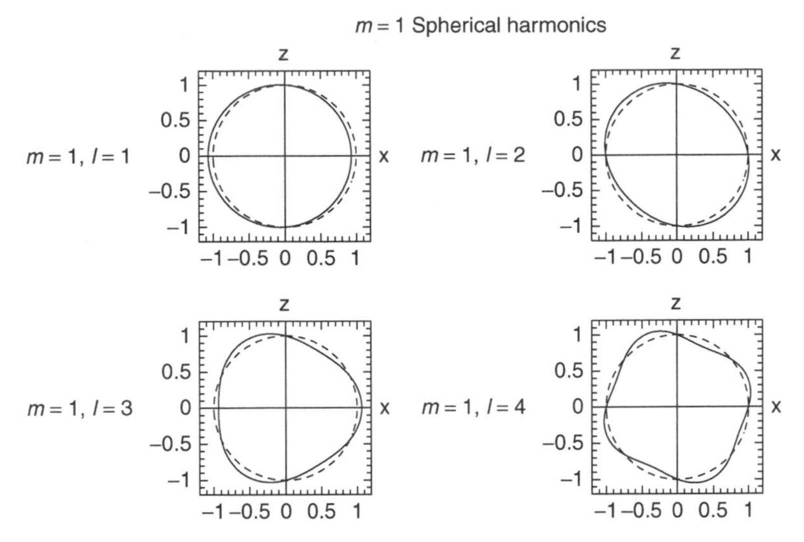

Fig. 3.12 $m = 1$ spherical harmonics.

One can see that the $l = 0$, $m = 0$ harmonic corresponds to a spherically symmetric expansion (or contraction) of the surface; the $l = 1$, $m = 0$ harmonic is a displacement of the sphere along the z-axis; the $l = 2$, $m = 0$ harmonic is an elliptical distortion of the surface; and higher-order harmonics correspond to more complicated distortions. The higher-order distortions have the appearance of sinusoidal oscillations along the surface of the sphere. As one travels from pole to pole along the sphere, the number of times each oscillation passes through zero equals l.

The $m \neq 0$ spherical harmonics are similar to the $m = 0$ harmonics with the same l, except that the distortions are tilted away from the z-axis. For a given value of m, a larger l corresponds to a smaller tilt. For $l = |m|$, the tilt is 90°, that is, the maximum distortion is in the x-y plane. In Fig. 3.12 are some pictures of the real part of the $m = 1$ modes. The reader is invited to modify the commands in Cell 3.21 and display other modes.

One can see that the $l = 1$, $m = \pm 1$ distortions correspond to displacements of the ball in the x-y plane; the $l = 2$, $m = \pm 1$ modes correspond to a tilted elliptical distortion. The distortion can be rotated about the z-axis through any angle ϕ_0 by multiplying the spherical harmonic by a complex number $e^{-i\phi_0}$ before taking the real part. Three-dimensional visualizations of some of these modes can also be found in Sec. 4.4.2.

Example Consider a hollow conducting sphere where the upper half is grounded and the lower half is held at potential V_0:

$$V(\theta, \phi) = \begin{cases} V_0 & \text{for} \quad 0 < \theta < \pi/2, \\ 0 & \text{otherwise}. \end{cases}$$

The potential within the sphere is found by first dropping terms in Eq. (3.2.35) proportional to $1/r^{l+1}$:

$$\Psi(r, \theta, \phi) = \sum_{l=0}^{\infty} \sum_{m=-1}^{l} B_{lm} r^l Y_{l,m}(\theta, \phi). \tag{3.2.41}$$

Matching this solution to $V(\theta, \phi)$, we obtain $\sum_{l=0}^{\infty} \sum_{m=-1}^{l} B_{lm} a^l Y_{l,m}(\theta, \phi) = V(\theta, \phi)$. Multiplying both sides by the complex conjugate of a spherical harmonic and integrating over the unit sphere, we pick out a single term in the sum according to Eq. (3.2.40):

$$B_{lm} a^l = \int_0^{2\pi} d\phi \int_0^{\pi} \sin\theta \, d\theta \, Y_{l,m}^*(\theta, \phi) V(\theta, \phi). \tag{3.2.42}$$

This result is equivalent to Eq. (3.2.38) but is more compact, thanks to the orthonormality of the spherical harmonics. Evaluating the integral over ϕ for our example, we obtain

$$B_{lm} a^l = 2\pi V_0 \delta_{m0} \int_0^{\pi/2} \sin\theta \, d\theta \, Y_{l,0}(\theta, \phi).$$

Thanks to the cylindrical symmetry of the boundary condition, only $m = 0$ spherical harmonics enter in the expansion of the potential. The integral over θ can be evaluated analytically by *Mathematica* for given values of l:

Cell 3.22

```
vlist = 2 Pi V0 Table[
   Integrate[Sin[θ] SphericalHarmonicY[l, 0, θ, ϕ],
    {θ, 0, Pi/2}], {l, 0, 20}]
```

$$\{\sqrt{\pi}V_0, \frac{1}{2}\sqrt{3\pi}V_0, 0, 0\frac{1}{8}\sqrt{7\pi}V_0, 0, \frac{1}{16}\sqrt{11\pi}V_0, 0,$$
$$-\frac{5}{128}\sqrt{15\pi}V_0, 0, \frac{7}{256}\sqrt{19\pi}V_0, 0, -\frac{21\sqrt{23\pi}V_0}{1024}, 0, \frac{99\sqrt{3\pi}V_0}{2048},$$
$$0, -\frac{429\sqrt{31\pi}V_0}{32768}, 0, \frac{715\sqrt{35\pi}V_0}{65536}, 0, -\frac{2431\sqrt{39\pi}V_0}{262144}, 0\}$$

This list of integrals can then be used to construct the solution for Ψ using Eq. (3.2.41), keeping terms up to $l = 20$:

Cell 3.23

```
Ψ[r_, θ_, ϕ_] =
  Sum[vlist[[l + 1]] SphericalHarmonicY[l, 0, θ, ϕ] (r/a)^l,
   {l, 0, 20}];
```

In Cell 3.24 we plot the solution as a surface plot in the x-z plane, taking $V_0 = 1$ volt:

Cell 3.24

```
a = 1;
V0 = 1;
ParametricPlot3D[{r Sin[θ], r Cos[θ], Evaluate[Ψ[r, θ, 0]]},
  {r, 0, a}, {θ, -Pi, Pi}, AxesLabel→{"x", "z", ""},
PlotLabel→"Potential inside a sphere",
  PlotPoints→{20, 100}, ViewPoint -> {2.557, -0.680, 1.414}];
```

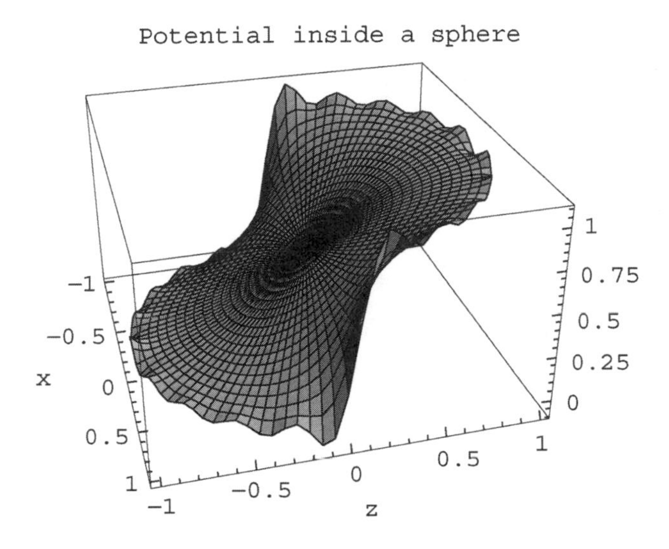

The solution varies smoothly throughout most of the spherical volume. However, near the surface at $r = a$ there is a Gibbs phenomenon due to the discontinuity in the boundary conditions. Fortunately this phenomenon vanishes as one moves inward, away from the surface.

3.2.5 3D Cylindrical Geometry

Introduction Consider a hollow cylindrical tube of length L and radius a, with closed ends at $z = 0$ and $z = L$. We will describe the solution in cylindrical coordinates (r, θ, z), where $x = r\cos\theta$ and $y = r\sin\theta$. The potential on the sides of the tube at $r = a$ is specified by a function $\phi_A(\theta, z)$:

$$\phi(a, \theta, z) = \phi_A(\theta, z),$$

and the potentials on the bottom and top of the tube are specified by functions $\phi_B(r, \theta)$ and $\phi_C(r, \theta)$ respectively:

$$\phi(r, \theta, 0) = \phi_B(r, \theta),$$
$$\phi(r, \theta, L) = \phi_C(r, \theta).$$

In order to solve Laplace's equation for the potential with the cylinder, it is best to follow the strategy discussed in the previous section on rectangular geometry, and break the problem into three separate problems, each of which has a nonzero potential only on one surface.

Nonzero Potential on One Cylinder End: Bessel Functions One of these three problems has boundary conditions $\phi(r,\theta,L)=\phi_C(r,\theta)$ and $\phi=0$ on the rest of the cylinder. Anticipating that the eigenmodes in θ vary as $e^{im\theta}$, we look for a solution of the form $\phi(r,\theta,z)=R(r)\,e^{im\theta}Z(z)$. Applying ∇^2 to this form using Eq. (3.2.16), one finds that

$$\frac{\nabla^2\phi}{\phi}=\frac{1}{rR(r)}\frac{\partial}{\partial r}\left(r\frac{\partial R}{\partial r}\right)-\frac{m^2}{r^2}+\frac{1}{Z(z)}\frac{\partial^2 Z}{\partial z^2}=0. \tag{3.2.43}$$

Separating variables results in the following equation for $Z(z)$:

$$\frac{\partial^2 Z}{\partial z^2}=\kappa z, \tag{3.2.44}$$

where κ is a separation constant. Thus, the general solution for $Z(z)$ is

$$Z(z)=A\,e^{\sqrt{\kappa}z}+B\,e^{-\sqrt{\kappa}z} \tag{3.2.45}$$

and can be either exponential or oscillatory, depending on the sign of κ. (We will find that $\kappa>0$.) The boundary condition that $\phi=0$ at $z=0$ is satisfied by taking $B=-A$. Thus, the z-solution is $Z(z)=2A\sinh(\sqrt{\kappa}z)$.

Bessel Functions. The separated equation for $R(r)$ is

$$\frac{1}{r}\frac{\partial}{\partial r}\left(r\frac{\partial R}{\partial r}\right)-\frac{m^2}{r^2}R+\kappa R=0. \tag{3.2.46}$$

This is a second-order ODE with a regular singular point at the origin. The boundary conditions on this problem are that $R=0$ at $r=a$, and that R is finite at $r=0$ (the regular singular point at $r=0$ implies that the solution can blow up there.) With these boundary conditions, one possible solution is $R=0$. Thus, Eq. (3.2.46) is another eigenvalue problem, where in this case the eigenvalue is κ.

The dependence of R on κ can be taken into account through a simple change in variables,

$$\bar{r}=\sqrt{\kappa}\,r, \tag{3.2.47}$$

yielding

$$\frac{1}{\bar{r}}\frac{\partial}{\partial \bar{r}}\left(\bar{r}\frac{\partial R}{\partial r}\right)-\left(\frac{m^2}{\bar{r}^2}-1\right)R(\bar{r})=0. \tag{3.2.48}$$

This ODE is *Bessel's equation.* The general solution is in terms of two independent functions, called *Bessel functions of the first kind,* $J_m(\bar{r})$ and $Y_m(\bar{r})$:

$$R(\bar{r})=AJ_m(\bar{r})+BY_m(\bar{r}). \tag{3.2.49}$$

The subscript m on these functions is the *order* of the Bessel function. The functions depend on m through its appearance in Bessel's equations, Eq. (3.2.48). *Mathematica* calls the Bessel functions `BesselJ[m,r̄]` and `BesselY[m,r̄]` respectively. We plot these functions in Cells 3.25 and 3.26 for several integer values of m. The reader is invited to reevaluate these plots for different values of m and over different ranges of r in order to get a feel for the behavior of these functions.

Cell 3.25

```
<<Graphics`;

Plot[{BesselJ[0, r], BesselJ[1, r], BesselJ[2, r]},
  {r, 0, 10}, PlotStyle→{Red, Blue, Green},
  PlotLabel→"J_m(r) for m=0,1,2", AxesLabel→{"r", " "}];
```

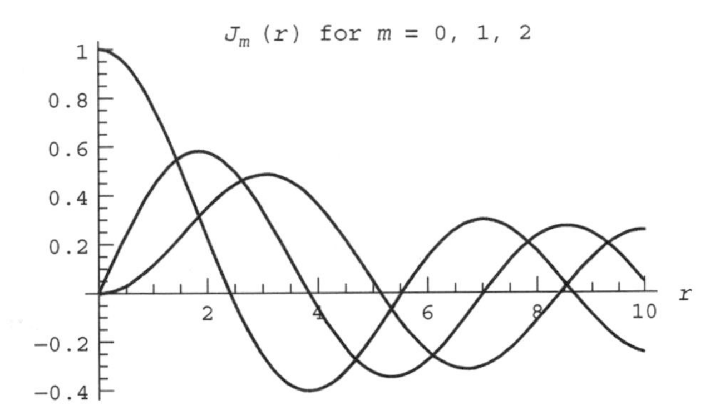

Cell 3.26

```
Plot[{BesselY[0, r], BesselY[1, r], BesselY[2, r]},
  {r, 0, 10}, PlotStyle→{Red, Blue, Green},
  PlotRange→{-2, 1},
  PlotLabel→"Y_m(r) for m=0,1,2", AxesLabel→{"r", " "}];
```

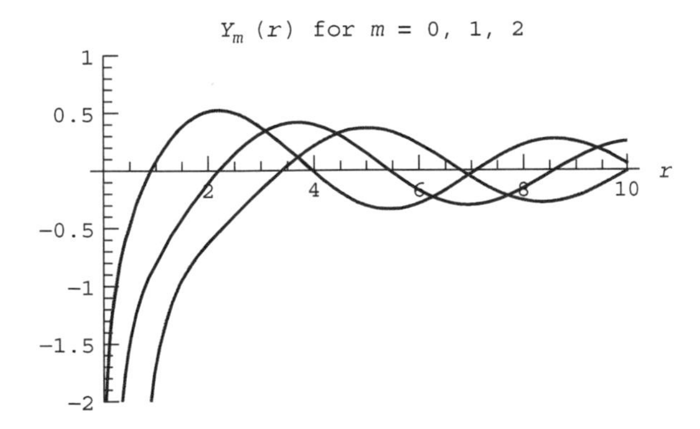

One thing that is immediately apparent is that the Y_m's are all singular at the origin, due to the regular singular point there. Therefore, we can rule out these functions for the eigenmodes in our problem, and write

$$R(\bar{r}) = AJ_m(\bar{r}). \tag{3.2.50}$$

Zeros of Bessel Functions. Another feature of Bessel functions also jumps out from the previous plots: like sines and cosines, these functions oscillate. In fact, one can think of $J_m(\bar{r})$ and $Y_m(\bar{r})$ as cylindrical coordinate versions of trigonometric functions. Each function crosses through zero an infinite number of times in the range $0 < r < \infty$. However, unlike trigonometric functions, the location of the zero crossings of the Bessel functions cannot be written down with any simple formula (except in certain limiting cases such as the zeros at large r: see the exercises for Sec. 5.2).

Starting with the smallest zero and counting upwards, we can formally refer to the nth zero of $J_m(\bar{r})$ as $j_{m,n}$; that is, $J_m(j_{m,n}) = 0$, $n = 1, 2, 3, \ldots$. Similarly, the nth zero of $Y_m(\bar{r})$ is referred to as $y_{m,n}$, and satisfies $Y_m(y_{m,n}) = 0$.

Although these zero crossings cannot be determined analytically, they can be found numerically. In fact, *Mathematica* has several intrinsic functions whose purpose is to evaluate the zeros of Bessel functions. They must be loaded from the add-on package **NumericalMath**:

Cell 3.27

```
<<NumericalMath`;
```

To obtain the first 10 consecutive zeros of $J_0(r)$, the syntax is as follows:

Cell 3.28

```
j0 = BesselJZeros[0, 10]

{2.40483, 5.52008, 8.65373, 11.7915,
 14.9309, 18.0711, 21.2116, 24.3525, 27.4935, 30.6346}
```

Thus, the smallest zero of $J_0(r)$, $j_{0,1}$, takes on the value $j_{0,1} = 2.40483\ldots$, while the next is at $j_{0,2} = 5.52008\ldots$, and so on. Similarly, the first four consecutive zeros of $J_1(r)$, $\{j_{1,1}, j_{1,2}, j_{1,3}, j_{1,4}\}$, are obtained via

Cell 3.29

```
BesselJZeros[1, 4]

{3.83171, 7.01559, 10.1735, 13.3237}
```

Although we do not need them here, the first M zeros of the Y_m Bessel function can also be obtained numerically, with the intrinsic function **BesselYZeros[m,M]**. For instance,

Cell 3.30

```
BesselYZeros[2, 6]

{3.38424, 6.79381, 10.0235, 13.21, 16.379, 19.539}
```

Radial Eigenfunctions and Eigenvalues. The potential is zero on the tube sides at $r = a$. We can use our knowledge of the zeros of the Bessel function in order to match the resulting boundary condition $R(a)=0$. According to Eqs. (3.2.50) and (3.2.47), $R(r)=AJ_m(\sqrt{\kappa}\,r)$. Therefore, $R(a)=AJ_m(\sqrt{\kappa}\,a)$, so we must specify

$$\kappa = (j_{m,n}/a)^2, \qquad n=1,2,3,\dots, \tag{3.2.51}$$

where $j_{m,n}$ is the nth zero of the mth Bessel function. This implies that the radial eigenfunctions $R(r)$ are

$$R(r) = AJ_m(j_{m,n}r/a). \tag{3.2.52}$$

A few of the $m=0$ radial eigenmodes are plotted in Cell 3.31. Eigenfunctions for other values of m can be plotted in the same way. This is left as an exercise for the reader.

Cell 3.31

```
Plot[Evaluate[Table[BesselJ[0, j0[[n]] r], {n, 1, 4}]],
  {r, 0, 1}, PlotStyle→{Red, Blue, Green, Purple},
  PlotLabel→"J0(j0,n r/a) for n=1 to 4",
  AxesLabel→{"r/a", " "}];
```

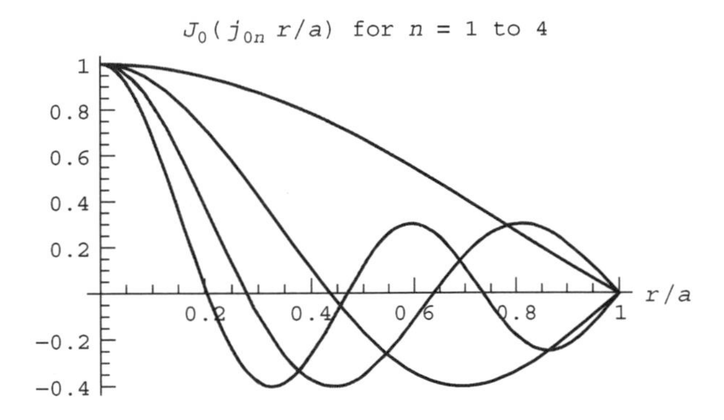

The General Solution for ϕ. The full solution for ϕ within the cylinder is obtained by summing over all radial eigenmodes and all θ-eigenmodes:

$$\phi(r,\theta,z) = \sum_{n=1}^{\infty}\sum_{m=-\infty}^{\infty} A_{mn}J_m(j_{m,n}r/a)\,e^{im\theta}\sinh(j_{m,n}z/a), \tag{3.2.53}$$

where the Fourier coefficients A_{mn} remain to be determined. This solution matches the homogenous boundary conditions on $r=a$ and $z=0$. To satisfy the inhomogeneous boundary condition at $z=L$, namely $\phi(r,\theta,L)=\phi_C(r,\theta)$, we choose the A_{mn}'s so that

$$\phi_C(r,\theta) = \sum_{n=1}^{\infty}\sum_{m=-\infty}^{\infty} A_{mn}J_m(j_{m,n}r/a)\,e^{im\theta}\sinh(j_{m,n}L/a).$$

Putting aside the radial dependence for a moment, note that this is a Fourier series in θ. Application of the orthogonality relation for the Fourier modes $e^{im\theta}$, Eq. (2.1.29), allows us to extract a single value of m from the sum in the usual way:

$$\sum_{n=1}^{\infty} A_{mn} \sinh(j_{m,n}L/a)\, J_m(j_{m,n}r/a) = \frac{1}{2\pi}\int_0^{2\pi} d\theta\, e^{-im\theta}\phi_C(r,\theta). \quad (3.2.54)$$

However, this is not enough to determine A_{mn}. It would be nice if there were some equivalent orthogonality relation that we could apply to the Bessel functions. Amazingly, such an orthogonality relation exists:

$$\int_0^a J_m(j_{m,\bar{n}}r/a) J_m(j_{m,n}r/a)\, r\, dr = \delta_{n\bar{n}} \frac{a^2}{2} J_{m+1}^2(j_{m,n}), \quad (3.2.55)$$

where $\delta_{n\bar{n}}$ is a Kronecker delta function. Thus, for $n \neq \bar{n}$, the different radial eigenmodes are orthogonal with respect to the radial integral $\int_0^a r\,dr$. This result can be checked using *Mathematica*:

Cell 3.32

```
Simplify[Integrate[BesselJ[m, j_{m,n} r/a]
  BesselJ[m, j_{m,n̄} r/a] r, {r, 0, a}] /.
  BesselJ[m, j_{m,n_}] → 0]
```

0

Cell 3.33

```
FullSimplify[Integrate[BesselJ[m, j_{m,n} r/a]^2 r, {r, 0, a}]-
  a^2/2 BesselJ[m + 1, j_{m,n}]^2, BesselJ[m, j_{m,n}] == 0 ]
```

0

We can use Eq. (3.2.55) to extract a single term from the sum over n in Eq. (3.2.55). Multiplying both sides of the equation by $J_m(j_{m,\bar{n}}r/a)$ and applying the integral $\int_0^a r\,dr$, we find

$$\sum_{n=1}^{\infty} A_{mn} \sinh(j_{mn}L/a) \int_0^a r\,dr J_m(j_{m,\bar{n}}r/a) J_m(j_{mn}r/a)$$
$$= \frac{1}{2\pi}\int_0^a r\,dr J_m(j_{m,\bar{n}}r/a) \int_0^{2\pi} d\theta\, e^{-im\theta}\phi_C(r,\theta).$$

Substituting for the radial integral on the left-hand side using Eq. (3.2.55), we see that only the term with $n = \bar{n}$ contributes, leaving us with the equation

$$A_{m\bar{n}} \sinh(j_{m\bar{n}}L/a) \frac{a^2}{2} J_{m+1}^2(j_{m,\bar{n}}) = \frac{1}{2\pi}\int_0^a r\,dr J_m(j_{m,\bar{n}}r/a) \int_0^{2\pi} d\theta\, e^{-im\theta}\phi_C(r,\theta). \quad (3.2.56)$$

This equation provides us with the Fourier coefficients A_{mn}.

It is really quite surprising that the Bessel functions form an orthogonal set. It appears that every time we solve an eigenvalue problem, we obtain an orthogonal set of eigenmodes. The reasons for this will be discussed in Chapter 4.

A trivial extension of the same method used here could also be used to determine the form of the potential due to the nonzero potential at $z=0$, $\phi(r,\theta,0)=\phi_B(r,\theta)$. This part of the problem is left as an exercise.

Example Say the potential on the top of the cylinder is fixed at $\phi_C(r,\theta)=V_0$, and the other sides of the cylinder are grounded. Then according to Eq. (3.2.56), only the $m=0$ Fourier coefficients A_{0n} are nonzero, and the potential within the cylinder is given by Eq. (3.2.56) by a sum over the zeros of a single Bessel function, J_0:

$$\phi(r,\theta,z)=\sum_{n=1}^{\infty} A_{0n} J_0(j_{0,n} r/a)\sinh(j_{0,n} z/a). \qquad (3.2.57)$$

As expected from the symmetry of the problem, the potential is cylindrically symmetric. The Fourier coefficients are given by Eq. (3.2.56):

$$A_{0n}=\frac{V_0}{(a^2/2)J_1^2(j_{0,n})\sinh(j_{0,n}L/a)}\int_0^a r\,dr\,J_0(j_{0,n}r/a).$$

The required integral can be performed analytically using *Mathematica*:

Cell 3.34

```
A[n_] = V0/(a^2 BesselJ[1, j[n]]^2 Sinh[j[n] L/a])
Integrate[r BesselJ[0, j[n] r/a], {r, 0, a}]

V0 Csch[L j[n]/a]
-------------------------
BesselJ[1, j[n]] j[n]
```

Here we have introduced the notation `j[n]` for the nth zero of the Bessel function J_0. For the first 20 zeros, this function can be defined in the following way:

Cell 3.35

```
<< NumericalMath`;

j0 = BesselJZeros[0, 20]; j[n_] := j0[[n]]
```

Now the potential can be evaluated numerically, keeping the first 20 terms in the sum in Eq. (3.2.57):

Cell 3.36

```
ϕ[r_, z_] = Sum[A[n] BesselJ[0, j[n] r/a] Sinh[j[n] z/a],
  {n, 1, 20}];
```

This potential is plotted in Cell 3.37 as a contour plot, taking $V_0=1$ volt and $L/a=2$. There is a Gibbs phenomenon near the top of the cylinder, because of the discontinuity in the potential between the top and the grounded sides. However, this phenomenon dies away rapidly with distance from the top, leaving a well-behaved solution for the potential in the cylinder.

Cell 3.37

```
L = 2a; a = 1; V0 = 1;
ContourPlot[φ[r, z], {r, 0, a}, {z, 0, L},
  AspectRatio→2, PlotPoints→40, FrameLabel→{"r", "z"},
  PlotLabel→"φ(r,z) in a cylinder \n with grounded sides"];
```

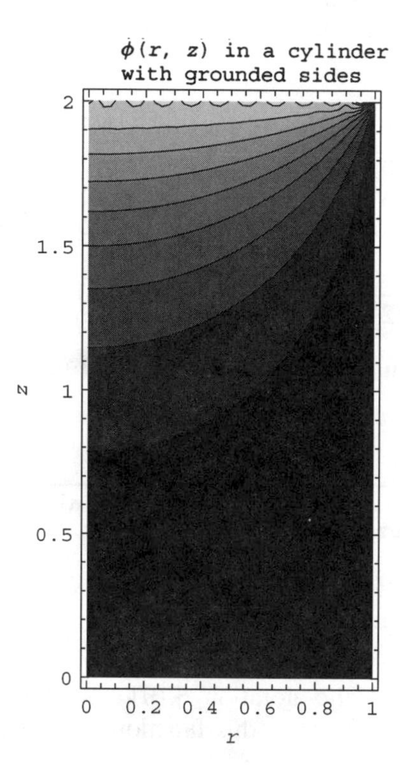

Nonzero Potential on the Cylinder Sides: Modified Bessel Functions We now consider the solution to the Laplace equation for a potential applied only to the sides of a cylinder of finite length L and radius a, $\phi(a, \theta, z) = \phi_A(\theta, z)$. On the ends of the cylinder at $z = 0$ and $z = L$, the boundary conditions are $\phi = 0$.

This problem is still solved by a sum of terms of the form $R(r)\,e^{im\theta}Z(z)$. Furthermore, separation of variables implies that $R(r)$ and $Z(z)$ are still governed by the same ODEs, Eqs. (3.2.43) and (3.2.45), with the same general solutions, Eqs. (3.2.45) and (3.2.49). However, the boundary conditions dictate that the coefficients in these general solutions must be chosen differently than in the previous case.

In order to satisfy $\phi(r, \theta, 0) = 0$, we require that $B = -A$ in Eq. (3.2.45), and in order to satisfy $\phi(r, \theta, L) = 0$, we require that

$$\kappa^2 = -(n\pi/L)^2, \tag{3.2.58}$$

so that $Z(z) = 2Ai\sin(n\pi z/L)$. Now the z-equation has provided us with eigenmodes and eigenvalues, and they are standard trigonometric Fourier modes.

The radial solution $R(r)$ is still given by Eq. (3.2.49) but with the imaginary value of κ given Eq. (3.2.58):

$$R(r) = CJ_m(in\pi r/L) + DY_m(in\pi r/L) \tag{3.2.59}$$

Bessel functions of an imaginary argument are called *modified Bessel functions*, I_m and K_m. These functions are defined below for integer m:

$$\begin{aligned} I_m(x) &= i^{-m} J_m(ix) \qquad \text{(for integer } m), \\ K_m(x) &= \frac{\pi\, i^{m+1}}{2} \left[J_m(ix) + iY_m(ix) \right] \qquad \text{(for integer } m). \end{aligned} \tag{3.2.60}$$

The modified Bessel functions bear a similar relation to J_m and Y_m to the one the hyperbolic functions sinh and cosh bear to the trigonometric functions sin and cos. In *Mathematica*, these functions are called **BesselI[m,x]** and **BesselK[m,x]**. The first few modified Bessel functions are plotted in Cells 3.38 and 3.39. The I_m's are finite at the origin, but become exponentially large at large x. The K_m's are singular at the origin, but approach zero (with exponential rapidity) at large x.

Cell 3.38

```
<< Graphics`;
Plot[{BesselI[0, x], BesselI[1, x], BesselI[2, x]},
  {x, 0, 4}, PlotStyle→{Red, Green, Blue},
  PlotLabel→"I_m(r) for m=0,1,2", AxesLabel→{"x", ""}];
```

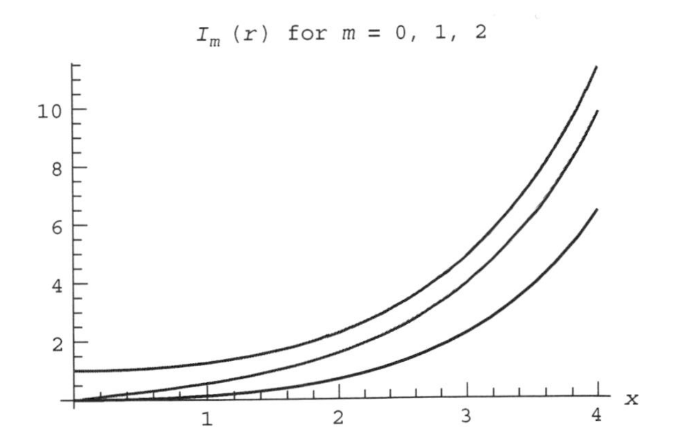

Cell 3.39

```
Plot[{BesselK[0, x], BesselK[1, x], BesselK[2, x]},
  {x, 0, 4}, PlotStyle→{Red, Green, Blue},
  PlotLabel→"K_m(r) for m=0,1,2", AxesLabel→ ["x", ""}];
```

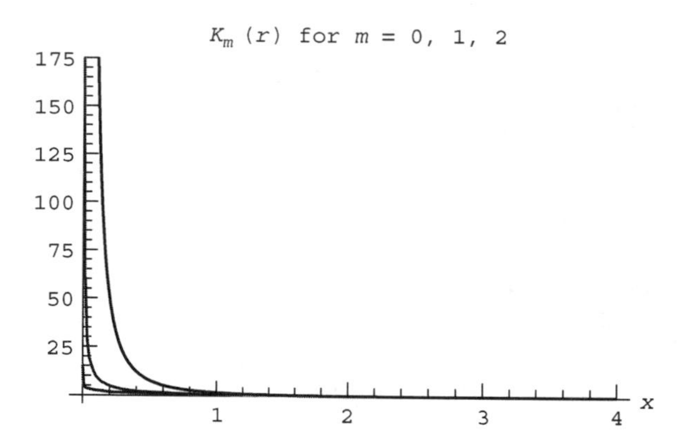

The general solultion for $R(r)$ is, according to Eq. (3.2.59), a sum of I_m and K_m. However, the previous plots show that only the I_m term should be kept, in order that the solution be finite at $r = 0$. Therefore, we find that the radial function is $R(r) = I_m(n\pi r/L)$, and the full solution for $\phi(r, \theta, z)$ is a linear combination of these functions:

$$\phi(r, \theta, z) = \sum_{m=-\infty}^{\infty} \sum_{n=1}^{\infty} B_{nm} I_m(n\pi r/L)\, e^{im\theta} \sin(n\pi z/L). \tag{3.2.61}$$

The Fourier coefficients a_{mn} are determined by matching to the inhomogeneous boundary condition at $r = a$:

$$\phi(a, \theta, z) = \phi_A(\theta, z) = \sum_{m=-\infty}^{\infty} \sum_{n=1}^{\infty} B_{nm} I_m(n\pi a/L)\, e^{im\theta} \sin(n\pi z/L).$$

Since this sum is a regular Fourier series in both θ and z, we can use orthogonality of the trigonometric functions to find the Fourier coefficients:

$$B_{nm} I_m(n\pi a/L) = \frac{1}{\pi L} \int_0^{2\pi} d\theta\, e^{-im\theta} \int_0^L dz \sin(n\pi z/L)\, \phi_A(\theta, z).$$

This completes the problem of determining the potential inside a cylindrical tube of finite length. More examples of such problems are left to the exercises.

EXERCISES FOR SEC. 3.2

(1) Solve $\nabla^2\phi(x, y) = 0$ for the following boundary conditions. Plot the solutions using **Plot3D**.

(a) $\phi(0, y) = \phi(2, y) = \phi(x, 0) = 0$; $\phi(x, 1) = 1$.

(b) $\frac{\partial \phi}{\partial x}(0, y) = \frac{\partial \phi}{\partial x}(1, y) = \frac{\partial \phi}{\partial y}(x, 0) = 0$; $\frac{\partial \phi}{\partial y}(x, 1) = \cos 2\pi n x$, n an integer. (What happens for $n = 0$?)

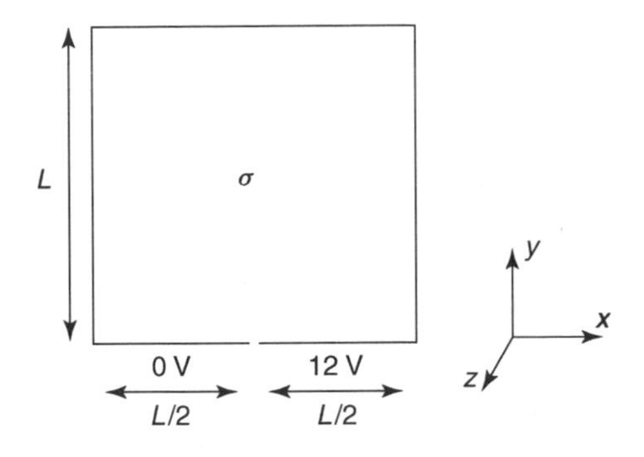

Fig. 3.13 Exercise (2).

(c) $\phi(0, y) = 2;\ \phi(1, y) = \phi(x, 0) = 0;\ \dfrac{\partial \phi}{\partial y}(x, 1) = 1.$

(d) $\dfrac{\partial \phi}{\partial x}(0, y) = \dfrac{\partial \phi}{\partial x}(1, y) = \dfrac{\partial \phi}{\partial y}(x, 0) = 0,\ \dfrac{\partial \phi}{\partial y}(x, 1) = 1.$

(2) A battery consists of a cube of side L filled with fluid of conductivity σ. The electrodes in the battery consist of two plates on the base at $y = 0$, one grounded and one at potential $V = 12$ volts (see Fig. 3.13). The other sides of the battery casing are not conductive. Find the potential ϕ everywhere inside the battery.

(3) Find the solution to $\nabla^2\phi(r, \theta) = 0$ inside a 2D cylinder for the following boundary conditions. Plot the solutions using contour plots.

(a) $\phi(1, \theta) = \cos n\theta$, n an integer.

(b) $\phi(1, \theta) = 0$ for $x > 0$; $\phi(1, \theta) = 1$ for $x < 0$.

(c) $\phi(1, \theta) = 0$, $\phi(2, \theta) = h(\theta)$ (find ϕ between the concentric cylinders; h is a Heaviside step function, $-\pi < \theta < \pi$ assumed).

(d) $\dfrac{\partial \phi}{\partial r}(1, \theta) = \sin 2\theta$, $\phi(2, \theta) = \cos\theta$ (find ϕ between the concentric cylinders).

(4) A long conducting cylinder of radius $a = 5$ cm has a sector of opening angle $\alpha = 20°$ that is electrically isolated from the rest of the cylinder by small gaps (see Fig. 3.14). The sector is placed at potential $V_0 = 1$ volt. Find the potential and plot it throughout the cylinder.

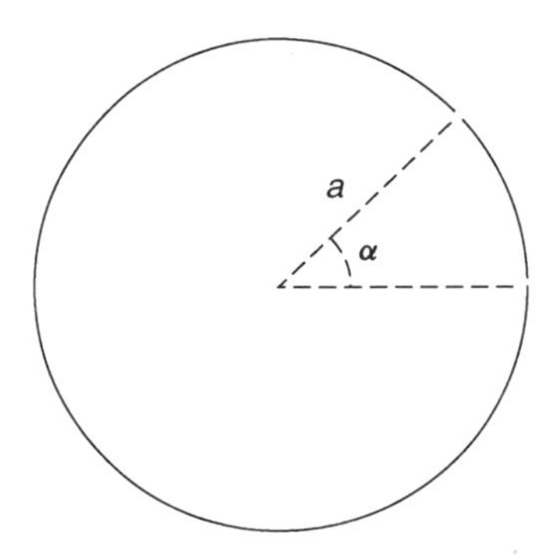

Fig. 3.14 Exercise (4).

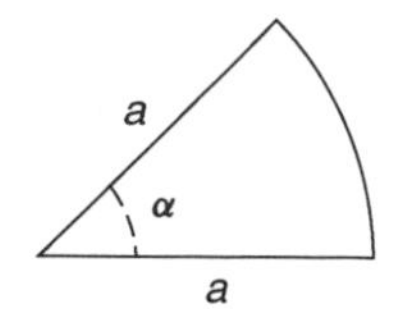

Fig. 3.15 Exercise (5).

(5) **(a)** The bottom and top of the wedge shown in Fig. 3.15 are grounded, but the end at $r = a$ is held at $V = 5$ volts. Find the potential everywhere inside the wedge by using separation of variables, and plot it as a surface plot for $a = 30°$.

(b) Show that the radial electric field near $r = 0$ is singular near $r = 0$ if $\alpha > \pi$, and find the form of the singularity. [Answer: $E_r \propto r^{\pi/\alpha - 1} \sin(\pi\theta/\alpha)$ as $r \to 0$.]

(6) An electrolytic cell consists of two plastic concentric cylinders of radii a and b, $a < b$, and length L. Between the cylinders is an acid with conductivity σ. Two conducting vanes of width $b - a$ and length L are placed in the cell between the cylinders, parallel to the axis of the cell, one along the positive x-axis and one along the negative x-axis. If the vanes are held at a fixed potential difference V_0, find the total current running between the vanes. Plot the contours of constant potential. [Hint: The potential satisfies Laplace's equation with von Neumann boundary conditions at the nonconducting surfaces and Dirichlet conditions at the conductors: see Exercise (2) above and Example 2 in Sec. 3.2.2. Be careful to keep *all* of the radial eigenmodes, including the $n = 0$ (constant) mode.]

(7) Find the solution to $\nabla^2\Psi(r, \theta, \phi) = 0$ inside a sphere with the following boundary conditions.

(a) $\Psi(1, \theta, \phi) = \sin^2 \theta$.

(b) $\Psi(1, \theta, \phi) = V_0 \theta^2(\pi - \theta)$.

(c) $\dfrac{\partial \Psi}{\partial r}(1, \theta, \phi) = \sin 2\theta \cos \phi$.

(8) A conducting sphere of radius $a = 2$ cm is cut in the half (see Fig. 3.16). The left half is grounded, and the right half is at 10 V. Find the electrostatic potential $\phi(r, \theta)$ everywhere outside the sphere, assuming that the potential goes to zero at infinity. Plot the potential for $a < r < 6$ cm.

(9) Two concentric spheres have radii a and b ($b > a$). Each is divided into two hemispheres by the same horizontal plane. The upper hemisphere of the

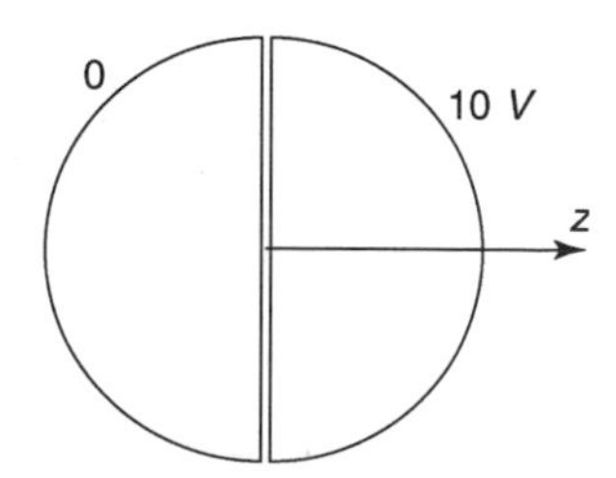

Fig. 3.16 Exercise (8).

outer sphere and the lower hemisphere of the inner sphere are maintained at potential V. The other hemispheres are grounded. Determine the potential in the region $a \le r \le b$ in terms of Legendre polynomials. Plot the result for $b = 3a/2$. Check your result against known solutions in the limiting cases $b \to \infty$ and $a \to 0$.

(10) A grounded conducting sphere of radius a is placed with its center at the origin, in a uniform electric field $\mathbf{E} = E\hat{\mathbf{z}}$ that is created by an external potential $\phi_e = -Ez$. Find the total potential $\phi(\mathbf{r})$ outside the sphere. (Hint: $\phi = 0$ at $r = a$; write ϕ_e in spherical coordinates.)

(11) By applying the Taylor expansion function `Series` to the Bessel functions, investigate the small-x form of $J_m(x)$ and $Y_m(x)$. Answer the following questions:

(a) Find a simple analytic expression for $J_m(x)$ (m an integer) as x approaches zero. Does your expression apply for $m < 0$ as well as $m > 0$?

(b) The Y_m's are singular near the origin. Find the form of the singularity for $m = 0, 1, 2, \dots$.

(12) Repeat the previous exercise for the modified Bessel functions $I_m(x)$ and $K_m(x)$ (m an integer).

(13) A cylinder has length $L = 1$ meter and radius $a = \frac{1}{2}$ meter. The sides and the top of the cylinder are grounded, but the base at $z = 0$ is held at potential $V = 50$ volts. Find the potential $\phi(r, z)$ throughout the interior. Plot the result using `ContourPlot`.

(14) A cylinder of unit height and radius has grounded ends, at $z = -\frac{1}{2}$ and $z = \frac{1}{2}$. The cylindrical wall at $r = 1$ is split in half lengthwise along the $x = 0$ plane. The half with $x > 0$ is held at 10 volts, and the half for which $x < 0$ is grounded. Find the potential and obtain the value of the electric field, $-\nabla\phi$ (magnitude and direction), at the origin. Plot the potential in the $z = 0$ plane using a surface plot.

(15) In a *Penning-Malmberg trap*, used to trap charged particles with static electric and magnetic fields, potentials are applied to coaxial cylindrical electrodes of radius a, as shown in Fig. 3.17 in order to provide an axial trapping potential.

(a) Find the potential $\phi(r, z)$ in the central region around $\mathbf{r} = \mathbf{0}$ as a Fourier series involving the Bessel function I_0.

(b) Taylor-expand this expression in r and z to show that the potential near $\mathbf{r} = \mathbf{0}$ has the form

$$\phi(r, z) = Ar^2 + Bz^2 + C. \tag{3.2.62}$$

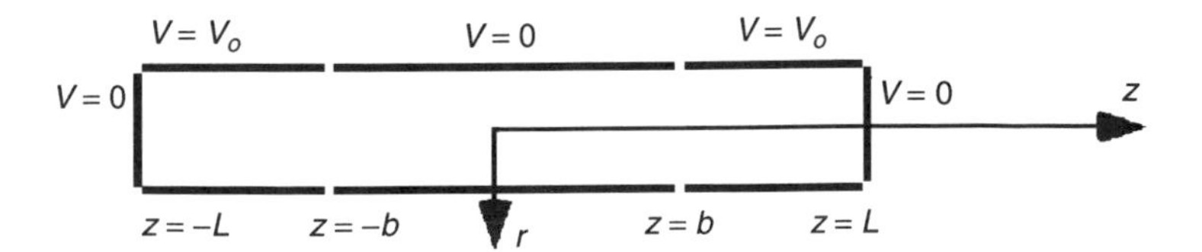

Fig. 3.17 Exercise (15).

This form of ϕ is simply an even Taylor expansion in r and z, as required by the symmetry of the system. Find values for A, B, and C when $a=5$ cm, $b=5$ cm, and $L=10$ cm.

(c) By direct substitution of Eq. (3.2.62) into Laplace's equation, show that $A=-B/2$. This implies that the center of the trap is a saddlepoint of the potential. Does your solution from part (a) satisfy this?

(d) By carefully taking a limit as $L \to \infty$, show that your result from part (a) can be written as the following Fourier integral:

$$\phi(r,z) = V_0\left(1 - \frac{2}{\pi}\int_0^\infty \frac{dk}{k}\sin kb \cos kz \frac{I_0(kr)}{I_0(ka)}\right).$$

[Hint: *Mathematica* can do the following sum analytically: $\sum_{n=0}^{\infty}(-1)^n/(2n+1)$.] Use the integral form to plot $\phi(0,z)$ for $-2b \le z \le 2b$, taking $a=b/2$.

(16) A cylindrical copper bar of conductivity σ has radius a and length L, with $0<z<L$. The surface of the bar is insulated, except on the ends at $r=0$, where wires are attached. The wire at $z=r=0$ injects current I, and the wire at $r=0$, $z=L$ removes the same current. Therefore, the form of the current density at $z=0$ and L is $\mathbf{j}(r,z=0)=\mathbf{j}(r,z=L)=I[\delta(r)/2\pi r]\hat{\mathbf{z}}$. Find the electrostatic potential $\phi(r,z)$ throughout the bar, and plot it as a contour plot, in suitably normalized coordinates, assuming that $\phi(0,L/2)=0$. Show that your solution satisfies $2\pi\int_0^a r\,dr\,j_z(r,z)=I$, and plot $j_z(r,L/2)$ vs. r. (Hint: keep *all* eigenmodes, including one with a zero eigenvalue. See Example 2 in Sec. 3.2.2.)

REFERENCES

Arthur C. Clarke, *The Fountains of Paradise* (Harcourt Brace Jovanovitch, New York, 1979).

D. G. Zill and M. R. Cullen, *Advanced Engineering Mathematics*, 2nd ed. (Jones and Bartlett, Sudbury, Mass., 2000).

CHAPTER 4

EIGENMODE ANALYSIS

4.1 GENERALIZED FOURIER SERIES

In Chapter 3 we constructed solutions to several linear partial differential equations using the method of separation of variables. The solutions were written in terms of an infinite sum of eigenmodes arising from an associated eigenvalue problem. Sometimes the eigenmodes were trigonometric functions, and the sums formed a Fourier series. In other examples the eigenmodes were Bessel functions, or associated Legendre functions. Nevertheless, in every case studied, these eigenmodes were orthogonal to one another, and it was only for this reason that they were useful in finding an analytic solution.

Why did the eigenmodes in each of these problems form an orthogonal set? In this section we answer this important question. In short, the answer is that these eigenmodes spring from a particular type of eigenvalue problem: a *Sturm–Liouville problem*. The differential operators in Sturm–Liouville problems have the property that they are *Hermitian*, and this property implies that the eigenmodes of the operators form a complete orthogonal set.

First, however, we need to generalize the idea of a Fourier series. There is nothing particularly special about trigonometric functions. One can describe a given function using an infinite series constructed from many orthogonal function sets, not just trigonometric Fourier modes. These series expansions are called *generalized Fourier series*. We will examine the workings of these series, and discuss a general way to create orthogonal sets of functions for use in these series: the *Gram–Schmidt method.*

4.1.1 Inner Products and Orthogonal Functions

Definition of an Inner Product In order to discuss generalized Fourier series, we must first extend our notion of orthogonality. This requires that we introduce the concept of an *inner product*.

We are all familiar with inner products through their use in linear algebra. The dot product of two N-dimensional real vectors $\mathbf{f}$ and $\mathbf{g}$, $\mathbf{f}\cdot\mathbf{g} = \sum_{i=1}^{N} f_i g_i$, is a type of inner product. For complex functions, an inner product acts on two functions defined on a given interval $a \le x \le b$, in general returning a complex number. The notation that we use for the inner product of the functions f and g is (f, g). One example of an inner product for complex functions f and g is

$$(f, g) = \int_a^b f^*(x) g(x)\, dx. \tag{4.1.1}$$

In inner product notation, two functions are orthogonal when their inner product vanishes: $(f, g) = 0$. This is like two vectors being perpendicular to one another. Using Eq. (4.1.1) for our inner product, the equation $(f, g) = 0$ is equivalent to the definition of orthogonality used in complex exponential Fourier series, Eq. (2.1.28).

It is also possible to write down other inner products. However, all inner products must satisfy certain rules. First,

$$(f, g) = (g, f)^*. \tag{4.1.2}$$

Equation (4.1.2) implies that $(f, f) = (f, f)^*$, so the inner product of a function with itself must be a real number. Another requirement for an inner product is that

$$(f, f) \ge 0 \tag{4.1.3}$$

with equality only if $f = 0$ on $a \le x \le b$. Also, the inner product must be linear in the second argument, so that

$$(f, g + Ch) = (f, g) + C(f, h), \tag{4.1.4}$$

where C is a constant. This implies that the inner product is *antilinear* in the first argument:

$$(f + Ch, g) = (g, f + Ch)^* = (g, f)^* + C^*(g, h)^* = (f, g) + C^*(h, g), \tag{4.1.5}$$

where in the first and last steps we used Eq. (4.1.2).

The inner product of Eq. (4.1.1) clearly satisfies these rules. Another inner product that does so is

$$(f, g) = \int_a^b f^*(x) g(x) p(x)\, dx, \tag{4.1.6}$$

for some real function $p(x)$ that has the property that $p(x) > 0$ on $a < x < b$.

Obviously, functions that are orthogonal with respect to the inner product given by Eq. (4.1.1) will generally not be orthogonal with respect to the inner product given in Eq. (4.1.6).

Sets of Orthogonal Functions. The Gram–Schmidt Method A set of functions $\{\psi_n(x)\}$ forms an orthogonal set with respect to some inner product if $(\psi_n, \psi_m) = 0$ for $n \ne m$. The trigonometric Fourier modes used in Fourier series are an example of an orthogonal set: $\{e^{i2\pi nx/(b-a)}\}$ forms an orthogonal set with respect to the inner product of Eq. (4.1.1).

It is also possible to find completely different *nontrigonometric* sets of functions that are orthogonal with respect to some inner product. We have already seen several examples of this in Chapter 3. For instance, Eq. (3.2.55) implies that the set of Bessel functions $\{J_m(j_{mn}r/a)\}$ form an orthogonal set with respect to the inner product defined by the integral $\int_0^a r\,dr$.

For a given inner product, one can in fact find an infinite number of different orthogonal sets of functions. One way to create such sets is via the *Gram–Schmidt* method.

The Gram–Schmidt method allows one to construct a set of orthogonal functions $\{\psi_n(x)\}$ out of a given set of functions $\{v_n(x)\}$, $n = 0, 1, 2, 3, \ldots$. There is almost no restriction on the functions chosen for the latter set, except that each function must be different than the previous functions. (More precisely, they must be linearly-independent functions; one function cannot be written as a sum of the others. Thus, $\{1, x, x^2, x^3, \ldots\}$ is a good set, but $\{x, 2x, 3x, 4x, \ldots\}$ is not.) Also, the inner products of these functions with one another must not be singular.

The method is analogous to the method of the same name used to create orthogonal sets of vectors in linear algebra. We will construct an orthogonal set of functions, $\{\psi_n(x)\}$, $n = 0, 1, 2, 3, \ldots$, by taking sums of the functions v_n. To start, we choose $\psi_0(x) = v_0(x)$. Next, we choose $\psi_1(x) = a_0 v_0(x) + v_1(x)$, where the constant a_0 is determined by imposing the requirement that ψ_0 and ψ_1 be orthogonal with respect to the given inner product:

$$(\psi_0, \psi_1) = 0 = a_0(v_0, v_0) + (v_0, v_1).$$

This implies that $a_0 = -(v_0, v_1)/(v_0, v_0)$. Next, we choose $\psi_2(x) = b_0 v_0(x) + b_1 v_1(x) + v_2(x)$, and we determine the constants b_0 and b_1 by the requirement that ψ_2 be orthogonal to both ψ_0 and ψ_1. This gives us two equations in the two unknowns b_0 and b_1:

$$\begin{aligned}
(\psi_0, \psi_2) &= 0 = b_0(v_0, v_0) + b_1(v_0, v_1) + (v_0, v_2),\\
(\psi_1, \psi_2) &= 0 = a_0 b_0(v_0, v_0) + a_0 b_1(v_0, v_1) + a_0(v_0, v_2)\\
&\qquad + b_0(v_1, v_0) + b_1(v_1, v_1) + (v_1, v_2).
\end{aligned}$$

After solving these coupled linear equations for b_0 and b_1, we repeat the procedure, defining $\psi_3(x) = c_0 v_0(x) + c_1 v_1(x) + c_2 v_2(x) + v_3(x)$, and so on.

In fact, it is possible to automate this process in *Mathematica*, constructing any set of orthogonal functions that we wish, for any given inner product. (See the exercises.)

As an example of the Gram–Schmidt method, we will take the following inner product:

$$(f, g) = \int_{-1}^{1} f^*(x) g(x)\,dx, \tag{4.1.7}$$

and we will construct an orthogonal set of functions using the set $\{1, x, x^2, x^3, \ldots\}$ as our starting point. According to the method outlined previously, we take

$\psi_0(x) = 1$, and $\psi_1(x) = a_0 + x$. In order to find a_0 we require that $(\psi_1, \psi_0) = 0$, which implies that $0 = a_0 \int_{-1}^{1} dx + \int_{-1}^{1} x\,dx$. Therefore, we find $a_0 = 0$, so $\psi_1(x) = x$.

Next, we set $\psi_2 = b_0 + b_1 x + x^2$. The conditions that $(\psi_2, \psi_1) = 0$ and $(\psi_2, \psi_0) = 0$ lead to two equations for b_0 and b_1, which can be solved using *Mathematica*.

Cell 4.1

```
ψ[0, x_] = 1;
ψ[1, x_] = x;
ψ[2, x_] = b0 + b1 x + x^2;
sol = Solve[{Integrate[ψ[2, x] ψ[1, x], {x, -1, 1}] == 0,
     Integrate[ψ[2, x] ψ[0, x], {x, -1, 1}] == 0},
       {b0, b1}][[1]];
ψ[2, x] = ψ[2, x]/.sol
```

$$-\frac{1}{3} + x^2$$

Thus, $\psi_2(x) = x^2 - \frac{1}{3}$.

Next, we set $\psi_3 = c_0 + c_1 x + c_2 x^2 + c_3 x^3$, and solve the three coupled equaitons $(\psi_3, \psi_2) = (\psi_3, \psi_1) = (\psi_3, \psi_0) = 0$:

Cell 4.2

```
ψ[3, x_] = c0 + c1 x + c2 x^2 + x^3;
sol = Solve[Table[Integrate[ψ[3, x] ψ[n, x],
              {x, -1, 1}] == 0, {n, 0, 2}],
    {c0, c1, c2}][[1]];
ψ[3, x] = ψ[3, x]/.
  sol
```

$$-\frac{3x}{5} + x^3$$

Thus, $\psi_3(x) = -3x/5 + x^3$.

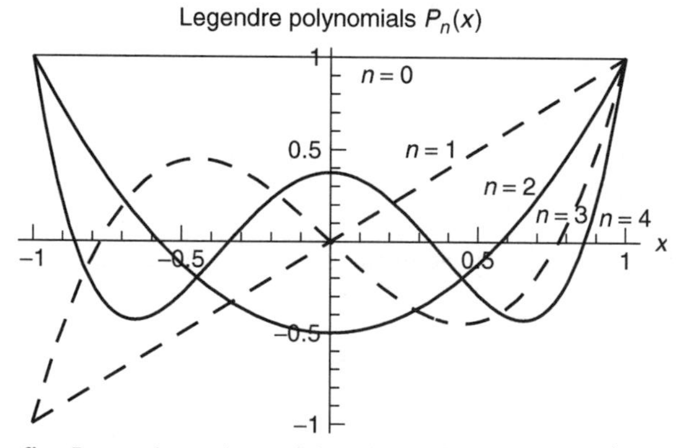

Fig. 4.1 The first five Legendre polynomials. The odd polynomials (n odd) are shown with dashed lines, the even polynomials (n even) with solid lines.

If we continue this process, the set of functions that we construct are proportional to *Legendre polynomials* $\{P_n(x)\}$, $n = 0, 1, 2, \ldots$. *Mathematica* refers to these functions as **LegendreP[n,x]**. We have already run across these polynomials, which are the $m = 0$ cases of the associated Legendre functions $P_l^m(x)$ discussed in connection with solution of the Laplace equation in spherical coordinates. A few of these polynomials are listed below, and are plotted in Fig. 4.1.

Cell 4.3

```
Do[Print[Subscript[P, n], "(x) = ", LegendreP[n, x]],
  {n, 0, 4}]
```

$P_0(x) = 1$

$P_1(x) = x$

$P_2(x) = -\frac{1}{2} + \frac{3x^2}{2}$

$P_3(x) = -\frac{3x}{2} + \frac{5x^3}{2}$

$P_4(x) = \frac{3}{8} - \frac{15x^2}{4} + \frac{35x^4}{8}$

One can verify that these polynomials form an orthogonal set with respect to the inner product of Eq. (4.1.7). In fact, these polynomials satisfy

$$(P_n(x), P_m(x)) = \delta_{nm} \frac{2}{2n+1}, \tag{4.1.8}$$

where δ_{nm}, the Kronecker δ-function, is defined by Eq. (1.6.22). The following is a matrix of inner products over a set of the first five Legendre polynomials:

Cell 4.4

```
MatrixForm[Table[
  Integrate[LegendreP[n, x] LegendreP[m, x], {x, -1, 1}],
   {n, 0, 4}, {m, 0, 4}]]
```

$$\begin{pmatrix} 2 & 0 & 0 & 0 & 0 \\ 0 & \frac{2}{3} & 0 & 0 & 0 \\ 0 & 0 & \frac{2}{5} & 0 & 0 \\ 0 & 0 & 0 & \frac{2}{7} & 0 \\ 0 & 0 & 0 & 0 & \frac{2}{9} \end{pmatrix}$$

The matrix is diagonal, as expected from orthogonality, and the values on the diagonal agree with Eq. (4.1.8).

Orthogonal polynomials such as the Legendre polynomials are useful because of their simplicity. One can easily integrate over them, or take their derivatives analytically.

Using the Gram–Schmidt method, we can construct other orthogonal sets by starting with a different set of functions $\{v_n(x)\}$. We will see examples of this in the exercises.

4.1.2 Series of Orthogonal Functions

Now that we have a set of orthogonal functions $\{\psi_n(x)\}$, we will create a *generalized Fourier series* in order to represent some function $f(x)$ defined on a given interval, $a \le x \le b$. To do so, we write

$$f(x) = \sum_{n}^{\infty} c_n \psi_n(x), \qquad (4.1.9)$$

where the c_n's are constant coefficients that need to be determined. These coefficients are called *generalized Fourier coefficients*; but to save space we will usually refer to them as just Fourier coefficients. To find these coefficients, we take an inner product of any one of the orthogonal functions, ψ_m, with respect to both sides of Eq. (4.1.9):

$$(\psi_m, f) = \sum_{n}^{\infty} c_n (\psi_m, \psi_n). \qquad (4.1.10)$$

Orthogonality then implies that all terms in the sum are zero except for the term $n = m$, so we obtain an equation for the mth coefficient:

$$c_m = \frac{(\psi_m, f)}{(\psi_m, \psi_m)}. \qquad (4.1.11)$$

By calculating the required inner products with respect to each function ψ_m in the set, Eq. (4.1.11) provides us with all of the Fourier coefficients required to construct the generalized Fourier series representation of f.

Say, for example, we wish to represent the function $f(x) = e^{-x} \sin 3x$ on the interval $-1 < x < 1$. We can do so using the Legendre polynomials, since they form an orthogonal set on this interval with respect to the inner product of Eq. (4.1.7). First we evaluate the Fourier coefficients c_n using Eq. (4.1.11). The required integrals could be found analytically, but the results are quite complicated. It is better just to evaluate the integrals numerically:

Cell 4.5

```
c[n_] := c[n] = NIntegrate[LegendreP[n, x] Exp[-x] Sin [3 x],
  {x, -1, 1}] /
   Integrate[LegendreP[n, x]^2, {x, -1, 1}]
```

Here we have used the trick of applying two sets of equal signs, so as to cause *Mathematica* to remember these integrals, evaluating each one only once. Next, we construct an approximation to the full series, keeping only the first M terms:

Cell 4.6

```
f_approx[x_, M_] := Sum[c[n] LegendreP[n, x], {n, 0, M}]
```

We can plot this series and compare it with the exact function, keeping increasing numbers of terms, as shown in Cell 4.7. Only six or so terms are required in order to achieve excellent convergence to the exact function (shown by the thin line). Of course, just as with trigonometric series, the more rapidly varying the function is, the more terms are needed in the series to obtain good convergence.

Cell 4.7

```
<< Graphics`;

Table[Plot[{Evaluate[f_approx[x, M]], Sin[3 x] Exp[-x]},
  {x, -1, 1},
   PlotStyle→{Red, Thickness[0.008]}, Blue},
   PlotLabel→"M = " <>ToString[M]], {M, 2, 8, 2}];
```

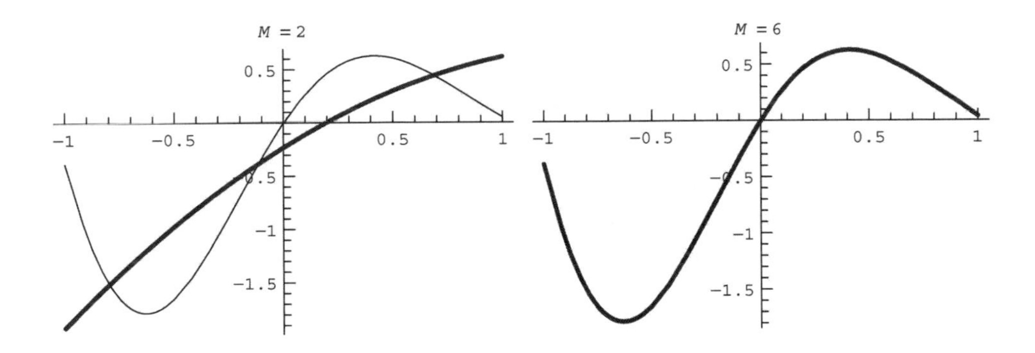

Let's try the same problem of constructing a generalized Fourier series for the function $f(x) = e^{-x} \sin 3x$, but with a different set of orthogonal functions on $[-1, 1]$: the set of *even* Legendre polynomials, $\psi_n(x) = P_{2n}(x)$. If we now try to expand the function $e^{-x} \sin 3x$ in this set, the expansion does not work, as seen in Cell 4.8. Only the even part of the function is properly represented by this set, because the orthogonal functions used in the series are all even in x. The odd part of the function cannot be represented by these even polynomials.

Cell 4.8

```
ψ[n_, x_] = LegendreP[2n, x];

Clear[c]; c[n_] := c[n] = NIntegrate[ψ[n, x] Exp[-x] Sin[3 x],
  {x, -1, 1}]/
   Integrate[ψ[n, x]^2, {x, -1, 1}]

f_approx[x_, M_] := Sum[c[n] ψ[n, x], {n, 0, M}]
```

```
Table[Plot[{Evaluate[f_approx[x, M]], Sin[3 x] Exp[-x]},
  {x, -1, 1},
   PlotStyle→{Red, Thickness[0.008]}, Blue},
   PlotLabel→"M = " <>ToString[M]], {M, 2, 6, 2}];
```

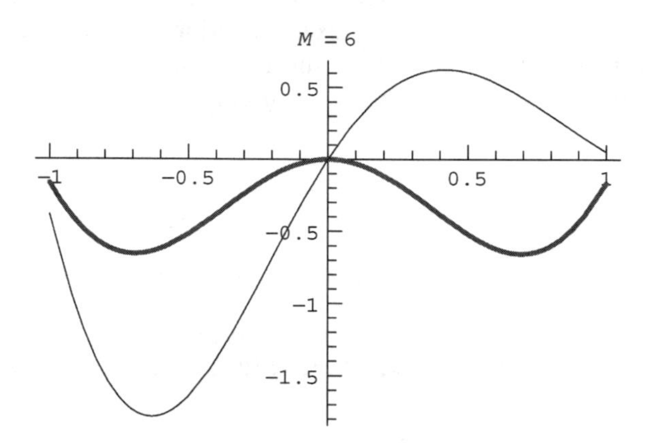

Thus, we cannot choose *any* set of orthogonal functions that we wish when making a generalized Fourier series expansion. The set of functions must be *complete*. That is, a linear combination of these functions must be able to represent any given function in the range of interest, $a < x < b$.

In the next section, we will discuss how to find sets of orthogonal functions that are guaranteed to be complete. These functions are eigenmodes of the spatial operators that appear in the linear PDEs we studied in Chapter 3. As such, they are just the thing for describing solutions to the PDEs in terms of generalized Fourier series.

4.1.3 Eigenmodes of Hermitian Operators

Hermitian Operators and Sturm–Liouville Problems An operator $\hat{L}$ is defined to be *Hermitian* with respect to a given inner product on the interval $a \le x \le b$, and with respect to some set of functions, if, for any two functions f and g taken from this set, $\hat{L}$ satisfies the following equation:

$$(f, \hat{L}g) = (\hat{L}f, g). \tag{4.1.12}$$

An operator can be Hermitian only with respect to a given set of functions and a given inner product.

As an example of a Hermitian operator consider the following second-order linear differential operator:

$$\hat{L} = \frac{1}{p(x)} \frac{d}{dx}\left(r(x) \frac{d}{dx}\right) + q(x), \tag{4.1.13}$$

on the interval $a \le x \le b$. The functions $p(x)$, $q(x)$, and $r(x)$ are assumed to be real, and also $p(x)$ and $r(x)$ are required to be positive-definite on the interval

$a < x < b$. This operator is called a Sturm–Liouville operator. Second-order operators of this type crop up regularly in mathematical physics. In fact, this kind of operator appeared in every eigenvalue problem that we encountered in Chapter 3. These eigenvalue problems are called *Sturm–Liouville problems*.

The Sturm–Liouville operator is Hermitian with respect to the inner product

$$(f,g) = \int_a^b f^*(x)g(x)p(x)\,dx, \tag{4.1.14}$$

[where the weight function $p(x)$ is the same as that which appears in Eq. (4.1.13)] and with respect to functions that satisfy a broad class of homogeneous boundary conditions. Recall that homogeneous conditions are such that one solution to the eigenvalue problem $\hat{L}\psi = \lambda\psi$ is $\psi = 0$.

For a Sturm–Liouville operator to be Hermitian, these homogeneous boundary conditions can take several forms:

If $r \neq 0$ at the end points a and b, the boundary conditions on the functions can be either homogeneous mixed, von Neumann, or Dirichlet. The boundary conditions can be different types at each end, e.g., Dirichlet at one end and von Neumann at the other.

If $r = 0$ at one or both ends of the interval, then at these ends the functions and their first derivatives need merely be finite.

The functions can also satisfy *periodic boundary conditions*, provided that $r(a) = r(b)$ (see the exercises).

For these classes of functions, it is easy to show that the Sturm–Liouville operator is Hermitian with respect to the inner product of Eq. (4.1.14). Starting with the inner product $(f, \hat{L}g)$, two integrations by parts yield

$$\begin{aligned}
(f,\hat{L}g) &= \int_a^b f^*\left(\frac{1}{p(x)}\frac{d}{dx}\left[r(x)\frac{d}{dx}g(x)\right] + q(x)g(x)\right)p(x)\,dx \\
&= rf^*\frac{\partial}{\partial x}g\bigg|_{x=a}^{x=b} - \int_a^b\left(r\frac{df^*}{dx}\frac{dg}{dx} + qpf^*g\right)dx \\
&= r\left(f^*\frac{\partial}{\partial x}g - g\frac{\partial}{\partial x}f^*\right)\bigg|_{x=a}^{x=b} + \int_a^b g\left[\frac{d}{dx}\left(r\frac{d}{dx}f^*\right) + qpf^*\right]dx \\
&= r\left(f^*\frac{\partial}{\partial x}g - g\frac{\partial}{\partial x}f^*\right)\bigg|_{x=a}^{x=b} + (\hat{L}f, g).
\end{aligned} \tag{4.1.15}$$

However, the boundary terms vanish because of the restriction to the sets of functions that satisfy the listed boundary conditions. Therefore, $(f, \hat{L}g) = (\hat{L}f, g)$, so the Sturm–Liouville operator is Hermitian for these sets of functions and this inner product.

Many second-order operators that do not appear to be of Sturm–Liouville form can be put in this form. For instance, the operator for damped harmonic motion,

$\hat{L} = d^2/dx^2 + v\,d/dx + \omega_0^2$, can be written as

$$\hat{L} = \frac{1}{e^{vx}} \frac{d}{dx}\left(e^{vx} \frac{d}{dx}\right) + \omega_0^2. \tag{4.1.16}$$

Since this operator is of Sturm–Liouville form with $p(x) = e^{vx}$, it is Hermitian with respect to the inner product $(f, g) = \int_a^b e^{vx} f^* g\,dx$, and with respect to functions that satisfy any of the homogeneous boundary conditions discussed with respect to Eqs. (4.1.13) and (4.1.14).

More generally, the operator $\hat{L} = d^2/dx^2 + u_1(x)\,d/dx + u_0(x)$ can also be put in Sturm–Liouville form:

$$\hat{L} = \frac{1}{e^{\int^x u_1(y)\,dy}} \frac{d}{dx}\left(e^{\int^x u_1(y)\,dy} \frac{d}{dx}\right) + u_0(x). \tag{4.1.17}$$

Some higher-order operators are also Hermitian. For instance, the operator $\hat{L} = d^4/dx^4$ can be shown to be Hermitian with respect to the inner product $(f, g) = \int_a^b f^* g\,dx$ for functions that vanish, along with their first derivatives, at the ends of the interval. (See the exercises.) However, other operators are not Hermitian. One simple example is $\hat{L} = d^3/dx^3$. Another is $d^2/dx^2 + v\,d/dx + b$ for *complex constants* v or b.

Eigenmodes Why should we care that an operator is Hermitian? Because the eigenmodes of Hermitian operators form an orthogonal set. Also, the eigenvalues of such operators are real. Consider the set of eigenfunctions $\{\psi_n\}$ of an operator $\hat{L}$. Each eigenfunction satisfies the ODE

$$\hat{L}\psi_n = \lambda_n \psi_n \tag{4.1.18}$$

on the interval $a \le x \le b$. Let us assume that this operator is Hermitian with respect to a set of functions that includes these eigenfunctions, and with respect to some inner product. Then the following theorem holds:

Theorem 4.1 Any two eigenfunctions ψ_n and ψ_m of a Hermitian operator $\hat{L}$ are orthogonal provided that the associated eigenvalues λ_n and λ_m are not equal. Furthermore, all the eigenvalues of $\hat{L}$ are real.

The proof is as follows. Consider the inner product $(\psi_m, \hat{L}\psi_n)$. According to Eq. (4.1.18) we can write this quantity as

$$(\psi_m, \hat{L}\psi_n) = (\psi_m, \lambda_n \psi_n) = \lambda_n(\psi_m, \psi_n), \tag{4.1.19}$$

where the last step follows from linearity of the inner product, Eq. (4.1.4). However, according to the Hermitian property, Eq. (4.1.12), we can also write this quantity as

$$(\psi_m, \hat{L}\psi_n) = (\psi_n, \hat{L}\psi_m)^*.$$

If we then apply Eq. (4.1.18), we obtain

$$(\psi_m, \hat{L}\psi_n) = (\psi_n, \lambda_m \psi_m)^* = \lambda_m^*(\psi_n, \psi_m)^* = \lambda_m^*(\psi_m, \psi_n), \qquad (4.1.20)$$

where in the last step we used Eq. (4.1.2) and in the next to last step we used Eq. (4.1.4).

Finally, equating Eq. (4.1.19) to Eq. (4.1.20) yields

$$(\lambda_n - \lambda_m^*)(\psi_m, \psi_n) = 0. \qquad (4.1.21)$$

Now, if $n = m$, then Eq. (4.1.21) becomes $(\lambda_n - \lambda_n^*)(\psi_n, \psi_n) = 0$. But $(\psi_n, \psi_n) > 0$ for nontrivial ψ_n, so we find that $\lambda_n - \lambda_n^* = 0$. Therefore, eigenvalues of $\hat{L}$ must be real numbers.

Since the eigenvalues are real, we can drop the complex conjugation from λ_m in Eq. (4.1.21). Then we have $(\lambda_n - \lambda_m)(\psi_m, \psi_n) = 0$. Therefore, $(\psi_m, \psi_n) = 0$ if $\lambda_m \neq \lambda_n$, proving that eigenfunctions associated with distinct eigenvalues are orthogonal.

We have finally solved the mystery encountered in the PDE problems of Chapter 3, of why the eigenmodes in these problems always formed orthogonal sets. We can now see that this occurred because in each case, the operators were of Sturm–Liouville form, so that the operators were Hermitian with respect to the inner product of Eq. (4.1.6), and with respect to functions that satisfied the homogeneous boundary conditions of the associated eigenvalue problem. In fact, Hermitian operators, and Sturm–Liouville operators in particular, dominate mathematical physics (especially at the introductory level).

For example, the set of Bessel eigenfucntions $\{J_m(j_{m,n} r/a)\}$, encountered in Sec. 4.2.5, satisfied the Sturm–Liouville problem on $0 < r < a$ given by Eq. (3.2.46),

$$\frac{1}{r}\frac{\partial}{\partial r}\left(r\frac{\partial \psi_n}{\partial r}\right) - \frac{m^2}{r^2}\psi_n - \kappa\psi_n = 0,$$

with Dirichlet boundary conditions at $r = a$. At $r = 0$ the eigenfunctions need only be finite. According to Eq. (4.1.14), these eigenmodes must form an orthogonal set with respect to the inner product $(f, g) = \int_0^a f^*(r) g(r) p(r)\, dr$, with $p(r) = r$. This corresponds to our previous result, Eq. (3.2.55), that $\int_0^a J_m(j_{m,\bar{n}} r/a) J_m(j_{m,n} r/a) r\, dr = 0$ for $n \neq \bar{n}$.

Completeness The fact that the eigenmodes of a Hermitian operator $\hat{L}$ form an orthogonal set means that they can be used in the generalized Fourier series representation of a function, Eq. (4.1.9). However, there is still the question whether these eigenmodes are complete. The answer to this question is that eigenfunctions of a Hermitian operator do form a complete set, under very general conditions. The proof can be found in Courant and Hilbert (1953, Chapters 2 and 4).

To be precise, the following completeness theorem holds for eigenmodes of a Hermitian operator:

Theorem 4.2 Given any function $f(x)$, described by a generalized Fourier series of eigenmodes of a Hermitian operator, the error between the function and the generalized Fourier series, $E_M(x) = f(x) - \Sigma_{n=0}^{M} c_n \psi_n(x)$, approaches zero as $M \to \infty$ in the following average sense:

$$\lim_{M\to\infty} (E_M(x), E_M(x)) = 0. \tag{4.1.22}$$

This is called *convergence in the mean*. If we write out the inner product using Eq. (4.1.6), we see that the error is averaged over x, weighted by the function $p(x)$:

$$\lim_{M\to\infty} \int_a^b |E_M(x)|^2 p(x)\,dx = 0. \tag{4.1.23}$$

Convergence in the mean is less restrictive than the uniform convergence discussed previously for Fourier series of functions that satisfy the conditions of Theorem 2.1. For example, Eq. (4.1.23) still holds for series that exhibit the Gibbs phenomenon. Also, for series using weight functions $p(x)$ that are small over certain ranges of the interval [such as for the Laguerre and Hermite polynomials at large x; see Exercise (5)(a) and (b)], there can be large differences between the series and $f(x)$ that are not revealed by Eq. (4.1.23). Nevertheless, this sort of convergence is usually all that is needed in applications.

4.1.4 Eigenmodes of Non-Hermitian Operators

From time to time, one will run into a problem where a linear operator is not Hermitian with respect to some given inner product and/or set of functions. One example that we already mentioned is the operator $\hat{L} = \partial^3/\partial x^3$ on the interval $a < x < b$. Eigenfunctions ψ_n of this operator do not form an orthogonal set with respect to the inner product defined by $(f, g) = \int_a^b f^* g p(x)\,dx$, for any $p(x)$. Nevertheless, we may want to expand some function in terms of these eigenmodes. For instance, we may need to solve a PDE involving $\hat{L}$, such as $\partial z/\partial t = \partial^3 z/\partial x^3$.

Fortunately, we can generalize our Fourier expansion techniques to allow series expansions in terms of eigenfunctions of non-Hermitian operators. To make the expansion work, we must first introduce the notion of the *adjoint* of an operator.

The adjoint of an operator $\hat{L}$ is another operator $\hat{L}^\dagger$ that is defined by the following equation:

$$(f, \hat{L}g) = (\hat{L}^\dagger f, g) \tag{4.1.24}$$

for some given inner product, where f and g are any two functions from some given set of functions. For instance, for $\hat{L} = \partial^3/\partial x^3$, the adjoint with respect to the inner product $(f, g) = \int_a^b f^* g\,dx$, and with respect to functions that satisfy homogeneous boundary conditions of various types, is simply $\hat{L}^\dagger = -\partial^3/\partial x^3$. This follows from three applications of integration by parts to Eq. (4.1.24), dropping the boundary terms because of the homogeneous boundary conditions.

Comparing Eq. (4.1.24) to Eq. (4.1.12), we see that a Hermitian operator satisfies $\hat{L}^\dagger = \hat{L}$: a Hermitian operator is its own adjoint. For this reason, Hermitian operators are also referred to as *self-adjoint*.

We will expand a function $f(x)$ in terms of the eigenmodes $\psi_n(x)$ of the non-Hermitian operator $\hat{L}$, writing

$$f(x) = \sum_n c_n \psi_n(x). \tag{4.1.25}$$

These eigenmodes satisfy the usual equation,

$$\hat{L}\psi_n = \lambda_n \psi_n. \tag{4.1.26}$$

However, as we have already stated, $(\psi_m, \psi_n) \neq 0$ for $m \neq n$, so our previous technique for finding the c_n's does not work. What to do?

Consider the eigenmodes $\psi_n^\dagger(x)$ of the *adjoint* operator. These eigenmodes satisfy the equation

$$\hat{L}^\dagger \psi_n^\dagger = \lambda_n^\dagger \psi_n^\dagger, \tag{4.1.27}$$

where $\lambda_n^\dagger$ is the associated eigenvalue. One can then prove the following:

$$\lambda_n^\dagger = \lambda_n^* \tag{4.1.28}$$

and

$$\left(\psi_n^\dagger, \psi_m\right) = 0 \qquad \text{if} \quad \lambda_n^* \neq \lambda_m. \tag{4.1.29}$$

The proof is almost identical to that given for Theorem 4.1, and is left to the exercises.

Since the adjoint eigenmodes form an orthogonal set with respect to the set $\{\psi_n\}$, we now take an inner product of Eq. (4.1.25) with respect to $\psi_n^\dagger$. This kills all terms in the sum except for the one involving c_n, and yields the result

$$c_n = \frac{\left(\psi_n^\dagger, f\right)}{\left(\psi_n^\dagger, \psi_n\right)}. \tag{4.1.30}$$

However, there is no guarantee that $(\psi_n^\dagger, \psi_n)$ is nonzero, because $\psi_n^\dagger \neq \psi_n$ in general. In fact, if this inner product vanishes for some value(s) of n, then Eq. (4.1.30) implies that an expansion of f in terms of eigenmodes of $\hat{L}$ is not possible, unless $(\psi_n^\dagger, f)$ also happens to equal zero for these n-values.

Also, even if $(\psi_n^\dagger, \psi_n)$ is nonzero for all n, there is generally no guarantee that the eigenmodes form a complete set, as there is with Hermitian operators. Nevertheless, this kind of eigenmode expansion can still be useful for those rare cases where non-Hermitian operators arise in a problem.

EXERCISES FOR SEC. 4.1

(1) Perform Gram–Schmidt orthogonalization by hand for the first three orthogonal polynomials extracted from the set $\{x^n\}$, $n \geq 0$, for the given inner products:

(a) $(f, g) = \int_0^\infty f^* g e^{-x}\, dx$ (these will be proportional to Laguerre polynomials).

(b) $(f, g) = \int_{-\infty}^\infty f^* g e^{-x^2}\, dx$ (these will be proportional to Hermite polynomials).

(c) $(f, g) = \int_{-1}^1 f^* g \dfrac{dx}{(1-x^2)^{1/2}}$ (these will be proportional to Chebyshev polynomials of the first kind).

(d) $(f, g) = \int_{-1}^{1} f^* g (1-x^2)^{1/2}\, dx$ (these will be proportional to Chebyshev polynomials of the second kind).

(2) Perform Gram–Schmidt orthogonalization by hand for the first three orthogonal functions from the set $\{e^{-nx}\}$, $n = 0, 1, 2, \ldots$. Take for the inner product $(f, g) = \int_0^\infty e^{-x} f^* g\, dx$.

(3) Create a *Mathematica* module called **gschmidt[M]** that automatically performs Gram–Schmidt orthogonalization for the first M orthogonal functions taken from a given set of predefined functions $v(n, x)$ and for a given predefined inner product. Run this *Mathematica* module for the orthogonal functions of Exercises (1) and (2), determining the first six orthogonal functions in each set.

(4) Find a generalized Fourier series representation of $x^2 e^{-2x}$ using the orthogonal functions derived in Exercise (3). Plot the result along with the exact function, keeping $M = 2, 4, 6$ terms.

(5) Find a generalized Fourier series representation for the following functions using the given orthogonal polynomials. Plot the resulting series for $M = 5$, 10, and 15 along with the functions. In each case, evaluate Eq. (4.1.23) for the different M-values to see whether convergence in the mean is being achieved.

(a) $f(x) = x/(1 + x^2)$ on $0 < x < \infty$, using Laguerre polynomials. Plot on $0 < x < 5$.

(b) $f(x) = (\sin x)/x$ on $-\infty < x < \infty$, using Hermite polynomials. Plot on $-8 < x < 8$.

(c) $f(t) = \sin \pi t$ on $-1 \leq t \leq 1$, using Legendre polynomials.

(d) $f(t) = t(1-t)/(2-t)$ on $-1 \leq t \leq 1$, using Chebyshev polynomials of the first kind.

(e) $f(t) = e^{-t}\sqrt{t+1}$ on $-1 \leq t \leq 1$, using Chebyshev polynomials of the second kind.

(Hint 1: These polynomials are already *Mathematica* intrinsic functions. You can find their definition and syntax in the help browser. Hint 2: The series representation may not converge well in every case.)

(6) **(a)** Prove that the generalized Fourier expansion of a polynomial of order N in terms of orthogonal polynomials is a finite series, involving only orthogonal polynomials of order N and lower.

(b) Expand x^4 in terms of Hermite polynomials.

(c) Expand x^2 in terms of Legendre polynomials.

(d) Expand x^6 in terms of Chebyshev polynomials of the second kind.

(7) Prove that the Sturm–Liouville operator (4.1.13) is Hermitian with respect to the inner product given by Eq. (4.1.14) and with respect to functions that satisfy mixed boundary conditions.

(8) Find an inner product for which the following operators are Hermitian with respect to functions that satisfy the given conditions:

(a) $\hat{L} = d^2/dx^2 - 2x\, d/dx$, for functions on $[-\infty, \infty]$ that go to ∞ more slowly than $e^{x^2/2}$ as $x \to \infty$.

(b) $\hat{L} = x\,d^2/dx^2 + (1-x)\,d/dx$, for functions on $[0, \infty]$ that go to ∞ more slowly than $e^{x/2}$ and that arc finite at the origin.

(c) $\hat{L} = (d/dx)(1-x^2)\,d/dx$, for functions on $[-1,\ 1]$ that are finite at the end points.

(d) $\hat{L} = d^2/dx^2 + d/dx + x^2$ on $[-2, 2]$, for functions that are zero at $x = \pm 2$.

(e) $\hat{L} = d^4/dx^4 + d^2/dx^2 + 1$ on $[0,\ 3]$, for functions that satisfy $f = 0$ and $f'' = 0$ at the end points.

(f) $\hat{L} = d^2/dx^2 + h(x)\,d/dx$, on $[-1, 1]$ for functions that satisfy $f = 0$ at the end points, and where $h(x)$ is the Heaviside step function.

(9) Show by substitution, using *Mathematica*, that the first five Hermite polynomials $H_n(x)$ are eigenfunctions of the operator $\hat{L}$ given in Exercise (8)(a), with eigenvalues $\lambda_n = -2n$.

(10) Show by substitution, using *Mathematica*, that the first five Laguerre polynomials $L_n(x)$ are eigenfunctions of the operator $\hat{L}$ given in Exercise (8)(b), with eigenvalues $\lambda_n = -n$.

(11) Show by substitution, using *Mathematica*, that the first five Legendre polynomials $P_n(x)$ are eigenfunctions of the operator $\hat{L}$ given in Exercise (8)(c), with eigenvalues $\lambda_n = -n(n+1)$.

(12) Prove that the Sturm–Liouville operator (4.1.13) is Hermitian with respect to the inner product given by Eq. (4.1.14) and with respect to functions $f(x)$ defined on $a \le x \le b$ that satisfy periodic boundary conditions $f(x) = f(x + b - a)$, provided that the function $r(x)$ satisfies $r(a) = r(b)$.

(13) Find the eigenfunctions and eigenvalues for the following operators. If the operators are Hermitian with respect to some inner product, show directly that the eigenfunctions are orthogonal with respect to that inner product.

(a) $\hat{L} = d^2/dx^2 + d/dx + 1$ with boundary conditions $\phi'(-1) = \phi'(1) = 0$.

(b) $\hat{L}$ and boundary conditions given in Exercise (8)(e).

(c) $\hat{L}$ and boundary conditions given in Exercise (8)(f). (Hint: Match solutions for $x < 0$ to those for $x > 0$.)

(d) $\hat{L} = d^2/dx^2 - 1$ with periodic boundary conditions on $[0, 1]$. [See Exercise (12).]

(14) Use the eigenfunctions obtained from the following operators in order to create a generalized Fourier series expansion of the following functions $f(x)$ on the given interval. In each case, plot the series and calculate the average error $(E_M(x), E_M(x))$ for $M = 5, 10, 15, 20$:

(a) $f(x) = x \sin 5x$ on $[-1, 1]$, using the eigenfunctions from Exercise (13)(a).

(b) $f(x) = e^{-x/3} \sin x$ on $[0, 3]$, using the eigenfunctions from the operator of Exercise (8)(e).

(c) $f(x) = x$ on $[-1, 1]$, using the eigenfunctions of the operator of Exercise (8)(f).

(15) A quantum particle of mass m is confined in a one-dimensional box with potential

$$V(x) = \begin{cases} 0, & |x| < a, \\ \infty, & |x| > a. \end{cases}$$

The energy levels E_n of the particle can be found by solving the time-independent *Schrödinger equation*

$$\hat{H}\psi_n = E_n \psi_n, \tag{4.1.31}$$

where $\hat{H}$ is the Hamiltonian operator $\hat{H} = -\hbar^2/2m(d^2/dx^2) + V(x)$, and $\hbar = 1.055 \times 10^{-34}$ J s is Planck's constant divided by 2π. For this problem the infinite potential at $\pm a$ implies that the wave functions vanish at $x = \pm a$. This provides homogeneous boundary conditions for the equation, which can be seen to be an eigenvalue problem of Sturm–Liouville form. The eigenvalues are the quantum energy levels E_n and the eigenfunctions $\psi_n(x)$ are the quantum wave functions corresponding to each energy level. Solve for the energy levels and energy eigenfunctions for this potential. Plot the three eigenfunctions with the lowest energies. (See problem 13 of Sec. 3.1, which solves an initial-value problem using these eigenmodes.)

(16) **(a)** A quantum particle of mass m is confined in a one-dimensional harmonic potential,

$$V(x) = \tfrac{1}{2} m \omega_0^2 x^2.$$

Now the boundary conditions are that $\psi_n \to 0$ as $x \to \pm\infty$. Show that the energy levels are given by $E_n = \hbar\omega_0(n + \frac{1}{2})$, and that the eigenfunctions are given by

$$\psi_n(x) = e^{-x^2/a^2} H_n(x/a), \tag{4.1.32}$$

where H_n is a Hermite polynomial and $a = \sqrt{\hbar/m\omega_0}$ is the spatial scale of the eigenfunctions. [Hint: Substitute the form $\psi_n(x) = e^{-x^2/2a^2} f_n(x/a)$ into the eigenmode equation (4.1.31), and show that f_n satisfies the Hermite polynomial ODE; see Exercises (8)(a) and (9).] Plot the three eigenfunctions with lowest energies, $n = 0$, 1, and 2.

(b) Use the eigenfunctions found in part (a) to solve the following initial value problem: $\Psi(x,0) = e^{-2(x-1)^2}$. Take $\hbar = m = \omega_0 = 1$ and animate the solution for $|\Psi(x,t)|^2$ using a table of plots for $0 < t < 4\pi$. [Hint: Recall that the wavefunction Ψ satisfies the time-dependent Schrödinger equation (3.1.81).]

(17) A quantum particle of mass m moves in a one-dimensional periodic harmonic potential with period $2a$, given by the following periodic δ-function:

$$V(x) = aV_0\,\delta(x), \qquad |x| < a,$$
$$V(x + 2a) = V(x).$$

Find the energy levels of a particle in this periodic potential (a model for the interaction of an electron with a periodic lattice of ions). Show that the modes break into two classes: those that are even in x, and those that are odd. For the odd modes, show that the energy levels are given by $E_n = \hbar^2 k_n^2/2m$, where $k_n = n\pi/a$ is the wavenumber, and n is a positive integer.

For the even modes, show that the wavenumbers satisfy the transcendental equation $k_n a \tan k_n a = mVa^2/\hbar^2$, where the energy E_n is still related to the $E_n = \hbar^2 k_n^2/2m$. For $mVa^2/\hbar^2 = 1$, find the smallest three values of $k_n a$ numerically and plot the corresponding eigenmodes on $-a < x < a$.

(18) A quantum particle is confined to $x > 0$, where x is now the vertical direction. The particle moves under the influence of a gravitational potential $V(x) = mgx$. Find an equation for the energy levels (which must be solved numerically) and find the corresponding energy eigenfunctions. Plot the lowest three energy eigenfunctions and find numerical values (up to the unknown constants) for their energies. Find the mean height of the particle above the surface $x = 0$ in each of the three energy levels. The mean height is given by

$$\langle x \rangle = \frac{\int x|\Psi|^2\,dx}{\int |\Psi|^2\,dx}.$$

(Hint: The eigenfunctions will be in terms of Airy functions.)

(19) Prove Eqs. (4.1.28) and (4.1.29).

(20) Find the adjoint operator for the given operator $\hat{L}$, inner product, and set of functions:

(a) $\hat{L} = \partial^2/\partial x^2$ with inner product $(f, g) = \int_a^b x f^* g\,dx$ and functions that obey $f(a) = f(b) = 0$.

(b) $\hat{L} = (1/x)(\partial/\partial x)x\,\partial/\partial x$ with inner product $(f, g) = \int_0^b x^2 f^* g\,dx$ and functions that obey $f(b) = 0$, $f(0)$ finite.

(c) $\hat{L} = p(x)\,\partial^2/\partial x^2 + q(x)\,\partial/\partial x + r(x)$ with inner product $(f, g) = \int_a^b f^* g\,dx$ and functions that obey $f(a) = f(b) = 0$.

(21) Find the eigenfunctions and eigenvalues of the adjoint operator for

(a) Exercise (20)(a),

(b) Exercise (20)(b).

Show that these eigenfunctions form an orthogonal set with respect to the eigenmodes of $\hat{L}$ and the given inner products.

(22) Find the first three eigenfunctions and eigenvalues of the operator $\hat{L} = \partial^3/\partial x^3$ and its adjoint with respect to the inner product $(f, g) = \int_0^1 f^* g\,dx$ and with respect to functions that satisfy $f(0) = f(1) = f'(0) = 0$. Show directly that the eigenfunctions of $\hat{L}$ are orthogonal with respect to the adjoint eigenfunctions. (Hint: For both the operator and its adjoint, the eigenvalues satisfy a transcendental equation that must be solved numerically.)

4.2 BEYOND SEPARATION OF VARIABLES: THE GENERAL SOLUTION OF THE 1D WAVE AND HEAT EQUATIONS

In this section we will obtain general solutions to the 1D wave and heat equations for *arbitrary* boundary conditions and *arbitrary* source functions. The analysis involves two steps: first, the PDE is put into *standard form*, by transforming away the inhomogeneous boundary, conditions, turning them into an extra source term in the PDE. The standard-form PDE then has homogeneous boundary conditions, and the solution to this PDE is obtained in terms of a generalized Fourier series of eigenmodes.

In later sections we will apply the same methods to other linear PDEs, including Poisson's equation and the wave and heat equations in more than one dimension.

4.2.1 Standard Form for the PDE

Let us consider the heat equation in one dimension on the interval $0<x<L$,

$$\frac{\partial T}{\partial t} = \frac{1}{C(x)} \frac{\partial}{\partial x}\left(\kappa(x) \frac{\partial T}{\partial x} \right) + S(x,t), \tag{4.2.1}$$

subject to general, possibly time-dependent boundary conditions of either the Dirichlet, von Neumann, or mixed type, as given by Eqs. (3.1.45)–(3.1.47). In order to solve this PDE, we will first put it into *standard form* with *homogeneous* boundary conditions. To do so, we write the solution for the temperature $T(x,t)$ as

$$T(x,t) = u(x,t) + \Delta T(x,t). \tag{4.2.2}$$

The function $u(x,t)$ is chosen to satisfy the inhomogeneous boundary conditions, but it is otherwise arbitrary. For example, if the boundary conditions are of the Dirichlet form (3.1.45), we choose *any function* that satisfies

$$u(0,t) = T_1(t),$$
$$u(L,t) = T_2(t).$$

One choice might be

$$u(x,t) = T_1(t) + [T_2(t) - T_1(t)]\, x/L, \tag{4.2.3}$$

but many others can also be used. For example, $u(x,t) = T_1(t) + [T_2(t) - T_1(t)](x/L)^n$ for any $n>0$ also works. However, we will soon see that it is best to choose a function with the slowest possible spatial variation, and especially try to avoid spatial discontinuities if at all possible.

If, on the other hand, the boundary conditions are of the von Neumann form (3.1.46), we choose some function $u(x,t)$ that satisfies

$$\frac{\partial u}{\partial x}(0,t) = -\frac{\Gamma_{x1}(t)}{\kappa},$$
$$\frac{\partial u}{\partial x}(L,t) = -\frac{\Gamma_{x2}(t)}{\kappa}.$$

Similarly, for mixed boundary conditions (3.1.47), we choose a function u that satisfies

$$\kappa \frac{\partial u}{\partial x}(0,t) = a\left[u(0,t) - T_1(t)\right],$$

$$\kappa \frac{\partial u}{\partial x}(L,t) = -b\left[u(L,t) - T_2(t)\right].$$

The remainder, $\Delta T(x,t)$, then satisfies *homogeneous* boundary conditions that are either Dirichlet,

$$\Delta T(0,t) = \Delta T(L,t) = 0, \tag{4.2.4}$$

von Neumann,

$$\frac{\partial \Delta T}{\partial x}(0,t) = \frac{\partial \Delta T}{\partial x}(L,t) = 0, \tag{4.2.5}$$

or mixed,

$$\kappa \frac{\partial \Delta T}{\partial x}(0,t) - a\,\Delta T(0,t) = \kappa \frac{\partial \Delta T}{\partial x}(L,t) + b\,\Delta T(L,t) = 0. \tag{4.2.6}$$

The function ΔT satisfies an inhomogeneous heat equation PDE that follows from applying Eq. (4.2.2) to Eq. (4.2.1):

$$\frac{\partial \Delta T}{\partial t} = \frac{1}{C}\frac{\partial}{\partial x}\left(\kappa \frac{\partial \Delta T}{\partial x}\right) + \bar{S}(x,t), \tag{4.2.7}$$

where the new source function $\bar{S}$ is given by

$$\bar{S}(x,t) = S(x,t) + \frac{1}{C}\frac{\partial}{\partial x}\left(\kappa \frac{\partial u}{\partial x}\right) - \frac{\partial u}{\partial t}. \tag{4.2.8}$$

Equation (4.2.7), with (4.2.8) and homogeneous boundary conditions, is called the *standard form* for the PDE. This approach to the problem bears some resemblence to the method of subtracting out the equilibrium solution, discussed in Sec. 3.1.2. Here, however, there need be no equilibrium solution, and we do not necessarily remove the source term in the equation by this technique. The main point is that we have made the boundary conditions homogeneous, so as to allow a generalized Fourier series expansion of eigenmodes to determine $\Delta T(x,t)$.

The same technique can be applied to the general wave equation with arbitrary boundary conditions and arbitrary time-dependent external transverse forces,

$$\frac{\partial^2}{\partial t^2} y(x,t) = \frac{1}{\rho(x)}\frac{\partial}{\partial x}\left(T(x)\frac{\partial}{\partial x} y(x,t)\right) + S(x,t). \tag{4.2.9}$$

We write

$$y(x,t) = \Delta y(x,t) + u(x,t), \tag{4.2.10}$$

where $u(x,t)$ is any function chosen to satisfy the inhomogeneous boundary conditions, and where $\Delta y(x,t)$ satisfies

$$\frac{\partial^2}{\partial t^2}\Delta y(x,t) = \frac{1}{\rho(x)}\frac{\partial}{\partial x}\left(T(x)\frac{\partial}{\partial x}\Delta y(x,t)\right) + \bar{S}(x,t) \tag{4.2.11}$$

with homogeneous boundary conditions, and the new source function $\bar{S}(x,t)$ is given by

$$\bar{S}(x,t) = S(x,t) + \frac{1}{\rho(x)}\frac{\partial}{\partial x}\left(T(x)\frac{\partial}{\partial x}u(x,t)\right) - \frac{\partial^2}{\partial t^2}u(x,t). \tag{4.2.12}$$

By putting the wave and heat equations into standard form, we have once again shown that inhomogeneous boundary conditions are equivalent to a source term in the differential equation, just as in discussion of ODE boundary-value problems in Sec. 1.4.5.

4.2.2 Generalized Fourier Series Expansion for the Solution

General Solution for the Wave Equation The general solutions for the standard form of the wave or heat equations follow the same route, so we will consider only the solution to the wave equation. The standard form of this equation, Eq. (4.2.11), can be written as

$$\frac{\partial^2}{\partial t^2}\Delta y(x,t) = \hat{L}\,\Delta y + \bar{S}(x,t). \tag{4.2.13}$$

where the operator $\hat{L}$ is

$$\hat{L}\,\Delta y = \frac{1}{\rho(x)}\frac{\partial}{\partial x}\left(T(x)\frac{\partial}{\partial x}\Delta y(x,t)\right). \tag{4.2.14}$$

The eigenmodes ψ_n of this operator satisfy

$$\hat{L}\psi_n(x) = -\omega_n^2\psi_n(x), \tag{4.2.15}$$

where ω_n is the corresponding eigenfrequency. A generalized Fourier series solution for $\Delta y(x,t)$ can then be constructed from these eigenmodes:

$$\Delta y(x,t) = \sum_{n=1}^{\infty} c_n(t)\,\psi_n(x). \tag{4.2.16}$$

Since $\rho(x) \ge 0$ and $T(x) \ge 0$ on $0 < x < L$, $\hat{L}$ is a Sturm–Liouville operator with eigenmodes that are orthogonal with respect to the inner product

$$(f,g) = \int_0^L \rho(x) f^*(x) g(x)\, dx. \tag{4.2.17}$$

Therefore, the time dependence of the Fourier amplitudes, $c_n(t)$, can be easily determined in the usual way. Substitution of Eq. (4.2.16) into Eq. (4.2.11) yields

$$\sum_{n=1}^{\infty} \psi_n(x) \frac{d^2}{dt^2} c_n(t) = \sum_{n=1}^{\infty} c_n(t) \hat{L} \psi_n(x) + \bar{S}(x,t) = -\sum_{n=1}^{\infty} c_n(t)\, \omega_n^2 \psi_n(x) + \bar{S}(x,t), \tag{4.2.18}$$

where in the last step we have applied Eq. (4.2.15). We then extract a single ODE for $c_n(t)$ by taking an inner product of both sides of Eq. (4.2.18) with respect to ψ_n. The result is

$$\frac{d^2}{dt^2} c_n(t) = -\omega_n^2 c_n(t) + \frac{\left(\psi_n, \bar{S}(x,t)\right)}{(\psi_n, \psi_n)}. \tag{4.2.19}$$

The general solution to this equation is

$$c_n(t) = A_n \cos \omega_n t + B_n \sin \omega_n t + c_{pn}(t), \tag{4.2.20}$$

where $c_{pn}(t)$ is a particular solution to Eq. (4.2.19).

A particular solution can be obtained in terms of the Green's function for the equation,

$$g(t) = \begin{cases} (\sin \omega_n t)/\omega_n, & t > 0, \\ 0, & t \le 0 \end{cases} \tag{4.2.21}$$

[see Eq. (2.3.77)]. The particular solution is

$$c_{pn}(t) = \int_0^t g(t - t_0) \frac{\left(\psi_n(x), S(x,t_0)\right)}{(\psi_n, \psi_n)}\, dt_0. \tag{4.2.22}$$

Combining Eqs. (4.2.16), (4.2.20), (4.2.22), and (4.2.10), we arrive at the general solution to the wave equation in one spatial dimension, with arbitrary boundary and initial conditions and an arbitrary source function:

$$y(x,t) = u(x,t) + \sum_{n=1}^{\infty} \left[A_n \cos \omega_n t + B_n \sin \omega_n t + c_{pn}(t) \right] \psi_n(x). \tag{4.2.23}$$

Any initial condition can be realized through appropriate choices of the constants A_n and B_n. In order to determine the A_n's, we apply the initial condition that $y(x,0) = y_0(x)$ for some function y_0, and we evaluate Eq. (4.2.23) at the initial time,

$$y(x,0) = u(x,0) + \sum_{n=1}^{\infty} A_n \psi_n(x) = y_0(x), \tag{4.2.24}$$

where we have recognized that $c_{pn}(0) = 0$ according to Eq. (4.2.22). The Fourier coefficients are determined in the usual fashion, by taking an inner product of both sides of the equation with respect to ψ_n and using orthogonality of the eigenmodes:

$$A_n = \frac{(\psi_n, y_0 - u(x,0))}{(\psi_n, \psi_n)}. \tag{4.2.25}$$

The coefficients B_n can be found in a similar way, using the second initial condition on $y(x,t)$, that $\frac{\partial y}{\partial t}(x,0) = v_0(x)$. According to Eq. (4.2.23), the initial time rate of change of y is given by

$$\frac{\partial y}{\partial t}(x,0) = \frac{\partial u}{\partial t}(x,0) + \sum_{n=1}^{\infty} \left[\omega_n B_n + c'_{pn}(0) \right] \psi_n(x) = v_0(x). \tag{4.2.26}$$

Taking the inner product with respect to ψ_n yields

$$B_n = \frac{(\psi_n, v_0)}{\omega_n(\psi_n, \psi_n)} - \frac{c'_{pn}(0)}{\omega_n}. \tag{4.2.27}$$

Our general solution to the wave equation, Eq. (4.2.23), satisfies the boundary conditions, because $u(x,t)$ is specifically chosen to satisfy these conditions and the eigenmodes satisfy homogeneous conditions. The solution also satisfies the initial conditions, since we have chosen the A_n's and B_n's to create a generalized Fourier series that sums to the correct initial conditions.

Wave Energy There is no dissipation in the wave equation: the system oscillates forever when excited. Therefore, we expect that energy is a conserved quantity, provided that there are no time-dependent sources or boundary conditions.

One can separate out static sources or boundary conditions by subtracting out the equilibrium solution to the string shape. The remaining perturbation satisfies the general wave equation (4.2.9) with homogeneous boundary conditions and no source. In this case energy conservation can be proven using the following argument. Multiply both sides of this equation by $\rho(x)\,\partial y/\partial t$, and integrate over the string length:

$$\int_0^L dx\, \rho(x) \frac{\partial y}{\partial t} \frac{\partial^2 y}{\partial t^2} = \int_0^L dx\, \rho(x) \frac{\partial y}{\partial t} \frac{1}{\rho(x)} \frac{\partial}{\partial x}\left(T \frac{\partial y}{\partial x} \right). \tag{4.2.28}$$

The integrals in Eq. (4.2.28) can be written as time derivatives. The left-hand integral is

$$\int_0^L dx\, \rho(x) \frac{\partial y}{\partial t} \frac{\partial^2 y}{\partial t^2} = \frac{\partial}{\partial t} \int_0^L dx \tfrac{1}{2} \rho(x) \left(\frac{\partial y}{\partial t} \right)^2 = \frac{\partial K(t)}{\partial t},$$

where $K(t) = \int_0^L dx \frac{1}{2}\rho(x)(\partial y/\partial t)^2$ is the kinetic energy associated with the vibrations. Similarly, one can cancel the ρ's in the right-hand side and integrate by parts, using the homogeneous boundary conditions, to obtain

$$\int_0^L dx\, \rho(x) \frac{\partial y}{\partial t} \frac{1}{\rho(x)} \frac{\partial}{\partial x}\left(T(x) \frac{\partial y}{\partial x}\right) = -\int_0^L dx\, T(x) \frac{\partial y}{\partial x} \frac{\partial}{\partial x}\left(\frac{\partial y}{\partial t}\right)$$

$$= -\frac{\partial}{\partial t} \int_0^L dx\, T(x)\left(\frac{\partial y}{\partial x}\right)^2 = -\frac{\partial U(t)}{\partial t},$$

where $U(t) = \int_0^L dx\, T(x)(\partial y/\partial x)^2$. Then Eq. (4.2.28) implies that

$$\frac{\partial}{\partial t}[K(t) + U(t)] = 0, \tag{4.2.29}$$

so $K + U$ is a constant of the string motion. Since K is the kinetic energy, we identify $U(t)$ as the potential energy of the system, and the quantity $E = K + U$ as the total energy.

The energy can also be written in terms of the normal modes of oscillation as follows. Since $y(x,t)$ is real, $y = y^*$ and we can write $\dot{y}^2 = \dot{y}^* \dot{y}$, and similarly $(\partial y/\partial x)^2 = (\partial y^*/\partial x)(\partial y/\partial x)$. If we substitute the Fourier expansion $y(x,t) = \sum_{n=1}^{\infty} c_n(t)\psi_n(x)$ into the expressions for K and U, we find that E can be written as

$$E = \sum_n \sum_m \left(\frac{1}{2} \dot{c}_n^*(t) \dot{c}_m(t)(\psi_n, \psi_m) + \frac{1}{2} c_n^*(t) c_m(t) \int_0^L dx\, T(x) \frac{\partial \psi_n^*}{\partial x} \frac{\partial \psi_m}{\partial x} \right),$$

where we have written the kinetic energy integral in terms of an inner product, using Eq. (4.2.17). Integrating by parts in the second term, we find that the integral can be written as $-\int_0^L dx\, \psi_n^*(\partial/\partial x)[T(x)\, \partial\psi_m/\partial x] = \omega_m^2(\psi_n, \psi_m)$, where we have used Eqs. (4.2.14), (4.2.15), and (4.2.17). Then orthogonality of the eigenmodes implies that the energy is

$$E = \sum_n \left[\tfrac{1}{2}|\dot{c}_n(t)|^2 + \tfrac{1}{2}\omega_n^2 |c_n(t)|^2 \right](\psi_n, \psi_n). \tag{4.2.30}$$

Equation (4.2.30) shows that the total energy E is a sum over the energies E_n of each normal mode, where

$$E_n = \left[\tfrac{1}{2}|\dot{c}_n(t)|^2 + \tfrac{1}{2}\omega_n^2 |c_n(t)|^2 \right](\psi_n, \psi_n).$$

One can see that E_n is the energy of a harmonic oscillator of frequency ω_n and effective mass (ψ_n, ψ_n). Therefore, E_n is also a conserved quantity. This follows from the fact that each mode amplitude satisfies the harmonic oscillator equation (4.2.19) (assuming no forcing). Thus, the total energy E is a sum of the energies E_n, of each normal mode, each of which is separately conserved.

Example 1: Temperature Oscillations Consider a slab of material of thickness L, and with uniform thermal diffusivity χ, initially at uniform temperature, $T(x,0) = T_0$. The left-hand side of the slab at $x = 0$ has fixed temperature, $T(0,t) = T_0$; but the right-hand side of the slab at $x = L$ has an oscillating temperature, $T(L,t) =$

$T_0 + T_1 \sin \omega_0 t$. Our task is to determine the evolution of the temperature within the slab of material, $T(x,t)$.

First, we write down the equation for T: it is the diffusion equation

$$\frac{\partial T}{\partial t} = \chi \frac{\partial^2 T}{\partial x^2}. \tag{4.2.31}$$

The boundary conditions on the equation are of Dirichlet form,

$$T(0,t) = T_0,$$
$$T(L,t) = T_0 + T_1 \sin \omega_0 t,$$

and the initial condition is $T(x,0) = T_0$. This information provides us with all we need to solve the problem. First, we put Eq. (4.2.31) into standard form by choosing a function $u(x,t)$ that satisfies the boundary conditions. We choose

$$u(x,t) = T_0 + \frac{x}{L} T_1 \sin \omega_0 t. \tag{4.2.32}$$

We next write the solution for $T(x,t)$ as $T(x,t) = u(x,t) + \Delta T(x,t)$, which implies that $\Delta T(x,t)$ is determined by the PDE

$$\frac{\partial \Delta T}{\partial t} = \chi \frac{\partial^2 \Delta T}{\partial x^2} + \bar{S}(x,t), \tag{4.2.33}$$

with homogeneous Dirichlet boundary conditions $\Delta T(0,t) = \Delta T(L,t) = 0$, and where the source term is

$$\bar{S}(x,t) = \chi \frac{\partial^2 u}{\partial x^2} - \frac{\partial u}{\partial t} = -\frac{x}{L} \omega_0 T_1 \cos \omega_0 t. \tag{4.2.34}$$

Next, we determine the eigenmodes of the spatial operator appearing in Eq. (4.2.33), $\hat{L} = \chi \partial^2/\partial x^2$. These eigenmodes satisfy

$$\chi \frac{\partial^2}{\partial x^2} \psi_n(x) = \lambda_n \psi_n(x), \tag{4.2.35}$$

with homogeneous Dirichlet boundary conditions $\psi_n(0) = \psi_n(L) = 0$. We have already seen this eigenvalue problem several times, in Chapter 3. The eigenmodes are

$$\psi_n(x) = \sin \frac{n\pi x}{L}, \tag{4.2.36}$$

and the eigenvalues are

$$\lambda_n = -\chi (n\pi/L)^2, \qquad n = 1, 2, 3, \ldots . \tag{4.2.37}$$

Next, we construct a generalized Fourier series solution for the function $\Delta T(x,t)$:

$$\Delta T(x,t) = \sum_n c_n(t)\,\psi_n(x). \tag{4.2.38}$$

Equations for each Fourier coefficient $c_n(t)$ are determined by substituting Eq. (4.2.38) into Eq. (4.2.33), then taking an inner product with respect to ψ_n. Using Eqs. (4.2.34) and (4.2.32) for the source function, we obtain

$$\frac{\partial}{\partial t}c_n(t) = \lambda_n c_n(t) - \frac{(\psi_n, x/L)}{(\psi_n, \psi_n)}\,\omega_0 T_1 \cos\omega_0 t. \tag{4.2.39}$$

The inner products appearing in this equation can be evaluated analytically. Since the medium is uniform, according to Eq. (4.2.15) these inner products are simply integrals from $x=0$ to $x=L$:

Cell 4.9

```
ψ[n_, x_] = Sin[n Pi x/L];

Integrate[ψ[n, x] x/L, {x, 0, L}]/Integrate[ψ[n, x]^2,
  {x, 0, L}];

Simplify[%, n ∈ Integers]
```

$$-\frac{2\,(-1)^n}{n\pi}$$

Thus, Eq. (4.2.39) becomes

$$\frac{\partial}{\partial t}c_n(t) = \lambda_n c_n(t) + \frac{2(-1)^n}{n\pi}\,\omega_0 T_1 \cos\omega_0 t. \tag{4.2.40}$$

The general solution to this ODE is given by a linear combination of a homogeneous and a particular solution.

There are two ways to proceed now. We can either simply apply **DSolve** to find the solution to Eq. (4.2.40), or we can write the particular solution in integral form using the Green's function for this first-order ODE:

$$c_m(t) = A_m\,e^{\lambda_m t} + \frac{2(-1)^n}{n\pi}\,\omega_0 T_1 \int_0^t e^{\lambda_m(t-\bar{t})}\cos\omega_0 \bar{t}\,d\bar{t}.$$

The integration can be performed using *Mathematica*:

Cell 4.10

```
c[n_, t_] = A[n] e^(λ[n] t) +

  FullSimplify[(2 (-1)^n)/(n π) ω0 T1 Integrate[e^(λ[n] (t-t̄)) Cos[ω0 t̄],
  {t̄, 0, t}]]
```

$$e^{t\lambda[n]}\,A[n] + \frac{2\,(-1)^n\,T_1\omega_0\,(\mathrm{Sin}[t\omega_0]\,\omega_0 + (e^{t\lambda[n]} - \mathrm{Cos}[t\omega_0])\,\lambda[n])}{n\pi(\omega_0^2 + \lambda[n]^2)}$$

With this result, the solution for ΔT is obtained using Eq. (4.2.38), keeping 20 terms in the sum over the Fourier modes:

Cell 4.11

```
M = 20;
ΔT[x_, t_] = Sum[c[n, t] ψ[n, x], {n, 1, M}];
```

However, $\Delta T(x,t)$ still depends on the Fourier coefficients A_n, which are determined by matching to the initial condition. These initial conditions are $\Delta T(x,0) = T(x,0) - u(x,0) = T_0 - T_0 = 0$. Thus, at $t=0$, we have that $\Delta T(x,0) = \Sigma_n A_n \psi_n(x) = 0$, so the solution for the A_n's is simply $A_n = 0$:

Cell 4.12

```
A[n_] = 0;
```

By adding the function $u(x,t)$ to $\Delta T(x,t)$, we obtain the full solution for the temperature evolution:

Cell 4.13

```
u[x_, t_] = T0 + T1 Sin[ω0 t]x/L;
T[x_, t_] = u[x, t] + ΔT[x, t];
```

The resulting function is displayed in Cell 4.14 as a series of plots. To evaluate this function numerically, we must choose values for L, χ, ω_0, T_0, and T_1. We must also define the eigenvalues λ_n. From these plots we can observe an interesting aspect of solutions to the heat equation with oscillatory boundary conditions: the temperature oscillations at the boundary only penetrate a short distance into the material, depending on the thermal diffusivity. The larger the diffusivity, the larger the penetration distance of the oscillations (try increasing χ by a factor of 4 in the above plots).

One can easily understand this intuitively: As the boundary condition oscillates, heat flows in and out of the slab though the surface. In an oscillation period $\tau = 2\pi/\omega_0$, the heat flows into the material only a certain distance, which grows larger as χ increases. (We will see in Chapter 5 that this distance scales as $\sqrt{\chi\tau}$.) In one half period, heat flows into the system, and in the next half period heat flows back out. The net effect is to average out the oscillations, and so produce a nearly time-independent temperature far from the surface.

Cell 4.14

```
λ[n_] = -χ (n Pi/L)^2;

L = 4; χ = 1/8; ω0 = 1; T0= 2; T1 = 1;

Table[Plot[T[x, t], {x, 0, L}, PlotRange→{{0, L}, {0, 3}},
   AxesLabel→{"x", ""},
   PlotLabel→"T[x, t], t=" <>ToString[t]], {t, 0, 15, .25}];
```

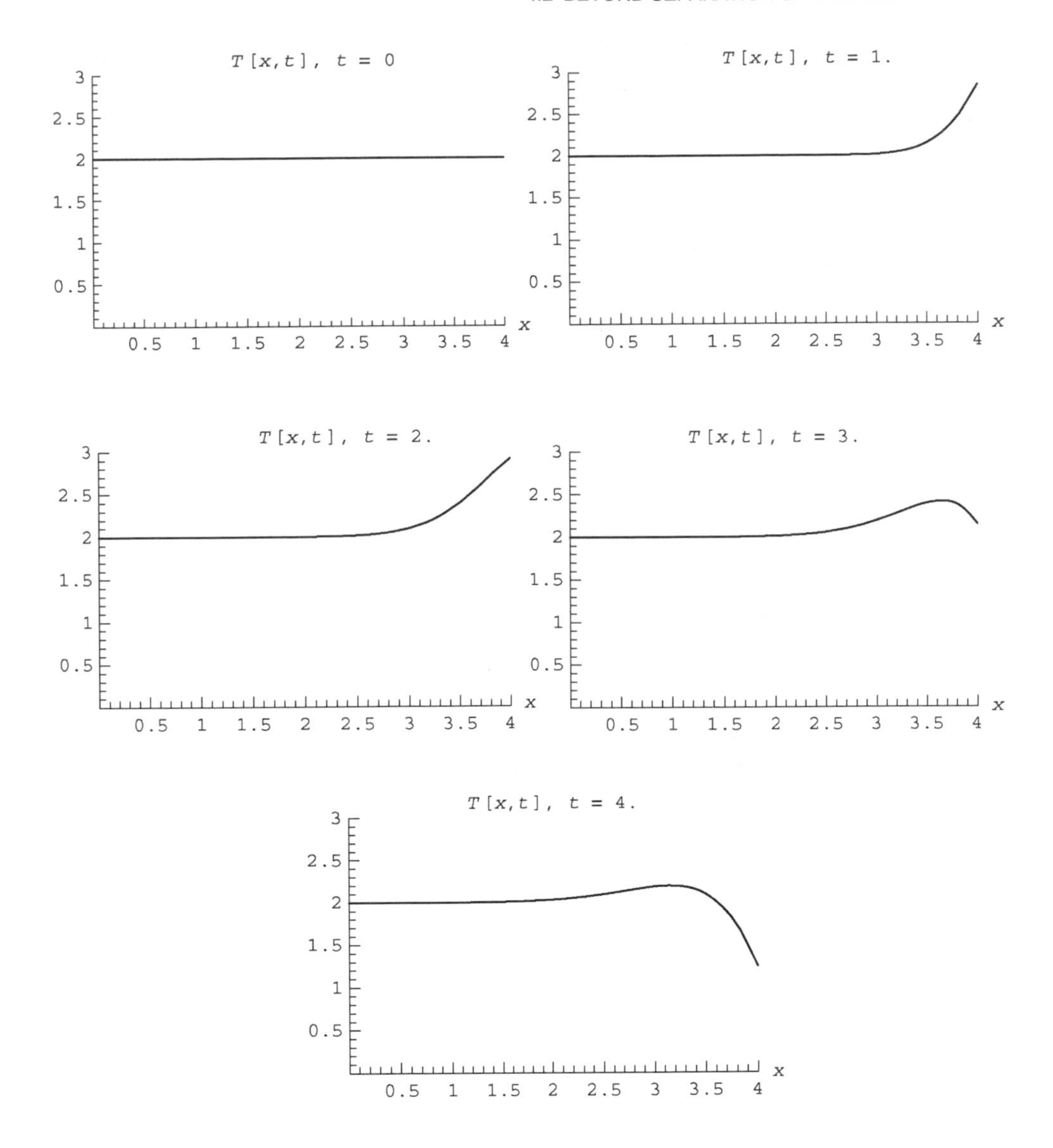

This is why wine cellars are buried underground. Rock and dirt have a rather low thermal diffusivity, so the temperature in a wine cellar remains nearly constant from day to day, although the earth's surface temperature oscillates considerably from day to night and from season to season.

Example 2: Cracking the Whip In the previous example, the system was uniform, so the eigenmodes were simply the usual trigonometric Fourier modes. Let us now consider a nonuniform system for which the eigenmodes are not trigonometric. This time we take an example from the wave equation, and consider a hanging string of uniform mass density ρ. The string is fixed to the ceiling at $z = L$, but the bottom of the string at $z = 0$ can move freely.

According to Eq. (3.1.1) the tension in such a string increases with height according to $T(z) = \rho g z$. Therefore, transverse perturbations on the string satisfy the nonuniform wave equation (3.1.7):

$$\frac{\partial^2}{\partial t^2} y(z,t) = g \frac{\partial}{\partial z}\left(z \frac{\partial}{\partial z} y(z,t) \right). \tag{4.2.41}$$

There is a regular singular point in this equation at $z = 0$, because the tension vanishes at the free end of the string. For this reason, as a boundary condition we need only specify that $y(0,t)$ is finite. The other boundary condition is that the string is fixed to the ceiling at $z = L$, so $y(L,t) = 0$. These boundary conditions are homogeneous, so the problem is already in standard form. Therefore, we can take $u = 0$ and find the evolution of $y(z,t)$ directly as a generalized Fourier series of eigenmodes,

$$y(z,t) = \sum_n c_n(t)\, \psi_n(z). \tag{4.2.42}$$

These eigenmodes satisfy

$$g \frac{\partial}{\partial z}\left(z \frac{\partial}{\partial z} \psi_n(z) \right) = -\omega_n^2 \psi_n(z), \tag{4.2.43}$$

with boundary conditions $\psi(L) = 0$ and $\psi(0)$ finite, where ω_n is the eigenfrequency. The general solution to Eq. (4.2.43) is in terms of Bessel functions:

$$\psi_n(z) = A J_0\left(2\,\omega_n \sqrt{\frac{z}{g}} \right) + B Y_0\left(2\,\omega_n \sqrt{\frac{z}{g}} \right). \tag{4.2.44}$$

This can be verified using *Mathematica*:

Cell 4.15

```
DSolve[g D[z D[ψ[z], z], z] == -ω²ψ[z], ψ[z], z]

{{ψ[z] → BesselJ[0, 2√z ω/√g] C[1] + BesselY[0, 2√z ω/√g] C[2]}}
```

Since the function Y_0 is singular at the origin, it cannot enter into the eigenmode. The boundary condition $\psi(L) = 0$ implies that $\psi_n(z) = A J_0(2\,\omega_n \sqrt{L/g}) = 0$. Therefore, $2\,\omega_n \sqrt{L/g} = j_{0,n}$, the nth zero of J_0, so the eigenfrequencies are

$$\omega_n = \frac{j_{0,n}}{2} \sqrt{\frac{g}{L}}, \tag{4.2.45}$$

and the eigenfunctions are

$$\psi_n(x) = A J_0\left(j_{0,n} \sqrt{z/L} \right). \tag{4.2.46}$$

A few of these eigenmodes are displayed in Cell 4.16.

Cell 4.16

```
<< NumericalMath`;
j0 = BesselJZeros[0, 5];
L = 3;

plts = Table[ParametricPlot[{BesselJ[0, j0[[n]]√(z / L)], z},
  {z, 0, L},
    PlotRange→{{-1, 1}, {0, L}}, AspectRatio→L/2,
      Axes→False,
    DisplayFunction→Identity, PlotStyle→Hue[1/n]],
      {n, 1, 5}];
Show[plts, DisplayFunction→$DisplayFunction]:
```

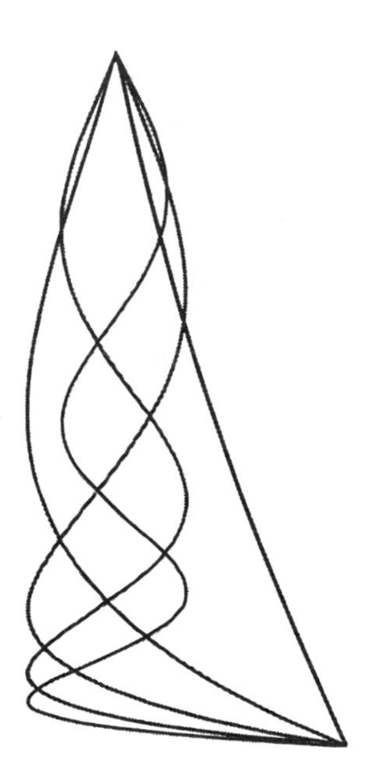

The lowest-frequency eigenmode is a sort of pendulum oscillation of the string from side to side, which can be easily set up in a real piece of string. The higher-order modes takes more work to view in a real string; one must oscillate the top of the string at just the right frequency to set up one of these modes. However, in *Mathematica* it is easy to see the modes oscillate, using an animation. In Cell 4.17 we show the $n = 2$ mode. To make the display realistic-looking, we must try to allow for the change in height of the bottom of the string during an oscillation. This change in height occurs because the length of the string is fixed. When a mode is present, the length is determined by an integral over length elements

$$ds = \sqrt{dy^2 + dz^2} = dz\sqrt{1 + \left(\frac{\partial y}{\partial z}\right)^2} \sim dz\left[1 + \frac{1}{2}\left(\frac{\partial y}{\partial z}\right)^2\right].$$

Thus, the fixed length L of the string determines the height $z_0(t)$ of the bottom of the string, according to

$$L = \int_{z_0(t)}^{L} ds \simeq \int_{z_0(t)}^{L} \left[1 + \frac{1}{2} \left(\frac{\partial y}{\partial z}(z,t) \right)^2 \right] dz$$

$$= L - z_0(t) + \frac{1}{2} \int_{z_0(t)}^{L} \frac{1}{2} \left(\frac{\partial y}{\partial z}(z,t) \right)^2 dz.$$

For small z_0, we can replace the lower bound of the last integral by zero, obtaining

$$z_0(t) = \frac{1}{2} \int_0^L \frac{1}{2} \left(\frac{\partial y}{\partial z}(z,t) \right)^2 dz.$$

The change in height of the string end is a nonlinear effect: $z_0(t)$ varies as the square of the mode amplitude. Strictly speaking this effect goes beyond our discussion of the linear wave equation. However, for finite-amplitude modes it can be important to allow for this effect; otherwise the modes simply don't look right.

The *Mathematica* commands in Cell 4.17 determine the height of the bottom of the string, $z_0(t)$, for a given mode n with amplitude a. We then animate the mode, *assuming* that the change in height of the bottom of the string corresponds to a mode on a slightly shortened string, of the form $y(z,t) \simeq a \cos \omega t J_0(j_{0,n}\{[z - z_0(t)]/[L - z_0(t)]\}^{1/2})$, $z_0(t) < z < L$. Of course, this is just a guess. To do this problem properly, we need to look for solutions to a nonlinear wave equation, which takes us beyond the bounds of this chapter. Some aspects of nonlinear waves will be considered in Chapter 7.

Cell 4.17

```
n = 2; (* mode number *)
L = 3; (* length of string *)
a = 1/2; (* mode amplitude *)

j0 = BesselJZeros[0, n];

z0[t_] =
 Integrate[1/2 a^2 Cos[t]^2 D[BesselJ[0, j0[[n]] Sqrt[z / L]], z]^2,
   {z, 0, L}];

Table[ParametricPlot[
   Evaluate[{a Cos[t] BesselJ[0, j0[[n]]
    Sqrt[(z - z0[t])/(L - z0[t])]], z}],
   {z, z0[t], L}, AspectRatio→L/2,
   PlotRange→{{-1, 1}, {0, L}}, PlotStyle→Thickness[0.02],
   PlotLabel->"n = "<>ToString[n]], {t, 0, 1.9 Pi, .1 Pi}];
```

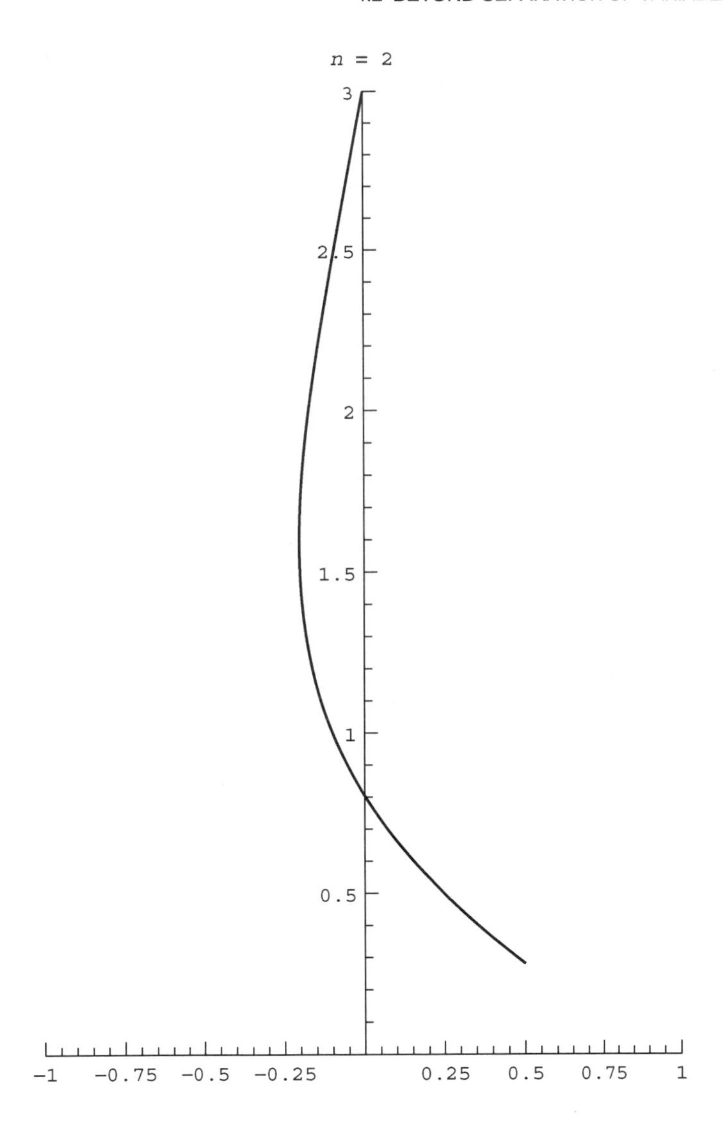

The reader is invited to reevaluate this cell for different values of the mode number n.

We can use these modes to determine the evolution of an initial disturbance on the string. Let's assume that initially the string has a sharply peaked Gaussian pulse shape $y(0,t) = y_0(x) = e^{-40(z-L/2)^2}$. We will also choose an initial velocity of the string consistent with this pulse traveling down the string. On a uniform string, such a pulse would have the form $y = y_0(z+ct)$ [see Eq. (3.1.26)], and this implies that $\partial y/\partial t = c\,\partial y/\partial z$. We will use this form for the initial condition of the nonuniform string, taking

$$\frac{\partial y}{\partial t}(z,0) = c(z)\frac{\partial y_0}{\partial z},$$

where $c(z) = \sqrt{gz}$ is the (nonuniform) propagation speed on the string. Then we know that the solution of Eq. (4.2.41) is

$$y(z,t) = \sum_{n=1}^{\infty} (A_n \cos \omega_n t + B_n \sin \omega_n t)\, \psi_n(z),$$

with ω_n and ψ_n given by Eqs. (4.2.45) and (4.2.46). According to the Sturm–Liouville form of Eq. (4.2.43), different modes are orthogonal with respect to the inner product $(f, g) = \int_0^L f^* g\, dz$. Therefore, A_n and B_n are given in terms of this inner product by Eqs. (4.2.25) and (4.2.27). We evaluate the required inner products in Cell 4.18 keeping $M = 50$ terms in the series solution, and we then plot the result. Here we do not bother to try to allow for the change in height of the string end during the evolution. The pulse travels toward the free end of the string, where a whipcrack occurs. The propagation speed $c(x)$ of the pulse decreases as it approaches the free end, due to the decrease in the tension. Thus, the back of the pulse catches up with the front, and as the pulse compresses, there is a buildup of the amplitude that causes extremely rapid motion of the string tip.

The speed of the string tip can actually exceed the speed of sound in air (roughly 700 miles per hour, or 340 m/s), causing a distinctive whipcrack sound as the string tip breaks the sound barrier. The velocity of the end of the string is plotted in Cell 4.19 in the vicinity of the first whipcrack. This velocity is in units of the maximum initial pulse amplitude per second.

Cell 4.18

```
<< NumericalMath`;
M = 50;
g = 9.8;
L = 3;
j0 = BesselJZeros[0, M];
ω[n_] := j0[[n]] Sqrt[g/L]/2;
ψ[n_, z_] := BesselJ[0, j0[[n]] Sqrt[z/L]];

y0[z_] = Exp[-(z - L/2)^2 40];

dy0[z_] = D[y0[z], z];

a = Table[NIntegrate[ψ[n, z] y0[z], {z, 0, L}]/
  Integrate[ψ[n, z]^2, {z, 0, L}], {n, 1, M}];
b = Table[NIntegrate[ψ[n, z] Sqrt[g z] dy0[z], {z, 0, L}]/
  (Integrate[ψ[n, z]^2, {z, 0, L}] ω[n]), {n, 1, M}];

y[z_, t_] = Sum[(a[[n]] Cos[ω[n] t] +
   b[[n]] Sin[ω[n] t]) ψ[n, z], {n, 1, M}];

Table[ParametricPlot[Evaluate[{y[z, t], z}], {z, 0, L},
  AspectRatio→1/2,
  PlotRange→{{-L, L}, {0, L}}, Axes→False, PlotPoints→60,
  PlotLabel→"y[z, t], t = "<>ToString[t] <>" sec"],
   {t, 0, 2, .05}];
```

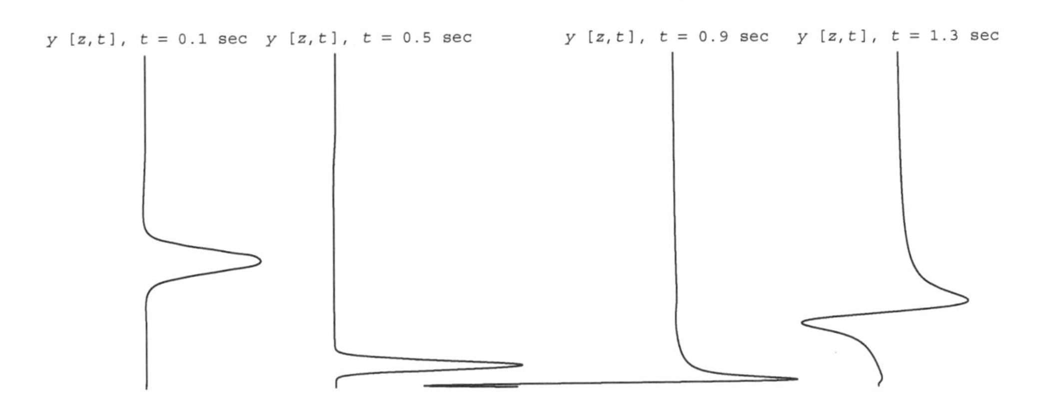

Cell 4.19

```
Plot[Evaluate[D[y[0, t], t]], {t, 0.5, 1.}, PlotRange→All,
  PlotLabel->"string tip velocity",
   AxesLabel→{"t (sec)", ""}];
```

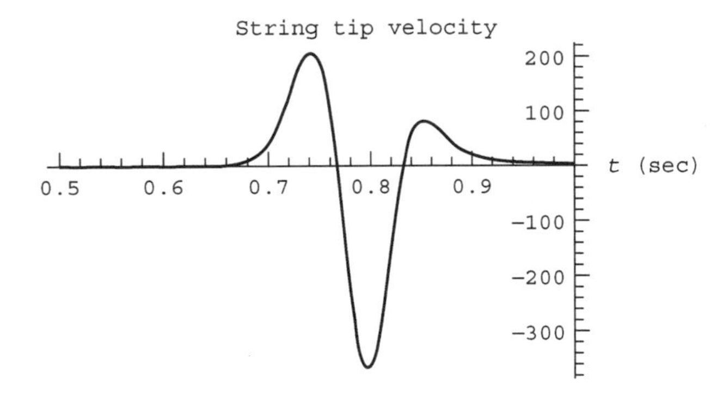

The plot shows that, for our Gaussian initial condition, an initial amplitude of 1 meter would theoretically produce a maximum tip speed that exceeds 300 m/s, which would break the sound barrier. However, such large-amplitude disturbances cannot be properly treated with our linear wave equation, which assumes small pulse amplitudes. We have also neglected many effects of importance in the motion of real whips, such as the effect of the pulse itself on the tension in the whip, the elasticity and plasticity of the whip material, and the effect of tapering the whip to small diameter at the tip end. The physics of whipcracks is, believe it or not, still an active area of research. Interested readers can find several recent references in the very clear paper by Goriely and MacMillan (2002).

EXERCISES FOR SEC. 4.2

(1) Put the following problems defined on $0 < x < 1$ into standard form (if necessary) and find a solution via a sum of eigenmodes:

(a) $\frac{\partial T}{\partial t} = 2\frac{\partial^2 T}{\partial x^2}$, $\frac{\partial T}{\partial x}(0,\mathrm{t}) = 0$, $\frac{\partial T}{\partial x}(1,\mathrm{t}) = 1$, $T(x,0) = \frac{x^4}{4}$. Animate the solution with a table of plots for $0 < t < 1$.

(b) $\frac{\partial T}{\partial t} = \frac{\partial T}{\partial x} + \frac{\partial^2 T}{\partial x^2}$, $T(0,t) = t e^{-4t}$, $T(1,t) = 0$, $T(x,0) = 0$. Animate the solution with a table of plots for $0 < t < 1$. (Hint: You must put the spatial operator in Sturm–Liouville form.)

(c) $\frac{\partial^2}{\partial t^2} y(z,t) = \frac{\partial}{\partial z}\left(z \frac{\partial}{\partial z} y(z,t)\right)$, $y(z,0) = 0.25(1-z)$, $\dot{y}(z,0) = 0$, $y(0,t) = y(1,t) = 0$. Animate the solution with a table of plots for $0 < t < 4$.

(2) When one cooks using radiant heat (under a broiler, for example), there is a heat flux Γ_r due to the radiation, incident on the surface of the food. On the other hand, the food is typically suspended (on a grill or spit for example) in such a way that it cannot conduct heat very well to the environment, so that little heat is reradiated. Under these conditions, find the time required to raise the internal temperature of a slab of meat of thickness $L = 5$ cm from $T = 20°\mathrm{C}$ to $T = 90°\mathrm{C}$. Animate the solution for $T(x,t)$ up to this time. Take $\chi = 3 \times 10^{-7}$ m^2/s, $C = 3 \times 10^6$ J/(m^3 K), and assume that both faces of the meat are subjected to the same flux of heat, equal to 10 kW/m^2 (a typical value in an oven).

(3) In a microwave oven, the microwave power P (in watts) is dissipated near the food surface, in a *skin depth* λ on the order of the wavelength of the microwaves. The power density in the food falls off as $S_0 e^{-2x/\lambda}$, where x is the distance from the food's surface.

(a) Assuming that all microwave power P is dissipated in a slab of meat, that the meat has surface area A on each side of the slab, and the slab thickness is everywhere much larger than λ, find S_0 in terms of P.

(b) A slab of cold roast beef, thickness 10 cm and surface area per side $A = 5000$ cm^2, with initially uniform temperature $T = 10°\mathrm{C}$, is placed in the microwave. The microwave is turned on high, with a power $P = 5$ kW. Taking $\lambda = 1$ cm, and assuming that both faces of the meat are heated equally, find $T(x,t)$ and determine the time required to heat the roast to at least $T = 50$ °C. Animate the solution for $T(x,t)$ up to this time. Take $\chi = 2 \times 10^{-7}$ m^2/s, and assume insulating boundary conditions at the faces of the meat.

(4) A child launches a wave on a skipping rope by flicking one end up and down. The other end is held fixed by a friend. The speed of waves on the rope is $c = 2$ m/s, and the rope is 2 m long. The end held by the child moves according to $u(x=0,t) = e^{-50(t-0.2)^2}$, and at $t = 0$ the rope is stationary, $u(x,0) = \frac{\partial u}{\partial t}(x,0) = 0$. Solve the wave equation for this problem using eigenmodes, and make an animation of the resulting motion of the rope for $0 < t < 1.5$ s.

(5) A child rotates a skipping rope with a frequency of $f = 1$ hertz. She does so by applying a displacement to the end of the rope (at $z = 0$) of the form $\mathbf{r}(0, t) = a(\hat{\mathbf{x}} \cos 2\pi ft + \hat{\mathbf{y}} \sin 2\pi ft)$, where $a = 0.05$ m. The tension in the rope is $T = 2$ newtons. The mass per unit length is $\rho = 0.5$ kg/m, and the length is $L = 3$ m. The other end of the rope is tied to a door handle: $\mathbf{r}(L, t) = 0$. Find the solution for $\mathbf{r}(t, z)$, assuming that initially the rope satisfies $\mathbf{r}(z, 0) = a(1 - z/L)\hat{\mathbf{x}}, \dot{\mathbf{r}}(z, 0) = 0$, and neglecting gravity. Animate the solution (using a table of `ParametricPlot3D`'s) over the time range $0 < t < 5$ s). (Careful: there is an exact resonance.)

(6) A stiff wooden rod is fixed to the wall at $x = 0$ and is free at the other end, at $x = L$. The rod vibrates in the x-direction (these are compressional vibrations, or sound waves, in the rod). These compressional vibrations satisfy the wave equation,

$$\frac{\partial^2}{\partial t^2} \eta(x, t) = c^2 \frac{\partial^2}{\partial x^2} \eta(x, t),$$

where η is the displacement from equilibrium of an element of the rod in the x-direction, and where c is the speed of sound in the rod. [This speed is given in terms of Young's modulus E, the mass density per unit length (ρ), and the cross-sectional area A by $c^2 = EA/\rho$.] The boundary condition at the free end is $\frac{\partial \eta}{\partial x}(L, t) = 0$, and at the fixed end it is $\eta(0, t) = 0$. Find the eigenmodes and their frequencies. Plot the first three eigenmodes.

(7) The horizontal wooden rod of the previous problem also supports transverse displacements (in the y-direction). However, these transverse displacements $y(x, t)$ satisfy a *biharmonic wave equation*,

$$\frac{\partial^2}{\partial t^2} y(x, t) = -\frac{D}{\rho} \frac{\partial^4}{\partial x^4} y(x, t), \qquad (4.2.47)$$

where $D = \pi a^4 E/4$ for a cylindrical rod of radius a, and E is Young's modulus. [See Landau and Lifshitz (1986, Sec. 25).] The boundary condition at the fixed end, $x = 0$, is $y = \partial y/\partial x = 0$. At the free end, $x = L$, the correct boundary condition is

$$\frac{\partial^2 y}{\partial x^2} = \frac{\partial^3 y}{\partial x^3} = 0.$$

(a) Find the first three eigenfrequencies of the rod, and plot the corresponding eigenmodes. (Hint: The eigenfrequencies satisfy a transcendental equation. Solve this equation numerically using `FindRoot`, after choosing suitable dimensionless variables.)

(b) Show numerically that these three eigenmodes are orthogonal with respect to the integral $\int_0^L dx$.

(c) Find and plot the equilibrium shape of the rod if it is subjected to a uniform gravitational acceleration g in the $-y$-direction. How does the maximum sag scale with the length L of the rod? [You may wish to compare this result with that obtained in Exercise (6), Sec. 9.10.]

(8) Sound waves are compressional waves that satisfy the wave equation

$$\frac{\partial^2}{\partial t^2}\eta(x,t) = c^2 \frac{\partial^2}{\partial x^2}\eta(x,t), \tag{4.2.48}$$

where $\eta(x,t)$ is the displacement of a fluid element from its equilibrium position (this displacement is in the x-direction, along the direction of the wave). In a gas the sound speed is given by the equation

$$c = \sqrt{\gamma p/\mathscr{M}}, \tag{4.2.49}$$

where p is the pressure of the equilibrium fluid, $\mathscr{M}$ is the mass density, and γ is the ratio of specific heats (equal to $\frac{5}{3}$ for an ideal gas of point particles).

(a) Find the speed of sound in an ideal gas consisting of helium atoms at 1 atmosphere and a temperature of 300 K. (Recall that $p = nk_BT$, where n is the number density.)

(b) A simple gas-filled piston consists of two flat parallel plates. One is fixed at $x = L$, and the other oscillates according to $x = a \sin \omega_0 t$. The boundary conditions are determined by the fact that the gas adjacent to the plates must move with the plates. Therefore,

$$\eta(0,t) = a \sin \omega_0 t \quad \text{and} \quad \eta(L,t) = 0$$

(provided that a is small). Solve for the motion of the gas between the plates, assuming that the gas is initially stationary.

(c) Find the conditions on ω_0 for which secular growth of the sound wave occurs (i.e., determine when there is an exact resonance).

(9) Water collects in a long straight channel with a sloping bottom and a vertical wall at $x = a$ (see Fig. 4.2). The water depth as function of transverse position x is $h(x) = \alpha x$, $0 < x < a$, where α is a dimensionless constant giving the slope of the channel bottom.

(a) Assuming that we can use the shallow-water equations, show that the horizontal fluid displacement $\eta(x,t)$ and wave height $z(x,t)$ are related by

$$z = -\frac{\partial[h(x)\eta]}{\partial x} \tag{4.2.50}$$

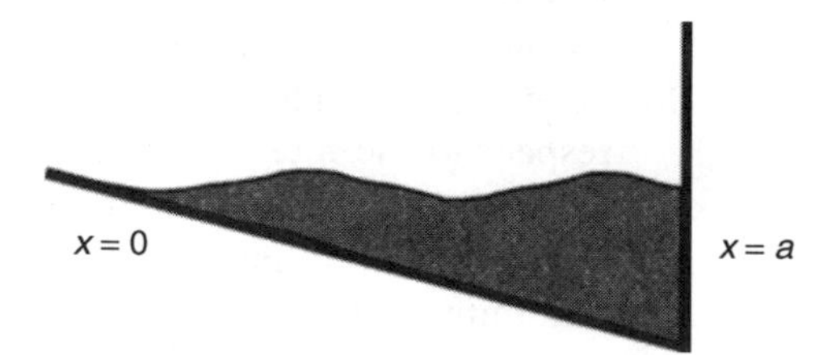

Fig. 4.2 Exercise (9).

and are determined by the wave equation

$$\frac{\partial^2 z(x,t)}{\partial t^2} = \frac{\partial}{\partial x}\left(c^2(x)\frac{\partial z}{\partial x}\right), \tag{4.2.51}$$

where $c(x) = \sqrt{gh(x)}$.

(b) Identify the proper boundary conditions for this problem, determine the eigenmodes, and show that the frequencies ω_n are given by solutions to

$$J_1\left(2\omega_n\sqrt{\alpha a/g}\right) = 0.$$

Find these frequencies for $a = 10$ m, $\alpha = \frac{1}{4}$, and $g = 9.8$ m/s^2. Plot the wave height and horizontal fluid displacement associated with the first three eigenmodes of z. (Hint: The horizontal fluid displacement does *not* vanish at $x = 0$: waves can move up and down the "beach" where the wave depth vanishes. You can, of course, use *Mathematica* to help solve the required differential equation for the spatial dependence of the modes.)

(c) Find the inner product with respect to which these eigenmodes are orthogonal, and solve the following initial-value problem, animating the solution for wave height z for $0 < t < 3$ sec:

$$z(x,0) = 0.3\, e^{-3(x-5)^2}, \qquad \dot{z}(x,0) = 0.$$

(10) (a) Suppose that a tidal estuary extends from $r = 0$ to $r = a$, where it meets the open sea. Suppose the floor of the estuary is level, but its width is proportional to a radial distance r (a wedge-shaped estuary, like Moray Firth in Scotland; see Fig. 4.3). Then using the same method as that which led to Eq. (3.1.78), show that the water depth $z(r,t)$ satisfies the following wave equation:

$$\frac{\partial^2}{\partial t^2} z = gh\frac{1}{r}\frac{\partial}{\partial r}\left(r\frac{\partial}{\partial r}z\right),$$

where g is the acceleration of gravity and h is the equilibrium depth. (Note that no θ-dependence is assumed: the boundary conditions along the upper and lower sides of the estuary are von Neumann, so this θ-independent solution is allowed.)

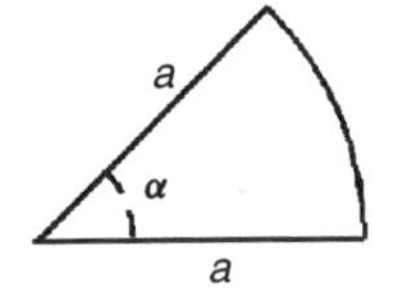

Fig. 4.3 Exercise (10).

(b) The tidal motion of the open sea is represented by the following Dirichlet boundary condition at the end of the estuary:

$$z(a,t) = h + d \cos \omega t.$$

Find a bounded solution of the PDE that satisfies this boundary condition, along with the initial conditions

$$y(r,0) = h + d, \qquad \frac{\partial y}{\partial t}(r,0) = 0.$$

Assume that $\omega \neq \omega_n$, where ω_n are the eigenfrequencies of the normal modes.

(c) Repeat part (b), assuming that $\omega = \omega_0$, the lowest eigenfrequency.

(11) **(a)** Find how much energy it takes to pluck a uniform string of mass density ρ and tension T, giving it a deformation of the form

$$y_0(x) = \begin{cases} ax, & 0 < x < L/2, \\ a(L-x), & L/2 < x < L. \end{cases}$$

(b) What fraction of the energy goes into each normal mode?

(12) A quantum particle is confined in a harmonic well of the form $V(x) = \frac{1}{2} m \omega_0^2 x^2$. Using the results of Eq. (4.1.32) for the eigenfunctions and the energy levels of the quantum harmonic oscillator, determine the evolution of the particle wavefunction $\psi(x,t)$, starting with $\psi(x,0) = \delta(x - x_0)$. Animate this evolution with a table of plots of $(|\psi|)^2$ for $0.01 < t < 6$, taking dimensionless units $m = \omega_0 = \hbar = 1$.

(13) A thick rope of length L and with mass per unit length ρ is spliced at $x = 0$ to a thin rope of the same length L with mass per length ρ'. The ropes are tied to posts at $x = \pm L$ and subjected to uniform tension T. Analytically find the form and the frequency of the first three eigenmodes of this composite rope. Plot the eigenmodes, assuming that $\rho' = \rho/4$. (Hint: Match separate trigonometric solutions for the modes across the splice. To do so, consider the mass elements in the two ropes that are adjacent to one another at the splice. According to Newton's third law, the forces of each element on the other must be equal and opposite. What does this say about the angle θ each element makes with the horizontal? [See Fig. 3.3 and Eq. (3.1.5).])

(14) A nonuniform rope is stretched between posts at $x = 0$ and $x = L$, and is subjected to uniform tension T. The mass density of the rope varies as $\rho = \rho_0 (L/x)^4$. Find the eigenmodes and eigenfrequencies for this rope. Plot the first three eigenmodes.

(15) A hanging string is attached to the ceiling at $z = L$ and has uniform mass density ρ and nonuniform tension due to the acceleration of gravity g. To the end of the rope at $z = 0$ is attached a mass m. The string motion is described by the function $y(z, t)$ as it moves in the y-z plane, and the mass also moves in y with position $Y(t) = y(0, t)$.

(a) Find the equations of motion of this coupled mass–string system, and find the eigenmodes of the system. Determine and plot the first three eigenmodes numerically for the case where $m = 2\rho L$. (Hint: The string eigenmodes satisfy a *mixed* boundary condition at $z = 0$, obtained by considering the horizontal acceleration of the mass due to the string tension.)

(b) Show numerically that the eigenmodes are orthogonal with respect to the inner product $\int_0^L \rho(z)\, dz$, where $\rho(z) = \rho + m\delta(z)$.

(16) An initially straight, horizontal rope of mass M, length $2L$, and tension T runs from $x = -L$ to $x = L$. A mass m is placed on the center of the rope at $x = 0$. The gravitational force causes the rope to sag, and then bounce up and down. Call the vertical position of the mass $Y(t)$, and of the rope $y(x, t)$. The point of the problem is to study the motion of this coupled mass–string system.

(a) Assuming that the rope takes on a triangular shape as it is depressed by the mass, and neglecting the mass of the rope itself, find the restoring force on the mass and show that the mass oscillates sinusoidally about an equilibrium position $-y_{eq}$ at a frequency of $\sqrt{2T/mL}$. [We found y_{eq} in Exercise (11) of Sec. 3.1.]

(b) In fact, the rope does not have a triangular shape during the motion. We will now do this problem properly, expanding in the eigenmodes of the system. Using symmetry, we expect that $y(-x, t) = y(x, t)$ during the motion, so we solve only for the motion in the range $0 < x < L$. Show that the string is described by $y = y_{eq}(x) + \Delta y(x, t)$, where y_{eq} is the equilibrium string displacement due to gravity (including the effect of the mass), and Δy is the deviation from equilibrium, described by a superposition of eigenmodes of the form $\psi_n(x) = \sin[\omega_n(L - x)/c]$, where the eigenfrequencies satisfy the equation $(\omega_n L/c)\tan(\omega_n L/c) = M/m$.

(c) For $M/m = 2$ find the first 10 eigenmodes numerically, and show that they are orthogonal with respect to the inner product $\int_0^L \rho(x)\, dx$, where $\rho(x) = \rho + m\delta(x)/2$ (the factor of two arises because only half the mass is supported by the right half of the string).

(d) Using these eigenmodes, find and plot $Y(t)$ for $0 < t < 20$ s, assuming that the string is initially straight and that the mass starts from rest at $Y = 0$. Take $L = 1$ m, $m = M = 0.5$ kg, $g = 0.3$ m/s^2, and $T = 0.25$ N.

(e) Sometimes the mass does not quite make it back to $Y = 0$ during its motion, and sometimes it actually gets to $y > 0$. This is in contrast with the sinusoidal oscillation found in part (a). Why is the energy of the mass not conserved?

(f) Write an expression for the total energy as a sum over the eigenmodes and the energy of the mass, involving its position $Y(t)$ and velocity $\dot{Y}(t)$. Show directly that this energy is conserved in your simulation by evaluating it as a function of time.

(17) In a bow, a bowstring under tension T with length $2L$ carries a mass m at its center. The mass is pulled back a distance $d \ll L$, and released, starting from rest. Use energy conservation to determine the speed of the mass as it leaves the string, assuming that the string takes on a triangular shape at all times during its motion. (Hint: At some point the mass comes off the string. You will need to identify this point.)

(18) Repeat the previous exercise, but do it properly, using the eigenmode approach of Exercise (16). Solve the equations of motion numerically, keeping 20 modes, using the same parameters in Exercise (16), and taking $d = 10$ cm. Compare the final energy of the mass with that obtained in Exercise (17).

(19) (a) A sound oscillation in a cubic enclosure of length $L = 1m$ and volume $V = 1\ \text{m}^3$ has the form $\eta(x,t) = \eta_0 \sin(\pi x/L)\cos(\omega t)$, where $\omega = c\pi/L$, $c = 340$ m/sec, and the maximum displacement η_0 of the air is 0.1 mm. Find the kinetic energy $K(t)$ (in Joules), where $K(t) = \int_V d^3r\, \mathcal{M}\dot{\eta}^2(x,t)$, and $\mathcal{M}$ is the mass density of air at atmospheric pressure.

(b) Find the potential energy $U(t)$ in Joules, and find the total energy in this sound oscillation. [Hint: Sound waves satisfy Eq. (4.2.48).]

4.3 POISSON'S EQUATION IN TWO AND THREE DIMENSIONS

4.3.1 Introduction. Uniqueness and Standard Form

Poisson's equation is the following partial differential equation:

$$\nabla^2\phi = -\rho/\epsilon_0. \tag{4.3.1}$$

The constant $\epsilon_0 = 8.85\ldots \times 10^{-12}$ F/m is the permittivity of free space. This PDE determines the electrostatic potential $\phi(\mathbf{r})$ (measured in volts) within a specified volume, given a charge density $\rho(\mathbf{r})$ (in coulombs per cubic meter) and boundary conditions on ϕ at the surface S of the volume (see Fig. 3.8). The boundary conditions are either Dirichlet, von Neumann, or mixed (see the introduction to Sec. 3.2 for a description of these boundary conditions).

For Dirichlet and mixed boundary conditions, the solution of Eq. (4.3.1) exists and is unique. The proof of uniqueness is the same as that given for the Laplace equation in Sec. 3.2.1. Existence will be shown by construction in Sec. 4.3.2.

However, for the von Neumann conditions that $\hat{\mathbf{n}}\cdot\nabla\phi|_S = E_0(\mathbf{r})$, the solution for ϕ is unique only up to an additive constant: if ϕ satisfies eq. (4.3.1) with von Neumann conditions, then $\phi + C$ also satisfies it. Also, for von Neumann boundary conditions, a solution only exists if the boundary conditions are consistent with Gauss's law,

$$\int_S \hat{\mathbf{n}}\cdot\nabla\phi\, d^2r = -Q_{\text{enc}}/\epsilon_0, \tag{4.3.2}$$

where $Q_{\text{enc}} = \int_V \rho\, d^3r$ is the charge enclosed by the domain. Gauss's law is merely the integral form of Eq. (4.3.1), obtained by applying the divergence theorem to this equation. An analogous result was obtained for the Laplace equation [see Eq. (3.2.3)].

In order to solve Eq. (4.3.1), we first put the PDE in standard form, by converting the inhomogeneous boundary conditions to a source function, just as we did for the wave and heat equations. That is, we write

$$\phi(\mathbf{r}) = \Delta\phi(\mathbf{r}) + u(\mathbf{r}), \tag{4.3.3}$$

where $u(\mathbf{r})$ is a function chosen to match the boundary conditions. The remainder, $\Delta\phi(\mathbf{r})$, then satisfies Poisson's equation with a new source,

$$\nabla^2 \Delta\phi = -\bar{\rho}/\epsilon_0, \tag{4.3.4}$$

where

$$\bar{\rho} = \rho + \epsilon_0 \nabla^2 u. \tag{4.3.5}$$

The boundary conditions on $\Delta\phi$ are homogeneous conditions of either Dirichlet, von Neumann, or mixed type.

The choice for $u(\mathbf{r})$ is arbitrary, but, as always, the simpler the choice, the better. Sometimes it is convenient to choose u to satisfy Laplace's equation $\nabla^2 u = 0$. As we will see in Sec. 4.3.4, this choice for u is particularly useful if $\rho = 0$, or if the inhomogeneous boundary conditions are rapidly varying. Techniques specific to the solution of Laplace's equation were developed in Sec. 3.2.

4.3.2 Green's Function

Equation (4.3.4) can be solved using a Green's function. The Green's function $g(\mathbf{r}, \mathbf{r}_0)$ satisfies

$$\nabla_{\mathbf{r}}^2 g(\mathbf{r}, \mathbf{r}_0) = \delta(\mathbf{r} - \mathbf{r}_0), \tag{4.3.6}$$

where the subscript $\mathbf{r}$ on $\nabla_{\mathbf{r}}^2$ is placed there to remind us that the derivatives in the Laplacian are with respect to $\mathbf{r}$ rather than $\mathbf{r}_0$. The boundary conditions on g are the same homogeneous boundary conditions required for $\Delta\phi$. The vector δ-function in Eq. (4.3.6) $\delta(\mathbf{r} - \mathbf{r}_0)$, is a δ-function at a point in space, i.e.,

$$\delta(\mathbf{r} - \mathbf{r}_0) = \delta(x - x_0)\,\delta(y - y_0)\,\delta(z - z_0). \tag{4.3.7}$$

One can see that this Green's function $g(\mathbf{r}, \mathbf{r}_0)$ is simply the potential at position $\mathbf{r}$ produced by a point charge at position $\mathbf{r}_0$ with "charge" $-\epsilon_0$. In free space, with the boundary condition that the potential equals zero at infinity, we know that this Green's function is simply the $1/r$ Coulomb potential:

$$g(\mathbf{r}, \mathbf{r}_0) = -\frac{1}{4\pi|\mathbf{r} - \mathbf{r}_0|}. \tag{4.3.8}$$

However, when the potential and/or its normal derivative is specified on a finite bounding surface S, the Green's function is more complicated because of image charges induced in the surface.

Assuming that the Green's function has been determined, the potential $\Delta\phi(\mathbf{r})$ can then be obtained using a multidimensional version of Eq. (2.4.30):

$$\Delta\phi(\mathbf{r}) = -\frac{1}{\epsilon_0}\int_V g(\mathbf{r},\mathbf{r}_0)\,\bar{\rho}(\mathbf{r}_0)\,d^3\mathbf{r}_0, \tag{4.3.9}$$

where the volume integral extends over the volume V. To prove Eq. (4.3.9), we simply apply the Laplacian $\nabla_{\mathbf{r}}^2$ to each side, and using Eq. (4.3.6) we have

$$\nabla_{\mathbf{r}}^2\Delta\phi(\mathbf{r}) = -\frac{1}{\epsilon_0}\int_V \nabla_{\mathbf{r}}^2 g(\mathbf{r},\mathbf{r}_0)\,\bar{\rho}(\mathbf{r}_0)\,d^3\mathbf{r}_0 = -\frac{1}{\epsilon_0}\int_V \delta(\mathbf{r}-\mathbf{r}_0)\,\bar{\rho}(\mathbf{r}_0)\,d^3\mathbf{r}_0 = -\frac{\bar{\rho}(\mathbf{r})}{\epsilon_0}. \tag{4.3.10}$$

Also, the homogeneous boundary conditions for $\Delta\phi$ are satisfied, because these same boundary conditions apply to g. For example, if the boundary conditions are Dirichlet, then $g=0$ for any point $\mathbf{r}$ on the surface S, and then Eq. (4.3.9) implies that $\Delta\phi=0$ on S as well.

Equation (4.3.10) has a simple physical interpretation: since $-g(\mathbf{r},\mathbf{r}_0)/\epsilon_0$ is the potential at $\mathbf{r}$ due to a unit charge at position $\mathbf{r}_0$, we can use the superposition principle to determine the total potential at $\mathbf{r}$ by superimposing the potentials due to all of the charges. When the charges form a continuous distribution $\bar{\rho}$, this sum becomes the integral given in Eq. (4.3.10). If the charges are discrete, at positions $\mathbf{r}_j$, each with charge e_j, then the charge density $\bar{\rho}$ is a sum of δ-functions,

$$\bar{\rho}(\mathbf{r}) = \sum_{j=1}^{N} e_j\,\delta(\mathbf{r}-\mathbf{r}_j),$$

and Eq. (4.3.9) implies that this collection of discrete charges produces the following potential:

$$\Delta\phi(\mathbf{r}) = -\sum_{j=1}^{N}\frac{e_j}{\epsilon_0}g(\mathbf{r},\mathbf{r}_j). \tag{4.3.11}$$

4.3.3 Expansion of g and ϕ in Eigenmodes of the Laplacian Operator

The Green's function can be determined as an expansion in eigenmodes $\psi_\alpha(\mathbf{r})$ of the Laplacian operator. These eigenmodes are functions of the vector position $\mathbf{r}$, and are defined by the eigenvalue problem

$$\nabla^2\psi_\alpha(\mathbf{r}) = \lambda_\alpha\,\psi_\alpha(\mathbf{r}). \tag{4.3.12}$$

This PDE, the *Helmholtz equation*, is subject to the previously described homogeneous boundary conditions. Each eigenfunction $\psi_\alpha(\mathbf{r})$ has an associated eigenvalue λ_α. The subscript α is merely a counter that enumerates all the different modes. (We will see presently that this counter can be represented as a list of integers that take on different values for the different modes.)

These eigenmodes form a complete orthogonal set with respect to the following inner product:

$$(f, g) = \int_V f^*(\mathbf{r}) g(\mathbf{r})\, d^3r. \tag{4.3.13}$$

The proof is straightforward, given what we already know about eigenmodes of Hermitian operators. All we need do is show that the Laplacian operator is Hermitian with respect to the above inner product and with respect to functions that satisfy homogeneous boundary conditions. We may then apply Theorems 4.1 and 4.2 (there is nothing in the proofs of these theorems that limits the operators in question to ODE operators, as opposed to PDE operators).

In order to show that ∇^2 is Hermitian, consider the following quantity:

$$(f, \nabla^2 g) = \int_V f^*(\mathbf{r}) \nabla^2 g(\mathbf{r})\, \mathbf{d}^3 r.$$

By application of Green's theorem, this inner product can be written as

$$(f, \nabla^2 g) = \hat{\mathbf{n}} \cdot (f^* \nabla g - g \nabla f^*)|_S + \int_V g(\mathbf{r}) \nabla^2 f^*(\mathbf{r})\, \mathbf{d}^3 r.$$

However, the surface term on the right-hand side vanishes for homogeneous boundary conditions of the Dirichlet, von Neumann, or mixed type, and the volume term is $(g, \nabla^2 f)^*$. This proves that ∇^2 is Hermitian, and therefore the eigenmodes of ∇^2 form a complete, orthogonal set with respect to the inner product of Eq. (4.3.13).

We can use the eigenmodes to express the Green's function as a generalized Fourier series:

$$g(\mathbf{r}, \mathbf{r}_0) = \sum_\alpha c_\alpha \psi_\alpha(\mathbf{r}). \tag{4.3.14}$$

The Fourier coefficients c_α are obtained by substituting Eq. (4.3.14) into Eq. (4.3.6):

$$\sum_\alpha c_\alpha \nabla_{\mathbf{r}}^2 \psi_\alpha(\mathbf{r}) = \delta(\mathbf{r} - \mathbf{r}_0).$$

If we then take an inner product of the equation with respect to one of the eigenmodes, $\psi_\beta(\mathbf{r})$, and apply Eq. (4.3.12), we obtain

$$\sum_\alpha c_\alpha \lambda_\alpha (\psi_\beta, \psi_\alpha) = \left(\psi_\beta(\mathbf{r}), \delta(\mathbf{r} - \mathbf{r}_0)\right) = \psi_\beta^*(\mathbf{r}_0),$$

where in the last step we used Eqs. (4.3.7), (4.3.13), and (2.3.29). However, orthogonality of the eigenmodes implies that the only term in the sum that survives is the one for which $\alpha = \beta$, which allows us to extract a single Fourier coefficient:

$$c_\beta = \frac{\psi_\beta^*(\mathbf{r}_0)}{\lambda_\beta (\psi_\beta, \psi_\beta)}. \tag{4.3.15}$$

Applying Eq. (4.3.15) to (4.3.14), we arrive at an eigenmode expansion for the Green's function:

$$g(\mathbf{r},\mathbf{r}_0) = \sum_\alpha \frac{\psi_\alpha^*(\mathbf{r}_0)\,\psi_\alpha(\mathbf{r})}{\lambda_\alpha(\psi_\alpha,\psi_\alpha)}. \tag{4.3.16}$$

Equation (4.3.16) is a general expression for the Green's function of the Poisson equation, and is called the bilinear equation. A similar expression can be obtained for the Green's function associated with any linear boundary-value problem. This equation can be used in Eq. (4.3.9) to determine the potential $\Delta\phi$ from an arbitrary charge distribution $\bar{\rho}(\mathbf{r})$:

$$\Delta\phi(\mathbf{r}) = -\frac{1}{\epsilon_0}\sum_\alpha \frac{(\psi_\alpha,\bar{\rho})}{\lambda_\alpha(\psi_\alpha,\psi_\alpha)}\,\psi_\alpha(\mathbf{r}), \tag{4.3.17}$$

where we have converted the volume integral over $\mathbf{r}_0$ to an inner product using Eq. (4.3.13).

Equation (4.3.17) is a generalized Fourier series for the potential $\Delta\phi$ due to a charge density $\bar{\rho}$. It applies to any geometry, with arbitrary homogeneous boundary conditions. Inhomogeneous boundary conditions can be easily accommodated using Eqs. (4.3.3) and (4.3.5). The only outstanding issue is the form of the eigenmodes $\psi_\alpha(\mathbf{r})$ and their associated eigenvalues λ_α.

It appears from Eq. (4.3.17) that a solution for the potential can always be constructed. On the other hand, we already know that the solution does not necessarily exist; boundary conditions must satisfy Gauss's law, Eq. (4.3.2). In fact, Eq. (4.3.17) only works if the eigenvalues λ_α are not equal to zero. For Dirichlet and mixed boundary conditions, it can be proven that this is actually true: $\lambda_\alpha \neq 0$ for all modes. The proof is simple: if some $\lambda_\alpha = 0$, then the corresponding eigenmode satisfies $\nabla^2\psi_\alpha = 0$, with *homogeneous* Dirichlet or mixed boundary conditions. However, we proved in Sec. 3.2.1 that this problem only has the trivial solution $\psi_\alpha = 0$. Therefore, the solution to Poisson's equation with Dirichlet or mixed boundary conditions always exists.

On the other hand, for the homogeneous von Neumann boundary conditions $\hat{\mathbf{n}}\cdot\nabla\psi_\alpha = 0$, the following eigenfunction satisfies the boundary conditions: $\psi_0 = 1$. This eigenfunction also satisfies $\nabla^2\psi_0 = 0$, so the corresponding eigenvalue is $\lambda_0 = 0$.

For von Neumann boundary conditions, a solution can only be obtained if the function $\bar{\rho}$ satisfies $(\psi_0, \bar{\rho}) = 0$, so that the ψ_0 term in Eq. (4.3.17) can be dropped and division by zero can be avoided. This inner product implies $\int_V \bar{\rho}\, d^3r = 0$. Using Eq. (4.3.5) and applying the divergence theorem, this equation can be shown to be the same as our previous condition for the existence of a solution, namely, Gauss's law, Eq. (4.3.2).

4.3.4 Eigenmodes of ∇^2 in Separable Geometries

Introduction In order to apply the generalized Fourier series, Eq. (4.3.17), for the potential within a specified domain V due to a given charge density and boundary condition on the surface of V, we require the eigenmodes of ∇^2 in this domain. For certain domain geometries, these eigenmodes can be determined analytically, using the method of separation of variables. We will consider three such geometries in the following sections: rectangular, cylindrical, and spherical domains. These by no means exhaust the possibilities: using separation of variables, analytic eigenmodes can be found in 11 different coordinate systems. [See Morse and Feshbach (1953, Chapter 5) for a full accounting.]

Rectangular Geometry

Rectangular Eigenmodes. We first solve for the eigenmodes of ∇^2 in a rectangular domain, as shown in Fig. 4.4. We will assume that the eigenmodes satisfy homogeneous Dirichlet boundary conditions at the surface of the domain, $\phi|_S = 0$. In physical terms, we are considering the z-independent eigenmodes inside a long grounded conducting tube with rectangular cross section.

Applying the method of separation of variables, we assume that a solution for an eigenmode $\psi_\alpha(x, y)$ can be found in the form

$$\psi_\alpha(x, y) = X(x)Y(y) \tag{4.3.18}$$

for some functions X and Y. If we substitute Eq. (4.3.18) into Eq. (4.3.12) and divide the result by ψ_α, we obtain

$$\frac{1}{X(x)}\frac{d^2X}{dx^2} + \frac{1}{Y(y)}\frac{d^2Y}{dy^2} = \lambda_\alpha. \tag{4.3.19}$$

Introducing a separation constant $-k^2$, this equation is separated into the two ODEs

$$\frac{1}{X(x)}\frac{d^2X}{dx^2} = -k^2, \tag{4.3.20}$$

$$\frac{1}{Y(y)}\frac{d^2Y}{dy^2} = \lambda_\alpha + k^2. \tag{4.3.21}$$

Boundary conditions for each ODE follow from the Dirichlet condition for each

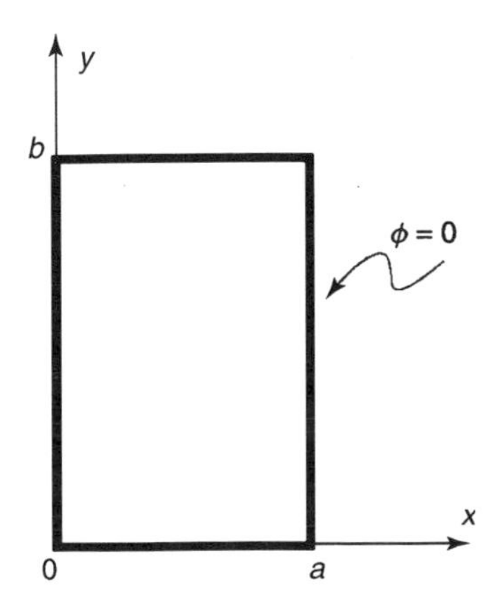

Fig. 4.4 Domain for eigenmodes in rectangular geometry with Dirichlet boundary conditions.

eigenmode, $\psi_\alpha|_S = 0$:

$$X(0) = X(a) = Y(0) = Y(b) = 0. \tag{4.3.22}$$

With these homogeneous boundary conditions, Eqs. (4.3.20) and (4.3.21) can be recognized to be separate eigenvalue problems. The solution to Eq. (4.3.20) is

$$k = n\pi/a, \tag{4.3.23}$$

$$X(x) = \sin\frac{n\pi x}{a}, \qquad n = 1, 2, 3, \ldots, \tag{4.3.24}$$

and the solution to Eq. (4.3.21) is

$$Y(y) = \sin\frac{m\pi y}{b}, \qquad m = 1, 2, 3, \ldots, \tag{4.3.25}$$

with eigenvalues given by $\lambda_\alpha + k^2 = -(m\pi/b)^2$. Using Eq. (4.3.23), this implies

$$\lambda_\alpha = -\left(\frac{n\pi}{a}\right)^2 - \left(\frac{m\pi}{b}\right)^2, \qquad m = 1, 2, 3, \ldots, \quad n = 1, 2, 3, \ldots. \tag{4.3.26}$$

Thus, the Dirichlet eigenfunction $\psi_\alpha(x, y)$ in rectangular geometry is a product of sine functions:

$$\psi_\alpha(x, y) = \sin\frac{n\pi x}{a}\sin\frac{m\pi y}{b}, \qquad m = 1, 2, 3, \ldots, \quad n = 1, 2, 3, \ldots, \tag{4.3.27}$$

with an eigenvalue given by Eq. (4.3.26). Different eigenfunctions and eigenvalues are selected by choosing different values for the positive integers m and n. Therefore, the counter α, which we have used to enumerate the eigenmodes, is actually a list: $\alpha = (m, n)$.

The eigenmodes clearly form a complete orthogonal set with respect to the inner product $\int_0^a dx \int_0^b dy$, as a consequence of the known orthogonality and completeness properties of the sine functions in a Fourier sine series. This is expected from the general arguments made in Sec. 4.3.3.

Example 1: Solution of Poisson's Equation with Smooth Boundary Conditions As an example, consider the case where the charge density $\rho = \rho_0$, a constant, and the walls of the enclosure are grounded, except for the base at $y = 0$, where $\phi(x, 0) = \beta x(a - x)$, where β is a constant.

In order to put Poisson's equation into standard form, Eqs. (4.3.3) and (4.3.4), we simply take $u = \beta x(a - x)(1 - y/b)$. This function satisfies the boundary conditions on all four sides of the rectangle.

According to Eq. (4.3.17) and (4.3.4), we require the inner product

$$(\psi_{mn}, \rho_0/\epsilon_0 + \nabla^2 u) = \int_0^a dx \int_0^b dy\, \psi_{mn}^*(\rho_0/\epsilon_0 + \nabla^2 u).$$

We can use *Mathematica* to evaluate this inner product:

Cell 4.20

```
u[x_, y_] = βx(a-x) (1-y/b);
ρ[m_, n_] = Integrate[Sin[mπy/b] Sin[nπx/a]
    (ρ0/ε0 + D[u[x, y], {x, 2}] + D[u[x, y], {y, 2}]),
     {x, 0, a}, {y, 0, b}];
ρ[m_, n_] = Simplify[ρ[m, n], m∈Integers&& n∈Integers]
```

$$-\frac{2\ a\ b\ \mathrm{Sin}\left[\frac{n\pi}{2}\right]^2\ (2\ \beta\ \epsilon_0\ +\ (-1\ +\ (-1)^m)\ \rho_0)}{mn\pi^2\ \epsilon_0}$$

Also, the inner product $(\psi_\alpha, \psi_\alpha)$ equals $ab/4$. The solution for $\phi(x, y)$, Eqs. (4.3.17) and (4.3.3), is plotted in Cell 4.21, taking $a = b = 1$ meter, $\rho_0/\epsilon_0 = 3\beta = 1$ $\mathrm{V/m^2}$. This corresponds to a charge density of $\rho_0 = 8.85 \times 10^{-12}$ $\mathrm{C/m^3}$. We keep only the first nine terms in each sine series, since each series converges quickly. The potential matches the boundary conditions, and has a maximum in the interior of the domain due to the uniform charge density.

Cell 4.21

```
a = 1; b = 1;
ρ0 = ε0; β = 1/3;
λ[m_, n_] = -(nπ/a)^2 - (mπ/b)^2;
φ[x_, y_] = - Sum[Sum[ρ[m, n] Sin[nπx/a] Sin[mπy/b] / (λ[m, n] 1/4 a b), {n, 1, 9}], {m, 1, 9}] + u[x, y];
Plot3D[φ[x, y], {x, 0, a}, {y, 0, b},
  AxesLabel→{"x", "y", ""},
  PlotLabel→"φ in a charge-filled square enclosure"];
```

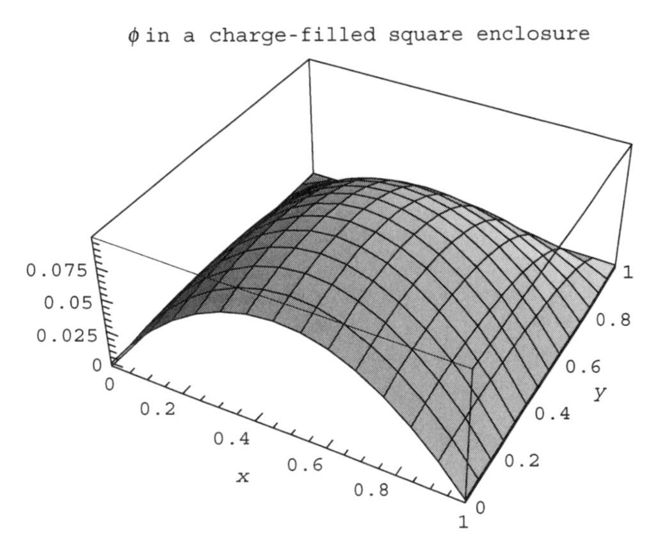

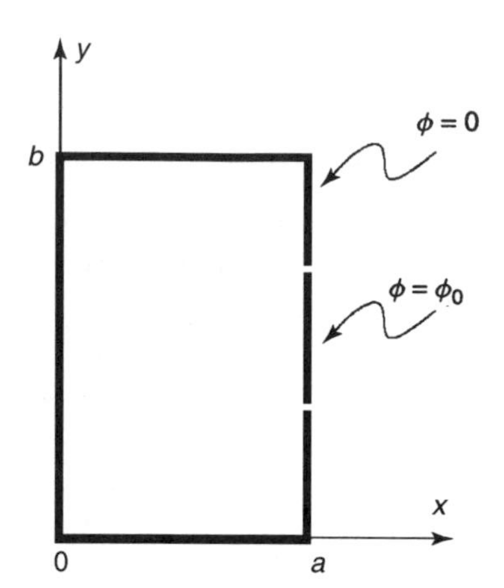

Fig. 4.5

Example 2: Solution of Laplace's Equation with Rapidly Varying Boundary Conditions In this example, we choose a case with no charge density, so that we are looking for the solution of Laplace's equation $\nabla^2\phi = 0$. For boundary conditions, we take $\phi = 0$ on three sides, but on the right side of the box at $x = a$, we take $\phi(a, y) = \phi_0 h(y - b/3)h(2b/3 - y)$, where $h(y)$ is a Heaviside step function. In other words, the center third of the right side of the box wall is at potential ϕ_0, but the rest of the box is grounded. (See Fig. 4.5.)

This is a case where it is best to solve the Laplace equation directly using the methods outlined in Sec. 3.2.2. To see why this is so, let us instead put the problem into standard form by choosing some function $u(x, y)$ that matches the boundary conditions. There are many choices we could make [see Exercise (6) at the end of this section]; one simple choice (which does not work very well) is

$$u(x, y) = \phi_0 h\left(y - \frac{b}{3}\right) h\left(\frac{2b}{3} - y\right) \frac{x}{a}. \tag{4.3.28}$$

Then, according to Eqs. (4.3.2) and (4.3.17), the solution to the problem is given by the following eigenmode expansion:

$$\phi(x, y) = u(x, y) - \sum_{m,n=1}^{\infty} \frac{(\psi_{mn}, \nabla^2 u)}{\lambda_{mn}(\psi_{mn}, \psi_{mn})} \psi_{mn}(x, y), \tag{4.3.29}$$

where the eigenfunctions and eigenvalues are given by Eqs. (4.3.27) and (4.3.26). The inner product $\psi_{mn}, \nabla^2 u)$ can be evaluated directly, but because u is discontinuous it is best to do so by first applying two integrations by parts in the y-integral:

$$(\psi_{mn}, \nabla^2 u) = \int_0^a dx \int_0^b dy\, \psi_{mn} \frac{\partial^2 u}{\partial y^2} = \int_0^a dx \int_0^b dy\, u \frac{\partial^2 \psi_{mn}}{\partial y^2}. \tag{4.3.30}$$

In the first step, we used the fact that Eqs. (4.3.28) implies $\partial^2 u/\partial x^2 = 0$, and in the second step we integrated by parts twice, and dropped the boundary terms because u and its derivatives vanish at $y = 0$ and $y = b$. The integrals in Eq. (4.3.30) can then easily be performed using *Mathematica*:

Cell 4.22

```
Clear["Global`*"];

u[x_, y_] = φ0 UnitStep[y-b/3] UnitStep[2 b/3-y] x/a;
ψ[m_, n_, x_, y_] = Sin[n Pi x/a] Sin[m Pi y/b];
ρ[m_, n_] =
  FullSimplify[Integrate[u[x, y] D[ψ[m, n, x, y], {y, 2}],
    {x, 0, a}, {y, 0, b}],
  b > 0&&m> 0&& n > 0&& m∈Integers && n∈Integers]
```

$$\frac{(-1)^{n}\ a\ m\ \left(\mathrm{Cos}\left[\frac{m\pi}{3}\right]\ -\ \mathrm{Cos}\left[\frac{2m\pi}{3}\right]\right)\ \phi_0}{b\ n}$$

In Cell 4.23, we plot the solution for a 1-meter-square box and $\phi_0 = 1$ volt, keeping $M = 30$ terms in the sums. There is a rather large Gibbs phenomenon caused by the discontinuities in our choice for $u(x, y)$, Eq. (4.3.28). This Gibbs phenomenon extends all the way into the interior of the box, as can be seen in Ccll 4.24 by looking at ϕ along the line $x = \frac{1}{2}$.

Cell 4.23

```
a = b = φ0 = 1;

M = 30;
```

$$\lambda[\mathrm{m\_},\ \mathrm{n\_}]\ =\ -\left(\frac{\mathrm{n}\pi}{\mathrm{a}}\right)^2\ -\ \left(\frac{\mathrm{m}\pi}{\mathrm{b}}\right)^2;$$

$$\phi[\mathrm{x\_},\ \mathrm{y\_}]\ =\ -\sum_{m=1}^{M}\sum_{n=1}^{M}\frac{\rho[\mathrm{m},\ \mathrm{n}]\ \psi[\mathrm{m},\ \mathrm{n},\ \mathrm{x},\ \mathrm{y}]}{\lambda[\mathrm{m},\ \mathrm{n}]\ \frac{1}{4}\mathrm{a\,b}}\ +\ \mathrm{u}[\mathrm{x},\ \mathrm{y}];$$

```
Plot3D[φ[x, y], {x, 0, a}, {y, 0, b},
  AxesLabel→{"x", "y", ""},
  PlotLabel→"φ in a square enclosure", PlotRange→All,
    PlotPoints→30];
```

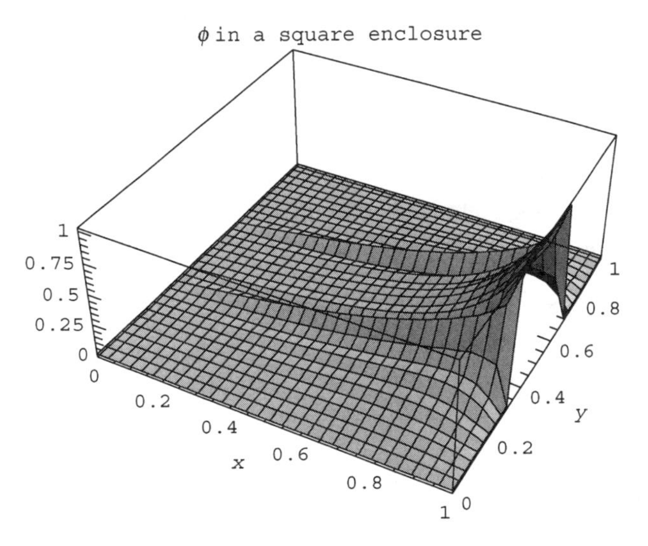

Cell 4.24

```
Plot[ϕ[1/2, y], {y, 0, b}];
```

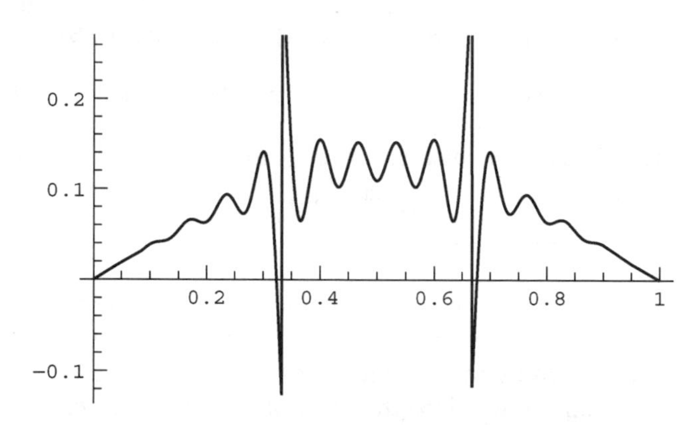

This is not a very good solution. One can see that the solution is "trying" to be smooth, but the discontinuities in $u(x, y)$ are creating problems. [A better choice for u can be found in Exercise (6) at the end of the section.]

Let's now compare this solution with the direct solution of Laplace's equation for this problem, as given by Eq. (3.2.10):

$$\phi(x, y) = \sum_n A_n \sinh\frac{n\pi x}{b} \sin\frac{n\pi y}{b}. \tag{4.3.31}$$

We can already observe one striking difference between Eqs. (4.3.31) and (4.3.29). The eigenmode expansion, Eq. (4.3.29), involves two sums, one over linearly independent eigenmodes in x and the other over independent modes in y. However, the direct solution of Laplace's equation, Eq. (4.3.31), involves only one sum. This is because in the direct solution, the $\sin(n\pi y/b)$ and $\sinh(n\pi x/b)$ functions are not independent; they are connected by the fact that the product of this pair of functions directly satisfies Laplace's equation.

Although Eqs. (4.3.31) and (4.3.29) are formally identical, the fact that one fewer sum is required in Eq. (4.3.31) is often a great practical advantage. Furthermore, for rapidly varying boundary conditions, we have seen that the eigenmode expansion does not converge well. We will now see that the direct solution works nicely.

The Fourier coefficients A_n in Eq. (4.3.31) are given by

$$\begin{aligned} A_n \sinh\frac{n\pi a}{b} &= \frac{2}{b}\int_0^b \phi_0 h\left(y - \frac{b}{3}\right) h\left(\frac{2b}{3} - y\right) \sin\frac{n\pi y}{b}\, dy \\ &= \frac{2}{b}\phi_0 \int_{b/3}^{2b/3} \sin\frac{n\pi y}{b}\, dy. \end{aligned} \tag{4.3.32}$$

In Cell 4.25, we evaluate A_n using *Mathematica* and calculate the direct solution for the potential, again taking $M = 30$ (this involves a sum of 30 terms, as opposed to the 900 terms in the sums for the previous eigenmode method). This solution is much better behaved than the previous eigenmode expansion. There is still a Gibbs phenomenon near the wall due to the discontinuity in the potential there, but by the time one reaches the middle of the box these oscillations are no longer apparent, as seen in Cell 4.26.

Cell 4.25

```
a = b = φ0 = 1;

M = 30;
A[n_] =
  Simplify[2/b φ0 Integrate[Sin[n Pi y/b], {y, b/3, 2 b/3}],
   n ∈ Integers] Sinh[n Pi a/b];

            M
φ[x_, y_] = Σ A[n] Sinh[n Pi x/b] Sin[n Pi y/b];
           n=1

Plot3D[φ[x, y], {x, 0, a}, {y, 0, b}, AxesLabel→{"x", "y", ""},
  PlotLabel→"φ in a square enclosure", PlotRange→All,
   PlotPoints→30];
```

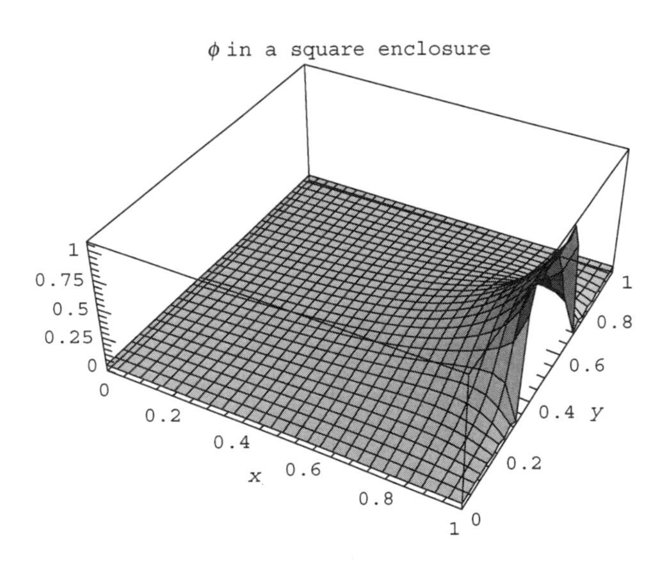

Cell 4.26

```
Plot[Evaluate[{φ[1, y], φ[1/2, y]}], {y, 0, 1},
  PlotLabel→"Potential along x=1/2 and x=1",
   AxesLabel→{"y", ""}];
```

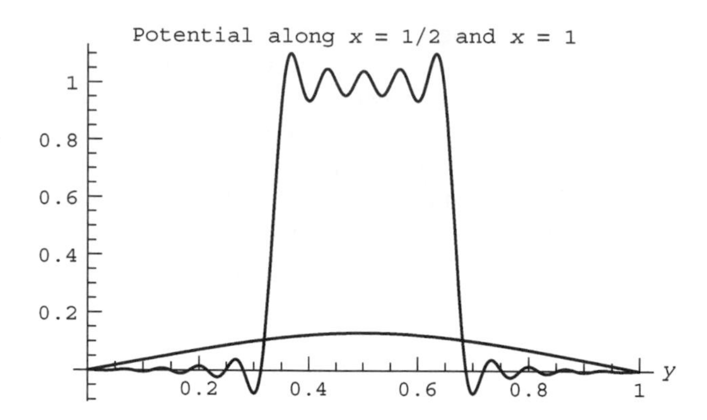

In summary, we have found that for potential problems with sharply varying boundary conditions, choosing an arbitrary function u that satisfies the boundary conditions and expanding the remainder in eigenmodes does not work very well. Rather, we found that it is better to use the direct solution of Laplace's equation, discussed in Sec. 3.2, to allow for the inhomogeneous boundary conditions.

It is possible to circumvent some of the problems with the eigenmode expansion method by careful choice of u [see Exercise (6) at the end of the section]. Nevertheless, we still recommend the direct solution for most applications with discontinuous boundary conditions, in view of its relative simplicity compared to the eigenmode expansion method.

On the other hand, for smoothly varying boundary conditions, the eigenmode expansion technique works quite well, as saw in our first example. When nonzero charge density ρ is present, this method has the important advantage that a separate solution to the Laplace equation need not be generated; rather, an arbitrary (but smoothly varying) function u can be chosen to match the boundary conditions.

The situation is summarized in Table 4.1.

Cylindrical Geometry

Cylindrical Eigenmodes. Eigenmodes of ∇^2 can also be determined analytically inside a cylindrical tube. The tube has radius a and length L, and is closed at the ends. We again assume Dirichlet boundary conditions for the eigenmodes, $\psi|_S = 0$.

Table 4.1. Pros and Cons of Different Choices for u

	Arbitrary Choice of u	u Satisfies $\nabla^2 u = 0$
Pro	No need to solve Laplace equation	Always works, efficient if $\rho = 0$
Con	Does not work well for rapidly varying boundary conditions unless care is taken in choice of u	If $\rho \neq 0$, must solve both Laplace equation for u and Poisson equation for $\Delta\phi$

We look for eigenmodes of the form

$$\psi_\alpha(r, \theta, z) = R(r)\Theta(\theta)Z(z). \tag{4.3.33}$$

Applying Eq. (4.3.33) to the eigenmode equation $(\nabla^2\psi_\alpha)/\psi_\alpha = \lambda_\alpha$ yields

$$\frac{1}{rR(r)}\frac{\partial}{\partial r}\left(r\frac{\partial R}{\partial r}\right) + \frac{1}{r^2\Theta(\theta)}\frac{\partial^2\Theta}{\partial\theta^2} + \frac{1}{Z(z)}\frac{\partial^2 Z}{\partial z^2} = \lambda_\alpha. \tag{4.3.34}$$

This equation can be separated into three ODEs in the usual way, using two separation constants $-k^2$ and $-m^2$:

$$\frac{1}{r}\frac{\partial}{\partial r}\left(r\frac{\partial R}{\partial r}\right) - \left(\frac{m^2}{r^2} + k^2\right)R(r) = \lambda_\alpha R(r), \tag{4.3.35}$$

$$\frac{\partial^2\Theta}{\partial\theta^2} = -m^2\Theta(\theta), \tag{4.3.36}$$

$$\frac{d^2Z}{dz^2} = -k^2 Z(z). \tag{4.3.37}$$

Each equation provides a separate eigenvalue problem. Starting with the last equation first, the solution with Dirichlet boundary conditions $Z(0) = Z(L) = 0$ is our standard trigonometric eigenfunctions

$$Z(z) = \sin\frac{l\pi z}{L}, \qquad l = 1, 2, 3, \ldots, \tag{4.3.38}$$

with eigenvalues

$$k = l\pi/L. \tag{4.3.39}$$

Next, we consider the θ-dependence of the eigenmodes. This is also a familiar problem, given the periodic boundary conditions $\Theta(\theta + 2\pi) = \Theta(\theta)$ required for a single-valued solution. This eigenvalue problem has the solution

$$\Theta(\theta) = e^{im\theta}, \qquad m \in \text{Integers}. \tag{4.3.40}$$

Finally, we turn to Eq. (4.3.35), which describes the radial dependence of the eigenmodes. The dependence of the solution on the parameter $k^2 - \lambda_\alpha$ can be accommodated by a simple change of variables to

$$\bar{r} = \sqrt{-\lambda_\alpha - k^2}\, r. \tag{4.3.41}$$

In this variable, Eq. (4.3.35) becomes Bessel's equation,

$$\frac{1}{\bar{r}}\frac{\partial}{\partial\bar{r}}\left(\bar{r}\frac{\partial R}{\partial\bar{r}}\right) - \left(\frac{m^2}{\bar{r}^2} - 1\right)R(\bar{r}) = 0. \tag{4.3.42}$$

Thus, the general solution for the radial eigenmodes is

$$R(r) = AJ_m\left(\sqrt{-\lambda_\alpha - k^2}\,r\right) + BY_m\left(\sqrt{-\lambda_\alpha - k^2}\,r\right), \tag{4.3.43}$$

where J_m and Y_m are the Bessel functions encountered previously in Sec. 3.2.5. To find the eigenmodes, we match to the boundary conditions. The fact that the solution must be finite at the origin implies that $B = 0$, because the Y_m's are singular at the origin. The fact that $R(a) = 0$ in order to satisfy the Dirichlet boundary condition at the wall implies that

$$\sqrt{-\lambda_\alpha - k^2}\,a = j_{m,n}, \qquad n = 1, 2, 3, \ldots, \tag{4.3.44}$$

where $j_{m,n}$ is the nth zero of $J_m(r)$, satisfying $J_m(j_{m,n}) = 0$.

Thus, the radial eigenmode takes the form

$$R(r) = J_m(j_{m,n} r/a). \tag{4.3.45}$$

Therefore, the cylindrical geometry Dirichlet eigenmodes of ∇^2 are

$$\psi_\alpha = J_m(j_{m,n} r/a)\, e^{im\theta} \sin(l\pi z/L), \tag{4.3.46}$$

and the corresponding eigenvalues follow from Eqs. (4.3.44) and (4.3.39):

$$\lambda_\alpha = -\frac{j_{m,n}^2}{a^2} - \left(\frac{l\pi}{L}\right)^2. \tag{4.3.47}$$

The counter α is now a list of three integers, $\alpha = (l, m, n)$, with $l > 0$, $n > 0$, and m taking on any integer value.

According to our previous general arguments, these eigenmodes form an orthogonal set with respect to the inner product given by Eq. (4.3.13). In fact, writing out the inner product as $\int_0^L dz \int_0^{2\pi} d\theta \int_0^a r\,dr$ and using Eq. (3.2.55), we find that the eigenmodes satisfy

$$(\psi_{\bar{\alpha}}, \psi_\alpha) = \delta_{l\bar{l}}\,\delta_{m\bar{m}}\,\delta_{n\bar{n}}\,\frac{\pi a^2 L}{2} J_{m+1}^2(j_{m,n}). \tag{4.3.48}$$

Example We now have all that we need to construct generalized Fourier series solutions to Poisson's equation inside a cylindrical tube with closed ends, via our general solution, Eqs. (4.3.17) and (4.3.3). As an example, let's take the case where the charge density inside the tube is linearly increasing with z: $\rho(r, \theta, z) = Az$. Also, let's assume that the base of the container has a potential $V(r) = V_0[1 - (r/a)^2]$, but that the other walls are grounded. This boundary condition is continuous, so we can simply choose an arbitrary function $u(r, z)$ that matches these boundary conditions. A suitable choice is

$$u(r,z) = V_0\left[1 - \left(\frac{r}{a}\right)^2\right]\left(1 - \frac{z}{L}\right). \tag{4.3.49}$$

Then according to Eqs. (4.3.4) and (4.3.5), we require the solution to Poisson's equation with homogeneous Dirichlet boundary conditions and a new charge density,

$$\bar{\rho} = \alpha + \beta z, \tag{4.3.50}$$

where $\alpha = -4V_0\epsilon_0/a^2$ and $\beta = A + 4V_0\epsilon_0/(La^2)$. The inner product $(\psi_\alpha, \bar{\rho})$ can then be worked out analytically:

$$(\psi_\alpha, \bar{\rho}) = \int_0^a J_m(j_{m,n}r/a)\,r\,dr \int_0^{2\pi} e^{-im\theta}\,d\theta \int_0^L (\alpha + \beta z)\sin\frac{l\pi z}{L}\,dz. \tag{4.3.51}$$

The θ-integral implies that we only require the $m = 0$ term. This is because the charge density is cylindrically-symmetric, so the potential is also cylindrically symmetric. For $m = 0$, the integrals over r and z can be done by *Mathematica*:

Cell 4.27

```
ρ[l_, n_] = Simplify[2 Pi Integrate[BesselJ[0, j0[n] r/a] r,
  {r, 0, a}]
   Integrate[(α + β z) Sin[l Pi z/L], {z, 0, L}], l ∈ Integers]

  2 a^2 L ((-1 + (-1^l)) α + (-1)^l Lβ) BesselJ[1, j0[n]]
- ------------------------------------------------------
                      l j0[n]
```

Here, we have introduced the function `j0[n]`, the nth zero of the Bessel function J_0. It can be defined as follows, up to the 20th zero:

Cell 4.28

```
<< NumericalMath`;

zeros = BesselJZeros[0, 20];
j0[n_] := zeros[[n]];
```

Finally, the solution can be constructed using the general form for the Fourier expansion, Eq. (4.3.17), along with our defined functions u, `ρ[l, n]`, and `j0[n]`:

Cell 4.29

```
u[r_, z_] = V0 (1-(r/a)^2) (1-z/L);

ψ[l_, n_, r_, z_] := BesselJ[0, j0[n] r / a] Sin[l π z / L];

λ[l_, n_] := -(l π / L)^2 - j0[n]^2 / a^2;
```

```
ψψ[l_, n_] := 1/2 πa^2 L BesselJ[1, j0[n]]^2;

(* :the inner product of ψ with itself; see Eq. (4.3.48) *)

ϕ[r_, z_] = u[r, z] - 1/ϵ0 Sum[Sum[ρ[l, n] ψ[l, n, r, z]/(λ[l, n] ψψ[l, n]), {l, 1, 20}], {n, 1, 5}];
```

Cell 4.30

```
a = 1; L = 2; A = ϵ0; V0=0.3;
α = -4 V0 ϵ0/a^2;
β = A + 4V0 ϵ0/(L a^2);
ContourPlot[ϕ[r, z], {r, 0, a}, {z, 0, L}, AspectRatio→L/a,
  PlotLabel→"ϕ in a charge-filled \ncylindrical tube",
  FrameLabel→{"r", "z"}, PlotPoints→25];
```

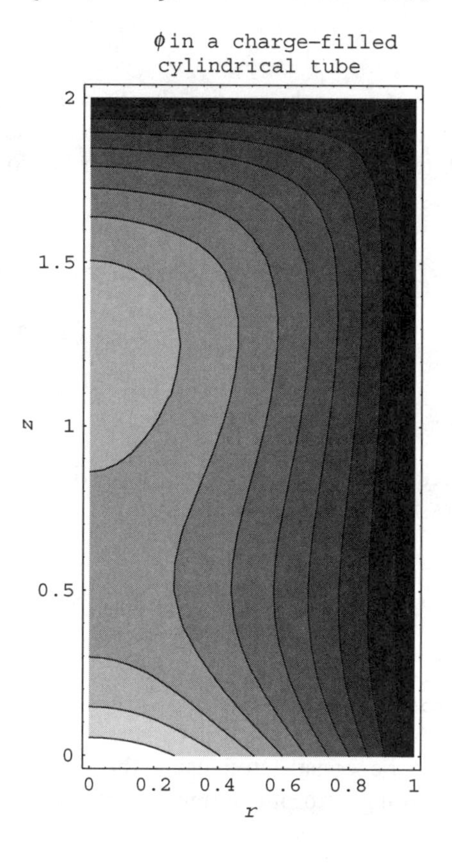

Here, `ψ[l, n, r, z]` is the eigenmode for given l and n for $m = 0$, `λ[l, n]` is the eigenvalue, and `ψψ[l, n]` is the inner product $(\psi_\alpha, \psi_\alpha)$. Only the first five Bessel function zeros and the first 20 axial modes are kept in the sum, because this achieves reasonable convergence to within a few percent. [This can be verified by evaluating ϕ at a few (r, z) points keeping different numbers of terms in the sum, which is left as an exercise for the reader.] In Cell 4.30 we plot the solution as a contour plot for the case $a = 1$ meter, $L = 2$ meter, and $A/\epsilon_0 = 1$ V/m, $V_0 = 0.3$ volt.

The potential is zero on the upper and side walls, as required by the boundary conditions. Along the bottom, the potential equals $V(r)$.

Spherical Geometry

Spherical Eigenmodes. Eigenmodes of ∇^2 can also be determined analytically in spherical coordinates (r, θ, ϕ). Here, we consider the Dirichlet eigenmodes inside a grounded spherical conducting shell of radius a. These eigenmodes satisfy the boundary condition $\psi_\alpha(a, \theta, \phi) = 0$. The eigenmodes are separable in spherical coordinates:

$$\psi_\alpha(r, \theta, \phi) = R(r)\Theta(\theta)\Phi(\phi). \tag{4.3.52}$$

The eigenmode equation, $\nabla^2\psi_\alpha/\psi_\alpha = \lambda_\alpha$, yields

$$\frac{1}{r^2 R(r)}\frac{\partial}{\partial r}\left(r^2\frac{\partial R}{\partial r}\right) + \frac{1}{r^2 \sin\theta\,\Theta(\theta)}\frac{\partial}{\partial\theta}\left(\sin\theta\frac{\partial\Theta}{\partial\theta}\right) + \frac{1}{r^2\sin^2\theta\,\Phi(\phi)}\frac{\partial^2\Phi}{\partial\phi^2} = \lambda_\alpha. \tag{4.3.53}$$

Following the approach of previous sub-subsections, one finds that this equation separates into three ODEs for R, Θ, and Φ:

$$\frac{\partial^2\Phi}{\partial\phi^2} = -m^2\Phi(\phi), \tag{4.3.54}$$

$$\frac{1}{\sin\theta}\frac{\partial}{\partial\theta}\left(\sin\theta\frac{\partial\Theta}{\partial\theta}\right) - \frac{m^2}{\sin^2\theta}\Theta(\theta) = -l(l+1)\Theta(\theta), \tag{4.3.55}$$

$$\frac{1}{r^2}\frac{\partial}{\partial r}\left(r^2\frac{\partial R}{\partial r}\right) - \frac{l(l+1)}{r^2}R(r) = \lambda_\alpha R(r). \tag{4.3.56}$$

Here, we have introduced the separation constants m and l, anticipating the form of the eigenvalues.

As usual, Eqs. (4.3.54)–(4.3.56) are separate eigenvalue problems. The solutions of the θ and ϕ equations were discussed in Sec. 3.2.4:

$$\Theta(\theta)\Phi(\phi) = Y_{l,m}(\theta,\phi), \qquad l, m \in \text{Integers and } l \geq |m|,$$

where

$$Y_{l,m}(\theta,\phi) = \sqrt{\frac{2l+1}{4\pi}\frac{(l-m)!}{(l+m)!}}\,P_l^m(\cos\theta)\,e^{im\phi}$$

is a spherical harmonic, and P_l^m is an associated Legendre function, discussed in relation to Eqs. (3.2.33) and (3.2.34), and given in Table 3.2.

Spherical Bessel Functions. We now determine the radial eigenmodes defined by Eq. (4.3.56) with boundary conditions that $R(a) = 0$ and $R(0)$ is finite. The general solution of the ODE is given by *Mathematica* (we have added a negative

Table 4.2. Spherical Bessel Functions That Are Finite at the Origin

l	$J_{l+1/2}(r)/\sqrt{r}$
0	$\dfrac{\sqrt{\frac{2}{\pi}}\sin r}{r}$
1	$\dfrac{\sqrt{\frac{2}{\pi}}\left(-\cos r+\frac{\sin r}{r}\right)}{r}$
2	$\dfrac{\sqrt{\frac{2}{\pi}}\left(-\frac{3\cos r}{r}-\sin r+\frac{3\sin r}{r^2}\right)}{r}$
3	$\dfrac{\sqrt{\frac{2}{\pi}}\left(\cos r-\frac{15\cos r}{r^2}+\frac{15\sin r}{r^3}-\frac{6\sin r}{r}\right)}{r}$

sign to λ_α, anticipating that the eigenvalue will be negative):

Cell 4.31

```
DSolve[1/r^2 D[r^2 D[R[r], r], r]-l (l + 1)/r^2
 R[r] == - λα R[r], R[r], r]
```

$$\left\{\left\{R[r]\to\frac{\text{BesselJ}[\frac{1}{2}\ (-1-2\ l),\ r\sqrt{\lambda_\alpha}]\ C[1]}{\sqrt{r}}+\frac{\text{BesselJ}\ [\frac{1}{2}\ (1+2\ l),\ r\sqrt{\lambda_\alpha}]\ C[2]}{\sqrt{r}}\right\}\right\}$$

The solution is in terms of Bessel functions of order $l+\frac{1}{2}$ and $-l-\frac{1}{2}$. Since the ODE has a regular singular point at the origin, one of the two solutions is singular there. The singular solution is $J_{-l-1/2}(r\sqrt{-\lambda_\alpha})/\sqrt{r}$. The other solution $J_{l+1/2}(r\sqrt{-\lambda_\alpha})/\sqrt{r}$, is well behaved at the origin. These functions are called spherical Bessel functions. Both sets of functions can be written in terms of trigonometric functions, as shown in Tables 4.2 and 4.3. Examples from both sets of spherical Bessel functions are plotted in Cells 4.32 and 4.33. Both sets of functions oscillate and have a similar form to the Bessel functions of integer order encountered in cylindrical geometry.

Cell 4.32

```
<<Graphics`;

Plot[Evaluate[Table[BesselJ[1/2 + l, r]/√r, {l, 0, 2}]],
  {r, 0, 15}, PlotStyle→{Red, Blue, Green},
  PlotLabel→"J_{l+1/2} (r)/√r for l=0,1,2",
   AxesLabel→{"r", " "}];
```

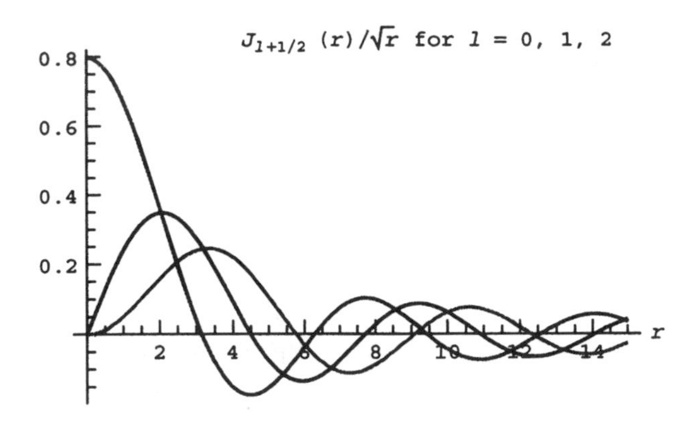

Table 4.3. Spherical Bessel Functions That Are Singular at the Origin

l	$J_{-l-1/2}(r)/\sqrt{r}$
0	$\dfrac{\sqrt{\frac{2}{\pi}}\cos r}{r}$
1	$\dfrac{\sqrt{\frac{2}{\pi}}\left(-\frac{\cos r}{r}-\sin r\right)}{r}$
2	$\dfrac{\sqrt{\frac{2}{\pi}}\left(-\cos r+\frac{3\cos r}{r^2}+\frac{3\sin r}{r}\right)}{r}$
3	$\dfrac{\sqrt{\frac{2}{\pi}}\left(-\frac{15\cos r}{r^3}+\frac{6\cos r}{r}+\sin r-\frac{15\sin r}{r^2}\right)}{r}$

Cell 4.33

```
Plot[Evaluate[Table[BesselJ[-1/2 -l, r]/√r, {l, 0, 2}]],
   {r, 0, 15},
  PlotStyle→{Red, Blue, Green}, PlotLabel→"J_{-l-1/2}(r)/√r  /
   for l=0,1,2",
  AxesLabel→{"r", " "}, PlotRange→{-1, 1}];
```

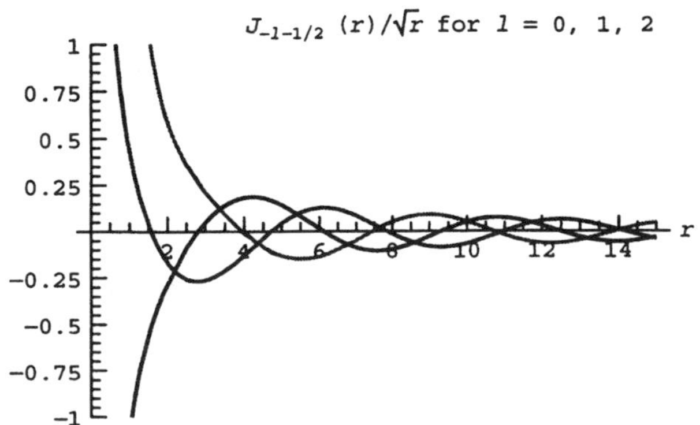

Since we require that the potential be finite throughout the spherical domain, we need not consider the singular spherical Bessel functions further. (However, they are required for problems where the origin is not included in the domain of interest.) The radial part of the spherical eigenfunction is then

$$R(r) = \frac{J_{l+1/2}\left(\sqrt{-\lambda_\alpha}\, r\right)}{\sqrt{r}}, \qquad l = 0, 1, 2, \dots . \tag{4.3.57}$$

We are finally in a position to determine the eigenvalue λ_α. The eigenvalue is determined by the boundary condition that the eigenmode ψ_α vanishes at $r = a$. Thus, $R(a) = 0$, and when applied to Eq. (4.3.57) this condition implies that

$$\lambda_\alpha = -\left(j_{l+1/2,\,n}/a\right)^2, \tag{4.3.58}$$

where $j_{l+1/2,\,n}$ is the nth zero of $J_{l+1/2}(x)$, satisfying $J_{l+1/2}(j_{l+1/2,\,n}) = 0$. For $l = 0$ the zeros can be determined analytically using the trigonometric form of the spherical Bessel function given in Table 4.2:

$$j_{1/2,\,n} = n\pi, \qquad n = 1, 2, 3, \dots . \tag{4.3.59}$$

However, for $l = 1$ or larger, the zeros must be found numerically. The intrinsic function **BesselJZeros** still works to determine lists of these zeros:

Cell 4.34

```
<<NumericalMath`;
BesselJZeros[3/2, 10]

{4.49341, 7.72525, 10.9041, 14.0662,
 17.2208, 20.3713, 23.5195, 26.6661, 29.8116, 32.9564}
```

For a given value of l these radial eigenfunctions are orthogonal with respect to the radial inner product:

$$\int_0^R \frac{J_{l+1/2}\left(j_{l+1/2,\,n} r/a\right)}{\sqrt{r}} \frac{J_{l+1/2}\left(j_{l+1/2,\,n} r/a\right)}{\sqrt{r}} r^2\, dr = 0 \qquad \text{if} \quad n \neq \bar{n}. \tag{4.3.60}$$

In fact, using Eq. (3.2.55) one can show that

$$\int_0^R \frac{J_{l+1/2}\left(j_{l+1/2,\,n} r/a\right)}{\sqrt{r}} \frac{J_{l+1/2}\left(j_{l+1/2,\,\bar{n}} r/a\right)}{\sqrt{r}} r^2\, dr = \delta_{n\bar{n}} \frac{a^2}{2} J_{l+3/2}^2\left(j_{l+1/2,\,n}\right). \tag{4.3.61}$$

We can now combine our results for the radial and angular eigenmodes to obtain the full spherical eigenmode $\psi_\alpha(r,\theta,\phi)$:

$$\psi_\alpha(r,\theta,\phi) = \frac{Y_{l,m}(\theta,\phi)J_{l+1/2}(j_{l+1/2,n}r/a)}{\sqrt{r}}. \tag{4.3.62}$$

The parameter α, used to enumerate the eigenmodes, can now be seen to be a list of three integers, $\alpha=(l,m,n)$. The integer l runs from 0 to infinity, determining the θ dependence of the mode, while $-l \le m \le l$ determines the ϕ-dependence, and $n=1,2,3,\dots$ counts the zeros in the radial mode.

According to Eqs. (4.3.61) and (4.2.40), these spherical eigenmodes are orthogonal with respect to the combined radial and angular inner products:

$$\begin{aligned}(\psi^*_{lmn},\psi_{\bar{l}\bar{m}\bar{n}}) &= \int_0^R r^2\,dr\int_0^\pi \sin\theta\,d\theta\int_0^{2\pi} d\phi\,\psi^*_{lmn}(r,\theta,\phi)\,\psi_{\bar{l}\bar{m}\bar{n}}(r,\theta,\phi)\\ &= \delta_{l\bar{l}}\,\delta_{m\bar{m}}\,\delta_{n\bar{n}}\frac{a^2}{2}J^2_{l+3/2}(j_{l+1/2,n}).\end{aligned} \tag{4.3.63}$$

However, this combined inner product is simply the three-dimensional integral over the volume V interior to the sphere. This is as expected from the general arguments given in Sec. 4.3.3.

Example We now use the spherical eigenmodes to solve the following potential problem: the upper half of a hollow sphere of radius a is filled with charge density ρ_0. The sphere is a conducting shell, cut in half at the equator. The upper half is grounded, and the lower half is held at potential V_0. (See Fig. 4.6.)

As in the previous examples, we take account of the inhomogeneous boundary condition by using a function $u(\mathbf{r})$ that matches this boundary condition:

$$u(a,\theta,\phi) = \begin{cases} 0, & 0 \le \theta \le \pi/2,\\ V_0, & \pi/2 < \theta < \pi.\end{cases} \tag{4.3.64}$$

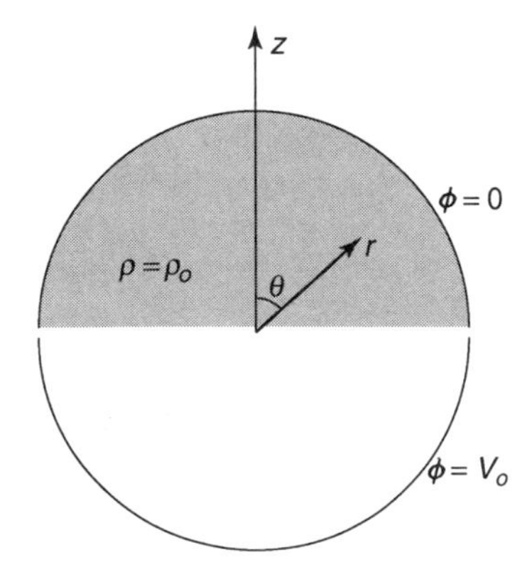

Fig. 4.6

However, this boundary condition is rapidly varying. Therefore, it is best to choose a function $u(r, \theta, \phi)$ that satisfies the Laplace equation with these boundary conditions, $\nabla^2 u = 0$.

We considered this part of the problem previously, in Sec. 3.2.4. The solution is a sum of spherical harmonics, given by Eqs. (3.2.41) and (3.2.42):

$$u(r, \theta, \phi) = \sum_{l=0}^{\infty} \sum_{m=-l}^{l} B_{lm} r^l Y_{l,m}(\theta, \phi), \tag{4.3.65}$$

where the Fourier coefficients B_{lm} are determined by matching to the boundary conditions:

$$B_{lm} a^l = V_0 \int_0^{2\pi} d\phi \int_{\pi/2}^{\pi} \sin\theta \, d\theta \, Y_{l,m}^*(\theta, \phi) = \delta_{m0} \, 2\pi V_0 \int_{\pi/2}^{\pi} \sin\theta \, d\theta \, Y_{l,0}^*(\theta). \tag{4.3.66}$$

As expected for cylindrically symmetric boundary conditions, only spherical harmonics with $m = 0$ enter in the expansion for u.

We then write the potential ϕ as $\phi = \Delta\phi + u$, and solve for $\Delta\phi$ using Eq. (4.3.17). The fact that $\nabla^2 u = 0$ implies that $\bar{\rho} = \rho$. Therefore the solution for $\Delta\phi$ is

$$\Delta\phi(r, \theta, \phi) = -\frac{1}{\epsilon_0} \sum_{\alpha} \frac{(\psi_\alpha, \rho)}{\lambda_\alpha (\psi_\alpha, \psi_\alpha)} \psi_\alpha(r, \theta, \phi), \tag{4.3.67}$$

where $\alpha = (l, m, n)$, λ_α is given by Eq. (4.3.58), and ψ_α is given by Eq. (4.3.62). The inner product (ψ_α, ρ) is

$$(\psi_{lmn}, \rho) = \int_0^a r^2 \, dr \frac{J_{l+1/2}(j_{l+1/2,n} r/a)}{\sqrt{r}} \int_0^{2\pi} d\phi \int_0^{\pi/2} \sin\theta \, d\theta \, \rho_0 Y_{l,m}(\theta, \phi). \tag{4.3.68}$$

The ϕ-integral picks out only the $m = 0$ eigenmodes, since the charge distribution is cylindrically symmetric. The r and θ integrals can be performed analytically, although the results are rather messy. It is best to simply leave the work to *Mathematica* by defining this inner product as a function **ρ[l,n]**. It is fastest and easiest to perform the radial integrals numerically using **NIntegrate**. This requires scaling the radius to a so that the radial integral runs from 0 to 1:

Cell 4.35

```
ρ[l_, n_] := ρ[l, n] =
ρ0 a^(5/2) NIntegrate[r^(3/2) BesselJ[l + 1/2, j[l, n] r], {r, 0, 1}] *
2 π Integrate[Sin[θ] SphericalHarmonicY[l, 0, θ, ϕ], {θ, 0, Pi/2}]
```

Here, we have also introduced another function, **j[l,n]**, which is $j_{l+1/2,\,n}$, the nth zero of the $J_{l+1/2}$. This can be defined as follows:

Cell 4.36

```
<<NumericalMath`;

zeros = Table[BesselJZeros[l + 1/2, 10], {l, 0, 9}];
j[l_, n_] := zeros [[l + 1, n]]
```

For example, the inner product (ψ_α, ρ) for $\alpha = (l, m, n) = (3, 0, 4)$ is given by

Cell 4.37

```
ρ[3, 4]

0.00375939 a^(5/2) ρ0
```

The solution for $\Delta\phi$ is given in terms of the spherical eigenmodes by Eq. (4.3.67). This generalized Fourier series is evaluated below:

Cell 4.38

```
(* define the spherical eigenmodes *)ψ[l_, m_, n_, r_, θ_, ϕ_] :=
BesselJ[l + 1/2, j[l, n] r/a]/Sqrt[r]
SphericalHarmonicY[l, m, θ, ϕ];

(* define the eigenvalues *)
λ[l_, n_] := -j[l, n]^2/a^2;

(* define the inner product (ψα, ψα) *)

ψψ[l_, n_] := a^2/2 BesselJ[l + 3/2, j[l, n]]^2;

(* sum the series to determine the potential, using Eq.
   (4.3.67) *)

Δϕ[r_, θ_] =
 -1/ϵ0 Sum[ρ[l, n]/(λ[l, n] ψψ[l, n]) ψ[l, 0, n, r, θ, ϕ], {l, 0, 4},
  {n, 1, 5}];
```

Now we must add to this the potential $u(r, \theta, \phi)$ arising from the boundary conditions, using Eqs. (4.3.65) and (4.3.66):

Cell 4.39

```
b[l_] =
  2 π V0 Integrate[Sin[θ] SphericalHarmonicY[l, 0, θ, ϕ],
   {θ, Pi/2, Pi}]/a^l;

u[r_, θ_] = Sum[b[l] r^l SphericalHarmonicY[l, 0, θ, ϕ],
  {l, 0, 40}];
```

Finally, we plot in Cell 4.40 the resulting potential on a cut throught the center of the sphere, along the $y = 0$ plane, taking the potential on the boundary to be

$V_0 = 1$ volt and a charge density $\rho_0 = 3\epsilon_0$. In the lower half sphere, $z < 0$, the potential rises to meet the boundary condition that $\phi(a, \theta, \phi) = 1$ volt; while in the upper half the potential on the surface of the sphere is zero. The convex curvature of the potential in the interior of the sphere reflects the positive charge density ρ_0 in the upper half of the sphere. Note the barely discernible Gibbs phenomenon near the discontinuities in the potential.

Cell 4.40

```
(* parameters *)
V0 = 1;
ρ0 = 3 ϵ0 ;
a = 1;
(* construct the total potential *)
Φ[r_, θ_] = Δϕ[r, θ] + u[r, θ];

ParametricPlot3D[{r Sin[θ], r Cos[θ], Φ[r, θ]},
  {r, 0.001, 1}, {θ, 0, 2 Pi}, PlotPoints→{20, 80},
  BoxRatios→{1, 1, 1/2}, ViewPoint -> {2.899, 0.307, 1.718},
  AxesLabel→{"x", "z", "ϕ"}, PlotLabel→"Potential inside
   a sphere"];
```

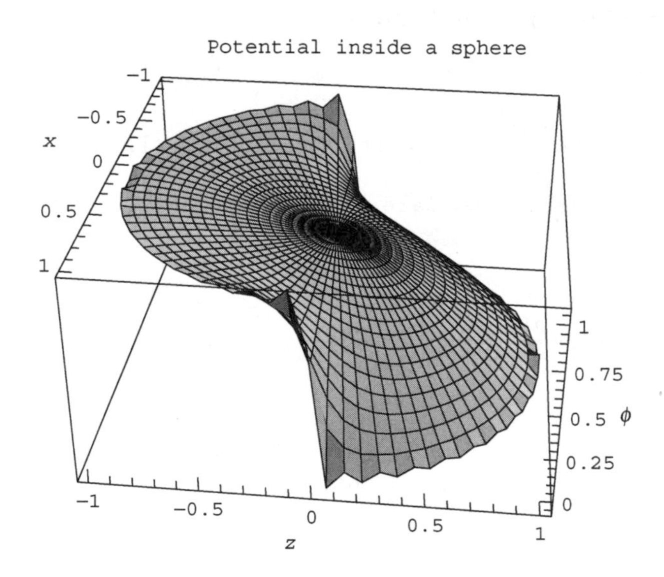

EXERCISES FOR SEC. 4.3

(1) Solve the following potential problems in rectangular geometry. Plot the solutions using `Plot3D`.

(a) $\nabla^2\phi(x, y) = x$, $\phi(0, y) = \phi(1, y) = \phi(x, 0) = 0$, $\phi(1, y) = \sin 2\pi y$.

(b) $\nabla^2\phi(x, y, z) = 10$, $\phi(x, y, 0) = \phi(x, y, 1) = \phi(0, y, z) = \phi(1, y, z) = \phi(x, 0, z) = 0$, $\phi(x, 1, z) = h(x - \frac{1}{2})$, where h is a Heaviside step function. Plot the solution in the $z = \frac{1}{2}$ plane.

(c) $\nabla^2\phi(x,y) = -xy$, $\phi(0,y) = \phi(2,y) = \phi(x,0) = 0$, $\phi(x,1) = 1$.

(d) $\nabla^2\phi(x,y,z) = 1$, $\phi(x,y,0) = \phi(x,y,1) = 0$, $\phi(0,y,z) = \phi(1,y,z) = z(1-z)$, $\phi(x,0,z) = \phi(x,1,z) = 0$. Plot the solution in the $z = \frac{1}{2}$ plane.

(e) $\nabla^2\phi(x,y) = \cos n\pi x$, n an integer, $\frac{\partial\phi}{\partial x}(0,y) = \frac{\partial\phi}{\partial x}(1,y) = \frac{\partial\phi}{\partial y}(x,0) = \frac{\partial\phi}{\partial y}(x,1) = 0$. (What happens for $n = 0$?)

(f) $\nabla^2\phi(x,y) = e^{x+y}$, $\phi(0,y) = \phi(1,y) = \phi(x,0) = 0$, $\frac{\partial\phi}{\partial y}(x,1) = 1$.

(2) For the problem

$$\nabla^2\phi(x,y) = 1, \quad \frac{\partial\phi}{\partial x}(0,y) = \frac{\partial\phi}{\partial x}(1,y) = \frac{\partial\phi}{\partial y}(x,0) = 0, \quad \frac{\partial\phi}{\partial y}(x,1) = ah(x-\tfrac{1}{2}),$$

where h is a Heaviside step function, find the value of a for which a solution exists, and find the solution for $\phi(x,y)$. [Hint: For the function u, choose $u = u_0(x,y) + f(x,y)$, where f is an arbitrary function that satisfies homogeneous von Neumann conditions on all sides except at $y = 1$, and that takes account of the charge in the box (i.e., $\int_V \nabla^2 f\,dx\,dy = 1$), and where $\nabla^2 u_0 = 0$.]

(3) **(a)** Show, using Gauss's law, that for Poisson's equation in a square box with *periodic* boundary conditions $\phi(x+L,y) = \phi(x,y+L) = \phi(x,y)$, a solution for $\phi(x,y)$ exists *only if* the net charge density in the square is zero:

$$\int_0^L dx \int_0^L dy\, \rho(x,y) = 0.$$

(b) For a given charge density $\rho(x,y)$ find a general form for the solution in terms of an exponential Fourier series.

(4) Find the Green's function $g(\mathbf{r},\mathbf{r}_0)$ as a generalized Fourier series for the potential inside a grounded rectangular cube with walls at $x = 0$, $x = a$, $y = 0$, $y = a$, $z = 0$, $z = L$.

(5) It is sometimes useful to write the Green's function for Poisson's equation in a different manner than the eigenmode expansion used in the bilinear equation. For the rectangular cube of Exercise (4), now employ only the x and y eigenmodes, writing

$$g(\mathbf{r},\mathbf{r}_0) = \sum_{m,n=1}^{\infty} f_{mn}(z,\mathbf{r}_0) \sin\frac{n\pi x}{a} \sin\frac{m\pi y}{a}. \tag{4.3.69}$$

(a) Show that f_{mn} solves the following boundary-value problem:

$$\frac{\partial^2}{\partial z^2} f_{mn} - \kappa_{mn}^2 f_{mn} = \sin\frac{n\pi x_0}{a} \sin\frac{m\pi y_0}{a}\, \delta(z - z_0), \quad f_{mn} = 0 \quad \text{at } z = 0 \text{ and } L,$$

where $\kappa_{mn}^2 = (m\pi/a)^2 + (n\pi/a)^2$.

(b) Solve this boundary-value problem, using the technique discussed in Sec. 3.4.4 to show that

$$f_{mn}(z,\mathbf{r}_0) = -4\sin\frac{n\pi x_0}{a}\sin\frac{m\pi y_0}{a}\sinh\kappa_{mn}z_<\frac{\sinh\kappa_{mn}(L-z_>)}{a^2\kappa_{mn}\sinh\kappa_{mn}L}, \quad (4.3.70)$$

where $z_{<(>)}$ is the lesser (greater) of z and z_0.

(c) By carefully taking the limit $L \gg z, z_0$ and $z, z_0 \gg 0$, show that the Green's function in an infinitely long grounded conducting tube of square cross section is

$$g(\mathbf{r},\mathbf{r}_0) = -2\sum_{m,n=1}^{\infty}\sin\frac{n\pi x}{a}\sin\frac{m\pi y}{a}\sin\frac{n\pi x_0}{a}\sin\frac{m\pi y_0}{a}\frac{e^{-\kappa_{mn}|z-z_0|}}{a^2\kappa_{mn}}. \quad (4.3.71)$$

(Hint: As $x \to \infty$, $\sinh x \sim \cosh x \sim e^x/2$.)

(d) Find the force in the z-direction on a charge q, located at position (x_0, y_0, z_0) in a rectangular box of length L with square cross section. Plot this force (scaled to $q^2/\epsilon_0 L^2$) vs. z for $0 < z_0 < L$, $y_0 = a/2 = x_0$, for $a = L/2$. [Hint: This force arises from the image charges in the walls of the grounded box. To determine the force, one needs to calculate the electric field $\mathbf{E}_0$ at the position $\mathbf{r}_0$ of the charge. However, one cannot simply evaluate $\mathbf{E}_0 = (q/\epsilon_0)\nabla_{\mathbf{r}} g(\mathbf{r},\mathbf{r}_0)|_{\mathbf{r}=\mathbf{r}_0}$ using Eq. (4.3.69), since the self-field of the charges is infinite at $\mathbf{r} = \mathbf{r}_0$. This is manifested in Eq. (4.3.69) and in Eq. (4.3.71)) by the fact that the series does not converge if $\mathbf{r} = \mathbf{r}_0$. (Try it if you like!) One must somehow subtract out this self-field term, and determine only the effect of the field due to the images in the walls of the tube. One way to do this is to note that for an infinite tube, there is no force in the z-direction on the charge, due to symmetry in z. Therefore, we can subtract Eq. (4.3.71) from Eq. (4.3.69) to obtain the effect on g of the tube's finite length. The resulting series converges when one takes $\mathbf{r} = \mathbf{r}_0$.]

(6) The following problem relates to the eigenmode expansion of the solution to Laplace's equation, Eq. (4.3.29).

(a) Use Green's theorem to show that, for Dirichlet boundary conditions in a two-dimensional domain,

$$(\psi_{nm}, \nabla^2 u) = \lambda_{nm}(\psi_{nm}, u) - \int_S u\hat{\mathbf{n}}\cdot\nabla\psi_{nm}\, dl,$$

where the surface integral runs over the boundary S of the domain, dl is a line element of this boundary, and ψ_{nm} and u have the same definitions as in Eq. (4.3.29).

(b) Determine the eigenmode expansion for the solution of Laplace's equation, $\nabla^2\phi = 0$, using the result of part (a) and Eq. (4.3.29). In particular, show that

$$\phi(x,y) = u(x,y) + \sum_{nm} b_{nm}\psi_{nm}(x,y), \quad (4.3.72)$$

where $b_{nm} = \int_S u\hat{\mathbf{n}} \cdot \nabla\psi_{nm}\, dl/[\lambda_{nm}(\psi_{nm}, \psi_{nm})] - (\psi_{nm}, u)/(\psi_{nm}, \psi_{nm})$. Note that in the surface integral the function u is uniquely determined by the boundary conditions. Equation (4.3.72) is equivalent to Eq. (4.3.29), but avoids taking derivatives of u. This is an advantage when u varies rapidly.

(c) Redo the Laplace equation problem associated with Fig. 4.5, taking $a = b = 1$ and $\phi_0 = 1$, and using an eigenmode expansion as given by Eq. (4.3.72). Note that on the boundary u is nonzero only on the right side, in the range $\frac{1}{3} < y < \frac{2}{3}$, so the required surface integral becomes $\int_{1/3}^{2/3} \hat{\mathbf{x}} \cdot \nabla\psi_{nm}(1, y)\, dy$. For $u(x, y)$ use the following function:

$$u(x,y) = xf(y - \tfrac{1}{3}, 1 - x, -\tfrac{1}{3}) f(\tfrac{2}{3} - y, 1 - x, -\tfrac{1}{3}),$$

where $f(x, y, z) = \frac{1}{2}[\tanh(x/y) - \tanh(z/y)]$. Plot $u(x, y)$ to convince yourself that it matches the boundary conditions. This function is chosen because it is continuous everywhere except on the right boundary. Therefore, its generalized Fourier series expansion in terms of ψ_{mn} has better convergence properties than our previous choice for u, Eq. (4.3.28). You will have to find the inner product (ψ_{nm}, u) via numerical integration. Keep $1 \le m \le 6$ and $1 \le n \le 6$. Avoid integrating in y all the way to $y = 1$, because of the singularity in u; rather, integrate only up to $y = 0.9999$. Compare your solution to the solution found via Eq. (4.3.31) (for $M = 36$) by plotting $\phi(0.9, y)$ and $\phi(x, \frac{1}{2})$ for both solutions. Which solution works better? [Answer: Eq. (4.3.31).]

(d) Show that in the limit that an infinite number of terms are kept, Eq. (4.3.72) becomes

$$\phi(x, y) = \sum_{nm} c_{nm} \psi_{nm}(x, y)$$

for (x, y) not on the boundary, where $c_{nm} = \int_S u\hat{\mathbf{n}} \cdot \nabla\psi_{nm}\, dl/[\lambda_{nm}(\psi_{nm}, \psi_{nm})]$. This is yet another form for the solution to Laplace's equation with Dirichlet boundary conditions, valid only in the interior of the domain. Repeat the calculation and plots of part (b) using this series, taking $1 \le m \le 40$ and $1 \le n \le 40$ (the coefficients c_{nm} can be determined analytically).

(7) Find the solution to the following potential problems inside a cylinder. Write the solution in terms of an eigenmode expansion. Put the equation into standard form, if necessary. Plot the solutions.

(a) $\nabla^2\phi(r, \theta) = xy$, $\phi(1, \theta) = 0$.

(b) $\nabla^2\phi(r, \theta, z) = z \sin\theta$, $\phi(1, \theta, z) = \phi(r, \theta, 0) = \phi(r, \theta, 2) = 0$. Plot in the $x = 0$ plane vs. y and z.

(c) $\nabla^2\phi(r, \theta, z) = 1$, $\phi(1, \theta, z) = \phi(r, \theta, -1) = \phi(r, \theta, 1) = 0$, $\phi(2, \theta, z) = h(z)h(\theta)$ (concentric cylinders, h is a Heaviside step function, $-\pi < \theta < \pi$ assumed). Plot in the $x = 0$ plane vs. y and z.

(d) $\nabla^2\phi(r, \theta) = y$, $\frac{\partial\phi}{\partial r}(1, \theta) = \sin\theta$.

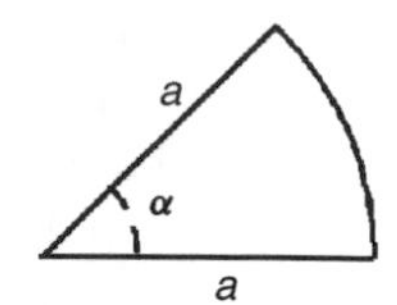

Fig. 4.7 Exercise (8).

(8) A wedge, shown in Fig. 4.7, has opening angle α. The wedge is filled with uniform charge, $\rho/\epsilon_0 = 1\ \text{V/m}^2$. The walls of the wedge are grounded, at zero potential.

(a) Find the eigenmodes for this geometry.

(b) Use these eigenmodes to solve for the potential $\phi(r, \theta)$ inside the wedge. Plot the solution using a contour plot for $\alpha = 65°$.

(9) A wedge, shown in Fig. 4.8, has opening angle α and radii a and b, $b < a$. The edges of the wedge have constant potentials as shown. Find the solution to Laplace's equation using separation of variables rather than eigenmodes. (Hint: You will still need to determine radial eigenfunctions, and the correct radial inner product with respect to which these functions are orthogonal.) Plot the solution using **ParametricPlot3D** for $\alpha = 135°$, $a = 1$, $b = \frac{1}{10}$, and $V_0 = 1$ volt. Answer: $\phi(r, \theta) = \sum_{n=1}^{\infty} A_n \sin\left(\frac{n\pi}{\log(b/a)} \log\frac{r}{a}\right) \sinh\left(\frac{n\pi}{\log(b/a)}\theta\right)$.

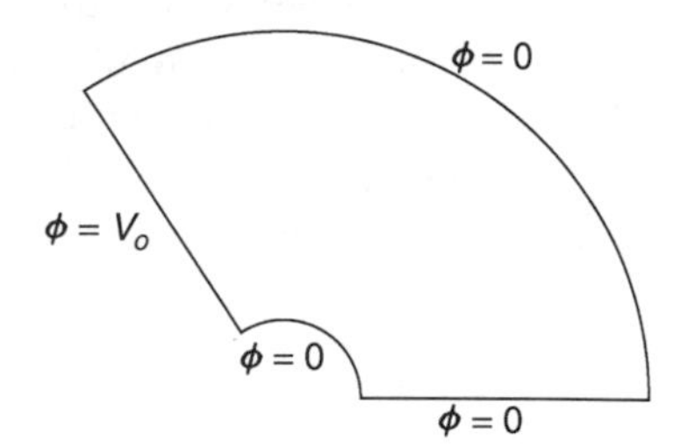

Fig. 4.8 Exercise (9).

(10) Find the solution to the following potential problems inside a sphere. Write the solution in terms of an eigenmode expansion. Convert inhomogeneous boundary conditions, if any, to a source term.

(a) $\nabla^2\Psi(r, \theta, \phi) = xyz$, $\Psi(1, \theta, \phi) = 0$.

(b) $\nabla^2\Psi(r, \theta, \phi) = 1$, $\Psi(1, \theta, \phi) = \sin^2\theta$.

(c) $\nabla^2\Psi(r, \theta, \phi) = (\cos\theta)/r$, $\Psi(1, \theta, \phi) = 0$, $\Psi(2, \theta, \phi) = 0$ (concentric spheres).

(d) $\nabla^2\Psi(r, \theta, \phi) = 1$, $\frac{\partial\Psi}{\partial r}(1, \theta, \phi) = a\sin^2\theta\cos^2\phi$. Find the value of a for which a solution exists, and find the solution.

(11) A hemispherical shell of radius a has a flat base, forming half of a sphere. This half sphere is filled with a uniform charge density, $\rho/\epsilon_0 = 10\ \text{V/m}^2$. The surface of the half sphere, including the base, is grounded. Find the potential inside the shell, and plot it. What and where is the maximum of the potential?

(12) In a plasma the potential due to a charge density ρ satisfies the linearized Poisson–Boltzmann equation $\nabla^2\phi = \phi/\lambda^2 - \rho/\epsilon_0$, where λ is the Debye length of the plasma. A spherical charge of radius a, uniform charge density, and total charge Q is placed in the plasma. Find the potential, assuming that it vanishes at infinity. [Hint: Solve the radial boundary-value problem directly in terms of homogeneous solutions, using boundary conditions that $\phi \to 0$ as $r \to \infty$ and $E_r = -\partial\phi/\partial r = Q/(4\pi\epsilon_0 a^2)$ at $r = a$.]

(13) **(a)** Repeat the analysis of the Green's function in Exercise (5), but for the inside of a spherical conducting shell of radius a. Now write

$$g(\mathbf{r}, \mathbf{r}_0) = \sum_{l=0}^{\infty} \sum_{m=-l}^{l} Y_{l,m}(\theta, \phi) f_{lm}(r, \mathbf{r}_0) \tag{4.3.73}$$

and find an ODE boundary-value problem for $f_{lm}(r, \mathbf{r}_0)$. Solve this boundary-value problem to show that

$$f_{lm}(r, \mathbf{r}_0) = -Y^*_{l,m}(\theta_0, \phi_0) \frac{1}{2l+1} \left(\frac{r_<^l}{r_>^{l+1}} - \frac{r_<^l r_>^l}{a^{2l+1}} \right), \tag{4.3.74}$$

where $r_{<(>)}$ is the lesser (greater) of r and r_0. Hint: In spherical coordinates the δ-function $\delta(\mathbf{r} - \mathbf{r}_0)$ is given by

$$\delta(\mathbf{r} - \mathbf{r}_0) = \frac{\delta(r - r_0)\,\delta(\theta - \theta_0)\,\delta(\phi - \phi_0)}{r^2 \sin\theta}. \tag{4.3.75}$$

(b) In the limit as $a \to \infty$, Eqs. (4.3.73) and (4.3.74) can be used to represent the potential at point $\mathbf{r}$ due to an arbitrary charge density $\rho(\mathbf{r}_0)$ in free space. Assume that this charge density is concentrated near the origin; that is, it is completely contained inside an imaginary sphere centered at the origin and of radius R. (See Fig. 4.9.) Then, using the Green's function, show that the electrostatic potential at locations far from this charge density is given by

$$\phi(\mathbf{r}) = \sum_{l=0}^{\infty} \sum_{m=-l}^{l} \frac{1}{2l+1} \rho_{lm} \frac{Y_{l,m}(\theta, \phi)}{\epsilon_0 r^{l+1}}, \qquad \text{provided that } r > R. \tag{4.3.76}$$

Here, $\rho_{lm} = \int d^3 r_0\, \rho(\mathbf{r}_0) r_0^l Y^*_{l,m}(\theta_0, \phi_0)$ is the *multipole moment* of the charge distribution. Equation (4.3.76) is called a *multipole expansion of the potential.*

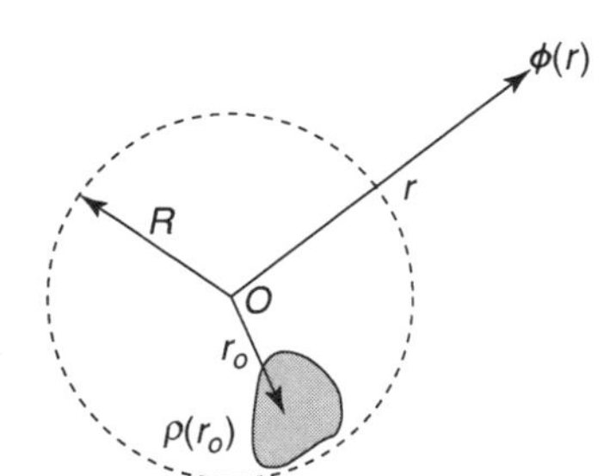

Fig. 4.9 Geometry assumed for the multipole expansion of Eq. (4.3.76).

(14) **(a)** The multipole moment ρ_{00} is called the monopole moment of the charge distribution. It is proportional to the total charge Q. The potential produced by the monopole moment is simply that given by Coulomb's law, $\phi = Q/4\pi\epsilon_0 r$. The moment ρ_{1m} is called a dipole moment, and ρ_{2m} is called a quadrupole moment. Plot contours of constant $\phi(x, y, z)$ in the x-z plane, assuming that

(i) only ρ_{10} is nonzero;

(ii) only ρ_{20} is nonzero.

(b) Show that ρ_{10} and ρ_{20} can be written in Cartesian coordinates as

$$\rho_{10} = \sqrt{\frac{3}{4\pi}} \int z_0 \, \rho(\mathbf{r}_0) \, d^3 r_0,$$

$$\rho_{20} = \sqrt{\frac{5}{16\pi}} \int (2z_0^2 - x_0^2 - y_0^2) \rho(\mathbf{r}_0) \, d^3 r_0.$$

(c) Evaluate the monopole, dipole, and quadrupole moments of two charges located on the z-axis: one at $+z_0$ with charge $+q$, and one at $-z_0$ with charge $-q$. Plot the potential $\phi(z)$ along the z-axis arising from the dipole and quadrupole terms for $0 < z < 10z_0$, and compare it with a plot of the exact potential $(q/\epsilon_0)(1/|z - z_0| - 1/|z + z_0|)$. Where does the multipole expansion work?

(15) A uniform density ellipsoid of total charge Q has a surface determined by the equation $x^2/a^2 + y^2/b^2 + z^2/c^2 = 1$. Find the quadrupole moments of this charge distribution, and show that

$$\rho_{20} = \sqrt{\frac{1}{80\pi}} \, Q(2c^2 - a^2 - b^2),$$

$$\rho_{21} = 0,$$

$$\rho_{22} = \sqrt{\frac{3}{160\pi}} \, Q(a^2 - b^2).$$

(16) A second form of multipole expansion is useful when we want to know the potential at a point near the origin due to charge density that is concentrated far from the origin, *outside* an imaginary sphere of radius R. (See Fig. 4.10.)

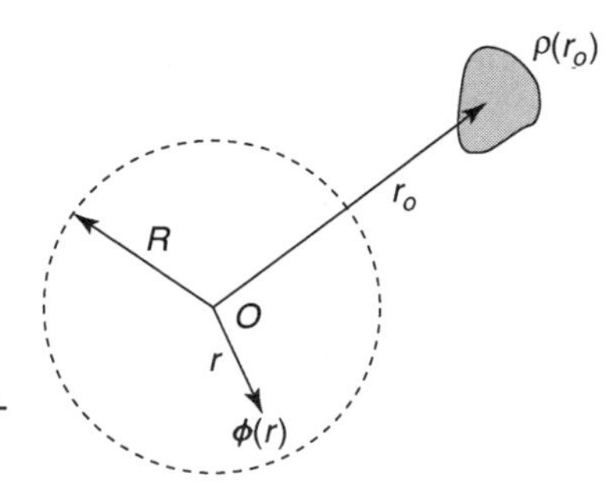

Fig. 4.10 Geometry assumed for the multipole expansion of Eq. (4.3.77).

For such a charge density, use the Green's function to show that

$$\phi(\mathbf{r}) = \sum_{l=0}^{\infty} \sum_{m=-l}^{l} \frac{1}{2l+1} \bar{\rho}_{lm} \frac{Y_{l,m}(\theta,\phi)}{\epsilon_0} r^l, \qquad \text{provided that} \quad r < R, \quad (4.3.77)$$

where $\bar{\rho}_{lm} = \int d^3r_0 \; \rho(\mathbf{r}_0) Y^*_{l,m}(\theta_0, \phi_0)/r_0^{l+1}$.

(17) Find the electrostatic potential near the origin due to a hemispherical shell of charge, total charge q, and radius a. The shell is oriented above the origin of coordinates, with its origin coincident with the origin of coordinates, and its axis of symmetry along the z-axis. Keep terms up to and including the quadrupole moments, and write the resulting potential $\phi(\mathbf{r})$ in terms of Cartesian coordinates (x, y, z).

(18) **(a)** Consider an object of mass m moving in gravitational free fall around a fixed mass M. At a given instant, the mass M is located a distance r_0 along the z-axis of a coordinate system whose origin is located near (or within) the object. Using the fact that the gravitational potential ϕ_G also satisfies Poisson's equation,

$$\nabla^2 \phi_G = 4\pi G \rho, \qquad (4.3.78)$$

where ρ is the *mass* density and G is the gravitational constant, find a multipole expansion of the gravitational potential due to the mass M that is valid near the origin within the object. Keep terms in the expansion up to and including the quadrupole terms, and show that the force in the z-direction on a mass element dm of the object, $dF_z = -dm \; \partial\phi_G/\partial z$, equals

$$dF_z = dm \; G\left(2\sqrt{\pi/3}\,\bar{\rho}_{10} + 4\sqrt{\pi/5}\,\bar{\rho}_{20}\, z\right). \qquad (4.3.79)$$

(b) If one assumes that the mass M is a point mass, then using Eq. (4.3.79), show that the total force in the z-direction on the object, $F_z = \int dF_z$, is

$$F_z = GMm/r_0^2,$$

provided that the coordinate system used to determine the multipole moments has its origin located at the center of mass of the object. The center-of-mass position $\mathbf{R}_{\mathrm{cm}}$ is defined as $\mathbf{R}_{\mathrm{cm}} = (1/m)\Sigma_i \, dm_i \, \mathbf{r}_i$, where the sum runs over all the mass elements dm_i of the object, each located at position $\mathbf{r}_i$. [Hint: You will need to use Eq. (4.3.75) to help determine the multipole moments of the point mass M.]

(c) The object in question has a spatial extent in the z-direction that runs from $-z_1$ to z_1, $z_1 \ll r_0$. Using Eq. (4.3.79), show that the acceleration of the point at $+z_1$ relative to that at $-z_1$ is given by

$$a_t = 4MGz_1/r_0^3.$$

This relative acceleration is called *tidal acceleration*. [Hint: Equation (4.3.75) will be needed to help determine the multipole moments of the point mass M.]

(d) Determine the tidal acceleration caused by the moon, calculated for the two points on the earth nearest and farthest from the moon. Treat the moon as a point mass.

(e) Determine the tidal acceleration due to the sun, in the same manner as was used for the moon in part (d).

(19) A neutron star has a mass $M = 2M_{\text{sun}}$, but a radius of only around 10 km. A rocket ship approaches the star in free fall, to a distance $r_0 = 3000$ km. Using Eq. (4.3.79), calculate the tension force (the tidal force) in a man floating inside the ship. Assume for simplicity that the mass distribution of the man is a uniform cylinder of total mass $m = 70$ kg and length $L = 2$ m, oriented with the cylinder axis pointing toward the star, and treat the star as a point mass. The tension force is defined here as the force between the halves of the man nearest and furthest from the star as they are pulled toward and away from the star by the tidal acceleration. Evaluate the tension force in pounds (1 pound = 4.45 newtons). [This problem is inspired by the science fiction novel *Neutron Star*, by Larry Niven (1968). (Answer: $T = MmGL/r_0^3$.)]

(20) A deformable incompressible body, in the presence of another gravitating body (both bodies at fixed positions) will deform until it is in equilibrium, in such a way that its volume remains unchanged. The equilibrium shape of the deformable body can be determined using the fact that, in equilibrium, the gravitational potential ϕ_G at the surface of the body is independent of position along the surface (i.e., the surface of the body is an equipotential). Take, for example, the earth–moon system. The earth will deform, attempting to come to equilibrium with the moon's gravitational attraction. (Actually, the earth's oceans deform. The rigidity of the solid part suppresses the response to the weak lunar tidal acceleration.) This is the basic effect responsible for the earth's tides. Assuming that the earth is a deformable incompressible body of uniform mass density, that the moon is located a distance r_0 from the earth along the z-axis of a coordinate system used to calculate the deformation (see Fig. 4.11), that the moon can be treated as a point mass, and that the resulting deformation is small and in equilibrium with the moon's attraction, show that the height $h(\theta)$ of the deformation of the earth's surface is

$$h(\theta) = \sqrt{\frac{5\pi}{4}}\,\frac{M_m}{M_e}\,\frac{R^4}{r_0^3}\,Y_{2,0}(\theta), \tag{4.3.80}$$

where M_e and R are the mass and radius of the earth respectively, and M_m is the mass of the moon. For the parameters of the earth–moon system, plot this deformation vs. θ, to show that the deformation is largest on the z-axis of our coordinate system at $\theta = 0$ and π, stretching the earth along the earth–moon axis by an amount equal to roughly 0.5 meter at each end. [Hint: Remember to allow for the effect of the deformation on the earth on its own

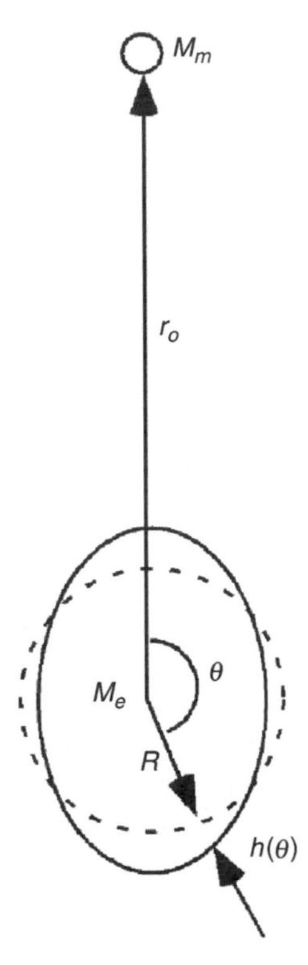

Fig. 4.11 Tidal effect in the earth–moon system (greatly exaggerated for clarity).

gravitational potential, using the multipole expansion of Eq. (4.3.76). Keep only up to quadrupole terms in the expansions. Along these lines, the results given in Exercise (15) may be useful. The total gravitational potential evaluated at the earth's (deformed) surface $r(\theta)$ will be of the form

$$\phi_{\text{tot}} = \frac{-GM_e}{r(\theta)} - \alpha Y_{2,0}(\theta) - \beta Y_{2,0}(\theta),$$

where small deformation is assumed, α is a constant proportional to the mass of the earth, and β is a constant proportional to the mass of the moon. The α-term is caused by the gravitational potential of the deformed earth, and the β-term is caused by the moon. The surface of the earth is deformed so as to be described by the equation $r(\theta) = R + h_0 Y_{2,0}(\theta)$, where $h_0 \ll R$ is a constant to be determined by making sure that ϕ_{tot} is independent of θ. But be careful: α is also proportional to h_0, since α arises from the distortion of the earth.]

4.4 THE WAVE AND HEAT EQUATIONS IN TWO AND THREE DIMENSIONS

In a uniform medium, the wave and heat equations in one dimension have the form $\partial^2 z/\partial t^2 = c^2 \partial^2 z/\partial x^2$ and $\partial T/\partial t = \chi \partial^2 T/\partial x^2$ respectively. In multiple spatial dimensions, the obvious generalization of these equations is

$$\frac{\partial^2 z}{\partial t^2} = c^2 \nabla^2 z(\mathbf{r}, t) \tag{4.4.1}$$

and

$$\frac{\partial T}{\partial t} = \chi \nabla^2 T(\mathbf{r}, t), \tag{4.4.2}$$

where ∇^2 is the Laplacian operator. This generalization allows the evolution of disturbances without any distinction between different spatial directions: the equations are *isotropic* in space.

As in the solution of Poisson's equation, we now consider solutions of the heat and wave equations within some specified volume V, which has a surface S. Boundary conditions of either Dirichlet, von Neumann, or mixed form are specified on this surface. Initial conditions must also be provided throughout the volume. As in the one-dimensional case, two initial conditions (on z and $\partial z/\partial t$) are required for the wave equation, but only one initial condition specifying $T(\mathbf{r}, t=0)$ is required for the heat equation.

General solutions for these equations can be found. The form of the solutions is a generalized Fourier series of eigenmodes of the Laplacian operator, just as for the one-dimensional case discussed in Sec. 4.2. However, for most boundary conditions, the eigenmodes cannot be determined analytically. Analytically tractable solutions can be obtained only for those special geometries in which the eigenmodes are separable. (Numerical solutions can be found using methods to be discussed in Chapter 6.) In the following sections we consider several analytically tractable examples.

4.4.1 Oscillations of a Circular Drumhead

General Solution Consider a drum consisting of a 2D membrane stretched tightly over a circular ring in the x-y plane, of radius a. The membrane is free to vibrate in the transverse (z) direction, with an amplitude $z(r, \theta, t)$, where r and θ are polar coordinates in the x-y plane. These vibrations satisfy the wave equation (4.4.1). The wave propagation speed is $c = \sqrt{T/\sigma}$, where T is the tension force per unit length applied to the edge of the membrane, and σ is the mass per unit area of the membrane. Since the membrane is fixed to the ring at $r = a$, the boundary condition on z is

$$z(a, \theta, t) = 0. \tag{4.4.3}$$

The initial conditions are

$$\begin{aligned} z(r, \theta, 0) &= z_0(r, \theta), \\ \frac{\partial z}{\partial t}(r, \theta, 0) &= v_0(r, \theta) \end{aligned} \tag{4.4.4}$$

for some initial transverse displacement and velocity, z_0 and v_0 respectively.

To solve for the evolution of z, we use a generalized Fourier series:

$$z(r, \theta, t) = \sum_{\alpha} c_\alpha(t) \psi_\alpha(r, \theta). \tag{4.4.5}$$

The functions $\psi_\alpha(r, \theta)$ are chosen to be eigenmodes of the Laplacian operator,

$$\nabla^2 \psi_\alpha(r, \theta) = \lambda_\alpha \psi_\alpha(r, \theta), \tag{4.4.6}$$

with boundary conditions identical to those on z,

$$\psi_\alpha(a, \theta) = 0. \tag{4.4.7}$$

From our study of Poisson's equation, we already know the form of these eigenmodes:

$$\psi_\alpha(r, \theta) = e^{im\theta} J_m(j_{m,n} r/a), \tag{4.4.8}$$

where m is any integer, J_m is a Bessel function, and $j_{m,n}$ is the nth zero of J_m. The corresponding eigenvalues are

$$\lambda_\alpha = -(j_{m,n}/a)^2. \tag{4.4.9}$$

Also, we know that these eigenmodes are orthogonal with respect to the inner product $(f, g) = \int_{r<a} f^*(\mathbf{r}) g(\mathbf{r}) d^2\mathbf{r}$. We can therefore extract an ODE for the time evolution of the Fourier amplitude $c_\alpha(t)$ in the usual manner. Substitution of Eq. (4.4.5) into the wave equation, together with Eq. (4.4.6), implies that

$$\sum_\alpha \psi_\alpha(r, \theta) \frac{d^2}{dt^2} c_\alpha(t) = c^2 \sum_\alpha \lambda_\alpha c_\alpha(t) \psi_\alpha(r, \theta). \tag{4.4.10}$$

Then an inner product with respect to ψ_α yields the harmonic oscillator equation,

$$\frac{d^2}{dt^2} c_\alpha(t) = c^2 \lambda_\alpha c_\alpha(t). \tag{4.4.11}$$

Using Eq. (4.4.9) we find that the general solution is

$$c_\alpha(t) = A_\alpha \cos \omega_\alpha t + B_\alpha \sin \omega_\alpha t, \tag{4.4.12}$$

where ω_α is the frequency associated with a given eigenmode,

$$\omega_\alpha = j_{m,n} c/a, \tag{4.4.13}$$

and A_α and B_α are constants determined by the initial conditions. To determine A_α, we evaluate Eq. (4.4.5) at time $t = 0$, and using Eq. (4.4.5) we find

$$z(r, \theta, 0) = \sum_\alpha A_\alpha \psi_\alpha(r, \theta) = z_0(r, \theta). \tag{4.4.14}$$

The usual inner product argument then yields

$$A_\alpha = \frac{(\psi_\alpha, z_0)}{(\psi_\alpha, \psi_\alpha)}. \tag{4.4.15}$$

Similarly, one finds that

$$\omega_\alpha B_\alpha = \frac{(\psi_\alpha, v_0)}{(\psi_\alpha, \psi_\alpha)}. \tag{4.4.16}$$

Thus, the solution takes the form

$$z(r,\theta,t) = \sum_{m=-\infty}^{\infty} \sum_{n=1}^{\infty} (A_{mn} \cos \omega_{mn} t + B_{mn} \sin \omega_{mn} t)\, e^{im\theta} J_m(j_{m,n} r/a). \tag{4.4.17}$$

This completes the solution of the problem. One can see that, aside from the higher dimensionality of the eigenmodes, the solution procedure is identical to that for the one-dimensional string.

Although the eigenmodes are complex, the coefficients A_{mn} and B_{mn} are also complex, so that the series sums to a real quantity. In particular, Eqs. (4.4.15) and (4.4.16) imply that the coefficients satisfy $A_{-mn} = A^*_{mn}$ and $B_{-mn} = B^*_{mn}$. Also, Eq. (4.4.13) implies that $\omega_{-mn} = \omega_{mn}$. If we use these results in Eq. (4.4.17), we can write the solution as a sum only over nonnegative m as

$$\begin{aligned} z(r,\theta,t) = {} & \sum_{n=1}^{\infty} (A_{0n} \cos \omega_{0n} t + B_{0n} \sin \omega_{0n} t) J_0(j_{0,n} r/a) \\ & + \sum_{m=1}^{\infty} \sum_{n=1}^{\infty} \left(\left[A_{mn} e^{im\theta} + A^*_{mn} e^{-im\theta} \right] \cos \omega_{mn} t \right. \\ & \left. + \left[B_{mn} e^{im\theta} + B^*_{mn} e^{-im\theta} \right] \sin \omega_{mn} t \right) J_m(j_{m,n} r/a). \end{aligned} \tag{4.4.18}$$

The quantities in the square brackets are real; for example, $A_{mn} e^{im\theta} + A^*_{mn} e^{-im\theta} = 2\,\mathrm{Re}(A_{mn} e^{im\theta})$. In fact, if we write $A_{mn} = |A_{mn}| e^{i\phi^A_{mn}}$, and $B_{mn} = |B_{mn}| e^{i\phi^B_{mn}}$, where ϕ^A_{mn} and ϕ^B_{mn} are the complex phases of the amplitudes, then Eq. (4.4.18) becomes

$$\begin{aligned} z(r,\theta,t) = {} & \sum_{n=1}^{\infty} (A_{0n} \cos \omega_{0n} t + B_{0n} \sin \omega_{0n} t) J_0(j_{0,n} r/a) \\ & + 2 \sum_{n=1}^{\infty} \sum_{m=1}^{\infty} \left[|A_{mn}| \cos(\theta + \phi^A_{mn}) \cos \omega_{mn} t \right. \\ & \left. + |B_{mn}| \cos(\theta + \phi^B_{mn}) \sin \omega_{mn} t \right] J_m(j_{m,n} r/a). \end{aligned} \tag{4.4.19}$$

This result is manifestly real, and shows directly that the complex part of the Fourier amplitudes merely produces a phase shift in the θ-dependence of the Fourier modes.

Drumhead Eigenmodes The cylindrically symmetric modes of the drumhead correspond to $m = 0$ and have frequencies $\omega_{0n} = j_{0,n} c/a$. Unlike the modes of a uniform string, these frequencies are not commensurate:

$$\omega_{01} = 2.40483c/a, \qquad \omega_{02} = 5.52008c/a, \qquad \omega_{03} = 8.65373c/a, \ldots .$$

These $m = 0$ modes have no θ-dependence. Like string modes, they are standing waves with stationary nodes at specific radial locations. In the lowest-order mode, the entire drumhead oscillates up and down, while in higher order modes different sections of the drumhead are oscillating 180° out of phase with the center. One of these modes, the $m = 0$, $n = 3$ mode, is shown in Cell 4.41.

Cell 4.41

```
<<NumericalMath`;
j0 = BesselJZeros[0, 3];
ψ[r_, θ_] := BesselJ[0, j0[[3]] r];

Table[ParametricPlot3D[{r Cos[θ], r Sin[θ], Cos[t] ψ[r, θ]},
   {r, 0, 1}, {θ, 0, 2 Pi}, PlotRange→{-1, 1}, PlotPoints→25,
   BoxRatios→{1, 1, 1/2}], {t, 0, 1.8 Pi, .2 Pi}];
```

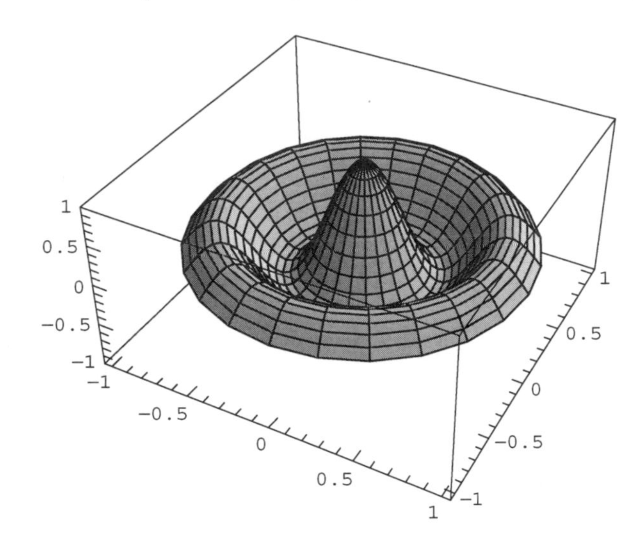

For $m \neq 0$, the eigenmodes obviously have θ-dependence $e^{im\theta} = \cos m\theta + i \sin m\theta$. As discussed above in relation to Eq. (4.4.19), the real and imaginary parts simply correspond to oscillations that are shifted in θ by 90°. Just as for the cylindrically symmetric ($m = 0$) modes, there are radial nodes where the drumhead is stationary. However, there are also locations in θ where nodes occur. This can be seen in Cell 4.42, which plots the real part of the $m = 1$, $n = 2$ mode. For this mode, the line $\theta = \pi/2$ (the y-axis) is stationary. The reader is invited to plot some of the other modes in this manner, so as to get a feeling for their behavior.

Cell 4.42

```
j1 = BesselJZeros[1, 3];
ψ[r_, θ_] := BesselJ[1, j1[[2]] r] Cos[θ];

Table[ParametricPlot3D[{r Cos[θ], r Sin[θ], Cos[t] ψ[r, θ]},
   {r, 0, 1}, {θ, 0, 2 Pi}, PlotRange→{-1, 1}, PlotPoints→
25,
   BoxRatios→{1, 1, 1/2}], {t, 0, 1.8 Pi, .2 Pi}];
```

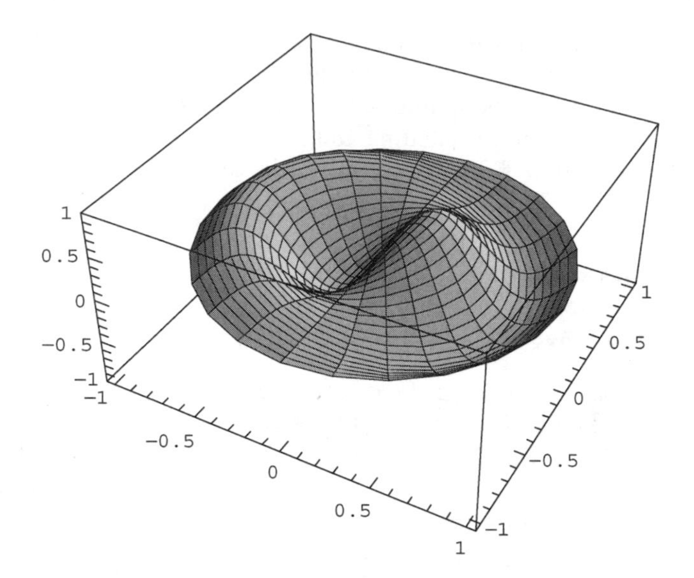

Traveling Waves in θ We have found that the general solution to the 2D wave equation is a sum of eigenmodes with oscillatory time dependence, given by Eq. (4.4.17). Each term in the sum has the form

$$(A_{mn} \cos \omega_{mn} t + B_{mn} \sin \omega_{mn} t)\, e^{im\theta} J_m(j_{m,n} r/a). \tag{4.4.20}$$

Let's consider a specific case, where the complex amplitude B_{mn} equals $-iA_{mn}$ for some given m and n. For this mode, Eq. (4.4.20) can be written as

$$\begin{aligned} A_{mn}(\cos \omega_{mn} t - i \sin \omega_{mn} t)\, e^{im\theta} J_m(j_{m,n} r/a) &= A_{mn}\, e^{-i\omega_{mn} t}\, e^{im\theta} J_m(j_{m,n} r/a) \\ &= A_{mn}\, e^{i(m\theta - \omega_{mn} t)} J_m(j_{m,n} r/a). \end{aligned}$$

This mode is a *traveling wave* in the θ-direction. The real part of the mode has a θ-variation of the form $\cos(m\theta - \omega_{mn} t + \phi_{mn}^A)$, so this wave moves, unlike a standing wave. For example, at $t = 0$, there is a maximum in the real part of the wave at $m\theta + \phi_{mn}^A = 0$; but as time progresses this maximum moves according to the equation $m\theta - \omega_{mn} t \phi_{mn}^A = 0$, or $\theta = -\phi_{mn}^A/m + (\omega_{mn}/m)t$.

The angular velocity ω_{mn}/m is also called the phase velocity c_ϕ of this wave. Since the wave is moving in θ, this phase velocity has units of radians/per second. In Chapter 6, we will consider traveling waves moving linearly in **r**. There, the phase velocity has units of meters per second.

We could also choose $B_{mn} = +iA_{mn}$ in Eq. (4.4.20). This results in a traveling wave proportional to $e^{i(m\theta + \omega_{mn} t)}$. This wave travels in the $-\theta$ direction.

In Cell 4.43 we exhibit a drumhead traveling wave for $m = 1$, $n = 1$, traveling in the positive θ-direction.

Cell 4.43

```
m = 1; ω = j1[[1]];

ψ[r_, θ_, t_] := BesselJ[1, j1[[1]] r] Cos[mθ - ω t] /; r≤1;
ψ[r_, θ_, t_] := 0/; r>1;
```

```
Table[ParametricPlot3D[{r Cos[θ], r Sin[θ], ψ[r, θ, t]},
 {r, 0, 1}, {θ, 0, 2 Pi},
   PlotPoints→25, BoxRatios→{1, 1, 1/2}],
    {t, 0, 1.9 Pi/ω, .1 Pi/ω}];
```

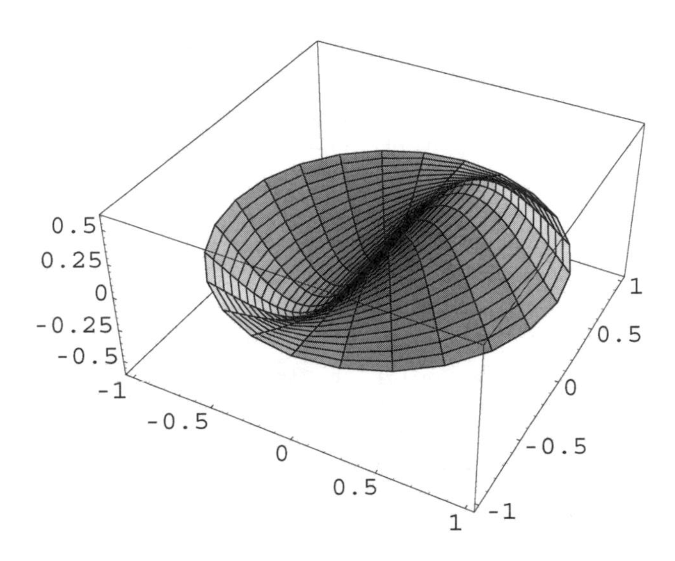

Sources and Inhomogeneous Boundary Conditions Traveling waves such as that shown above are often created by moving disturbances (time-dependent sources). For example, a boat traveling across the surface of a lake creates a wake of traveling waves. Mathematically, these sources enter as a function $S(\mathbf{r}, t)$ on the right-hand side of the wave equation (4.4.1). The response to time-dependent sources is found using an eigenmode expansion, in a manner that is completely identical to that used for the one-dimensional wave equation. Problems such as this will be left to the exercises.

Inhomogeneous boundary conditions on the wave equation can also be handled in an analogous manner to the methods used for the one-dimensional wave equation. For example, on a circular drumhead the rim might be warped, with a height that is given by some function $z_0(\theta)$. This implies a Dirichlet boundary condition

$$z(a, \theta) = z_0(\theta). \tag{4.4.21}$$

The wave equation is then solved by breaking the solution into two pieces, $z(r, \theta, t) = \Delta z(r, \theta, t) + u(r, \theta)$. One can now use two approaches to finding the solution. In the eigenmode approach, one chooses the function $u(r, \theta)$ to be any function that matches the boundary condition, $u(a, \theta) = z_0(\theta)$, and the remainder Δz then satisfies homogeneous boundary conditions, and also satisfies the wave equation with a source created by u:

$$\frac{\partial^2 \Delta z}{\partial t^2} = c^2 \nabla^2 \Delta z - \frac{\partial^2 u}{\partial t^2} + c^2 \nabla^2 u. \tag{4.4.22}$$

However, for this time-independent boundary condition it is easier to use a second approach, by choosing a form for u which satisfies the Laplace equation, $\nabla^2 u = 0$. The solution for u is the equilibrium shape of the drumhead: $z = u(r,\theta)$ is a time-independent solution to the wave equation (4.4.1). The remainder term Δz then satisfies the wave equation without sources, subject to whatever initial conditions are given in the problem.

The equilibrium solution for the shape of the drumhead can be found using the Laplace solution methods discussed in Sec. 3.2.3. For instance, if the warp follows the equation $z(a,\theta) = a\sin 2\theta$, the solution to Laplace's equation that matches this boundary condition is simply $u(r,\theta) = (r^2/a)\sin 2\theta$. [This follows from Eq. (3.1.24), and can be verified by direct substitution into $\nabla^2 u = 0$.] This equilibrium shape is displayed in Cell 4.44.

Cell 4.44

```
u[r_, θ_] = r^2 Sin[2 θ];

ParametricPlot3D[{r Cos[θ], r Sin[θ], u[r, θ]},
  {r, 0, 1}, {θ, 0, 2 Pi}, PlotPoints→25
  BoxRatios→{1, 1, 1/2},
  PlotLabel→"equilibrium of a warped circular drumhead"];
```

On the other hand, for time-dependent boundary conditions, the eigenmode approach must be used. Even if we choose u to satisfy the Laplace equation, a source function will still appear in Eq. (4.4.16), because the time-dependent boundary conditions imply that $\partial^2 u/\partial t^2 \neq 0$. Problems of this sort follow an identical path to solution as for the one-dimensional wave equation with a source function, and examples are left to the exercises.

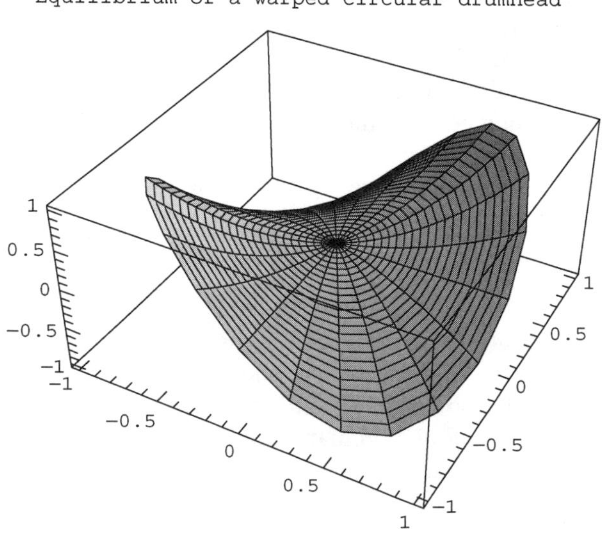

4.4.2 Large-Scale Ocean Modes

The large-scale waves of the ocean provide an instructive example of waves in spherical geometry. Here, for simplicity we take the idealized case of a "water world," where the ocean completely covers the earth's surface, with uniform depth $d_0 \ll R$, where R is the radius of the earth. Also, we will neglect the effects of earth's rotation on the wave dynamics.

As we saw in Exercise (4) of Sec. 3.1, waves in shallow water (for which the wavelength $\gg d_0$) satisfy a wave equation with speed $c = \sqrt{gd_0}$, where $g = 9.8$ m/s^2 is the acceleration of gravity. On the spherical surface of the earth, these waves will also satisfy the wave equation, written in spherical coordinates. That is, the depth of the ocean now varies according to $d = d_0 + h(\theta, \phi, t)$, where h is the change in height of the surface due to the waves, at latitude and longitude specified by the spherical polar angles θ and ϕ. The function h satisfies the wave equation,

$$\frac{\partial^2 h}{\partial t^2} = c^2 \nabla^2 h(\theta, \phi, t) = \frac{c^2}{R^2}\left[\frac{1}{\sin\theta}\frac{\partial}{\partial\theta}\left(\sin\theta\frac{\partial h}{\partial\theta}\right) + \frac{1}{\sin^2\theta}\frac{\partial^2 h}{\partial\phi^2}\right], \quad (4.4.23)$$

where we have used the spherical form of ∇^2, Eq. (3.2.27) [see Lamb (1932, pg. 301)].

From our work on Poisson's equation in spherical coordinates, we know that the eigenmodes of the operator on the right-hand side of Eq. (4.4.23) are spherical harmonics $Y_{l,m}(\theta, \phi)$. Furthermore, we know that the eigenvalues of this operator are $-c^2 l(l+1)/R^2$ [see Eq. (4.3.55)]. Therefore, the amplitude of each spherical harmonic oscillates in time at the frequency

$$\omega_l = \frac{c\sqrt{l(l+1)}}{R}. \quad (4.4.24)$$

The solution is a sum of these modes,

$$h(\theta, \phi, t) = \sum_{l=0}^{\infty} \sum_{m=-l}^{l} (A_{lm} \cos\omega_l t + B_{lm} \sin\omega_l t) Y_{l,m}(\theta, \phi). \quad (4.4.25)$$

It is entertaining to work out the frequencies and shapes of some of the low-order modes, for earthlike parameters. Taking an average ocean depth of roughly $d_0 = 4$ km, the wave speed is $c = \sqrt{9.8 \times 4000}$ m/s $= 198$ m/s. The earth's radius is approximately $R = 6400$ km, so the lowest-frequency modes, with $l = 1$, have frequency $\omega_1 = \sqrt{2}\,c/R = 4.4 \times 10^{-5}$ s^{-1}, corresponding to a period of $2\pi/\omega_1 = 1.4 \times 10^5$ s, or 40 hours. The $l = 2$ modes have a frequency that is larger by the factor $\sqrt{3}$, giving a period of 23 hours. The $l = 1$ and 2 modes are shown in Cells 4.45 and 4.46 for $m = 0$. In the $l = 1$, $m = 0$ mode, water moves from pole to pole, with the result that the center of mass of the water oscillates axially. Such motions could actually only occur if there were a time-dependent force acting to accelerate the water with respect to the solid portion of the earth, such as an undersea earthquake or a meteor impact (see the exercises). Of course, such events also excite other modes.

The next azimuthally symmetric mode has $l = 2$, and is shown in Cell 4.46. In this $l = 2$, $m = 0$ mode, water flows from the equator to the poles and back, producing elliptical perturbations. The reader is invited to explore the behavior of other modes by reevaluating these animations for different l-values.

Cell 4.45

```
l = 1; m = 0;
Table[ParametricPlot3D[(1 + .4 Cos[t]
SphericalHarmonicY[l, m, θ, ϕ])
     {Sin[θ] Cos[ϕ], Sin[θ], Sin[ϕ], Cos[θ]}, {θ, 0, Pi},
   {ϕ, 0, 2 Pi}, PlotRange→{{-1.2, 1.2}, {-1.2, 1.2},
     {-1.2, 1.2}}],
 {t, 0, 2 Pi-0.2 Pi, .2 Pi}];
```

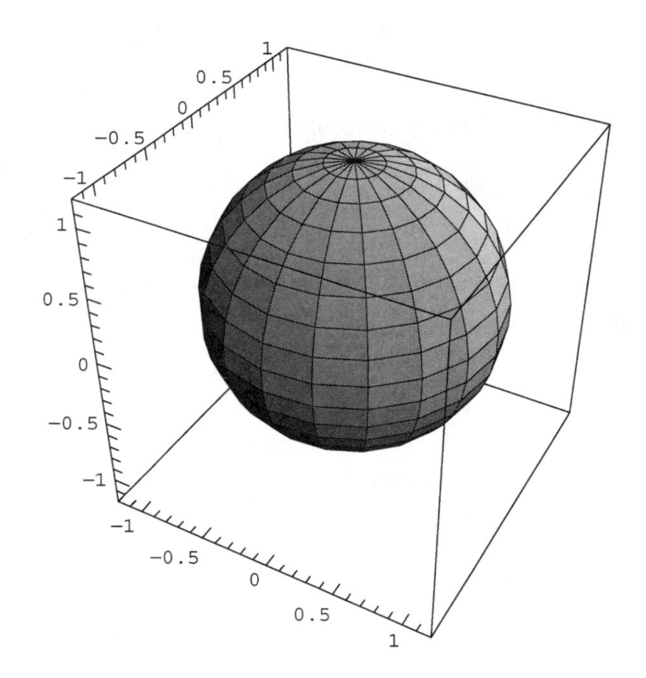

Cell 4.46

```
l = 2; m = 0;
Table[ParametricPlot3D[1 + .2 Cos[t]
SphericalHarmonicY[l, m, θ, ϕ])
     {Sin[θ] Cos[ϕ], Sin[θ], Sin[ϕ], Cos[θ]}, {θ, 0, Pi},
   {ϕ, 0, 2 Pi}, PlotRange→{{-1.2, 1.2}, {-1.2, 1.2},
     {-1.2, 1.2}}], {t, 0, 2 Pi - 0.2 Pi, .2 Pi}];
```

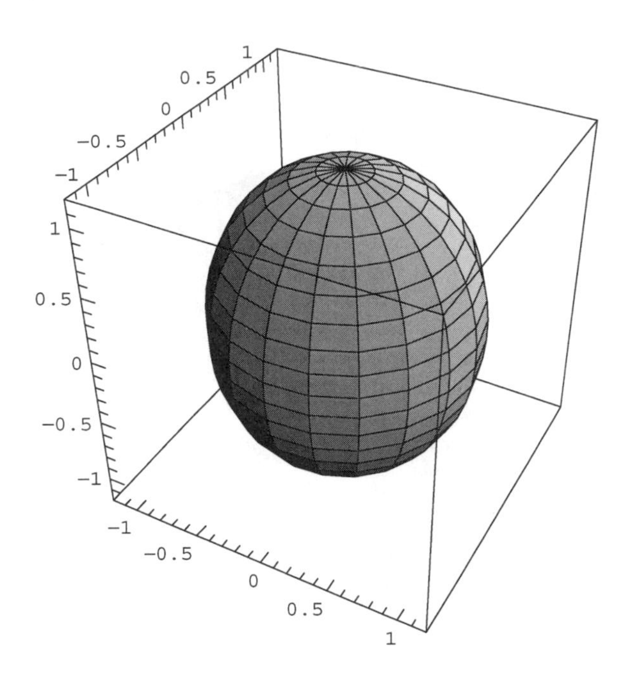

Oscillations with $m \neq 0$ are also of great importance. Of particular interest is the traveling-wave form of the $l = m = 2$ perturbation, which is an elliptical distortion of the ocean surface that travels around the equator. This mode is actually a linear combination of the $l = 2$, $m = \pm 2$ spherical harmonics, and can most easily be expressed in terms of the associated Legendre functions making up these harmonics: $Y_{2,\pm 2} \propto P_2^2(\cos\theta)\, e^{i(\pm 2\phi - \omega_2 t)}$. [Here we have used the fact that $P_2^{-2}(\cos\theta)$ is proportional to $P_2^2(\cos\theta)$; see Table 3.2 in Sec. 3.2.4.] The resulting disturbance is shown in Cell 4.47.

This type of elliptical traveling wave can be excited by the gravitational attraction of the earth to the moon. The moon appears to revolve about the earth daily in the earth's rest frame. This revolution, in concert with the earth–moon attraction, causes an elliptical distortion that follows the moon's apparent position and is responsible for the earth's tides [see Eq. (4.3.80)]. It is interesting that the natural frequency of this mode is 23 hours for the parameters that we chose; this means that the mode is almost resonant with the gravitational force caused by the moon (in our simple model, that is—on the real earth, there are many effects neglected here, not least of which are the continents, which tend to get in the way of this mode).

Cell 4.47

```
l = 2; m =2;
Table[ParametricPlot3D[(1 + .05 LegendreP[l, m, Cos[θ]]
  Cos[m φ -t])
    {Sin[θ] Cos[φ], Sin[θ] Sin[φ], Cos[θ]}, {θ, 0, Pi},
   {φ, 0, 2 Pi}, PlotRange→{{-1.2, 1.2}, {-1.2, 1.2},
    {-1.2, 1.2}}],  {t, 0, 2 Pi - 0.2 Pi, .2 Pi}];
```

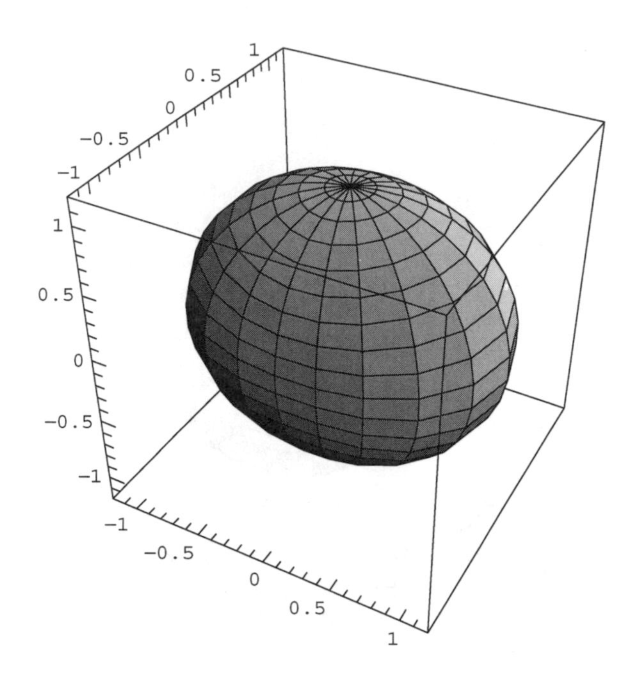

4.4.3 The Rate of Cooling of the Earth

We now consider a classic heat equation problem in spherical coordinates, in which the following question is addressed: a sphere of radius R, with thermal diffusivity χ, is initially at uniform temperature T_0, and the surroundings are at lower temperature T_1. What is the rate at which the sphere cools to temperature T_1?

Since the problem has spherical symmetry, the spherical harmonics are not needed, and the temperature $T(r,t)$ evolves according to the spherically symmetric diffusion equation,

$$\frac{\partial T}{\partial t} = \chi \frac{1}{r^2} \frac{\partial}{\partial r}\left(r^2 \frac{\partial T}{\partial r} \right), \tag{4.4.26}$$

with boundary condition $T(R,t) = T_1$, and initial condition $T(r,0) = T_0$.

As usual, we remove the inhomogeneous boundary condition by writing $T(r,t) = \Delta T(r,t) + u(r)$, where u is chosen to match the boundary condition. A simple choice is $u(r) = T_1$. Then ΔT evolves according to Eq. (4.4.26) with boundary condition $\Delta T(R,t) = 0$ and initial condition

$$\Delta T(r,0) = T_0 - T_1. \tag{4.4.27}$$

The solution for $\Delta T(r,t)$ follows a by now familiar path: we expand ΔT in the spherically symmetric Dirichlet eigenmodes of the spatial operator in Eq. (4.4.26):

$$\Delta T(r,t) = \sum_{n=1}^{\infty} c_n(t)\, \psi_n(r), \tag{4.4.28}$$

then recognize that these eigenmodes are the $l=0$ (spherically symmetric) spherical Bessel functions studied in previous examples:

$$\psi_n(r) = \frac{\sin(n\pi r/R)}{r}, \tag{4.4.29}$$

with eigenvalues

$$\lambda_n = -\chi(n\pi/R)^2, \qquad n = 1,2,3,\dots \tag{4.4.30}$$

[see Eq. (4.3.58) and Table 4.2]. These modes are orthogonal with respect to the radial inner product given in Eq. (4.3.61). The evolution of $c_n(t)$ then follows by taking the inner product with ψ_n, yielding $(d/dt)c_n(t) = -\chi(n\pi/R)^2 c_n$, which has the solution

$$c_n(t) = A_n\, e^{-(n\pi/R)^2 \chi t}. \tag{4.4.31}$$

Finally, the constants A_n are determined by the initial condition (4.4.27):

$$A_n = \frac{(\psi_n(r), T_0 - T_1)}{(\psi_n, \psi_n)}. \tag{4.4.32}$$

This completes the formulation of the problem. Before we exhibit the full solution, however, it is instructive to examine the behavior of $c_n(t)$ for different mode numbers. Equation (4.4.32) implies that an infinite number of eigenmodes are excited by the initial condition. However, eigenmodes with large n decay away very rapidly according to Eq. (4.4.31). Therefore, at late times, the evolution is determined by the $n=1$ mode alone, with an exponential decay of the form $A_1\, e^{-(\pi/R)^2 \chi t}$.

Let's determine this exponential rate of thermal decay for the earth, a sphere with radius $R = 6400$ km. The thermal diffusivity of the earth has been estimated to be roughly $\chi \sim 2\times 10^{-6}$ m^2/s. The rate constant for the $n=1$ mode is then $(\pi/R)^2\ \chi \sim 5\times 10^{-19}$ s^{-1}. The reciprocal of this rate is the time for the temperature to drop by a factor of $e = 2.71\dots$, and equals 60 billion years! This is rather satisfying, since it is much longer than the age of the earth, currently estimated at around 4 billion years. Thus, we would expect from this argument that the earth's core would still be hot, as in fact it is.

Looked at more carefully, however, there is a contradiction. The average temperature gradient at the surface of the earth has been measured to be about 0.03K/m (or 30 K/km, measured in mineshafts and boreholes). Below, we plot this surface gradient using our solution, Eq. (4.4.28), and assuming that the initial temperature of the earth is uniform, and around the melting temperature of rock, $T_0 = 2000$ K.

Cell 4.48

```
A[n_] = (T0 - T1) Simplify[Integrate[r^2 Sin[n Pi r/R] /r,
  {r, 0, R}] /
    Integrate[r^2 Sin[n Pi r/R]^2/r^2, {r, 0, R}],
     n ∈ Integers]
```

$$-\frac{2\ (-1)^n\ R\ (-T_1 + T_0)}{n\pi}$$

Cell 4.49

```
T0 = 2000; T1 = 300; R = 6.4 * 10^6; χ = 2 * 10^-6;
M = 700;
year = 60*60*24*365;
(* the temperature gradient: *)

                 M
dT[r_, t_] =  Σ A[n] D[Sin[n Pi r/R]/r, r] e^(-χ(n Pi/R)^2 t);
                n=1

Plot[dT[R, t 10^6 year], {t, 1, 50},
   AxesLabel → {"t (10^6 years)","dT/dr|r=R (°K/m)"}],
```

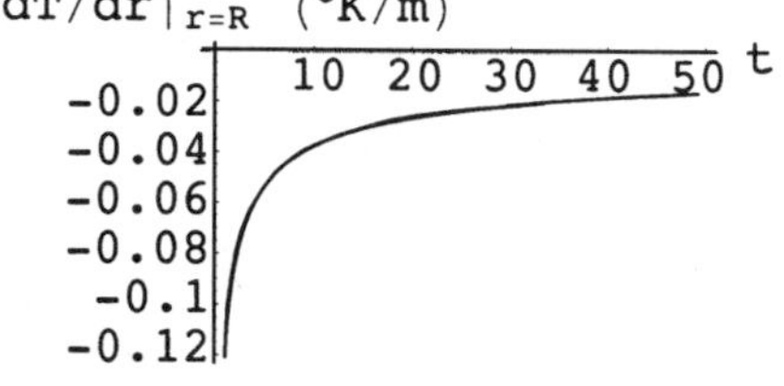

(Note the large number of radial modes needed in order to obtain a converged solution.) From the plot of the temperature gradient, we can see that its magnitude drops to the present value of 0.03K/m after only 20 million years or so, This is much too short a time compared to the age of the earth.

The resolution of this paradox lies in the fact that the earth contains trace amounts of naturally occurring radioactive elements. The radioactive decay of these elements is a source of heat. The heat flux caused by this source creates a temperature gradient at the surface through Fick's law, Eq. (3.1.37). It is currently believed that there is sufficient natural radioactivity in the earth's interior to explain the large surface temperature gradient observed in present experiments [Garland (1979)].

EXERCISES FOR SEC. 4.4

(1) A drumhead has uniform mass density σ per unit area, uniform wave speed c, and a fixed boundary of arbitrary shape. Starting with the equations of motion, show that the energy E of transverse perturbations $z(\mathbf{r}, t)$ is a conserved quantity, where

$$E = \int d^2r \left[\frac{\sigma}{2}\left(\frac{\partial z}{\partial t}\right)^2 + \frac{\sigma c^2}{2} \nabla z \cdot \nabla z \right]. \tag{4.4.33}$$

(2) **(a)** Find the frequencies and spatial form of the normal modes of oscillation of a rectangular trampoline with length a, width b, and propagation speed c.

(b) Find and plot (as a `Plot3D` graphics object) the equilibrium shape of the trampoline under the action of gravity, $g = 9.8\ \text{m/s}^2$, assuming that

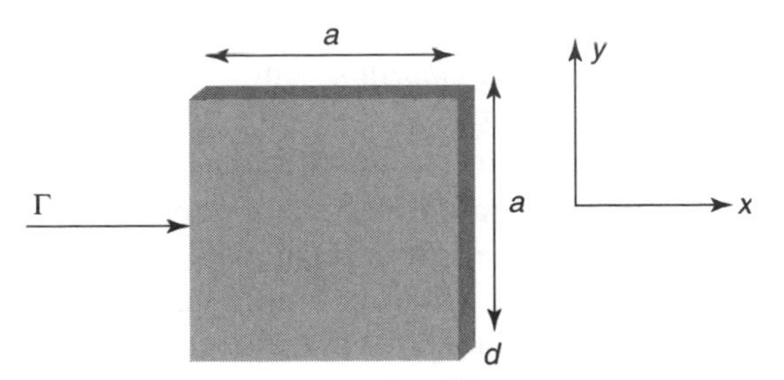

Fig. 4.12 Exercise (5).

$a = b = 3$ m and $c = 3$ m/s. What is the maximum displacement of the trampoline from the horizontal, to three significant figures?

(c) Determine the evoluation of a square trampoline with fixed edges of length $L = 3$ m and wave speed $c = 3$ m/s. Neglect gravity. The initial condition is $z(x, y, 0) = t(x)t(y)$, where $t(x)$ is a triangular shape,

$$t(x) = \begin{cases} x, & x < L/2, \\ L - x, & x > L/2. \end{cases}$$

Animate this evolution for $0 < t < 2$ s using **`Plot3D`**.

(3) One edge of the frame of a square drumhead with unit sides is warped, following $z(0, y) = \frac{1}{4}\sin 2\pi y$. The other edges are straight, satisfying $z = 0$. Find $z(x, y)$ in equilibrium, and plot it as a surface plot using **`Plot3D`**.

(4) A french fry has a square cross section with sides of length $a = 1$ cm and an overall length $L = 5$ cm. The fry is initially at a uniform temperature $T = 0°\text{C}$. It is tossed into boiling fat at $T = 150°\text{C}$. How long does it take for the center of the fry to reach $100°\text{C}$? Take $\chi = 2 \times 10^{-7}$ m^2/s.

(5) A cooling vane on a motor has the shape of a thin square with sides of length $a = 0.7$ cm with thickness $d = 0.2$ cm (see Fig. 4.12). Initially, the motor is off and the vane is at uniform temperature $T = 300$ K. When the motor is turned on, a constant heat flux $\Gamma = 500$ W/cm^2 enters one thin edge of the vane, and the other five faces are all kept at $T = 300$ K. The thermal conductivity of the vane is $\kappa = 0.1$ W/cm K, and the specific heat is $C = 2$ J/cm^3 K.

(a) Find the equilibrium temperature distribution, and determine what is the maximum temperature in the vane, and where it occurs.

(b) Plot the temperature vs. time as a sequence of contour plots in x and y at $z = d/2$ for $0 < t < 0.5$ s.

(6) Solve for the motion of a circular drumhead with a fixed boundary of radius $a = 1$ and sound speed $c = 1$, subject to the following initial conditions:

(a) $z(r, \theta, 0) = r^2(1 - r)^3 \cos 4\theta$, $\partial_t z(r, \theta, 0) = 0$. Animate the solution for $0 < t < 2$ using **`Plot3D`**.

(b) $z(r, \theta, 0) = 0$, $\partial_t z(r, \theta, 0) = \delta(r)/r$. Animate the solution for $0 < t < 2$ using **`Plot`**.

(7) **(a)** Assuming that the measured temperature gradient at the earth's surface, 0.03 K/m, is due to an *equilibrium* temperature profile $T_{\text{eq}}(r)$, find the required mean heat source $\langle S \rangle$, in W/m^3, averaged over the earth's interior (presumably due to radioactivity). Take $\kappa = 2$ W/(m K).

(b) Plot $T_{\text{eq}}(r)$, assuming that the heat source is distributed *uniformly* throughout the earth's interior, and that κ is uniform, given in part (a). Show that the temperature of the core is of order 10^5 K. (This huge temperature is 30 times larger than current estimates. Evidently, the radioactive heat source cannot be uniform, but instead must be concentrated mostly near the earth's surface where the heat can more easily escape [Garland, (1976), p. 356)].)

(8) Solve the following heat equation problems in cylindrical coordinates:

(a) $T(r, \theta, 0) = \delta(r)/r$ in an insulated cylinder of radius $a = 1$ and thermal diffusivity $\chi = 1$. Animate the solution for $0 < t < 0.5$.

(b) $T(r, \theta, 0) = 0$, in a cylinder of radius $a = 1$, thermal diffusivity $\chi = 1$, and thermal conductivity $\kappa = 1$. There is an incident heat flux $\Gamma = -\cos\theta\,\hat{\mathbf{r}}$.

(9) Damped waves on a circular drumhead, radius a, satisfy the following PDE:

$$\gamma \frac{\partial z(r, \theta, t)}{\partial t} + \frac{\partial^2 z(r, \theta, t)}{\partial t^2} = c^2 \nabla^2 z[r, \theta, t],$$

where $\gamma > 0$ is a damping rate.

(a) Find the eigenmodes and eigenfrequencies for this wave equation, assuming that the edge of the drumhead at $r = a$ is fixed.

(b) Solve for the motion for the initial conditions $z(r, \theta, 0) = (1 - r)r^2 \sin 2\theta$, $\dot{z}(r, \theta, 0) = 0$, and boundary condition $z(1, \theta, t) = 0$. Animate the solution for $\gamma = 0.3$, $c = 1$, $0 < t < 2$.

(10) A can of beer, radius $a = 3$ cm and height $L = 11$ cm, is initially at room temperature $T = 25°\text{C}$. The beer is placed in a cooler of ice, at 0°C. Solve the heat equation to determine how long it takes the beer to cool to less than 5°C. Assume that the thermal diffusivity is that of water, $\chi = 1.4 \times 10^{-7}$ m^2/s.

(11) A drumhead has mass σ per unit area (units kg/m^2) and radius R. The speed of propagation of transverse waves is c. A force per unit area, $F(r, \theta, t)$, is applied to the drumhead in the z-direction. The wave equation then becomes

$$\frac{\partial^2 z(r, \theta, t)}{\partial t^2} = c^2 \nabla^2 z(r, \theta, t) + \frac{F(r, \theta, t)}{\sigma}. \tag{4.4.34}$$

A ring of radius $a = 3$ cm and mass $m = 5$ kg is placed in the center of a circular drumhead of radius $R = 1$ m. The speed of propagation is $c = 100$ m/s, and $\sigma = 0.1$ kg/m^2. Including the effect of gravity on the drumhead itself, find the equilibrium shape of the drumhead. [Hint: The force per unit area due to the ring is proportional to $\delta(r - a)$.]

(12) A marble rolls in a circle of radius a around the center of a drumhead of radius R, with mass per unit area σ and wave propagation speed c. The

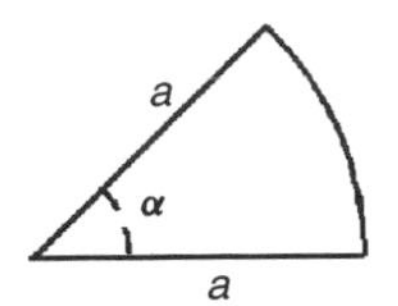

Fig. 4.13 Exercise (14).

marble creates a force per unit area $F = F_0\, e^{-\alpha[(r-a)^2+(\theta-\omega t)^2]}$, where ω is the angular speed of the marble. Find the response of the drumhead to the marble, assuming that the drumhead is initially motionless, and neglecting the effect of gravity on the drumhead itself. Animate this response for two rotations of the marble using contour plots, as a function of time, taking $F_0 = \sigma = a = 1$, $R = 2$, $c = 1$, $\alpha = 20$, and

(a) $\omega = \frac{1}{2}$,

(b) $\omega = 4$.

(13) The edge of a circular drumhead of radius $a = 2$ m and $c = \frac{1}{2}$ m/s is flicked up and down, following $z(a, \theta, t) = t\exp(-t)$ (t in seconds). Find the evolution of the drumhead, assuming that it is initially at rest at $z = 0$. Animate $z(r, t)$ as a series of plots for $0 < t < 10$.

(14) A drumhead has the shape of a wedge, shown in Fig. 4.13, with opening angle α. The edges of the drumhead are fixed at $z = 0$. Find analytic expressions for the frequencies and spatial forms of the normal modes for this drumhead. Find the lowest-frequency mode for $\alpha = 27°$, and plot its form as a surface plot. Plot the frequency of this mode as a function of α for $0 < \alpha < 360°$.

(15) A wheel of radius $R = 10$ cm rotates on a bearing of radius $a = 1$ cm, with angular frequency $\omega = 100$ rad/s. The surface of the wheel is insulated, except at the bearing. At time $t = 0$ the wheel has uniform temperature $T_0 = 300$ K, but due to friction on the bearing it begins to heat. Taking the torque due to friction as $\tau = 1$ newton per meter of length of the bearing, the heating power per unit length is $\tau\omega = 100$ W/m. Assuming that this power is dissipated into the metal wheel, with the thermal conductivity and heat capacity of copper, find $T(r, t)$ in the wheel.

(16) A meteor strikes the north pole of a spherical planet of radius $a = 5000$ km, covered with water of uniform depth $d_0 = 1$ km. The acceleration of gravity on the planet is $g = 10$ m/s^2. Initially, the perturbed ocean height satisfies $h(\theta, \phi, 0) = h_0 \exp(-50\theta^2)$, $\partial_t h(\theta, \phi, 0) = 0$, where $h_0 = 100$ m. Find the evolution $h(\theta, \phi, t)$ of the subsequent tsunami,and animate it vs. time using `Plot` (as a function of θ only) for $0 \le t \le 50$ hours.

(17) A hemispherical chunk of fat has radius $a = 2$ cm. The flat side of the hemisphere sits on a stove, at height $z = 0$. At $t = 0$ the temperature of the fat is $T_0 = 300$ K. At this time, the stove is turned on and the surface at $z = 0$ heats according to $T = T_0 + (T_1 - T_0)\tanh(t/60)$, where $T_1 = 400$ K, and times are in seconds. Assuming that the rest of the fat surface exposed to air has an insulating boundary condition, and that the fat has the thermal properties of water, find $T(r, \theta, t)$. Plot the temperature vs. time at the point farthest from the stove.

(18) A copper sphere of radius $a = 10$ cm is placed in sunlight, with an incident thermal flux of $\Gamma = 1000$ W/m^2, in the $-z$ direction (on the upper side of the sphere only). The sphere also radiates away this energy like a blackbody, with a flux $\Gamma_r = \sigma T(a, \theta)^4$ in the radial direction, where σ is the Stefan–Boltzmann constant.

(a) Assuming that the temperature distribution is spherically symmetric, what is the temperature of the sphere in equilibrium?

(b) Find the actual equilibrium temperature distribution $T_{\text{eq}}(r, \theta)$ in the sphere. [Hint: Copper is a very good heat conductor, so the temperature distribution is *nearly* spherically symmetric. Therefore, you only need to keep two or three terms in the generalized Fourier series for $T_{\text{eq}}(r, \theta)$, and you can Taylor-expand the Stefan–Boltzmann law around the spherically symmetric solution.]

(19) **(a)** The energy levels E_{lmn} for an electron confined in a spherical cavity of radius a (a *quantum dot*) are described by the time-independent Schrödinger equation,

$$\hat{H}\psi_{lmn} = E_{lmn}\psi_{lmn}, \tag{4.4.35}$$

where $\hat{H} = -(\hbar^2/2m)\nabla^2 + V(\mathbf{r})$ is the energy operator, m is the electron mass, V is the potential of the cavity ($V = 0$ for $r < a$, $V = \infty$ for $r \geq a$), $\psi_{lmn}(r, \theta, \phi)$ are the associated energy eigenfunctions, and l, m, n are quantum numbers enumerating the energy levels. Apply separation of variables to this problem in order to find the energy levels and the energy eigenfunctions. (Hint: In this potential, the boundary condition is $\psi = 0$ at $r = a$.) What is the lowest energy level in electron volts for a dot of radius $a = 5$ Å? (1Å $= 10^{-10}$ m; 1 eV $= 1.60 \times 10^{-19}$ J.)

(20) The electron energy levels in a hydrogen atom also satisfy Eq. (4.4.35), with Hamiltonian operator $\hat{H} = -(\hbar^2/2m)\nabla^2 - e^2/(4\pi\epsilon_0 r)$, where m is the reduced mass of the system, roughly equaling the electron mass, and e is the electron charge. Show by substitution of the following solutions into Eq. (4.4.35) that the energy levels are given by

$$E_n = \frac{e^2}{8\pi\epsilon_0 a n^2}, \qquad n = 1, 2, 3, \ldots, \tag{4.4.36}$$

and that the eigenfunctions are

$$\psi_{lmn} = Y_{lm}(\theta, \phi) r^l L_{n-l-1}^{2l+1}(2r/na)\, e^{-r/(na)}, \qquad 0 < l < n, \quad |m| \leq l, \tag{4.4.37}$$

where $a = 4\pi\epsilon_0\hbar^2/me^2$ is the Bohr radius, and where $L_\alpha^\beta(x)$ are generalized Laguerre polynomials, equal to the ordinary Laguerre polynomials for $\beta = 0$. (In *Mathematica* these polynomials are referred to as `LaguerreL[`$\alpha, \beta,$ `x]`.) In your solution you may use the fact that these polynomials satisfy the ODE

$$xL_\alpha^{\beta''} + (\beta + 1 - x)L_\alpha^{\beta'} + \alpha L_\alpha^\beta = 0.$$

(Hint: In the Schrödinger equation scale distances to a and energies to $e^2/4\pi\epsilon_0 a$.)

(21) Rubber supports two types of bulk wave motions: compressional modes and shear modes. The compressional modes have displacements $\delta\mathbf{r}$ in the direction of propagation, creating compressions in the rubber. The shear modes have displacements transverse to the propagation direction, so that $\nabla\cdot\delta\mathbf{r}=0$, and therefore no compressions occur in the shear modes. Consider the shear modes in a spherical rubber ball of radius R. These modes satisfy the following vector wave equation:

$$\frac{\partial^2}{\partial t^2}\delta\mathbf{r}=c_s^2\nabla^2\delta\mathbf{r},$$

where c_s is the speed of shear wave propagation [see Love (1944)]. Assuming that $\delta\mathbf{r}=\hat{\boldsymbol{\phi}}\,\delta r_\phi(r,\theta,\phi)$ [i.e., the motion is in the (x,y) plane, in the ϕ-direction], the boundary condition at the free surface of the sphere, $r=R$, can be shown to be

$$R\frac{\partial}{\partial r}\delta r_\phi=\delta r_\phi.$$

(a) By applying separation of variables to this wave equation, show that the modes have the form

$$\delta r_\phi(r,\theta,\phi)=\frac{J_{l+1/2}(kr)}{\sqrt{r}}e^{im\theta}\frac{\partial}{\partial\theta}P_l^m(\cos\theta),$$

and that the frequency ω of normal modes satisfy

$$R\frac{\partial}{\partial R}\frac{J_{l+1/2}(kR)}{\sqrt{R}}=\frac{J_{l+1/2}(kR)}{\sqrt{R}},$$

where $k=\omega/c_s$ and where J_l is a Bessel function, P_l^m is an associated Legendre function, $l=0,1,2,\dots$ and $m=-l,-l+1,\dots,l-1,l$. Solve numerically for the lowest four modes. Plot the dependence of these four modes on r.

(b) Choose 50 points distributed evenly along the great circle defined by $\phi=0$ and π, $r=R$, and follow the motion of these points as they are carried along by the lowest normal mode for one period of oscillation.

(22) Spherically symmetric compressional waves in a rubber ball of radius R satisfy the following elastic wave equation:

$$\frac{\partial^2}{\partial t^2}\delta r=c_p^2\left(\frac{\partial^2}{\partial r^2}\delta r+\frac{2}{r}\frac{\partial}{\partial r}\delta r-\frac{2}{r^2}\delta r\right),$$

where δr is the radial position change of a mass element, and c_p is the speed of compressional waves. These waves satisfy the following boundary condition at the free surface $r=R$:

$$\left(2c_p^2-4c_s^2\right)\frac{\delta r}{r}+c_p^2\frac{\partial}{\partial r}\delta r=0,$$

where c_s is the speed of shear waves.

(a) Show that the dispersion relation for these waves is

$$4kRc_s^2 \cos kR = \left(4c_s^2 - \omega^2 R^2\right) \sin kR, \qquad \text{where} \quad k = \omega/c_p.$$

(b) Solve for the lowest three frequencies, and in each case plot the radial dependence of δr, for (i) $c_s = c_p$, (ii) $c_s = c_p/\sqrt{2}$, (iii) $c_s = 0$.

(23) A child drops a pebble into the center of a shallow pool of water of radius $a = 1$ m and depth h. The wave height $z(r,t)$ satisfies the wave equation with wave propagation speed $c = \sqrt{gh} = 1$ m/s. The initial disturbance has the form $z(r,0) = -z_0 \exp(-30r^2)(1 - 30r^2)$, $\partial z(r,0) = 0$, where $z_0 = 1$ cm (and r is measured in meters). The boundary condition at the pool edge is not Dirichlet: the water surface at the pool edge can move up and down. To find the correct boundary condition, we must consider the radial displacement η of fluid during the wave motion.

(a) Using the same techniques as were used in the derivation of Eq. (3.1.78), show that the radial fluid displacement η is related to the wave height z by

$$z = -\frac{h}{r}\frac{\partial(r\eta)}{\partial r}.$$

(b) The boundary conditions are $\eta = 0$ at $r = a$, and η finite at $r = 0$. Find the eigenmodes and eigenfrequencies, and plot the wave height z vs. r for the first three eigenmodes.

(c) Solve the initial-value problem stated previously, and animate the solution using `Plot` for $0 < t < 2$.

(24) For bounded pools of shallow water with waves moving in one dimension only, we saw in the previous problem that it is necessary to consider the horizontal fluid displacement η. For general wave motion in more than one dimension, this displacement is a vector, $\boldsymbol{\eta}(\mathbf{r},t)$. Wave height z is related to fluid displacement according to $z = -h\nabla \cdot \boldsymbol{\eta}$. Using the same method as that which led to Eq. (3.1.78), one can show that $\boldsymbol{\eta}$ satisfies the vector equation

$$\frac{\partial^2}{\partial t^2}\boldsymbol{\eta}(\mathbf{r},t) = c^2 \nabla\nabla \cdot \boldsymbol{\eta}(\mathbf{r},t), \tag{4.4.38}$$

where $c = \sqrt{gh}$, for a pool with constant equilibrium depth h. The boundary condition on this equation is $\hat{\mathbf{n}} \cdot \boldsymbol{\eta} = 0$, where $\hat{\mathbf{n}}$ is a unit vector normal to the edge of the pool.

(a) Using Eq. (4.4.38), show that wave height z satisfies a wave equation.

(b) Assume that $\boldsymbol{\eta} = \nabla\phi(\mathbf{r},t)$. This is called *potential flow* [see Landau and Lifshitz (1959)]. Show that ϕ also satisfies the wave equation

$$\frac{\partial^2}{\partial t^2}\phi = c^2 \nabla^2 \phi \tag{4.4.39}$$

with von Neumann boundary condition $\hat{\mathbf{n}} \cdot \nabla \phi = 0$. Show that ϕ and z are related according to $z = -h\nabla^2\phi$.

(25) **(a)** A coffee cup has radius a and is filled with coffee of depth h. Assuming potential flow and using Eq. (4.4.39), with $\mathbf{r} = (r, \theta)$, find analytic expressions for frequency and spatial form of the normal modes of the coffee. What is the lowest-frequency mode, and what is its frequency? (Hint: This mode is not cylindrically symmetric.) Plot the spatial form of $z(r, \theta, t)$ for this mode.

(b) Find the frequency of this lowest mode for a cup of radius $a = 3$ cm, filled to a depth of $h = 1$ cm. See if you can set this mode up in your coffee. Measure its frequency with your watch, and use a ruler to determine the depth of the coffee and the cup diameter. Compare to the theory, and discuss any errors.

(26) A wave machine excites waves at one end of a shallow square pool of side $L = 1$ by injecting and removing water sinusoidally in time along the left edge. Along this edge ($x = 0$), the boundary conditions are time-dependent, with fluid being displaced according to $\eta_x(0, y, t) = y^2(1-y)^2 \sin \omega_0 t$. Along the other three edges, $\hat{\mathbf{n}} \cdot \boldsymbol{\eta} = 0$. Assume potential flow.

(a) Solve for the wave height $z(x, y, t)$, assuming that $\phi = 0$ and $\partial_t \phi = \omega_0 y^2(1-y)^2 x$ initially. Create an animation of the wave height using `Plot3D` for the case $\omega_0 = 1$ and $c = 0.2$ for $0 < t < 6\pi$.

(b) There are values of ω_0 for which the wave response becomes large. Find these values, and explain their meaning physically.

(27) Sound waves in a fluid such as air or water satisfy equations of the same form as those for gravity waves on a fluid surface, Eqs. (4.4.38) and (4.4.39), except written in three dimensions. Assuming potential flow, the displacement of a fluid element in the gas is given by $\boldsymbol{\eta}(x, y, z, t) = \nabla\phi$, where ϕ satisfies Eq. (4.4.39), and the sound speed is $c = \sqrt{\gamma p_0/M}$, where M is the mass density of the fluid, p_0 is the equilibrium pressure, and γ is the ratio of specific heats. Also, the perturbed pressure δp in the wave is related to $\boldsymbol{\eta}$ by

$$\delta p = -p_0 \nabla \cdot \boldsymbol{\eta} = -p_0 \nabla^2 \phi. \tag{4.4.40}$$

(a) A closed tube of length 2 with circular cross section supports normal modes in the enclosed gas. The tube volume lies in the domain $0 < r < 1$, $0 < z < 2$. Find the frequencies of the cylindrically symmetric normal modes in this tube and the spatial form of the modes. (Hint: First, show that the function ϕ satisfies von Neumann boundary conditions.)

(b) Call the (r, z) mode numbers the integers (l, n), with $(l, n) = (0, 0)$ denoting the lowest mode (with zero frequency). Plot the displacement field $\boldsymbol{\eta}$ for the $(l, n) = (1, 2)$ mode in the (x, z) plane using `PlotVectorField`.

(c) A circular portion of the end of the cylinder at $z = 0$ is a loudspeaker: an elastic membrane, which is driven by an external force to oscillate as $\delta z(r, 0, t) = z_0(1 - 4r^2) \sin \omega_0 t$, for $r < \frac{1}{2}$, and $\delta z(r, 0, t) = 0$ otherwise. Find the response of the pressure $\delta p(r, z, t)$ in the gas inside the piston as a Fourier series, assuming that $\delta p = 0$ for all $t < 0$. [Hint: The potential ϕ at $z = 0$ has a discontinuous derivative as a function of r. Therefore, to put the equation in standard form it is important to find a function $u(r, z, t)$ that matches these boundary conditions but remains smooth in the interior of the domain. Use the solution of $\nabla^2 u = \rho(t)$ with the appropriate discontinuous boundary conditions, choosing $\rho(t)$ so as to satisfy the Gauss's law constraint condition on problems with von Neumann boundary conditions.]

(d) Taking $c = 1$, $\omega_0 = 10$, and keeping 15 normal modes in the r-dimension and 6 in the z-dimension, animate δp vs. x and z in the $y = 0$ plane for $0 \le t \le 2$.

(28) A solid sphere of radius a is concentric with a larger hollow sphere of radius R. At $t = 0$, the smaller sphere begins to move up and down along the z-axis, according to $\delta z(t) = z_0 \sin \omega_0 t$. Assume potential flow, and also assume a free-slip boundary condition at the surface of the spheres, so that the boundary condition at $r = a$ is $\hat{\mathbf{r}} \cdot \boldsymbol{\eta}|_{r=a} = \delta z(t) \cos\theta$, and at $r = R$ is $\hat{\mathbf{r}} \cdot \boldsymbol{\eta}|_{r=R} = 0$. (The fluid is allowed to slip along the sphere's surface, but cannot go through the surface.) Find the resulting pressure distribution in the cavity between the spheres, $\delta p(r, \theta, t)$, in spherical coordinates by solving for $\phi(r, \theta, t)$ with the appropriate von Neumann boundary conditions and applying Eq. (4.4.40). Plot δp vs. time in the (x, z) plane for $0 < t < 4$, taking $c = 1$, $a = 1$, $R = 3$, and $\omega_0 = 2\pi$, and keeping the first 10 radial eigenmodes. (The mode frequencies must be determined numerically using `FindRoot`.)

REFERENCES

R. Courant and D. Hilbert, *Methods of Mathematical Physics, Vol. 1* (Wiley-Interscience, New York, 1953).

G. D. Garland, *Introduction to Geophysics*, 2nd ed. (W. B. Saunders, Philadelphia, 1979).

A. Goriely and T. MacMillan, *Shape of a cracking whip*, Phys. Rev. Lett. **88**, 244301 (2002).

H. Lamb, *Hydrodynamics*, 6th ed. (Dover, New York, 1932).

L. D. Landau and E. M. Lifshitz, *Theory of Elasticity* (Pergamon, Oxford, 1986).

L. D. Landau and E. M. Lifshitz, *Fluid Mechanics* (Pergamon, London, 1959).

A. E. H. Love, *A Treatise on the Mathematical Theory of Elasticity*, 4th ed. (Dover, New York, 1944).

P. M. Morse and H. Feshbach, *Methods of Theoretical Physics*, (McGraw-Hill, New York, 1953).

Larry Niven, *Neutron Star* (Ballantine Books, New York, 1968).

CHAPTER 5

PARTIAL DIFFERENTIAL EQUATIONS IN INFINITE DOMAINS

In Chapters 3 and 4, we found solutions to various partial differential equations in finite spatial domains. In order to determine the solution, it was necessary to specify boundary conditions. In this chapter, we explore solutions to PDEs in infinite spatial domains. One often encounters effectively infinite domains, where the boundaries are sufficiently far from the region of interest to have negligible influence on the local behavior of the solution.

Take, for example, a large object, inside which a localized temperature perturbation $T(\mathbf{r}, t)$ evolves, far from the edges of the object. (See Fig. 5.1.) This perturbation could be expanded as a sum of eigenmodes for the system, but then boundary conditions would have to be specified, and this ought to be unnecessary. It should make no difference to the local evolution of the temperature whether the boundaries are insulated or are conducting, given that they are far from the perturbation. Also, if the boundary has some complicated shape, it could be difficult as a practical matter to determine the eigenmodes.

In Sec. 5.1 we will see that problems such as this can be solved using Fourier transforms, provided that the system in question is uniform in space and time over the region of interest (i.e., the system is without spatial or temporal variation in intrinsic parameters such as the wave speed or the thermal conductivity). Recall that in Chapter 2 Fourier transforms were used to describe arbitrary functions defined on the entire real line. Thus, they are just what is needed to describe the evolution of solutions to PDEs in infinite domains. Furthermore, we will see that these Fourier transform solutions can be determined without imposing specific boundary conditions.

In Sec. 5.2 (and Sec. 5.3 in the electronic version) we consider systems that are nonuniform but still effectively infinite, in the sense that perturbations are localized far from boundaries, so that boundary conditions need not be specified. Now parameters such as the sound speed or conductivity vary with position or time in

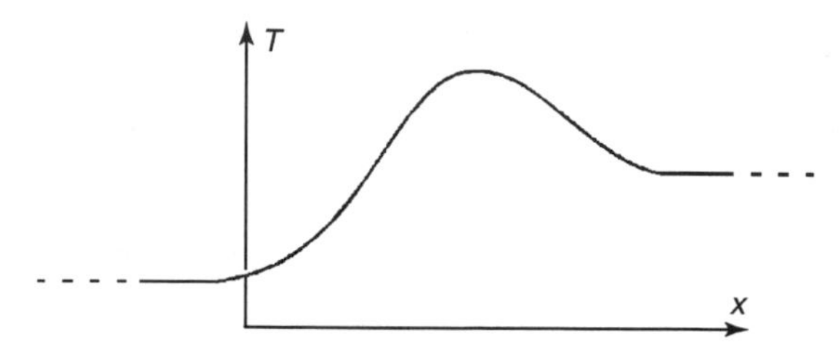

Fig. 5.1 Localized temperature perturbation.

the region of interest. We will discuss an analytic approximation technique for determining the local evolution of the solution, called WKB theory.

5.1 FOURIER TRANSFORM METHODS

5.1.1 The Wave Equation in One Dimension

Traveling Waves We first consider solutions to the wave equation in one dimension, on an infinite uniform string running from $-\infty < x < \infty$:

$$\frac{\partial^2 y}{\partial t^2} = c^2 \frac{\partial^2 y}{\partial x^2}, \tag{5.1.1}$$

subject to the initial conditions

$$\begin{aligned} y(x,0) &= y_0(x), \\ \frac{\partial y}{\partial t}(x,0) &= v_0(x). \end{aligned} \tag{5.1.2}$$

This problem can be solved by applying the spatial Fourier transform operator $\hat{F} = \int_{-\infty}^{\infty} dx\, e^{-ikx}$ to both sides of Eq. (5.1.1):

$$\hat{F} \frac{\partial^2 y}{\partial t^2} = \hat{F} c^2 \frac{\partial^2 y}{\partial x^2}. \tag{5.1.3}$$

On the left-hand side, we can exchange $\hat{F}$ with $\partial^2/\partial t^2$, writing

$$\hat{F} \frac{\partial^2 y}{\partial t^2} = \frac{\partial^2}{\partial t^2} \hat{F} y = \frac{\partial^2}{\partial t^2} \tilde{y},$$

where $\tilde{y}(k,t)$ is the spatial Fourier transform of $y(x,t)$, given by

$$\tilde{y}(k,t) = \int_{-\infty}^{\infty} dx\, e^{-ikx} y(x,t). \tag{5.1.4}$$

On the right-hand side of Eq. (5.1.3), we use Eq. (2.3.18) to write $\hat{F}c^2\,\partial^2 y/\partial x^2 = -c^2k^2\hat{F}y = -c^2k^2\tilde{y}$. Thus, Eq. (5.1.3) becomes a second-order ODE in time,

$$\frac{\partial^2}{\partial t^2}\tilde{y}(k,t) = -c^2k^2\tilde{y}(k,t). \tag{5.1.5}$$

This ODE is merely the harmonic oscillator equation, with an oscillation frequency that depends on the wavenumber k according to

$$\omega(k) = ck. \tag{5.1.6}$$

Such a relation between frequency and wavenumber is called a *dispersion relation*.

It is convenient to write the general solution to Eq. (5.1.5) in exponential notation,

$$\tilde{y}(k,t) = C(k)\,e^{-i\,\omega(k)t} + D(k)\,e^{i\,\omega(k)t}, \tag{5.1.7}$$

where $C(k)$ and $D(k)$ are the two undetermined coefficients that appear in this second-order ODE. These coefficients can take on different values for different k's and so are written as functions of k. We can then transform the solution back to x-space by applying an inverse Fourier transformation, $\hat{F}^{-1} = \int_{-\infty}^{\infty}(dk/2\pi)\,e^{ikx}$, to Eq. (5.1.7):

$$y(x,t) = \hat{F}^{-1}\tilde{y}(k,t) = \int_{-\infty}^{\infty}\frac{dk}{2\pi}C(k)\,e^{i[kx-\omega(k)t]} + \int_{-\infty}^{\infty}\frac{dk}{2\pi}D(k)\,e^{i[kx+\omega(k)t]}. \tag{5.1.8}$$

Equation (5.1.8) is the general solution to the wave equation on a one-dimensional infinite uniform string. It looks very similar to the eigenmode expansions encountered in Chapter 4, except that now we integrate over a continuous spectrum of modes rather than summing over a countably infinite set. In fact, one can think of the functions e^{ikx} and e^{-ikx} in Eq. (5.1.8) as eigenmodes of the operator $c^2\,\partial^2/\partial x^2$, in that this operator returns these functions unchanged except for a multiplicative constant (the eigenvalue, $-c^2k^2$). However, unlike regular eigenvalue problems, there are no boundary conditions associated with these functions at $\pm\infty$.

Equation (5.1.8) can be further simplified by substituting the dispersion relation, Eq. (5.1.6):

$$y(x,t) = \int_{-\infty}^{\infty}\frac{dk}{2\pi}C(k)\,e^{ik(x-ct)} + \int_{-\infty}^{\infty}\frac{dk}{2\pi}D(k)\,e^{ik(x+ct)}. \tag{5.1.9}$$

The functions $e^{ik(x+ct)}$ and $e^{ik(x-ct)}$ represent *traveling waves*, propagating to the right and left respectively. The real parts of these two functions are shown in Cells 5.1 and 5.2 taking $c = 1$ and $k = 1$. (The plots in Cells 5.1 and 5.2 differ only when animated, so only the former plot is included in the printed version of the book.) These traveling waves extend over the whole real line, and are periodic in both time and space. The spatial period of the waves is the wavelength λ, and is related to the wavenumber k by $\lambda = 2\pi/k$. The temporal period T is related to the frequency $\omega(k)$ by $T = 2\pi/\omega(k)$.

Cell 5.1

```
c = 1; k = 1;
Table[Plot[Re[e^(ik (x-ct))], {x, 0, 6 Pi},
   PlotLabel → "Re[e^(ik (x-ct))]", AxesLabel → {"x", ""}],
   {t, 0, 1.8 Pi, .2 Pi}];
```

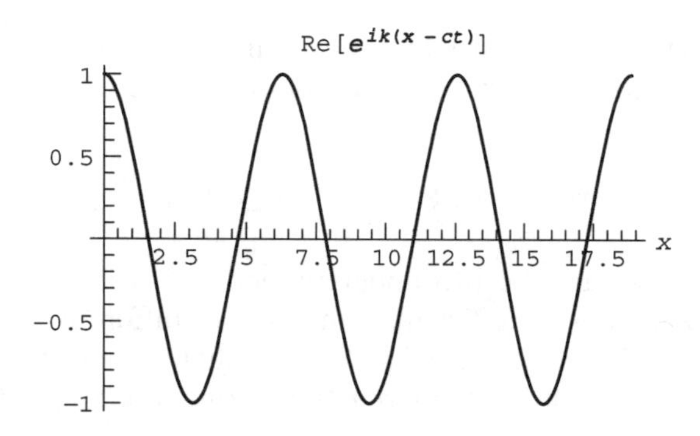

Cell 5.2

```
c =1; k = 1;
 Table[Plot[Re[e^(ik (x+ct))], {x, 0, 6 Pi},
   PlotLabel → "Re[e^(ik (x+ct))]", AxesLabel → {"x", ""}],
   {t, 0, 1.8 Pi, .2 Pi}];
```

The speed at which the waves propagate can be seen to equal c, since the waves can be written as $e^{ik(x \pm ct)}$. For instance, the motion of a wave maximum in the previous animations is determined by the condition that the argument of $\cos k(x \pm ct)$ equals 0. Therefore, the location of the maximum satisfies $x \pm ct = 0$, which implies that the maximum moves with velocity $\mp c$.

General Solution The general solution to the wave equation (5.1.9) is a linear superposition of these traveling waves, propagating to the left and the right. This solution can also be written in another, illuminating form. If we define two functions $f(x)$ and $g(x)$ according to

$$f(x) = \int_{-\infty}^{\infty} \frac{dk}{2\pi} C(k)\, e^{ikx}, \qquad g(x) = \int_{-\infty}^{\infty} \frac{dk}{2\pi} D(k)\, e^{ikx}, \tag{5.1.10}$$

then Eq. (5.1.9) becomes

$$y(x,t) = f(x - ct) + g(x + ct). \tag{5.1.11}$$

Equation (5.1.11) is another way to represent the general solution to the wave equation in one dimension. It is called D'Alembert's solution. The functions

$f(x-ct)$ and $g(x+ct)$ are arbitrary disturbances traveling to the right and left respectively with constant speed c, and without changing their shape. We discussed this behavior previously; see Eq. (3.1.26).

In order to determine the connection between these propagating disturbances and the initial conditions, we evaluate Eq. (5.1.11) and its time derivative at $t=0$, and use Eq. (5.1.2):

$$y(x,0) = f(x) + g(x) = y_0(x),$$

$$\frac{\partial y}{\partial t}(x,0) = -c\frac{\partial f}{\partial x}(x) + c\frac{\partial g}{\partial x}(x) = v_0(x).$$

In the second line we have used the chain rule to write $(\partial f/\partial t)(x-ct) = -c(\partial f/\partial x)(x-ct)$, and similarly for the time derivative of g. These two equations are sufficient to determine f and g in terms of y_0 and v_0. To do so, operate on both equations with a Fourier transform. Then, using Eq. (5.1.10), we obtain

$$\begin{aligned} C(k) + D(k) &= \tilde{y}_0(k), \\ -i\omega(k)[C(k) - D(k)] &= \tilde{v}_0(k), \end{aligned} \tag{5.1.12}$$

where we have also used Eq. (2.3.18) and (5.1.6), and where $\tilde{y}_0$ and $\tilde{v}_0$ are the Fourier transforms of y_0 and v_0 respectively. Solving these coupled equations for $A(k)$ and $B(k)$ yields

$$\begin{aligned} C(k) &= \frac{1}{2}\left(\tilde{y}_0(k) + i\frac{\tilde{v}_0(k)}{\omega(k)}\right), \\ D(k) &= \frac{1}{2}\left(\tilde{y}_0(k) - i\frac{\tilde{v}_0(k)}{\omega(k)}\right). \end{aligned} \tag{5.1.13}$$

The functions f and g then follow from Eq. (5.1.10).

For example, if the initial perturbation is stationary, so that $v_0(x)=0$ but $y_0(x) \neq 0$, then equations (5.1.13) imply that $C(k) = D(k) = \tilde{y}_0(k)/2$. Equations (5.1.10) and (5.1.11) then yield

$$y(x,t) = \frac{y_0(x+ct) + y_0(x-ct)}{2}. \tag{5.1.14}$$

The initial perturbation breaks into two equal pulses, traveling in opposite directions. Recall that this was the behavior observed in Example 2 of Sec. 3.1.2, up to the time where the pulses encountered the boundaries.

5.1.2 Dispersion; Phase and Group Velocities

The Schrödinger Equation for a Free Particle Moving in One Dimension

Let's now apply the Fourier transform analysis to another wave equation: the Schrödinger equation for the evolution of the wave function $\psi(x,t)$ of a free particle of mass m moving in one dimension,

$$i\hbar \frac{\partial}{\partial t} \psi(x,t) = -\frac{\hbar^2}{2m} \frac{\partial^2}{\partial x^2} \psi(x,t). \tag{5.1.15}$$

This ODE is first-order in time, and so is supplemented by a single initial condition on the wave function at $t = 0$:

$$\psi(x,0) = \psi_0(x). \tag{5.1.16}$$

Application of the spatial Fourier transformation operator $\hat{F} = \int_{-\infty}^{\infty} dx\, e^{-ikx}$ to both sides of Eq. (5.1.15) yields the following ODE:

$$i\hbar \frac{\partial}{\partial t} \tilde{\psi}(k,t) = \frac{\hbar^2 k^2}{2m} \tilde{\psi}(k,t), \tag{5.1.17}$$

where $\tilde{\psi} = \hat{F}\psi$ is the spatial Fourier transform of ψ. This first-order ODE has the general solution

$$\tilde{\psi}(k,t) = C(k)\, e^{-i\omega(k)t}, \tag{5.1.18}$$

where $C(k)$ is the undetermined constant, a function of the wavenumber k, and $\omega(k)$ is given by

$$\omega(k) = \frac{\hbar k^2}{2m}. \tag{5.1.19}$$

An inverse transformation of Eq. (5.1.18) then yields the general solution for $\psi(x,t)$:

$$\psi(x,t) = \int_{-\infty}^{\infty} \frac{dk}{2\pi} C(k)\, e^{i[kx - \omega(k)t]}. \tag{5.1.20}$$

Equation (5.1.20) implies that the wave function for a free particle can be written as a superposition of traveling waves, each of the form $e^{i[kx-\omega(k)t]}$. These waves propagate with a velocity that depends on the wavenumber, termed the *phase velocity* $v_\phi(k)$. The phase velocity refers to the velocity of a point of given phase ϕ on the wave, where $\phi = kx - \omega(k)t$. For example, the real part of the wave has a maximum at $\phi = 0$. The maximum moves according to the equation $0 = kx - \omega(k)t$. Therefore, the speed at which this point moves is given by

$$v_\phi(k) = \omega(k)/k. \tag{5.1.21}$$

For a string, with dispersion relation $\omega(k) = ck$, Eq. (5.1.21) implies that $v_\phi(k) = c$, so that waves of any wavenumber propagate at the same speed. Here however, Eqs. (5.1.21) and (5.1.19) imply that $v_\phi(k) = \hbar k/2m$, so long waves propagate more slowly than short waves. This variation of phase velocity with wavenumber is called *dispersion*. Dispersion has important consequences for the behavior of the free-particle wave function, as we will now see.

The effects of dispersion are most easily addressed with an example. Consider the evolution of a wave function from the following initial condition:

$$\psi_0(x) = A\, e^{-x^2/2a^2 + ik_0 x}. \tag{5.1.22}$$

The functions $|\psi_0(x)|$ and Re $\psi_0(x)$ are displayed in Cell 5.3. This initial condition is a wave with wavenumber k_0 and varying amplitude that vanishes as $|x| \to \infty$. This sort of wave function is often referred to as a *wave packet*. According to the tenets of quantum theory, the probability of finding the particle in the range from x to $x + dx$ is $|\psi(x)|^2\, dx$, so Eq. (5.1.22) describes a particle localized to within a distance of roughly a from the origin. The constant A in Eq. (5.1.22) is determined by the condition that

$$\int_{-\infty}^{\infty} |\psi(x)|^2\, dx = 1;$$

or in words, there is a unit probability of finding the particle somewhere on the real line. For $\psi_0(x)$ given by Eq. (5.1.22), this implies that $A = 1/\sqrt{2\pi a^2}$.

Cell 5.3

```
A = 1/√(2Pi a²);
ψ₀[x_] = A Exp[-x ^2/(2 a^2) + I k₀ x]; a = 1/Sqrt[2];
  k₀ = 12;
Plot[{Abs[ψ₀[x]], Re[ψ₀[x]]}, {x, -4, 4}, PlotRange→All,
  PlotLabel→TableForm[{{StyleForm["|ψ₀(x)|", FontColor→
   RGBColor[0, 0, 0]],
      ",", StyleForm["Re[ψ₀[x]]", FontColor→
       RGBColor[1, 0, 1]]}},
    TableSpacing→0], AxesLabel→{"x/√2 a", ""},
  PlotStyle→{RGBColor[0, 0, 0], RGBColor[1, 0, 1]}];
```

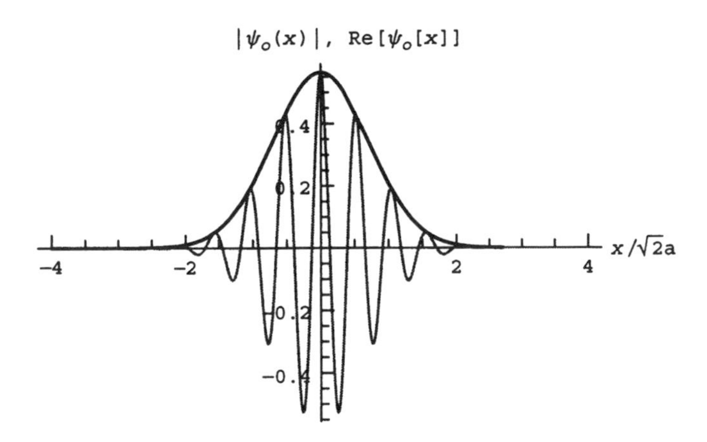

To find how this initial condition evolves, we can use Eq. (5.1.20). But we need the function $C(k)$. This function is determined using Eqs. (5.1.20) and (5.1.16):

$$\psi_0(x) = \int_{-\infty}^{\infty} \frac{dk}{2\pi} C(k)\, e^{ikx}. \tag{5.1.23}$$

Thus, $C(k)$ can be recognized as the Fourier transform of the initial condition:

$$C(k) = \int_{-\infty}^{\infty} dx\, e^{-ikx} \frac{1}{\sqrt{2\pi a^2}} e^{-x^2/2a^2 + ik_0 x}. \tag{5.1.24}$$

This integral can be evaluated analytically:

Cell 5.4

```
Clear["Global`*"];

Integrate[1/(Sqrt[2 π] a) Exp[-I k x + I k0 x - x^2/(2a^2)],

{x, -Infinity, Infinity}, Assumptions → a> 0]

e^(-1/2 a^2 (k-k0)^2)
```

Therefore,

$$C(k) = e^{-a^2(k-k_0)^2/2}. \tag{5.1.25}$$

The Fourier transform of the initial condition is another Gaussian, peaked around the central wavenumber of the packet, k_0. The width in k of this spectrum is of order $1/a$. Thus, as the width a of the inital condition increases, the Fourier transform (5.1.25) of the packet decreases in width, becoming more sharply peaked around k_0. This is expected from the uncertainty principle of Fourier analysis, discussed in Chapter 2.

Equation (5.1.20) implies that the wave function evolves according to

$$\psi(x,t) = \int_{-\infty}^{\infty} \frac{dk}{2\pi} e^{-\frac{1}{2}a^2(k-k_0)^2} e^{i[kx-(hk^2/2m)t]}.$$

This integral can also be evaluated analytically, although the precise form of the result is not of importance at this point:

Cell 5.5

```
ψ[x_, t_] = Integrate[

    e^(-1/2 a^2 (k-k0)^2) Exp[I (k x - ħ k^2 t/(2 m))],
    {k, -Infinity, Infinity}]/(2 Pi)

e^(-(m x^2 + i a^2 k0 (-2m x + k0 t ħ))/(2 a^2 m + 2 i t ħ)) / (Sqrt[2 π] Sqrt[a^2 + i t ħ/m])
```

The probability density $|\psi|^2$ is shown in Cell 5.6 for an electron with $k_0 = 100$ Å^{-1} and $a = 700$ Å, where 1 angstrom (Å) is 10^{-10} m.

Cell 5.6

```
m = 9.11 10^-31 ; ℏ = 1.05 10^-34;
Å = 10^-10; a = 700 Å ; k0 = 10^-2 Å^-1;

Table[
  Plot[Abs[ψ[x, t]]^2, {x, -2000 Å, 20000 Å},
    AxesLabel→{"x (meters)", ""},
   PlotLabel→"(|ψ|)^2, "<>ToString[t/10^-11]<>"x10^-11sec",
   PlotRange→{0, 1/(2 Pi a^2)}], {t, 0., 15 10^-11,
    1. 10^-11}];
```

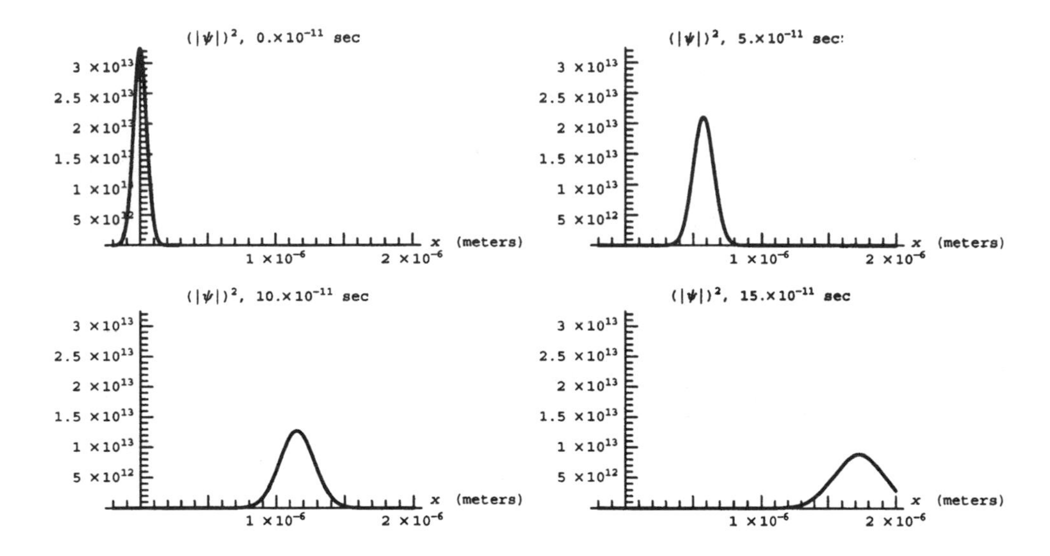

Our choice of initial condition leads to an electron that travels to the right with constant velocity. As the electron propagates, its probability density spreads: the electron becomes *delocalized.* First, we will consider what sets the speed of the wave function as it moves to the right, then we will discuss the spreading of the wave function.

Group Velocity The previous calculation shows that the electron traveled roughly 10^{-6} m in 10^{-10} s, implying a speed of about 10^4 m/s. What sets this speed? One might think that the speed is determined by the phase velocity v_ϕ of waves with the central wavenumber k_0 of the wave packet. Using Eq. (5.1.19) and (5.1.21), $v_\phi(k_0)$ is evaluated as

Cell 5.7

```
ℏk0/(2m)
5762.9
```

This phase velocity, measured in meters per second, is less than the observed electron speed by roughly a factor of two. In fact, we will now show that the electron wave packet travels at the *group velocity*.

In this wave packet, the spectrum of wavenumbers $C(k)$ is sharply peaked around k_0. We can then approximate the wavenumber integral in Eq. (5.1.20) by Taylor-expanding $\omega(k)$ about $k = k_0$:

$$\psi(x,t) = \int_{-\infty}^{\infty} \frac{dk}{2\pi} C(k) \exp\left\{ i \left[kx - \left(\omega(k_0) + (k - k_0) \frac{\partial \omega}{\partial k}\bigg|_{k=k_0} + \cdots \right) t \right] \right\}. \tag{5.1.26}$$

If we neglect the higher-order terms in the Taylor expansion, and keep only those terms shown, then we can write the wave packet as

$$\psi(x,t) = \exp\left[i \left(k_0 \frac{\partial \omega}{\partial k}\bigg|_{k=k_0} - \omega(k_0) \right) t \right] \int_{-\infty}^{\infty} \frac{dk}{2\pi} C(k) \exp\left[ik \left(x - \frac{\partial \omega}{\partial k}\bigg|_{k=k_0} t \right) \right]. \tag{5.1.27}$$

The Fourier integral in Eq. (5.1.27) can now be evaluated using Eq. (5.1.23):

$$\psi(x,t) = \exp\left[i \left(k_0 \frac{\partial \omega}{\partial k}\bigg|_{k=k_0} - \omega(k_0) \right) t \right] \psi_0\left(x - \frac{\partial \omega}{\partial k}\bigg|_{k=k_0} t \right).$$

Thus, aside from a phase factor that is unimportant in determining $|\psi|^2$, we see that the initial condition $\psi_0(x)$ simply moves with a velocity equal to $\partial\omega/\partial k|_{k=k_0}$. This is the group velocity of the wave packet:

$$v_g(k_0) = \frac{\partial \omega}{\partial k}\bigg|_{k=k_0}. \tag{5.1.28}$$

Evaluating this derivative using the dispersion relation for a free particle, Eq. (5.1.19), yields the following result for the group velocity:

$$v_g(k_0) = \hbar k_0/m = 2v_\phi(k_0). \tag{5.1.29}$$

For the parameters of our example, $v_g \simeq 11{,}500$ m/s, which is in rough agreement with the observed speed of the packet.

The fact that the phase velocity of waves at $k = k_0$ is unequal to the group velocity means that waves making up the packet move with respect to the packet itself. This cannot be seen when we plot $|\psi|^2$, but is very clear if we instead plot $\text{Re}\,\psi$, as shown in Cell 5.8. In the animation one can clearly observe the peaks and troughs falling behind in the packet, which is what we should expect given that the phase velocity is half the group velocity.

Cell 5.8

```
Table[Plot[Re[ψ[x, t]], {x, -2000 Å, 20000 Å},
   PlotRange→{-1, 1}/Sqrt[2 Pi a^2], AxesLabel→
    {"x (meters)", ""},
   PlotLabel→"Re[ψ], t=" <>ToString[t/10^-11] <>"x10^-11sec",
   PlotPoints→300], {t, 0., 1.5 10^-10, .2 10^-11}];
```

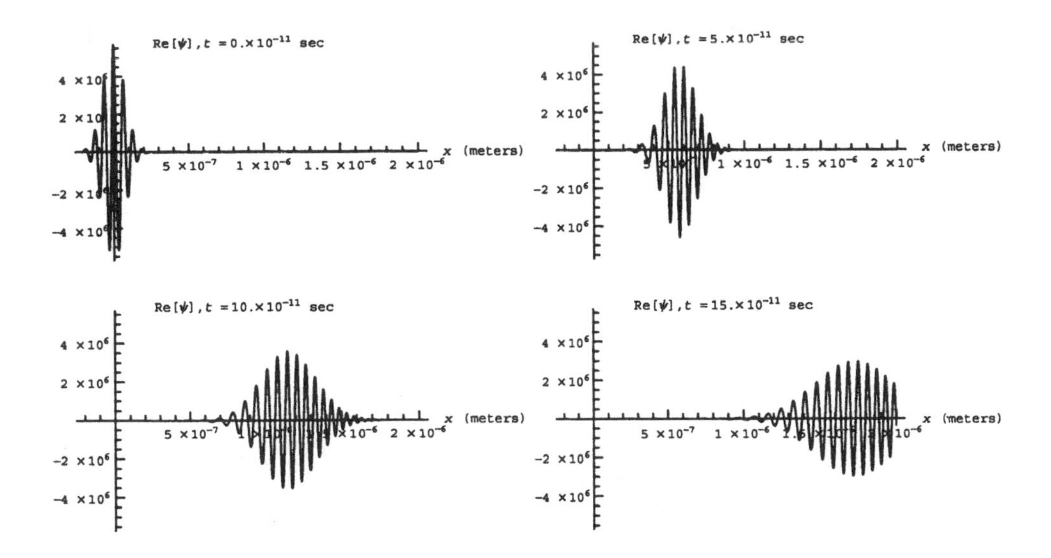

Dispersion We now consider the spreading of the wave packet. This spreading is a consequence of the dispersion in phase velocities of the separate wave making up this wave function. Each wave, with form $e^{i[kx-\omega(k)t]}$, travels with its own distinct phase velocity, and as a result, the initial condition *disperses* as slow waves fall behind and fast waves outpace the packet.

In order to understand this dispersion, it is necessary to go beyond the first-order Taylor expansion in Eq. (5.1.26), and keep second-order terms as well, writing $\omega(k) \simeq \omega(k_0) + (k-k_0)v_g(k_0) + \alpha(k-k_0)^2$, where the constant $\alpha = \frac{1}{2}(\partial/\partial k_0)v_g(k_0)$. [For the free-particle group velocity given by Eq. (5.1.29), this constant is independent of k_0: $\alpha = \hbar/2m$.] Then Eq. (5.1.20) becomes

$$\psi(x,t) \simeq \int_{-\infty}^{\infty} \frac{dk}{2\pi} C(k)\, e^{i\{kx-[\omega(k_0)+(k-k_0)v_g(k_0)+\alpha(k-k_0)^2]t\}}.$$

The wavenumber integration in this equation cannot be performed analytically in general, but can be for various special cases, including the Gaussian wave packet given by the spectrum of Eq. (5.1.24). The result for this case is

$$\psi(x,t) \simeq \frac{e^{i[k_0x-\omega(k_0)t]}\, e^{-[x-v_g(k_0)t]^2/(2a^2+4i\alpha t)}}{\sqrt{2\pi(a^2+2i\alpha t)}}. \tag{5.1.30}$$

This shows that the envelope of the wave packet remains a Gaussian during its evolution, but the width of the Gaussian increases with time. This is what we observed in the previous animations.

We define the width in the following (somewhat arbitrary) way. For a Gaussian of the form $e^{-x^2/2a^2}$, the width is defined as a (the value of x where the Gaussian falls to $1/2\,e = 0.606\ldots$ of its maximum). For a Gaussian of the form $e^{-x^2/2(A+iB)}$, the width is defined as the width of the envelope function. Writing the Gaussian as $e^{-x^2(A-iB)/2(A^2+B^2)}$, the envelope is the nonoscillatory part of the exponential, $e^{-x^2A/2(A^2+B^2)}$. Thus, the width w is $w = \sqrt{(A^2+B^2)/A}$. Applying this definition to Eq. (5.1.30), we find that the width of the wave packet increases with time according to

$$w(t) = \sqrt{a^2 + \frac{4\alpha^2 t^2}{a^2}} . \tag{5.1.31}$$

This result applies to Gaussian wave packet traveling through any dispersive medium, not just to solutions of Schrödinger's equation. We will see other examples in the exercises and in future sections. Note that for times $t \ll a^2/2\alpha$, no spreading occurs. For these early times we can neglect dispersion and use Eq. (5.1.27) to describe the wave packet.

The rate of increase in the width is controlled by the parameter α. For a nondispersive system, where $v_g = c =$ constant, we have $\alpha = 0$ and packets do not disperse. This is in line with our previous understanding of solutions to the wave equation on a uniform string.

5.1.3 Waves in Two and Three Dimensions

Schrödinger Equation in Three Dimensions When a free particle of mass m propagates in three dimensions, the wave function $\psi(\mathbf{r}, t)$ of the particle is described by the Schrödinger equation,

$$i\hbar \frac{\partial}{\partial t} \psi(\mathbf{r}, t) = -\frac{\hbar^2}{2m} \nabla^2 \psi(\mathbf{r}, t). \tag{5.1.32}$$

As in the previous one-dimensional case, this equation is first-order in time, and must therefore be supplemented with an initial condition,

$$\psi(\mathbf{r}, 0) = \psi_0(\mathbf{r}). \tag{5.1.33}$$

The evolution of $\psi(\mathbf{r}, t)$ from this initial condition can again be obtained using Fourier transform methods. We apply three separate transforms consecutively, in x, y, and z, defining

$$\tilde{\psi}(k_x, k_y, k_z, t) = \int_{-\infty}^{\infty} dx\, e^{-ik_x x} \int_{-\infty}^{\infty} dy\, e^{-ik_y y} \int_{-\infty}^{\infty} dz\, e^{-ik_z z} \psi(x, y, z, t). \tag{5.1.34}$$

This Fourier transform can be written more compactly as

$$\tilde{\psi}(\mathbf{k}, t) = \int d^3 r\, e^{-i\mathbf{k}\cdot\mathbf{r}} \psi(\mathbf{r}, t), \tag{5.1.35}$$

where $\mathbf{k} = (k_x, k_y, k_z)$ is called the *wave vector*. Then by Fourier transforming Eq. (5.1.32) in x, y, and z we find that

$$i\hbar \frac{\partial}{\partial t} \tilde{\psi}(\mathbf{k}, t) = \frac{\hbar^2 k^2}{2m} \tilde{\psi}(\mathbf{k}, t), \tag{5.1.36}$$

where $k = \sqrt{k_x^2 + k_y^2 + k_z^2}$ is the magnitude of the wave vector. This equation is identical to that for propagation in one dimension, and has the solution

$$\tilde{\psi}(\mathbf{k}, t) = C(\mathbf{k})\, e^{-i\,\omega(k)t}, \tag{5.1.37}$$

where the constant of integration $C(\mathbf{k})$ can depend on all three components of the wave vector, and where $\omega(k)$ depends only on the magnitude of $\mathbf{k}$, and is given by Eq. (5.1.19). Taking an inverse transform in all three dimensions then yields the solution for $\psi(\mathbf{r}, t)$:

$$\psi(\mathbf{r}, t) = \int \frac{d^3k}{(2\pi)^3} C(\mathbf{k})\, e^{i\mathbf{k}\cdot\mathbf{r} - i\,\omega(k)t}. \tag{5.1.38}$$

This is a wave packet consisting of a superposition of traveling waves, each of the form $e^{i\mathbf{k}\cdot\mathbf{r} - i\,\omega(k)t}$. A given phase ϕ of this wave satisfies

$$\phi = \mathbf{k}\cdot\mathbf{r} - \omega(k)t. \tag{5.1.39}$$

The meaning of Eq. (5.1.39) can be understood by defining coordinates $(\bar{x}, \bar{y}, \bar{z})$ with the $\bar{z}$-axis aligned along $\mathbf{k}$. (See Fig. 5.2.) Then Eq. (5.1.39) becomes

$$\phi = k\bar{z} - \omega(k)t, \tag{5.1.40}$$

or in other words, $\bar{z} = \phi/k + \omega(k)t/k$. But $\bar{x}$ and $\bar{y}$ are not determined, and can take on any values. This implies that Eq. (5.1.39) defines a plane of constant phase that moves with speed $\omega(k)/k$ in the direction of $\mathbf{k}$. Therefore, the phase velocity

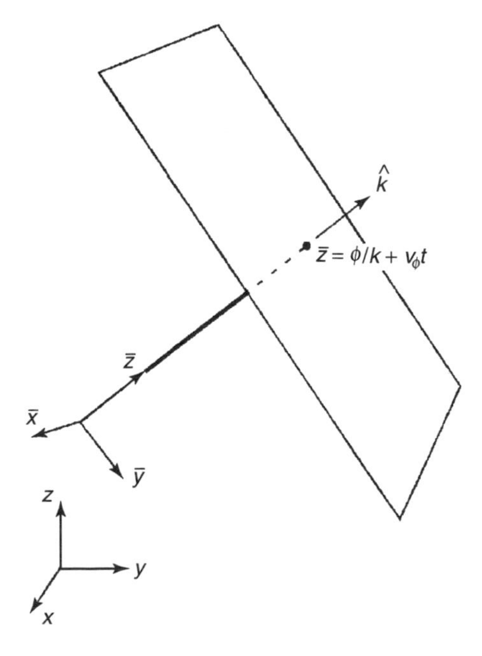

Fig. 5.2 Location of a plane of constant phase ϕ at time t.

of a wave with wave vector $\mathbf{k}$ is

$$\mathbf{v}_{\phi}(\mathbf{k}) = \frac{\omega(k)}{k}\hat{\mathbf{k}}, \tag{5.1.41}$$

where $\hat{\mathbf{k}}$ is a unit vector in the direction of $\mathbf{k}$. This generalizes our previous one-dimensional result, Eq. (5.1.21). On the other hand, a wave packet centered at wavenumber $\mathbf{k}$ will move at the group velocity,

$$\mathbf{v}_g(\mathbf{k}) = \left(\frac{\partial \omega}{\partial k_x}, \frac{\partial \omega}{\partial k_y}, \frac{\partial \omega}{\partial k_z} \right). \tag{5.1.42}$$

This follows from an argument that is analogous to that given previously for the one-dimensional case. One can write Eq. (5.1.42) more compactly as $\mathbf{v}_g(\mathbf{k}) = \partial\omega/\partial\mathbf{k}$, where $\partial/\partial\mathbf{k} \equiv (\partial/\partial k_x, \partial/\partial k_y, \partial/\partial k_z)$.

Group Velocity in an Isotropic Medium If a system is isotropic, so that ω depends only on the magnitude of $\mathbf{k}$ and not its direction, then the group velocity simplifies somewhat. The definition of the group velocity, Equation (5.1.42), implies that

$$\mathbf{v}_g(\mathbf{k}) = \frac{\partial}{\partial \mathbf{k}}\omega(k) = \frac{\partial k}{\partial \mathbf{k}}\frac{\partial \omega}{\partial k},$$

where the second equality follows from the chain rule. However,

$$\frac{\partial k}{\partial \mathbf{k}} = \left(\frac{\partial}{\partial k_x}, \frac{\partial}{\partial k_y}, \frac{\partial}{\partial k_z} \right)\sqrt{k_x^2 + k_y^2 + k_z^2} = \left(\frac{k_x}{k}, \frac{k_y}{k}, \frac{k_z}{k} \right) = \hat{\mathbf{k}},$$

where $\hat{\mathbf{k}}$ is a unit vector in the direction of $\mathbf{k}$. Thus, the group velocity is

$$\mathbf{v}_g(\mathbf{k}) = \hat{\mathbf{k}}\frac{\partial \omega}{\partial k}. \tag{5.1.43}$$

In an isotropic system, the group and phase velocities are both along $\mathbf{k}$, although they are generally of different magnitudes.

The Wave Equation in Three Dimensions

The Initial-Value Problem. Let's now consider the wave equation for propagation in two or three dimensions,

$$\frac{\partial^2}{\partial t^2}p(\mathbf{r},t) = c^2\nabla^2 p(\mathbf{r},t). \tag{5.1.44}$$

Here we are thinking of p as the pressure in a compressional sound wave, and c as the speed of sound. However, this same equation applies to the propagation of light waves, and in two dimensions it applies to waves on a drumhead, or on the surface of water (with a suitable reinterpretation of the meaning of the symbols p and c).

Equation (5.1.44) must be supplemented by two initial conditions,

$$\begin{aligned} p(\mathbf{r},0) &= p_0(\mathbf{r}), \\ \frac{\partial p}{\partial t}(\mathbf{r},0) &= v_0(\mathbf{r}). \end{aligned} \tag{5.1.45}$$

The solution for p follows from Fourier transformation with respect to $\mathbf{r}$. The Fourier transform of p, $\tilde{p}(\mathbf{k},t)$, satisfies the second-order ODE

$$\frac{\partial^2}{\partial t^2}\tilde{p}(\mathbf{k},t) = -c^2k^2\tilde{p}(\mathbf{k},t). \tag{5.1.46}$$

Therefore,

$$p(\mathbf{r},t) = \int \frac{d^3k}{(2\pi)^3} C(\mathbf{k})\, e^{i[\mathbf{k}\cdot\mathbf{r}-\omega(k)t]} + \int \frac{d^3k}{(2\pi)^3} D(\mathbf{k})\, e^{i[\mathbf{k}\cdot\mathbf{r}+\omega(k)t]}, \tag{5.1.47}$$

where $C(\mathbf{k})$ and $D(\mathbf{k})$ are undetermined coefficients, and $\omega(k) = ck$ is the dispersion relation for the waves, k being the magnitude of the wave vector $\mathbf{k}$. The functions $C(\mathbf{k})$ and $D(\mathbf{k})$ are found by matching to the two initial conditions (5.1.45). Evaluation of Eq. (5.1.47) and its time derivative at time $t = 0$ yields, after Fourier transformation, two equations for $C(\mathbf{k})$ and $D(\mathbf{k})$. Following the same approach as in the 1D case studied in Sec. 5.1.1, we obtain

$$\begin{aligned} C(\mathbf{k}) &= \frac{1}{2}\left(\tilde{p}_0(\mathbf{k}) + i\frac{\tilde{v}_0(\mathbf{k})}{\omega(k)}\right), \\ D(\mathbf{k}) &= \frac{1}{2}\left(\tilde{p}_0(\mathbf{k}) - i\frac{\tilde{v}_0(\mathbf{k})}{\omega(k)}\right), \end{aligned} \tag{5.1.48}$$

where $\tilde{p}_0$ and $\tilde{v}_0$ are the Fourier transforms of the initial conditions. Equations (5.1.47) and (5.1.48) are the solution to the problem.

However, the form of Eq. (5.1.47) can be inconvenient in some applications, for the following reason: for given wavenumber k, there are *two* possible frequencies appearing in Eq. (5.1.47), $\pm\omega(k)$. These two choices merely correspond to wave propagation with and against the direction of $\mathbf{k}$. However, this implies that there are now *two* group velocities and *two* phase velocities for every value of $\mathbf{k}$, each pair with equal magnitude but opposite direction. This occurs because the wave equation is second-order in time, so the dispersion relation is a polynomial in ω of order 2, with two solutions. This was not the case for the solution to Schrödinger's equation (5.1.38), because that equation is first-order in time.

In fact, for more general wave equations (such as vector wave equations, or scalar wave equations of order 3 or higher in time), there can be multiple roots to the dispersion relation, $\omega_i(\mathbf{k})$, $i = 1, 2, \dots$. Each of the roots gets its own Fourier integral in Eq. (5.1.47). Such equations describe wave propagation through media such as plasmas or solids. The multiple dispersion relations correspond to physically different modes of propagation, such as shear and compressional waves in the case of a solid.

However, Eq. (5.1.44) describes only one type of wave, and it should be possible to simplify the solution to reflect this fact. Indeed, we will now show that Eq. (5.1.47) can be simplified, because our system is isotropic. Recall that an isotropic system is one that looks the same in all directions, and in particular, the dispersion relation depends only on the magnitude of **k**, $\omega = \omega(k)$.

Because of isotropy, we can write Eq. (5.1.47) in a form where there is only one frequency for each value of **k**. First, we transform the integral involving $D(\mathbf{k})$ in Eq. (5.1.47), taking $\mathbf{k} \to -\mathbf{k}$:

$$p(\mathbf{r},t) = \int \frac{d^3k}{(2\pi)^3} C(\mathbf{k})\, e^{i[\mathbf{k}\cdot\mathbf{r} - \omega(k)t]} + \int \frac{d^3k}{(2\pi)^3} D(-\mathbf{k})\, e^{i[-\mathbf{k}\cdot\mathbf{r} + \omega(k)t]}. \quad (5.1.49)$$

Note that the frequency $\omega(k)$ remains the same in this transformation, since ω depends only on the magnitude of k. Next, we observe that Eq. (5.1.48) implies that $D(-\mathbf{k}) = \frac{1}{2}[\tilde{p}_0(-\mathbf{k}) - i\tilde{v}_0(-\mathbf{k})/\omega(k)]$. However, the functions $p_0(\mathbf{r})$ and $v_0(\mathbf{r})$ are real, so according to Eq. (2.1.7), their Fourier transforms have the property that $\tilde{p}_0(-\mathbf{k}) = \tilde{p}_0(\mathbf{k})^*$ and $\tilde{v}_0(-\mathbf{k}) = \tilde{v}_0(\mathbf{k})^*$. Therefore, we find that $D(-\mathbf{k}) = C(\mathbf{k})^*$, from which it follows that

$$p(\mathbf{r},t) = \int \frac{d^3k}{(2\pi)^3} \left[C(\mathbf{k})\, e^{i[\mathbf{k}\cdot\mathbf{r} - \omega(k)t]} + \left(C(k)\, e^{i[\mathbf{k}\cdot\mathbf{r} - \omega(k)t]} \right)^* \right].$$

Since the second term is the complex conjugate of the first, we arrive at

$$p(\mathbf{r},t) = 2\,\mathrm{Re} \int \frac{d^3k}{(2\pi)^3} C(\mathbf{k})\, e^{i[\mathbf{k}\cdot\mathbf{r} - \omega(k)t]}. \quad (5.1.50)$$

For a given value of **k**, the argument of the integral in Eq. (5.1.50) describes a wave that propagates only in one direction, along **k**. Furthermore, these waves only have positive frequencies. This is as opposed to the solution of Eq. (5.1.47), where for given **k** waves propagate both with and against **k**, with both positive and negative frequencies. Thanks to isotropy, we do not lose anything in Eq. (5.1.50): a wave with negative frequency and wave vector **k**, propagating in the $-\mathbf{k}$ direction, can equally well be described by a wave with positive frequency but with the wave vector $-\mathbf{k}$.

How can we match to the two initial conditions of Eq. (5.1.45) when there is only a single function $C(\mathbf{k})$ in Eq. (5.1.50), rather than the two functions in Eq. (5.1.47)? The answer is that $C(\mathbf{k})$ has both a real and an imaginary part, and this gives us the two functions we need: see Eq. (5.1.48). The fact that we take the real part in Eq. (5.1.50) allows us latitude in choosing these functions that is absent in Eq. (5.1.47), where the integrals must by themselves provide a real result. In fact, one can regard the function $D(\mathbf{k})$ as merely a way to obtain a real result in Eq. (5.1.47), which is obviated in Eq. (5.1.50) by our directly taking the real part.

Also, it is important to realize that no part of the argument leading to Eq. (5.1.50) relied on the specific form of the dispersion relation, $\omega = ck$; rather, only the isotropy of the dispersion was required. In fact, Eq. (5.1.50) also applies to more general dispersive isotropic systems where $\omega = \omega(k)$ for some general function $\omega(k)$. Examples of such systems include electromagnetic waves propagating in isotropic dispersive media with an index of refraction that depends on frequency, $n = n(\omega)$; phonons propagating though amorphous materials such as rubber or glass; waves in unmagnetized plasmas; water waves in deep water; and sound waves in air. These examples, and others, will be taken up in the exercises.

Wave Packets. One advantage of Eq. (5.1.50) over Eq. (5.1.47), aside from its compact form, is the ease with which it can be applied to describe the evolution of wave packets. If we consider an initial condition for which $C(\mathbf{k})$ is sharply peaked around a wave vector $\mathbf{k}_0$, then we can Taylor-expand $\omega(\mathbf{k})$ about $\mathbf{k}_0$, obtaining $\omega(\mathbf{k}) \simeq \omega(\mathbf{k}_0) + \mathbf{v}_g(\mathbf{k}_0)\cdot(\mathbf{k}-\mathbf{k}_0)$, where the group velocity $\mathbf{v}_g$ is given by Eq. (5.1.42). Using this approximation in Eq. (5.1.49) allows us to write

$$p(\mathbf{r},t) = \operatorname{Re}\left(e^{i[\mathbf{k}_0\cdot\mathbf{r}-\omega(k_0)t]} \int \frac{d^3k}{(2\pi)^3} 2C(\mathbf{k})\, e^{i(k-\mathbf{k}_0)\cdot[\mathbf{r}-\mathbf{v}_g(\mathbf{k}_0)t]} \right).$$

The integral over $\mathbf{k}$ results in a function only of $\mathbf{r} - \mathbf{v}_g(\mathbf{k}_0)t$, which we write as $A(\mathbf{r} - \mathbf{v}_g(\mathbf{k}_0)t)$. Furthermore, since $C(\mathbf{k})$ is peaked around $\mathbf{k} = \mathbf{k}_0$, the Fourier components $\mathbf{k} - \mathbf{k}_0$ that contribute to A are small, so $A(\mathbf{r})$ is slowly varying in space. The function $A(\mathbf{r})$ can be recognized as the slowly varying amplitude function that is superimposed on the wave train $e^{i[\mathbf{k}_0\cdot\mathbf{r}-\omega(k_0)t]}$ in order to create a finite wave packet:

$$p(\mathbf{r},t) = \operatorname{Re}\left[A(\mathbf{r} - \mathbf{v}_g(\mathbf{k}_0)t)\, e^{i[\mathbf{k}_0\cdot\mathbf{r}-\omega(k_0)t]} \right]. \tag{5.1.51}$$

The amplitude function travels with the group velocity of the wave packet, while planes of constant phase within the packet move with the average phase velocity of the packet. The packet travels without changing shape, because in deriving Eq. (5.1.51) we neglected dispersion.

Dispersion and Diffraction. We can take account of dispersion in much the same manner as was used to analyze the 1D wave equation. Keeping the next-order term in the Taylor expansion of $\omega(k)$ about the central wave vector $\mathbf{k}_0$ of the wave packet, we obtain

$$\omega(k) = \omega(\mathbf{k}_0) + \mathbf{v}_g(\mathbf{k}_0)\cdot(\mathbf{k}-\mathbf{k}_0) + \tfrac{1}{2}(\mathbf{k}-\mathbf{k}_0)\cdot\boldsymbol{\gamma}\cdot(\mathbf{k}-\mathbf{k}_0), \tag{5.1.52}$$

where $\boldsymbol{\gamma}$ is a symmetric tensor, given in component form by $\gamma_{ij} = \partial^2\omega/\partial k_i\,\partial k_j|_{\mathbf{k}=\mathbf{k}_0}$. Here the indices i and j each run over the x, y, and z components of the vector $\mathbf{k}$. The tensor $\boldsymbol{\gamma}$ describes the spreading of the wave packet. For an isotropic system where $\omega = \omega(k)$, we can write this tensor as

$$\gamma_{ij} = \frac{\partial}{\partial k_j} v_{g_i}(k)\bigg|_{\mathbf{k}=\mathbf{k}_0} = \frac{\partial}{\partial k_j}\left(v_g(k)\frac{k_i}{k} \right)\bigg|_{\mathbf{k}=\mathbf{k}_0},$$

where we have used Eq. (5.1.43). Then applying the identities $\partial k_i/\partial k_j = \delta_{ij}$, $\partial k/\partial k_j = k_j/k$, and using the chain rule, we obtain

$$\gamma_{ij} = \frac{v_g(k)}{k}\left(\delta_{ij} - \frac{k_i k_j}{k^2}\right) + \frac{\partial v_g}{\partial k}\frac{k_i k_j}{k^2}\bigg|_{\mathbf{k}=\mathbf{k}_0}, \tag{5.1.53}$$

where δ_{ij} is a Kronecker delta function (the component form of the unit tensor).

The tensor $\boldsymbol{\gamma}$ has a simple diagonal form when written in components with the z-axis chosen along the direction of $\mathbf{k}_0$:

$$\boldsymbol{\gamma} = \begin{pmatrix} \dfrac{v_g(k_0)}{k_0} & 0 & 0 \\ 0 & \dfrac{v_g(k_0)}{k_0} & 0 \\ 0 & 0 & \dfrac{\partial v_g}{\partial k_0} \end{pmatrix}. \tag{5.1.54}$$

The zz component of the tensor, $\partial v_g/\partial k_0$, causes dispersion along the direction of propagation (*longitudinal* dispersion). The xx and yy components, $v_g(k_0)/k_0$, cause spreading of the wave packet *transverse* to the direction of propagation (*transverse* dispersion). The fact that these latter two components are identical implies that the wave packet spreads isotropically in the transverse directions, as expected for a homogeneous isotropic medium. However, the transverse dispersion is not the same as the longitudinal dispersion. Transverse dispersion is also known as *diffraction*.

We can work out the dispersion quantitatively for a Gaussian wave packet with central wave vector $k_0\hat{\mathbf{z}}$. When Eqs. (5.1.52) and (5.1.54) are employed in Eq. (5.1.50), the result is

$$p(\mathbf{r},t) = 2\,\mathrm{Re}\left\{ e^{i[\mathbf{k}_0\cdot\mathbf{v}_g(\mathbf{k}_0) - \omega(k_0)]t} \int \frac{d^3k}{(2\pi)^3} C(\mathbf{k}) \right.$$
$$\left. \times \exp\left[i\mathbf{k}\cdot[\mathbf{r} - \mathbf{v}_g(\mathbf{k}_0)t] - i\left(\frac{v_g(k_0)}{2k_0}(k_x^2 + k_y^2) + \frac{1}{2}\frac{\partial v_g}{\partial k_0}(k_z - k_0)^2 \right)t \right] \right\}.$$

Taking a packet with a spectrum of the form

$$C(\mathbf{k}) = A e^{-(k_x^2 + k_y^2)b^2/2} e^{-(k_z - k_0)^2 a^2/2}$$

allows us to perform the required integrations over k_x, k_y and k_z analytically. The result is given below, defining $\alpha = \frac{1}{2}\partial v_g/\partial k_0$ and $\beta = v_g(k_0)/2k_0$:

$$p(\mathbf{r},t) = 2\,\mathrm{Re}\left[\frac{4\pi^{3/2}A}{(b^2 + 2i\beta t)\sqrt{2a^2 + 4i\alpha t}} e^{i(k_0 z - \omega(k_0)t)} \right.$$
$$\left. \times \exp\left(-\frac{x^2 + y^2}{2b^2 + 4i\beta t} - \frac{(z - v_g(k_0)t)^2}{2a^2 + 4i\alpha t} \right) \right]. \tag{5.1.55}$$

This packet has an initial transverse width of b and an initial length of a, but spreads over time. The rate of longitudinal spreading is the same as for a 1D packet, Eq. (5.1.31), and is controlled by the same parameter $\alpha = \frac{1}{2}\,\partial v_g/\partial k_0$ as for the 1D case.

However, the transverse spreading (diffraction) is controlled by the parameter $\beta = v_g(k_0)/2k_0$. This parameter is nonzero even if the medium is dispersionless. This implies that diffraction occurs even in vacuum.

The transverse width $w(t)$ of the Gaussian wave packet, defined in the same way as in Eq. (5.1.31), is $w(t) = (b^2 + 4\beta^2 t^2/b^2)^{1/2}$. Thus, the width of the packet increases by a factor $\sqrt{2}$ in a time $t_2 = b^2/2\beta = b^2 k_0/v_g(k_0)$. In this time, the packet has moved a distance

$$d = v_g(k_0)t_2 = b^2 k_0. \tag{5.1.56}$$

This distance is called the *Rayleigh length*. It is the distance over which a Gaussian wave packet increases in transverse width from b to $\sqrt{2}\,b$.

This diffraction is simply a consequence of the uncertainty principle of Fourier analysis. According to the uncertainty principle, any packet with finite transverse extent must contain a finite range of transverse wavenumbers. This necessarily implies a range of directions for the phase velocities of the different waves making up the packet. These waves then propagate off in separate directions, causing the packet to spread.

Even tightly focused laser beams must eventually diffract. In fact, Eq. (5.1.56) shows that the narrower the beam, the faster it spreads. However, the higher the wavenumber k_0 of the beam, the further it can propagate before diffracting appreciably.

For example, for a Gaussian light beam with a wavelength in the visible, $2\pi/k_0 = 5000$ Å, and an initial radius of $b = 1$ cm $= 0.01$ m, Eq. (5.1.56) implies that the beam will propagate $0.01^2 \times 2\pi/(5000 \times 10^{-10})$ m $= 1.2$ km before spreading by a factor of $\sqrt{2}$. However, for a 1-cm-radius ultraviolet beam, with a wavelength of 40 Å, the beam propagates 160 km before spreading by the same factor.

Wave-Packet Energy and Momentum. In this section we derive general expressions for the energy and momentum of a wave packet for the wave equation, and for a more general class of wave equations that we term *classical* wave equations. Quantum wave equations, such as Schrödinger's equation, are considered in the exercises. We begin with the energy conservation relation for the simple nondispersive wave equation, $\partial^2 z/\partial t^2 = c^2\nabla^2 z$. Multiplying both sides of this equation by $\rho\,\partial z/\partial t$, and differentiating by parts using, for example, $(\partial z/\partial t)(\partial^2 z/\partial t^2) = \frac{1}{2}(\partial/\partial t)(\partial z/\partial t)^2$, we obtain

$$\frac{\partial}{\partial t}\epsilon + \nabla\cdot\boldsymbol{\Gamma} = 0, \tag{5.1.57}$$

where the energy density ϵ is

$$\epsilon = \frac{\rho}{2}\left(\frac{\partial z}{\partial t}\right)^2 + \frac{\rho c^2}{2}\nabla z\cdot\nabla z, \tag{5.1.58}$$

and the energy flux $\boldsymbol{\Gamma}$ is

$$\boldsymbol{\Gamma} = -\rho c^2 \frac{\partial z}{\partial t} \nabla z. \tag{5.1.59}$$

Recall that Eq. (5.1.57) also described energy conservation for the heat equation, although the form of the energy density and flux differed: see Eq. (3.1.38). By integrating Eq. (5.1.57) over all space, applying the divergence theorem, and assuming that the energy flux vanishes at infinity, we can regain the integral form of energy conservation,

$$\frac{d}{dt} \int \epsilon(\mathbf{r}, t)\, d^3 r = 0,$$

as previously discussed for 1D or 2D wave propagation [see Eqs. (4.2.29) and (4.4.33)]. [Note: The interpretation of ϵ and $\boldsymbol{\Gamma}$ as energy density and energy flux is easiest to understand when z is the wave displacement (with units of distance). Then ρ is mass density, and ϵ is the sum of kinetic and potential energies. If z is some other quantity such as wave pressure or potential, ρ must be reinterpreted.]

Now consider the energy density and energy flux for a wave packet. According to Eq. (5.1.51), such packets can be described as traveling waves with a slowly varying amplitude A and phase ϕ:

$$z(\mathbf{r}, t) = A(\mathbf{r} - \mathbf{v}_g t) \cos\left[\mathbf{k}_0 \cdot \mathbf{r} - \omega_0 t + \phi(\mathbf{r} - \mathbf{v}_g t)\right], \tag{5.1.60}$$

where for our simple wave equation $\omega_0 = c k_0$ and $\mathbf{v}_g = c\hat{\mathbf{k}}_0$. We will now substitute this expression into Eq. (5.1.58), and note that derivatives of the amplitude and phase are small compared to derivatives of the cosine function, since by assumption the packet has slowly varying amplitude and phase. Then $\partial z/\partial t \approx \omega_0 A \sin(\mathbf{k}_0 \cdot \mathbf{r} - \omega_0 t + \phi)$ and $\nabla z \simeq -\mathbf{k}_0 A \sin(\mathbf{k}_0 \cdot \mathbf{r} - \omega_0 t + \phi)$, and the energy density and flux become

$$\begin{aligned} \epsilon &= \frac{\rho}{2} \left(\omega_0^2 + c^2 k_0^2 \right) A^2 \sin^2(\mathbf{k}_0 \cdot \mathbf{r} - \omega_0 t + \phi), \\ \boldsymbol{\Gamma} &= \rho c^2 \omega_0 \mathbf{k}_0 A^2 \sin^2(\mathbf{k}_0 \cdot \mathbf{r} - \omega_0 t + \phi). \end{aligned} \tag{5.1.61}$$

The energy density varies in time because energy is traded back and forth between different parts of the packet as the wave oscillates, and also because the packet moves at the group velocity. However, if we take a time average over the oscillations, using $\langle \sin^2 x \rangle = \frac{1}{2}$, we arrive at an expression for the average energy and flux associated with the packet:

$$\begin{aligned} \bar{\epsilon}(\mathbf{r}, t) &= \frac{\rho}{2} \omega_0^2 A\left(\mathbf{r} - c\hat{\mathbf{k}}_0 t\right)^2, \\ \bar{\boldsymbol{\Gamma}}(\mathbf{r}, t) &= \bar{\epsilon}(\mathbf{r}, t) c\hat{\mathbf{k}}_0, \end{aligned} \tag{5.1.62}$$

where we have used the dispersion relation $\omega_0 = c k_0$. Both energy density and flux are proportional to amplitude squared, and propagate at the wave velocity $c\hat{\mathbf{k}}_0$.

Also, we see that the energy flux consists of the energy density transported at the wave velocity $c\hat{\mathbf{k}}_0$. This is to be expected, since energy is neither created nor destroyed in this wave system, but is simply moved from place to place by the waves. The Poynting flux $\mathbf{S}$ for electromagnetic waves traveling in a dispersionless medium has the same form, $\mathbf{S} = \epsilon c\hat{\mathbf{k}}$, where ϵ is the wave energy density.

However, for more general wave equations we might expect the average energy flux to involve the group velocity $\mathbf{v}_g$. We will now show that this is in fact the case. We will also find a simple expression for the energy density in terms of the dispersion relation for the waves.

Consider a wave equation of the form

$$\sum_{n=0,2,4,6,\ldots} a_n \frac{\partial^n}{\partial t^n} z = \sum_{n=2,4,6,\ldots} \sum_{j=1,2,3} b_{nj} \frac{\partial^n}{\partial r_j^n} z, \tag{5.1.63}$$

where r_j, $j = 1,2,3$, are the x, y, and z components of $\mathbf{r}$, and a_n and b_{nj} are constant coefficients. Even more general wave equations could be constructed, involving cross-derivatives between different components of r_j and t, but Eq. (5.1.63) is sufficient for our purposes. We refer to wave equations of the form of Eq. (5.1.63) as *classical wave equations*. Note that wave equations such as Schrödinger's equation, with odd powers of the time derivative, are not included in this form. Conservation laws for Schrödinger's equation and its cousins will be considered in the exercises.

The dispersion relation associated with this wave equation takes the form of a polynomial, found by replacing $\partial/\partial t$ by $-i\omega$ and $\partial/\partial r_j$ by ik_j. We write this dispersion relation in the compact form $D(\mathbf{k}, \omega) = 0$, where the dispersion *function* D is given by

$$D(\mathbf{k}, \omega) = \sum_n (-1)^{n/2-1} a_n \omega^n + \sum_{nj} (-1)^{n/2} b_{nj} k_j^n, \tag{5.1.64}$$

and we have used the fact that n is even. For example, for electromagnetic waves traveling through a medium with refractive index $n(\omega)$, the dispersion function is $D = \omega^2 n^2(\omega) - c^2 k^2 = 0$, and we can think of Eq. (5.1.64) as a polynomial expansion of this equation.

We now construct an energy conservation law for Eq. (5.1.63) by following the same approach as for the wave equation. We multiply both sides of the equation by $\rho\, \partial z/\partial t$, and differentiate by parts repeatedly. For example,

$$\begin{aligned} \frac{\partial z}{\partial t}\frac{\partial^n z}{\partial t^2} &= \frac{\partial}{\partial t}\left(\frac{\partial z}{\partial t}\frac{\partial^{n-1} z}{\partial t^{n-1}}\right) - \frac{\partial^2 z}{\partial t^2}\frac{\partial^{n-1} z}{\partial t^{n-1}} \\ &= \frac{\partial}{\partial t}\left(\frac{\partial z}{\partial t}\frac{\partial^{n-1} z}{\partial t^{n-1}} - \frac{\partial^2 z}{\partial t^2}\frac{\partial^{n-2} z}{\partial t^{n-2}}\right) + \frac{\partial^3 z}{\partial t^3}\frac{\partial^{n-2} z}{\partial t^{n-2}} = \cdots \\ &= \frac{\partial}{\partial t}\left(\sum_{m=1}^{n-1} (-1)^{m-1} \frac{\partial^m z}{\partial t^m}\frac{\partial^{n-m} z}{\partial t^{n-m}}\right) + (-1)^{n-1}\frac{\partial z}{\partial t}\frac{\partial^n z}{\partial t^n}. \end{aligned}$$

However, if n is even, $(-1)^{n-1}(\partial z/\partial t)\,\partial^n z/\partial t^n = -(\partial z/\partial t)\,\partial^n z/\partial t^n$. This allows us to bring this term over to the left-hand side of the equation, and so obtain

$$\frac{\partial z}{\partial t}\frac{\partial^n z}{\partial t^n} = \frac{1}{2}\frac{\partial}{\partial t}\left(\sum_{m=1}^{n-1}(-1)^{m-1}\frac{\partial^m z}{\partial t^m}\frac{\partial^{n-m} z}{\partial t^{n-m}}\right), \qquad n = 2,4,6,\dots. \tag{5.1.65}$$

For the case $n = 0$, we can simply write

$$\frac{\partial z}{\partial t} z = \frac{1}{2}\frac{\partial z^2}{\partial t}. \tag{5.1.66}$$

Similarly, we require terms of the form $(\partial z/\partial t)\,\partial^n z/\partial r_j^n$ on the right-hand side of the energy equation. These terms can also be rewritten via differentiation by parts:

$$\frac{\partial z}{\partial t}\frac{\partial^n z}{\partial r_j^n} = \frac{\partial}{\partial r_j}\left[\sum_{m=1}^{n/2}(-1)^{m-1}\frac{\partial^m}{\partial r_j^m}\left(\frac{\partial z}{\partial t}\right)\frac{\partial^{n-m-1} z}{\partial r_j^{n-m-1}}\right] - \frac{1}{2}\frac{\partial}{\partial t}\left(\frac{\partial^{n/2} z}{\partial r_j^{n/2}}\right)^2, \tag{5.1.67}$$

where again we have used the fact that n is even. Combining Eqs. (5.1.65)–(5.1.67), we obtain the general energy conservation relation in the form of Eq. (5.1.57), with

$$\epsilon = \frac{\rho}{2}a_0 z^2 + \frac{\rho}{2}\sum_{n=2,4,\dots}\left[a_n\sum_{m=1}^{n-1}(-1)^{m-1}\frac{\partial^m z}{\partial t^m}\frac{\partial^{n-m} z}{\partial t^{n-m}} + \sum_j b_{nj}\left(\frac{\partial^{n/2} z}{\partial r_j^{n/2}}\right)^2\right] \tag{5.1.68}$$

and

$$\Gamma_j = -\rho\sum_{n=2,4,\dots} b_{nj}\sum_{m=1}^{n/2}(-1)^{m-1}\frac{\partial^m}{\partial r_j^m}\left(\frac{\partial z}{\partial t}\right)\frac{\partial^{n-m-1} z}{\partial r_j^{n-m-1}}. \tag{5.1.69}$$

In their present form, these expressions for energy density and flux are of such complexity as to be of limited usefulness. However, when applied to wave packets, the expressions simplify considerably. On substituting Eq. (5.1.60) into Eqs. (5.1.68) and (5.1.69), again dropping derivatives of the slowing varying amplitude and phase, and averaging over a cycle, each term in the sum over m in Eq. (5.1.68) yields the same result, $(-1)^{n/2-1}\omega_0^n A^2/2$; the same is true for each term in the sum over m in Eq. (5.1.69), which becomes $(-1)^{n/2}\omega_0(k_{0j})^n A^2/2$. We then obtain the following expressions for the averaged energy density and flux:

$$\bar{\epsilon} = \frac{\rho}{4}A^2\left[a_0 + \sum_{n=2,4,\dots}(-1)^{n/2-1}\left(a_n\,\omega_0^n(n-1) + \sum_j b_{nj}(k_{0j})^n\right)\right],$$

$$\bar{\Gamma}_j = \frac{\rho}{4}\sum_{n=2,4,\dots}(-1)^{n/2-1}b_{nj}A^2\omega_0 n(k_{0j})^{n-1}.$$

These sums collapse into amazingly simple and general expressions if they are written in terms of the dispersion function $D(\mathbf{k}, \omega)$, defined in Eq. (5.1.64). By applying differentiation formulae such as $\partial\omega_0^n/\partial\omega_0 = n\,\omega_0^{n-1}$, we obtain

$$\bar{\epsilon} = \frac{\rho}{4} A^2 \omega_0^2 \frac{\partial}{\partial\omega_0} \frac{D(\mathbf{k}_0, \omega_0)}{\omega_0} \tag{5.1.70}$$

and

$$\bar{\Gamma}_j = -\frac{\rho}{4} A^2 \omega_0 \frac{\partial}{\partial k_{0j}} D(\mathbf{k}_0, \omega_0). \tag{5.1.71}$$

Two more simplification can be made. First, since $D(\mathbf{k}_0, \omega_0) = 0$, Eq. (5.1.70) can also be written as

$$\bar{\epsilon} = \frac{\rho}{4} A^2 \omega_0 \frac{\partial}{\partial\omega_0} D(\mathbf{k}_0, \omega_0). \tag{5.1.72}$$

Second, we note that the group velocity $\mathbf{v}_g$ can be written as

$$\mathbf{v}_g = \frac{\partial\omega_0}{\partial\mathbf{k}_0} = -\frac{\partial D/\partial\mathbf{k}_0}{\partial D/\partial\omega_0}.$$

This allows us to write the energy flux in terms of the energy density as

$$\bar{\mathbf{\Gamma}} = \mathbf{v}_g \bar{\epsilon}. \tag{5.1.73}$$

Thus, the energy density propagates in the direction of the group velocity, as we expected.

Equations (5.1.72) and (5.1.73) provide simple general expressions for the energy density and flux associated with a wave-packet. Similar expressions can also be obtained for the momentum density and momentum flux. Here, the conservation law is obtained by multiplying the wave equation by $\rho\nabla z$ rather than by $\rho\,\partial z/\partial t$, and then applying the same differentiation-by-parts regimen. The resulting conservation law is written as follows:

$$\frac{\partial}{\partial t}\mathbf{p} + \nabla\cdot\mathbf{T} = \mathbf{0}, \tag{5.1.74}$$

where $\mathbf{p}$ is the momentum density, and $\mathbf{T}$ is a tensor (the stress tensor), which is the flux of momentum. (Since momentum is a vector, the flux of momentum is a tensor. The components of $\mathbf{T}$, T_{ij}, provide the flux of p_i in direction j.) The units of $\mathbf{T}$ are pressure, and in fact $\mathbf{T}$ is the pressure created by the waves. For a wave packet, the expressions for $\mathbf{p}$ and $\mathbf{T}$, averaged over a cycle, are

$$\bar{\mathbf{p}} = \mathbf{k}_0 \bar{\epsilon}/\omega_0 \tag{5.1.75}$$

and

$$T_{ij} = \bar{p}_i v_{gj}. \tag{5.1.76}$$

The proof of Eqs. (5.1.75) and (5.1.76) is left as an exercise.

Thus, just as with energy density, the flux of momentum is given by the momentum density, transported at the group velocity. The wave momentum itself is in the direction of the wave vector $\mathbf{k}_0$ (rather than the group velocity), since the phase fronts responsible for pushing on an object in the path of the wave travel in this direction. For waves traveling in the x-direction though a nondispersive medium, the only nonzero component of $\mathbf{T}$ is $T_{xx} = \bar{\epsilon}$, indicating that a pressure equal to the wave energy density will be exerted on an absorbing barrier placed in the path of the wave packet.

Spherically Symmetric Disturbance. We now consider the evolution of waves in a dispersionless medium, starting from a spherically symmetric initial condition, $p_0(\mathbf{r}) = P\,e^{-r^2/a^2}$, $u_0(\mathbf{r}) = 0$. The Fourier transform of $p_0(\mathbf{r})$ is $\tilde{p}_0(\mathbf{k}) = (\pi a^2)^{3/2} P e^{-a^2k^2/4}$. Then according to Eq. (5.1.48) and (5.1.50), the response to this initial condition is

$$p(\mathbf{r},t) = 2\,\mathrm{Re}\int \frac{d^3k}{(2\pi)^3}\tilde{p}_0(\mathbf{k})\, e^{i(\mathbf{k}\cdot\mathbf{r}-ckt)} = (\pi a^2)^{3/2} P\,\mathrm{Re}\int \frac{d^3k}{(2\pi)^3} e^{i(\mathbf{k}\cdot\mathbf{r}-ckt)-a^2k^2/4}. \tag{5.1.77}$$

It is best to perform this integration using spherical coordinates in $\mathbf{k}$-space, (k, θ, ϕ). Here θ and ϕ are polar angles used to determine the direction of $\mathbf{k}$ with respect to fixed axes. We choose the vector $\mathbf{r}$ to be along the k_z-axis of this coordinate system, so that $\mathbf{k}\cdot\mathbf{r} = k\cos\theta$, where $r = |\mathbf{r}|$. Also, $d^3k = k^2\,dk\sin\theta\,d\theta\,d\phi$. Equation (5.1.77) then becomes

$$p(\mathbf{r},t) = \frac{(\pi a^2)^{3/2} P}{(2\pi)^3}\mathrm{Re}\int_0^\infty k^2\,dk\int_0^{2\pi} d\phi\int_0^\pi \sin\theta\,d\theta\, e^{-a^2k^2/4}\, e^{ik(r\cos\theta - ct)}.$$

The ϕ-integral is trivial, and the θ-integral can be calculated by changing variables to $x = \cos\theta$. This implies that $dx = -\sin\theta\,d\theta$, and the range of integration runs from $x = 1$ to $x = -1$. The result is

$$p(\mathbf{r},t) = \frac{(\pi a^2)^{3/2} P}{(2\pi)^2}\mathrm{Re}\int_0^\infty k^2\,dk\,e^{-ikct}\,e^{-a^2k^2/4}\,\frac{2\sin kr}{kr}.$$

The k-integral can also be calculated analytically, yielding

$$p(r,t) = \frac{P}{2r}\left[e^{-(r+ct)^2/a^2}(r+ct) + e^{-(r-ct)^2/a^2}(r-ct)\right]. \tag{5.1.78}$$

This solution is displayed in Cell 5.9, taking $P = c = 1$, $a = 0.02$.

Cell 5.9

```
p[r_, t_] = P/(2r) (e^(-(r+ct)^2/a^2) (r + ct) + e^(-(r-ct)^2/a^2) (r - ct))/.
  {P → 1, a → 0.02, c → 1};

Table[Plot[p[r, t], {r, 0, .3},
   PlotRange → {{0, .3}, {-.5, 1}}, AxesLabel → {"r", "p"},
   PlotLabel → "t = "<>ToString[t]], {t, 0, .3, .01}];
```

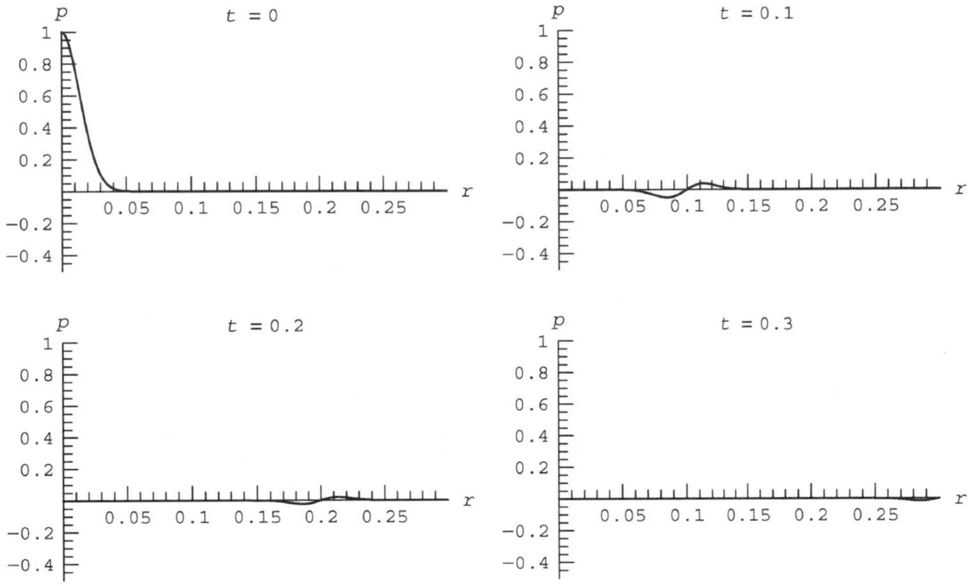

This spherical wave decreases in amplitude as it expands in r. This is to be expected, given that the surface area of the pulse is increasing, yet the overall energy of the pulse is conserved. The general considerations of the previous section imply that the total wave energy should scale as p^2V, where V is the volume occupied by the pulse. The radial width of the pulse is roughly constant, but the surface area of the pulse is proportional to r^2. We therefore expect p to be proportional to $1/r$ at the peak of the pulse. This is in fact what Eq. (5.1.78) implies, for large t. Interestingly, there is also a negative section to the pulse (a pressure rarefaction), trailing the forward peak, even though the initial pulse was entirely positive. This does not occur in propagation in one dimension along a string, and is a consequence of the spherical geometry.

Response to a Source The wave equation with a source $S(\mathbf{r}, t)$ is

$$\frac{\partial^2}{\partial t^2} p(\mathbf{r}, t) = c^2 \nabla^2 p(\mathbf{r}, t) + S(\mathbf{r}, t). \tag{5.1.79}$$

A particular solution for the response to the source can be obtained by Fourier transformation in both time and space. We denote the time–space Fourier transform of $p(\mathbf{r}, t)$ as $\tilde{p}(\mathbf{k}, \omega)$, where

$$\tilde{p}(\mathbf{k}, \omega) = \int dt \int d^3r p(\mathbf{r}, t)\, e^{-i(\mathbf{k}\cdot\mathbf{r} - \omega t)}, \tag{5.1.80}$$

and similarly for the time–space transform of the source, $\tilde{S}(\mathbf{k}, \omega)$. Note the use of differing transform conventions for the time and space parts of the transform, as discussed in Sec. 2.3.1.

Application of the time–space transform to Eq. (5.1.79) yields the algebraic equation

$$-\omega^2 \tilde{p}(\mathbf{k}, \omega) = -c^2 k^2 \tilde{p}(\mathbf{k}, \omega) + \tilde{S}(\mathbf{k}, \omega),$$

which has the solution $\tilde{p}(\mathbf{k}, \omega) = \tilde{S}(\mathbf{k}, \omega)/(c^2k^2 - \omega^2)$. Taking an inverse time–space transform then yields the particular solution,

$$p(\mathbf{r}, t) = \int \frac{d\omega\, d^3k}{(2\pi)^4} \frac{\tilde{S}(\mathbf{k}, \omega)}{c^2k^2 - \omega^2} e^{i(\mathbf{k}\cdot\mathbf{r} - \omega t)}. \tag{5.1.81}$$

As an example, let us determine the response to a point source in time and space, $S(\mathbf{r}, t) = \delta(\mathbf{r})\delta(t)$. Then $\tilde{S}(\mathbf{k}, \omega) = 1$. We first perform the inverse frequency transformation. This is easily done with the **InverseFourierTransform function:**

Cell 5.10

```
FullSimplify[InverseFourierTransform[1/(c^2 k^2 - ω^2),
 ω, t]/Sqrt[2 Pi]]

Sign[t] Sin[c k t]
------------------
      2 c k
```

Thus, we obtain

$$p(\mathbf{r}, t) = \mathrm{Sign}(t) \int \frac{d^3k}{(2\pi)^3} \frac{\sin ckt}{2ck} e^{i\mathbf{k}\cdot\mathbf{r}} \tag{5.1.82}$$

This solution is nonzero for $t < 0$, which is inconvenient. But it is only one particular solution to the PDE. To obtain a solution that is zero for $t < 0$, we must add an appropriate solution to the homogeneous equation: Eq. (5.1.50), or equivalently, Eq. (5.1.47). By choosing $C(\mathbf{k}) = 1/(4cki) = -D(\mathbf{k})$, Eq. (5.1.47) becomes

$$\int \frac{d^3k}{(2\pi)^3} \frac{\sin ckt}{2ck} e^{i\mathbf{k}\cdot\mathbf{r}}.$$

Adding this homogeneous solution to Eq. (5.1.82), we cancel out the solution for $t < 0$ and are left with

$$p(\mathbf{r},t) = h(t) \int \frac{d^3k}{(2\pi)^3} \frac{\sin ckt}{ck} e^{i\mathbf{k}\cdot\mathbf{r}},$$

where $h(t)$ is the Heaviside step function. [We could also have obtained this result directly from Eq. (5.1.79) by Fourier-transforming only in space, and then using the Green's function for the harmonic oscillator equation, Eq. (2.3.77).]

The remaining Fourier integrals are best evaluated in spherical coordinates (k, θ, ϕ). As in the derivation of Eq. (5.1.78), we choose $\mathbf{r}$ to lie along the z-axis of the spherical coordinate system. Then $e^{i\mathbf{k}\cdot\mathbf{r}} = e^{ikr\cos\theta}$, and the θ and ϕ integrals yield

$$p(\mathbf{r},t) = \frac{h(t)}{rc} \int_0^\infty \frac{dk}{2\pi^2} \sin ckt \sin kr = \frac{h(t)}{rc} \int_0^\infty \frac{dk}{4\pi^2} [\cos k(r-ct) - \cos k(r+ct)]. \tag{5.1.83}$$

The integral over k can be recognized as a generalized Fourier integral of the type considered in Sec. 2.3.4. The result is $p(\mathbf{r},t) = [h(t)/4\pi cr][\delta(r-ct) - \delta(r+ct)]$ (see the exercises). However, since $r \geq 0$ and we only require the solution for $t > 0$ thanks to the Heaviside function, we can neglect $\delta(r+ct)$ and drop $h(t)$ (since the δ-function makes it redundant), and so we obtain

$$p(\mathbf{r},t) = \frac{1}{4\pi cr} \delta(r-ct). \tag{5.1.84}$$

This solution, the response to a δ-function impulse at the origin at time $t = 0$, is a spherical δ-function wave that moves with speed c, decreasing in amplitude like $1/r$ as it expands radially. This response function also arises in the theory of the retarded potentials for electromagnetic radiation from a point source. [See, for example, Griffiths (1999).]

Equation (5.1.84) may be compared with the response to an initially localized spherically symmetric perturbation, derived in the previous section [see Eq. (5.1.78) and Cell 5.9]. In that solution, there was a negative portion to the response, trailing the forward wave. Here, there is no trailing negative response, merely an outward-propagating δ-function peak, because here the source function creates an initial *rate of change* to the pressure, $\partial_t p(\mathbf{r}, t = 0^+) = \delta(\mathbf{r})$; but the pressure itself is initially zero: $p(\mathbf{r}, t = 0^+) = 0$. In electromagnetism, this sort of initial condition is the most physically relevant: it describes the response to a source in the wave equation (in electromagnetism, a current or moving charge) rather than the evolution of an initial field distribution.

Equation (5.1.84) can also be thought of as a Green's function for the response to a general source of the form $S(\mathbf{r}, t)$. Since any such source can be constructed from a series of δ-function sources in space and time, the superposition of the response to each of these point sources is

$$p(\mathbf{r},t) = \int dt_0\, d^3r_0 \frac{1}{4\pi c|\mathbf{r}-\mathbf{r}_0|} \delta(|\mathbf{r}-\mathbf{r}_0| - c(t-t_0)) S(\mathbf{r}_0,t_0). \quad (5.1.85)$$

In this equation, the time t_0 is generally referred to as the *retarded time*, because it is the time at which a source at $\mathbf{r}_0$ must emit a wave in order for the response to reach position $\mathbf{r}$ at time t.

Wave Propagation in an Anisotropic Medium When waves travel through some medium that is moving with velocity $\mathbf{U}$ with respect to the observer, the dispersion relation is affected by the motion of the medium. This can be easily understood by making a Galilean transformation to the frame of reference of the medium, where the position of some point on the wave is $\bar{\mathbf{r}}$. The same point observed in the lab frame has position $\mathbf{r}$, where

$$\mathbf{r} = \bar{\mathbf{r}} + \mathbf{U}t. \quad (5.1.86)$$

(Here we consider nonrelativistic wave propagation, such as that of sound or water waves.) In the moving frame, the medium is stationary, and the waves obey the usual wave equation (5.1.44). The solution is given by Eq. (5.1.50) with $\mathbf{r}$ replaced by $\bar{\mathbf{r}}$. If we now substitute for $\bar{\mathbf{r}}$ using Eq. (5.1.86), we obtain

$$\begin{aligned} p(\mathbf{r},t) &= 2\,\mathrm{Re} \int \frac{d^3k}{(2\pi)^3} C(\mathbf{k})\, e^{i[\mathbf{k}\cdot\bar{\mathbf{r}} - \bar{\omega}(k)t]} \\ &= 2\,\mathrm{Re} \int \frac{d^3k}{(2\pi)^3} C(\mathbf{k})\, e^{i[\mathbf{k}\cdot\mathbf{r} - Ut - \bar{\omega}(k)t]} = 2\,\mathrm{Re} \int \frac{d^3k}{(2\pi)^3} C(\mathbf{k})\, e^{i[\mathbf{k}\cdot\mathbf{r} - \omega(\mathbf{k})t]}, \end{aligned} \quad (5.1.87)$$

where $\omega(\mathbf{k})$ is the frequency as seen in the lab frame and $\bar{\omega}(k)$ is the wave frequency as seen in the frame moving with the medium. According to Eq. (5.1.87) the frequencies in the lab and moving frames are related by the equation

$$\omega(\mathbf{k}) = \bar{\omega}(k) + \mathbf{k}\cdot\mathbf{U}. \quad (5.1.88)$$

The term $\mathbf{k}\cdot\mathbf{U}$ is the *Doppler shift* of the frequency, caused by the motion of the medium. Equation (5.1.88) is an example of *anisotropic* dispersion: the frequency depends on the direction of propagation as well as the magnitude of the wave vector. As a result, we will now see that the group and phase velocities are no longer in the same directions.

The phase velocity is

$$\mathbf{v}_\phi = \frac{\omega(\mathbf{k})\hat{\mathbf{k}}}{k} = \left(\frac{\bar{\omega}(k)}{k} + \hat{\mathbf{k}}\cdot\mathbf{U} \right)\hat{\mathbf{k}}, \quad (5.1.89)$$

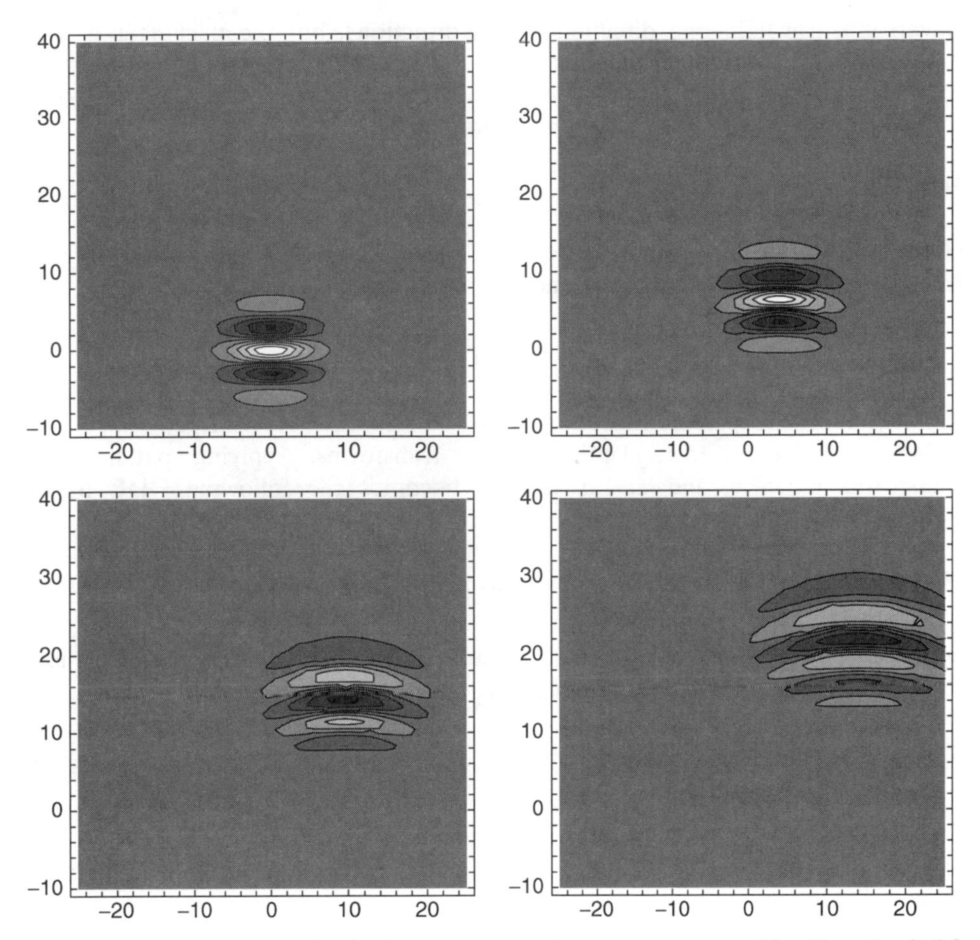

Fig. 5.3 Gaussian wave packet in deep water, moving across a current. [See Exercise (26).]

where $\hat{\mathbf{k}}$ is a unit vector in the direction of $\mathbf{k}$; but the group velocity is

$$\mathbf{v}_g = \frac{\partial \overline{\omega}}{\partial k} \hat{\mathbf{k}} + \boldsymbol{U}. \tag{5.1.90}$$

Consequently, the phase fronts within the packet propagate along $\mathbf{k}$, but the packet itself propagates in any entirely different direction as it is carried along by the flowing medium.

Waves in a moving medium are a particularly simple example of how anisotropy can modify the character of wave propagation. In these systems, phase and group velocities are completely independent of one another; they not only differ in magnitude, as in an isotropic dispersive medium; they also differ in direction.

An example of the propagation of a wave packet in an anisotropic system is displayed in Fig. 5.3. [This wave packet is the solution to Exercise (26) at the end of this section, on the propagation of deep-water waves in moving water. The water is moving to the right, and carries the packet with it.] Note how the wavefronts travel in one direction (the y-direction), but the packet as a whole propagates in a different direction. Also note that the phase velocity is greater than the group velocity for these waves, and that the wave packet disperses both in the axial and transverse directions as it propagates.

Heat Equation

General Solution. Consider the heat equation for the evolution of the temperature $T(\mathbf{r},t)$ in an isotropic homogeneous material:

$$\frac{\partial}{\partial t}T(\mathbf{r},t)=\chi\,\nabla^2 T(\mathbf{r},t)+S(\mathbf{r},t). \tag{5.1.91}$$

This equation is supplemented by an initial condition, $T(\mathbf{r},t)=T_0(\mathbf{r})$.

We can solve Eq. (5.1.91) using Fourier transforms. Applying spatial Fourier transforms in x, y, and z, Eq. (5.1.91) becomes the following ODE for the Fourier-transformed temperature $\tilde{T}(\mathbf{k},t)$:

$$\frac{\partial}{\partial t}\tilde{T}(\mathbf{k},t)=-\chi k^2\tilde{T}(\mathbf{k},t)+\tilde{S}(\mathbf{k},t),$$

where $\tilde{S}(\mathbf{k},t)$ is the spatial Fourier transform of $S(\mathbf{r},t)$. This inhomogeneous first-order ODE has the general solution

$$\tilde{T}(\mathbf{k},t)=A(\mathbf{k})\,e^{-\chi k^2 t}+\int_0^t e^{-\chi k^2(t-\bar{t})}\tilde{S}(\mathbf{k},\bar{t})\,d\bar{t}. \tag{5.1.92}$$

The coefficient $A(\mathbf{k})$ is determined by the initial condition, $\tilde{T}(\mathbf{k},0)=\tilde{T}_0(\mathbf{k})$. This implies that $A(\mathbf{k})=\tilde{T}_0(\mathbf{k})$. Taking the inverse Fourier transform of Eq. (5.1.92) then yields the solution,

$$T(\mathbf{r},t)=\int\frac{d^3k}{(2\pi)^3}e^{i\mathbf{k}\cdot\mathbf{r}-\chi k^2 t}\left(\tilde{T}_0(\mathbf{k})+\int_0^t e^{\chi k^2\bar{t}}\tilde{S}(\mathbf{k},\bar{t})\,d\bar{t}\right). \tag{5.1.93}$$

The exponential that appears in this solution, $e^{i\mathbf{k}\cdot\mathbf{r}-\chi k^2 t}$, looks something like the traveling waves $e^{i[\mathbf{k}\cdot\mathbf{r}-\omega(k)t]}$ that appear in the solutions of the wave and Schrödinger equations. In fact, by comparing with that form for traveling waves, we can identify a dispersion relation:

$$\omega(k)=-i\chi k^2. \tag{5.1.94}$$

The "frequency" of these temperature waves is imaginary, indicating that the waves do not propagate, but rather decay with time.

It is often the case that dispersion relations yield *complex* frequencies. Waves on a real string damp as they propagate due to a variety of effects, and a mathematical description of the motion leads to a dispersion relation with complex frequencies. Only the real part of the frequency describes propagation. Thus, for damped waves, the phase velocity is defined as

$$\mathbf{v}_\phi=(\mathrm{Re}\,\omega)\hat{\mathbf{k}}/k, \tag{5.1.95}$$

and the group velocity is

$$\mathbf{v}_g = \frac{\partial}{\partial \mathbf{k}} \mathrm{Re}\, \omega. \tag{5.1.96}$$

For the heat equation, however, the waves do not propagate at all; the phase and group velocities are zero. However, this does not mean that heat does not propagate through a material. Rather, it merely implies that heat does not propagate like a wave, with constant phase and group velocities. Rather, heat *diffuses* through a material. This diffusion is described by the general solution to the heat equation, Eq. (5.1.93). To understand this diffusion, it suffices to consider an example.

Green's Function for the Heat Equation. Let us take as an example a system that is initially at zero temperature, $T_0 = 0$, and that is subjected to a δ-function heat source at time t_0:

$$S(\mathbf{r},t) = \delta(\mathbf{r}-\mathbf{r}_0)\,\delta(t-t_0). \tag{5.1.97}$$

The evolution of the temperature resulting from this heat pulse is given by Eq. (5.1.93). The initial condition $T_0 = 0$ implies that $\tilde{T}_0(\mathbf{k}) = 0$, and Eq. (5.1.97) implies that $\tilde{S}(\mathbf{k},\bar{t}) = e^{-i\mathbf{k}\cdot\mathbf{r}_0}\,\delta(\bar{t}-t_0)$. Applying these results to Eq. (5.1.93) and performing the integration over $\bar{t}$ yields

$$T(\mathbf{r},t) = \begin{cases} \displaystyle\int \frac{d^3k}{(2\pi)^3} e^{i\mathbf{k}\cdot(\mathbf{r}-\mathbf{r}_0) - \chi k^2 (t-t_0)}, & t > t_0, \\ 0, & t < t_0. \end{cases}$$

The integral can be evaluated in the usual way, by writing $\mathbf{k}$ in spherical coordinates with the z-axis chosen in the direction of $\mathbf{r}-\mathbf{r}_0$. Then $\mathbf{k}\cdot(\mathbf{r}-\mathbf{r}_0) = k\,\Delta r \cos\theta$, where $\Delta r = |\mathbf{r}-\mathbf{r}_0|$, and $d^3k = k^2\,dk \sin\theta\, d\theta\, d\phi$. The θ and ϕ integrals then yield

$$T(\mathbf{r},t) = \begin{cases} \displaystyle\int_0^\infty \frac{k\,dk}{(2\pi)^2} 2\frac{\sin(k\,\Delta r)}{\Delta r} e^{-\chi k^2 (t-t_0)}, & t > t_0, \\ 0, & t < t_0. \end{cases}$$

The k-integral can also be evaluated analytically:

Cell 5.11

```
FullSimplify[
 Integrate[2 Exp[-χk² (t - t₀)] Sin[k Δr]/Δr k,
   {k, 0, Infinity}] / (2 Pi)²,
 Δr> 0 && χ (t - t₀) > 0]
```

$$\frac{e^{\frac{\Delta r^2}{-4t\chi+4\chi t_0}}}{8\pi^{3/2}\,(\chi(t-t_0))^{3/2}}$$

Thus, we find that the temperature pulse is described by

$$T(\mathbf{r},t) = \frac{h(t-t_0)}{8\pi^{3/2}[\chi(t-t_0)]^{3/2}} e^{-\Delta r^2/4\chi(t-t_0)}, \tag{5.1.98}$$

where h is a Heaviside step function. This is a Gaussian pulse that spreads in time. Initially the width of the pulse is zero because it begins as a δ-function. In fact, Eq. (5.1.98) can be regarded as a Green's function for the heat equation.

The evolution from any source $S(\mathbf{r},t)$ can be written as

$$T(\mathbf{r},t) = \int d^3r_0\, dt_0\, g(\mathbf{r}-\mathbf{r}_0, t-t_0) S(\mathbf{r}_0, t_0), \tag{5.1.99}$$

where g is given by Eq. (5.1.98). The proof of this equation is left as an exercise. However, this is easy to understand physically. Breaking the source S into a collection of δ-function sources at different positions and times, the resulting temperature is a superposition of the responses to all of them. Of course, Eq. (5.1.99) is only a particular solution, because it does not necessarily satisfy the initial conditions.

The width w of the Gaussian heat pulse given by Eq. (5.1.98), defined in the same manner as in Eq. (5.1.31), increases with time according to

$$w = \sqrt{2\chi(t-t_0)}, \qquad t > t_0. \tag{5.1.100}$$

Thus, the heat pulse does not propagate outward with a constant speed as in the wave equation. Rather, the pulse width increases as the square root of time. This is a signature of diffusion, a subject to which we will return in Chapter 8.

EXERCISES FOR SEC. 5.1

(1) Using Fourier transforms, find the solution of the 1D wave equation in an infinite system with $c = 1$, for the following initial conditions. Animate the solution for $0 < t < 2$.

- **(a)** $\dfrac{\partial y}{\partial t}(x,0) = xe^{-|x|}$, $y(x,0) = 0$.
- **(b)** $y(x,0) = h(x)$, $\dfrac{\partial y}{\partial t}(x,0) = 0$, where $h(x)$ is a Heaviside step function.
- **(c)** $y(x,0) = 0$, with boundary condition $y(0,t) = te^{-t}$, and initial condition that $y = 0$ for $t < 0$. Solve for $y(x,t)$ on $0 < x < \infty$. (Hint: Put the equation in standard form, and use a Fourier sine integral.)

(2) A quantum particle moving freely in one dimension has an initial wave packet given in k-space by $\tilde{\psi}_0(k) = \sqrt{\pi/\Delta k}$, $|k - k_0| \le \Delta k$, and zero otherwise. Find the subsequent evolution of the wave packet. Animate the evolution of $|\psi|^2$ for $\hbar = m = 1$ and $k_0 = 5$, $\Delta k = \frac{1}{2}$, for $0 < t < 4$. How far does the center of the wave packet move in this time? Compare the speed of the wave packet with the mean group velocity and the mean phase velocity, based on the central wavenumber k_0. Which velocity best fits the actual motion of the packet?

(3) Water waves in water of depth h have a dispersion relation given by the equation $\omega = (gk \tanh kh)^{1/2}$, where g is the acceleration of gravity, $g = 9.8$ m/s^2. [See Kundu (1990, pg. 94).] Plot the phase and group velocities of these waves in meters per second, as a function of kh. Show analytically that in the limit that $kh \gg 1$ (deep-water waves) the phase velocity is twice the group velocity.

(4) Water waves in deep water, including surface tension, have the following dispersion relation: $\omega = (gk + \sigma k^3)^{1/2}$, where σ is the surface tension coefficient, equal to 71.4 cm^3/s^2 for fresh water.

(a) Find the group and phase velocities for a wave packet with central wavenumber of magnitude k. Plot each as a function of k. Find the critical value of k, k^*, at which the phase and group velocities coincide. Waves with wavenumbers above k^* are dominated by surface tension; waves below k^* are gravity-dominated.

(b) Show that k^* also corresponds to a minimum possible phase velocity as a function of k, and find its value in centimeters per second. As one moves a stick through water, the speed of the stick must be larger than this minimum phase velocity in order to excite waves, since the stick only excites waves which move at the same phase velocity. (You can test this in your sink or bathtub.)

(c) Using Eq. (5.1.30), construct a 1D Gaussian wave packet consisting of ripples traveling in the positive x-direction, with central wavenumber $k_0 = 5$ cm^{-1}, and initial shape in k-space $C(k) = e^{-(k-k_0)^2/2\,\Delta k^2}$ with $\Delta k = 1$ cm^{-1}. Keeping the effect of dispersion to the lowest possible nontrivial order, make a movie of the propagation of the real part of this wave packet for $0 < t < 1$ s, in increments of 0.025 s, as seen in the lab frame. How many cm does the packet travel in 1 s? Compare the mean speed of the packet with the group and phase velocities at wavenumber k_0.

(5) Find the solution to the heat equation with $\chi = 1$ for the following initial conditions:

(a) $T(x,0) = e^{-|x|}$. Animate the solution on $-5 < x < 5$ for $0 < t < 2$.

(b) $T(x,0) = h(x)$, where $h(x)$ is a Heaviside step function. Animate the solution on $-5 < x < 5$ for $0 < t < 2$.

(6) The biharmonic wave equation,

$$\frac{\partial^2}{\partial t^2} y(x,t) = -\alpha \frac{\partial^4}{\partial x^4} y(x,t),$$

where α is a constant with units of m^4/s^2, describes transverse waves on a long rod [see Eq. (4.2.47)].

(a) What is the dispersion relation for waves on this rod?

(b) Without making approximations, find the evolution of the following initial perturbation on the rod:

$$y(x,0) = \left(\sin\frac{10x}{a}\right) e^{-x^2/a^2}, \qquad \frac{\partial y}{\partial t}(x,0) = 0.$$

Animate the evolution, taking $\alpha = a = 1$, for $0 < t < 0.2$.

(7) A laser beam of frequency ω_0 and wavenumber k_0 in the z-direction is fired into a vacuum. The form of the beam is $p(r,z,t) = e^{i(k_0 z - \omega_0 t)} p_0(r)$. Using the wave equation, show that $p_0(r) = AJ_0(k_\perp r)$ provided that $\omega_0 > ck_0$, and find $k_\perp$. Note that, unlike a Gaussian wave packet, this beam does not spread as it propagates. This does not violate our expression for the Rayleigh length, because the Bessel function already has an effectively infinite transverse extent, falling off slowly with increasing r. This type of beam is called a *Bessel beam.*

(8) The starship Enterprise fires its phaser at a Klingon warship 10^5 km away. Assuming that the beam is initially 1 m in radius and consists of electromagnetic waves that have a frequency centered in the visible ($\lambda = 5 \times 10^{-7}$ m), how wide is the beam by the time it strikes the warship?

(9) Prove Eqs. (5.1.75) and (5.1.76).

(10) **(a)** Schrödinger's equation, $-i\hbar\, \partial\psi/\partial t = -(\hbar^2/2m)\nabla^2\psi + V\psi$, ($V$ assumed real), does not fit the pattern of the classical wave equations described by Eq. (5.1.63), because it has a first-order time derivative. Now, the conserved density is the probability density $\rho(r,t) = |\psi|^2$. Prove that ρ satisfies a conservation law of the form of Eq. (5.1.57), and find the form of the flux Γ. Hint: Multiply Schrödinger's equation by ψ^*, then subtract the result from its complex conjugate.

(b) By applying the results of part (a) to a wave packet of the form $\psi = \psi_0(\mathbf{r} - \mathbf{v}_g t)\, e^{i(\mathbf{k}\cdot\mathbf{r} - \omega t)}$, show that the flux of probability density is $\boldsymbol{\Gamma} = \rho\hbar\mathbf{k}/m$.

(11) Schrödinger's equation is just one member of a class of wave equations with odd-order time derivatives, describing the evolution of a complex function $\psi(\mathbf{r},t)$:

$$i \sum_{n=1,3,5,\ldots} a_n \frac{\partial^n}{\partial t^n}\psi = \sum_{n=0,2,4,\ldots} \sum_{j=1,2,3} b_{nj} \frac{\partial^n}{\partial r_j^n}\psi, \tag{5.1.101}$$

where a_n and b_{nj} are real coefficients. Using the same manipulations as described in the previous problem, show that a conservation law of the form of Eq. (5.1.57) exists for a generalized probability density $N(\mathbf{r},t)$, and that for a wave packet $N(\mathbf{r},t)$ is given by

$$N(\mathbf{r},t) = |\psi|^2 \frac{\partial D(\mathbf{k},\omega)}{\partial\omega}. \tag{5.1.102}$$

and the flux of N is $\boldsymbol{\Gamma} = N\mathbf{v}_g$, where $D(\mathbf{k},\omega)$ is the dispersion function associated with Eq. (5.1.101), and where $\mathbf{v}_g$ is the group velocity.

(12) Sunlight creates an energy flux (Poynting flux) at the earth with intensity $\Gamma = 1300\ \mathrm{W/m^2}$. Use this result to determine the pressure per unit area on a solar sail in earth orbit, via Eqs. (5.1.73) and (5.1.76). The sail consists of an ultrathin Mylar surface oriented normal to the direction of the sun's rays. For simplicity, assume the sail absorbs all incoming radiation. For a sail with area $A = 1\ \mathrm{km^2}$ and mass 1000 kg, find the acceleration due to the light pressure. (It's not very large, but it lasts 24 hours a day, and it's free.) Neglecting gravitational effects, how fast (in kilometers per second) is the sail moving after one month of constant acceleration?

(13) Prove that Eq. (5.1.84) follows from Eq. (5.1.83) by calculating the required generalized Fourier integral by hand.

(14) The intensity of sound is measured in a logarithmic scale in units of *bels*, where x bels is an energy flux of $10^x \overline{\Gamma}_0$ and $\overline{\Gamma}_0 = 10^{-12}\ \mathrm{W/m^2}$. A decibel is 0.1 bel. Assume that the sound travels in air at atmospheric pressure and a temperature of 20°C, the frequency is 440 hz, and the intensity is 20 decibels.

(a) Find the mean energy density in the sound wave, in Joules per cubic meter.

(b) Find the maximum displacement of the air, in microns (1 micron $= 10^{-6}$ m). The displacement $\eta(x,t)$ of air in a compressional sound wave follows the wave equation, Eq. (4.2.48).

(c) Find the maximum pressure change due to the wave, in atmospheres. The pressure change in the air is related to the displacement by Eq. (4.4.40).

(15) A particle of mass m is described initially by a spherically symmetric Gaussian wave function, $\psi(r, t=0) = A e^{-\beta r^2}$. Find the subsequent evolution in three dimensions, using Schrödinger's equation. Animate the solution for $|\psi(r,t)|^2$, taking $\beta = \frac{1}{10}$, for $0 < t < 20$.

(16) In an exploding-wire experiment, a long thin wire is subjected to an extremely large current, causing the wire to vaporize and explode. (Such explosions are used as x-ray sources, and to study the properties of hot dense materials.) Assuming that the wire creates a pressure source of the form $S(\mathbf{r}, t) = s h(t)\delta(x)\delta(y)$ where $h(t)$ is a Heaviside function, solve Eq. (5.1.79) for the resulting pressure wave, assuming that $p = 0$ for $t < 0$. Show that

$$p(r,t) = \begin{cases} \dfrac{s}{2\pi c^2} \log\left(\dfrac{ct}{r} + \sqrt{\dfrac{c^2 t^2}{r^2} - 1} \right), & r \le ct, \\ 0, & r > ct, \end{cases}$$

where r is the *cylindrical* radial coordinate. Taking $c = s = 1$, animate $p(r,t)$ for $0 < t < 10$. [Hint: Find the solution for p in terms of an inverse Fourier transform involving $\mathbf{k} = (k_x, k_y)$. Then write $\mathbf{k}\cdot(x,y) = kr\cos\theta$, where θ is the angle between $\mathbf{k}$ and $\mathbf{r} = (x,y)$, and perform the integral over $\mathbf{k}$ in polar coordinates (k, θ) where $k_x = k\cos\theta$, $k_y = k\sin\theta$. In these coordinates, $d^2k = k\,dk\,d\theta$.]

(17) Waves in deep water, neglecting surface tension, follow the dispersion relation $\omega(k) = \sqrt{gk}$, where g is the acceleration of gravity.

(a) Show that a wave packet with cylindrical symmetry will evolve according to

$$z(r,t) = 2\,\mathrm{Re}\int_0^\infty kA(k)\,J_0(kr)\,e^{-i\,\omega(k)t}\,\frac{dk}{2\pi},$$

where $A(k)$ is the Fourier amplitude of the packet at time $t=0$.

(b) Assuming that $A(k) = e^{-k^2/36}$ (distances in meters), evaluate $z(r,t)$ numerically, and create a sequence of **Plot3D** graphics objects plotting $z(\sqrt{x^2+y^2}, t)$ for $-3<x<3$, $-3<y<3$ in steps of 0.1 s for $0<t<3$ s. [Hint: These plots would take a long time to evaluate if done directly. Therefore, use the following technique. First, define $z(r,t)$ using **NIntegrate**, and do the k-integral only over the range $0<k<15$. Then, evaluate a table of values of $z(r_i, t)$, and interpolate them via the command **zint[t_]:= zint[t] = Interpolation[Table[{r,z[r,t]},{r,0,5,.2}]]**. Using **Plot3D** the interpolation function **zint[t][$\sqrt{x^2+y^2}$]** can now be plotted.] The result is shown in Fig. 5.4 for the case of shallow-water waves, where $\omega = \sqrt{gh}\,k$, taking $\sqrt{gh} = 1$ m/s. The reader is encouraged to vary the parameters of the problem,

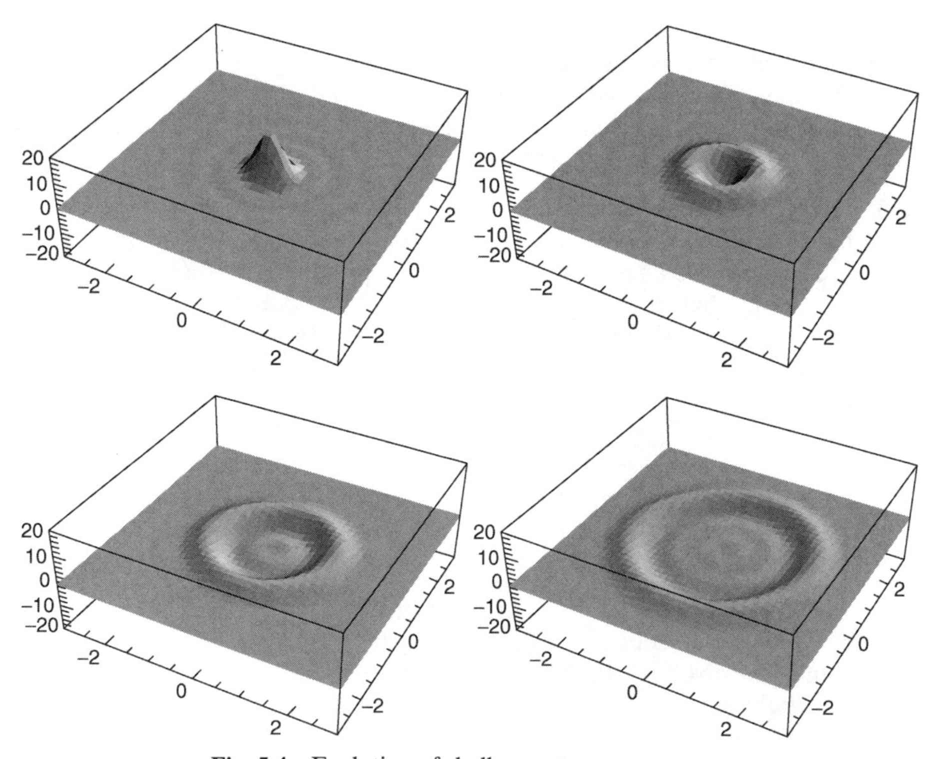

Fig. 5.4 Evolution of shallow-water waves.

and to investigate the behavior of other dispersion relations. (For example, what happens if surface tension is added?)

(18) A general spherically symmetric function $z(r)$ has a Fourier transform representation of the form

$$z(r) = \int \tilde{z}(k)\, e^{i\mathbf{k}\cdot\mathbf{r}}\, d^3k/(2\pi)^3.$$

By comparing the result written in cylindrical coordinates (ρ, ϕ, z) with that written in terms of the spherical radial coordinate r, prove the following identity:

$$\int_0^{\pi} J_0(\rho \sin\theta)\, e^{iz\cos\theta} \sin\theta\, d\theta = 2\frac{\sin\sqrt{\rho^2+z^2}}{\sqrt{\rho^2+z^2}}. \tag{5.1.103}$$

(19) A stationary point sound emits waves of frequency ω_0: $S(\mathbf{r},t) = s\,\delta(\mathbf{r})\, e^{-i\omega_0 t}$. Using Eq. (5.1.85), show that the pressure response to this source is

$$p(\mathbf{r},t) = \frac{s}{4\pi c^2 r} e^{-i\omega_0(t-r/c)}.$$

(20) An explosion occurs at a height $z = H$ above a flat surface. The explosion can be modeled as a point pressure source in the three-dimensional wave equation, $\partial^2 p/\partial t^2 = c^2\nabla^2 p + s\delta(x)\delta(y)\delta(z-H)h(t)te^{-\gamma t}$. At ground level, the boundary condition is that $\partial p/\partial z = 0$. Using either Fourier methods or Eq. (5.1.85), determine the evolution of the pressure p in the resulting blast wave. (Hint: Show that the boundary condition can be met by adding an appropriate image source below the plane. Then solve for the pressure from a source in free space, and add in the image to match the boundary condition.) For $H = 5$, $\gamma = 3$, $c = s = 1$, make an animation of surface plots of the evolution of $p(r, z, t)$ in the (x, z) plane for $0 < t < 8$.

(21) Fourier transforms can also be used to solve certain potential problems. Solve the following problem in the domain $z > 0$: $\nabla^2\phi(x, y, z) = 0$ with boundary conditions in the (x, y) plane that $\phi(x, y, 0) = V_0$ in a circular patch of radius a, and $\phi(x, y, 0) = 0$ otherwise. Another boundary condition is $\phi \to 0$ as $z \to \infty$. Solve this problem by Fourier-transforming in x and y and solving the resulting ODE for $\tilde{\phi}(k_x, k_y, z)$ with the appropriate boundary conditions. Show that the result can be written as the following integral:

$$\phi(r,z) = V_0 a \int_0^{\infty} J_0(k_r r) J_1(k_r a)\, e^{-k_r z}\, dk_r.$$

Plot the solution for ϕ/V_0 in the $y = 0$ plane as a `Plot3D` graphics object taking $a = 1$, for $-3 < x < 3$, $0 < z < 6$, by numerically evaluating this integral.

(22) A circular speaker sits in the x-y plane, in air with equilibrium pressure p_0. The speaker is an elastic membrane that moves according to $\delta z(r,0,t) = z_0 \sin\omega_0 t$ for $r < a$, and $\delta z(r,0,t) = 0$ otherwise. The perturbed air pressure

δp around the speaker satisfies potential flow with fluid displacement $\boldsymbol{\eta}(r, z, t) = \nabla\phi$, where ϕ satisfies Eq. (4.4.39) and the sound speed is c. Also, the perturbed pressure is related to $\boldsymbol{\eta}$ by Eq. (4.4.40).

(a) Using Fourier cosine transforms, write the solution for $\phi(r, z, t)$ as

$$\phi(r,z,t) = u(r,z,t) + \int \frac{d^2k_\perp}{(2\pi)^2} \int_0^\infty dk_z A(k_\perp, k_z, t)\, e^{i\mathbf{k}_\perp \cdot \mathbf{r}} \cos(k_z z),$$

where $\mathbf{k}_\perp = k_r(\cos\theta\,\hat{\mathbf{x}} + \sin\theta\,\hat{\mathbf{y}})$ is the perpendicular component of the wavenumber, and u satisfies the boundary conditions. Solve for $A(k_\perp, k_z, t)$ to show that

$$\frac{\delta p(r,z,t)}{p_0} = \frac{az_0\omega_0^2}{\pi} \int_0^\infty dk_r\, J_0(k_r r) J_1(k_r a) \int_0^\infty dk_z$$
$$\times \frac{\omega_0 \sin ckt - ck \sin \omega_0 t}{ck(\omega_0^2 - c^2k^2)} \cos k_z z,$$

where $k = \sqrt{k_r^2 + k_z^2}$. [Hint: To satisfy the von Neumann boundary conditions in the x-y plane, put the wave equation for ϕ in standard form by employing a function $u(r, z, t)$ that matches the boundary conditions, but that also satisfies the Laplace equation $\nabla^2 u = 0$.]

(b) Although the integrals in the solution for δp are well defined, they cannot be evaluated analytically, and are nontrivial to evaluate numerically. We will use the following method to evaluate δp in the $z = \frac{1}{2}$ plane at $t = 1$, taking $a = z_0 = p_0 = c = 1$ and $\omega_0 = 10$. First, define a numerical function

$$f(k_r, z, t) = \int_0^{1000} dk_z \frac{\omega_0 \sin ckt - ck \sin \omega_0 t}{ck(\omega_0^2 - c^2k^2)} \cos k_z z,$$

using `NIntegrate`. Next, make a table of values of $\{k_r, f(k_r, \frac{1}{2}, 1)\}$ for $k_r = i$, $i = 0, 1, 2, \ldots, 50$. Next, create an interpolating function $g(k_r)$ by interpolating this data. Finally, define a numerical function $\delta p(r) = (az_0 p_0 \omega_0^2/\pi)\int_0^{50} dk_r\, J_0(k_r r) J_1(k_r a) g(k_r)$, and plot the result for $0 < r < 2$ to obtain the pressure field in the $z = \frac{1}{2}$ plane at $t = 1$. [The solution for $\delta p(r, t = 1)$ is shown in Fig. 5.5 for the case of the $z = 0$ plane. Waves of wavelength $2\pi c/\omega_0 = 0.63\ldots$ are being excited. Note the kink at the edge of the speaker, at $r = 1$. Also note that the solution vanishes for $r > 2$, as expected for $t = 1$, since the signal has not yet propagated into this region.]

(23) A jet aircraft moves with constant speed v in the z-direction. The jet acts as a point pressure source, and the air pressure responds according to the three-dimensional wave equation, $\partial^2 p/\partial t^2 = c^2\nabla^2 p + s\delta(z - vt)\delta(x)\delta(y)$.

(a) Using Fourier transforms, find the particular solution for the wake behind this moving source of the form $p(z - vt, r)$ in cylindrical coordi-

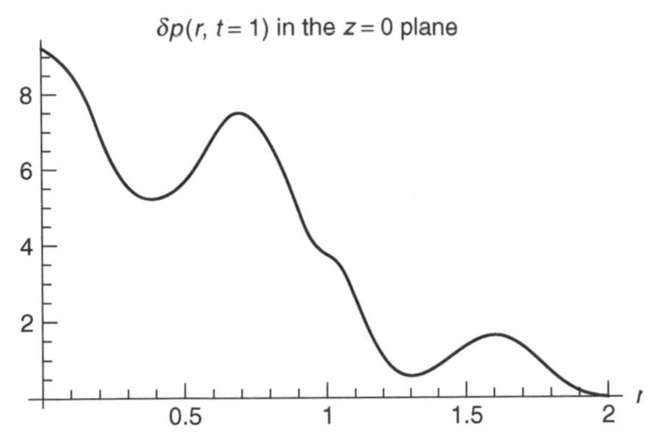

Fig. 5.5 $\delta p(r, t = 1)$ in the $z = 0$ plane.

nates (r, θ, z), *assuming that* $v < c$. (As seen in a frame moving with the source, this wake is stationary.)

(b) In the moving frame $\bar{z} = z - vt$, plot contours of constant p in the x-z plane for (i) $v = 0$, (ii) $v = c/2$, (iii) $v = 9c/10$.

(c) Show that for $v > c$, your solution breaks down along lines defined by $(\bar{z}/r)^2 = (v/c)^2 - 1$. What is the physical significance of these lines? (Actually, your answer is incorrect if $v > c$. See the next exercise.)

(24) Redo the calculation in the previous problem, taking $v > c$, and using Eq. (5.1.85) rather than Fourier methods.

(a) Show that the retarded time t_0 in Eq. (5.1.85) satisfies

$$t_0 = t - \frac{v\bar{z} \pm \sqrt{r^2(c^2 - v^2) - \bar{z}^2 c^2}}{c^2 - v^2}, \qquad \text{where} \quad \bar{z} = z - vt,$$

with the added requirement $t_0 < t$ used to determine the choice of the $\pm$ sign, and with no solution if the argument of the square root is negative.

(b) Show that for $c > v$ the requirement that $t_0 < t$ implies that only the $+$ sign should be kept.

(c) Show that for $c < v$ the requirement that $t_0 < t$ implies that $\bar{z} < 0$ and that both signs can be kept.

(d) Using Eq. (2.3.33), show that $p(\mathbf{r}, t)$ can be written as

$$p(\mathbf{r}, t) = \frac{s}{4\pi c} \sum_{t_0} \frac{1}{|t_0(v^2 - c^2) + c^2 t - vz|},$$

where the sum is over the solution(s) for the retarded time.

(e) Using the previous results, show that, for $v > c$,

$$p(\mathbf{r}, t) = \frac{s}{2\pi c} \frac{1}{\sqrt{r^2(c^2 - v^2) + \bar{z}^2 c^2}} h\left(-r\sqrt{c^2 - v^2} - \bar{z}c\right),$$

where h is a Heaviside function, and for $v < c$,

$$p(\mathbf{r},t) = \frac{s}{4\pi c} \frac{1}{\sqrt{r^2(c^2 - v^2) + \bar{z}^2 c^2}}.$$

(This is the solution to the previous problem.) Make a surface plot of $p(r, \bar{z})$ for (i) $v/c = 2$, (ii) $v/c = 4$. (The cone structure visible behind the source is called a *Mach cone*. The large amplitude just behind the cone implies that linear theory breaks down and a nonlinear *shock wave* forms. Shocks will be discussed in Chapter 7.)

(25) A long straight wire of radius a and resistivity ρ is initially at temperature $T = 0°C$. The wire is clad with an infinitely thick shell of material, also at $T = 0°C$. The cladding is not electrically conductive, but has the same thermal diffusivity χ as the conductor in the wire. At time $t = 0$, the wire begins to carry a uniform current density j, creating a heating power density $S = \rho j h(a - r)$, where h is a Heaviside function.

(a) Find the evolution of the temperature, and show that in cylindrical coordinates,

$$T(r,t) = aS \int_0^\infty J_0(kr) J_1(ka) \left(\frac{1 - e^{-\chi k^2 t}}{\chi k^2} \right) dk.$$

(b) Perform the integral to show that the temperature along the axis of the wire satisfies

$$T(0,t) = S \left[t(1 - e^{-a^2/4t\chi}) + \frac{a^2 \Gamma(0, a^2/4t\chi)}{4\chi} \right],$$

where $\Gamma(n, x)$ is a special function called an *incomplete Gamma function*. Plot this result for $\chi = a = S = 1$ and for $0 < t \le 10^3$. Show that at large times the temperature diverges, and find the form of this divergence. (The material cannot conduct heat away from the source rapidly enough to equilibrate its temperature.)

(26) A Gaussian wave packet of water waves propagates across a river in the $+y$ direction. The form of the packet is $z(x, y, t) = 2\,\mathrm{Re} \int [d^2k/(2\pi)^2] A(\mathbf{k})\, e^{i(\mathbf{k}\cdot\mathbf{r} - \omega(\mathbf{k})t)}$, where $A(\mathbf{k}) = 25\pi\, e^{-\frac{25}{4}[k_x^2 + (-1 + k_y)^2]}$, with distances measured in centimeters. Thus, the wave packet is peaked at $\mathbf{k}_0 = (0, 1)$. In still water, the dispersion relation for the waves is $\omega(k) = \sqrt{gk}$. However, there is a current in the $+x$ direction, with speed $U = 10$ cm/s. Keeping dispersion in both x and y directions, plot the contours of constant z for $0 < t < 2$ s. (Solution: See Fig. 5.3.)

(27) Magnetized plasmas carry many different types of waves that have anisotropic dispersion due to the magnetic field. Assuming a uniform magnetic field in the z-direction, magnetized electron plasma waves have the following dispersion relation:

$$\omega = \omega_p k_z / k = \omega_p \cos\theta,$$

where θ is the angle between **k** and the magnetic field, $\omega_p = \sqrt{e^2 n/\epsilon_0 m}$ is the electron plasma frequency, n is the electron density in the plasma, e is the electron charge, and m is the electron mass.

(a) Find the group and phase velocities. Show that $v_g = 0$ if $\theta = 0$, and for fixed ω_p and k find the angle θ that provides the maximum value of v_g.

(b) A disturbance that is initially spherically symmetric propagates in the $+z$ direction according to $E(\mathbf{r},t) = 2\,\mathrm{Re} \int d^3k/(2\pi)^3 A(k)\, e^{i[\mathbf{k}\cdot\mathbf{r} - \omega(\mathbf{k})t]}$. Show that this disturbance can be written as a single integral over the magnitude of **k**,

$$E(\mathbf{r},t) = 8\pi\ \mathrm{Re} \int \frac{k^2\, dk}{(2\pi)^3} A(k) \frac{\sin\left(\sqrt{k^2 r^2 + (kz - \omega_p t)^2}\right)}{\sqrt{k^2 r^2 + (kz - \omega_p t)^2}},$$

where (r, z) are cylindrical coordinates. [Hint: Use Eq. (5.1.103).] Evaluating this integral numerically over the range $0 < k < 8$, create an animation of a pulse with a spectrum of the form $A(k) = e^{-k^2/4}$. Plot the pulse evolution in the r-z plane for $0 < \omega_p t < 20$, via a series of `Plot3D` graphics functions.

(28) Whistler waves are low-frequency electromagnetic waves that propagate along the magnetic field in a magnetized plasma. They have the following dispersion relation:

$$\omega = \Omega_c c^2 k^2 / \omega_p^2, \tag{5.1.104}$$

where $\Omega_c = eB/m$ is the electron cyclotron frequency, we have assumed that **k** is parallel to the magnetic field, and we have also assumed that $ck \ll \omega \ll \Omega_c$. [See Stix (1962, pg. 55).]

(a) A group of whistler waves is created by a lightning flash in the upper atmosphere. The waves propagate along the earth's magnetic field in the ionospheric plasma, according to $E(x,t) = 2\,\mathrm{Re} \int \frac{dk}{2\pi} A(k)\, e^{i[kx - \omega(k)t]}$, where x is the distance along the magnetic field. Taking $A(k) = e^{-k^2/k_0^2}$, find the evolution of the wave packet analytically, without approximation. Animate the evolution for the following parameters: $k_0 = 0.01\ \mathrm{m}^{-1}$, $\Omega_c = 10^5\ \mathrm{s}^{-1}$, and $\omega_p = 10^7\ \mathrm{s}^{-1}$, and for $0 < t < 0.01$ s.

(b) A radio receiver is 3000 kilometers away from the initial location of the lightning flash, as measured along the magnetic field. Using the `Play` function, play the sound that the whistler waves make as they are picked up by the receiver. (A good time range is from zero to 5 seconds.) Explain qualitatively why you hear a descending tone (this is why they are called whistler waves). (Hint: How does the phase velocity depend on frequency?) To hear the sound of real whistler waves, picked up by the University of Iowa plasma wave instrument on NASA's POLAR space craft, go to the following link:

http://www-pw.physics.uiowa.edu/plasma-wave/
istp/polar/magnetosound.html

(c) For propagation that is not aligned with the magnetic field, the whistler-wave dispersion relation is $\omega = (\Omega_c c^2 k^2 \cos\theta)/\omega_p^2$, where θ is the angle between **k** and the magnetic field. Show that this dispersion relation implies that the angle between the group velocity and the magnetic field is always less than 19.5°. Hence, the earth's magnetic field guides these waves.

(29) Using Fourier transforms find the Green's function for the following heat equation with anisotropic diffusion, in 2 dimensions:

$$\frac{\partial T}{\partial t} = A\frac{\partial^2 T}{\partial x^2} + B\frac{\partial^2 T}{\partial x\,\partial y} + C\frac{\partial^2 T}{\partial y^2}.$$

Show that a bounded solution only exists provided that the diffusion coefficients A, B, and C satisfy $4AC - B^2 > 0$. Such anisotropic diffusion can occur in crystals and in magnetized plasmas, for example.

(30) (a) A semiinfinite slab of material, with thermal diffusivity χ, runs from $0 < x < \infty$ and is initially at uniform temperature T_0. At time $t = 0$, it is put in contact with a heat reservoir, setting the temperature at $x = 0$ to be $T(0,t) = T_1$. Use a Fourier sine transform to show that the subsequent evolution of the temperature is given by

$$T(x,t) = (T_0 - T_1)\operatorname{erf}\left(x/2\sqrt{\pi\chi t}\right) + T_1,$$

where erf(x) is an error function. Plot in the ranges $0 < x < 2$ and $0.0001 < t < 1$, taking $\chi = T_0 = 1$ and $T_1 = 0$.

(b) Show that the temperature gradient at $x = 0$ is given by $\partial T/\partial x|_{x=0} = (T_0 - T_1)/\sqrt{\pi\chi t}$. For $T_0 = 2000$ K, $T_1 = 300$ K and $\chi = 2\times10^{-6}$ m^2/s, find the time in years required for the gradient to relax to 0.03 K/m. Compare to the result obtained for the cooling of the earth in Chapter 4, Cell 4.49. Explain in a few words why the results are almost identical.

5.2 THE WKB METHOD

5.2.1 WKB Analysis without Dispersion

The Eikonal So far in this chapter we have considered systems that are *homogeneous in space and time*: their intrinsic properties such as wave speed c or conductivity κ do not vary from place to place, or from instant to instant. As a consequence, Fourier transform techniques work to determine the solution. In this section we consider systems that are inhomogeneous, but slowly varying, in space or time. The inhomogeneity is assumed to be slowly varying in space on the scale of the wavelength of the waves that make up the solution, and slowly varying in time compared to the frequency of these waves.

For such systems, the technique of WKB analysis allows us to determine analytic approximations to the solution. (WKB stands for G. Wentzel, H. Kramers, and L. Brillouin, who more or less independently discovered the theory. Several other researchers also made important contributions to its development, including H. Jeffreys and Lord Rayleigh.)

As a simple example, consider a string under uniform tension T, but for which the mass per unit length varies, $\rho = \rho(x)$. Then the wave equation for this system is given by Eq. (3.1.7):

$$\frac{\partial^2 y}{\partial t^2} = c^2(x)\frac{\partial^2 y}{\partial x^2}, \tag{5.2.1}$$

where $c(x) = \sqrt{T/\rho(x)}$ is the propagation speed. This speed now varies with position on the string, so the solution for the string motion can no longer be written as a superposition of traveling waves of the form $e^{i(kx-\omega t)}$. Rather, we will look for traveling-wave solutions of the more general form

$$y(x,t) = e^{-i\omega t}\psi(x), \tag{5.2.2}$$

where the function $\psi(x)$ provides the spatial dependence in the traveling wave. Substituting Eq. (5.2.2) into Eq. (5.2.1) provides an ODE for $\psi(x)$:

$$\frac{\partial^2 \psi}{\partial x^2} = -k^2(x)\psi(x), \tag{5.2.3}$$

where $k(x) = \omega/c(x)$.

If $k(x)$ were constant, then the independent solutions of Eq. (5.2.3) would be $e^{\pm ikx}$ and we would refer to k as the wavenumber of the waves. We will continue with this nomenclature, and refer to the function $k(x)$ as the wavenumber, but now the solution will no longer be of the form $e^{\pm ikx}$.

To find the solution, we introduce the *eikonal* $S(x)$, writing

$$\psi(x) = e^{S(x)}. \tag{5.2.4}$$

The eikonal has real and imaginary parts:

$$S(x) = \ln A(x) + i\phi(x). \tag{5.2.5}$$

Here, $A(x)$ is the amplitude of the wave, and $\phi(x)$ is the phase, as may be seen when this form is substituted into Eq. (5.2.4): $\psi(x) = A(x)\,e^{i\phi(x)}$. Substituting Eq. (5.2.4) into Eq. (5.2.3) yields a nonlinear ODE for $S(x)$,

$$\frac{\partial^2 S}{\partial x^2} + \left(\frac{\partial S}{\partial x}\right)^2 = -k^2(x). \tag{5.2.6}$$

So far no approximations have been made. However, it doesn't appear that we have made much progress: we have replaced a linear equation, Eq. (5.2.3), by a *nonlinear* equation, Eq. (5.2.6). Usually, nonlinear ODEs are much harder to solve than linear ones. But it turns out to be relatively easy to find an *approximate* solution to Eq. (5.2.6), by dropping one of the terms in the equation because it is smaller than the others. To determine the size of the terms, let us define a length scale $L(x)$ for the spatial variation of $k(x)$:

$$L(x) = \left| \frac{k(x)}{dk/dx} \right| = \left| \left(\frac{d \ln k(x)}{dx} \right)^{-1} \right|. \tag{5.2.7}$$

We will assume that the wavelength of the wave in question, $2\pi/k(x)$, is small compared $L(x)$: $1/k(x) \ll L(x)$. Let us define a small dimensionless parameter ϵ,

$$\epsilon = \frac{1}{k(x)L(x)} \ll 1. \tag{5.2.8}$$

We will refer to the limit where the ϵ is small as the *WKB limit*.

We now determine the relative size of the terms in Eq. (5.2.6). To do so, divide the equation by $k^2(x)$, and replace $\partial^n/\partial x^n$ by $1/L^n$. The result is

$$\epsilon^2 S + \epsilon^2 S^2 = -1.$$

This equation is no longer strictly correct, and must be regarded as only a qualitative indicator of the relative size of terms. The idea is that *$\partial/\partial x$ is on the order of $1/L$*. After determining the small term(s), we can go back to using the correct equation and then drop the small term(s) to get an approximate solution.

Which term(s) are small? Sometimes that is obvious, but not in this case. It appears that both terms on the left-hand side are small because they are multiplied by ϵ^2, but it is best not to make any assumptions about the size of S. Instead, we will systematically assume that each term in the equation is small, drop that term, solve for S, and then see if our assumption is consistent with the solution for S. This trial-and-error method is called *finding the dominant balance* between terms in the equation.

If we assume that -1 is the small term, and drop -1, then we are left with the equation $S^2 = -S$, which has the "solutions" $S = 0$ and $S = -1$. [These are not the actual solutions for $S(x)$, which must follow from Eq. (5.2.6). Rather, they only provide the order of magnitude of S.] However, these solutions aren't consistent with dropping -1, because $\epsilon^2 S + \epsilon^2 S^2$ is then small compared to -1.

If we instead assume $\epsilon^2 S^2$ is the small term, we obtain the equation $S = -1/\epsilon^2$. However, this implies that $\epsilon^2 S^2 = 1/\epsilon^2$, which is even larger than the terms we kept, so this also is not a consistent approximation.

We are therefore left with $\epsilon^2 S$ as the small term. Dropping this term, we obtain $\epsilon^2 S^2 = -1$ which yields $S = \pm i/\epsilon$. Then the term we dropped, $\epsilon^2 S$, equals $\pm i\epsilon$, which is in fact small compared to either of the other terms that we kept. Therefore, *this* is a consistent approximation. We have found the dominant balance between the terms in the equation: $\epsilon^2 S^2$ balances -1, and $\epsilon^2 S$ is a small correction to this dominant balance.

We have found that the eikonal S is large, scaling as $1/\epsilon$. This reflects the fact that in the WKB limit, over a distance of order L there must be many wavelengths in ψ. The phase ϕ must vary by a large amount over this distance.

Returning now to the equation, Eq. (5.2.6), we drop the small term $\partial^2 S/\partial x^2$ to obtain an approximate solution for S. Let us call the result the zeroth approximation for S, $S = S_0$, where

$$\left(\frac{\partial S_0}{\partial x}\right)^2 = -k^2(x).$$

This equation has the solution

$$S_0(x) = B \pm i \int^x k(x)\, dx, \tag{5.2.9}$$

where B is a constant of integration. Comparing with Eq. (5.2.5), we have found that to lowest order in ϵ the amplitude is constant and the phase is $\phi(x) = \pm \int^x k(x)\, dx$. Since the wavenumber $k(x)$ now varies with position, phase accumulates according to Eq. (5.2.9) rather than merely as $\pm kx$.

We will now confirm that $|\partial^2 S/\partial x^2| \ll |\partial S/\partial x|^2$ in the WKB limit. Comparing $\partial^2 S_0/\partial x^2$ and $(\partial S_0/\partial x)^2$ using Eq. (5.2.9), we have $|\partial S_0/\partial x|^2 = k^2$ and $|\partial^2 S_0/\partial x^2| = |\partial k/\partial x| \sim k/L$, so

$$\left|\frac{\partial^2 S_0}{\partial x^2} \bigg/ \left(\frac{\partial S_0}{\partial x}\right)^2\right| \sim \left|\frac{1}{k^2}\frac{k}{L}\right| = \epsilon \ll 1,$$

where the inequality follows directly from Eq. (5.2.8). Therefore, Eq. (5.2.9) is a consistent approximate solution to Eq. (5.2.3) in the WKB limit.

Equation (5.2.9), along with Eq. (5.2.4), provides us with (approximations to) the two independent solutions to Eq. (5.2.3), which when superimposed provide an approximate general solution for $\psi(x)$:

$$\psi(x) = C e^{i \int^x k(x)\, dx} + D e^{-i \int^x k(x)\, dx}. \tag{5.2.10}$$

Using Eq. (5.2.10) in Eq. (5.2.2), we have for $y(x,t)$ a left- and a right-propagating wave,

$$y(x,t) \simeq C e^{-i\omega t + i \int^x k(x)\, dx} + D e^{-i\omega t - i \int^x k(x)\, dx}. \tag{5.2.11}$$

These waves are generalizations of traveling waves of the form $e^{-i(\omega t \pm kx)}$, encountered previously in the description of waves in a homogeneous system.

However, Eq. (5.2.10) is a rather crude approximation to the exact solution of Eq. (5.2.3). We will now improve on the solution by first rewriting the exact equation for S, Eq. (5.2.6), as

$$\left(\frac{\partial S}{\partial x}\right)^2 = -k^2(x) - \lambda \frac{\partial^2 S}{\partial x^2}. \tag{5.2.12}$$

We have placed an *ordering parameter* λ in front of the small term $\partial^2 S/\partial x^2$ in order to remind us that this term is small, of order ϵ, compared to the other terms. This allows us to keep track of small terms in the following calculation. We can even expand our solutions in powers of λ, since this is equivalent to expanding in powers of ϵ. However, when we are finished, we will set $\lambda = 1$. This is a bookkeeping device that allows us to identify small terms at a glance, and easily perform Taylor expansions in powers of ϵ.

Previously, we found S_0 by dropping the λ-term altogether. Let us call an improvement to this solution $S = S_1$. To obtain an equation for S_1 we rewrite $\lambda\,\partial^2 S/\partial x^2$ as $\lambda\,\partial^2 S/\partial x^2 = \lambda\,\partial^2 S_0/\partial x^2 + \lambda\,\partial^2(S - S_0)/\partial x^2$. However, in the WKB limit we expect that $S - S_0$ will be small, so that $\lambda\,\partial^2(S - S_0)/\partial x^2$ is a small correction to the small correction, and so we drop this term. Then Eq. (5.2.12) becomes

$$\left(\frac{\partial S_1}{\partial x}\right)^2 = -k^2(x) - \lambda\frac{\partial^2 S_0}{\partial x^2}. \tag{5.2.13}$$

Furthermore, since we already know S_0 from Eq. (5.2.9), this equation can be solved for S_1. The result is

$$S_1(x) = B \pm i\int^x k(x)\sqrt{1 + \frac{\lambda}{k^2(x)}\frac{\partial^2 S_0}{\partial x^2}}\,dx.$$

We can simplify the square root by noting that the λ-term is small, of order ϵ. A first-order Taylor expansion in λ yields

$$S_1(x) = B \pm i\int^x k(x)\,dx \pm i\frac{\lambda}{2}\int^x \frac{1}{k(x)}\frac{\partial^2 S_0}{\partial x^2}dx + O(\lambda^2).$$

Also, we can perform the second integral analytically by substituting for S_0 using Eq. (5.2.9), yielding

$$\int^x \frac{1}{k(x)}\frac{\partial^2 S_0}{\partial x^2}dx = \int^x \frac{\pm i}{k(x)}\frac{\partial k}{\partial x}dx = \pm i\log k(x).$$

Therefore, we obtain

$$S_1(x) = B \pm i\int^x k(x)\,dx - \frac{\lambda}{2}\log k(x) + O(\lambda^2). \tag{5.2.14}$$

Setting $\lambda = 1$, and using this improved approximation for S in (5.2.4), we find that the term $\frac{1}{2}\log k(x)$ has the effect of causing the amplitude of the wave to vary with position, according to

$$\psi(x) = e^{\pm i\int^x k(x)\,dx + B + \log\, k(x)^{-1/2}} = \frac{e^B\, e^{\pm i\int^x k(x)\,dx}}{\sqrt{k(x)}}. \tag{5.2.15}$$

Therefore, in the WKB limit wc have the following traveling-wave solution to Eq. (5.2.1):

$$y(x,t) \simeq \frac{C}{\sqrt{k(x)}} e^{-i\omega t + i\int^x k(x)\,dx} + \frac{D}{\sqrt{k(x)}} e^{-i\omega t - i\int^x k(x)\,dx}. \tag{5.2.16}$$

This is a much better approximation to the exact solution than Eq. (5.2.11), because it has amplitude variation that is missing in Eq. (5.2.11).

However, Eq. (5.2.16) is also only an approximate solution to Eq. (5.2.1), valid only in the WKB limit $\epsilon \ll 1$. One could, if one wished, further improve the solution. Defining the nth approximation to the solution as S_n, we replace Eq. (5.2.13) by a *recursion relation*,

$$\left(\frac{\partial S_n}{\partial x}\right)^2 = -k^2(x) - \lambda \frac{\partial^2 S_{n-1}}{\partial x^2}. \tag{5.2.17}$$

For instance, we could obtain S_2 by using S_1 in the right-hand side of Eq. (5.2.15). We could then use that form on the right-hand side to obtain S_3, and so on. The hope is that the resulting approximants converge, so that for n large, $\partial^2 S_{n-1}/\partial x^2 \to \partial^2 S_n/\partial x^2$ and Eq. (5.2.17) becomes the same as Eq. (5.2.12), so that we are obtaining the exact solution for S in the limit.

At each stage in the recursion, we can Taylor-expand S_n in λ to the appropriate order so as to obtain the simplest possible analytic result. In fact, one can automate this procedure quite easily, since evaluating recursion relations and performing Taylor expansions are just the sort of things at which *Mathematica* excels. We do so below, defining a function **Sp[n,x]**, which is $\partial S_n/\partial x$. According to Eq. (5.2.17),

Cell 5.12

```
Sp[n_, x_] := Sqrt[-k[x]^2 - λD[Sp[n - 1, x], x]];
```

(One could also choose a negative sign in front of the square root to obtain a second solution.) For the initial condition in this recursion relation, we take

Cell 5.13

```
Sp[0, x_] := Sqrt[-k[x]^2];
```

Then we can ask for the result at any order. For instance, at second order,

Cell 5.14

```
Sp[2, x]
```

$$\sqrt{-k[x]^2 - \frac{\lambda\left(-2k[x]\;k'[x] + \frac{\lambda\,k[x]^2 k'[x]^2}{(-k[x]^2)^{3/2}} + \frac{\lambda\,k'[x]^2}{\sqrt{-k[x]^2}} + \frac{\lambda k[x]\;k''[x]}{\sqrt{-k[x]^2}}\right)}{2\sqrt{-k[x]^2 + \frac{\lambda k[x]\;k'[x]}{\sqrt{-k[x]^2}}}}}$$

This is very messy, but can be simplified by noting that it is correct only to second order in λ. Therefore, we can Taylor-expand in λ:

Cell 5.15

```
Expand[Simplify[Normal[Series[%, {λ, 0, 2}]], k[x]>0]]
```

$$\text{i k[x]} - \frac{\lambda \text{k}'[\text{x}]}{2\ \text{k[x]}} + \frac{3\ \text{i}\ \lambda^2 \text{k}'[\text{x}]^2}{8\ \text{k[x]}^3} - \frac{\text{i}\ \lambda^2\ \text{k}''[\text{x}]}{4\ \text{k[x]}^2}$$

This is the derivative of the second-order approximant, $\partial S_2/\partial x$. It contains the contributions from S_1 given by Eq. (5.2.14) as well as two correction terms of order λ^2. It is easy to continue this process and obtain even higher-order corrections.

More often than not, however, the approximants do *not* converge as $n \to \infty$; instead, they converge for a time, and then begin to diverge as n increases past some value that depends on the size of ϵ. This is typical behavior for *asymptotic expansions*, of which WKB solutions are an important case. [Some examples of this behavior are considered in the exercises. Readers interested in learning more about asymptotic expansions are referred to the book by Bender and Orzag (1978).] Nevertheless, if ϵ is sufficiently small, the WKB solution S_1 is usually a reasonably good approximation. We will not bother with the higher-order approximants for S here, because they become more and more complicated in form, and make only minor corrections to Eq. (5.2.16).

Example: Waves on a String with Varying Mass Density As a simple example of traveling-wave solutions in an inhomogeneous system, let us consider the case where the string mass varies as $\rho(x) = \rho_0\, e^{x/L}$ (i.e., the string gets exponentially heavier as one moves to the right.) Then the wave speed obeys $c(x) = \sqrt{T/\rho(x)} = c_0\, e^{-x/(2L)}$, where $c_0 = \sqrt{T/\rho_0}$ is the wave speed at $x = 0$, and the local wavenumber varies with position as $k(x) = \omega\, e^{x/(2L)}/c_0$. This implies that phase in the wave accumulates according to $\int^x k(x)\, dx = (2|\omega|L/c_0)\, e^{x/(2L)}$ and so the traveling wave solution is

$$y(x,t) \approx \frac{C}{e^{x/(4L)}} \exp\left[-i\left(\omega t - \frac{2|\omega|L}{c_0} e^{x/(2L)}\right)\right] + \frac{D}{e^{x/(4L)}} \exp\left[-i\left(\omega t + \frac{2|\omega|L}{c_0} e^{x/(2L)}\right)\right]. \tag{5.2.18}$$

The real part of this solution is plotted in Cell 5.16, taking $D = 0$ (i.e., the wave propagates to the right) and $c_0 = \omega = 1$ and $L = 5$. This plot shows that the wave decreases in amplitude as it propagates to the right, because the string gets heavier as one moves in this direction. Also, since the wave slows down as it propagates, wavefronts pile up, making the wavelength smaller.

Cell 5.16

```
y[x_, t_] = 1/e^(x/(4L)) exp[-i(ωt- (2ωL)/c0 e^(x/(2L))/. {c0→1, L→5,ω →1};

Table[Plot[Re[y[x, t]], {x, -4, 20},
   PlotRange→{-1, 1}, AxesLabel→{"x", "y"}],
  {t, 0, 1.8 Pi, .2 Pi}];
```

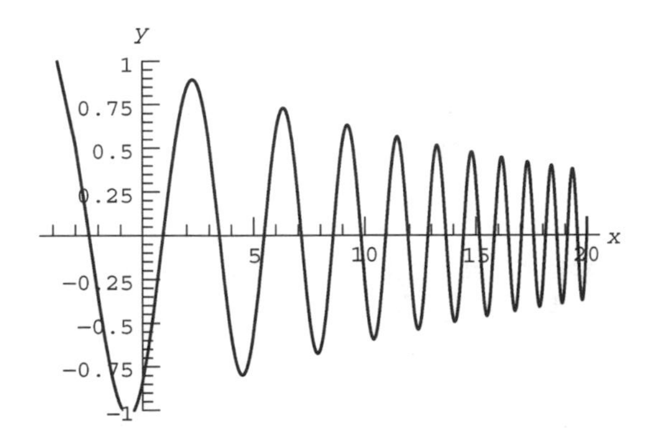

This solution propagates to the right. As an exercise, reevaluate this cell, modifying the definition of y so that it is a traveling wave moving to the left. What does the animation look like now? How does it relate to the one in Cell 5.16?

Another question is: how close is this WKB solution to the exact solution for this traveling wave? One can answer this question by solving Eq. (5.2.3) for ψ using **DSolve**:

Cell 5.17

```
ψ[x_] = ψ[x]/. DSolve[ψ''[x] == - ω^2/c0^2 Exp[x/L] ψ[x], ψ[x], x][[1]]

BesselJ[0, 2 Sqrt[e^(x/L)] Lω/c0] C[1] + BesselY[0, Sqrt[e^(x/L)] Lω/c0]C[2]
```

The result is a superposition of two Bessel functions. In order to compare this exact solution with our WKB approximation, we must determine the constants **C[1]** and **C[2]**. To do so, we will match $\psi(x)$ to $y(x,0)$ at $x=0$, and $\partial\psi/\partial x$ to $\partial y/\partial x|_{t=0}$ at $x=0$. Therefore, the solutions will match at $x=0$, but not necessarily at any other x:

Cell 5.18

```
Solve[
{ψ[0] == y[0, 0], (D[ψ[x], x]/. x→0) == (D[y[x, 0],
 x]/. x→0)}, {C[1], C[2]}];

ψ[x_] = ψ[x]/. %[[1]];

yexact[x_, t_] = ψ[x] Exp[-I ω t]/. {ω → 1., L→5., c0→1.};
```

The command for plotting the exact solution is given in Cell 5.19. The result is almost identical to that shown in Cell 5.16, so the actual plot is not shown in order to save space. However, there is a small difference with the WKB solution, plotted in Cell 5.20. This error in the WKB solution grows as the length scale L decreases, for fixed ω and c_0. This is because the WKB limit depends on L being large. The reader is invited to reevaluate these plots for smaller L, in order to investigate the error in the WKB approach.

Cell 5.19

```
Plot[Re[yexact[x, 0]], {x, -4, 20}, PlotRange→{-1, 1},
  AxesLabel→{"x", "y"}];
```

Cell 5.20

```
Plot[Re[yexact[x, 0] -y[x, 0]], {x, -4, 20},
  AxesLabel→{"x", "Δy"}];
```

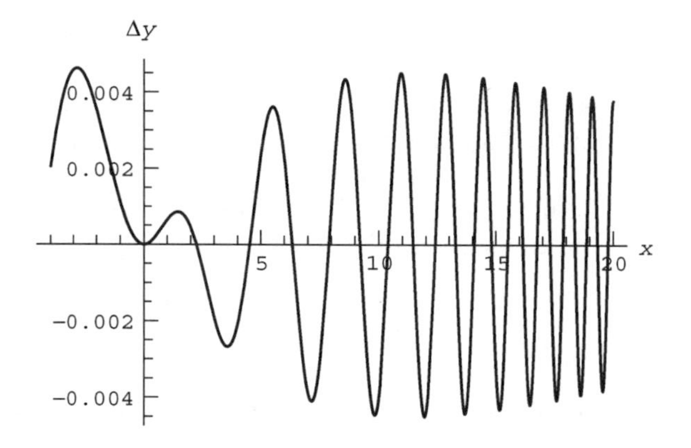

WKB Method for Systems That Vary Slowly in Time: Adiabatic Invariance

Consider a quantum-mechanical system, such as an atom, that is in a particular energy eigenstate described by some set of quantum numbers. When the system is put in an external electric or magnetic field that varies rapidly in time, it is well known that this can cause the system to jump to other energy eigenstates, with different quantum numbers.

On the other hand, if the external fields vary slowly, it is observed that the system remains in the original quantum state. For example, hydrogen atoms remain in the ground state if they are slowly transported from a field-free region to a region where magnetic or electric fields exist. Thus, the quantum numbers are *constants of the motion* when the external fields vary slowly, but not when the fields vary quickly.

In 1911, at the first Solvay conference, the question was raised as to whether there is analogous behavior in classical systems. That is, when a classical system is placed in slowly varying external fields, is there some quantity that remains time-independent, in analogy to the quantum numbers of the quantum system? Albert Einstein, a conference attendee, showed that such a quantity does in fact exist. It is called an adiabatic invariant, because it is only invariant if the fields vary adiabatically (slowly).

To illustrate the concept of an adiabatic invariant, Einstein chose the following model system: a pendulum whose length is varied slowly compared to its oscillation frequency. A slightly simpler version of this problem is a harmonic oscillator for which the oscillator frequency varies slowly in time in some prescribed manner, $\omega_0 = \omega_0(t)$ (the pendulum problem is considered in the exercises). The Hamiltonian of the oscillator is

$$H = \tfrac{1}{2}m\dot{x}^2 + \tfrac{1}{2}m\,\omega_0^2(t)\,x^2. \tag{5.2.19}$$

Since H depends explicitly on time, the energy of the system is not a conserved quantity. To find the adiabatic invariant for this system, one can examine the equation of motion for the oscillator:

$$\ddot{x}(t) = -\omega_0^2(t)\,x(t). \tag{5.2.20}$$

This equation has the same form as Eq. (5.2.3), so we can apply WKB theory to find an approximate solution. This solution will be valid only if the frequency $\omega_0(t)$ is slowly varying in time. Specifically, the requirement on ω_0 is $\dot{\omega}_0/\omega_0^2 \ll 1$ [this is analogous to Eq. (5.2.8)]. Comparing with the WKB solution of Eq. (5.2.3), we can write the WKB solution to Eq. (5.2.20) as

$$x(t) = \frac{A}{\sqrt{\omega_0(t)}}\cos\left(\int^t \omega_0(t)\,dt + \phi\right), \tag{5.2.21}$$

where ϕ is a constant phase factor and A and ϕ are determined by the initial conditions on Eq. (5.2.10). Equation (5.2.21) shows that when the frequency of the oscillator increases with time, the amplitude of the oscillations decreases like $1/\sqrt{\omega_0(t)}$.

We can see this behavior if we solve Eq. (5.2.20) numerically for a given initial condition and a given $\omega_0(t)$. For example, take $\omega_0(t) = -\alpha t + e^{\alpha t}$, $x(0) = 1$, $x'(0) = 0$. Then the WKB solution for these initial conditions is $x(t) = \cos\int_0^t \omega_0(t)\,dt/\sqrt{\omega_0(t)}$. This is compared with the numerical solution of Eq. (5.2.20) in Cell 5.21. The WKB solution (dashed line) works reasonably well, considering that the time variation of $\omega_0(t)$ is rather rapid. By changing the value of α, the reader can observe how the error in the WKB solution depends on rate of change of $\omega_0(t)$.

Cell 5.21

```
Clear["Global`*"];
<<Graphics`;
α = 1/3;
ω0[t_] = -αt + e^(αt);
```

```
xWKB[t_] = Cos[Integrate[ω0[t1], {t1, 0, t}]]/Sqrt[ω0[t]];
NDSolve[{x"[t] == -ω0[t]^2 x[t], x[0] == 1, x'[0] == 0},
  x[t], {t, 0, 15}, MaxSteps→5000];
x1[t_] = x[t]/. %[[1]];
Plot[{xWKB[t], x1[t]}, {t, 0, 15}, PlotStyle→
  {Red, Dashing[{0.01, 0.02}], Thickness[0.006]}, Blue},
AxesLabel→{"t", ""}, PlotLabel→"Oscillator with varying
    frequency"];
```

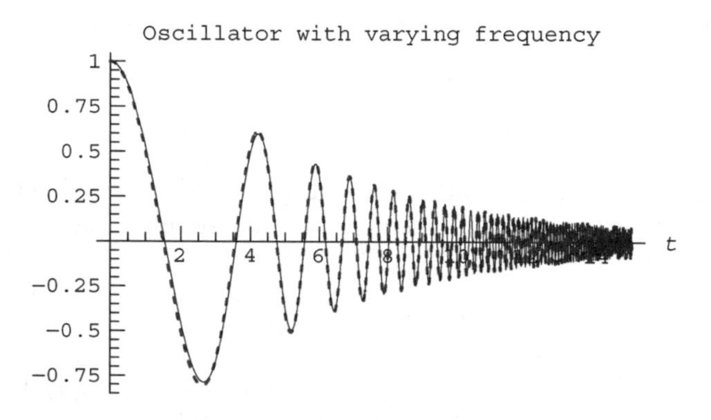

Can we find a constant of the motion for this problem in the WKB limit, where $\dot{\omega}_0/\omega_0^2 \ll 1$? We know that energy is not conserved. However, consider the following quantity:

$$J = E(t)/\omega_0(t). \tag{5.2.22}$$

We will show that the quantity J is constant in the WKB approximation, and is the adiabatic invariant we are looking for. To show this, we will substitute Eq. (5.2.21) into Eq. (5.2.19). This requires that we first evaluate $\dot{x}(t) = dx/dt$:

$$\dot{x}(t) = \frac{A}{\sqrt{\omega_0(t)}}\left[-\frac{\dot{\omega}_0}{2\omega_0}\cos\left(\int^t \omega_0(t)\,dt + \phi\right) + \omega_0 \sin\left(\int^t \omega_0(t)\,dt + \phi\right)\right].$$

However, the first term is negligible compared to the second term in the WKB limit where $\dot{\omega}_0/\omega_0^2 \ll 1$, so we will drop it. Then we have

$$H = E \simeq \frac{1}{2}m\frac{A^2}{\omega_0(t)}\omega_0^2 \sin^2\left(\int^t \omega_0(t)\,dt + \phi\right) + \frac{1}{2}m\omega_0^2\frac{A^2}{\omega_0(t)}\cos^2\left(\int^t \omega_0(t)\,dt + \phi\right)$$
$$= \frac{1}{2}mA^2\omega_0(t).$$

This proves that $J = E/\omega_0(t)$ is a constant of the motion in the WKB limit. However, it is only an approximate constant. For example, if we evaluate $E/\omega_0(t)$ using the previous numerical solution, we find (Cell 5.22) that this quantity does vary slightly. Initially, J is not very well conserved. But as time goes by, $\dot{\omega}_0/\omega_0^2 \sim e^{-\alpha t}$ becomes smaller, we approach the WKB limit, and the adiabatic invariant is almost a constant of the motion.

Cell 5.22

```
J[t_] = (1/2 x1'[t]^2 + 1/2 ω0[t]^2 x1[t]^2)/ω0[t];
Plot[J[t], {t, 0, 15}, PlotPoints→2000,
  PlotRange→All, PlotLabel→"adiabatic invariant vs time"];
```

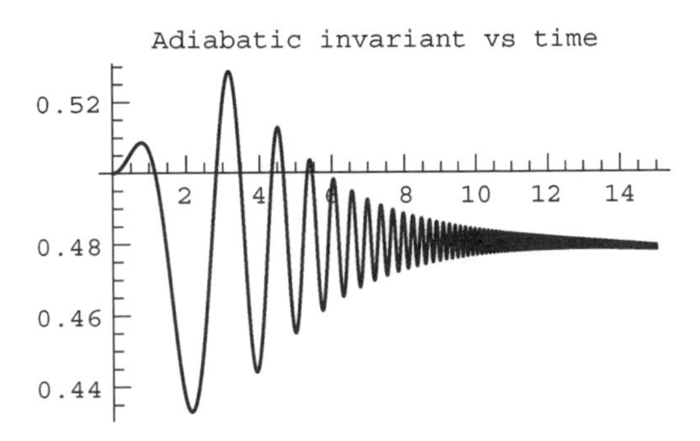

Adiabatic invariants can also be defined for other systems where a degree of freedom exhibits high-frequency oscillations. These approximate invariants can be related to action integrals. For in-depth discussions of this important subject, see Goldstein (1980, pg. 431), and Landau and Lifshitz (1976, pg. 154).

WKB Wave Packets Let us return now to the problem of waves on an inhomogeneous string. General solutions to the wave equation can be constructed out of superpositions of these traveling waves. In order to see how this is accomplished, perform a temporal Fourier transform of the wave equation (5.2.1): $-\omega^2\tilde{y}(x,\omega) = c^2(x)\,\partial^2\tilde{y}/\partial x^2$, where $\tilde{y}(x,\omega)$ is the Fourier transform in time of $y(x,t)$. Dividing through by c^2, we obtain

$$\frac{\partial^2 \tilde{y}}{\partial x^2} = k^2(x,\omega)\,\tilde{y}(x,\omega), \tag{5.2.23}$$

where $k(x,\omega) = \omega/c(x)$. The general solution to this ODE is of the form

$$\tilde{y}(x,\omega) = A(\omega)\,\psi_1(x,\omega) + B(\omega)\,\psi_2(x,\omega), \tag{5.2.24}$$

where $\psi_1(x,\omega)$ and $\psi_2(x,\omega)$ are two independent solutions to this second-order ODE. Taking the inverse transform, we find that the general solution to Eq. (5.2.23) may be expressed as

$$y(x,t) = \int_{-\infty}^{\infty} \frac{d\omega}{2\pi}\left[A(\omega)\,e^{-i\omega t}\,\psi_1(x,\omega) + B(\omega)\,e^{-i\omega t}\,\psi_2(x,\omega)\right]. \tag{5.2.25}$$

The functions $A(\omega)$ and $B(\omega)$ can be used to help match to given initial conditions and/or boundary conditions (if any).

We found WKB approximations for $\psi_1(x,\omega)$ and $\psi_2(x,\omega)$ previously; see Eq. (5.2.15). Assuming that the WKB limit holds, we can use these solutions in Eq. (5.2.25):

$$y(x,t) = \int_{-\infty}^{\infty} \frac{d\omega}{2\pi}\,\frac{A(\omega)\,e^{-i\omega t + i\int^x k(x,\omega)\,dx} + B(\omega)\,e^{-i\omega t - i\int^x k(x,\omega)\,dx}}{\sqrt{k(x,\omega)}}. \tag{5.2.26}$$

For the case of a nonuniform string, where $k(x, \omega) = \omega/c(x)$, Eq. (5.2.26) can be further simplified:

$$y(x,t) = \sqrt{c(x)} \int_{-\infty}^{\infty} \frac{d\omega}{2\pi} \left[\bar{A}(\omega)\, e^{-i\omega(t - \int^x dx/c(x))} + \bar{B}(\omega)\, e^{-i\omega(t + \int^x dx/c(x))} \right], \tag{5.2.27}$$

where $\bar{A} = A/\sqrt{\omega}$ and $\bar{B} = B/\sqrt{\omega}$. Let us now define a function $\tau(x) = \int^x dx/c(x)$; this is the time required to propagate to point x. [It is convenient to leave the integral in $\tau(x)$ with an indeterminate lower bound, so we have not specified an initial position. To be precise, we should say that $\tau(x_2) - \tau(x_1) = \int_{x_1}^{x_2} dx/c(x)$ is the time required to propagate from x_1 to x_2.]

Using this definition for $\tau(x)$ in Eq. (5.2.26), we can write the solution as

$$y(x,t) = \sqrt{c(x)}\, [f(t - \tau(x)) + g(t + \tau(x))], \tag{5.2.28}$$

where $f(t) = \int_{-\infty}^{\infty}(d\omega/2\pi)\bar{A}(\omega)\, e^{i\omega t}$ and $g(t) = \int_{-\infty}^{\infty}(d\omega/2\pi)\bar{B}(\omega)\, e^{-i\omega t}$ are Fourier transforms of $\bar{A}$ and $\bar{B}$. This WKB form of the general solution to the wave equation for an inhomogeneous medium has much in common with the expression for the d'Alembert solution for a homogeneous medium, Eq. (5.1.11). Like Eq. (5.1.11), Eq. (5.2.28) describes two counterpropagating wave packets. But now the time required to propagate to position x is $\tau(x)$ rather than x/c, and the shape and amplitude of the packets vary.

For example, let us consider an initially Gaussian packet, $y(x,0) = y_0(x) = e^{-x^2}$, moving to the right. Then we can take $g = 0$, since this part of the solution describes a packet moving to the left. Also, $f(t)$ is determined by the initial condition,

$$y(x,0) = \sqrt{c(x)}\, f(-\tau(x)) = y_0(x). \tag{5.2.29}$$

We can solve this equation for $f(t)$ by inverting the function $\tau(x)$, given that τ increases monotonically with x. Let $X(t)$ be the solution for x to the equation $t = \tau(x)$, for given t. If we let $t = \tau(x)$ in Eq. (5.2.29), then $x = X(t)$, and the equation becomes

$$f(-t) = \frac{y_0(X(t))}{\sqrt{c(X(t))}}. \tag{5.2.30}$$

This equation determines $f(t)$, and when used in Eq. (5.2.28) it provides us with the solution for the motion of the wave packet.

We will work out the WKB solution for the case $c(x) = c_0\, e^{-x/2L}$ discussed previously. Then $\tau(x) = \int^x dx/c(x) = (2L/c_0)\, e^{x/2L}$. This shows that it takes exponentially longer to travel to larger x, because the wave is slowing down as x increases. The solution for x to the equation $\tau(x) = t$ is $X(t) = 2L \ln[c_0 t/(2L)]$. Then Eq. (5.2.30) implies $f(-t) = e^{-X^2(t)}\sqrt{t/(2L)}$. Using this expression in Eq. (5.2.28), together with our expression for $T(x)$, provides us with the required solution for the propagation of the packet in the WKB approximation. This is shown in Cell 5.23, taking $L = 2$, $c_0 = 1$. As the wave packet propagates into the heavier region of the string where $c(x)$ decreases, the packet slows and becomes smaller. It also narrows, because the back of the packet catches up with the front as the packet slows down.

Cell 5.23

```
c[x_] = c0 e^(-x/(2L));
τ[x_] = (2L/c0) e^(x/(2L));
X[t_] = 2 L Log[c0 t/(2L)];
f[t_] = e^(-X[-t]^2) Sqrt[-t/(2L)];
y[x_, t_] = Sqrt[c[x]] f[t - τ[x]]/. {c0→1, L→2};

Table[Plot[y[x, t], {x, -3, 15},
   PlotRange→{{-3, 10}, {-0.2, 1}}, AxesLabel→{"x", ""},
   PlotLabel→"y[x, t], t =" <>ToString[t]], {t, 0, 30, 1}];
```

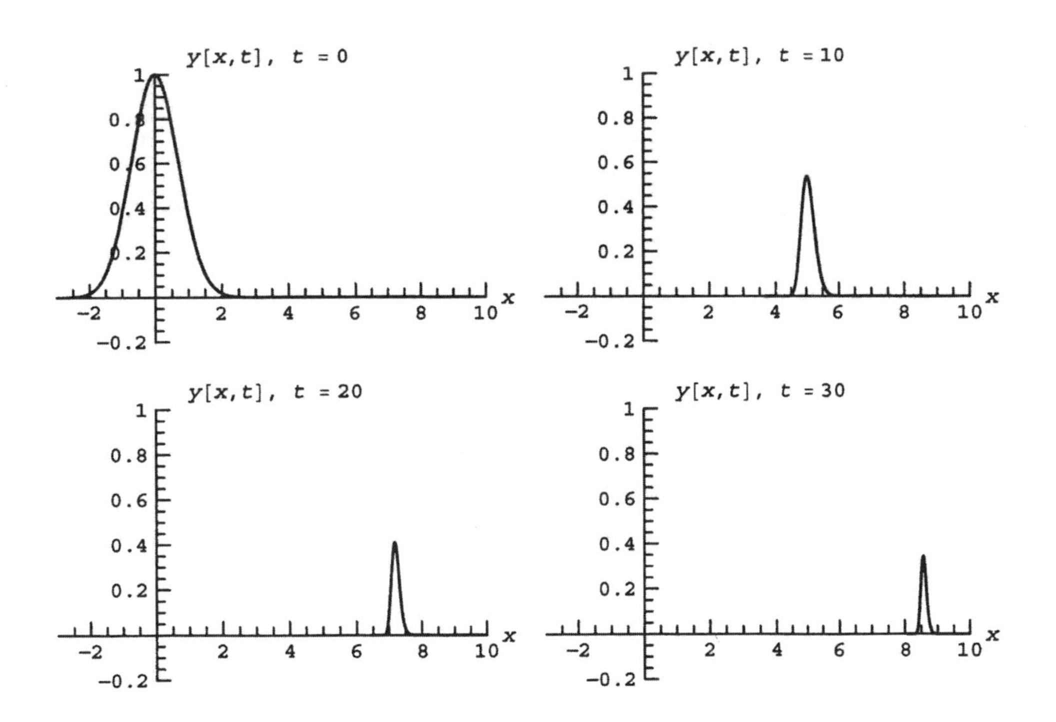

We also observed this narrowing of the packet when we analzyed the motion of a hanging string in Chapter 4, Sec. 4.2. There, however, it was the tension that varied with position, not the mass. As a result, the amplitude of the packet *increased* as the packet slowed and narrowed, rather than decreasing as it does here. The reason for this difference in behavior between strings with nonuniform tension and nonuniform mass density will be taken up in the exercises.

WKB Waves in Two Dimensions: Water Waves Near the Shore Waves in shallow water of varying depth $h(\mathbf{r})$, with amplitude $z(\mathbf{r}, t)$, are described by the following wave equation:

$$\frac{\partial^2 z}{\partial t^2} = \nabla \cdot \left[c^2(\mathbf{r}) \nabla z \right], \tag{5.2.31}$$

where $c^2(\mathbf{r}) = gh(\mathbf{r})$, and g is the acceleration of gravity. Equation (5.2.31) is a generalization of Eqs. (3.1.78) and (3.1.79) for varying depth and propagation in two dimensions, with $\mathbf{r} = (x, y)$. This equation follows from the same arguments that led to Eq. (3.1.78). It provides a useful model for a discussion of wave propagation in inhomogeneous 2D systems.

For simplicity, we will assume that $h = h(x)$ only, so that there is no y-dependence to the water depth. In this case, traveling-wave solutions to Eq. (5.2.31) are of the form $z = e^{i(k_y y - \omega t)} \psi(x)$, where k_y is the y-component of the wavenumber, a constant. The function ψ satisfies the ODE

$$\frac{\partial}{\partial x} \left[\frac{1}{k_x^2(x, \omega, k_y)} \frac{\partial}{\partial x} \psi(x) \right] = -\psi(x), \tag{5.2.32}$$

where

$$k_x^2(x, \omega, k_y) = \frac{\omega^2}{c^2(x)} - k_y^2 \tag{5.2.33}$$

is the square of the x-component of the wave vector. Equation (5.2.33) can be rearranged to read

$$\omega^2 = c^2(x)\left(k_x^2 + k_y^2\right), \tag{5.2.34}$$

which is simply the dispersion relation for these waves. Since ω and k_y are constants, as $c(x)$ varies k_x must also vary in order to satisfy Eq. (5.2.34).

Equation (5.2.32) is a linear ODE that is similar in form to Eq. (5.2.3), so WKB analysis can be used to follow the behavior of these shallow-water waves, provided that the depth varies little over a wavelength. That is, we require that $k_x L \gg 1$, where $L = |1/(\partial \ln k_x / \partial x)|$ is the scale length due to the depth variation. If this is so, then we can use the eikonal method discussed in the previous section, first writing $\psi = e^{S(x)}$, and then approximating $S(x)$ in the WKB limit. The result is

$$z(x, y, t) = A\sqrt{k_x(x, \omega, k_y)}\, e^{i[k_y y \pm \int^x k_x(x, \omega, k_y)\, dx - \omega t]}, \tag{5.2.35}$$

(see the exercises). Equation (5.2.35) describes a wave propagating in either the $+x$ or the $-x$ direction, and at the same time propagating in the y-direction with wavenumber k_y.

For example, let's assume that the water depth $h(x)$ decreases as x increases, according to

$$h(x) = \begin{cases} h_0 a/(a + x), & x > 0, \\ h_0, & x < 0. \end{cases} \tag{5.2.36}$$

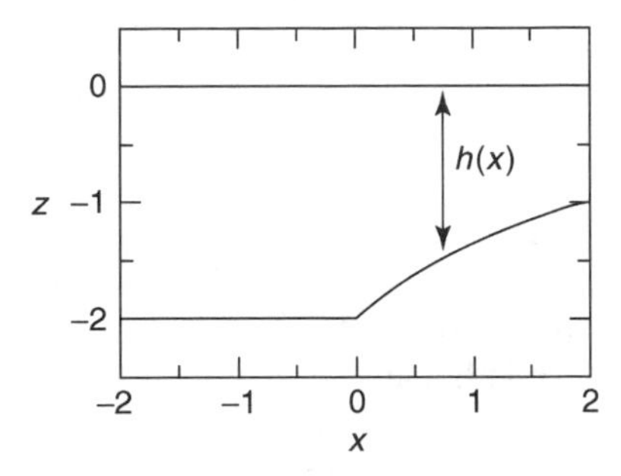

Fig. 5.6 Equilibrium water depth as a function of position.

In other words, as x increases we approach the shore. (See Fig. 5.6.) In Cell 5.24, we display $z(x, 0, t)$ for a wave propagating in the positive x and y directions. In this example we choose ω so that $k_x = k_y = 1\ \text{m}^{-1}$ for $x < 0$. According to Eq. (5.2.34) this requires $\omega = \sqrt{2gh_0} \times 1\ \text{m}^{-1}$. Also, for $x > 0$, $k_x = \sqrt{1 + 2x/a}\ \text{m}^{-1}$ according to Eqs. (5.2.33) and (3.1.79).

Cell 5.24

```
h[x_] := h0 a/(a + x)/; x > 0
h[x_] := h0/; x ≤ 0
c[x_] = √(g h[x]);

kx[x_] := √(ω²/c[x]² - ky²);

ω = √(2 g h0);
ky = 1;
g = 9.8;
a = 2;
h0 = 2;

z[x_, y_, t_] := √(kx[x]) Exp[I (ky y +NIntegrate[kx[x1],
  {x1, 0, x}] -ω t}]

Table[Plot[Re[z[x, 0, t]], {x, -5, 10}, PlotRange→{-3, 3},
  AxesLabel→{"x", "z[x, 0, t]"}], {t, 0, 1.8Pi/ω, .2Pi/ω}];
```

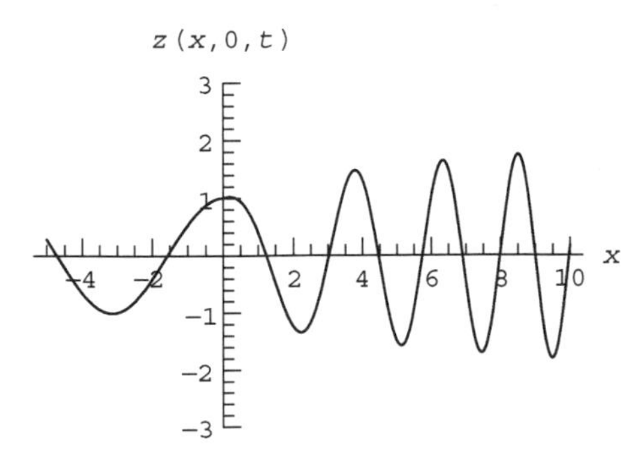

As the waves move toward the beach, they slow down [$c(x)$ decreases as the depth decreases]. As the waves slow, the wavefronts pile up, and this results in a shortening of the x-wavelength, $2\pi/k_x$ (that is, k_x increases). However, unlike the waves on a nonuniform string, the amplitude of these water waves *increases* as the waves slow. Eventually, if the amplitude becomes sufficiently large, nonlinear effects become important: the waves break. This will be discussed in Chapter 7.

While k_x increases as x increases, k_y remains constant, and this implies that the wave vector $\mathbf{k} = (k_x, k_y)$ changes direction as the waves propagate, increasingly pointing in the x-direction as x increases. This is shown in Cell 5.25 using a contour plot. The waves bend toward the shore. This is because the portion of a wavefront that is closer to the beach moves more slowly, falling behind. This behavior is familiar to anyone who has observed waves at the beach. The bending of the wavefronts is analogous to the refraction of light waves. We discuss this connection in Sec. 5.2.2.

Cell 5.25

```
waves = Table[
  ContourPlot[Re[z[x, y, t]], {x, -5, 10}, {y, -5, 10},
   PlotRange→{-3, 3},
   FrameLabel→{"x", "y"}, PlotPoints→100],
    {t, 0, 1.8Pi/ω, .2Pi/ω}];
```

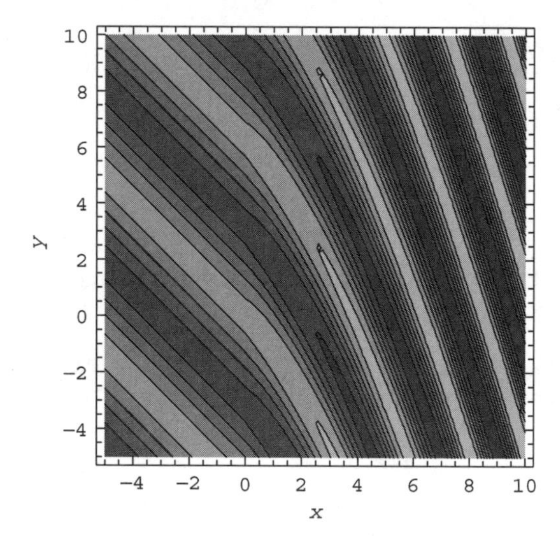

Ray Trajectories For waves propagating in two dimensions in a medium where $c = c(x)$, we have seen that the wave's phase $\phi(x, y, t)$ is given in the WKB approximation by

$$\phi(x, y, t) = k_y y + \int^x k_x \, dx - \omega t, \tag{5.2.37}$$

where $k_x = \sqrt{\omega^2/c^2(x) - k_y^2}$. This phase appears in the exponent of Eq. (5.2.35). Curves of constant ϕ define the phase fronts that are observed to propagate and bend in the previous animation.

Consider a point on a phase front at position (x_0, y_0) at time $t=0$. Let's assume that this point moves with the wave, and determine the path that it follows. One might think of this point as a surfer who is allowing the wave to push him/her along at the local wave speed, in a direction normal to the local wavefront.

The equations of motion for this point of constant phase can be obtained by taking a differential of Eq. (5.2.37) and setting $d\phi$ equal to zero:

$$d\phi = 0 = k_y\, dy + k_x(x, \omega, k_y)\, dx - \omega\, dt,$$

or in other words, $d\mathbf{r}\cdot\mathbf{k} = \omega\, dt$. Since the direction of motion is the **k**-direction, we obtain

$$\frac{d\mathbf{r}}{dt} = \frac{\omega}{k}\hat{\mathbf{k}} = c(x)\hat{\mathbf{k}}, \tag{5.2.38}$$

where $\hat{\mathbf{k}} = \hat{\mathbf{k}}(x)$ is a unit vector in the **k**-direction, and where we have used Eq. (5.2.34) in the second step.

Thus, the phase point travels with the local phase velocity $\mathbf{v}_\phi(x) = (\omega/k)\hat{\mathbf{k}}$, which is hardly surprising. This is the same equation for phase velocity as for waves in a homogeneous medium, Eq. (5.1.41), but now the velocity varies with position. For a given initial condition, $\mathbf{r}(t=0) = (x_0, y_0)$, the solution to Eq. (5.2.38) defines a curve $\mathbf{r}(t)$ in the x-y plane, which is the trajectory of the point of constant phase as time progresses. This curve is called a *ray trajectory*.

In component form, Eq. (5.2.38) becomes

$$\begin{aligned} \frac{dx}{dt} &= c(x)\frac{k_x}{k}, \\ \frac{dy}{dt} &= c(x)\frac{k_y}{k}. \end{aligned} \tag{5.2.39}$$

In Cell 5.26, we solve Eq. (5.2.39) for three separate initial conditions, and plot them, using our previous results for $k_x(x)$. The rays are curves $y(x)$ that are normal to each surface of constant phase. These surfaces are nearly identical to the contours of constant wave height shown in Cell 5.26, since the wave amplitude, $\sqrt{k_x(x)}$, is slowly varying compared to ϕ [see Eq. (5.2.35)].

Cell 5.26

```
Do[x0 = -5 + 2 n; y0 = -5 + 4n;
  ray = NDSolve[{x'[t] == c[x[t]] kx[x[t]]/
   Sqrt[ky^2 + kx[x[t]]^2],
     y'[t] == c[x[t]] ky/Sqrt[ky^2 + kx[x[t]]^2],
     x[0] == x0, y[0] == y0}, {x[t], y[t]}, {t, 0, 10}];
  xray[t_] = x[t]/. ray[[1]]; yray[t_] = y[t]/. ray[[1]];
 plot[n] = ParametricPlot[{xray[t], yray[t]}, {t, 0, 10},
   DisplayFunction→ Identity, PlotStyle→RGBColor[1, 0, 0]
   ];, {n, 0, 2}];
Show[Join[{waves[[1]]}, Table[plot[n], {n, 0, 2}]],
  PlotRange→{{-5, 10}, {-5, 10}}, DisplayFunction→
   $DisplayFunction];
```

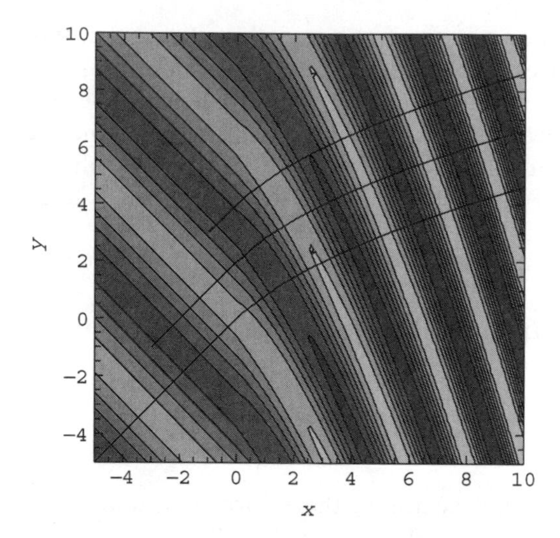

The bending of the rays can also be understood by dividing dy/dt by dx/dt in Eq. (5.2.39):

$$\frac{dy/dt}{dx/dt} = \frac{dy}{dx} = \frac{k_y}{k_x(x, \omega, k_y)}. \tag{5.2.40}$$

This equation has a simple physical interpretation. If we define $\theta(x)$ as the angle the rays make with the horizontal (see Fig. 5.7), Eq. (5.2.40) implies that $\tan\theta(x) = k_y/k_x$, or equivalently, $\sin\theta(x) = k_y/k(x) = k_y c(x)/\omega$, where in the second step we have used Eq. (5.2.34). Dividing through by $c(x)$ yields

$$\frac{\sin\theta(x)}{c(x)} = \frac{k_y}{\omega} = \text{constant}. \tag{5.2.41}$$

Equation (5.2.41) is the famous Snell's law for the propagation of waves through an inhomogeneous medium. As the phase speed $c(x)$ decreases, Snell's law implies that $\theta(x)$ also decreases. In other words, *rays bend toward regions of lower phase speed.* This same equation applies to the refraction of light rays as they propagate through an inhomogeneous medium.

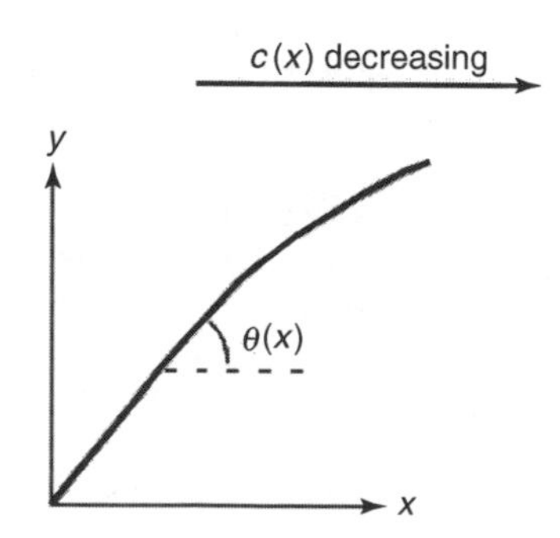

Fig. 5.7 Ray trajectory in a medium with wave speed $c = c(x)$ that decreases to the right.

It is interesting that Snell's law, derived here for a medium that varies slowly compared to the wavelength, also applies even for a medium with sharp interfaces, such as occurs for light propagating from air to glass. What is missing in the WKB theory of this problem is the *reflected* wave from the sharp interface. Reflections do not occur in the WKB approximation, except at turning points where total internal reflection occurs. (Turning points are discussed at the end of the next section.)

5.2.2 WKB with Dispersion: Geometrical Optics

Wave-Packet Trajectory When light travels through a dispersive medium such as glass, the phase velocity is given by

$$v_\phi = \omega/k = c/n(x,\omega), \tag{5.2.42}$$

where c is the speed of light in vacuum and $n(x, \omega)$ is the index of refraction of the medium. The index depends on the frequency of the light; for example, in the visible region, the index of window glass is an increasing function of frequency. We would like to describe the behavior of a wave packet traveling through such a medium, assuming that the refractive index varies little in space over a wavelength of the light, so that the WKB method is valid.

The previous discussion of WKB theory focused on dispersionless systems where the phase velocity $\omega/k = c(x)$ is independent of frequency ω. On the other hand, the Fourier-transformed wave equation (5.2.23) could also be applied to dispersive systems where $k(x, \omega)$ is a general function of ω, given by $k(x, \omega) = \omega n(x, \omega)/c$. Therefore, the WKB solution (5.2.26) applies to dispersive media. Such media have a general dispersion relation, found by inverting the equation $k = k(x, \omega)$ in order to obtain the frequency vs. x and k:

$$\omega = \omega(x,k). \tag{5.2.43}$$

Let us assume for simplicity that we are dealing only with a traveling wave packet moving to the right, so that we can take $B = 0$ in Eq. (5.2.26). Then one possible approach is to assume that $A(\omega)$ is sharply peaked around some frequency ω_0. (That is, we assume that we are dealing with a wave packet that is nearly a single wave of frequency ω_0, with an envelope that varies slowly in time compared to ω_0. This is analogous to the discussion of wave packet for homogeneous systems in Sec. 5.1.2.) In this case, we can Taylor-expand $k(x, \omega)$ about ω_0, writing $k(x, \omega) \simeq k(x, \omega_0) + (\omega - \omega_0)\, \partial k(x, \omega_0)/\partial\omega_0$. Using this expansion in Eq. (5.2.26), we obtain

$$y(x,t) = \frac{\exp\left[i\left(\int^x k(x,\omega_0)\,dx - \omega_0 \int^x \frac{\partial k}{\partial\omega_0}(x,\omega_0)\,dx\right)\right]}{\sqrt{k(x,\omega_0)}} \times \int_{-\infty}^{\infty} \frac{d\omega}{2\pi} A(\omega) \exp\left[-i\omega\left(t - \int^x \frac{\partial k(x,\omega_0)}{\partial\omega_0}\,dx\right)\right].$$

If we now introduce the function $f(t) = \int_{-\infty}^{\infty} (d\omega/2\pi) A(\omega) e^{-i\omega t}$, this equation becomes

$$y(x,t) = \frac{\exp\left[i\left(\int^x k(x,\omega_0)\,dx - \omega_0 \tau(x,\omega_0)\right)\right]}{\sqrt{k(x,\omega_0)}} f(t - \tau(x,\omega_0)), \quad (5.2.44)$$

where $\tau(x, \omega_0) = \int^x \partial k(x, \omega_0)/\partial \omega_0 \, dx$. The function $\tau(x, \omega_0)$ determines the time for the wave packet to propagate to position x, or more precisely, $\tau(x_2, \omega_0) - \tau(x_1, \omega_0)$ is the time required to propagate from position x_1 to x_2. This can be seen by noting that the function $k = k(x, \omega)$ can be inverted to obtain the dispersion relation for the waves, $\omega = \omega(x, k)$. Then the following identity holds:

$$\left.\frac{\partial k(x,\omega_0)}{\partial \omega_0} \frac{\partial \omega(x,k)}{\partial k}\right|_{k=k(x,\omega_0)} = 1.$$

However, this implies that $\partial k(x, \omega_0)/\partial \omega_0 = 1/v_g(x, \omega_0)$, where $v_g(x, \omega_0) = \partial \omega(x, k)/\partial k|_{k=k(x,\omega_0)}$ is the local group velocity of a wave packet with central frequency ω_0, given by Eq. (5.1.28). Thus, we can write

$$\tau(x,\omega_0) = \int^x \frac{dx}{v_g(x,\omega_0)}, \quad (5.2.45)$$

which is clearly the time required to travel to position x for a wave packet of constant frequency ω_0 traveling with the given group velocity $v_g(x, \omega_0)$.

Although the central frequency ω_0 of the wave packet is a constant parameter, determined by $A(\omega)$, the central wavenumber is not; the wavenumber varies slowly in space according to $k = k(x, \omega_0)$. It is useful to consider the time variation of k as one follows along with a wave packet with given central frequency ω. (We drop the subscript on ω_0 to simplify the notation in what follows.) Taking a time derivative and using the chain rule, we have $dk/dt = (dx/dt)\,\partial k(x, \omega)/\partial x$. However, we have seen that the packet moves at the group velocity, so its position satisfies

$$\frac{dx}{dt} = \frac{\partial \omega(x,k)}{\partial k}, \quad (5.2.46)$$

where $\omega(x, k)$ is the wave frequency, determined as a function of x and k by the dispersion relation (5.2.43). Therefore, the central wavenumber of the packet satisfies

$$\frac{dk}{dt} = \frac{\partial \omega(x,k)}{\partial k} \frac{\partial k(x,\omega)}{\partial x} = -\frac{\partial \omega(x,k)}{\partial x}, \quad (5.2.47)$$

where the second equality follows from a standard identity for differentials. Equations (5.2.46) and (5.2.47) are coupled first-order ODEs for the evolution of the central wavenumber and position of a WKB wave packet. If we know x and k at some initial time, these ODEs can be used to find $x(t)$ and $k(t)$ for all time.

These equations are called the equations of *geometrical optics*. We will see later, in Sec. 5.3, that the equations can be generalized to wave-packet motion in two or more dimensions, according to

$$\frac{d\mathbf{r}}{dt} = \frac{\partial \omega(\mathbf{r},\mathbf{k})}{\partial \mathbf{k}},$$
$$\frac{d\mathbf{k}}{dt} = -\frac{\partial \omega(\mathbf{r},\mathbf{k})}{\partial \mathbf{r}}. \tag{5.2.48}$$

If we compare this with Eqs. (1.2.15), we see that these ODEs have a Hamiltonian form, with $\mathbf{k}$ taking the role of the momentum $\mathbf{p}$ and the frequency $\omega(\mathbf{r},\mathbf{k})$ taking the role of the Hamiltonian $H(\mathbf{r},\mathbf{p})$. Thus, the fact that the central frequency ω of the wave packet is constant in time during the evolution of the wave packet corresponds to the fact that energy is conserved for time-independent Hamiltonians.

Also, for wave-packet motion in the x-y plane, if $\omega = \omega(x,\mathbf{k})$, then Eq. (5.2.48) implies that $dk_y/dt = 0$: the y-component of the central wavenumber for the packet is a constant of the motion. This is equivalent to the conservation of the y-component of momentum when the Hamiltonian is y-independent.

The trajectory of a wave packet moving in the x-y plane is given by the ratio of the x and y components of $d\mathbf{r}/dt = \partial\omega/\partial\mathbf{k}$:

$$\frac{dy/dt}{dx/dt} = \frac{dy}{dx} = \frac{k_y |v_g|/k}{k_x |v_g|/k} = \frac{k_y}{k_x}, \tag{5.2.49}$$

where in the second step we assumed that the medium is isotropic, and used Eq. (5.1.43) for the group velocity. This ODE is identical to the equation for ray trajectories, Eq. (5.2.40). (Recall that ray trajectories describe the propagation of phase fronts, as opposed to the propagation of the wave packet.) Therefore, in an isotropic medium the trajectory of the center of a wave packet with central frequency ω is identical to the trajectory of a ray with the same initial direction and the same frequency. This is because the phase and group velocities in an isotropic medium are in the same direction, along $\mathbf{k}$: see Eqs. (5.1.41) and (5.1.43). (However, the time required to propagate to a given point along the trajectory will generally differ in the two cases, because the magnitude of the phase and group velocities need not be the same.) Later in the chapter, we will consider anisotropic media where the ray and wave-packet trajectories are completely different.

Ray trajectories trace out the path followed by points of constant phase (points on phase fronts). Wave-packet trajectories trace out the path followed by wave packets. For a given frequency and initial propagation direction, in an isotropic medium for which $\omega = \omega(|\mathbf{k}|, \mathbf{r})$, the two trajectories are the same.

Of course, a wave packet contains a spread of frequencies, and this leads to dispersion: different frequency components travel with different phase velocities, in different directions. We will not consider wave-packet dispersion in an inhomogeneous medium in any detail here. For the most part, the qualitative features of the dispersion are the same as for a homogeneous medium. Around any given wave-packet trajectory [a solution to Eq. (5.2.48)], the wave packet spreads in both the transverse and longitudinal directions.

However, there is one novel dispersive effect in an inhomogeneous medium. Recall that transverse dispersion is weak for a collimated beam that is sufficiently broad compared to its wavelength: the Rayleigh length becomes very large [Eq. (5.1.56)]. In such a beam, different frequency components are all traveling in the same direction. However, in an inhomogeneous dispersive medium, transverse dispersion can occur even if all the rays in a packet are initially propagating in the same direction. This transverse dispersion is responsible for the rainbow spectrum that occurs when a ray of white light is bent through a prism.

One can see this effect qualitatively by following the ray trajectories for different-frequency rays as they propagate through an inhomogeneous medium with a frequency-dependent refractive index $n(x, \omega)$. According to Eq. (5.2.49), rays with different frequencies take different paths, even if they start in the same direction. This may be seen explicitly by noting that k_y is constant in Eq. (5.2.49) and k_x is determined by the equation $k = \sqrt{k_x^2 + k_y^2} = \omega n(x, \omega)/c$. The second equality follows from the dispersion relation for these waves, Eq. (5.2.42). Solving this equation for k_x yields $k_x = \{[\omega n(x\ \omega)/c]^2 - k_y^2\}^{1/2}$. Then the ray equation becomes

$$\frac{dy}{dx} = \frac{1}{\sqrt{\left[\omega n(x, \omega)/ck_y\right]^2 - 1}},$$

which has the solution

$$y(x) = y(0) + \int_0^x \frac{1}{\sqrt{\left[\omega n(x, \omega)/ck_y\right]^2 - 1}}\, dx. \tag{5.2.50}$$

For example, if the index of refraction is $n(x, \omega) = 1 - \omega x^2/a$ for some constant a, then we can find the ray trajectory for a given frequency ω by evaluating the above integral. Let us assume that all the rays are initially propagating at 45° with respect to the horizontal, so that $dy/dx = 1$ at $x = 0$, where $n = 1$; then Eq. (5.2.50) [or Eq. (5.2.49)] implies that $(\omega/ck_y)^2 = 2$, and we have

Cell 5.27

```
Clear["Global`*"];
n[x_, ω_] = 1-x^2 ω/a;

y[x_, ω_] = Integrate[1/Sqrt[2n[x1, ω]^2 - 1], {x1, 0, x}]
```

$$-\frac{i \sqrt{1-\frac{2x^2\omega}{2a-\sqrt{2}\,a}}\sqrt{1-\frac{2x^2\omega}{2a+\sqrt{2}\,a}}\ \text{EllipticF}\left[i\ \text{ArcSinh}\left[\sqrt{2}\,x\sqrt{-\frac{\omega}{2a-\sqrt{2}\,a}}\right], \frac{2a-\sqrt{2}\,a}{2a+\sqrt{2}\,a}\right]}{\sqrt{2}\sqrt{-\frac{\omega}{2a-\sqrt{2}\,a}}\sqrt{-1+2\left(1-\frac{x^2\omega}{a}\right)^2}}$$

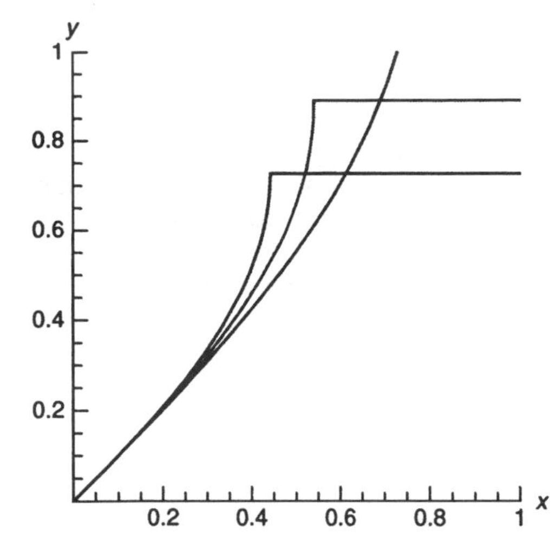

Fig. 5.8 Three ray trajectories in a dispersive medium where $n(x, \omega) = 1 - \omega x^2$.

This solution is plotted in Fig. 5.8, taking $a = 1$ for three rays with $\omega = 1, 2$, and 3. The rays bend toward the region of lower wave speed (more negative x) as expected from Snell's law. Although all three rays start out moving in the same direction, the refractive index for the blue ray (higher frequency) is larger than for the others, so this ray bends more. A light beam made up of these three frequencies would spread accordingly, revealing the spectral components of the beam.

Note that when $dy/dx = \infty$ on a given ray, our analytic solution for $y(x)$ fails. This is the location of a *turning point* where the ray trajectory is reflected back toward $-\hat{x}$. At the turning point, $k_x \to 0$, and we should have switched the sign of the square root in Eq. (5.2.50). One can avoid this problem with the analytic form of the ray trajectory by simply integrating Eqs. (5.2.48) numerically to obtain $\mathbf{r}(t)$. The appearance of a turning point in the rays is an example of *total internal reflection*, wherein rays are reflected back toward the region of higher refractive index (lower wave speed).

At a planar interface between two homogeneous media with refractive indices n_1 and n_2, $n_1 > n_2$, one can show that total internal reflection will occur for a ray propagating in medium 1 provided that the angle ψ between the interface and the ray is sufficiently small so that $\psi < \cos^{-1}(n_2/n_1)$ (see the exercises at the end of the section). We will have more to say about turning points later in the chapter.

As the spectral components of this beam spread, the beam amplitude would be expected to decrease as the wave energy is distributed over a larger volume. In fact, the behavior of the wave-packet amplitude near a turning point is rather more complicated than this simple picture suggests. The dynamics of the wave-packet amplitude are considered briefly in the sub-subsection entitled Wave-packet Amplitude, and in Sec. 5.3.

Hamiltonian Mechanics as the WKB Limit of Quantum Mechanics The evolution of the wave function $\psi(x, t)$ for a particle of mass m, moving in one dimension under the influence of a given potential $V(x)$, is determined by

Schrödinger's equation,

$$i\hbar \frac{\partial}{\partial t}\psi(x,t) = -\frac{\hbar^2}{2m}\frac{\partial^2}{\partial x^2}\psi(x,t) + V(x)\psi(x,t). \tag{5.2.51}$$

The WKB method is useful for understanding the solution to this equation, provided that the wave function varies rapidly in space compared with the spatial scale of $V(x)$, so that the WKB limit, Eq. (5.2.7), holds.

To put Eq. (5.2.51) into a form amenable to WKB analysis, we apply a Fourier transform in time, defining $\tilde{\psi}(x, \omega)$ to be the temporal Fourier transform of $\psi(x,t)$. Then Eq. (5.2.51) becomes

$$\hbar\,\omega\tilde{\psi}(x,\omega) = -\frac{\hbar^2}{2m}\frac{\partial^2}{\partial x^2}\tilde{\psi}(x,\omega) + V(x)\tilde{\psi}(x,\omega).$$

This equation can be rearranged to read

$$\frac{\partial^2}{\partial x^2}\tilde{\psi}(x,\omega) = -k^2(x,\omega)\tilde{\psi}(x,\omega), \tag{5.2.52}$$

where $k(x, \omega)$ is the local wavenumber of the wave function, defined by the equation

$$k^2(x,\omega) = \frac{2m}{\hbar^2}\left[\hbar\,\omega - V(x)\right]. \tag{5.2.53}$$

Let us assume for the moment that the energy of the packet, $\hbar\,\omega_0$, is greater than $V(x)$, so that $k(x, \omega_0)$ is a real number. Then Eq. (5.2.52) is identical to Eq. (5.2.3), and we already know the WKB solution to this problem. WKB wave packets propagating to the left or right are described by Eq. (5.2.26), and if these packets are narrow in frequency, a packet traveling to the right with frequency ω_0 is given by Eq. (5.2.44), with an analogous form for a packet traveling to the left.

The center of the packet travels according to the equations of geometrical optics, given by Eqs. (5.2.46) and (5.2.47). These equations can be written in a more recognizable form. The group velocity is related to the wavenumber k through $v_g = \hbar k/m = p/m$, where $p = \hbar k$ is the particle momentum. Also, the energy $H = \hbar\,\omega$ of the particle can be written in terms of p via Eq. (5.2.53): $H = \hbar\,\omega = \hbar^2 k^2/2m + V(x) = p^2/2m + V(x)$. Using these two results, Eqs. (5.2.46) and (5.2.47) can be written as

$$\begin{aligned} \frac{dx}{dt} &= \frac{\partial H(x,p)}{\partial p} = \frac{p}{m}, \\ \frac{dp}{dt} &= -\frac{\partial H(x,p)}{\partial x} = -\frac{\partial V}{\partial x}. \end{aligned} \tag{5.2.54}$$

Equations (5.2.54) are the Hamiltonian equations of motion for a classical particle moving in the potential $V(x)$. Thus, the equations of geometrical optics for the wave packet of a quantum particle lead directly to the equations of classical mechanics.

Turning Points Equations (5.2.53) and (5.2.54) exhibit a rather unsettling phenomenon from the point of view of WKB theory: there can be positions for which $k(x, \omega)$ vanishes. These positions occur where $E = \hbar\omega = V(x)$, which is merely the equation for the classical turning points in the motion of the particle. At these points the WKB approximation breaks down. One can see this in two ways: (1) the limit $k \to 0$ implies that $(d \ln k/dx)^{-1}$ is no longer small and (2) the amplitude of the WKB solution, Eq. (5.2.24), is singular where $k \to 0$.

In the optics example discussed in relation to Eq. (5.2.48), an analogous phenomenon occurred, referred to there as total internal reflection. The behavior of a wave packet near a turning point in either optics or in quantum theory can be treated with the same approach.

Near the turning point, we cannot use the WKB method. Instead, we will Taylor-expand $k(x, \omega)$ around the location of the turning point, $x = x_0$. At this point, $k(x_0, \omega) = 0$. Then we may write

$$k^2(x, \omega) \approx \alpha(x - x_0), \tag{5.2.55}$$

where the constant α is the slope of $k^2(x, \omega)$ at $x = x_0$. This expansion is valid provided that we do not stray too far from x_0: we need to keep $x - x_0 \ll L$, where L is the scale length for variation in k. Also, we assume here that $\alpha \neq 0$.

Let us further assume that $\alpha > 0$. This implies that the region $x < x_0$ is the region of classically allowed orbits, and $x > x_0$ is classically disallowed. If we use Eq. (5.2.55) in Eq. (5.2.52), and define $y = x - x_0$, we obtain the Airy equation:

$$\psi''(y) = \alpha y \psi(y). \tag{5.2.56}$$

The general solution is

$$\psi(y) = C_1\, Ai(\alpha^{1/3} y) + C_2\, Bi(\alpha^{1/3} y), \tag{5.2.57}$$

where $Ai(x)$ and $Bi(x)$ are Airy functions [see the discussion surrounding Eq. (1.3.2)]. However, for $\alpha > 0$, $Bi(\alpha^{1/3} y)$ blows up as $y \to \infty$, so we must take $C_2 = 0$ in order to obtain a finite wave function. Therefore, the wave function near the turning point has the form

$$\psi(x) = C_1\, Ai\big(\alpha^{1/3}(x - x_0)\big). \tag{5.2.58}$$

[This form also turns out to be correct when $\alpha < 0$, provided that when $\alpha < 0$ we take the branch of $\alpha^{1/3}$ given by $\alpha^{1/3} = -|\alpha|^{1/3}$.] The solution is plotted in Cell 5.28 (for $\alpha > 0$).

Cell 5.28

```
Plot[AiryAi[x], {x, -20, 3},
  PlotLabel → "Behavior of the wavefunction near a turning
   point",
  AxesLabel → {"α^(1/3) (x-x0)", ""}];
```

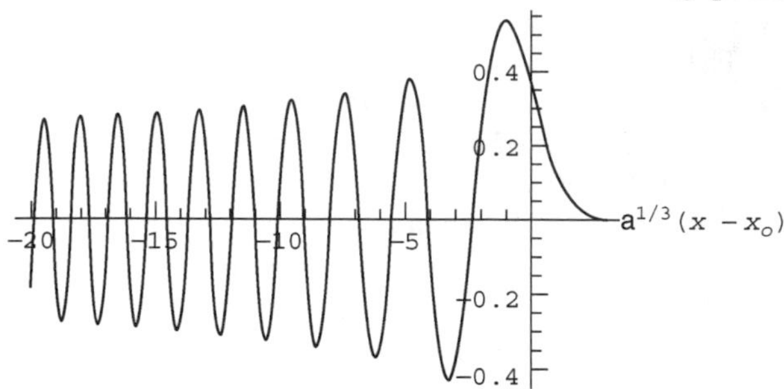

One can see that the amplitude and wavelength of the wave are increasing as one nears the turning point, as expected from WKB analysis as $k \to 0$. However, at the turning point the solution is not singular; rather, it decays exponentially as one passes through the turning point to the other side. The exponential decay of the wave function in the range of x that is classically disallowed is the well-known phenomenon of quantum tunneling; that is, quantum particles have a small but finite probability of being found in classically disallowed regions. (In optics, the exponential decay of a wave in the forbidden region is referred to as *evanescence*).

The solution shown above corresponds to a single wave component of frequency ω. If a traveling wave of this frequency impinges on the turning point from the left, it turns and creates a reflected wave, moving back to the left from whence it came. The interference between these two traveling waves makes the pattern of nodes observed in the Airy function plotted in Cell 5.28. Thus, the classical turning point at $x = x_0$, which reflects classical particles, also causes the quantum wave function to reflect.

In order to describe the reflection of a wave packet in detail, we would need to integrate over a range of frequencies in the packet. The turning-point location differs for the different frequency components and causes dispersion of the wave packet, as discussed previously in relation to Eq. (5.2.50). If we neglect dispersive spreading, this leads to the equations of geometrical optics for the wave packet, equivalent to the equations of motion of the classical particle as it is reflected at the turning point. To keep dispersive effects requires considerably more work, and will not be discussed further here. In Chapter 6 we will return to the behavior of wave functions near a turning point, when we examine the energy eigenfunctions of the Schrödinger equation in the WKB limit.

Wave-Packet Amplitude When a wave packet with frequency ω propagates through a nonuniform medium, the amplitude of the packet varies. We saw several examples of this for one-dimensional systems in previous sections, but here we discuss the general case of propagation in two or more dimensions.

One can determine the amplitude variation using the law of energy conservation for wave packets, discussed in Sec. 5.1.3 for classical wave equations with even-order time derivatives. (The following does not apply to Schrödinger's equation, which is taken up in the next section.) Although this energy conservation law was derived for uniform systems, it should also be possible to apply it to nonuniform systems, provided that the intrinsic parameters in the system (wave speed, refractive index, etc.) remain time-independent. Assuming that this is the case, we write the energy conservation law (5.1.57) as

$$\frac{\partial}{\partial t}\bar{\epsilon}(\mathbf{r},t) + \nabla\cdot\left[\mathbf{v}_g(\mathbf{r})\bar{\epsilon}(\mathbf{r},t)\right] = 0, \tag{5.2.59}$$

where $\bar{\epsilon}$ is the energy density of a packet with wavenumber $\mathbf{k}$ and frequency ω, time-averaged over an oscillation period. Following Eq. (5.1.62), the energy density can be expressed in terms of the wave-packet amplitude $A(\mathbf{r},t)$ as

$$\bar{\epsilon} = \frac{\rho}{4}A^2\omega\frac{\partial}{\partial\omega}D(\mathbf{r},\mathbf{k},\omega), \tag{5.2.60}$$

where D is the dispersion function. The frequency and wavenumber are related by this dispersion function through the equation $D(\mathbf{r},\mathbf{k},\omega)=0$, the solution of which provides the dispersion relation $\omega=\omega(\mathbf{r},\mathbf{k})$. The group velocity is $\mathbf{v}_g(\mathbf{r}) = \partial\omega(\mathbf{r},\mathbf{k})/\partial\mathbf{k}|_{\mathbf{k}=\mathbf{k}(\mathbf{r})}$, where $\mathbf{k}(\mathbf{r})$ is the local wavenumber of the packet, determined by solution of the equations of geometrical optics. For example, for the optics problems considered in the previous section, $D(\mathbf{r},\mathbf{k},\omega) = \omega^2 n^2(\mathbf{r},\omega) - c^2k^2$, where n is the refractive index.

Equation (5.2.59) is a first-order PDE that is amenable to the method of characteristics, to be discussed in Chapter 7. For a given initial energy density in the wave packet and given initial wavenumber and frequency, the solution for $\bar{\epsilon}(\mathbf{r},t)$ can be found and can then be used to determine the wave-packet amplitude A via Eq. (5.2.60). This is somewhat involved, and we will not go into detail here. Some examples will be considered in Chapter 7, and in Sec. 5.3.1.

However, the method of characteristics is not needed for the case of a time-independent beam in a medium that varies only in the x-direction. In this case, Eq. (5.2.59) reduces to

$$\frac{\partial}{\partial x}\left[v_{gx}(x)\bar{\epsilon}(x)\right] = 0,$$

which has solution $\bar{\epsilon}(x) = C/v_{gx}(x)$, where C is a constant. Using Eq. (5.2.60), we can then obtain the amplitude $A(x)$ of the beam: $A(x) \propto 1/[|v_{gx}(x)\,\partial D(x,\mathbf{k},\omega)/\partial\omega|]^{1/2}$. However, recalling that $v_{gx} = \partial\omega/\partial k_x$, and applying the identity $(\partial\omega/\partial k_x)(\partial D/\partial\omega) = -\partial D/\partial k_x$, we arrive at

$$A(x) \propto \frac{1}{\sqrt{|\partial D(x,\mathbf{k},\omega)/\partial k_x|}}. \tag{5.2.61}$$

For the optics example discussed previously, where $D(\mathbf{r},\mathbf{k},\omega) = \omega^2 n^2(x,\omega) - c^2k^2$, Eq. (5.2.61) implies that the beam amplitude varies as $1/\sqrt{k_x(x)}$, which is typical for WKB problems. In fact, this result could also be obtained directly from the WKB solution without consideration of the wave energy [see Exercise (9) at the end of the section].

In this optics example we assumed that $n(x,\omega) = 1 - x^2\omega$ (see Fig. 6.8 and the surrounding discussion). Then Eq. (5.2.48) implies that $dk_x/dt < 0$ along the wave-packet trajectory, which is another way to see why the rays bent away from positive x in Fig. 6.8. In turn, $A(x) \propto 1/\sqrt{k_x}$ then implies that the electric field *increases* as k_x decreases. However, this result contradicts our previous intuition that the spreading of the rays making up the packet should cause the amplitude of the packet to *decrease* as the turning point is approached. In fact, both effects occur at once as the turning point is approached: the amplitude of a given ray increases (because the x-component of the wave speed is decreasing, so wavefronts pile up), but the dispersal of the wave packet of rays works against this effect.

However, it is important to remember that Eqs. (5.2.59)–(5.2.61) *neglect dispersion*. Recall that Eq.(5.2.59) is based on Eq. (5.1.60), which does not include dispersive spreading of the wave packet.

One can see the limitations of Eq. (5.2.59) in another way: for a wave packet of plane waves moving in one dimension through a uniform medium, Eq. (5.2.59) implies that $\partial\bar{\epsilon}/\partial t + \mathbf{v}_g \cdot \nabla\bar{\epsilon} = 0$. This equation implies that the wave packet is transmitted without change in the energy density (see Sec. 7.1). However, we know that dispersion will cause such a wave packet to spread. Equation (5.2.59) is valid only for problems where such dispersive effects are negligible.

EXERCISES FOR SEC. 5.2

(1) Bessel's equation is the following ODE:

$$\frac{1}{r}\frac{\partial}{\partial r}\left(r\frac{\partial R}{\partial r}\right) - \left(\frac{m^2}{r^2} - 1\right)R(r) = 0.$$

(a) Using Eq. (5.2.8) show that WKB analysis can be applied to understand the solutions to the equation, provided that $r \gg 1$.

(b) The two independent solutions to Bessel's equation are the Bessel functions $J_m(r)$ and $Y_m(r)$. Show that these functions have the following form at large r:

$$J_m(r), Y_m(r) \sim Ar^{-1/2}\cos(r - \alpha_m), \qquad \text{where } A \text{ and } \alpha_m \text{ are constants.}$$

[By other means, it can be shown that $A = \sqrt{2/\pi}$, $\alpha_m = \pi(m/2 + \frac{1}{4})$ for J_m, and $\alpha_m = \pi(m/2 + \frac{3}{4})$ for Y_m.]

(2) **(a)** Using Eq. (5.2.8), show that WKB solutions of the Airy equation, $\partial^2 y(x)/\partial x^2 = xy(x)$, are valid provided that $|x| \gg 1$.

(b) Show that these solutions are of the form $y(x) \propto (1/\sqrt[4]{x})\,e^{\pm\frac{2}{3}x^{2/3}}$ for $x \gg 1$, and $y(x) \propto (1/\sqrt[4]{|x|})\,e^{\pm i\frac{2}{3}|x|^{2/3}}$ for $x \ll -1$.

(3) By setting up a recursion relation analogous to Eq. (5.2.17), show that the eikonal S_n for the nth approximation to the solution of the Airy equation

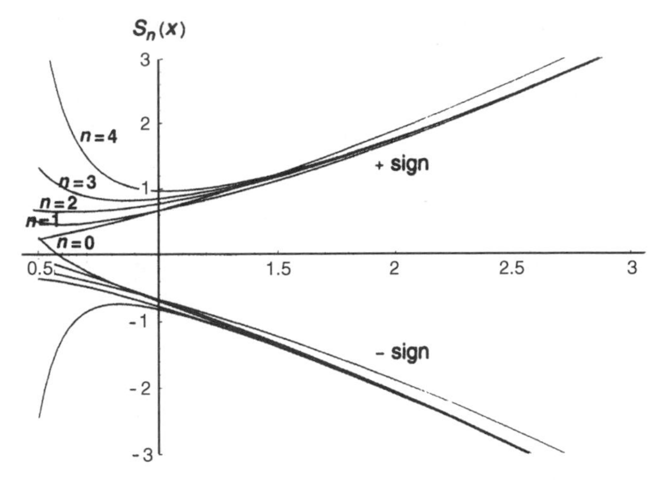

Fig. 5.9 Approximations to the eikonal for the Airy equation. Order of the approximant is shown for the + sign solution. Note the breakdown of convergence for $x \leq 1$, as expected for an asymptotic expansion that holds only for $x \gg 1$.

satisfies

$$S_n(x) = B \pm \int^x \sqrt{x - \lambda \frac{\partial^2 S_{n-1}}{\partial x^2}}\, dx \qquad \text{for} \quad x \gg 1$$

with $S_0 = \pm 2x^{3/2}/3$. Using *Mathematica*, solve this recursion relation and plot $S_n(x)$ for $0 = 1, 2, \ldots, 4$ in the range $x = 0.5$–3. (Drop the integration constant B.) (Hint: At each order n, Taylor-expand the square root to order n to simplify the integrals. Write a recursion relation for S_n that includes this Taylor expansion in the definition.) The solution is shown in Fig. 5.9.

(4) **(a)** Approximants found from recursion relations can also be used to solve certain nonlinear equations. For example, consider the equation $\lambda x^3 - x + 1 = 0$, where $\lambda \ll 1$. A dominant balance exists between x and 1: if we assume that $\lambda x^3 \ll 1$ and drop this term, the solution is $x = 1$. Then $\lambda x^3 = \lambda \ll 1$, which is consistent with our assumption. Set up a recursion relation of the form $x_n = 1 + \lambda x_{n-1}^3$, and solve for x_3, given $x_0 = 1$. (No need to Taylor-expand the solution.) Plot the result for $0 < \lambda < 1$.

(b) The cubic polynomial in part (a) has three roots, but we only found one approximate solution in part (a). Evidently there are two other solutions. Find another dominant balance between two terms in the equation that yields approximations to these solutions, set up a recursion relation, and solve it up to at least x_3. (Again, no Taylor expansion is necessary.)

(c) Plot the asymptotic expressions found in parts **(a)** and **(b)** together on the same graph as a function of λ for $0 < \lambda < 1$. Superimpose these results on the exact solutions of the cubic equation vs. λ.

(5) The general wave equation for a 1D string with mass density $\rho(x)$ and tension $T(x)$ is

$$\frac{\partial^2}{\partial t^2} y(x,t) = \frac{1}{\rho(x)} \frac{\partial}{\partial x}\left(T(x) \frac{\partial}{\partial x} y(x,t)\right).$$

Show that the WKB solution for traveling waves on this string is

$$y(x,t) = \frac{C}{[T(x)\rho(x)]^{1/4}} e^{-i\omega t + i\int_0^x k(x)\,dx} + \frac{D}{[T(x)\rho(x)]^{1/4}} e^{-i\omega t + -i\int_0^s k(x)\,dx}, \tag{5.2.62}$$

where $k(x) = \omega/\sqrt{T(x)/\rho(x)}$ is the local wavenumber.

(b) Use the result of part (a) to prove that the WKB solution to Eq. (5.2.32) is Eq. (5.2.35).

(6) A simple pendulum with varying length $l(t)$ satisfies the following differential equation:

$$\frac{d}{dt}\left[l^2(t)\dot{\theta}(t)\right] = -gl(t)\theta(t),$$

where $g = 9.8$ m/s^2 and $\theta \ll 1$ is assumed.

(a) Use WKB analysis to determine how $\theta(t)$ varies in time, assuming that $l(t)$ varies slowly. Plot $\theta(t)$ assuming that $\theta(0) = 10°$, $\dot{\theta}(0) = 0$, and $l(t) = 11 - t$, for $0 < t < 10$. Compare with the exact motion, found using `NDSolve`.

(b) Find an adiabatic invariant for the pendulum motion. (Hint: The energy of the pendulum is $E(t) = \frac{1}{2}ml^2\dot{\theta}^2 + \frac{1}{2}mgl\theta^2$, where m is the mass of the bob.)

(7) For the hanging string of length L described by Eq. (4.2.41), use WKB analysis to determine the evolution of a Gaussian wave packet of the initial form

$$y(z,0) = \exp\left[-40(z - L/2)^2\right],$$

moving down the string (in the $-z$-direction, toward the free end at $z = 0$). Make an animation taking $L = 3$, and compare this evolution with the exact evolution determined in Example 2 of Sec. 4.2.2, for $0 < t < 0.6$ s.

(8) A rod with varying mass density $\rho(x)$ supports transverse vibrations that satisfy the inhomogeneous biharmonic wave equation

$$\frac{\partial^2}{\partial t^2} y(x,t) = -\frac{D}{\rho(x)} \frac{\partial^4}{\partial x^4} y(x,t),$$

where D is a constant [see Eq. (4.2.48)]. Using WKB theory, analyze the behavior of traveling waves on this rod, and determine the manner in which the wave amplitude varies as $\rho(x)$ varies.

(9) **(a)** A beam of light of frequency ω has an electric field $E(x,t)$ that varies as $E(x)e^{-i\omega t}$, and that propagates in the $+x$ direction through a clear medium with varying refractive index $n = n(x)$. The electric field in the light wave satisfies

$$\frac{\partial^2 E(x,t)}{\partial t^2} = \frac{c^2}{n(x)^2} \frac{\partial^2 E(x,t)}{\partial x^2},$$

where c is the speed of light in vacuum. Analyze this equation using WKB theory, and determine how the electric field varies with x, assuming that the index varies little over a wavelength of the light.

(b) The average flux of energy (power per unit area) carried by the beam in the x-direction is $\frac{1}{2}n(x)|E[x,t]|^2 c\epsilon_0$ [Griffiths (1999, p. 383)]. According to the WKB approach, how does this energy flux vary with x? Comment.

(10) When a light beam of frequency ω propagates in an *arbitrary* direction through a (nonmagnetic) medium with index of refraction $n(\mathbf{r})$, Maxwell's equations imply that

$$\frac{\partial^2}{\partial t^2}\mathbf{E}(\mathbf{r},t) = \frac{c^2}{n(\mathbf{r})^2}\left[\nabla^2\mathbf{E} + 2\nabla\left(\mathbf{E}\cdot\frac{\nabla n}{n}\right)\right]. \tag{5.2.63}$$

Assume that $n = n(x)$, and that the beam propagates in the x-y plane. Take components of the equation, and use WKB theory to analyze the wave propagation in two cases.

(a) If $E_x = E_y = 0$ but $E_z \neq 0$, show that this remains true everywhere, so that $\hat{\mathbf{k}}\cdot\mathbf{E} = 0$ everywhere, and find the dependence of $|E_z|$ on x.

(b) If $E_z = 0$ but E_x and E_y are not equal to zero initially, show that E_z remains zero. Although the propagation direction varies with x, the waves must remain transverse, i.e., $\nabla\cdot(\epsilon\mathbf{E}) = 0$, where $\epsilon = n^2$ is the dielectric function. Solve Eq. (5.2.63) for E_x in the WKB limit, apply $\mathbf{k}\cdot\mathbf{E} \simeq 0$ to find E_y, and find the dependence of $|\mathbf{E}|$ on x.

(c) Show that for both cases (a) and (b), the results are compatible with the conservation of the x-component of the average Poynting flux:

$$\mathbf{S}\cdot\hat{\mathbf{x}} = \frac{1}{2\mu_0}(\mathbf{E}\times\mathbf{B}^*)\cdot\hat{\mathbf{x}} = \frac{1}{2}|\mathbf{E}|^2 c^2\epsilon_0 k_x/\omega = \text{constant}.$$

(This is the same result as obtained in the previous problem for a scalar wave equation.)

(11) An electromagnetic wave of frequency $\omega = 3$ and speed $c = 1$ is polarized in the z-direction and propagates in the x-y plane. The beam is incident on a medium with index of refraction $n(x)$. The electric field in these waves is described by Eq. (5.2.63). The angle of incidence is $\theta_0 = 45°$. The index varies in x according to

$$n(x) = \begin{cases} 1, & x < 0, \\ 1 + \frac{1}{4}\sin x, & x \geq 0. \end{cases}$$

(a) Plot contours of constant $\mathrm{Re}\, E_z$ at $t=0$ in the range $-5<x<15$, $-5<y<15$.

(b) Plot a ray trajectory $y(x)$ starting at $x=-5$, $y=-5$, for $-5<x<15$, using the previous parameters. Superimpose this ray on the previous contour plot.

(12) A surfer is a distance of 100 m from the shore, which runs north–south. Swells with a 17-m wavelength are moving past him with a period of 4 seconds. Use the full dispersion relation for water waves, $\omega^2 = g|\mathbf{k}| \tanh[h(x)|\mathbf{k}|]$.

(a) What is the depth of the water at this location?

(b) At the surfer's location, the waves travel in the direction 40° north of due east with period and wavenumber given above. The ocean depth varies as $h_0/(1+x^2/10^3)$, where x is in meters, measured from the surfer's position. The surfer lets the swells carry him in, eventually taking him 100 m in the x-direction into shore (where the depth is roughly one-tenth its previous value and the surfer can stand up). Assuming he rides directly transverse to the wave face, and that he travels at the same speed as the swells, plot the surfer's trajectory. How long does it take him to make this distance? How far north of his initial location does he get before he reaches shore?

(13) An undersea mountain causes the water depth to vary as $h(x,y)=1-0.7\exp[-2(x-\frac{1}{2})^2-5(y-\frac{1}{2})^2]$. Water waves have frequency $\omega(k,\mathbf{r}) = [gk \tanh kh(\mathbf{r})]^{1/2}$, where we will take $g=1$, and where $k=\sqrt{k_x^2+k_y^2}$ is the magnitude of the wavenumber. Use the equations of geometric optics to trace 11 wave-packet trajectories which initially have $k_x=0$, $k_y=1$, and start at $y=0$, $x=0.1n$, $n=0,1,2,\ldots,10$. Follow the trajectories for $0<t<2$, and plot them all on a single parametric plot of y vs. x. What happens to the rays? (The same principle is at work in any lens system.)

(14) In shortwave radio communication, the diffuse plasma in the ionosphere can be used as a mirror to bounce radio waves to receivers beyond the horizon. For radio waves of frequency ω, the index of refraction is a function of altitude z, and is approximately $n(z,\omega)=[1-\omega_p^2(z)/\omega^2]^{1/2}$, where $\omega_p = [e^2N(z)/\epsilon_0 m]^{1/2}$ is the *electron plasma frequency*, e and m are the electron charge and mass respectively, and $N(z)$ is the electron density in m^{-3}. At ground level $z=0$ the electron density is negligible, but at altitudes on the order of 50–100 km it grows to some maximum value due to the ionizing effect of UV radiation, and then decays again at still higher altitudes as the atmospheric density falls off.

(a) Show that electromagnetic waves travel through this plasma with the dispersion relation

$$\omega^2 = c^2k^2 + \omega_p^2(z).$$

(b) Show that waves shone directly upward, with $\mathbf{k}=k\hat{\mathbf{z}}$, will reflect back provided that $\omega < \omega_{p\,\max}$, where $\omega_{p\,\max}$ is the maximum plasma frequency. (Hint: Find the turning point for the rays.)

(c) More generally, show that for $\omega > \omega_{p\,\max}$, there is a maximum angle $\theta_{\max}$ with respect to the horizontal at which one can direct the waves and still have them bounce back. Find an expression for the angle $\theta_{\max}(\omega)$. [Answer: $\sin^2\theta_{\max} = (\omega_{p\,\max}/\omega)^2$.]

(d) Take the following simple model for the electron density in the ionosphere:

$$N(z) = \begin{cases} 0, & z < H, \\ (5 \times 10^{14}\ \mathrm{m}^{-3})(z/H - 1)^2\, e^{-10(z/H-1)}, & z \geq H, \end{cases}$$

with $H = 70$ km. Using this model and the results of part (c), trace several ray trajectories for values of θ in the range $0 < \theta \leq \theta_{\max}$ to show that communication with a receiver closer than a certain distance $d(\omega)$ to the transmitter (but beyond the horizon) is impossible when $\omega > \omega_{p\,\max}$. Find $d(\omega)$ graphically for (i) $\omega = 2\pi c/10$ (10-meter wavelength) and (ii) $\omega = 2\pi c/5$.

(15) A nebula consists of plasma with plasma frequency $\omega_p(x, y, z)$ that varies in space as $\omega_p(x, y, z) = (Ax^2 + By^2 + Cz^2)^{1/2}$, where a, b, and c are constants. A radio source at the center of the nebula emits waves of frequency ω in all directions. Using the dispersion relation from the previous problem, and the equations of geometrical optics, find the ray trajectories for the radio waves analytically, and show that the waves are trapped within the nebula. Plot a ray trajectory using `ParametricPlot3D` for the following conditions: $A = 1$ $\mathrm{s}^{-2}/\mathrm{km}^2$, $B = 4\ \mathrm{s}^{-2}/\mathrm{km}^2$, $C = 9\ \mathrm{s}^{-2}/\mathrm{km}^2$, $\omega = 10^{10}\ \mathrm{s}^{-1}$, and k initially in the direction $(\hat{\mathbf{x}} + \hat{\mathbf{y}} + \hat{\mathbf{z}})/\sqrt{3}$.

(16) **(a)** At a planar interface between two media with refractive indices n_1 and n_2, $n_1 > n_2$, show using Snell's law that a ray in medium 1 propagating toward the interface will be completely reflected provided that the angle ψ between the interface and the ray satisfies $\psi < \psi_c = \arccos(n_2/n_1)$. (Hint: Show that no real solution for the refracted-wave angle exists if this inequality is satisfied.)

(b) In an optical fiber, the index of refraction $n = n(r)$ is a decreasing function of cylindrical radius r. In typical fibers the index is of the form $n(r) = n_1$ $(r < a)$, $n = n_2$ $(r > a)$, with $a = 4$ μm, $n_1 = 1.451$, and $n_2 = 1.444$. The critical angle is then $\psi_c = 5.63°$. The point of an optical fiber is that it guides the rays, even if the fiber bends. Initially, a ray propagates along the axis of the fiber. Then the fiber is bent into a circle of radius $R \gg a$. Given that n_1 is nearly the same as n_2, so that $\psi_c \ll 1$, show that the ray will be trapped in the fiber provided that $R > 2a/\psi_c^2 = 830$ μm (ψ_c measured in radians).

(17) The speed c_p of compressional sound waves in the earth (P-waves) is an increasing function of depth d measured from the earth's surface, thanks to the fact that the earth's mean density increases with depth. In accordance with Snell's law, this causes waves that are initially propagating downward to bend back toward the surface. A theorist's model for the dependence of the wave speed on depth d into the Earth's upper crust is

$$c_p(d) = c_0 + (c_1 - c_0)d/d_0, \qquad d < d_0,$$

where $c_0 = 8$ km/s is the speed at the surface, $d_0 = 2900$ km is the depth of the earth's outer core, and $c_1 = 14$ km/s is the speed at $d = d_0$. Assuming that the waves travel along the great circle defined by $\phi = 0$, the equations for geometrical optics is spherical coordinates (r, θ, ϕ) are

$$\frac{dr}{dt} = \frac{\partial \omega}{\partial k_r}, \qquad r\frac{d\theta}{dt} = \frac{\partial \omega}{\partial k_\theta},$$

$$\frac{dk_r}{dt} = -\frac{\partial \omega}{\partial r}, \qquad \frac{dk_\theta}{dt} = -\frac{1}{r}\frac{\partial \omega}{\partial \theta},$$

where the wavevector $\mathbf{k} = k_r\hat{\mathbf{r}} + k_\theta\hat{\boldsymbol{\theta}}$, and $\omega = c_p(d = R_e - r)k$, and $R_e = 6400$ km is the earth's radius.

(a) Use these equations to plot the ray trajectories in the (x, z) plane (where $x = r\sin\theta$, $z = r\cos\theta$) followed by 10 sound rays until they reach the surface again, starting from $\theta = 0$, $r = R_e$ (i.e. $x = 0$, $z = R_e$), $k_r = -5$ m^{-1}, and $k_\theta = n$ m^{-1}, $n = 5, 6, \ldots, 15$. (This takes up to 1000 seconds for the longest ray trajectories.)

(b) Let θ_0 be the value of θ for which a given ray reappears at the earth's surface, $r = R_e$. Using `FindRoot`, determine the propagation time $t_p(\theta_0)$ to propagate from $\theta = 0$ to $\theta = \theta_0$, for the rays found in part (a). Plot this propagation time vs. θ_0. [Given the positions of three or more seismographs and the arrival time of the pulse at each, one can fix the location of the epicenter. See, for example, Garland (1979) in the reference list for Chapter 4.]

(18) A point source generates sound wave pulses at $\boldsymbol{r} = 0$. The sound speed in still air is c. The pulses initially travel in the $+x$ (horizontal) direction (i.e., $k_x > 0$, $k_y = 0$). However, a wind of velocity $\mathbf{v} = -ay\hat{\mathbf{x}}$ is blowing, where y is altitude and a is a positive constant. The shear in the wind causes the pulses to travel upward. Using the equations of geometrical optics, find an analytic expression for the time t required for the pulses to reach a height y. Plot the shape of the wave-packet trajectory $y(x)$ for $a = 1$, $c = 340$, and $0 < x < 100$ (in MKS units). [The shape of this trajectory explains why it is difficult to make someone hear you when they are upwind of your position: your shouts into the wind are refracted up over the person's head.] [Hint: see Eq. (5.1.88).]

(19) Whistler waves are electromagnetic waves that travel along the earth's magnetic field lines, and are described by the dispersion relation given by Eq. (5.1.104). The earth's magnetic field can be approximated as the field from a point magnetic dipole of order $M = 8 \times 10^{15}$ Tm^3. Therefore, in spherical coordinates the magnetic field has the form

$$\mathbf{B} = -\frac{M}{r^3}(2\hat{\mathbf{r}}\cos\theta + \hat{\boldsymbol{\theta}}\sin\theta).$$

The plasma frequency in the earth's magnetosphere varies roughly as $\omega_p(r) = \omega_{p0}[(R\sin^2\theta)/r]^2$, where $R = 6400$ km, and $\omega_{p0} = 3 \times 10^7$ s^{-1} (see Shulz and Lanzerotti, 1974). Follow the trajectories of four wave packets, that begin in the northern hemisphere at $\phi = 0$, $\theta = \pi/4$, $r = R$, $|\mathbf{k}| = 3 \times 10^{-5}$ m, but with random initial propagation directions. Plot the trajectories using `ParametricPlot3D`, and plot the earth's surface for reference, to determine visually

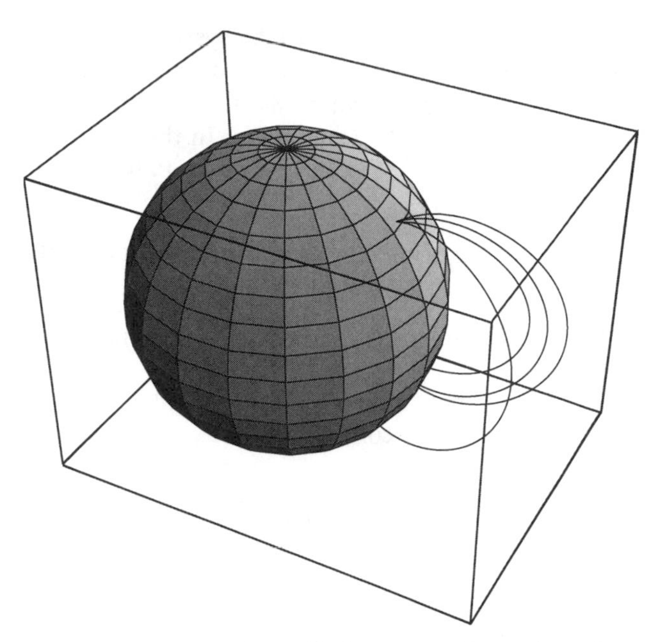

Fig. 5.10 Four random whistler-wave packet trajectories.

at what location the trajectories return to the earth's surface. (The solution for one set of random trajectories is shown in Fig. 5.10.

(20) Moving boundary conditions present special difficulties, but can often be put in standard form by means of a coordinate transformation. Consider the following example. A plucked uniform string, of initial length L_0, is initially vibrating in its lowest mode with $y(x,t) = \sin(\pi x/L_0)\cos\omega_0 t$, where $\omega_0 = \pi c/L_0$, and c is the propagation speed for waves on the string. For $t > 0$, the Dirichlet boundary condition at $x = L_0$ begins to move according to $y(L(t),t) = 0$, where $L(t) = L_0 - vt$. This can be accomplished by running one's finger along the string at speed v. Solve this problem by first putting it in standard form through the transformation $\bar{x} = x/L(t)$. In these coordinates the boundary conditions are stationary, $y(0,t) = y(1,t) = 0$, but:

(a) Show that the wave equation becomes

$$\hat{D}^2 y(\bar{x},t) = \frac{c^2}{L^2(t)} \frac{\partial^2}{\partial \bar{x}^2} y(\bar{x},t), \tag{5.2.64}$$

where the operator

$$\hat{D} \equiv \frac{\partial}{\partial t} - \frac{\bar{x}\dot{L}(t)}{L(t)} \frac{\partial}{\partial \bar{x}},$$

and the time derivative is at fixed $\bar{x}$. Hint: Use the following identities:

$$\frac{\partial}{\partial x} y(\bar{x},t)\bigg|_t = \frac{\partial \bar{x}}{\partial x}\bigg|_t \frac{\partial}{\partial \bar{x}} y(\bar{x},t)\bigg|_t,$$

$$\frac{\partial}{\partial t} y(\bar{x},t)\bigg|_x = \frac{\partial}{\partial t} y(\bar{x},t)\bigg|_{\bar{x}} + \frac{\partial \bar{x}}{\partial t}\bigg|_x \frac{\partial}{\partial \bar{x}} y(\bar{x},t)\bigg|_t.$$

(b) This problem is amenable to WKB analysis, provided that $v \ll c$. The solution can be written in the approximate form $y(\bar{x}, t) = \mathrm{Re}[C(\bar{x}, t)\, e^{-i\int_0^t \omega(t)\, dt}] \sin \pi \bar{x}$, where $\omega(t) = \pi c/L(t)$, and $C(\bar{x}, t)$ is a relatively slowly varying (complex) amplitude. In the limit that $v/c \ll 1$, show that $C(\bar{x}, t)$ satisfies the following first-order PDE:

$$(\tan \pi \bar{x}) \frac{\partial C}{\partial t} - i \frac{c}{L(t)} \frac{\partial C}{\partial \bar{x}} = \left(\pi \bar{x} + \frac{1}{2} \tan \pi \bar{x} \right) \frac{\dot{L}}{L} C.$$

(c) First-order PDEs of this type can be solved analytically using the method of characteristics. (See Chapter 7.) **DSolve** can also solve this PDE analytically. Show that, for $v/c \ll 1$, the solution from **DSolve** that matches the proper initial conditions is $C(x, t) \sim 1$, so that

$$y(x,t) \approx \cos \int_0^t \omega(t)\, dt \sin \frac{\pi x}{L(t)}. \tag{5.2.65}$$

(d) Use the **Play** function to listen to the sound made by $\cos \int_0^t \omega(t)\, dt$ for the case $L_0 = 1$ m, $c = 300$ m/s, and $v = 2$ m/s.

(e) Find the rate at which the string energy $E(t)$ increases, and using this show that $E(t)/\omega(t)$ is an adiabatic invariant.

(f) Use the result from part (e) to determine the horizontal force F that the string exerts on the boundary.

(21) Consider the following wave equation on the interval $0 < x < L$:

$$\frac{\partial^2 y}{\partial t^2} = c^2(t) \frac{\partial^2 y}{\partial x^2}.$$

This equation describes waves on a string for which the string tension T and/or the mass density ρ vary in time, but the length L is fixed. (For example, tightening a guitar string by turning the peg both increases the tension and reduces the mass density as more string becomes wrapped around the peg. Boundary conditions are taken to be $y = 0$ at $x = 0, L$, with an initial condition $y(x,0) = \sin(n\pi x/L)$, $\dot{y}(x,0) = 0$. In other words, like the previous problem, the system starts off in a single normal mode. Assume that $c(t)$ varies in time slowly compared to the frequencies of the normal modes, find the WKB solution for the string motion, and show that

(a) the system remains in mode n;

(b) mode frequency varies in time according to $\omega(t) = n\pi c(t)/L$;

(c) the adiabatic invariant for *this* system is $E(t)\omega(t)/T(t)$.

REFERENCES

C. Bender and S. Orzag, *Advanced Mathematical Methods for Scientists and Engineers* (McGraw-Hill, New York, 1978).

H. Goldstein, *Classical Mechanics*, 2nd ed. (Addison Wesley, Reading, MA, 1980).

D. J. Griffiths, *Introduction to Electrodynamics*, 3rd ed. (Prentice Hall, Englewood Cliffs, NJ, 1999).

I. Kay and J. B. Keller, *Asymptotic evaluation of the field at a caustic*, J. Appl. Phys. **25**, 879 (1954).

P. K. Kundu, *Fluid Mechanics* (Academic Press, San Diego, 1980).

L. D. Landau and E. M. Lifshitz, *Mechanics*, 3rd ed. (Pergamon Press, Oxford, 1976).

E. H. Lockwood, *A Book of Curves* (Cambridge University Press, Cambridge, 1967).

M. Shulz and L. Lanzerotti, *Particle Diffusion in the Radiation Belts* (Springer-Verlag, 1974).

T. Stix, *Theory of Plasma Waves*, (McGraw Hill, New York, 1962).

S. Weinberg, *Eikonal methods in geometrical optics*, Phys. Rev. **126**, 1899 (1962).

CHAPTER 6

NUMERICAL SOLUTION OF LINEAR PARTIAL DIFFERENTIAL EQUATIONS

6.1 THE GALERKIN METHOD

6.1.1 Introduction

The Galerkin method is a numerical method for the solution of differential equations that is based on the ideas of eigenmode analysis. The method can be applied to both linear and nonlinear PDEs of any order, for which solutions are required in some given spatial domain. The domain can be infinite, but if it is finite the method is most efficient when the boundary conditions are continuous functions of position along the boundary, and when any edges formed by the intersection of the bounding surfaces form convex cusps. For instance, the two-dimensional domain shown in Fig. 6.1(a) can be easily treated via the Galerkin method. However, the domain in Fig. 6.1(b) is considerably more difficult to deal with, because it has a concave cusp pointing into the domain. The reason for these restrictions will be discussed in the next section.

6.1.2 Boundary-Value Problems

Theory Consider a time-independent boundary-value problem in two dimensions,

$$\hat{L}\phi(x,y) = \rho(x,y), \tag{6.1.1}$$

where $\hat{L}$ is some second order linear spatial operator such as ∇^2. We assume for the moment that the boundary conditions are Dirichlet, specified on the boundary S of a domain V such as that shown in Fig. 6.1(a):

$$\phi|_S = \phi_0(x,y), \tag{6.1.2}$$

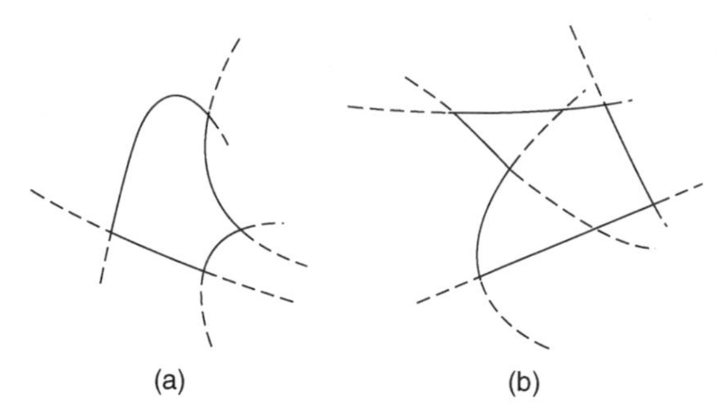

Fig. 6.1 Spatial domains (shown outlined by solid lines, and defined by the smooth dashed curves) for which (a) one can easily apply the Galerkin method; (b) difficulties arise.

for some function $\phi_0(x, y)$. This function must be smoothly varying on the boundary in order for the Galerkin method to work efficiently.

The boundary is itself determined by the intersection of M smooth curves, each of which is defined by an equation of the form

$$z_n(x, y) = 0, \qquad n = 1, \dots, M. \tag{6.1.3}$$

These curves are the lines shown schematically in Fig. 6.1(a).

To solve this problem numerically via the Galerkin method, we will expand $\phi(x,y)$ in a set of N different *basis functions* $\{v_\alpha\}$, chosen to satisfy homogeneous Dirichlet boundary conditions:

$$v_\alpha|_S = 0. \tag{6.1.4}$$

We write the solution in the standard form for an eigenmode expansion:

$$\phi(x, y) = u(x, y) + \sum_\alpha c_\alpha v_\alpha(x, y), \tag{6.1.5}$$

where $u(x, y)$ is any (smooth) function that satisfies the boundary condition $u|_S = \phi_0$, and where the c_α's are the Fourier coefficients that must be determined. In order to find these coefficients, we substitute Eq. (6.1.5) into Eq. (6.1.1), yielding

$$\sum_\alpha c_\alpha \hat{L} v_\alpha(x, y) = \rho(x, y) - \hat{L} u(x, y). \tag{6.1.6}$$

Next, we take an inner product with respect to one of the basis functions, v_β:

$$\sum_\alpha c_\alpha \left(v_\beta, \hat{L} v_\alpha \right) = \left(v_\beta, \rho - \hat{L} u \right). \tag{6.1.7}$$

So far, the method is identical to an eigenmode expansion. However, the v_α's need *not* be eigenmodes of $\hat{L}$; and they need not even form an orthogonal set. Rather, they can be any set of functions that we choose, provided that in the limit as

$N \to \infty$ they form a complete set, and provided that they satisfy Eq. (6.1.4). (Also, it is preferable that the functions be as analytically simple in form as possible.)

Furthermore, the inner product used in Eq. (6.1.7) can be *any inner product over the domain V*. (Again, however, the simpler the better.)

For example, one possible choice for the v_α's is the following: let the iterator α equal the set of integers (i, j), and take

$$v_{ij}(x, y) = x^i y^j \prod_{n=1}^{M} z_n(x, y) \qquad \text{for} \quad i = 0, 1, \ldots, P_x, \quad j = 0, 1, \ldots, P_y. \quad (6.1.8)$$

With this choice, the total number of functions is

$$N = (P_x + 1)(P_y + 1). \qquad (6.1.9)$$

The product of z_n's implies that the v_{ij}'s are zero on the boundary surface, and the polynomial form in x and y provides a sufficient range of variation so that in the limit as P_x and P_y approach infinity, the basis functions form a complete set.

For the inner product, one simple choice is

$$(f, g) = \int_V f^*(x, y) g(x, y)\, dx\, dy, \qquad (6.1.10)$$

where the integral runs over the domain V of the problem. One could also choose $(f, g) = \int_V f^*(x, y) g(x, y) p(x, y)\, dx\, dy$ for some choice of weighting function $p(x, y) > 0$. But it is preferable that p does not weight any one region of the domain disproportionately (unless there are regions of the domain where the solution is zero, or nearly zero). Otherwise, the solution may not converge well.

There is a certain art to finding good choices for the basis functions and the inner product. What we are looking for are choices that minimize the number of terms required to get a well-converged expansion for ϕ, and that are sufficiently simple so that inner products in Eq. (6.1.7) can be easily evaluated. In other words, we want a small number of v_α's to sum to something close to the exact solution.

The best way to learn how to make good choices is by gaining experience with the method. We will work through several examples in this and later sections.

We now solve for the Fourier coefficients c_α. To do so, we note that for different β-values, Eq. (6.1.7) forms a set of N linear equations in the N unknowns c_α. Taking $\alpha = (k, l)$ and $\beta = (i, j)$, these equations are rewritten explicitly in terms of (i, j) and (k, l):

$$\sum_{k=0}^{P_x} \sum_{l=0}^{P_y} \left(v_{ij}, \hat{L} v_{kl} \right) c_{kl} = \left(v_{ij}, \rho - \hat{L} u \right), \qquad i = 0, 1, \ldots, P_x, \quad j = 0, 1, \ldots, P_y. \qquad (6.1.11)$$

Since the inner products in Eq. (6.1.11) are merely some integrals that we can find once we have chosen the basis functions, we know everything that we need in order to solve these N coupled linear equations for the N c_{kl}'s.

From a theory perspective, it may be of use to note that the inner product $(v_\beta, \hat{L}v_\alpha)$ is the projection of the operator $\hat{L}$ onto the basis functions v_β and v_α, and that this projection is just another way to represent the operator numerically. We can think of this inner product as a square matrix $L_{\beta\alpha}$, so that Eq. (6.1.7) becomes the matrix equation

$$\sum_\alpha L_{\beta\alpha} c_\alpha = \bar{\rho}_\beta, \tag{6.1.12}$$

where $\bar{\rho}_\beta = (v_\beta, \rho - \hat{L}u)$. We need merely invert the matrix to solve for the c_α's: $c_\alpha = (\mathbf{L}^{-1})_{\alpha\beta}\,\bar{\rho}_\beta$. Thus, this method has much in common with the matrix inversion methods that we employed in solving linear boundary-value problems in Chapter 2.

We can now see why the boundary shown in Fig. 6.1(b) are more difficult to treat. The basis functions $v_\alpha(x, y)$ are zero along the smooth curves $z_n(x, y) = 0$ that define the bounding surface. These curves are displayed as the dashed lines in the figure. Because of the shape of the cusp in Fig. 6.1(b), two of the curves penetrate into the interior of the region, and so the functions v_α are all zero along these curves. Therefore, they cannot form a complete set, since they must be able to represent arbitrary functions within the domain.

While it is possible to define boundary curves $z_n(x, y) = 0$ that are not smooth, forming concave cusps as shown in Fig. 6.1(b) without continuing into the interior of the domain (using nonanalytic functions in the definition of z_n: see Example 3 below), such boundary curves introduce singular behavior into the v_α's through Eq. (6.1.8), and therefore the functions may not behave well when acted upon by $\hat{L}$. Unless the functions z_n are carefully chosen to minimize the singular behavior, derivatives of these basis functions are not well behaved, and the generalized Fourier series solution does not converge very well.

However, we will see in Example 3 that even if care is taken in choosing mathematical forms for the boundary curves so as to minimize the singular behavior, the Galerkin method still faces difficulties when concave cusps are present. The reason is simple: such cusps introduce rapid variation in the solution. Taking Poisson's equation as an example, we noted in Exercise (5)(b) of Sec. 3.2 that the electric field near the point of a concave cusp becomes infinite. Such rapid variation in the derivative of ϕ requires a Galerkin solution that keeps a large number of basis functions.

For essentially the same reason, it is difficult to deal with boundary conditions that are discontinuous. A situation where one electrode is held at potential V_1 and an adjacent electrode is held at potential V_2 is difficult to treat with the Galerkin method, because the discontinuous boundary condition causes rapid variation in the solution that requires many basis functions to resolve.

When dealing with boundary conditions of the sort that cause rapid variation in the solution, it is often easier to use other methods, such as the grid methods discussed in Sec. 6.2.

Example 1 As an example of the Galerkin method, we will solve a problem that we could also do analytically, so that we can compare the numerical solution with the exact solution. The problem is to find the solution to $\nabla^2\phi(x, y) = 0$ in a square box with the boundary condition that $\phi = 0$ on the edges, except along $y = 0$, where $\phi(x,0) = \sin \pi x$. The exact solution for ϕ, found by separation of variables, is

$$\phi_{\text{exact}}(x, y) = \sin \pi x \frac{\sinh \pi(1-y)}{\sinh \pi}. \tag{6.1.13}$$

To find the solution via the Galerkin method, we first choose functions v_{ij} according to Eq. (6.1.8):

Cell 6.1

```
v[i_, j_, x_, y_] = x^i y^j xy (1-x) (1-y);
```

These functions are zero along each edge of the box. Next, we define an inner product. Sometimes it is more convenient to evaluate these inner product integrals numerically rather than analytically, but in this case, analytic evaluation is faster. However, even if one performs the integrals analytically, it is usually best to convert any exact numbers to approximate numbers using **N**, so that the solutions to the equations are not too complicated:

Cell 6.2

```
norm[f_, g_] := Integrate[f g, {x, 0, 1}, {y, 0, 1}]//N
```

[Note the use of a delayed equal sign, since the integration cannot be performed until $f(x, y)$ and $g(x, y)$ are known.]

Next, we choose a function $u(x, y)$ that matches the boundary conditions:

Cell 6.3

```
u[x_, y_] = (1-y) Sin[Pi x];
```

This function is indeed zero on the three sides defined by $x = 0$, $x = 1$, and $y = 1$, and it equals $\sin \pi x$ on the remaining side $y = 0$.

It is also useful to define the operator $\hat{L} = \nabla^2$:

Cell 6.4

```
L[f_] := D[f, {x, 2}] + D[f, {y, 2}]
```

We can now define the inner product $L_{ijkl} = (v_{ij}, \hat{L} v_{kl})$,

Cell 6.5

```
Lmat[i_, j_, k_, l_] := norm[v[i, j, x, y], L[v[k, l, x, y]]]
```

and the inner product $\bar{\rho}_{ij} = (v_{ij}, \rho - \hat{L}u)$ (in this example, $\rho = 0$):

Cell 6.6

```
ρvec[i_, j_] := norm[v[i, j, x, y], -L[u[x, y]]];
```

Next, we create the coupled equations for the Fourier coefficients c_{kl}. We will refer to these coefficients as **c[k,l]**, and we will give each equation a name, **eq[i, j]**, so that we may refer to it later:

Cell 6.7

```
eq[i_, j_] :=
 eq[i, j] = Sum[Lmat[i, j, k, l] c[k, l], {k, 0, Px},
    {l, 0, Py}] == ρvec[i, j]
```

Here we made *Mathematica* remember each equation. Since each equation takes some time to create, we only want to do it once. At this point it is useful to check that everything is working, so we will use a **Do** statement that prints out each equation. We choose values of $P_x = 3$ and $P_y = 3$ so as to create 16 equations; see Eq. (6.1.9):

Cell 6.8

```
Px = 3; Py = 3;
Do[Print["eq", {i, j}, " = ", eq[i, j]], {i, 0, Px},
   {j, 0, Py}];
```

We omit the results in order to save space. But the reader can perform the evaluations to confirm that everything appears to be working properly. We now solve for the c_{kl}'s using **Solve**. **Solve** takes arguments of the form **Solve[{***equations***}, {***variables***}]** so we need to create a one-dimensional list of equations and another list of the variables. The list of equations can be created using a **Table** command:

Cell 6.9

```
eqns = Flatten[Table[eq[i, j], {i, 0, Px}, {j, 0, Py}]];
```

We flattened the two-dimensional table into a one-dimensional list using the **Flatten** command, in order for the list to obey the proper syntax for **Solve**.

To create the variable list, we use another flattened table:

Cell 6.10

```
vars = Flatten[Table[c[k, l], {k, 0, Px}, {l, 0, Py}]]

{c[0, 0], c[0, 1], c[0, 2], c[0, 3], c[1, 0], c[1, 1],
 c[1, 2], c[1, 3], c[2, 0], c[2, 1], c[2, 2], c[2, 3],
 c[3, 0], c[3, 1], c[3, 2], c[3, 3]}
```

Now we solve for the c_{mn}'s:

Cell 6.11

```
coefficients = Solve[eqns, vars][[1]]

{c[0, 0] → -6.67286, c[0, 1] → 8.2558, c[0, 2] → -5.74432,
 c[0, 3] → 1.91403, c[1, 0] → -7.6008, c[1, 1] → 9.47786,
```

```
c[1, 2] → -6.67373, c[1, 3] → 2.24739, c[2, 0] → 7.6008,
c[2, 1] → -9.47786, c[2, 2] → 6.67373, c[2, 3] → -2.24739,
c[3, 0] → -1.7257×10^-10, c[3, 1] → 1.13872×10^-9,
c[3, 2] → -2.3676×10^-9, c[3, 3] → 1.47698×10^-9}
```

This list of substitutions is used to construct $\phi(x, y)$ as follows:

Cell 6.12

```
φ[x_, y_] =
  u[x, y] + Sum[c[k, l]v[k, l, x, y], {k, 0, Px},
   {l, 0, Py}]/.coefficients;
```

The result is plotted in Cell 6.13. This solution appears to match the analytic solution given by Eq. (6.1.13). The error $\Delta\phi = \phi - \phi_{\text{exact}}$ is, in fact, quite small, as shown in Cell 6.14. The error could be reduced even further by increasing P_x and P_y. However, the more terms one keeps, the more time it takes to create and solve the equations.

Cell 6.13

```
Plot3D[φ[x, y], {x, 0, 1}, {y, 0, 1},
  AxesLabel → {"x", "y", "φ"}];
```

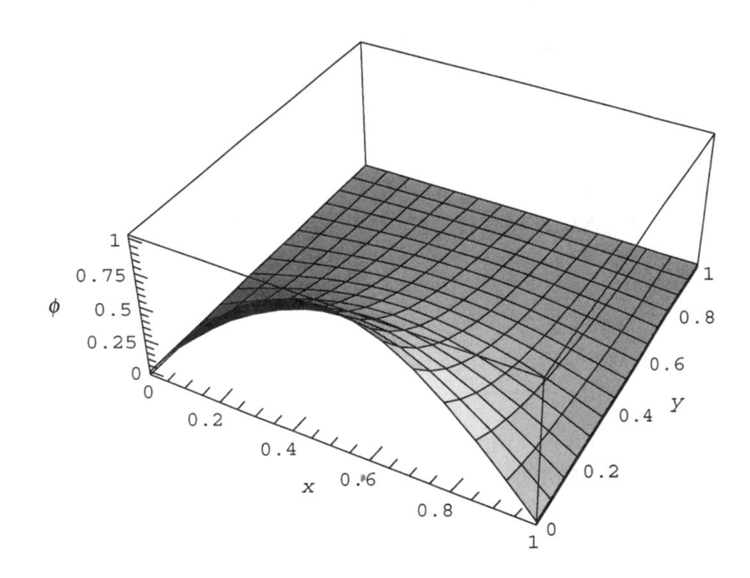

Cell 6.14

```
φexact[x_, y_] = Sin[Pi x] Sinh[Pi (1-y)]/Sinh[Pi];
Plot3D[φ[x, y] - φexact[x, y], {x, 0, 1},
 {y, 0, 1}, AxesLabel → {"x", "y", "Δφ"}, PlotPoints → 50];
```

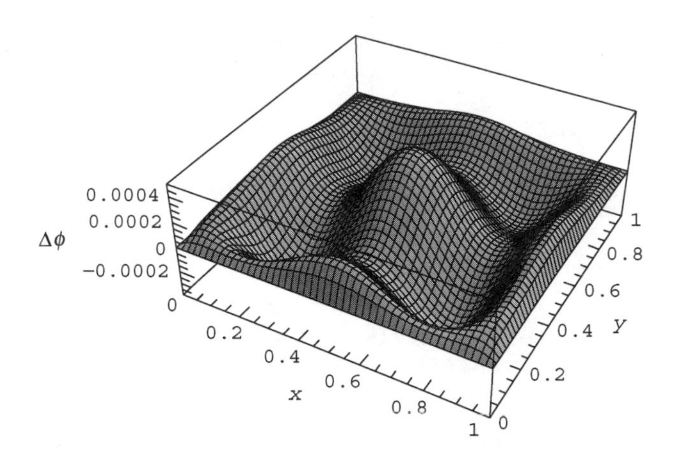

The computation time can be reduced significantly if a general analytic form for one or more of the inner products can be found. For example, the inner product L_{ijkl} in the above example can be determined analytically for general i, j, k and l:

Cell 6.15

```
Lmat[i_, j_, k_, l_] =
 Integrate[v[i, j, x, y], L[v[k, l, x, y]], {x, 0, 1},
   {y, 0, 1}]
```

(the result is suppressed to save space). Even though this expression for $L[i, j, k, l]$ is quite complicated, it takes less time to evaluate than the previous expression that contained a delayed evaluation of the inner product. Previously, when the equations `eq[i,j]` were constructed, the integrals in the inner product had to be performed separately for every value of i, j, k, and l. Now, the equations take much less time to create. (Try it.)

However, it is not always possible to work out the inner products analytically for general values of the indices. For instance, the integrals needed in $\bar{\rho}_{kl}$ are too complex to be done for general k and l.

Example 2 In the next example, we will solve Poisson's equation in a domain that does not allow for an analytic solution. The domain is defined by the equations $y \leq x/6+1$, $y \geq x^2$. This domain is plotted in Cell 6.17. Here we have used a technique involving the function `Boole`, available only in *Mathematica* version 4.1 or later, that allows one to plot or integrate over complex regions defined by inequalities. We will also use this technique in defining the inner product over this region. This technique is merely a convenience; it is also possible to write out the ranges of integration in the inner product explicitly.

Cell 6.16

```
<<Calculus`
```

Cell 6.17

```
ContourPlot[Boole[y≤x/6 + 1 && y≥x^2],
  {x, -1, 1.3}, {y, 0, 1.3}, PlotPoints→100];
```

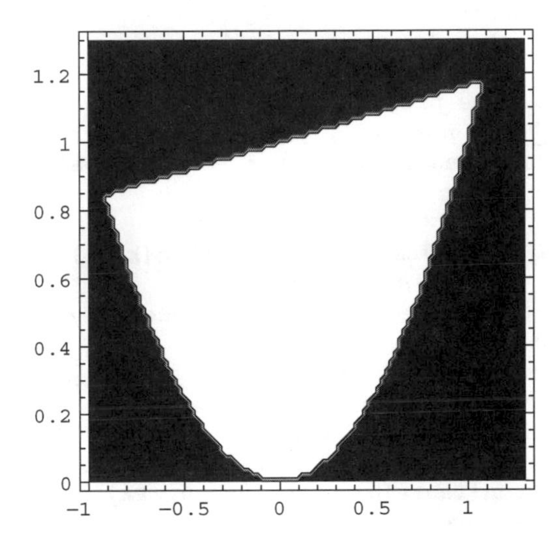

Within this domain, we will solve $\nabla^2\phi(x, y) = 1$, with boundary conditions that $\phi|_S = 0$.

The following module puts the whole Galerkin method together in one place:

Cell 6.18

```
(* define the curves that determine the boundary: *)
z[1, x_, y_] = y-(x/6 + 1);
z[2, x_, y_] = y-x^2;
(* define the charge density *)
ρ[x_, y_] = 1;

(* define the function u used to
   match inhomogeneous boundary conditions, if any *)
  u[x_, y_] =
    0;

  (* define the inner product
    using Boole notation (Mathematica 4.1 only) *)
norm[f_, g_] := Integrate[f g Boole[z[1, x, y] ≤ 0 && z[2, x,
  y] ≥ 0],
    {x, -Infinity, Infinity}, {y, -Infinity, Infinity}]//N;
```

Cell 6.19

```
galerkin[Px_, Py_, M_] := Module[{v, Lmat, ρvec, eq,
  eqns, vars},
  (* define the basis functions: *)
v[i_, j_, x_, y_] = x^i y^j Product[z[n, x, y], {n, 1, M}];
(* define the operator: *)
L[f_] := D[f, {x, 2}] + D[f, {y, 2}];
(* determine the projection of the operator onto the basis
   functions: *)
Lmat[i_, j_, k_, l_] := norm[v[i, j, x, y],
 L[v[k, l, x, y]]];
(* determine the projection of ρ̄ onto the basis functions: *)
ρvec[i_, j_] := norm[v[i, j, x, y], ρ[x, y]-L[u[x, y]]];
(* define the equations for the Fourier coefficients
   c[m,n] *)
eq[i_, j_] :=
   eq[i, j] = Sum[Lmat[i, j, k, l] c[k, l], {k, 0, Px},
    {l, 0, Py}] == ρvec[i, j];
  (* Print out the equations (not necessary, but a useful
    check) *)
Do[Print["eq", {i, j}, " = ", eq[i, j]], {i, 0, Px},
 {j, 0, Py}];
(* create lists of equations and variables *)
eqns = Flatten[Table[eq[i, j], {i, 0, Px}, {j, 0, Py}]];
vars = Flatten[Table[c[k, l], {k, 0, Px}, {l, 0, Py}]];
(* solve the equations *)
coefficients = Solve[eqns, vars][[1]];
(* define the solution *)
ϕ[x_, y_] =
   u[x, y] + Sum[c[k, l]v[k, l, x, y], {k, 0, Px},
   {l, 0, Py}]/. coefficients;]
```

We will run the module taking $P_x = P_y = 4$:

Cell 6.20

```
galerkin[4, 4, 2]
```

The module returns the approximate solution for the potential $\phi(x, y)$. We plot the solution in Cell 6.21 as a contour plot.

Cell 6.21

```
ContourPlot[ϕ[x, y] Boole[z[1, x, y] ≤ 0&&z[2, x, y] ≥ 0],
  {x, -1, 1.3}, {y, 0, 1.3}, PlotPoints→100,
   FrameLabel→{"x", "y"},
 Contours→{0, -.02, -0.04, -0.06, -0.08}];
```

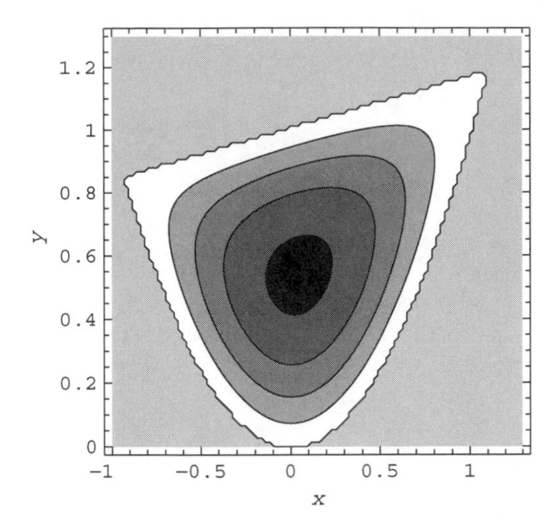

We can check how well our approximate solution solves the Poisson equation by substituting it into the PDE:

Cell 6.22

```
L[φ[x, y]]-1;
```

The result deviates only slightly from zero, as shown in Cell 6.23. The error is largest near the cusps in the domain, where the solution is expected to vary most rapidly. Of course, one could improve the solution by keeping more terms in the expansion.

Cell 6.23

```
Plot3D[Abs[%] Boole[z[1, x, y] ≤ 0&&z[2, x, y] ≥ 0],
  {x, -1, 1}, {y, 0, 1.3}, PlotPoints→50];
```

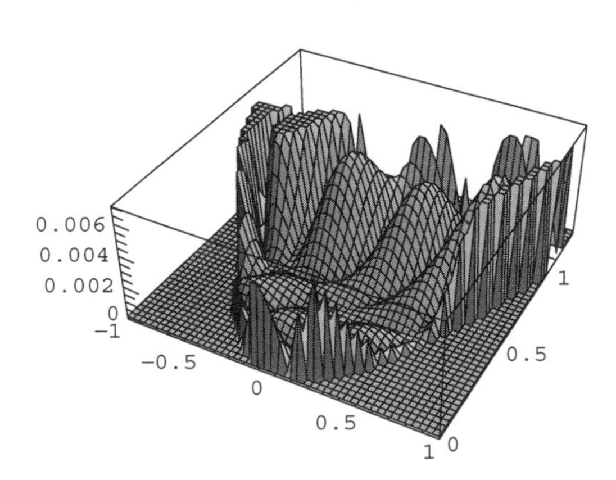

Example 3 In this example, we consider the solution of Laplaces equation, $\nabla^2\phi = 0$, in a domain that has a concave cusp of the type discussed in relation to Fig. 6.1(b). The domain boundary is defined by the following two equations:

$$z_1(x, y) = x^2 + y^2 - 1,$$
$$z_2(x, y) = y - |x|.$$

The domain is shown in Cell 6.24. For boundary conditions, on the edges of the cusp we take $\phi(x = y, y) = \phi(x = -y, y) = 1 - x^2 - y^2$. This boundary condition matches continuously onto the boundary condition $\phi = 0$, applied to the circular part of the boundary.

Cell 6.24

```
<<Calculus`;

z[1, x_, y_] = x^2 + y^2-1;
z[2, x_, y_] = y-Abs[x];
ContourPlot[ Boole[z[1, x, y] ≤ 0 && z[2, x, y] ≤ 0],
  {x, -1.3, 1.3}, {y, -1.3, 1.3}, PlotPoints→100];
```

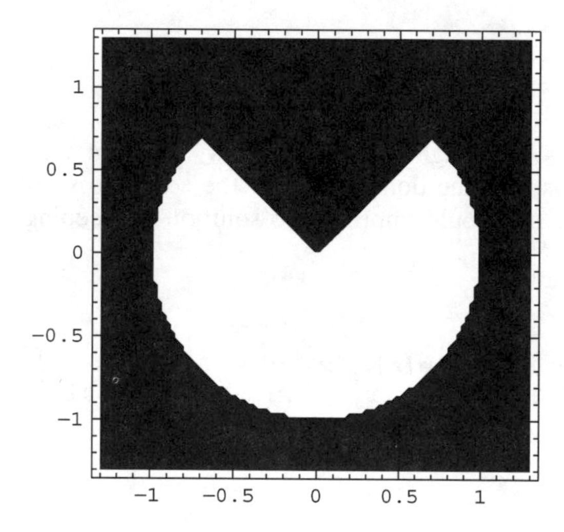

This problem is analytically tractable. The solution was developed as an exercise in Chapter 4 [Sec. 4.3, Exercise (9)] and is given below in cylindrical coordinates:

$$\phi(r, \theta) = \lim_{\epsilon \to 0} \sum_{n=1}^{\infty} A_n(\epsilon) \sin\left(\frac{n\pi}{\log \epsilon} \log r\right)$$
$$\times \left[\sinh\left(\frac{n\pi}{\log \epsilon}(\theta - \beta)\right) - \sinh\left(\frac{n\pi}{\log \epsilon}(\theta - \beta - \alpha)\right)\right], \quad (6.1.14)$$

where θ runs between $\beta = 3\pi/4$ and $\beta + \alpha$, where $\alpha = 3\pi/2$ is the opening angle of the domain. The Fourier coefficients A_n are determined according to the equation

$$\sinh\left(\frac{n\pi\alpha}{\log\epsilon}\right) A_n(\epsilon) = \frac{\int_\epsilon^1 \sin\left(\frac{n\pi}{\log\epsilon}\log r\right)(1-r^2)\frac{dr}{r}}{\int_\epsilon^1 \sin^2\left(\frac{n\pi}{\log\epsilon}\log r\right)\frac{dr}{r}}.$$

This analytic solution is constructed below:

Cell 6.25

```
A[n_] = Simplify[Integrate[Sin[n Pi Log[r]/Log[ε]] (1-r²)/r,
 {r, ε, 1}]/
  Integrate[Sin[n Pi Log[r]/Log[ε]]^2/r, {r, ε, 1}],
   n ∈ Integers]
```

$$\frac{2\ (-1)^n\ n^2\ \pi^2\ (-1 + \epsilon^2) - 8\ (-1 + (-1)^n)\ \mathrm{Log}[\epsilon]^2}{n^3\ \pi^3\ +\ 4\ n\pi\,\mathrm{Log}[\epsilon]^2}$$

Cell 6.26

```
ε = 10.^-8; α = 3 Pi/2.; β = 3 Pi/4;

φ[r_, θ_] = Sum[A[n] Sin[n Pi Log[r]/Log[ε]]
    (Sinh[n Pi (θ - β)/Log[ε]] + Sinh[n Pi (α + β - θ)/Log[ε]])/
     Sinh[nπα/Log[ε]], {n, 1, 150}];
```

Along the edges of the cusp at $\theta = \beta$ and $\theta = \beta + \alpha$, this analytic solution can be seen to match the boundary condition $\phi = 1 - r^2$, as shown in Cell 6.27. However, along $\theta = 3\pi/2$ (i.e. the $-y$ axis), the solution has a singular slope near the origin, as shown in Cell 6.28. The singularity in $\partial\phi/\partial r$ is expected near the point of a cusp, as surface charge accumulates at such points [see Sec. 3.2, Exercise (5)(b)]. This singular behavior will make it difficult to find a convergent numerical solution using the Galerkin method.

Cell 6.27

```
Plot[{φ[r, β], φ[r, β + α], 1-r^2}, {r, 0.0001, 1},
 PlotRange → All,
  AxesLabel → {"r", ""}, PlotLabel -> " φ along edges
   of cusp"];
```

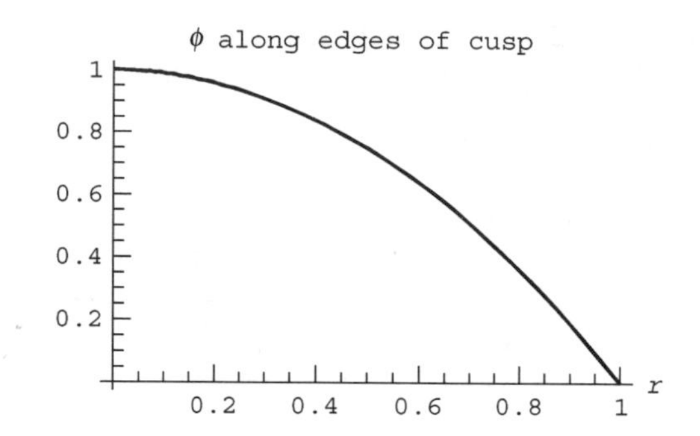

Cell 6.28

```
Plot[ϕ[r, β + α /2], {r, 0.0001, 1}, PlotRange→All];
```

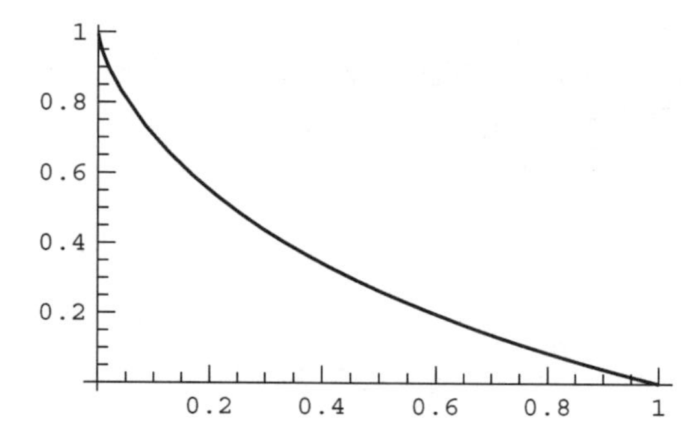

Nevertheless, we will now attempt a Galerkin solution. First, we must choose basis functions. These functions must be zero on the boundary. However, if we were to use the function $z_2(x, y) = y - |x|$ to define the basis functions via Eq. (6.1.8), grave problems would develop in the Galerkin method. Because gradients of the v_{mn} are required in the solution, the result would exhibit a singularity along $x = 0$ and strong Gibbs phenomena throughout the domain.

What we need instead is a different function $z_2(x, y)$ that equals zero on the domain boundary, but is not singular, except at $(0, 0)$. Such functions do exist (after all, the solution of Poisson's equation is one such function). One simple example is

$$z_2(x, y) = y - \sqrt{(1-a)x^2 + ay^2}, \qquad 0 < a < 1.$$

It is easy to see that this function equals zero only along the lines $y = |x|$, but the function is singular only at $(x, y) = (0, 0)$. If we use this choice in the Galerkin method, we can find a convergent solution for $\phi(x, y)$, although the convergence is slow, as we will see. The code is given below. Note the use of symmetry in x when defining the basis functions and the inner product: the solution is symmetric in x, so we choose basis functions that have this symmetry, and we integrate only over the right half of the domain.

Cell 6.29

```
(* define the curves that determine the boundary: *)
z[1, x_, y_] = x^2 + y^2-1;
z[2, x_, y_] = y-Sqrt[3 x^2/4 + y^2/4];
(* define the charge density *)
ρ[x_, y_] = 0;
(* define the function u used to
  match inhomogeneous boundary conditions, if any *)
u[x_, y_] = 1-(x^2 + y^2);
(* define the inner product
  using Boole notation (Mathematica 4.1 only) *)
norm[f_, g_] := 2NIntegrate[f g Boole[z[1, x, y] ≤
 0 && z[2, x, y] ≤ 0],
   {x, 0, Infinity}, {y, -Infinity, Infinity},
    MaxRecursion → 10];
```

Numerical integration is now the fastest approach to determining the inner product. Even so, the equations take some time to evaluate for the chosen case of $P_x = P_y = 3$. (Each equation requires 16 two-dimensional numerical integrals to be evaluated, and there are 16 equations for the 16 Fourier coefficients.) The result is plotted in Cell 6.31.

Cell 6.30

```
galerkin[3, 3, 2]
```

The solution is apparently well behaved throughout the domain of interest. However, closer examination shows that it is not particularly accurate near the origin. In Fig. 6.2, we compare the Galerkin solution with the exact solution found previously along the line $\theta = 3\pi/2$, for the case $P_x = P_y = 1, 3, 5$. The solid curve is the analytic solution, and the dashed, dotted, and dot-dashed curves are Galerkin solutions in order of increasing accuracy. The Galerkin solutions do not converge very rapidly near the origin, because of the singularity in the radial electric field caused by the cusp point. This can be seen directly in the plot (Fig. 6.3) of the radial electric field along the same line, $\theta = 3\pi/2$. As might be expected, our Galerkin solution fails to capture the singular behavior near the origin, although it is reasonably accurate for $r \gtrsim 0.5$. Many more terms are needed to roughly approximate the singularity, but calculation of the required inner products is computationally intensive. In general, numerical solutions near singularities are

Cell 6.31

```
Plot3D[(φ[x, y]) Boole[z[1, x, y] ≤ 0 && z[2, x, y] ≤ 0],
  {x, -1.3, 1.3}, {y, -1.3, 1.3}, PlotPoints → 60,
   PlotRange → All,
  Mesh → False, AxesLabel → {"x", "y", ""}, PlotLabel -> "φ"];
```

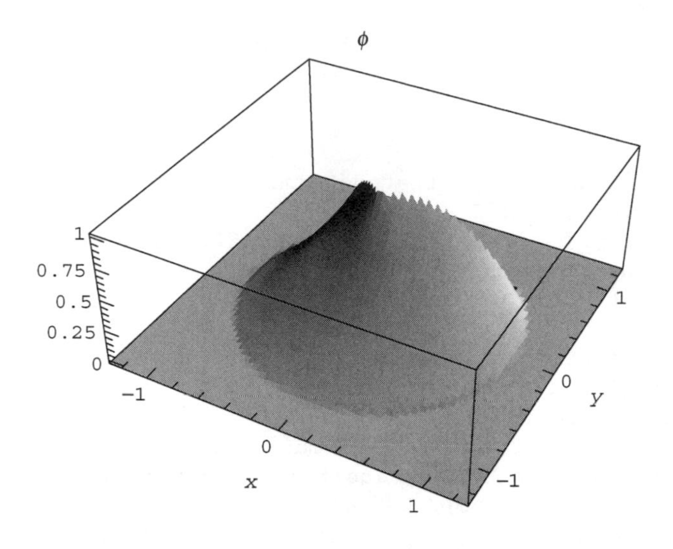

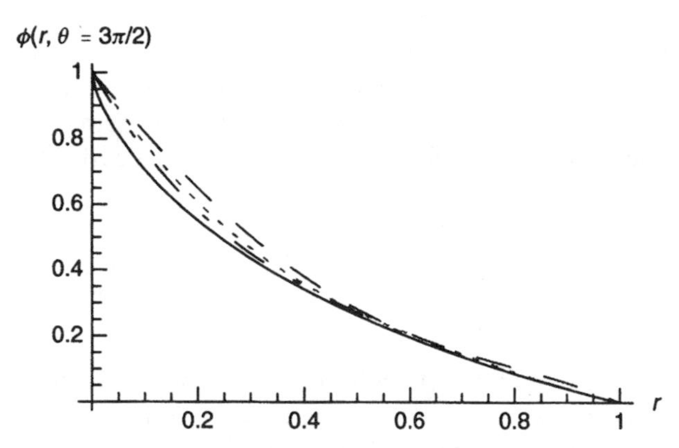

Fig. 6.2 Potential along $\theta = 3\pi/2$ for exact solution (solid) and Galerkin solutions with $P_x = P_y = 1$ (dashed), 3 (dotted), 5 (dot-dashed).

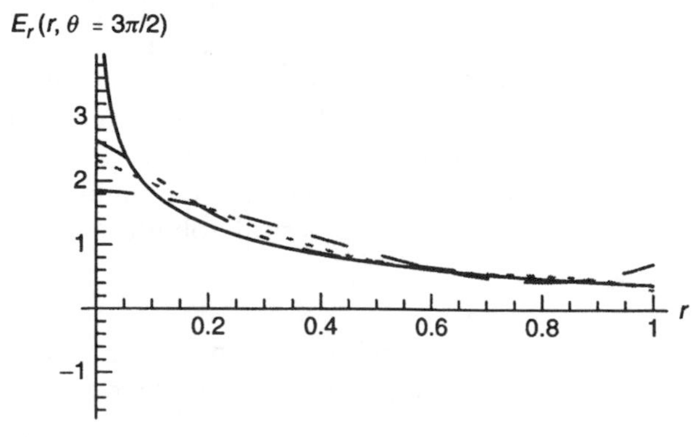

Fig. 6.3 Radial electric field along $\theta = 3\pi/2$ for exact solution (solid) and Galerkin solutions with $P_x = P_y = 1$ (dashed), 3 (dotted), 5 (dot-dashed).

problematic in any numerical method. However, the Galerkin method in particular is best suited to problems with smoothly varying solutions, unless the required inner products are easy to evaluate.

In Sec. 6.2, we will observe that grid methods can do a somewhat better job in resolving solutions near singularities, if the grid is chosen to be sufficiently fine.

6.1.3 Time-Dependent Problems

The Heat Equation The Galerkin method can also be used in the solution of PDEs with time dependence. As a simple example, we will use it to solve the heat equation in one dimension on the interval $0 \le x \le b$,

$$\frac{\partial}{\partial t} T(x,t) = \chi \frac{\partial^2}{\partial x^2} T(x,t) + S(x,t), \tag{6.1.15}$$

subject to mixed boundary conditions

$$\begin{aligned} T(0,t) &= T_1(t), \\ \frac{\partial T}{\partial x}(b,t) &= 0, \end{aligned} \tag{6.1.16}$$

and the initial condition

$$T(x,0) = T_0(x). \tag{6.1.17}$$

We write the solution in the usual manner as

$$T(x,t) = \sum_{n=0}^{M} c_n(t) v_n(x) + u(x,t), \tag{6.1.18}$$

where $u(x,t)$ is any smooth function that satisfies the boundary conditions. For this problem, the choice

$$u(x,t) = T_1(t) \tag{6.1.19}$$

works well.

The basis functions $v_n(x)$ are any set of M functions that form a complete set as $M \to \infty$, and that satisfy homogeneous mixed boundary conditions of the same type as specified in the problem. In this case, this implies homogeneous Dirichlet conditions on one end, and homogeneous von Neumann conditions on the other:

$$v_n(0) = \frac{\partial v_n}{\partial x}(b) = 0.$$

For example, we could choose these functions to be eigenmodes of the operator $\partial^2/\partial x^2$, and then we would have the usual eigenmode expansion, as discussed previously in Secs. 3.1.4 and 4.2.2. Here, however, we will choose some other set. For instance, one simple choice is

$$v_n(x) = x^n\left(1 - \frac{n}{n+1}\frac{x}{b}\right), \qquad n = 1, 2, \ldots, M. \tag{6.1.20}$$

Some of these functions are displayed in Cell 6.32, assuming $b = 2$.

Cell 6.32

```
v[n_, x_] = x^ n (1-n (x/b)/(n + 1));
b = 2; Plot[Evaluate[Table[v[n, x], {n, 1, 4}]], {x, 0, b}];
```

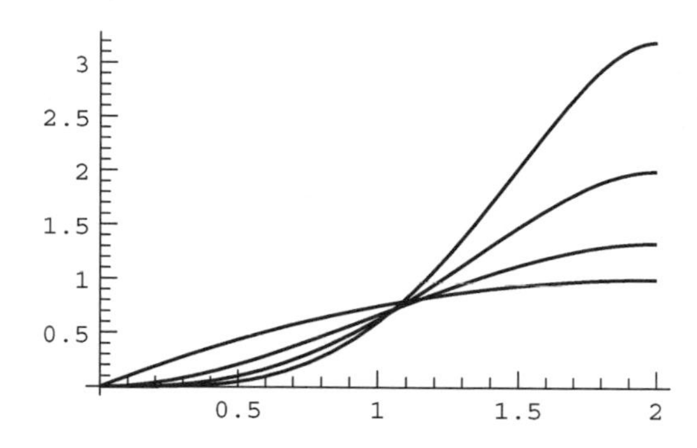

Next, we choose an inner product. We will make the simple choice

$$(f,g) = \int_0^b f^* g \, dx. \qquad (6.1.21)$$

The Fourier coefficients $c_n(t)$ in Eq. (6.1.18) must now be determined. To do so, we first substitute Eq. (6.1.18) into Eq. (6.1.15), obtaining

$$\sum_{n=1}^{M} \frac{\partial}{\partial t} c_n(t) v_n(x) = \chi \sum_{n=1}^{M} c_n(t) \frac{\partial^2}{\partial x^2} v_n(x) + \bar{S}(x,t), \qquad (6.1.22)$$

where the modified source function $\bar{S}(x,t)$ is

$$\bar{S}(x,t) = S(x,t) + \chi \frac{\partial^2}{\partial x^2} u(x,t) - \frac{\partial}{\partial t} u(x,t). \qquad (6.1.23)$$

Now, we take an inner product of Eq. (6.1.22) with respect to $v_m(x)$, yielding

$$\sum_{n=1}^{M} (v_m, v_n) \frac{\partial}{\partial t} c_n(t) = \sum_{n=1}^{M} c_n(t) \left(v_m, \chi \frac{\partial^2}{\partial x^2} v_n \right) + (v_m, \bar{S}), \qquad m = 1, 2, \ldots, M$$

$$(6.1.24)$$

Equation (6.1.24) provides a set of M coupled ODEs for the time variation of the Fourier coefficients $c_n(t)$. One can solve these ODEs either analytically or numerically (using **DSolve** or **NDSolve**). The initial conditions on the ODEs are determined by Eq. (6.1.17):

$$\sum_{n=1}^{M} c_n(0) v_n(x) + u(x,0) = T_0(x). \qquad (6.1.25)$$

Taking an inner product of this equation with respect to $v_m(x)$, we obtain a set of coupled equations for the initial values of the Fourier coefficients:

$$\sum_{n=1}^{M} c_n(0)(v_m, v_n) = (v_m, T_0 - u(x,0)), \qquad m = 1, 2, \ldots, M. \qquad (6.1.26)$$

These M equations, together with the M ODEs given by Eq. (6.1.24), are sufficient to specify $c_n(t)$.

The following *Mathematica* commands solve this problem on the time interval $0 \le t \le \tau$. We will use **NDSolve** to numerically determine the solutions to Eqs. (6.1.24), subject to the initial conditions of Eq. (6.1.26). The Fourier coefficients $c_n(t)$ are referred to as **c[n][t]**. This notation is equivalent to **c[n,t]**, and it works in **NDSolve**, whereas the notation **c[n,"t]** does not, because **NDSolve** interprets that as an instruction to solve a PDE for a function of two variables **c[n,t]**. (Recall that the same issue arose when we performed molecular dynamics simulations in Chapter 1.)

Since this is a numerical solution to the problem, we must choose specific forms for the initial and boundary conditions explicitly, and we must also choose a numerical value for the thermal diffusivity χ and the size of the interval, b. Below, we make these choices, keeping $M = 8$ terms in the series expansion of the solution, and solving over a time interval of $0 < t < 15$:

Cell 6.33

```
T0[x_] = 1; T1[t_] =1 + Sin[t]; S[x_, t_] = 0;

χ = 1/8; b = 2; M = 8;
τ = 15.;
```

The temperature on the left side of the slab oscillates in time, and on the right the temperature gradient is zero.

To construct the Galerkin solution to this problem, we first define the function $u(x, t)$, according to Eq. (6.1.19):

Cell 6.34

```
u[x_, t_] = T1[t];
```

We then define the inner product, as in Eq. (6.1.21). In this case, we will assume that the functions to be integrated are sufficiently straightforward that the integrals can be done analytically using **Integrate**:

Cell 6.35

```
norm[f_, g_] := Integrate[fg, {x, 0, b}]
```

It is also useful to define the spatial operator $\hat{L} = \chi \partial^2 / \partial x^2$:

Cell 6.36

```
L[f_] := χD[f, {x,2}];
```

Next, we define the projection of the operator onto the functions, $L_{mn} = (v_m, \hat{L} v_n)$:

Cell 6.37

```
Lmat[m_, n_] := norm[v[m, x], L[v[n, x]]]
```

We then define the inner product of the functions with each other, (v_m, v_n). Here we use symmetry of the inner product with respect to the interchange of n and m in order to reduce the number of integrations needed. Also, we use the double-equal-sign notation to store previously determined values of the inner product without redoing the integration:

Cell 6.38

```
nm[n_, m_] := (nm[n, m] = norm[v[m, x], v[n, x]] )/; n≥m;
nm[n_, m_] := nm[m, n]/; n < m
```

Next, we define the projection of the source function $\bar{S}(x,t)$ onto the $v_m(x)$:

Cell 6.39

```
Svec[m_] := norm[v[m, x], S[x, t]-D[u[x, t], t] + L[u[x, t]]];
```

We are ready to specify the differential equations for the Fourier coefficients:

Cell 6.40

```
eq[m_] := eq[m] = Sum[nm[m, n] D[c[n][t], t],
  {n, 1, M}] ==
    Sum[Lmat[m, n] c[n][t], {n, 1, M}] + Svec[m];
```

It is useful to point out the equations as a check:

Cell 6.41

```
Do[Print["eq[", m, "]= ", eq[m]], {m, 1, M}];
```

The initial conditions on the Fourier coefficients are specified according to Eq. (6.1.26). We print these out as well:

Cell 6.42

```
ic[m_] :=
  ic[m] = Sum[nm[m, n] c[n][0], {n, 1, M}] == norm[v[m, x],
   T0[x]-u[x, 0]];

Do[Print["ic[", m, "]=", ic[m]], {m, 1, M}];
```

In order to use **NDSolve** to determine the solution to the coupled ODEs, we must make a list of the equations and initial conditions:

Cell 6.43

```
eqns = Join[Table[eq[m], {m, 1, M}], Table[ic[m], {m, 1, M}]];
```

Also, we create a list of the variables for which we are solving:

Cell 6.44

```
vars = Table[c[m][t], {m, 1, M}];
```

Finally, we solve the equations over a time τ and define the solution:

Cell 6.45

```
coefficients = NDSolve[eqns, vars, {t, 0, τ},
 MaxSteps → 5000] [[1]];

T[x_, t_] = u[x, t] + Evaluate[Sum[c[n][t] v[n, x],
 {n, 1, M}]];
T[x_, t_] = T[x, t]/. coefficients;
```

Cell 6.46 displays our solution. This solution exhibits the same behavior as was observed in Example 1 of Sec. 4.2.2, where we solved a similar problem using the exact eigenmodes for the system. The temperature oscillations only penetrate a short distance into the material before they are averaged out.

Cell 6.46

```
Table[Plot[T[x, t], {x, 0, b}, PlotRange → {{0, 2}, {0, 2}},
 AxesLabel → {"x", ""},
   PlotLabel → "T[x, t], t=" <>ToString[t]], {t, 0, 15, .5}];
```

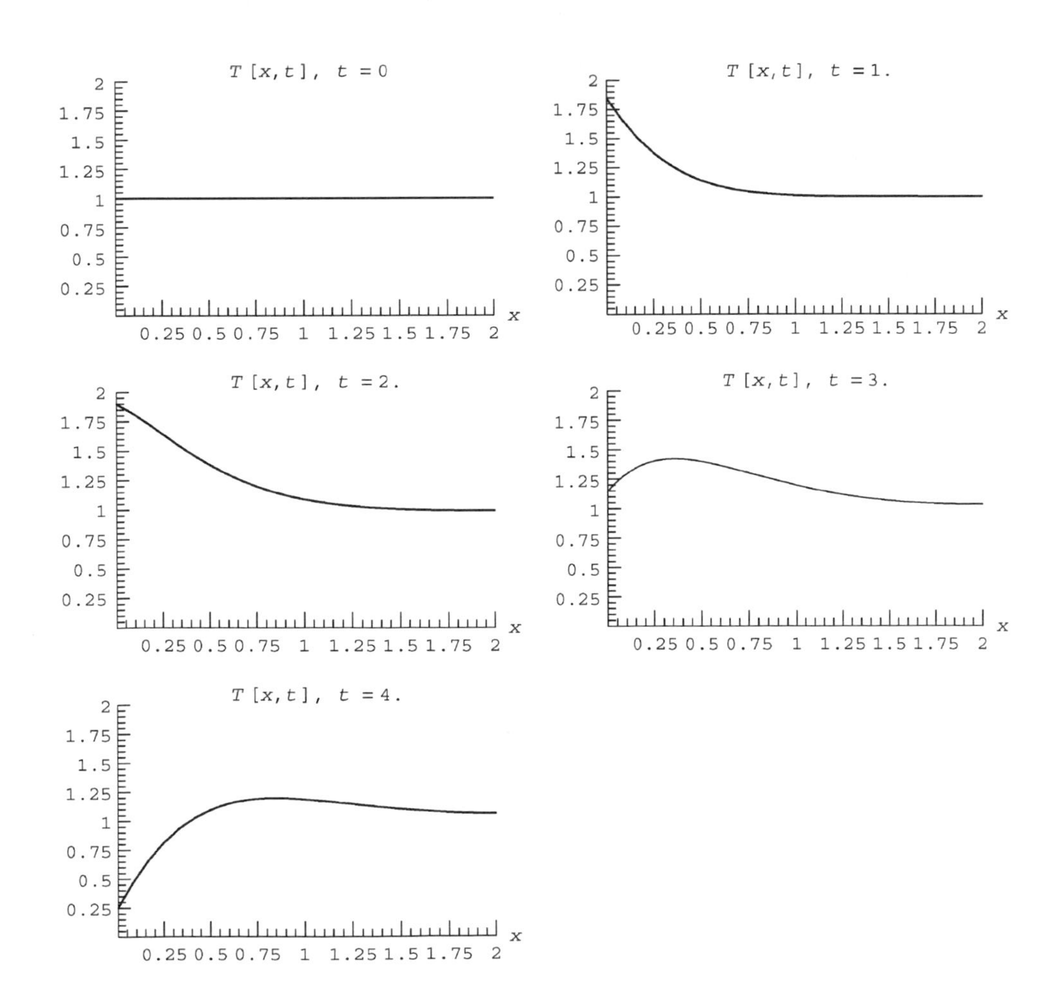

Below, we collect the Galerkin method commands in one module:

Cell 6.47

```
galerkint[M_, τ_] := Module[{u, v, norm, L, eq, ic,
 eqns, vars},
  (* define the function u used to
    match inhomogeneous boundary conditions, if any *)
  u[x_, t_] = T1[t];
  (* define the basis functions: *)
                          n
  v[n_, x_] = x^n (1- ───── (x/b));
                        n + 1
(* define the inner product *)
  norm[f_, g_] := Integrate[f g, {x, 0, b}];
  (* define the operator: *)
  L̂[f_] := χD[f, {x, 2}];
  (* determine the projection of the operator onto the basis
    functions: *)
  L[m_, n_] := norm[v[m, x], L̂[v[n, x]]];
  (* determine the norm of the basis functions with each
    other. Use symmetry of the norm to reduce the number of
    integrals that need be done; remember their values. *)
  nm[n_, m_] := (nm[n, m] = norm[v[m, x], v[n, x]])/; n ≥ m;
  nm[n_, m_] := nm[m, n]/; n < m ;
  (* determine the projection of S̄ onto the basis functions:
   *)
  S̄[m_] := norm[v[m, x], S[x, t]-D[u[x, t], t] +
   L̂[u[x, t]]];
  (* define the equations for the Fourier coefficients c[n][t]
    *)
  eq[m_] := eq[m] = Sum[nm[m, n] D[c[n][t], t], {n, 1, M}]
    ==Sum[L[m, n] c[n][t], {n, 1, M}] + S̄[m];
  (* Print out the equations (not necessary, but a useful
    check) *)
  Do[Print["eq[", m, "]= ", eq[m]], {m, 1, M}];
  (* define the initial conditions *)
  ic[m_] :=
   ic[m] = Sum[nm[m, n] c[n][0], {n, 1, M}] == norm[v[m, x],
    T0[x]-u[x, 0]];
  (* Print out the initial conditions (not necessary, but a
    useful check) *)
  Do[Print["ic[", m, "]= ", ic[m]], {m, 1, M}];
  (* create lists of equations, variables *)
  eqns = Join[Table[eq[m], {m, 1, M}], Table[ic[m],
   {m, 1, M}]];
  vars = Table[c[m][t], {m, 1, M}];
  (* solve the equations *)
  coefficients = NDSolve[eqns, vars, {t, 0, τ}][[1]];
  (* define the solution *)
  T[x_, t_] = u[x, t] + Evaluate[Sum[c[n][t] v[n, x],
   {n, 1, M}]];
  T[x_, t_] = T[x, t]/. coefficients;]
```

Reflection of a Quantum Particle from a Barrier The evolution of a quantum particle moving in an external potential is another case where the Galerkin method is useful. This is a case where all of the steps in the method can be performed analytically, except for the solution of the coupled ODEs.

We consider a particle that encounters a potential barrier $V(x)$ of finite height and width. The particle wave function $\psi(x,t)$ satisfies the time-dependent Schrödinger equation,

$$i\hbar \frac{\partial}{\partial t}\psi(x,t) = -\frac{\hbar^2}{2m}\frac{\partial^2}{\partial x^2}\psi(x,t) + V(x)\psi(x,t). \tag{6.1.27}$$

The initial condition on the wave function is

$$\psi(x,0) = \psi_0(x). \tag{6.1.28}$$

For this problem, we take the barrier potential $V(x)$ to be

$$V(x) = \begin{cases} V_0, & |x| < a, \\ 0, & |x| \geq a. \end{cases} \tag{6.1.29}$$

We will assume that the system has periodic boundary conditions, with period $2L$. (So this barrier, and the wave function itself, are periodically replicated in cells of width $2L$.) Then to match these boundary conditions we will expand the wave function in terms of a periodic exponential Fourier series,

$$\psi(x,t) = \sum_k c_k(t)\, e^{ikx}, \qquad k = n\pi/L, \quad n = -M, \dots, M. \tag{6.1.30}$$

Substituting this series into Eq. (6.1.27), we obtain

$$i\hbar \sum_k \frac{\partial}{\partial t} c_k(t)\, e^{ikx} = \sum_k \frac{\hbar^2 k^2}{2m} c_k(t)\, e^{ikx} + \sum_{\bar{k}} c_{\bar{k}}(t) V(x)\, e^{i\bar{k}x}. \tag{6.1.31}$$

In the last term, we replaced the dummy variable k by $\bar{k}$. For the inner product, we choose the one for which the Fourier modes are orthogonal:

$$(f,g) = \int_{-L}^{L} f^* g\, dx. \tag{6.1.32}$$

Then, taking an inner product of Eq. (6.1.31) with respect to e^{ikx} yields the following coupled equations for the Fourier coefficients $c_k(t)$:

$$i\hbar \frac{\partial}{\partial t} c_k(t) = \frac{\hbar^2 k^2}{2m} c_k(t) + \sum_{\bar{k}} V_{k\bar{k}} c_{\bar{k}}(t), \tag{6.1.33}$$

where $V_{k\bar{k}}$ is the projection of the potential onto the Fourier modes:

$$V_{k\bar{k}} = \frac{\left(e^{ikx}, V(x)\, e^{i\bar{k}x}\right)}{\left(e^{ikx}, e^{ikx}\right)}. \tag{6.1.34}$$

For the barrier potential of Eq. (6.1.29), this matrix can be evaluated analytically:

$$V_{k\bar{k}} = \frac{V_0}{(\bar{k}-k)L} \sin(\bar{k}-k)a. \tag{6.1.35}$$

We are then left with the task of solving the coupled ODEs given by Eq. (6.1.33), subject to the initial condition, Eq. (6.1.28). When Eq. (6.1.30) is used in Eq. (6.1.28) and an inner product is taken, we obtain

$$c_k(0) = \frac{\left(e^{ikx}, \psi_0(x)\right)}{\left(e^{ikx}, e^{ikx}\right)} = \frac{1}{2L}\int_{-L}^{L} e^{-ikx}\, \psi_0(x)\, dx, \tag{6.1.36}$$

which is simply the usual expression for the Fourier coefficient of the initial condition.

Let's solve this problem for specific (dimensionless) choices of the parameters:

Cell 6.48

```
ℏ = 1; mass = 1;
```

For the initial condition, we choose a wave function centered at $k_0 = 5$ and at position $x = -5$, moving toward the potential barrier from the left:

$$\psi_0(x) = e^{-2(x+5)^2}\, e^{5ix}.$$

This corresponds to a classical particle with energy $(\hbar k_0)^2/(2m) = 12.5$ in our dimensionless units. We take the potential barrier height equal to $V_0 = 15$, in order to reflect the classical particle.

The Fourier coefficients of the initial condition can be evaluated for general k:

$$c_k(0) = \frac{e^{-50-\frac{1}{8}(-5-20\,i+k)^2}\sqrt{\pi/2}}{2L}.$$

The following module integrates the equations of motion for this initial condition. The Fourier coefficients $c_k(t)$ are referred to as `c[n][t]`, where $k = 2\pi n/(2L)$, $n = -M, \dots, M$. The equations for the coefficients are integrated up to time τ. The module can be easily modified to solve for propagation of a quantum wave function starting with any initial condition and moving in any one-dimensional potential, simply by changing the definition of the initial conditions $c_k(0)$ and the potential matrix $V_{k\bar{k}}$.

Cell 6.49

```
schrödinger[L_, a_, V0_, M_, τ_] := Module[{k, v, ic, V, eq,
 eqns, vars},
  (* DEFINE THE WAVENUMBER *)
  k[n_] = Pi n/L;
  (* DEFINE THE BASIS FUNCTIONS *)
  v[n_, x_] = Exp[I k[n] x];
  (* DEFINE AND PRINT THE INITIAL CONDITIONS *)
```

$$\text{ic[m\_] := ic[m] = c[m][0] ==}\ \frac{E^{-50-\frac{1}{8}\,((-50-20\,I)+k[m])^2}\sqrt{\frac{\pi}{2}}}{2L};$$

```
  Do[Print["ic[", m, "]=", ic[m]], {m, -M, M}];

  (* DEFINE PROJECTION OF THE POTENTIAL ONTO THE MODES *)
  V[n_, m_] := V0 Sin[(k[n]-k[m]) a]/((k[n]-k[m]) L)/; n≠m;
  V[n_, n_] := V0 a/L;
  (* DEFINE AND PRINT THE ODES *)
  eq[n_] := eq[n] = I D[c[n][t], t] ==
    ℏ² k[n]²/(2 mass) c[n][t] + Sum[V[n, m] c[m][t],
    {m, -M, M}];
  Do[Print["eq[", m, "]=", eq[m]],
   {m, -M, M}];
  (* CREATE LISTS OF THE EQUATIONS AND VARIABLES FOR USE IN
    NDSOLVE*)
  eqns = Join[Table[eq[m], {m, -M, M}], Table[ic[m],
   {m, -M, M}]];
  vars = Table[c[n][t], {n, -M, M}];
  (* SOLVE THE EQUATIONS *)
  coefficients = NDSolve[eqns, vars, {t, 0, τ},
   MaxSteps→6000][[1]];
  (* THE SOLUTION *)
ψ[x_, t_] = Sum[c[n][t] v[n, x], {n, -M, M}];
ψ[x_, t_] = ψ[x, t]/. coefficients;]
```

In Cell 6.50 we run this module, taking $L = 15$ and a barrier half width $a = 1$. We integrate over a time $\tau = 2.4$ and take $M = 60$ (a large number of modes must be kept in order to represent the sharply peaked initial condition).

Cell 6.50

```
schrödinger[15., 1., 15., 60, 2.4];
```

Cell 6.51

```
Table[Plot[Abs[ψ[x, t]]^2, {x, -15, 15},
   PlotRange→{0, 1}, PlotLabel →" (|ψ[x, t]|)²,
    t= " <>ToString[t],
   AxesLabel→{"x", ""}], {t, 0, 2.4, .2}];
```

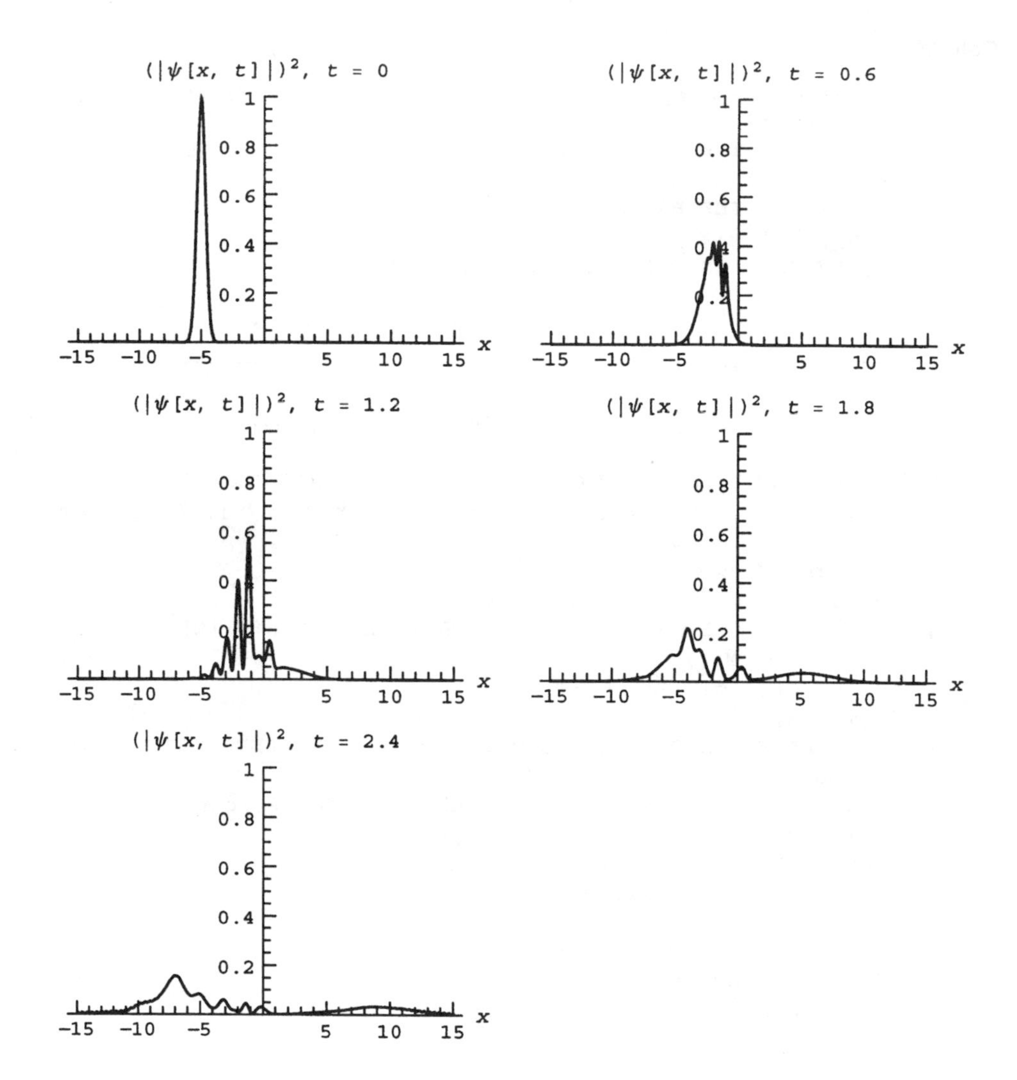

The plots of the wave packet in Cell 6.51 display several interesting effects. First, we can clearly see that dispersion causes the packet to spread, just as in the discussion of freely propagating packets in Sec. 5.1.2. Second, during the reflection, there is a complex interference pattern produced as the reflected packet travels back through the incident packet. Third, a fair fraction of the wave packet is transmitted through the barrier, which runs from -1 to 1. A classical particle with energy 12.5 would have been reflected from this barrier, which has potential 15. This is the well-known phenomenon of *quantum tunneling* through a classically forbidden region.

One can also see something else in this simulation: with a speed of 5 in our scaled units, the center of a freely propagating wave packet would have made it to position $x = -5 + \tau \times 5 = 7$ in time $\tau = 2.4$. However, the center of the packet in our simulation has actually made it *farther* than this: the animation clearly shows that the center of the packet is roughly at $x = 9$ at time $\tau = 2.4$. Adding a potential barrier that creates a classically forbidden region has actually *increased* the average group velocity of the transmitted part of the packet. Note that the width of the potential barrier was 2, which is just the observed increase in position compared to free propagation. It is as if the wave packet were instantaneously transported across the forbidden zone.

This phenomenon is referred to in the literature as *superluminal pulse propagation*. Actually, this is something of a misnomer. A detailed analysis of the pulse propagation shows that it does not violate the tenets of special relativity, because the wave packet has components that rapidly disperse (the leading edge of the packet is far in advance of the pulse center.) These forward components already signal the arrival of the packet, and they are causal, always traveling at speeds less than or equal to that of light.

Superluminal propagation can also occur in classical systems, such as light traveling in a dispersive medium. Although this phenomenon is well-known, it has recently received renewed attention [see, e.g., Winful (2003)].

The reader is invited to vary the parameters in this simulation in order to further study these phenomena. Questions that one might ask are: how does the mean group velocity of the transmitted packet vary with barrier height and width, and as a function of the initial packet speed and width? What fraction of the packet is transmitted as a function of these parameters? How is the dispersion of the packet affected by the barrier? (Caution: The simulation runs rather slowly on older machines, due to the large number of modes that must be kept.)

EXERCISES FOR SEC. 6.1

(1) Use the Galerkin method to solve the potential in the grounded charge-filled enclosure of Fig. 6.4. The potential satisfies $\nabla^2\phi(x, y) = x$, and is zero on the walls. Plot the resulting potential as a contour plot, and find the place where the potential is minimum.

(2) A grounded toroidal conducting shell has a shape given by the equation $z^2/2 + (r - R)^2 = a^2$, where (r, θ, z) are cylindrical coordinates, and $R = 1$ meter, $a = 0.75$ meter. Plot this shell using a **ParametricPlot3D** command. The shell is filled with a uniform charge density $\rho/\epsilon_0 = 10$ V. Find and plot the contours of constant electrostatic potential in the (r, z) plane. At what r and z is the potential maximum, and what is its value, to three significant figures? (Hint: Solve Poisson's equation in cylindrical coordinates.)

(3) A hemispherical shell of radius a sits in the base of a cylindrical tube of the same radius (see Fig. 6.5). The top of the tube at $z = L$ is flat, and the sides of the tube are insulated. The interior of the tube is filled with material of

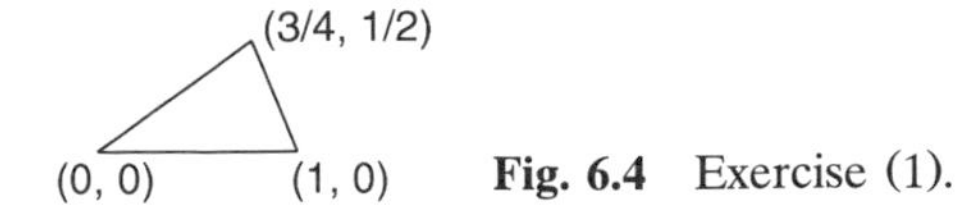

Fig. 6.4 Exercise (1).

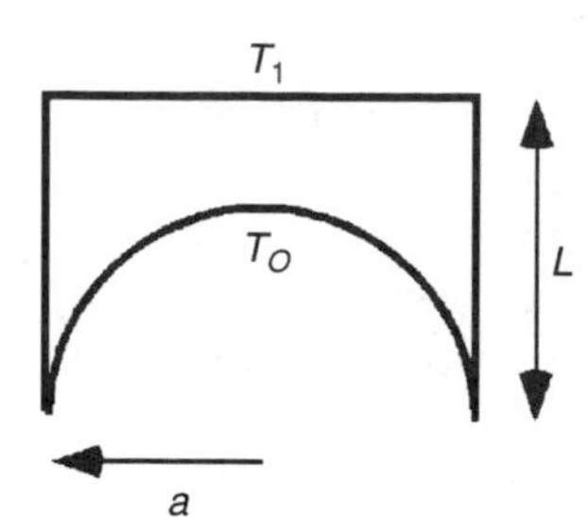

Fig. 6.5 Exercise (3).

some uniform thermal diffusivity. The hemisphere is held at fixed temperature T_0, and thc top end and of the tube is held at fixed temperature T_1. Use the Galerkin method to find the form of $T(r, z)$ in equilibrium within the tube, for $a = 1$, $L = 1.5$. (Hint: Try basis functions of the form $v_{nm}(r, z) = [r^n - A(z)](a^2 - r^2 - z^2)(L - z)z^m$, and find the form of $A(z)$ required to match the von Neumann boundary conditions at $r = a$.)

(4) A triaxial ellipsoid has a shape given by the equation $x^2/a^2 + y^2/b^2 + z^2/c^2 = 1$, where $a = 2$, $b = 3$, and $c = 4$. The ellipsoid surface is at potential V_0. The ellipsoid is filled with a uniform negative charge of density $-\rho_0$. Using the Galerkin method, find the electrostatic potential inside the ellipsoid as a function of (x, y, z). (Hint: The solution is a simple polynomial function of x, y, z.)

(5) Use the Galerkin method to solve the following heat equation problem in 1D, with inhomogeneous conductivity, on $0 < x < 1$, for $0 < t < \frac{1}{2}$. Choose your own modes, and test to make sure you have kept enough modes. Plot the result as an animation:

$$\frac{\partial}{\partial x}\left((4x + 1)\frac{\partial}{\partial x}T(x, t)\right) = \frac{\partial}{\partial t}T(x, t) \qquad \text{on} \quad 0 < x < 1$$

with

$$T(0, t) = 0, \qquad T(1, t) = 1, \quad \text{and} \quad T(x, 0) = x^6.$$

(6) Solve the time-dependent (dimensionless) Schrödinger equation for the wave function $\psi(x, t)$ of a particle in a potential $V(x)$ for $0 < t < 1$, subject to boundary conditions $\psi(\pm a, t) = 0$ and initial condition $\psi(x, 0) = \psi_0(x)$:

$$i\frac{\partial \psi(x, t)}{\partial t} + V(x)\psi(x, t) - \frac{1}{2}\frac{\partial^2 \psi(x, t)}{\partial x^2}.$$

Take $V(x) = x^4/4$, use trigonometric functions for the modes in the Galerkin method, and work out the matrix elements V_{nm} of $V(x)$ analytically for general n and m. (Note: Check to make sure the integrals for V_{nn} are evaluated properly.) Take $\psi_0(x) = \exp[-2(x - 1)^2]$, and $a = 4$. Keep as many basis functions as necessary to obtain a converged answer. Plot the solution in time increments of 0.01 for $0 < t < 4$.

(7) **(a)** Use the Galerkin method to solve for the dynamics of the string with the moving boundary condition, described in scaled coordinates by Eq. (5.2.64). Again take the case $L_0 = 1$ m, $c = 300$ m/s, and $v = 2$ m/s. For basis functions, use $\sin n\theta\bar{x}$. Plot $y(\bar{x} = \frac{1}{2}, t)$ and listen to the sound the string makes using the `Play` function, for $0 < t < 0.45$ s. Compare the plot and the sound with the WKB solution, Eq. (5.2.65). (Hint: You will need to increase `MaxSteps` in `NDSolve` to around 5000.)

(b) Redo the problem, and make an animation of $y(x,t)$, for a case where the WKB solution fails: $L_0 = 1$ m, $c = 300$ m/s, and $v = 200$ m/s. Plot the solution over the time interval $0 < t < 0.0045$ s.

(c) Evaluate the energy variation of the string dynamics found in part (b), and use this to determine the force on the moving boundary vs. time.

(8) A large sheet of cloth has an elliptical shape given by $x^2/a^2 + y^2/b^2 = 1$, where $a = 2$ m and $b = 1$ m. The edge of the sheet is fixed at $z = 0$, and the propagation speed of waves on the sheet is $c = 3$ m/s. Initially the sheet is flat and stationary, $z(x, y, 0) = \dot{z}(x, y, 0) = 0$. However, thanks to the acceleration of gravity $g = 9.8$ m/s^2, the sheet begins to sag. The equation of motion for the sheet is the damped wave equation,

$$\left(\frac{\partial^2}{\partial t^2} + \gamma\frac{\partial}{\partial t}\right) z(x,y,t) = c^2\left(\frac{\partial^2}{\partial x^2} + \frac{\partial^2}{\partial y^2}\right) z(x,y,t),$$

where $\gamma = 10$ s^{-1} due to air drag on the sheet. Solve for the dynamics of the sheet using the Galerkin method, and plot the resulting motion as a series of `ParametricPlot3D` graphics objects, for $0 < t < 1$ s.

(9) A circular wading pool of radius $a = 1$ m has nonuniform depth, $h(r,\theta) = h_0[1 + (2r^2\cos 2\theta)/(3a^2)]$, where $h_0 = 5$ cm and coordinates are measured from the center of the pool. Assuming potential flow, the horizontal fluid displacement vector can be expressed as $\boldsymbol{\eta} = \nabla\phi(r,\theta,t)$, with ϕ satisfying the PDE

$$\frac{\partial^2}{\partial t^2}\phi = \nabla\cdot c^2(r,\theta)\,\nabla\phi(r,\theta,t),$$

and where $c^2(r,\theta) = gh(r,\theta)$ is the square of the wave speed. The wave height is $z(r,\theta,t) = -\nabla\cdot[h(r,\theta)\boldsymbol{\eta}]$. The boundary conditions are $\boldsymbol{\eta}\cdot\hat{\mathbf{r}}|_{r=a} = 0$. [See Eqs. (4.2.50), (4.4.38) and (4.4.39) and the surrounding discussion.]

(a) We will use the Galerkin method to solve this problem. First, we must choose basis functions. Show that functions of the form $v_{mn}(r,\theta) = r^n(1 - b_n r^2)\cos m\theta$ match the von Neumann boundary conditions required at $r = a$, provided that b_n is chosen properly. (Why don't we need $\sin m\theta$ basis functions?)

(b) Boundary conditions at $r = 0$ must also be satisfied: the solution and its derivatives must be finite. By examining the form of ∇^2 in cylindrical coordinates, show that this leads to the requirement on the basis functions that

$$\lim_{r\to 0}\left(\frac{\partial v_{mn}}{\partial r} - \frac{m^2}{r}v_{mn}\right) = 0. \tag{6.1.37}$$

Show that this equation implies that we must choose n in the range $n \geq m$.

(c) Initially, $\phi(r,\theta,0) = -J_0(j_{1,1}r/a)/10$ and $\dot\phi(r,\theta,0) = 0$. Find the wave motion and plot $z(r,\theta,t)$ for $0 < t < 3$ s.

6.2 GRID METHODS

6.2.1 Time-Dependent Problems

FTCS Method for the Heat Equation In Chapters 1 and 2, we showed how boundary- and initial-value ODE problems can be solved using grid methods, wherein the linear ODE is converted into a difference equation by discretizing the independent variable. In this section, we apply analogous methods to partial differential equations.

As a first example of a grid method in the solution of a PDE, consider the heat equation in a slab running from $0 < x < L$,

$$\frac{\partial}{\partial t}T(x,t) = \chi\frac{\partial^2}{\partial x^2}T(x,t) + S(x,t). \tag{6.2.1}$$

We solve for $T(x,t)$ subject to the initial condition that $T(x,0) = T_0(x)$ and the Dirichlet boundary conditions $T(0,t) = T_1(t)$, $T(L,t) = T_2(t)$. To solve this problem, we discretize both space and time, according to

$$\begin{aligned} t_n &= n\,\Delta t, \qquad n = 0,1,2,\ldots, \\ x_j &= j\,\Delta x, \qquad j = 0,1,2,\ldots,M, \end{aligned} \tag{6.2.2}$$

where Δt is the step size in time, and $\Delta x = L/M$ is the step size in space. We will solve Eq. (6.2.1) for $T(x,t)$ only on this grid of points. It is useful to introduce the notation $T(x_j, t_n) = T_j^n$ (subscripts refer to the spatial grid point, and superscripts refer to the time grid point).

We must now discretize the space and time derivatives in Eq. (6.2.1). There are many ways to do so, some better than others. To start with, we will discuss the FTCS method, where we use a forward difference method for the time derivative and a centered difference method for the space derivative (hence the name: forward time, centered space). That is, for the time derivative, we write

$$\frac{\partial T}{\partial t}(x_j,t_n) \to \frac{T(x_j,t_{n+1}) - T(x_j,t_n)}{\Delta t} = \frac{T_j^{n+1} - T_j^n}{\Delta t}, \tag{6.2.3}$$

and for the space derivative,

$$\frac{\partial^2 T}{\partial x^2}(x_j,t_n) \to \frac{T(x_{j+1},t_n) - 2T(x_j,t_n) + T(x_{j-1},t_n)}{\Delta x^2} = \frac{T_{j+1}^n - 2T_j^n + T_{j-1}^n}{\Delta x^2}. \tag{6.2.4}$$

(See Sec. 2.4.5 for a discussion of finite-difference derivatives.) Substituting these expressions into Eq. (6.2.1), we collect all terms on the right-hand side except for the term involving T_j^{n+1}:

$$T_j^{n+1} = T_j^n + \Delta t\, S_j^n + \frac{\chi \Delta t}{\Delta x^2}\left(T_{j+1}^n - 2T_j^n + T_{j-1}^n\right). \tag{6.2.5}$$

Equation (6.2.5) is a recursion relation that bears a distinct resemblance to Euler's method. If, at the nth time step, we know T_j^n for all values of j on the spatial grid, then we can use Eq. (6.2.5) to determine T at the $n+1$st time step at each spatial gridpoint. The programming of this method in *Mathematica* is nearly as simple as for Euler's method. We refer to T_j^n as the function **T[j,n]**, and write the recursion relation as

Cell 6.52

```
Clear["Global`*"];

T[j_, n_] := T[j, n] =
  T[j, n-1] + Δt S[j, n-1] + (χ Δt/Δx^2) (T[j + 1, n-1]
   -2 T[j, n-1] + T[j-1, n-1])
```

Note the use of the double equality, in order to force *Mathematica* to remember previously evaluated values of T, just as was done in Euler's method.

However, Eq. (6.2.5) works *only for the interior points*. We cannot use it to determine T_0^n or T_M^n, because the equation would require T_{-1}^n and T_{M+1}^n, which are outside the range of the grid. Fortunately, these boundary values can be specified by the boundary conditions $T_0^n = T_1(t_n)$ and $T_M^n = T_2(t_n)$:

Cell 6.53

```
T[0, n_] := T1[t]/. t→n Δt;
M = 10; T[M, n_] := T2[t]/. t→n Δt
```

> In grid methods, we determine the unknown function at the boundary points of the grid using the boundary conditions, and use the finite-differenced differential equation only at the interior points.

In order to start the recursion, we must also specify the initial condition:

Cell 6.54

```
T[j_, 0] := T0[x]/. x→j Δx
```

To use this method, all we need do is choose Δt and M, and specify respectively the source function S, the initial and boundary conditions **T0**, **T1**, and **T2**, and the width L of the system and the thermal diffusivity χ. We do so below for a specific case:

Cell 6.55

```
L = 1; χ = 1/4; Δt = 0.01; Δx = L/M; S[j_, n_] = 0;
T0[x_] = x^2 (1-x);
T1[t_] = t/(1 + 3 t); T2[t_] = 0;
```

We now make a table of plots of the solution, evaluated at different times, as shown in Cell 6.56.

Cell 6.56

```
Table[ListPlot[Table[{Δx j, T[j, n]}, {j, 0, M}],
   PlotJoined→True, PlotRange→{0, .4},
    AxesLabel→{"x", ""},
   PlotLabel→"T[x, t],t=" <>ToString[Δt n]],
    {n, 0, 200, 10}];
```

T[x,t], t = 0 — T[x,t], t = 0.2 — T[x,t], t = 0.4 — T[x,t], t = 0.6

Note, however, the rather small time step size that we took, $\Delta t = 0.01$. In part, this is because the forward difference method used for the time derivative is only first-order accurate in Δt, so small steps are required to achieve reasonable accuracy. But there is also another reason. If we attempt to run the code with a larger time step size, we run into an immediate problem: *numerical instability*. This is shown in Cell 6.57 for $\Delta t = 0.05$. So that we need not reevaluate previous cells defining T, we use an **Unset** command to clear only specific values of **T[j, n]**.

Cell 6.57

```
Do[Unset[T[j, n]], {j, 1, M-1}, {n, 1, 200}];
Δt=0.05;
Table[ListPlot[Table[{Δx j, T[j, n]}, {j, 0, M}],
  PlotJoined→True, PlotRange→{0, .4}, AxesLabel→{"x", ""},
  PlotLabel→"T[x,t],t="<>ToString[Δt n]], {n, 0, 10, 2}];
```

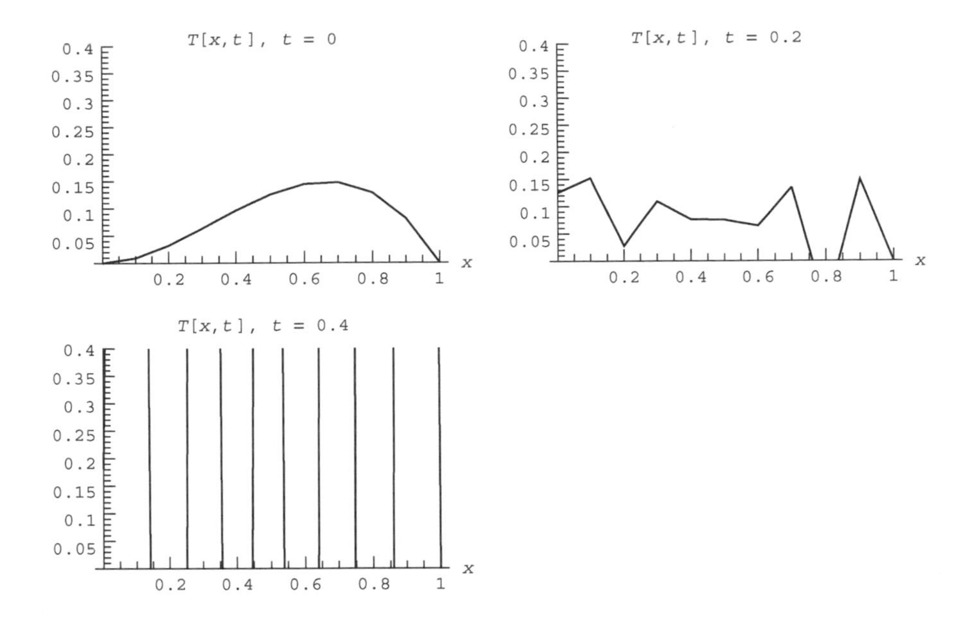

The result shows that thc solution immediately blows up. What happened?

von Neumann Stability Analysis The problem encountered in the previous example is one of numerical instability caused by an overly large step size Δt. *von Neumann stability analysis* provides us with the tools we need to understand and overcome this problem. The analysis relies on the idea that certain short-wavelength spatial modes in the numerical solution can be unstable. In the stability analysis for these modes, it suffices to neglect boundary conditions and source terms, if any, and solve the recursion relation analytically, using Fourier modes. That is, keeping only one mode, we write $T_j^n = a_n e^{ikx_j}$, where k is the wavenumber of the mode. Applying this ansatz to the FTCS method (6.2.5), and neglecting the source term, we find that we can divide out a factor of e^{ikx_j} from both sides, leaving us with the following difference equation for a_n:

$$a_{n+1} = a_n + \frac{\chi \Delta t}{\Delta x^2}\left(e^{ik\Delta x} a_n - 2a_n + e^{-ik\Delta x} a_n\right).$$

Using a trigonometric identity, this equation can be rewritten as

$$a_{n+1} = a_n\left(1 - 4\frac{\chi \Delta t}{\Delta x^2}\sin^2\frac{k\Delta x}{2}\right). \tag{6.2.6}$$

Linear homogeneous difference equations of this sort have a general solution of the form

$$a_n = A\xi^n, \tag{6.2.7}$$

where A is an undetermined constant, dependent on the initial condition. The parameter ξ is called the amplification factor, since it is the factor by which the

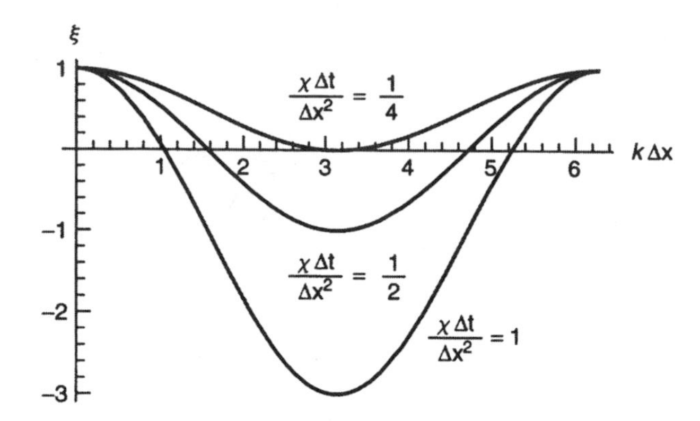

Fig. 6.6 von Neumann stability analysis for the heat equation.

solution is amplified in a single timestep. This parameter can be determined by substitution of Eq. (6.2.7) into Eq. (6.2.6):

$$\xi = 1 - 4\frac{\chi\,\Delta t}{\Delta x^2}\sin^2\frac{k\,\Delta x}{2}. \tag{6.2.8}$$

The amplification factor depends on the step size in both time and space, and on the wavenumber k of the mode in question. The important issue facing us is that, for certain choices of these parameters, $|\xi|$ can be larger that unity. If this is so, the solution is unstable, because as the timestep n increases, the solution blows up like $\xi^n\, e^{ikx_j}$. However, if $|\xi| < 1$ for all values of k, then each mode will decay in time rather than grow, and the solution will be well behaved. (The decay of these modes does not necessarily imply that the solution to the heat equation goes to zero, because boundary conditions and source terms neglected here can keep the solution finite. Rather, it implies that any short-wavelength noise in the solution will decay away. This is just what we want to happen in order to obtain a smooth solution.)

These considerations lead us to the condition that $|\xi| \le 1$ for all wavenumbers k in order for our numerical solution to be stable. In Fig. 6.6 we graph Eq. (6.2.8) vs. k for several values of the parameter $\chi\,\Delta t/\Delta x^2$. From the figure, one can see that the most unstable wavenumber satisfies $k\,\Delta x = \pi$. For this wavenumber, Eq. (6.2.8) becomes $\xi = 1 - 4\chi\,\Delta t/\Delta x^2$, and the solution is stable for $\xi \ge -1$. This implies

$$\frac{\chi\,\Delta t}{\Delta x^2} \le \frac{1}{2} \tag{6.2.9}$$

in order for the FTCS method to be stable. This inequality is known as a Courant condition. The Courant condition sets a rather stringent limit on the maximum time step size Δt for a given value of Δx. For example, in the problem solved in the previous section, we took $\chi = \frac{1}{4}$ and $\Delta x = 0.1$, implying a maximum time step size given by $\Delta t \le 0.02$. Thus, the first solution, with $\Delta t = 0.01$, worked well, but the second solution, with $\Delta t = 0.05$, violated the Courant condition and the method failed.

CTCS Method The FTCS method uses a forward difference approximation for the time derivative, which is only first-order accurate in Δt. A higher-accuracy difference approximation to the time derivative would be preferable. One simple approach is the CTCS method, where both the time and space derivatives are center-differenced. For the heat equation, this corresponds to the difference equation

$$T_j^{n+1} = T_j^{n-1} + 2\,\Delta t\, S_j^n + \frac{2\chi\,\Delta t}{\Delta x^2}\left(T_{j+1}^n - 2T_j^n + T_{j-1}^n\right). \tag{6.2.10}$$

Unfortunately, however, von Neumann stability analysis of this equation leads us to the conclusion that the CTCS method is unstable for *any* time step size. For, when we substitute in $T_j^n = \xi^n\, e^{ikx_j}$, we obtain a quadratic equation for ξ:

$$\xi^2 + \alpha\xi - 1 = 0, \tag{6.2.11}$$

where $\alpha = (8\chi\,\Delta t/\Delta x^2)\sin^2(k\,\Delta x/2)$. The two solutions for ξ are plotted in Cell 6.58. One of the two solutions has $|\xi| > 1$ for all possible α-values, corresponding to all possible Δt, so the CTCS method does not work for the heat equation. However, the CTCS method does work for other equations, such as the wave equation and the Schrödinger equation. (See the exercises.)

Cell 6.58

```
sol = Solve[ξ^2 + αξ - 1 == 0, ξ];
Plot[Evaluate[ξ/. sol], {α, 0, 1}, AxesLabel→{"α", "ξ"}];
```

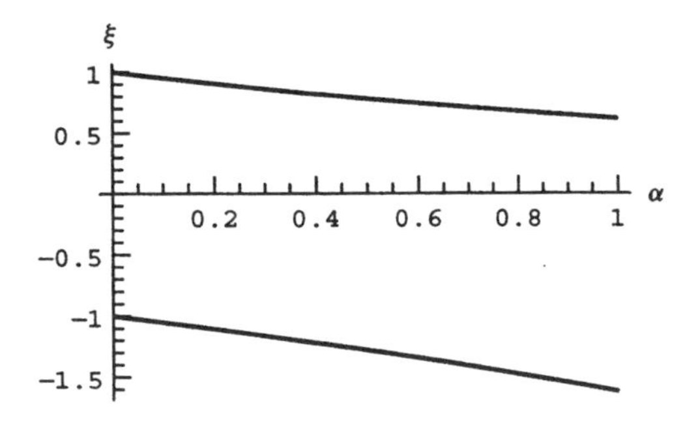

A way to modify the CTCS method so that it works for the heat equation is discussed in the next sub-subsection.

Lax Method Finding a stable, efficient algorithm for the solution of a PDE is "more of an art than a science," in the words of Press, Flannery, Teukolsky, and Vetterling (1992). Nowhere is this more evident than in the Lax method. In this method, we make a seemingly trivial modification to the previous CTCS method for the heat equation, and completely change the equation's stability properties. The change that we make is as follows: in the differenced time derivative, $(T_j^{n+1} - T_j^{n-1})/(2\,\Delta t)$, we replace T_j^{n-1} by its average value from surrounding grid points: $[T_j^{n+1} - \frac{1}{2}(T_{j+1}^{n-1} + T_{j-1}^{n-1})]/(2\,\Delta t)$. The resulting difference equation is

$$T_j^{n+1} = \frac{1}{2}\left(T_{j+1}^{n-1} + T_{j-1}^{n-1}\right) + 2\,\Delta t\, S_j^n + \frac{2\chi\,\Delta t}{\Delta x^2}\left(T_{j+1}^n - 2T_j^n + T_{j-1}^n\right). \quad (6.2.12)$$

von Neumann stability analysis of this equation leads to the following quadratic equation for the amplification factor ξ:

$$\xi^2 + 8\,\delta\left(\sin^2\frac{k\,\Delta x}{2}\right)\xi - \cos k\,\Delta x = 0,$$

where $\delta = \chi\,\Delta t/\Delta x^2$. This is the same as Eq. (6.2.11), except that $\cos(k\,\Delta x)$ replaces the constant term. But this makes a big difference. The magnitude of the solution to this quadratic is plotted in Cell 6.59, for $\delta = \frac{1}{5}$. For this value of δ, one can see that the solution is stable, since $|\xi| \le 1$. In fact, by varying the value of δ, one finds that the maximum possible value of δ is $\frac{1}{4}$. Thus, the Courant condition for Eq. (6.2.12) is

$$\chi\,\Delta t/\Delta x^2 \le \tfrac{1}{4}. \quad (6.2.13)$$

Cell 6.59

```
sol = Solve[ξ^2 + 8 δ Sin[(k/2)]^2 ξ-Cos[k] == 0, ξ];
Plot[Evaluate[Abs[(ξ/. sol)]/. δ→1/5, {k, 0, 2 Pi},
  AxesLabel→{"k", "|ξ|"}, PlotLabel→"|ξ|,
   for χ Δt/Δx^2 = 1/5"];
```

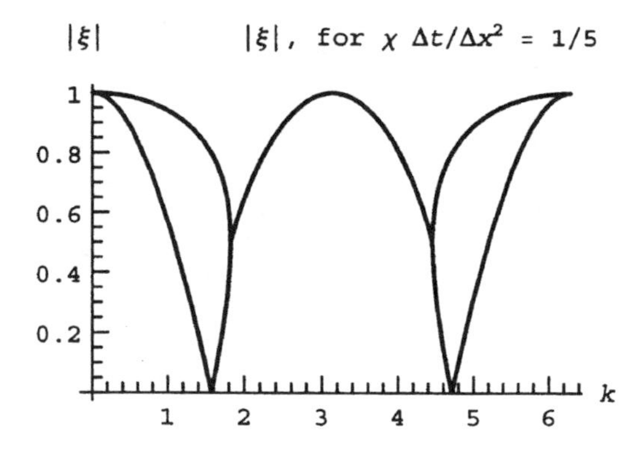

The Lax modification to the CTCS method has improved matters over the original FTCS method: the new method is now second-order accurate in both time and space. However, the Courant condition of Eq. (6.2.13) is still a rather stringent criterion for Δt, making this method impractical for real-world problems. It would be better if we could find a method that loosens this condition. One such method is discussed below.

Implicit Methods. The Crank Nicolson Algorithm Problems that exhibit diffusive behavior, such as the heat equation, are often best treated by *implicit* methods. As an example of an implicit method, take the FTCS method applied to the heat equation, Eq. (6.2.5), and replace the forward-differenced time derivative with a backward difference, $\partial T/\partial t|_{t_n} \to (T_j^n - T_j^{n-1})/\Delta t$. Then the heat equation becomes

$$T_j^n - T_j^{n-1} = \Delta t\, S_j^n + \frac{\chi\,\Delta t}{\Delta x^2}\left(T_{j+1}^n - 2T_j^n + T_{j-1}^n\right). \tag{6.2.14}$$

This algorithm could be called the BTCS (backward time, centered space) method but instead it is referred to as an *implicit* method. The reason is that Eq. (6.2.14) does not explicitly provide T_j^n, given T_j^{n-1}. Rather, since T_j^n now appears on both sides of Eq. (6.2.14), it provides a set of coupled linear equations for T_j^n which we must solve. This is as opposed to previous methods such as FTCS, where T_j^n is given explicitly without needing to solve any equations. These previous methods are called *explicit* methods.

The added complexity of Eq. (6.2.14) is offset by its superior stability properties. von Neumann stability analysis applied to Eq. (6.2.14) yields the following amplification factor:

$$\xi = \left(1 + 4\delta \sin^2 \frac{k\,\Delta x}{2}\right)^{-1}, \tag{6.2.15}$$

where $\delta = \chi\,\Delta t/\Delta x^2$. Thus, $|\xi| \le 1$ for any value of δ. This method is stable for *any* choice of Δt.

However, the method is still only first-order accurate in Δt, because of the backward time difference. A second-order method can now be easily obtained by taking the average of the right-hand side in Eq. (6.2.14) between time steps $n-1$ and n:

$$T_j^n - T_j^{n-1} = \frac{1}{2}\Bigg(\Delta t\left(S_j^n + S_j^{n-1}\right) + \frac{\chi\,\Delta t}{\Delta x^2}\left(T_{j+1}^n - 2T_j^n + T_{j-1}^n + T_{j+1}^{n-1} - 2T_j^{n-1} + T_{j-1}^{n-1}\right)\Bigg). \tag{6.2.16}$$

Now the right-hand and left-hand sides are both centered at timestep $n - \frac{1}{2}$, so the time derivative is now second-order accurate in Δt. Equation (6.2.16) is called the *Crank–Nicolson* scheme. von Neumann analysis of this equation leads to the following amplification factor:

$$\xi = \frac{1 - 2\delta \sin^2(k\,\Delta x/2)}{1 + 2\delta \sin^2(k\,\Delta x/2)}, \tag{6.2.17}$$

which also exhibits stability for any choice of Δt. The Crank–Nicolson method is an implicit scheme that is very useful for solving linear problems with diffusive behavior. Below, we implement the Crank–Nicholson scheme in *Mathematica*.

First, we choose values for the parameters:

Cell 6.60

```
Clear["Global`*"];

L = 1; M = 10; Δt = 0.05; Δx = L/M; χ = 1/4;
```

Note the large time step size relative to the Courant condition for the FTCS ($\Delta t \leq 0.02$) or the CTCS–Lax ($\Delta t \leq 0.01$) method.

Next, we choose initial and boundary conditions, and a source term:

Cell 6.61

```
T0[x_] = Sin[Pi x/L]//N;
T[0, n_] := 0;
T[M, n_] := 0;
S[j_, n_] = 0;
```

These initial and boundary conditions correspond to a single Fourier mode, which should decay away according to

$$T(x,t) = e^{-\chi\pi^2 t/L^2}\sin(\pi x/L). \tag{6.2.18}$$

Now we solve Eq. (6.2.16) for T_j^n. This equation couples the different j-values together, so we need to solve a set of linear equations. This can be easily done using **Solve**, particularly since the equations are *sparse* (most elements are zero), and the method of *Gaussian elimination* can be employed. [These details need not concern us; **Solve** takes care of them. Those who wish to learn about Gaussian elimination and sparse-matrix solution techniques are referred to Press et al. (1992), or any other book on basic numerical techniques.]

Since the output of **Solve** is a list of substitutions, it is more convenient to write the recursion relation in terms of this list of substitutions rather than T_j^n itself. To this end, we define **sol[n]** as this list of substitutions for the solution at time step n. For instance, the initial condition is given by the following list:

Cell 6.62

```
sol[0] = Table[T[j, 0] → T0[j Δx], {j, 1, M-1}]

{T[1, 0] → 0.309017, T[2, 0] → 0.587785, T[3, 0] → 0.809017,
 T[4, 0] → 0.951057, T[5, 0] → 1., T[6, 0] → 0.951057,
 T[7, 0] → 0.809017, T[8, 0] → 0.587785, T[9, 0] → 0.309017}
```

Note that we need not add the end points to the list, since they are already determined by the boundary conditions. We can then write the Crank–Nicolson method as the following recursion relation on **sol[n]**:

Cell 6.63

```
sol[n_] := sol[n] =
  Module[{vars, eqns},
   (*define the variables*)
   vars = Table[T[j, n], {j, 1, M-1}];
   (*define the equations (Eqs. 6.2.16),
   substituting for the values of T determined at the last
    timestep *)

   eqns = Table[T[j, n]-T[j, n-1] == 1/2 (Δt (S[j, n] +
          S[j, n-1]) + χΔt/Δx^2 (T[j + 1, n]-2T[j, n] +
```

```
            T[j-1, n] +T[j + 1, n-1] -2T[j, n-1] +
            T[j-1, n-1])), {j, 1, M-1}]/. sol[n-1];

    (*solve the equations *)
  Solve[eqns, vars] [[1]]]
```

Cell 6.64

```
CrankNicolson = Table[ListPlot[Table[{Δx j, T[j, n]},
    {j, 0, M]}/. sol[n],
    PlotJoined→True, PlotRange→{0, L}, {0, 1}},
    AxesLabel→{"x", ""}, PlotLabel→"T[x,t],t="
     <>ToString[Δt n],
    DisplayFunction→Identity], {n, 0, 40, 5}];
exact = Table[Plot[e^(-π²χt/L²) Sin[Pi x/L], {x, 0, L},
 DisplayFunction→Identity,
    PlotStyle→RGBColor[1, 0, 0]], {t, 0, 40 Δt, 5 Δt}];
Table[Show[CrankNicolson[[n]], exact[[n]],
   DisplayFunction→$DisplayFunction],
    {n, 1, Length[exact]}];
```

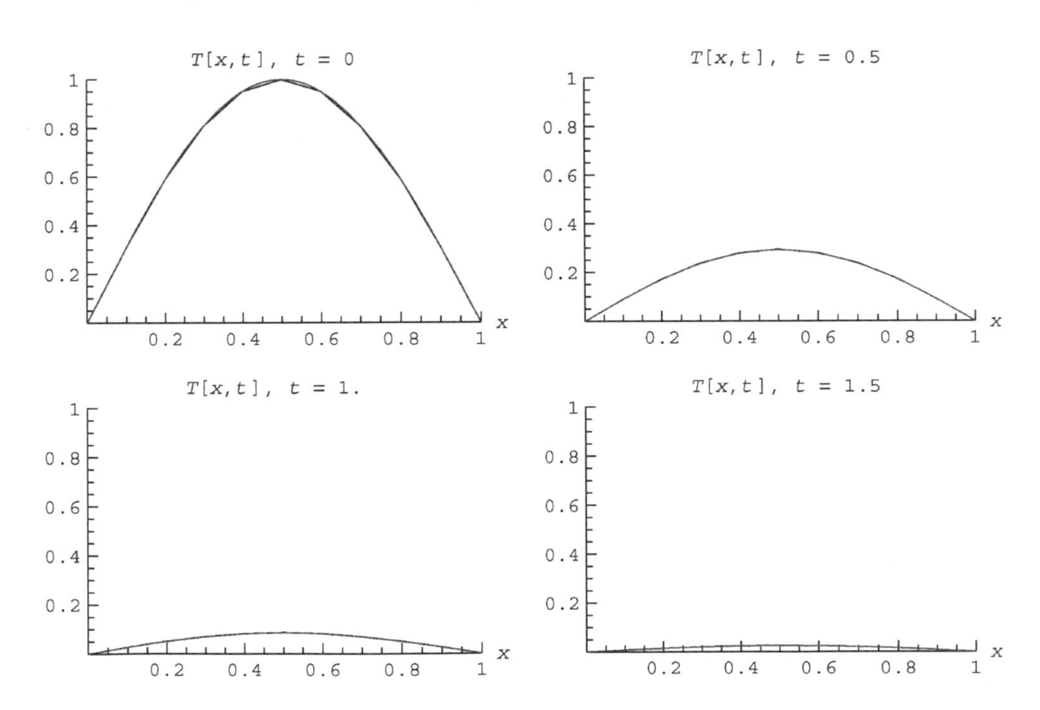

The recursion comes about because **sol[n]** calls **sol[n-1]** (see the substitution command in the definition of **eqns**). We can now evaluate the solution and animate the result. In Cell 6.64 we compare this solution with the exact solution, Eq. (6.2.18). The method works beautifully, following the exact result precisely even for these rather coarse grids in time and space.

Finally, we note that, just as it was possible to create an interpolation function for grid solutions to ODEs, it is possible to do the same for grid solutions to PDEs:

Cell 6.65

```
Tsol = Interpolation[
  Flatten[Table[Table[{j Δx, n Δt, T[j, n]},
   {j, 0, M}]/. sol[n], {n, 0, 25}], 1]]

InterpolatingFunction[{{0., 1.}, {0., 1.25}}, <>]
```

The **Flatten** command used here removes an extra set of brackets that the **Interpolation** function finds objectionable. We can plot the solution function **Tsol[x,t]** vs. both **x** and **t**, and we do so in Cell 6.66.

Cell 6.66

```
Plot3D[Tsol[x, t], {x, 0, 1}, {t, 0, 1.25},
 AxesLabel → {"x", "t", "T[x,t]"}];
```

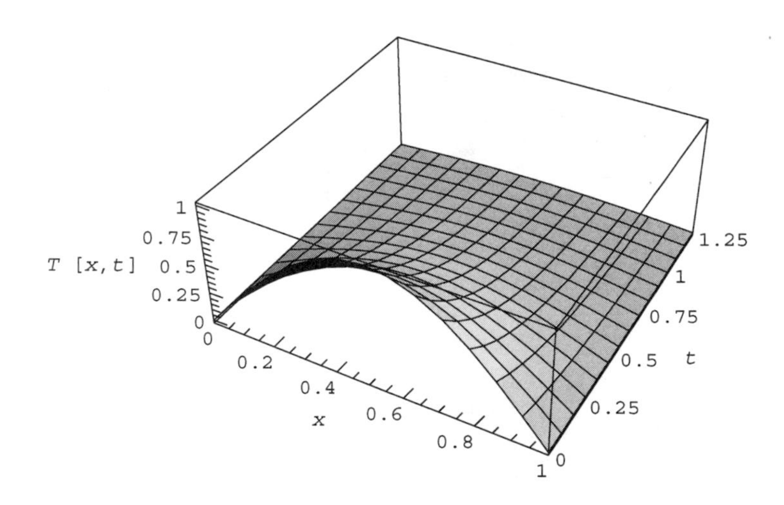

von Neumann and Mixed Boundary Conditions Grid methods can also be employed in problems involving von Neumann or mixed boundary conditions. Take, for example, a heat flow problem on $0 < x < L$ with a boundary condition that $\kappa\, \partial T/\partial x|_{x=0} = -\Gamma$. This von Neumann condition can be met by finite-differencing the derivative at the boundary $x = 0$:

$$\frac{T_1^n - T_0^n}{\Delta x} = -\frac{\Gamma}{\kappa}, \tag{6.2.19}$$

from which we can obtain the boundary value for the temperature in terms of the interior points:

$$T_0^n = T_1^n + \Delta x\, \Gamma/\kappa. \tag{6.2.20}$$

On the other hand, the left-hand side of Eq. (6.2.19) is a forward-difference form for the derivative at $x = 0$, and is only first-order accurate in Δx. Thus, the error in our numerical method will scale only as the first power of Δx. This is unfortunate, because the centered-difference equations used for the interior points are all of second-order accuracy in Δx. It would be better to use a more accurate approximation for $T'(x=0,t)$.

One simple improvement is to place the boundary at $x=0$ *halfway between* the grid points 0 and 1. Then we can use a *centered-difference* form for the von Neumann boundary condition, $T'(x=0,t) \to [T_1^n - T_0^n]/\Delta x = -\Gamma/\kappa$. The equation is the same as before, but now it is second-order rather than first-order accurate. The only change that we need make is to the location of the grid points. Now,

$$x_j = \Delta x\left(j - \tfrac{1}{2}\right), \qquad j = 0,1,2,\ldots,M. \tag{6.2.21}$$

Considering now the boundary condition at $x=L$, if this condition is also von Neumann, then $x=L$ must fall between the $M-1$st and Mth points in order to employ the same centered-difference scheme there. Therefore, Eq. (6.2.21) implies that $L = \Delta x(M - \frac{1}{2} - \frac{1}{2})$ and so $\Delta x = L/(M-1)$. On the other hand, if the boundary condition at $x=L$ is Dirichlet, then we need to have the Mth grid point fall at $x=L$, and this implies that $\Delta x = L/(M-\frac{1}{2})$.

For example, consider the heat equation with insulating boundary conditions at both ends, $\partial T/\partial x|_{x=0} = \partial T/\partial x|_{x=L} = 0$. For the grid we use Eq. (6.2.21) with $\Delta x = L/(M-1)$. For this example, we take $M=10$, $L=1$, $\chi = \frac{1}{4}$, $\Delta t = 0.05$ and set source terms equal to zero:

Cell 6.67

```
M = 10; L = 1.;
Δx = L/(M-1);
x[j_] = Δx (j-1/2);
χ = 1/4; Δt = 0.05;
S[j_, n_] = 0;
```

According to Eq. (6.2.20) the boundary conditions are satisfied by the equations

Cell 6.68

```
T[0, n_] = T[1, n];
T[M, n_] = T[M-1, n];
```

For the initial condition, we again choose a single Fourier mode as a test, but this time we choose a cosine mode so as to satisfy the boundary conditions:

Cell 6.69

```
T0[x_] = Cos[Pi x/L];
```

In theory this mode should decay away exponentially at the rate $\chi(\pi/L)^2$. We can now run any of our previously discussed methods to solve for the temperature evolution. We use the Crank–Nicolson module created in the previous section, and plot the result in Cell 6.71. The close agreement between the numerical result (solid) and the exact solution (dashed) speaks for itself.

Cell 6.70

```
sol[0] = Table[T[j, 0] → T0[x[j]], {j, 1, M-1}]

{T[1, 0] → 0.984808, T[2, 0] → 0.866025, T[3, 0] → 0.642788,
 T[4, 0] → 0.34302, T[5, 0] → 6.12323×10^-17,
 T[6, 0] → -0.34202, T[7, 0] → -0.642788,
 T[8, 0] → -0.866025, T[9, 0] → -0.984808}
```

Cell 6.71

```
CrankNicolson = Table[ListPlot[Table[{x[j], T[j, n]},
  {j, 0, M}] /. sol[n],
    PlotJoined → True, PlotRange → {{0, L}, {-1, 1}},
    AxesLabel → {"x", ""}, PlotLabel → "T[x,t],t="
     <>ToString[Δt n],
    DisplayFunction → Identity], {n, 0, 40, 5}];
exact = Table[Plot[e^(-π^2 χt/L^2) Cos[Pi x/L], {x, 0, L},
  DisplayFunction → Identity,
    PlotStyle → {RGBColor[1, 0, 0], Dashing[{0.05, 0.05}]}],
     t, 0, 40 Δt, 5 Δt}];
Table[Show[CrankNicolson[[n]], exact[[n]],
   DisplayFunction → $DisplayFunction],
    {n, 1, Length[exact]}];
```

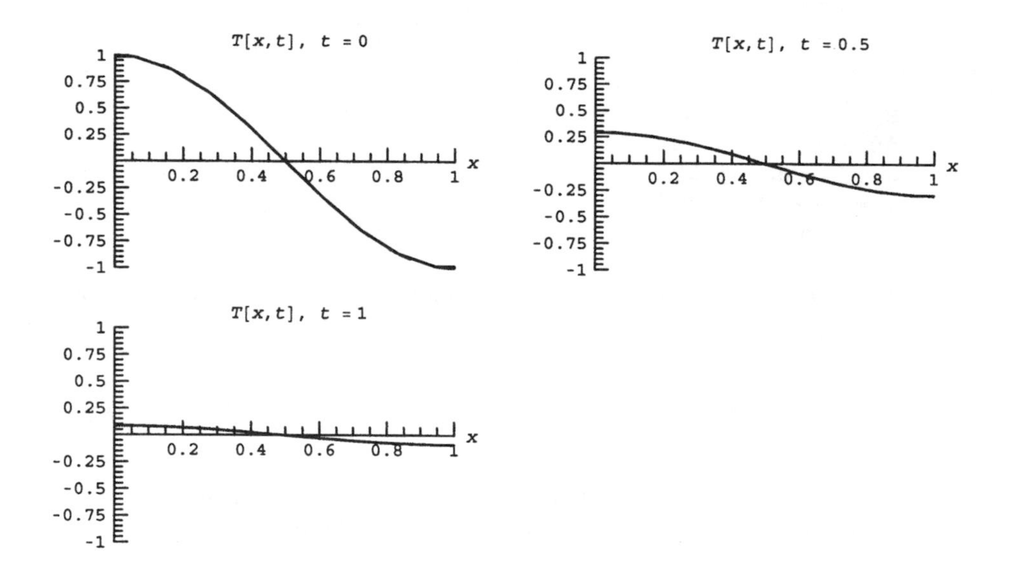

Mixed boundary conditions are more of a challenge. Now both the function and its derivative are required on the boundary. For example, at $x = 0$ the general mixed boundary condition is of the form $\kappa\, \partial T/\partial x|_{x=0} = a[T(0,t) - \tau]$, where κ, a, and τ are given constants. How do we obtain second-order-accurate values for both T and its derivative at the boundary point?

Since the boundary at $x = 0$ falls halfway between the $j = 0$ and $j = 1$ grid points, one simple approach is to use the average value, $\frac{1}{2}(T_0^n + T_1^n)$, to approximate T at $x = 0$. We still use the centered-difference form for the first derivative for $\partial T/\partial x|_{x=0}$ at $x = 0$, so the mixed boundary condition becomes

$$\kappa \frac{T_1^n - T_0^n}{\Delta x} = a\left[\frac{1}{2}(T_0^n + T_1^n) - \tau\right]. \tag{6.2.22}$$

Solving for T_0^n yields

$$T_0^n = \frac{2a\tau\,\Delta x + (2\kappa - a\,\Delta x)T_1^n}{2\kappa + a\,\Delta x}. \tag{6.2.23}$$

Similarly, for a mixed boundary condition at $x = L$, namely $\kappa\,\partial T/\partial x|_{x=L} = -b[T(L,t) - \tau]$, we take a grid for which $x = L$ falls halfway between gridpoints $M - 1$ and M, so that $T|_{x=L} = \frac{1}{2}(T_M^n + T_{M-1}^n)$. Then the boundary condition is

$$\kappa \frac{T_M^n - T_{M-1}^n}{\Delta x} = -b\left[\frac{1}{2}(T_{M-1}^n + T_M^n) - \tau\right], \tag{6.2.24}$$

which leads to

$$T_M^n = \frac{2b\tau\,\Delta x + (2\kappa - b\,\Delta x)T_{M-1}^n}{2\kappa + b\,\Delta x}. \tag{6.2.25}$$

Implementation of these boundary conditions follows the same course as in our previous examples, and is left to the exercises.

Multiple Spatial Dimensions

Explicit Method. The explicit methods discussed previously generalize easily to PDEs in multiple spatial dimensions. Continuing with our heat equation model, we now consider the problem in two dimensions:

$$\frac{\partial}{\partial t} T(x,y,t) = \chi\left(\frac{\partial^2}{\partial x^2} + \frac{\partial^2}{\partial y^2}\right) T(x,y,t) + S(x,y,t). \tag{6.2.26}$$

For simplicity, we take the boundary to be a rectangular box of sides L_x and L_y, and create a rectangular spatial grid in x and y according to

$$\begin{aligned} x_j &= j\,\Delta x, \qquad j = 0,1,2,\dots,M_x, \\ y_k &= k\,\Delta y, \qquad k = 0,1,2,\dots,M_y, \end{aligned} \tag{6.2.27}$$

where $\Delta x = L_x/M_x$ and $\Delta y = L_y/M_y$. (Nonrectangular boundaries are treated in Sec. 6.2.2.)

The FTCS method on this grid is a trivial extension of the method in one dimension:

$$T_{jk}^{n+1} = T_{jk}^{n} + \Delta t\, S_{jk}^{n} + \frac{\chi\,\Delta t}{\Delta x^2}\left(T_{j+1\,k}^{n} - 2T_{jk}^{n} + T_{j-1\,k}^{n}\right) + \frac{\chi\,\Delta t}{\Delta y^2}\left(T_{jk+1}^{n} - 2T_{jk}^{n} + T_{jk-1}^{n}\right). \tag{6.2.28}$$

Equation (6.2.28) is valid for the points interior to the rectangle, and must be supplemented by boundary conditions that determine the points on the boundary, as well as by initial conditions specifying T_{jk}^{0}. von Neumann stability analysis for this difference equation considers the stability of a Fourier mode of the form $\xi^m\, e^{ik_x x}\, e^{ik_y y}$, and leads to a Courant condition that

$$\chi\,\Delta t \le \frac{1}{2}\,\frac{\Delta x^2 \Delta y^2}{\Delta x^2 + \Delta y^2} \tag{6.2.29}$$

(see the exercises). The other explicit methods discussed previously, CTCS and Lax, also can be easily extended to multiple dimensions.

Example: Wave Equation with Nonuniform Wave Speed As an example of an explicit method in two dimensions, we construct the numerical solution to a wave equation in two dimensions with nonuniform wave speed,

$$\frac{\partial^2}{\partial t^2} z(x,y,t) = c^2(x,y)\left(\frac{\partial^2}{\partial x^2} + \frac{\partial^2}{\partial y^2}\right) z(x,y,t),$$

with initial conditions $z(x,y,0) = z_0(x,y)$, $\dot{z}(x,y,0) = v_0(x,y)$. Here we take the simple case $z_0 = v_0 = 0$. For boundary conditions, we choose a square domain with unit sides and take an oscillating boundary condition,

$$z(0,y,t) = z(1,y,t) = z(x,1,t) = 0, \qquad z(x,0,t) = x(1-x)\sin 2t.$$

For the wave speed, we choose

$$c(x,y) = \tfrac{1}{10}\left\{1 - 0.7\exp\left[-2(x-0.5)^2 - 5(y-0.5)^2\right]\right\}.$$

The wave speed, plotted in Cell 6.72, has a minimum at the center of the box.

Cell 6.72

```
cw[x_, y_] = 1/10 (1-.7 Exp[-2 (x-0.5)^2-5 (y-0.5)^2]);

Plot3D[cw[x, y], {x, 0, 1}, {y, 0, 1},
  AxesLabel→{"x", "y", ""}, PlotLabel -> "wave speed"];
```

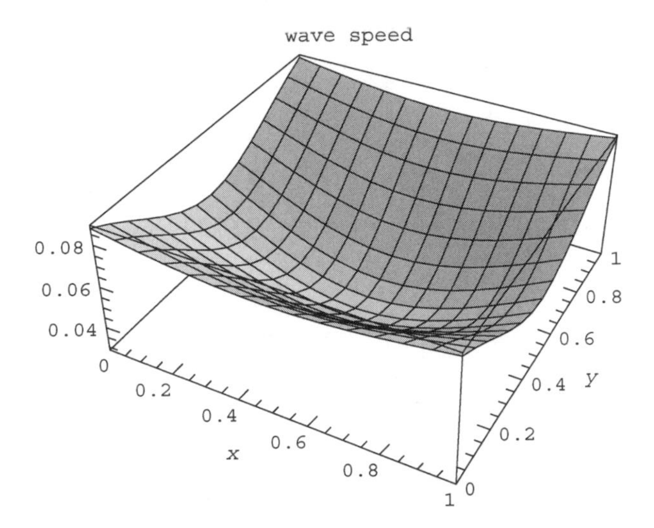

For the numerical method, we choose CTCS. This explicit method works well for the wave equation, being second-order in both the space and time step sizes, and allowing relatively large steps according to the von Neumann stability criterion: for $\Delta x = \Delta y = \Delta$, $c^2\,\Delta t^2/\Delta^2 \le \frac{1}{2}$ for stability (see the exercises). Therefore, if $c \le \frac{1}{10}$ and $\Delta = \frac{1}{30}$, $\Delta t \le 1/(3\sqrt{2}) = 0.24$. Below, we choose $\Delta t = 0.2$:

Cell 6.73

```
Clear["Global`*"];

(* define the grid *)
Mx = 30; My = 30;
Lx = Ly =1;
Δx = Lx/Mx; Δy = Ly/My;
Δt = 0.2;

x[j_] := j Δx;
y[k_] := k Δy;

(* initial conditions *)
z0[j_, k_] := 0;
v0[j_, k_] := 0;

(* zeroth step: *)
z[j_, k_, 0] := z0[j, k];

(* define the wave speed on the grid *)
c[j_, k_] := cw[x[j], y[k]];

(* boundary conditions *)
z[0, k_, n_] := 0;
z[Mx, k_, n_] := 0;
z[j_, 0, n_] := x[j] (1-x[j]) Sin[2 n Δt];
z[j_, My, n_] := 0;
```

The CTCS method for the wave equation is

$$z_{jk}^{n+1} = 2z_{jk}^{n} - z_{jk}^{n-1} + \frac{c_{jk}^2\,\Delta t^2}{\Delta x^2}\left(z_{j+1\,k}^{n} - 2z_{jk}^{n} + z_{j-1\,k}^{n}\right) + \frac{c_{jk}^2\,\Delta t^2}{\Delta y^2}\left(z_{j\,k+1}^{n} - 2z_{jk}^{n} + z_{j\,k-1}^{n}\right).$$

We construct the recursion relation for this method below:

Cell 6.74

```
z[j_, k_, n_] := (z[j, k, n] = 2 z[j, k, n-1]-z[j, k, n-2] +
   Δt² c[j, k]^2
   ───────────── ((z[j+1, k, n-1]-2 z[j, k, n-1] +
        Δx²
   z[j-1, k, n-1]) + (z[j, k + 1, n-1]-2 z[j, k, n-1] +
    z[j, k-1, n-1])))/; n≥2
```

The method can only be used for $n \geq 2$, because evaluation of the nth timestep refers back to the $n-2$nd step. Therefore, for the first step, we use the trick employed in Chapter 1 for second-order ODEs [see Eq. (1.4.28) and the surrounding discussion]:

$$z_{jk}^{1} = z_{jk}^{0} + v_{0jk}\,\Delta t + \frac{\Delta t^2}{2} a_{jk},$$

where v_{0jk} is the initial velocity (equal to zero for our choice of initial conditions), and a_{jk} is the initial acceleration, given by

$$a_{jk} = \frac{c_{jk}^2}{\Delta x^2}\left(z_{j+1\,k}^{0} - 2z_{jk}^{0} + z_{j-1\,k}^{0}\right) + \frac{c_{jk}^2\,\Delta t^2}{\Delta y^2}\left(z_{j\,k+1}^{0} - 2z_{jk}^{0} + z_{j\,k-1}^{0}\right)$$

(this is also equal to zero for our initial conditions). Therefore, we simply have $z_{jk}^{1} = z_{jk}^{0} = 0$:

Cell 6.75

```
z[j_, k_, 1] := 0;
```

(Of course, this cell must be modified if other initial conditions are used.) At each timestep, we construct an interpolation function of the solution in x and y:

Cell 6.76

```
zsol[n_] := zsol[n] = Interpolation[
  Flatten[Table[{j Δx, k Δy, z[j, k, n]}, {j, 0, Mx},
   {k, 0, My}], 1]]
```

In Cell 6.77 we plot the solution. The wavefronts curve toward the region of lower c according to Snell's law, and as they do so the amplitude and wavelength decrease, as expected from our WKB analysis of this equation in Chapter 5. However, note the formation of a focus at a point behind the region of low wave speed, where the amplitude increases as wave energy is concentrated. This region is acting as a lens for the waves (see Exercise 13 in Sec. 5.2 and the examples in Sec. 5.3.1 for similar problems, in which the ray trajectories for the waves are evaluated).

Implicit Methods and Operator Splitting. Putting implicit methods in multidimensional form takes a bit more thought. For example, if we try to implement the Crank–Nicolson scheme directly in two dimensions, we obtain

$$T_{jk}^{n} = T_{jk}^{n-1} + \tfrac{1}{2}\Delta t\left(S_{jk}^{n} + S_{jk}^{n-1}\right) + D_x T_{jk}^{n} + D_x T_{jk}^{n-1} + D_y T_{jk}^{n} + D_y T_{jk}^{n-1}, \quad (6.2.30)$$

where for convenience we have introduced the notation

$$D_x T_{jk}^{n} = \frac{1}{2}\frac{\chi\,\Delta t}{\Delta x^2}\left(T_{j+1\,k}^{n} - 2T_{jk}^{n} + T_{j-1\,k}^{n}\right),$$

$$D_y T_{j}^{n} = \frac{1}{2}\frac{\chi\,\Delta t}{\Delta y^2}\left(T_{jk+1}^{n} - 2T_{jk}^{n} + T_{jk-1}^{n}\right).$$

(D_x and D_y are difference operators for second derivatives in the x and y directions respectively, multiplied by $\chi\,\Delta t/2$.)

Solution of Eq. (6.2.30) requires some finesse, since the number of coupled equations for T_{jk}^{n} equals $(M_x - 1)(M_y - 1)$, which can easily exceed several hundred. The problem is even worse in three dimensions, where thousands of coupled equations must be solved simultaneously. However, the equations are sparse, and *Mathematica*'s **Solve** function is sufficiently adept at solving such problems so that even in three dimensions, the solution can be obtained directly without too much difficulty. We leave this brute-force approach to the problems at the end of the section.

Rather than using a direct solution of the coupled equations, it is often preferable to generalize the Crank–Nicolson scheme to multiple dimensions in a slightly different way. The idea is to split the spatial operator ∇^2 into its constituent operators $\partial^2/\partial x^2$ and $\partial^2/\partial y^2$, and also split the time step into two half steps, each of size $\Delta t/2$. We treat the $\partial^2/\partial x^2$ part of the operator implicitly

Cell 6.77

```
Table[Plot3D[Evaluate[zsol[n][x, y]], {x, 0, 1},
   {y, 0, 1}, PlotPoints→60, PlotRange→{-.5, .5},
   Mesh→False,
   PlotLabel -> "t= <>ToString[nΔt]], {n, 0, 140, 4}];
```

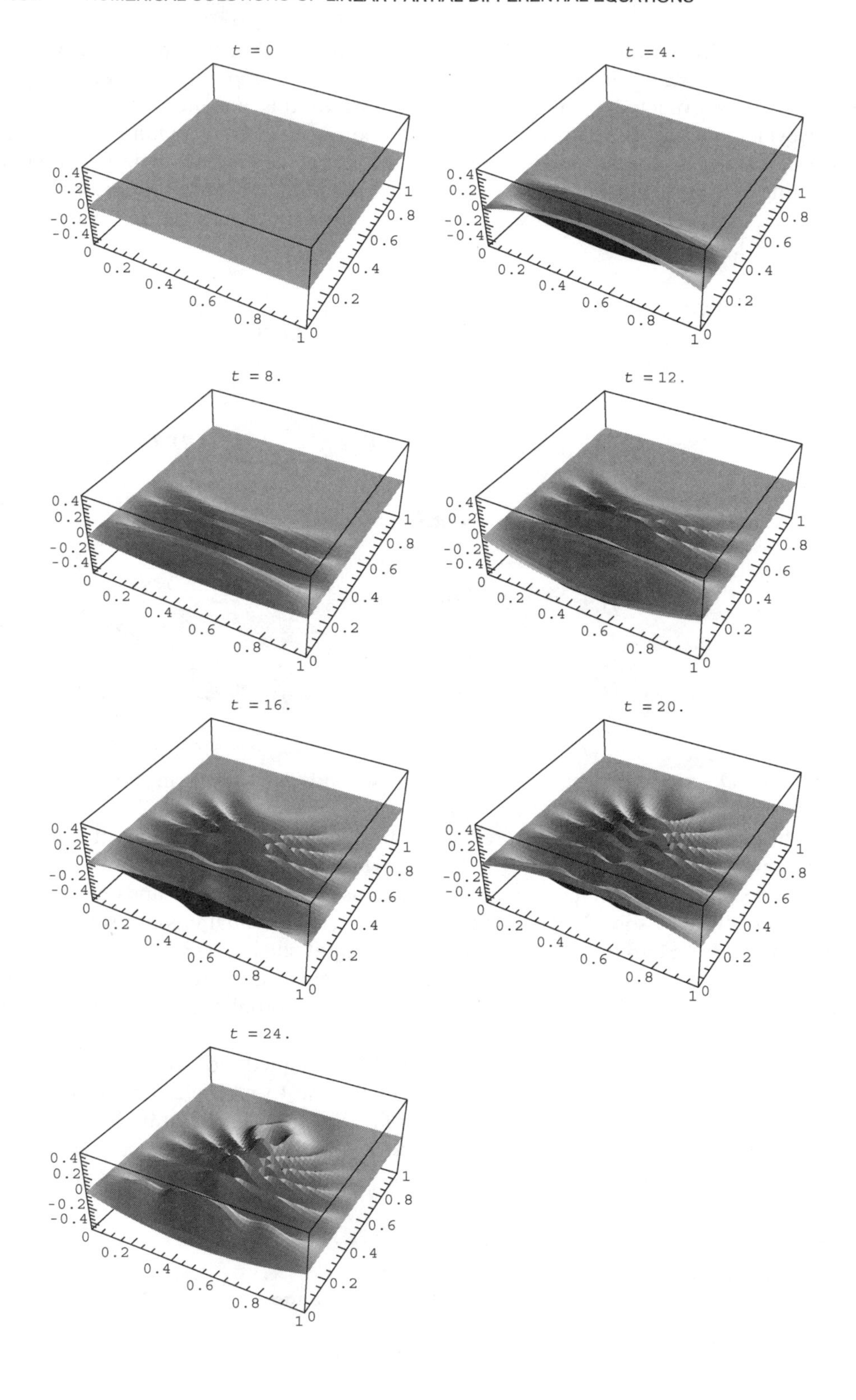

t = 0
t = 4.
t = 8.
t = 12.
t = 16.
t = 20.
t = 24.
0.4
0.2
0
-0.2
-0.4
0.6
0.8
1

in the first half step, and the $\partial^2/\partial y^2$ part implicitly in the next half step. [Note: this type of operator splitting is called the *alternating-direction implicit method*. Another type of operator splitting is discussed in Chapter 7 in relation to a nonlinear convection/diffusion equation: see Eqs. (7.2.29)–(7.2.31). For more on operator splitting, see Press et al. (1992).] The equations used for each half step are

$$\begin{aligned} T_{jk}^{n-1/2} &= T_{jk}^{n-1} + \tfrac{1}{2}\Delta t S_{jk}^{n-1} + D_x T_{jk}^{n-1/2} + D_y T_{jk}^{n-1}, \\ T_{jk}^{n} &= T_{jk}^{n-1/2} + \tfrac{1}{2}\Delta t S_{jk}^{n} + D_x T_{jk}^{n-1/2} + D_y T_{jk}^{n}. \end{aligned} \tag{6.2.31}$$

In the first half step, the source and $\partial^2/\partial y^2$ are treated with forward time differencing as in Euler's method, and $\partial^2/\partial x^2$ is treated with a backward-difference; but in the next half step this is reversed and $\partial^2/\partial x^2$ is treated with a forward-difference derivative, while the source and $\partial^2/\partial y^2$ are treated with a backward-difference derivative. As a result of this time symmetry, the method is still second-order accurate in time (and in space as well, due to the centered-difference form for D_x and D_y). Also, one can show that it is still absolutely stable. Now, however, we need only solve a one-dimensional set of coupled equations in each half step. In the first half step we must solve $M_x - 1$ coupled equations for each value of k, and in the second half step we solve $M_y - 1$ equations for each value of j. All in all, we still have to solve $(M_x - 1)(M_y - 1)$ equations, but they are only coupled in batches of $M_x - 1$ or $M_y - 1$ equations at a time, and this simplifies the task enormously. The method is implemented below.

Cell 6.78

```
Lx = 1; Ly = 2; Mx = 10; My = 10; Δt = 0.05; Δx = Lx/Mx;
Δy = Ly/My ; χ = 1/4;
```

We now must choose initial and boundary conditions on all four sides of the rectangle. For simplicity, we take homogeneous Dirichlet conditions, no source term, and an initial condition corresponding to the slowest-varying eigenmode of ∇^2:

Cell 6.79

```
(4 initial condition 4)
T0[x_, y_] = Sin[Pi y/Ly] Sin[Pi x/Lx]//N;

(* boundary conditions on the four sides *)
T[j_, 0, n_] := 0;
T[0, k_, n_] := 0;
T[j_, My, n_] := 0;
T[Mx, k_, n_] := 0;

(* source term *)
S[j_, k_, n_] = 0;
```

This initial condition should decay away according to

$$T(x,t) = e^{-\chi t(\pi^2/L_x^2 + \pi^2/L_y^2)} \sin(\pi x/L_x)\sin(\pi y/L_y). \tag{6.2.32}$$

Now we solve Eq. (6.2.31) for T_{jk}^n. Just as in the implementation of the 1D Crank–Nicolson method, we define **sol[n]** as a list of substitutions for the solution at time step n. The initial condition is given by the following list:

Cell 6.80

```
sol[0] =
   Flatten[Table[T[j, k, 0] → T0[j Δx, k Δy], {j, 1, Mx-1},
     {k, 1, My-1}]];
```

Also, we define the second-derivative difference functions to simplify the notation:

Cell 6.81

```
Dx[j_, k_, n_] := χΔt/(2 Δx^2) (T[j + 1, k, n]-2 T[j, k, n] +
T [j-1, k, n]);

Dy[j_, k_, n_] := χΔt/(2 Δy^2) (T[j, k + 1, n]-2 T[j, k, n] +
 T[j, k-1, n]);
```

Then we have

Cell 6.82

```
sol[n_] := sol[n] =
  Module[{},
   (*define the variables for the first half-step *)
   vars1[k_] := Table[T[j, k, n-1/2], {j, 1, Mx-1}];
   (*define the equations for the first half-step in Eqs.
     (6.2.31), substituting for the values of T determined
     at the last time step *)
  eqns1[k_] := Flatten[Table[T[j, k, n-1/2]-T[j, k, n-1] ==
     1/2 Δt S[j, k, n-1] + Dx[j, k, n-1/2] + Dy[j, k, n-1],
     {j, 1, Mx-1}] /. sol[n-1]];
   (*solve the equations *)
   soll = Flatten[Table[Solve[eqns1[k], vars1[k]],
    {k, 1, My-1}]];
   (*define the variables for the second half-step *)
   vars2[j_] := Table[T[j, k, n], {k, 1, My-1}];
   (*define the equations for the second half-step in Eqs.
     (6.2.31), substituting for the values of T determined at
     the last time step and the first half-step*)

   eqns2[j_] := Flatten[Table[T[j, k, n] -T[j, k n-1/2] ==
   1/2 Δt S[j, k, n] + Dx[j, k, n-1/2] + Dy[
            j, k, n], {k, 1, My-1}] /. Join[sol[n-1], soll]];
   (*solve the equations *)
   Flatten[Table[Solve[eqns2[j], vars2[j]], {m, 1, Mx-1}]]]
```

In Cell 6.84 we display the result, after creating an interpolation function **Tsol[x,y,t]**.

Cell 6.83

```
Tsol = Interpolation[
  Flatten[Table[Table[{j Δx, k Δy, n Δt, T[j, k, n]},
   {j, 0, Mx}, {k, 0, My}] /.
     sol[n], {n, 0, 20}], 2]]
```

```
InterpolatingFunction[{{0., 1.}, {0., 2.}, {0., 1.}}, <>]
```

Cell 6.84

```
Table[Plot3D[Tsol[x, y, t], {x, 0, Lx},
   {y, 0, Ly}, PlotRange → {0, 1}, AxesLabel → {"x", "y", ""},
   PlotLabel → "T[x,y,t],t=" <>ToString[t] <> "\n",
   BoxRatios → {Lx, Ly, 1}], {t, 0, 1, .1}];
```

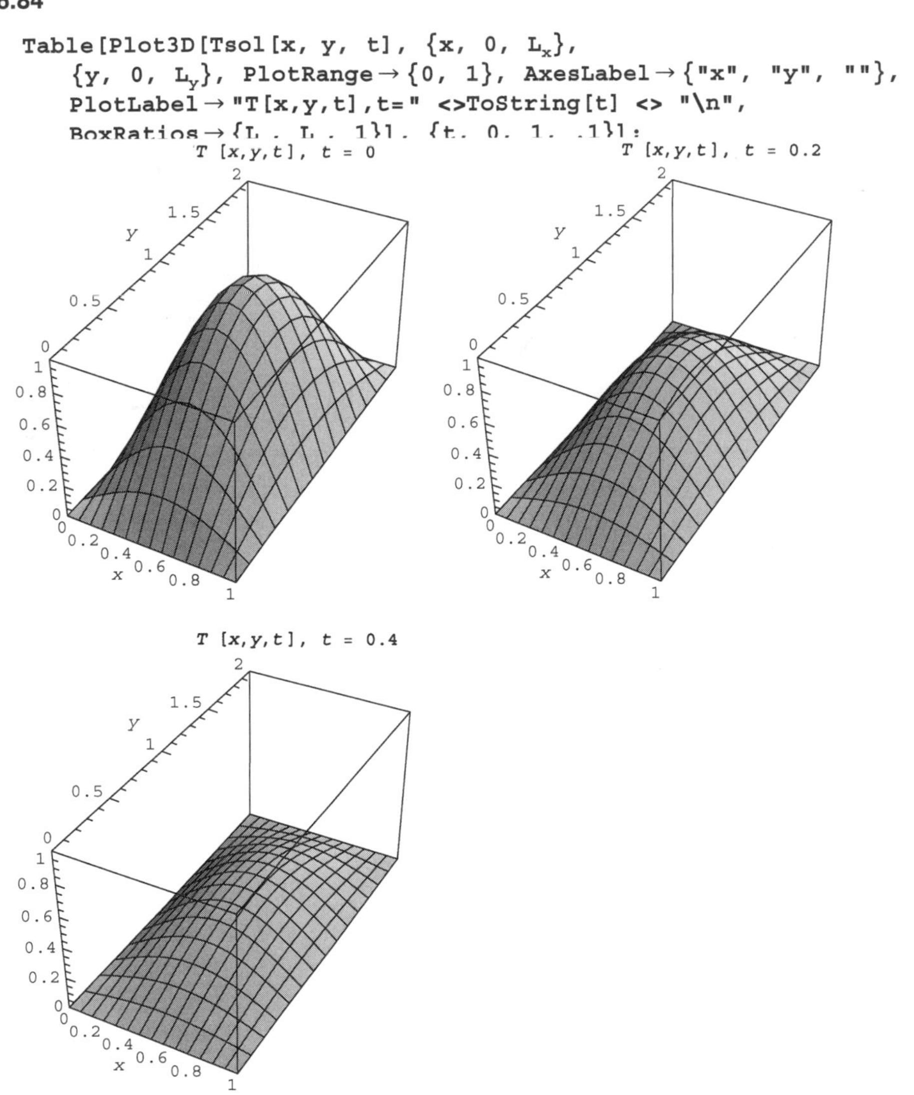

In Cell 6.85, we compare the Crank–Nicolson solution with the exact solution, Eq. (6.2.32), along a cut at $x = \frac{1}{2}$. The numerical solution (solid line) is so close to the exact solution (dashed line) that the two curves cannot be distinguished.

Cell 6.85

```
CrankNicolson = Table[Plot[Tsol[1/2, y, t], {y, 0, Ly},
    PlotRange → {0, 1}, AxesLabel → {"y", ""},
     DisplayFunction → Identity,
    PlotLabel → "T[1/2, y, t],t="<>ToString[t]],
    {t, 0, 1, .1}];

exact = Table[Plot[e^(-π^2 χ t (1/Lx^2+1/Ly^2)) Sin[Pi 5 Δx/Lx] Sin[
  Pi y/Ly], {y, 0, Ly},
    DisplayFunction → Identity, PlotStyle → {RGBColor[1, 0, 0],
    Dashing[{0.05, 0.05}]}, {t, 0, 1, .1}];

Table[Show[CrankNicolson[[n]], exact[[n]],
   DisplayFunction → $DisplayFunction],
   {n, 1, Length[exact]}];
```

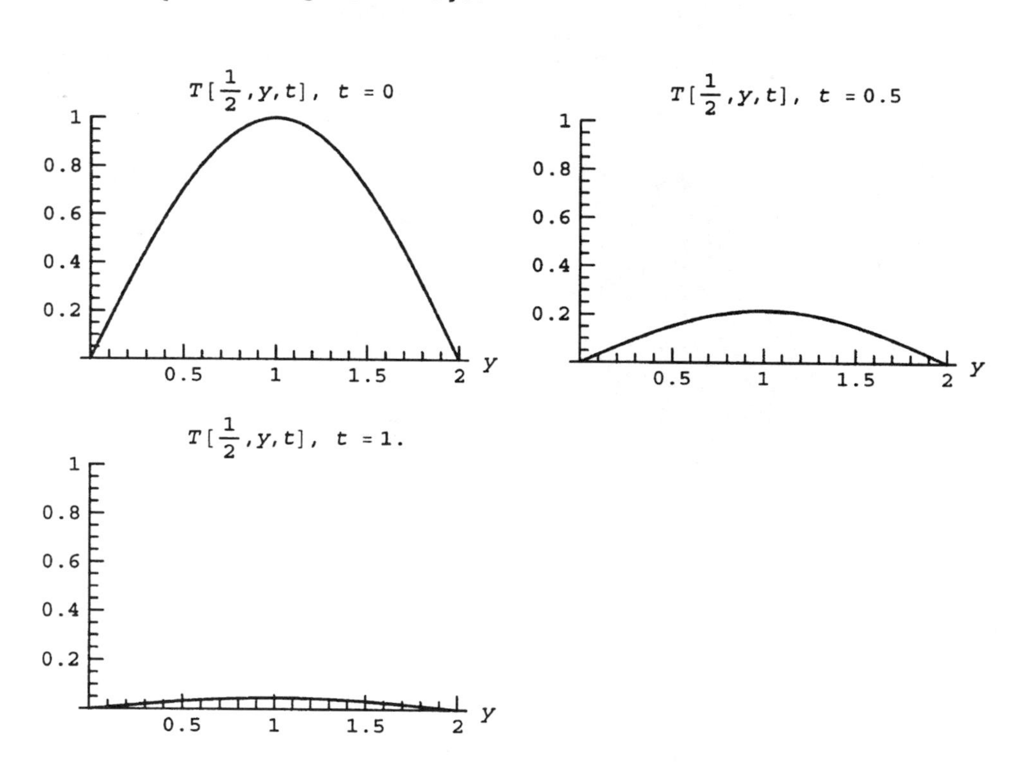

6.2.2 Boundary-Value Problems

Direct Solution We now turn to the solution of time-independent boundary-value problems using grid techniques. The canonical example of such a problem is Poisson's equation in two dimensions,

$$\left(\frac{\partial^2}{\partial x^2}+\frac{\partial^2}{\partial y^2}\right)\phi(x,y)=\rho(x,y). \tag{6.2.33}$$

We first consider the problem in a rectangular box with dimensions L_x and L_y, with ϕ specified on the boundaries (Dirichlet boundary conditions). As before, we construct a grid on which we will determine ϕ, at positions

$$x_j = j\,\Delta x, \qquad j = 0,1,2,\dots,M_x,$$

$$y_k = k\,\Delta y, \qquad k = 0,1,2,\dots,M_y,$$

with $\Delta x = L_x/M_x$ and $\Delta y = L_y/M_y$.

One way to solve this problem is to centered-difference the spatial derivatives, defining a spatial grid that turns Eq. (6.2.33) into a system of coupled linear equations:

$$\frac{\phi_{j+1\,k} - 2\phi_{jk} + \phi_{j-1\,k}}{\Delta x^2} + \frac{\phi_{j\,k+1} - 2\phi_{jk} + \phi_{j\,k-1}}{\Delta y^2} = \rho_{jk},$$

$$j = 1,2,\dots,M_x - 1, \quad k = 1,2,\dots,M_y - 1. \qquad (6.2.34)$$

These equations for the potential in the interior of the box are supplemented by the boundary conditions on the box edges, which directly determine ϕ_{jk} for $j = 0$, $j = M_x$, $k = 0$, and $k = M_y$.

We can solve this problem directly using **Solve**. This involves the solution of $(M_x - 1)(M_y - 1)$ simultaneous equations for the potential at the interior points. This is a daunting task, but let's try it anyway to see if *Mathematica* can handle it. First, we specify the dimensions of the box and the number of grid points:

Cell 6.86

```
Mx = 10; My = 10;
Lx = Ly = 1.;

Δx = Lx/Mx; Δy = Ly/My;
```

Next, we specify the boundary conditions. We take homogeneous Dirichlet conditions:

Cell 6.87

```
ϕ[0, k_] = 0;
ϕ[Mx, k_] = 0;
ϕ[j_, 0] = 0;
ϕ[j_, My] = 0;
```

For the source, we choose uniform charge density, $\rho = 1$:

Cell 6.88

```
ρ[j_, k_] = 1;
```

We are now ready to create the list of equations using Eq. (6.2.34) for the interior points:

Cell 6.89

```
eqns = Flatten[
  Table[(φ[j + 1, k] - 2 φ[j, k] + φ[j-1, k])/Δx^2 +
  (φ[j, k + 1] - 2 φ[j, k] + φ[j, k-1])/Δy^2 ==
  ρ[j, k], {j, 1, Mx-1}, {k, 1, My-1}]];
```

The list of variables for which we are solving is given by the following command:

Cell 6.90

```
vars = Flatten[Table[φ[j, k], {j, 1, Mx-1}, {k, 1, My-1}]];
```

Finally, we solve the equations:

Cell 6.91

```
sol = Solve[eqns, vars] [[1]];
```

Mathematica is able to solve this problem quite quickly. We can construct an interpolation function of the solution in the usual way:

Cell 6.92

```
φsol = Interpolation[
  Flatten[Table[{j Δx, k Δy, φ[j, k]}, {j, 0, Mx},
   {k, 0, My}] /. sol, 1]]

InterpolatingFunction[{{0., 1.}, {0., 1.}},<>]
```

A contour plot of the solution is shown in Cell 6.93.

Cell 6.93

```
ContourPlot[φsol[x, y], {x, 0, 1}, {y, 0, 1},
  FrameLabel → {"x", "y"},
  PlotLabel → "Potential inside a charge-filled square box"];
```

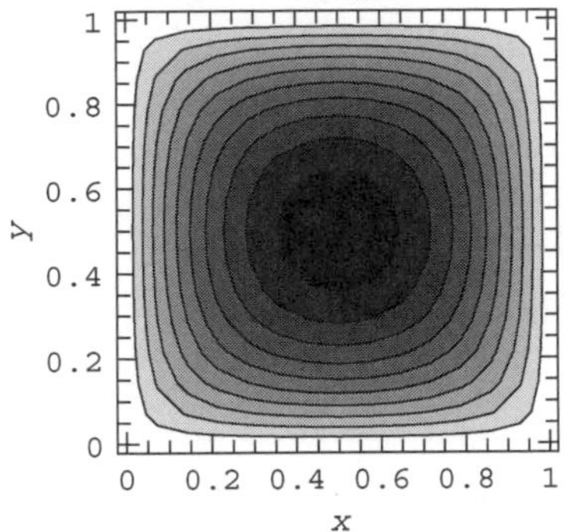

Nonrectangular Boundaries The same direct method of solution can be used for Poisson's equation in domains with nonrectangular boundaries. We will assume for now that the boundary can be matched to some rectangular grid, so that grid points fall on the boundary. The grid is chosen as

$$\begin{aligned} x_j &= x_0 + j\,\Delta x, \qquad j = f(k)+1,\dots,g(k), \\ y_k &= y_0 + k\,\Delta y, \qquad k = 0,1,2,\dots,M_y, \end{aligned} \tag{6.2.35}$$

where $\Delta x = L_x/M_x$ and $\Delta y = L_y/M_y$; $f(k)$ and $g(k)$ are two integer functions that determine the left and right edges of the domain, respectively; and L_x and L_y are the dimensions of a rectangular box that is sufficiently large to contain the domain of interest. (This procedure does not work very well for smoothly curved boundaries, as we will see.)

The equations for the interior points are the same as Eq. (6.2.34):

$$\frac{\phi_{j+1\,k} - 2\phi_{jk} + \phi_{j-1\,k}}{\Delta x^2} + \frac{\phi_{j\,k+1} - 2\phi_{jk} + \phi_{j\,k-1}}{\Delta y^2} = \rho_{jk},$$

$$j = f(k)+1,\dots,g(k)-1, \quad k = 1,2,\dots,M_y - 1. \tag{6.2.36}$$

These equations for the potential in the interior of the box are supplemented by the boundary conditions on the box edges, which directly determine ϕ_{jk} on the boundary points.

Cell 6.94

```
Clear["Global`*"];

Mx = 30; My = 30;
Lx = 2; Ly = 2;
Δx = Lx/Mx; Δy = Ly/My;
x0 =-1; y0 =-1;
x[j_] = x0 + j Δx;
y[k_] = y0 + k Δy;
```

The boundary conditions are specified by the following statements:

Cell 6.95

```
(* boundary conditions *)
ϕ[j_, k_] := V0[j, k]/; j ≤ f[k]
ϕ[j_, k_] := V0[j, k]/; j ≥ g[k]
ϕ[j_, 0] := V0[j, 0];
ϕ[j_, My] := V0[j, My];
```

Let's choose the following equations for the boundaries: $x^2+y^2=1$ and $x=-|y|$. This boundary, shown in Cell 6.98, is the same as that chosen in Example 3 of Sec. 7.1, except that it is rotated by 90 degrees.

We choose boundary conditions $V_0(\mathbf{r})=1-r^2$ on the cusp surfaces, and $V_0(\mathbf{r})=0$ on the circular part of the boundary:

Cell 6.96

```
V0[j_, k_] := 0/; Abs[y[k]] ≥ 1/Sqrt[2] || x[j] > 0
V0[j_, k_] := 1-x[j]^2-y[j]^2/; Abs[y[k]] <
  1/Sqrt[2] && x[j] ≤ 0
```

To determine the integer functions $f(k)$ and $g(k)$ that best fit this boundary, we will first break the boundary up into left and right boundaries, $x(y)=x_R(y)$ on the right and $x(y)=x_L(y)$ on the left. For our example, equations for the left and right boundaries are of the following form:

$$x_R(y)=\sqrt{1-y^2},$$

$$x_L(y)=\begin{cases}-\sqrt{1-y^2}, & |y|>1/\sqrt{2},\\ -|y|, & |y|<1/\sqrt{2}.\end{cases}$$

Then the integer functions $f(k)$ and $g(k)$ are determined by the values of $j(k)$ corresponding to these boundary equations. Using $x(j)=x_0+j\,\Delta x=x_R(y(k))$ on the right boundary yields $j=[x_R(y(k))-x_0]/\Delta x$. Rounding down to the nearest integer, we obtain the value of j corresponding to the right boundary at $y=y(k)$. This is our function $g(k)$:

$$g(k)=\text{Floor}\left(\frac{x_R(y(k))-x_0}{\Delta x}\right). \tag{6.2.37}$$

Similarly, on the left boundary,

$$f(k)=\text{Floor}\left(\frac{x_L(y(k))-x_0}{\Delta x}\right). \tag{6.2.38}$$

These integer functions are defined below using the above definitions of $x_R(y)$ and $x_L(y)$:

Cell 6.97

$$\text{g[k\_] := Floor}\left[\frac{1}{\Delta x}\text{ (-x0 + Sqrt[1-y[k]^2])}\right]$$

$$\text{f[k\_] := Floor}\left[\frac{1}{\Delta x}\text{ (-x0-Sqrt[1-y[k]^2])}\right]\text{/;}$$

```
Abs[y[k]] > 1/Sqrt[2]
```

$$\text{f[k\_] := Floor}\left[\frac{1}{\Delta x}\text{ (-x0-Abs[y[k]])}\right]\text{/; Abs[y[k]]} \le \text{1/Sqrt[2]}$$

The shape of this domain is shown in Cell 6.98.

Cell 6.98

```
ListPlot[
  Join[Table[{x[f[k]], y[k]}, {k, 0, My}],
   Table[{x[g[k]], y[k]}, {k, 0, My}]],
  PlotRange→{{-1.2, 1.2}, {-1.2, 1.2}}];
```

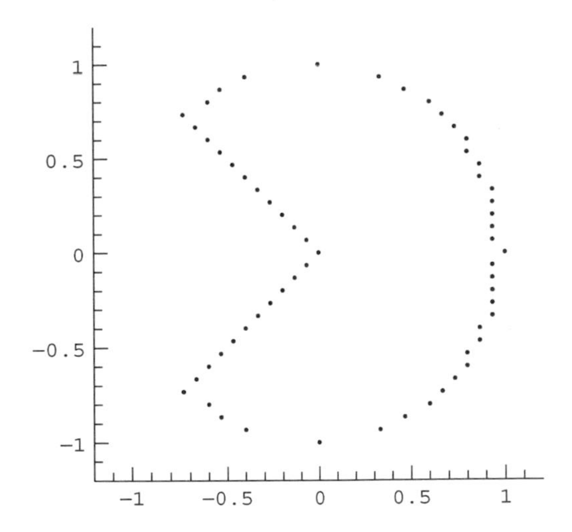

The irregularities in the boundary reflect the difficulty of fitting a rectangular grid to a curved boundary. They can be reduced by using a finer grid, but this increases the number of coupled equations that we need to solve. They can also be reduced by employing an averaging technique, determining the value of ϕ on the boundary as a weighted average of surrounding grid points. (See the exercises for an implementation of this technique.) Alternatively, a nonrectangular grid can be chosen that better matches the edge(s). However, this is rather nontrivial, and will not be pursued further here. An excellent introduction to irregular grids, finite element analysis and the like, can be found on the web in a set of course notes by Peter Hunter and Andrew Pullan (2002). A recent textbook on the subject is also listed in the chapter references.

Here we will make do with the coarse fit obtained above. Since we may be off by as much as a grid spacing Δx in the boundary location, we can only expect our solution to be first-order accurate in Δx. However, for distances from the boundary much greater than Δx, the effect of the irregularities in the boundary often average out, and the solution can be more accurate than this estimate suggests.

Continuing with our example, we take $\rho = 0$ for the source,

Cell 6.99

```
ρ[j_, k_] = 0.;
```

Then the list of equations, using Eq. (6.2.34) for the interior points, is given by

Cell 6.100

```
eqns = Flatten[Table[
   Table[(ϕ[j + 1, k] - 2 ϕ[j, k] + ϕ[j - 1, k])/Δx^2 +
   (ϕ[j, k + 1] - 2 ϕ[j, k] + ϕ[j, k - 1])/Δy^2 ==
   ρ[j, k], {j, f[k] + 1, g[k] - 1}], {k, 1, My - 1}]];
```

and the list of variables is

Cell 6.101

```
vars = Flatten[Table[Table[ϕ[j, k], {j, f[k] + 1, g[k] - 1}],
   {k, 1, My - 1}]];
```

Finally, we solve the equations:

Cell 6.102

```
sol = Solve[eqns, vars][[1]];
```

We can again construct an interpolation function of the solution:

Cell 6.103

```
ϕsol = Interpolation[
   Flatten[Table[{x[j], y[k], ϕ[j, k]}, {j, 0, Mx},
     {k, 0, My}] /. sol, 1]]
InterplatingFunction[{{-1., 1.}, {-1., 1.}}, <>]
```

A contour plot of the solution is shown in Cell 6.105.

Cell 6.104

```
<<Calculus`
```

Cell 6.105

```
ContourPlot[φsol[x, y] Boole[1-x^2-y^2 ≥
 0&& x ≥ -Abs[y]], {x, -1, 1.},
  {y, -1, 1}, FrameLabel → {"x", "y"}, PlotLabel → "φ(x,y)",
    PlotPoints → 60];
```

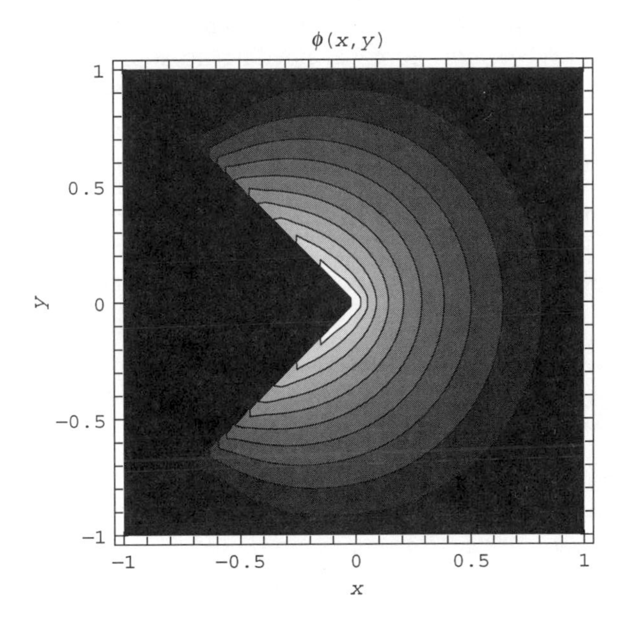

Since this problem is analytically tractable, we can compare our numerical solution to the analytic solution. The analytic solution is given by Eq. (6.1.14), with $\beta = 5\pi/4$. In Fig. 6.7, we compare the analytic solution (solid) to the grid solution along the x-axis, for $M_x = M_y = 30$ (dashed) and 60 (dotted). The solution is more accurate than the corresponding Galerkin solution (see Fig. 6.2) near the singularity at $x = y = 0$. However, the Galerkin solution is superior near the circular boundary, because of the aforementioned difficulty of fitting a curved edge to a grid.

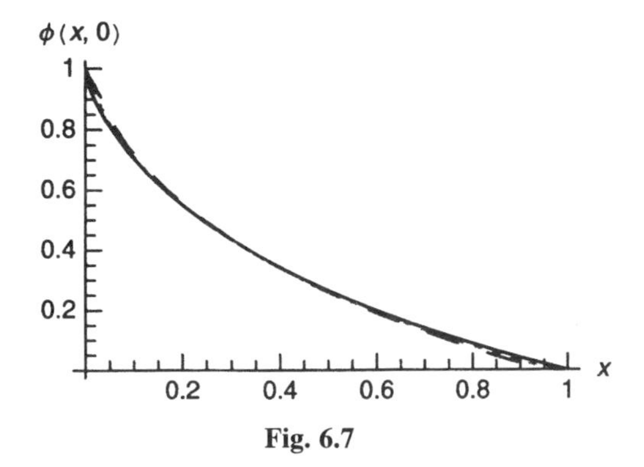

Fig. 6.7

Relaxation Methods

Jacobi's Method. Although direct solution of the coupled equations for the potential is not difficult using *Mathematica*'s **Solve** function, it can be much more difficult to program an efficient direct solution in other computing environments where access to sophisticated linear algebra packages may be limited. In these situations, relaxation methods are often useful. These methods have the advantage that they are relatively simple to code, needing no sophisticated external library routines. On the other hand, they are typically not as fast as the direct methods.

In the following relaxation method, one uses the following idea: solutions to Poisson's equation can be thought of as the equilibrium solution to the heat equation with a source, Eq. (6.2.26). Starting with any initial condition, the heat equation solution will eventually relax to a solultion of Poisson's equation.

For example, we can take the FTCS algorithm for the heat equation, (6.2.28), replace T by ϕ and S/χ by $-\rho$, and take the largest possible time step size that is stable according to the Courant condition (6.2.29):

$$\phi_{jk}^{n+1} = \phi_{jk}^{n} + \frac{1}{2}\frac{\Delta x^2\,\Delta y^2}{\Delta x^2+\Delta y^2}$$

$$\times\left(-\rho_{jk} + \frac{1}{\Delta x^2}\left(\phi_{j+1\,k}^{n} - 2\phi_{jk}^{n} + \phi_{j-1\,k}^{n}\right) + \frac{1}{\Delta y^2}\left(\phi_{j\,k+1}^{n} - 2\phi_{jk}^{n} + \phi_{j\,k-1}^{n}\right)\right).$$

If one expands out the expression, one finds that the terms involving ϕ_{jk}^{n} cancel, leaving

$$\phi_{jk}^{n+1} = \frac{1}{2}\frac{\Delta x^2\,\Delta y^2}{\Delta x^2+\Delta y^2}\left(-\rho_{jk} + \frac{1}{\Delta x^2}\left(\phi_{j+1\,k}^{n} + \phi_{j-1\,k}^{n}\right) + \frac{1}{\Delta y^2}\left(\phi_{j\,k+1}^{n} + \phi_{j\,k-1}^{n}\right)\right). \tag{6.2.39}$$

Equation (6.2.39) is called *Jacobi's method*. The equation provides us with a recursion relation for determining the solution to Poisson's equation. Starting with any initial choice for ϕ at timestep $n=0$, the solution eventually relaxes to a solution of the Poisson equation $\nabla^2\phi=\rho$. We stop the recursion when the difference between ϕ at adjacent time steps is less than some tolerance.

The following simple code implements the Jacobi method in a domain of arbitrary shape:

Cell 6.106

```
Clear["Global`*"];

Mx = 20; My = 20;
Lx = 1.5; Ly = 2.;
Δx = Lx/Mx; Δy = Ly/My;
```

The homogeneous boundary conditions are specified by the following statements:

Cell 6.107

```
ϕ[j_, k_, n_] := 0/; j≤f[k]
ϕ[j_, k_, n_] := 0/; j≥g[k]
ϕ[j_, 0, n_] = 0;
ϕ[j_, M_y, n_] = 0;
```

The main body of the code is the Jacobi recursion relation, Eq. (6.2.39):

Cell 6.108

```
ϕ[j_, k_, n_] :=
```

$$\phi[j,\ k,\ n] = \frac{1}{2}\ \frac{\Delta x^2\,\Delta y^2}{(\Delta x^2 + \Delta y^2)}$$

$$\left(-\rho[j,\ k] + \frac{1}{\Delta x^2}\ (\phi[j+1,\ k,\ n-1] + \phi[j-1,\ k,\ n-1]) + \frac{1}{\Delta y^2}\ (\phi[j,\ k+1,\ n-1] + \phi[j,\ k-1,\ n-1])\right)$$

We must start the recursion with an initial guess. Typically the closer the guess to the exact solution, the better the method works. However, we will simply choose $\phi = 0$ inside the domain:

Cell 6.109

```
ϕ[j_, k_, 0] := 0;
```

In order to determine how well the solution is converging to the correct answer, we take the difference between the solutions at the nth and $n-1$st steps:

Cell 6.110

```
error[n_] := Max[Table[
   Table[Abs[ϕ[j, k, n] - ϕ[j, k, n-1]],
    {j, f[k] + 1, g[k]-1}], {k, 1, M_y-1}]]
```

When this difference is sufficiently small, we can say that the solution has converged.

Let's now perform a solution of $\nabla^2\phi = 1$ in the irregular domain specified below:

Cell 6.111

```
ρ[j_, k_] = 1;

g[k_] := M_x; k≤5;
g[k_] := M_x-k + 5/; 5 <k<15;
g[k_] := M_x-10/; k ≥ 15;
f[k_] = Floor[12 (k/M_y-1)^2];
```

It requires many steps, but eventually the error falls to an acceptable level; as shown in Cells 6.112 and 6.113. We can again construct an interpolation function of the solution, taken from the 200th iteration:

Cell 6.112

```
error[200]
```

```
4.74156×10^-6
```

Cell 6.113

```
<<Graphics`;

LogLogListPlot[Table[{n, error[n]}, {n, 1, 200}],
AxesLabel→{"n", "error[n]"}];
```

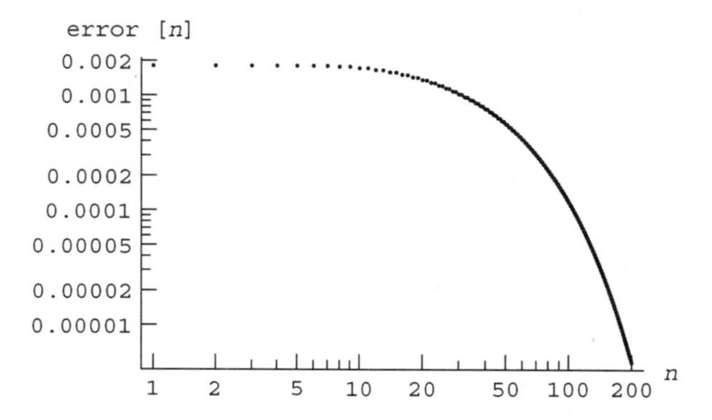

Cell 6.114

```
φsol = Interpolation[
  Flatten[Table[{j Δx, k Δy, φ[j, k, 200]}, {j, 0, Mx},
   {k, 0, My}], 1]]
```

```
InterpolatingFunction[{{0., 1.5}, {0., 2.}}, <>]
```

A contour plot of the solution is shown in Cell 6.115.

Cell 6.115

```
ContourPlot[φsol[x, y], {x, 0, 1.5}, {y, 0, 2},
 FrameLabel→{"x", "y"},
  PlotLabel→"Potential inside a\ncharge-filled irregular
     box",
  AspectRatio→4/3, PlotPoints→30];
```

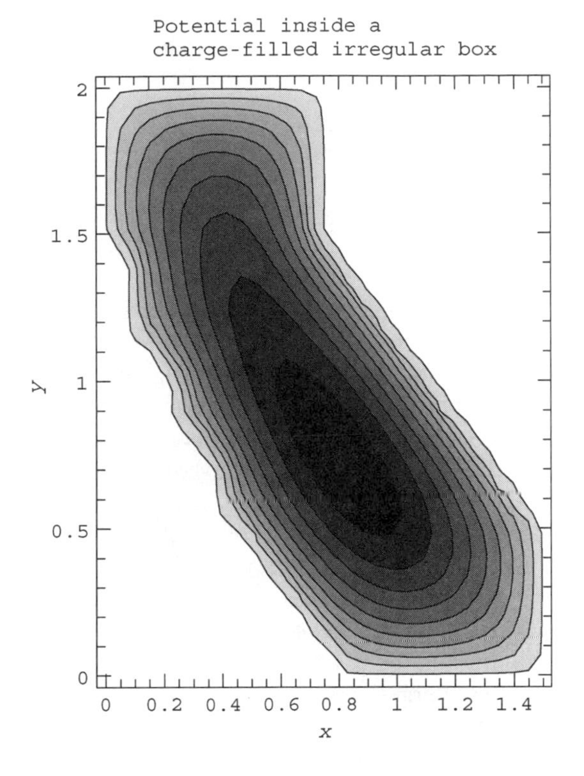

Simultaneous Overrelaxation The convergence of the Jacobi method is quite slow, rendering it impractical in most applications. Furthermore, the larger the system, the slower the convergence. This is simply because we are solving a heat equation problem, and we know that for a system of size L the solution to the heat equation decays toward equilibrium at a rate v of order χ/L^2 [the order of magnitude of the lowest eigenvalue: see Eq. (3.1.60)]. Multiplying this rate by our time step size Δt, and using the Courant condition $\Delta t \sim \Delta x^2/\chi$, where Δx is the grid spacing, we obtain

$$\frac{1}{v\,\Delta t} = \text{number of timesteps} \sim \left(\frac{L}{\Delta x}\right)^2.$$

The number of timesteps required to get a converged solution scales like the square of the number of grid points across the system.

In order to improve on this poor scaling, various recipes have been put forward. One of the most successful is the method of *simultaneous overrelaxation* (SOR). In this algorithm, the Jacobi method is modified in two ways, as pointed out by the arrows in Eq. (6.2.40):

$$\phi_{jk}^{n+1} = (1 - \omega)\,\phi_{jk}^{n} + \frac{\omega}{2}\,\frac{\Delta x^2\,\Delta y^2}{\Delta x^2 + \Delta y^2}$$
$$\times\left(-\rho_{jk} + \frac{1}{\Delta x^2}\left(\phi_{j+1\,k}^{n} + \phi_{j-1\,k}^{n+1}\right) + \frac{1}{\Delta y^2}\left(\phi_{j\,k+1}^{n} + \phi_{j\,k-1}^{n+1}\right)\right). \quad (6.2.40)$$

First, we use improved values for ϕ on the right-hand side as soon as they become available, in order to help improve the guess we are making for the next step. This looks like an implicit method with ϕ_{jk}^{n+1} appearing on both sides, but it is not: assuming that we are stepping through the equations from smaller to larger values of j and k, by the time we evaluate ϕ_{jk}^{n+1} we have already evaluated $\phi_{j-1\,k}^{n+1}$ and $\phi_{j\,k-1}^{n+1}$, so we can use these values on the right-hand side. (Stepping through the grid is performed when we evaluate the error in the solution, creating a table of the error at each grid point, just as in the Jacobi method. The steps in the table are performed from smaller to larger values of j and k.)

Also, we have added terms involving a *relaxation parameter* ω. It has been proven that this parameter must take on a value in the range $1 < \omega < 2$ for the method to work. This parameter is added in order to take advantage of the fact that the Jacobi method converges exponentially toward equilibrium (it is a solution to the heat equation), and tries to overshoot the exponential decay and go directly to the final result in one step. However, the rate of convergence of the SOR method depends sensitively on the choice of ω, and there is no way to know the correct choice beforehand (without doing considerable extra work). This is the main weakness of the method. Typically, one must simply make a guess for ω. However, even if the guess is not close to the optimum value, the SOR method usually works better than the Jacobi method, and is just as easy to program. We provide a module below, and use it to solve the same problem as we did previously:

Cell 6.116

```
Clear["Global`*"];
SOR[ω_, Nstep_] := Module[{ϕ},
  (* choose grid parameters and shape of region*)
  Mx = 20; My = 20;
  Lx = 1.5; Ly = 2.;
  Δx = Lx/Mx; Δy = Ly/My;

  g[k_] := Mx; k≤5;
  g[k_] := Mx-k + 5/; 5 < k < 15;
  g[k_] := Mx-10/; k ≥ 15;
  f[k_] = Floor[12 (k/My-1)^2];
  (* boundary conditions *)
  ϕ[j_, k_, n_] := 0/; j≤f[k];
  ϕ[j_, k_, n_] := 0/; j ≥ g[k];
  ϕ[j_, 0, n_] = 0;
  ϕ[j_, My, n_] = 0;
  (* charge density *)
```

```
ρ[j_, k_] = 1;
(* SOR recursion relation *)
ϕ[j_, k_, n_] := ϕ[j, k, n] = (1 - ω) ϕ[j, k, n-1] +
```

$$\omega\frac{1}{2}\frac{\Delta x^2\,\Delta y^2}{(\Delta x^2+\Delta y^2)}\left(-\rho[j,\,k] + \frac{1}{\Delta x^2}(\phi[j+1,\,k,\,n-1]) + \phi[j-1,\,k,\,n]) + \frac{1}{\Delta y^2}(\phi[j,\,k+1,\,n-1] + \phi[j,\,k-1,\,n])\right);$$

```
(* initial choice for ϕ to start the iteration *)
ϕ[j_, k_, 0] = 0;
(* define error at the nth step *)
error[n_] := Max[Table[Table[
   Abs[ϕ[j, k, n] - ϕ[j, k, n-1]], {j, f[k] + 1, g[k]-1}],
  {k, 1, My-1}]];
(*evaluate error at step NStep*)
Print["error at step NStep = ", error[Nstep]];
(* create interpolation function of solution *)
ϕsol = Interpolation[
  Flatten[Table[{j Δx, k Δy, ϕ[j, k, Nstep]}, {j, 0, Mx},
   {k, 0, My}], 1]]]
```

Cell 6.117

```
SOR[1.6, 30]

error at step NStep = 4.15829×10^-7

InterpolatingFunction[{{0., 1.5}, {0., 2.}}, <>]
```

With the guess $\omega = 1.6$, the result has already converged within 30 steps, as compared to the 200 steps needed in the Jacobi method. The error is plotted in Cell 6.118, showing faster convergence than the Jacobi method. A description of the theory behind SOR, and further references, can be found in Press et al. (1992). The reader is invited to vary the parameter ω, the shape of the region, and the boundary conditions in order to examine the convergence properties of the method.

FFT Method Solution of Poisson's equation within a rectangle (or a 3D box) can be efficiently performed using the analytic trigonometric Fourier series methods discussed in Chapter 4. However, for high-accuracy results, the number of terms kept in the series in each dimension must be large, and the total number of terms in the series is the square (in 2D) or cube (in 3D) of this large number. Evaluation of the enormous list of Fourier coefficients can be most rapidly accomplished using

Cell 6.118

```
<<Graphics`
LogLogListPlot[Table[{n, error[n]}, {n, 1, 30}],
  AxesLabel → {"n", "error[n]"}];
```

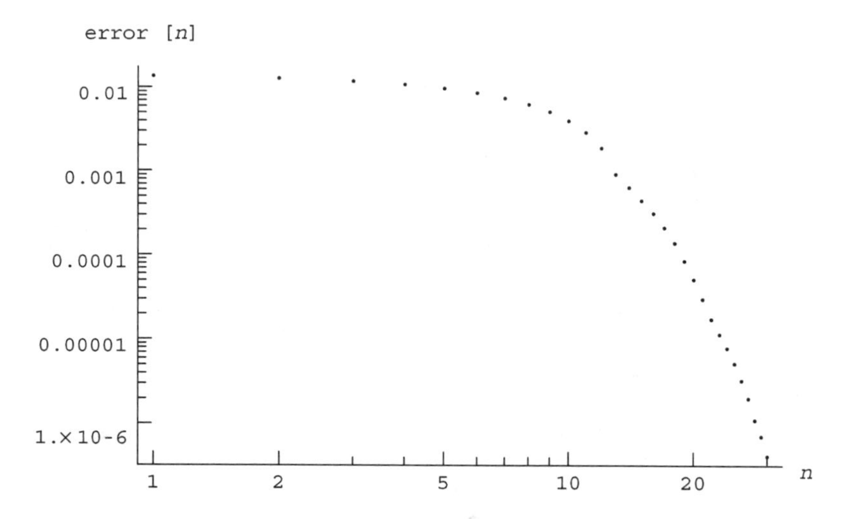

the FFT methods discussed in Chapter 2. However, use of the FFT requires that we discretize the potential on a grid.

As an example, consider the Poisson equation $\nabla^2\phi = \rho$ with periodic boundary conditions in a rectangular box with dimensions L_x and L_y. The density and potential are discretized on the usual rectangular grid,

$$x_j = j\,\Delta x, \qquad j = 0, 1, 2, \ldots, M_x,$$
$$y_k = k\,\Delta y, \qquad k = 0, 1, 2, \ldots, M_y,$$

where $\Delta x = L_x/M_x$ and $\Delta y = L_y/M_y$. Then the density and potential at the grid point (j, k) can be written in terms of a double discrete inverse Fourier transform in x and y, using Eq. (2.3.63) (changed to use Fourier conventions for spatial transforms):

$$\rho_{jk} = \sum_{m=0}^{M_x-1} \sum_{n=0}^{M_y-1} \tilde{\rho}_{mn}\, e^{2\pi imj/M_x}\, e^{2\pi ink/M_y} \tag{6.2.41}$$

and

$$\phi_{jk} = \sum_{m=0}^{M_x-1} \sum_{n=0}^{M_y-1} \tilde{\phi}_{mn}\, e^{2\pi imj/M_x}\, e^{2\pi ink/M_y}. \tag{6.2.42}$$

If we substitute these expressions into the discretized Poisson equation (6.2.34) and use orthogonality of the separate Fourier modes, we find that the Fourier coefficients $\tilde{\phi}_{mn}$ and $\tilde{\rho}_{mn}$ are related according to

$$\tilde{\phi}_{mn}\left(\frac{e^{2\pi im/M_x} - 2 + e^{-2\pi im/M_x}}{\Delta x^2} + \frac{e^{2\pi in/M_y} - 2 + e^{-2\pi in/M_y}}{\Delta y^2} \right) = \tilde{\rho}_{mn},$$

or in other words,

$$\tilde{\phi}_{mn} = -\frac{\tilde{\rho}_{mn}}{\kappa^2(m,n)} \tag{6.2.43}$$

where

$$\kappa^2(m,n) = \frac{4}{\Delta x^2}\sin^2\frac{\pi m}{M_x} + \frac{4}{\Delta y^2}\sin^2\frac{\pi n}{M_y}.$$

We can then use this expression in Eq. (6.2.42) to evaluate the potential at each grid point. Thus, our strategy will be to:

(a) Compute the Fourier coefficients $\tilde{\rho}_{mn}$ by a FFT.
(b) Determine the Fourier coefficients of the potential via Eq. (6.2.43).
(c) Perform an inverse FFT to obtain ϕ_{jk}. We can then interpolate, if we wish, to determine $\phi(x, y)$ at any point (x, y) in the rectangle.

Because of the centered-difference method used, the error in this method scales roughly as $k^2\,\Delta x^2$, where k is the dominant wavenumber in the density and potential. One can see this directly in Eq. (6.2.43). For low wavenumbers, $\kappa^2 \approx k^2$ and Eq. (6.2.43) correctly yields the Fourier-transformed Poisson equation, $\tilde{\phi} = -\tilde{\rho}/k^2$. However, if the density is dominated by wavenumbers on the order of the grid spacing, κ^2 is not close to k^2 and the FFT method fails.

Before we proceed, we note that Eq. (6.2.43) blows up for $n = m = 0$. This is because, for periodic boundary conditions, constant charge density (the $m = n = 0$ term) creates an infinite potential: the uniform charge density stretches to infinity in both x and y directions in the periodically replicated boxes, creating an infinite potential. Therefore, when using periodic boundary conditions, we require that this term in the charge density vanish, so that the net charge density in the box is zero.

First, we set the parameters of the grid, taking a rather fine grid in x and y to illustrate the speed of the method:

Cell 6.119

```
Clear["Global`*"];
Mx = My = 50;
Lx = Ly = 1;

Δx = Lx/Mx; Δy = Ly/My;
```

This corresponds to a discrete Fourier series with 2500 terms, which would be quite time-consuming to evaluate using a standard Fourier series expansion. For fine grids, the FFT method is also more efficient than the direct or relaxation methods discussed previously.

Next, we choose a charge density:

Cell 6.120

```
ρ[x_, y_] = x Sin[Pi x] Sin[Pi y]-2/Pi^2;
```

As required by our periodic boundary conditions, this charge density has a zero average, as can be seen by evaluating the integral over the box:

Cell 6.121

```
Integrate[ρ[x, y], {x, 0, Lx}, {y, 0, Ly}]

0
```

The charge density is plotted in Cell 6.122.

Cell 6.122

```
Plot3D[ρ[x, y], {x, 0, 1}, {y, 0, 1},
  PlotLabel → "ρ[x, y]", AxesLabel → {"x", "y", ""}];
```

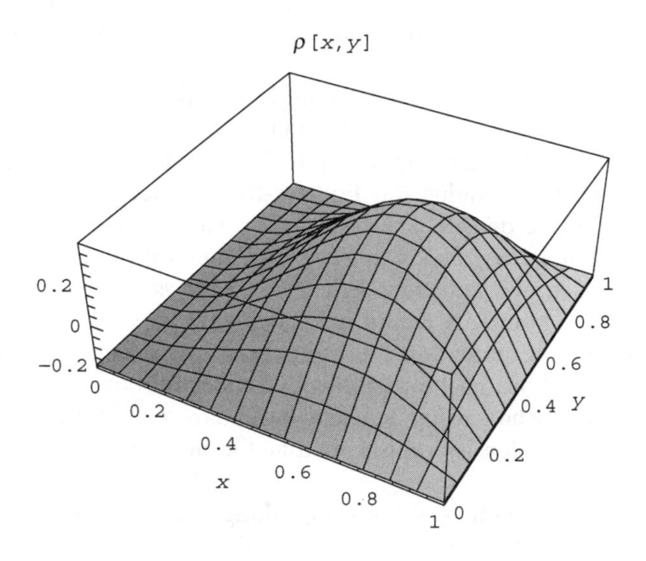

We now create a matrix with elements ρ_{jk}, and convert the matrix to approximate numbers in order to assure that no exact arithmetic is attempted in what follows:

Cell 6.123

```
ρmatrix = Table[ρ[j Δx, k Δy], {j, 0, Mx-1}, {k, 0, My-1}]//N;
```

We next determine the Fourier coefficient matrix, $\tilde{\rho}_{mn}$, using an FFT:

Cell 6.124

```
ρft = Fourier[ρmatrix];
```

The $(1, 1)$ element of this matrix corresponds to the $m = 0, n = 0$ term, and should be zero:

Cell 6.125

```
ρft[[1, 1]]

-0.00666601 + 0. i
```

This element is not quite zero because of the discretization of the charge density: the discrete Fourier transform is not exactly the same as the integral of the charge over the rectangular domain, although it is close for large M_x and M_y. Therefore, we simply set this term equal to zero directly:

Cell 6.126

```
ρft[[1, 1]] = 0;
```

Now we define the factor by which we are to divide $\tilde{\rho}_{mn}$ [see Eq. (6.2.43)]:

Cell 6.127

```
factor[m_, n_] :=-(4/Δx^2 Sin[πm/Mx]^2 + 4/Δy^2 Sin[π n/My]^2);

factor[0, 0] = 1;
```

This factor is defined so that we may safely divide by it without getting an error when $m = n = 0$:

Cell 6.128

```
φ̃ = Table[ρft[[m + 1, n + 1]]/factor[m, n], {m, 0, Mx-1},
  {n, 0, My-1}];
```

Finally, we obtain the potential ϕ_{jk} at the grid points $0 \leq j \leq M_x - 1, 0 \leq k \leq M_y - 1$:

Cell 6.129

```
φmatrix = InverseFourier[φ̃];
```

It is enlightening to compare the time required to complete these operations with that required by the direct solution method, or the relaxation methods for the same fine grid. The reader is invited to evaluate these methods, which should provide convincing proof of the efficacy of the FFT method for this rectangular geometry.

We can create an interpolation of the matrix `φmatrix` by first adding to the data the corresponding x, y positions for each grid point, and then interpolating the resulting data list:

Cell 6.130

```
φ = Table[{j Δx, k Δy, φmatrix[[j + 1, k + 1]]}, {j, 0, Mx-1},
  {k, 0, My-1}];

φsol = Interpolation[Flatten[φ, 1]]
InterpolatingFunction[{{0., 0.98}, {0., 0.98}}, <>]
```

This function is defined only in the square from 0 to 0.98, rather than from 0 to 1, because of the discretization. If we wished, we could use the periodicity of the system to extend the result to the entire *x*-*y* plane. The solution is plotted in Cell 6.131.

Cell 6.131

```
Plot3D[φsol[x, y], {x, 0, .98}, {y, 0, .98},
  PlotLabel→"φ(x,y)", AxesLabel→{"x", "y", ""}];
```

EXERCISES FOR SEC. 6.2

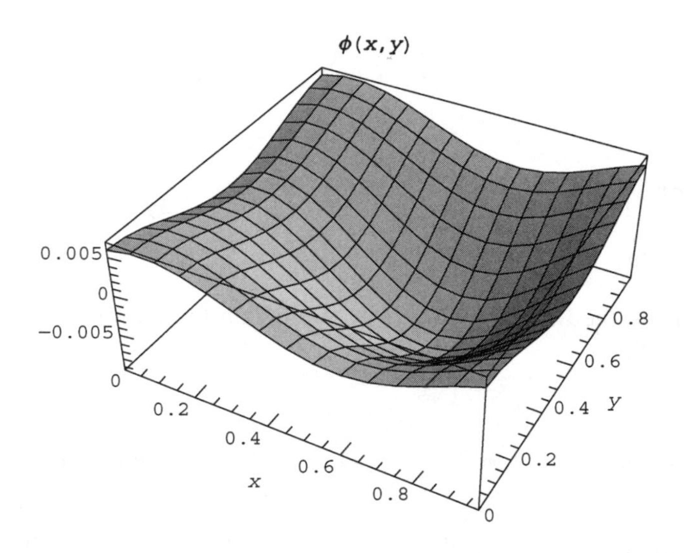

(1) Consider the following heat equation problem on $0 \le x \le 1$:

$$\frac{\partial T(x,t)}{\partial t} = \chi(x)\frac{\partial^2 T(x,t)}{\partial x^2} + 2x, \qquad T(0,t) = T(1,t) = 0, \quad T(x,0) = 0,$$

where $\chi(x) = 1$ for $0 < x < \frac{1}{2}$ and $\chi(x) = 2$ for $\frac{1}{2} < x < 1$.

(a) Using the FTCS method, and taking $M = 10$ (i.e., 11 grid points including the end points), what is the largest time step size one can use?

(b) Solve the equation with the largest possible time step size. Make an animation showing only every fifth timestep for $0 \le t \le \frac{1}{2}$.

(2) Using the Lax method with the largest possible time step size, solve the following heat equation problem on $0 \le x \le 1$:

$$\frac{\partial T(x,t)}{\partial t} = \frac{\partial}{\partial x}\chi(x)\frac{\partial T(x,t)}{\partial x}, \qquad T(0,t) = \sin(6\pi t),$$

$$T(1,t) = 0, \quad T(x,0) = 0,$$

where $\chi(x) = x$ for $0 < x < \frac{1}{2}$ and $\chi(x) = 2x - \frac{1}{2}$ for $\frac{1}{2} < x < 1$. Make an animation showing only every fifth timestep for $0 \le t \le \frac{1}{2}$.

(3) Consider the following first-order PDE:

$$\frac{\partial y(x,t)}{\partial t} + c\frac{\partial y(x,t)}{\partial x} = 0.$$

Using von Neumann stability analysis, show that the CTCS method applied to that equation is stable provided that the following Courant condition is satisfied:

$$c\,\Delta t/\Delta x \le 1. \tag{6.2.44}$$

(4) For the time-dependent Schrödinger equation in 1D [Eq. (3.1.81)], and for a particle of mass m in a constant potential $V(x) = V_0$, show that the CTCS method is stable only if $\Delta t[|V_0|/\hbar + 4\hbar/(m\,\Delta x^2)] < 2$. This sets a rather stringent condition on Δt, rendering CTCS of only academic interest for quantum problems.

(5) **(a)** Use the CTCS method to finite-difference the wave equation in D dimensions (on a uniform Cartesian grid). Perform a von Neumann stability analysis to show that this method is stable provided that $c\,\Delta t/\Delta x \le 1/\sqrt{D}$.

(b) For the 1D wave equation, $\partial^2 y/\partial t^2 = c^2(x)\,\partial^2 y(x,t)/\partial x^2$ on $-4 < x < 4$, where $c(x) = 1$ for $x < 0$ and $c(x) = \frac{1}{2}$ for $x \ge 0$, use the CTCS method to solve the following problem for $0 < t < 20$. Take $\Delta x = 0.25$, $\Delta t = 0.2$:

$$y(x,0) = e^{-2(x-2)^2}, \qquad \left.\frac{\partial y(x,t)}{\partial t}\right|_{t=0} = -2(x-2)\,e^{-2(x-2)^2},$$

$$y(-4,t) = y(4,t) = 0.$$

Make an animation of the result showing every fifth time step. Note: The CTCS scheme will require you to specify u_k^1 directly from the two initial conditions. To determine u_k^1 use the equation for uniform acceleration,

$$u_k^1 = u_k^0 + v_{0k}\,\Delta t + a_k\,\Delta t^2/2,$$

where v_{0k} is the initial velocity of the kth point and a_k is the initial acceleration, determined by

$$a_k = c^2 \left.\frac{\partial^2 u(x,0)}{\partial x^2}\right|_{x=x_k}.$$

(6) **(a)** Prove Eq. (6.2.15).

(b) Prove Eq. (6.2.17).

(7) Repeat Exercise (1) using the Crank–Nicolson method. Now take as a time step size $\Delta t = 0.02$ and show every frame of the animation for $0 < t < \frac{1}{2}$.

(8) **(a)** For the 1D Schrödinger equation and for a particle of mass m in a uniform potential $V(x)=V_0$, show that the Crank–Nicolson method is stable for any time step size.

(b) Use the Crank–Nicolson method to solve the Schrödinger equation for a particle of mass m, described by a wave function $\psi(x,t)$, moving in a potential $V(x)=8x$ on $-4<x<4$, over the time range $0<t<3$. Animate $|\psi|^2$ over this time interval. Take $\psi=0$ on the boundaries, and assume an initial condition $\psi(x,0)=e^{5ix-2(x+2)^2}$. Also, take $m=\hbar=1$.

(9) Consider the following heat equation problem on $0\le r\le 1$ in *cylindrical* geometry:

$$\frac{\partial T}{\partial t}=\frac{\partial^2 T}{\partial t^2}+\frac{1}{r}\frac{\partial T}{\partial r}+\frac{1}{r^2}\frac{\partial^2 T}{\partial \theta^2}+r^2\cos m\theta,$$

$$T(1,\theta,t)=-\left.\frac{\partial T(r,\theta,t)}{\partial r}\right|_{r=1},\quad T(r,\theta,0)=0.$$

(a) The solution will depend on θ as $\cos m\theta$, so only a radial grid is necessary. While this simplifies the numerics, we now require a boundary condition at $r=0$. Recall from Sec. 6.1 that the proper boundary condition is

$$\lim_{r\to 0}\frac{\partial T}{\partial r}-\frac{m^2}{r}T=0$$

[see Eq. (6.1.37)]. For $m=0$ this implies $\partial T/\partial r=0$ at $r=0$, but for $m\ne 0$ it implies $T=0$ at $r=0$. In both cases the grid must avoid $r=0$. The best choice is $r(j)=(j-\frac{1}{2})\Delta r$, $j=0,1,\dots,$ using a centered difference for $\partial T/\partial r|_{r=0}$ (if $m=0$), or the average value at grid points $j=0$ and 1 for $T|_{r=0}$ (if $m\ne 0$). Use the differencing techniques discussed in connection with Eqs. (6.2.23) and (6.2.25) to satisfy the mixed boundary condition at $r=1$.

(b) Use the Crank–Nicolson method to solve this problem for the case $m=1$. Take the time step size $\Delta t=0.02$, and take $M=10$ (i.e., 11 grid points in r). Make an animation of the solution as a function of r at $\theta=0$ for $0\le t\le\frac{1}{2}$.

(10) A sandbar causes the water depth in the ocean to vary as $h(x)=1-e^{-x^2}/2$. Water waves impinge on the sandbar at normal incidence, satisfying the 1D shallow-water wave equation $\partial^2 z(x,t)/\partial t^2=(\partial/\partial x)[c^2(x)\,\partial z(x,t)/\partial x]$, where $c^2(x)=gh(x)$. Solve using CTCS for $z(x,t)$ on a grid with $M=50$ points, with *periodic* boundary conditions on $-5<x<5$, and with initial conditions that $z(x,0)=e^{-2(x-2)^2}$ and $\dot z(x,0)=-4(x-2)e^{-2(x-2)^2}$. [Hint: Periodic boundary conditions imply that $z(-5,t)=z(5,t)$ and $z'(-5,t)=z'(5,t)$.] Animate the solution for $0<t<6$, taking $g=1$ and Δt as large as possible.

(11) A square pool with $0<x<a$, $0<y<a$, has a sloped bottom, becoming deeper as x or y increases: $h(x,y)=h_0(1+x/a+y/a)$. Assuming potential

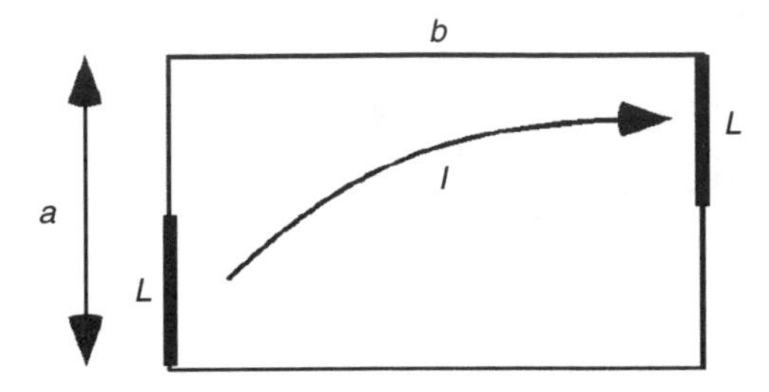

Fig. 6.8 Exercise (12).

flow, the horizontal fluid displacement satisfies $\boldsymbol{\eta} = \nabla\phi(x, y, t)$, and the wave height is given by $z(x, y, t) = -\nabla\cdot[h(x, y)\nabla\phi]$, with ϕ satisfying the following wave equation:

$$\frac{\partial^2\phi(x,y,t)}{\partial t^2} = \frac{\partial}{\partial x}\left(c^2(x,y)\frac{\partial\phi(x,y,t)}{\partial x}\right) + \frac{\partial}{\partial y}\left(c^2(x,y)\frac{\partial\phi(x,y,t)}{\partial y}\right), \qquad (6.2.45)$$

where $c^2 = gh$. Initially, the fluid is stationary, but an inlet at the center of the left-hand side, of length $a/2$, begins to add and remove water according to $\eta_x(0, y) = x_0 \sin \omega_0 t$, $a/4 < y < 3a/4$. Around the rest of the pool, the normal component of $\boldsymbol{\eta}$ is zero. Using CTCS, solve $\phi(x, y, t)$ on a 15-by-15 grid for $0 < t < 4$ s, assuming that $a = 5$ m, $h_0 = 10$ cm, $\omega_0 = 3$, $x_0 = 2$ cm. Make an animation of the resulting wave height $z(x, y, t)$ over this time period.

(12) An electrolytic cell consists of a glass tank of width b and height a, and length c in the z-direction (into the page). The tank is filled with fluid that has electrical conductivity σ. The tank has electrodes of length $L = a/2$, located as shown in Fig. 6.8, and attached on the left and right sides; the rest of the surface is electrically insulated. The right electrode is grounded, and the left electrode is at a potential V_0. Using the direct method, solve for the potential in the fluid, and use this result to find the total current I running between the two electrodes. (See Example 2 in Sec. 3.2.2 for help on setting up the equations for this problem.) Do this problem for $b = na/5$, $n = 1, 2, \ldots, 10$, and plot the result for I vs. b/a, with I suitably normalized in terms of the given parameters. [Hint: This is really a two-dimensional problem; current density and potential depend only on x and y. Part of the boundary has a von Neumann boundary condition, and part has a Dirichlet condition. For second-order accuracy, use a grid for which the boundary falls halfway between the grid points. For the von Neumann portion of the boundary, use the centered-difference method discussed in relation to Eq. (6.2.20), and for the Dirichlet part of the boundary, use the averaging technique discussed in relation to Eq. (6.2.22).]

(13) An enclosure in the shape of an isosceles triangle with base of length a and with height b is filled with a uniform charge density $\rho/\epsilon_0 = 2$ V. The surface of the triangle is grounded. Find the best grid to fit this problem, taking

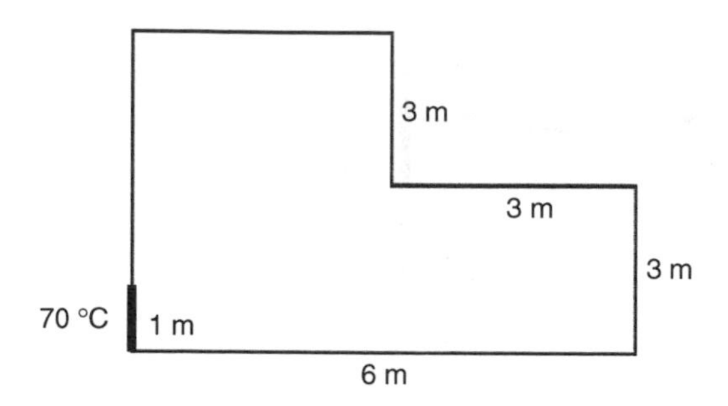

Fig. 6.9 Exercise (14).

$M_x = M_y = 20$, and solve it using the direct method, for the case $b = \sqrt{5}$ meter, $a = 1.5$ meter.

(14) A room is heated with a corner radiator, held at temperature $T_0 = 70°\text{C}$. As seen in Fig. 6.9, the bottom and left wall of the room are insulated, but the right and top walls are at constant temperature of 15°C. Treating the air in the room as a thermally conductive material, and for the dimensions of the room shown in the figure, find the temperature distribution $T(x, y)$ in the room using the direct method. What is the total heat flux entering the room through the radiator in watts per square meter, assuming that, for air, $\kappa = 0.1$ W/m K)?

(15) A grounded square box, with an interior defined by $0 < x < 1$ and $0 < y < 1$, is cut into two sections by a conducting plate at $x = \frac{1}{2}$. The plate is held at potential $V_0 = 1$ volt. The plate has a gap at the center, leaving an opening at $\frac{1}{4} < y < \frac{3}{4}$. Find the potential $\phi(x, y)$ in the box, using a 20-by-20 grid. Solve this problem using simultaneous overrelaxation, and add conditions to the recursion equation for the interior points that avoid the points that fall on the central plate. The solution for $\phi(x, y)$ is shown in Fig. 6.10.

(16) A wire with square cross section is defined by $-1 < x < 1$, $-1 < y < 1$. It is held at fixed temperature T_0, and is embedded in insulation with thermal diffusivity χ, also in the shape of a square defined by $-2 < x < 2$, $-2 < y < 2$. The surface of the insulation is at temperature T_2. Find and plot as a surface plot the equilibrium temperature $T(x, y)$ in the insulator, using simultaneous overrelaxation.

(17) The bottom of a cubical box has a pyramidal cusp. The box walls are defined by the equations $z = 1$, $y = 0$, $y = 1$, $x = 0$, $x = 1$, $z = x$ (for $x < \frac{1}{2}$), $z = y$ (for $y < \frac{1}{2}$), $z = 1 - x$ (for $x > \frac{1}{2}$), $z = 1 - y$ (for $y > \frac{1}{2}$). The potential on the pyramidal cusp is held fixed at $V_0 = 2$ volts, but the potential on the other box walls is zero. Using the direct method in three dimensions on a 10-by-10-by-10 grid, solve for the potential everywhere inside the box. Plot the result for $\phi(x, \frac{1}{2}, z)$ as a contour plot.

(18) It is easy to convince oneself that it is not possible to find any rectangular grid that will fit the grounded triangular enclosure shown in Fig. 6.11, for $a = 1$,

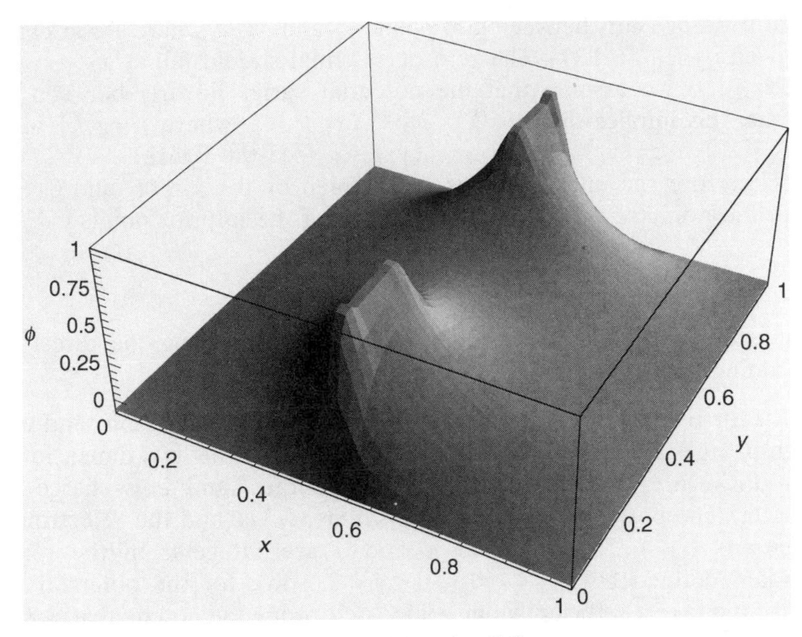

Fig. 6.10 Exercise (15).

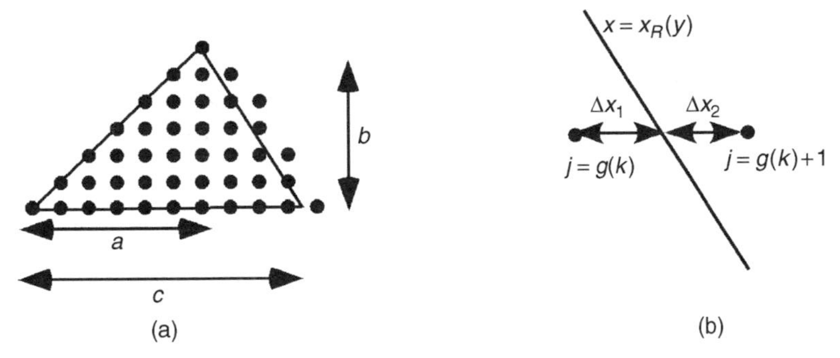

Fig. 6.11 (a) Triangular domain together with a possible choice of grid. The rightmost point in each row has index $j = g(k) + 1$. (b) Points $g(k)$ and $g(k) + 1$ at the end of row k.

$b = 1$, $c = \sqrt{3}$. One can either content oneself with an order-Δx approximation to the solution using the methods discussed in Sec. 6.2.2, or else attempt to find a better way of matching the boundary conditions. One way to do so is as follows. Choose the grid so that the bottom and left sides fall on the grid. This requires that $\Delta y = b/M_y$ and $\Delta x = a/P$ for integers P and M_y With these choices, the grid in x must run out to $x_j = j\,\Delta x$, $j = 0, 1, 2, \ldots, M_x$, where $M_x = \text{Ceiling}[c/\Delta x]$. However, the right side of the triangle does not fall on the grid.

Since we wish to set the potential on the right side of the triangle equal to zero, we will use the following averaging technique. The equation for the triangle's right edge is $x_R(y) = -(c-a)y/b + c$. For each value of y_k on the grid, this edge cuts between grid points x_j and x_{j+1}, where $j = g(k)$ and $g(k)$ is given by Eq. (6.2.37). The zero of potential should fall at $x_R(y_k)$, not at x_j or x_{j+1}. If we assume that the potential varies linearly between the grid points, this implies that $0 = \Delta x_2\, \phi_{jk} + \Delta x_1\, \phi_{j+1\,k}$, where $j = g(k)$, and where $\Delta x_2 = x_{j+1} - x_R(y_k)$ and $\Delta x_1 = x_R(y_k) - x_j$ (see the figure).

Therefore the interior points are defined by $0 < k < M_y$ and $0 < j \le g(k)$, with the points $\phi_{g(k)+1\,k}$ defined in terms of the interior points by

$$\phi_{g(k)+1\,k} = -\Delta x_1 \phi_{g(k)\,k} / \Delta x_2 .$$

Implement this procedure and solve for $\phi(x, y)$ using the direct method, assuming that $\rho/\epsilon_0 = x$ and taking $M_y = P = 10$.

(19) Solve for the potential created by a rectangular lattice of ions and electrons, with charges e and $-e$ respectively. (Actually, in this two-dimensional problem these are charged rods, not point charges, and e is charge per unit length.) The positive "ions" are at positions $(i, 2j)$, and the "electrons" are at positions $(i + \frac{1}{2}, 2j + \frac{1}{2})$, where i and j are integers. In the rectangular domain defined by $-\frac{1}{4} < x < \frac{3}{4}$, $0 < y < 2$, solve for the potential using an FFT, and take a sufficiently fine grid to determine the potential at $x = \frac{1}{2}$, $y = \frac{3}{4}$ to three significant figures.

(20) The following charge density $\rho(x, y, z)$, defined in a cube, is periodically replicated throughout space:

$$\rho(x, y, z) = Axyz^2(1 - x)(1 - y^2)(1 - z)^2 - \tfrac{A}{720},$$
$$0 < x < 1, \quad 0 < y < 1 \quad 0 < z < 1.$$

Find the potential at the center of the cube to three significant figures using the FFT method.

REFERENCES

P. Hunter and A. Pullan, *FEM/BEM Notes*, at http://www.esc.auckland.ac.nz/Academic/Texts/FEM-BEM-notes.html/.

Klaus-Jurgen Bathe, *Finite Element Procedures* (Prentice Hall, Englewood Cliffs, NJ, 1996).

L. D. Landau and E. M. Lifshitz, *Quantum Mechanics* (*Nonrelativistic Theory*) (Pergamon Press, Oxford, 1981).

W. H. Press, S. A. Teukolsky, W. T. Vetterling and B. P. Flannery, *Numerical Recipes*, 2nd ed. (Cambridge Press, Cambridge, 1992).

H. G. Winful, *Nature of "superluminal" barrier tunneling*, Phys. Rev. Lett. **90**, 023901-1 (2003).

CHAPTER 7

NONLINEAR PARTIAL DIFFERENTIAL EQUATIONS

Linear partial differential equations depend on only the first power of the unknown function. On the other hand, many natural phenomena require nonlinear partial differential equations for their description. These equations can depend on the unknown function in an arbitrary fashion, and consequently the principle of superposition no longer applies: the sum of two solutions to the homogeneous PDE is no longer itself a solution. Therefore, the eigenmode and Fourier expansion methods developed previously for linear PDEs will not work on nonlinear PDEs.

Generally, nonlinear PDEs require numerical methods for their solution. The Galerkin method and the grid methods discussed in Chapter 6 can still be applied, and we will see examples in this chapter. We will also examine an advanced simulation method, the particle-in-cell method, for the numerical solution of some classes of nonlinear PDEs.

First, however, we will discuss an analytic method that can be applied to solve first-order nonlinear PDEs: the method of characteristics.

7.1 THE METHOD OF CHARACTERISTICS FOR FIRST-ORDER PDEs

7.1.1 Characteristics

Consider the following PDE for a function $f(x,t)$:

$$\frac{\partial f}{\partial t} + v(x,t,f)\frac{\partial f}{\partial x} = s(x,t,f). \tag{7.1.1}$$

This equation is first-order in time, and so requires a single initial condition:

$$f(x,0) = f_0(x). \tag{7.1.2}$$

The equation is also first-order in space, and so in principle a single boundary condition could be imposed. However, this can lead to difficulties that are tangential to the discussion at hand. In what follows, for simplicity we will solve the PDE on the entire real line without imposing boundary conditions, as we did in Chapter 5.

First-order PDEs of this sort arise in a number of different physical contexts. For example, the energy and wave action equations (5.2.59) and (5.3.23) describing wave-packet dynamics are of this general form. We will also see that this equation and its generalizations describe dye in a fluid flow, nonlinear waves, ideal gases in an external force field, plasmas, and self-gravitating matter such as stars and galaxies.

We will solve Eq. (7.1.1) using the method of characteristics, which is specially suited for PDEs of this form, and which works whether or not the PDE is linear in f. In this method, by means of a coordinate transformation $x \to x_0(x,t)$ that takes us to a moving frame, we convert the PDE into an ODE. The moving frame that we choose is one that moves with velocity $v(x,t,f)$.

This is not a standard Galilean transformation, where the entire frame moves with the same velocity. Rather, the velocity varies from place to place and from time to time. Because of this, it is best to think of the transformation in terms of an infinite number of moving observers, spread along the x-axis. Each observer moves along a trajectory $x = x(t, x_0)$, where x_0 is the initial location of the observer:

$$x = x_0 \qquad \text{at} \quad t = 0. \tag{7.1.3}$$

The trajectory $x(t, x_0)$ of some given observer is called a *characteristic*.

Each observer is responsible for recording the behavior of f at his or her own (moving) position. Along this trajectory, $f = f(x(t, x_0), t)$. Taking a differential and applying the chain rule, we find that each observer sees a differential change in f given by

$$df = \frac{\partial f}{\partial t}\,dt + \frac{\partial f}{\partial x}\frac{dx}{dt}\,dt.$$

If we now divide by dt and substitute for $\partial f/\partial t$ from Eq. (7.1.1), we obtain the following equation for the *total time derivative* of f along the characteristic trajectory, df/dt:

$$\frac{df}{dt} = \frac{\partial f}{\partial x}\frac{dx}{dt} + s - v\frac{\partial f}{\partial x}. \tag{7.1.4}$$

This equation simplifies if we take $dx/dt = v$:

$$\frac{df}{dt} = s(x,t,f). \tag{7.1.5}$$

With this choice the original PDE, Eq. (7.1.1), is converted into an ODE involving the total rate of change of f as seen by an observer moving along the characteristic. The characteristic itself is defined by the ODE

$$\frac{dx}{dt} = v(x,t,f), \tag{7.1.6}$$

with initial condition $x = x_0$ at time $t = 0$. The solution to this ODE is our transformation $x = x(t, x_0)$, and we use this solution in solving Eq. (7.1.5) for f.

Before we go on to solve Eq. (7.1.5), it is instructive to compare it with Eq. (7.1.1). We can then immediately make the following identification:

$$\frac{df}{dt} = \frac{\partial f}{\partial t} + v\frac{\partial f}{\partial x}. \tag{7.1.7}$$

The total rate of change of f as seen by the moving observer consists of two terms. The origin of the first term, $\partial f/\partial t$, is obvious. This term would occur even if the observer were stationary. The second term, $v\,\partial f/\partial x$, would occur even if f were not a function of time. This term arises because the observer sees a change in f due to his or her own motion at velocity v.

There is another useful way to write the total derivative df/dt. Since $x = x(t, x_0)$ defines a transformation from coordinate x to x_0, we could write f in terms of x_0 and t rather than x and t. If we now consider a general variation of $f(x_0, t)$, as x_0 and t are varied by infinitesimal amounts, we obtain the differential relation

$$df = \left.\frac{\partial f}{\partial t}\right|_{x_0} dt + \left.\frac{\partial f}{\partial x_0}\right|_t dx_0. \tag{7.1.8}$$

However, a given characteristic is defined by a single initial condition x_0 that does not vary, so $dx_0 = 0$ along the characteristic. Dividing through by dt, we then obtain

$$\frac{df}{dt} = \left.\frac{\partial f}{\partial t}\right|_{x_0}. \tag{7.1.9}$$

Since the initial condition x_0 is a fixed parameter of the characteristic, it should hardly be surprising that the total time derivative along a characteristic is the same as a partial derivative, holding x_0 fixed.

7.1.2 Linear Cases

We now turn to the solution of Eqs. (7.1.5) and (7.1.6). These are coupled first-order ODEs. It is easiest to understand the solution to these ODEs by first considering a linear case, where $s = s(x, t)$ and $v = v(x, t)$. Then Eqs. (7.1.5) and (7.1.6) become

$$df/dt = s(x, t), \tag{7.1.10}$$

$$dx/dt = v(x, t). \tag{7.1.11}$$

We first solve Eq. (7.1.11) for the characteristics. This is a first-order ODE with initial condition $x = x_0$ at $t = 0$. We write the solution as $x = x(t, x_0)$.

Next, we use these characteristics in solving Eq. (7.1.10) for $f(x, t)$. Here, it must be remembered that df/dt is the *total* rate of change, as seen by the moving observer. Thus, x varies in time as $x(t, x_0)$ along the characteristic, and s varies in time as $s(x(t, x_0), t)$. The solution to $df/dt = s$ is therefore

$$f(t) = f(0) + \int_0^t s(x(t', x_0), t')\, dt', \tag{7.1.12}$$

where $f(0) = f_0(x_0)$ is the initial value for f at position x_0, and where $f(t) = f(x,t)|_{x=x(t,x_0)}$ is the value of f on the characteristic at time t.

The right-hand side is written in terms of x_0 rather than x, so we must invert $x = x(t, x_0)$ to obtain $x_0 = x_0(x,t)$. Using this equation in Eq. (7.1.12) yields

$$f(x,t) = f_0(x_0(x,t)) + \int_0^t s(x(t',x_0),t')\,dt'\bigg|_{x_0=x_0(x,t)}. \qquad (7.1.13)$$

This is our solution for f, written explicitly as a function of x and t. In order to clarify the meaning of this result, we consider several examples below. Other examples can be found in the exercises.

Example 1: $s = 0$ and $v(x,t,f) = c$ Now Eq. (7.1.1) is

$$\frac{\partial f}{\partial t} + c\frac{\partial f}{\partial x} = 0. \qquad (7.1.14)$$

We will see that the solutions to this equation are like those to the wave equation, except that propagation is now only in one direction. The solution is obtained using Eqs. (7.1.5) and (7.1.6):

$$df/dt = 0,$$
$$dx/dt = c.$$

The solution to $dx/dt = c$ with initial condition $x = x_0$ is $x = x_0 + ct$. These straight parallel lines are the characteristics for the PDE for this example, and are shown for three initial positions x_0 in Fig. 7.1.

The equation $df/dt = 0$ implies that $f =$ constant along these characteristics; that is, along each characteristic the value of f remains unchanged from its initial value. For the characteristic starting at $x = x_0$, this value is $f = f_0(x_0)$, according to Eq. (7.1.2). Therefore, the solution for f is

$$f(t) = f_0(x_0), \qquad (7.1.15)$$

where $f(t) = f(x,t)|_{x=x_0+ct}$ is the value of f at time t, evaluated along the characteristic.

This equation could equally well have been obtained by using Eq. (7.1.9), which when combined with $df/dt = 0$ yields

$$\partial f/\partial t|_{x_0} = 0.$$

The solution to this equation is clearly any function $f(x_0)$, since any such function can be differentiated with respect to t (holding x_0 fixed) to yield zero. To satisfy the initial condition, Eq. (7.1.2), we require that $f(x_0) = f_0(x_0)$ (because $x = x_0$ at $t = 0$), and so we return to Eq. (7.1.15).

To finish the solution, we now invert the characteristic equation $x = x_0 + ct$ to obtain $x_0 = x - ct$. Substituting into Eq. (7.1.14) yields

$$f(x,t) = f_0(x - ct).$$

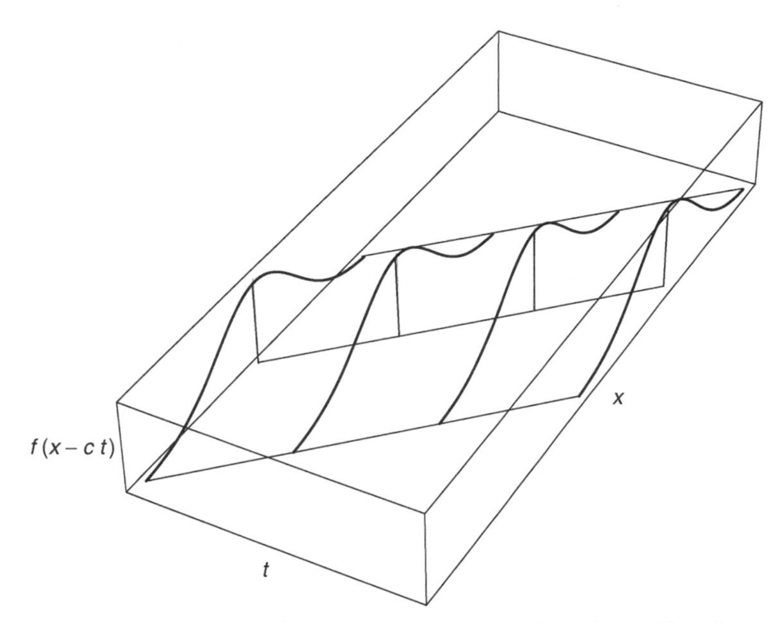

Fig. 7.1 The solution to Eq. (7.1.1) for $s = 0$ and $v = c$, at four times. Also shown are three characteristics.

This solution is simply a disturbance that moves at velocity c without varying from its initial shape $f_0(x)$. One can verify that this is the solution to Eq. (7.1.14) by direct substitution:

Cell 7.1

```
Simplify[D[f[x, t], t] + c D[f[x, t], x] == 0 /.f→
  Function[{x, t}, f0[x - ct]]]

True
```

This form of the solution is similar to d'Alembert's solution for the wave equation, Eq. (5.1.11). Here, however, the disturbance moves only in one direction, because (7.1.1) is first-order rather than second-order in time and space. The shape of the disturbance does not change, because f does not vary along each characteristic, and because the characteristics are parallel lines. Thus, the entire solution is transported without change along the characteristics, as depicted in Fig. 7.1.

Example 2: $s = ax^n$ and $v = bx$ Equations (7.1.5) and (7.1.6) are now

$$df/dt = ax^n,$$
$$dx/dt = bx.$$

The solution to $dx/dt = bx$ with initial condition $x = x_0$ provides us with the characteristics:

$$x(t, x_0) = x_0\, e^{bt}. \qquad (7.1.16)$$

These characteristics are shown in Fig. 7.2.

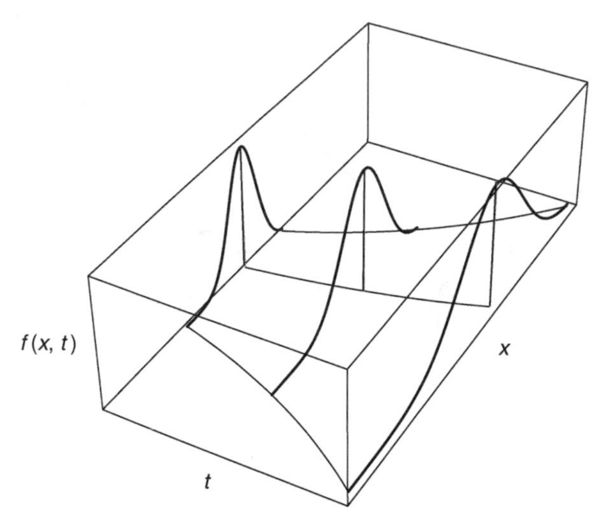

Fig. 7.2 The solution to Eq. (7.1.1) for $s = 0$ and $v = bx$, at three times. Also shown are three characteristics with initial conditions $x_0 = -2, 1, 3$.

The characteristics must be used when solving $df/dt = ax^n$. The solution to $df/dt = ax_0^n e^{nbt}$ is

$$f(t) = f(0) + \frac{ax_0^n}{nb}(e^{nbt} - 1), \tag{7.1.17}$$

where $f(0) = f_0(x_0)$ and $f(t) = f(x,t)|_{x=x(t,x_0)}$. Finally, we must invert Eq. (7.1.16) to obtain $x_0(x,t) = xe^{-bt}$. Substituting this result into Eq. (7.1.17) yields

$$f(x,t) = f_0(xe^{-bt}) + \frac{ax^n e^{-nbt}}{nb}(e^{nbt} - 1).$$

An example of the resulting evolution is shown in Cell 7.2 for $a = 0$, $b = 1$, and $f_0(x) = e^{-(x-1)^2}$. Although f is constant along each characteristic ($df/dt = 0$ when $a = 0$), the spreading of the characteristics results in a change in shape of the solution over time. The reader is invited to vary the parameters of the problem so as to investigate other solutions.

Example 3: Mixing Dye Next, we consider a generalization of Eq. (7.1.1) to two spatial dimensions:

$$\frac{\partial f}{\partial t} + v_x(x,y,t)\frac{\partial f}{\partial x} + v_y(x,y,t)\frac{\partial f}{\partial y} = 0. \tag{7.1.18}$$

Cell 7.2

```
f0[x_] = Exp[-(x - 1)^2];
f[x_, t_] = f0[x Exp[- t]];
Table[Plot[f[x, t], {x, -10, 20}, PlotRange→{0, 1}],
  {t, 0, 4, .2}];
```

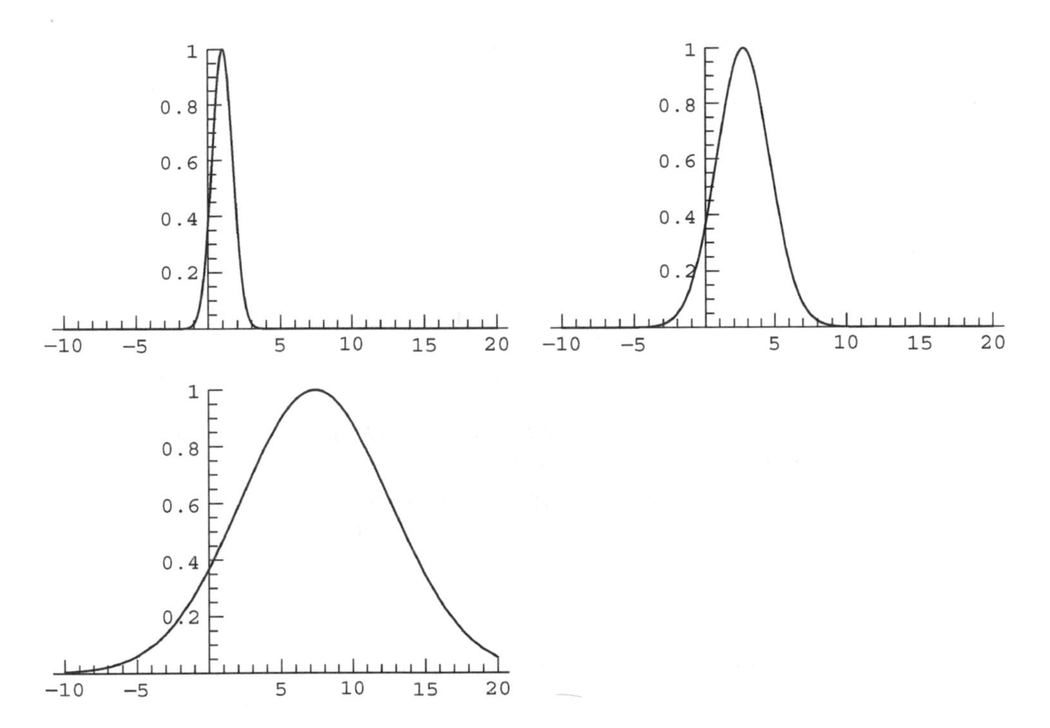

Here, the functions v_x and v_y provide the x and y components of the velocity. Equation (7.1.18) describes the behavior of dye placed in a fluid that moves with flow velocity $\mathbf{v} = (v_x, v_y)$. The density of the dye, $f(x, y, t)$, is carried along with the fluid, and so follows

$$\frac{df}{dt} = 0. \tag{7.1.19}$$

Equation (7.1.18) is merely an expression for this total time derivative, analogous to the one-dimensional total time derivative of Eq. (7.1.7). Equation (7.1.19) implies that $f =$ constant along the characteristics. This means that elements of the dye are carried along by the fluid without change in their density.

The characteristics for Eq. (7.1.18) are the coupled first-order ODEs,

$$\begin{aligned} dx/dt &= v_x(x, y, t), \\ dy/dt &= v_y(x, y, t), \end{aligned} \tag{7.1.20}$$

with initial conditions $x = x_0$ and $y = y_0$ at $t = 0$. For general functions v_x and v_y, these ODEs need not have an analytic solution and can even display chaotic behavior. That is, the dye can be mixed chaotically if the fluid flow is sufficiently complex. Thus, even though Eq. (7.1.18) is linear, it is not always possible to find a closed-form analytic solution.

Below, we show some characteristics for a fluid flow with

$$\begin{aligned} v_x &= y + x, \\ v_y &= -y - (x^3 - 1)(1 + \cos t). \end{aligned}$$

Note that $\nabla \cdot (v_x, v_y) = 0$, so this flow is incompressible [see Eq. (1.2.12)]. Therefore, the area of a patch of dye carried along by this particular flow will not change over time.

The characteristics for the flow can be computed using **NDSolve**, for given initial conditions (x_0, y_0):

Cell 7.3

```
c[x0_, y0_] := c[x0, y0] = {x[t], y[t]}/.
     NDSolve[{x'[t] == (y[t] + x[t]), y'[t] == -y[t]
      - (x[t]^3 - 1) (1 + Cos[t]),
        x[0] == x0, y[0] == y0}, {x, y}, {t, 0, 10}][[1]];
```

The characteristics can then be plotted in the (x, y) plane using **ParametricPlot**, as shown in Cell 7.4. Even this smooth flow produces a complex pattern of characteristics. By computing the Lyapunov exponent for these trajectories, one can show that in fact the trajectories are chaotic, exhibiting sensitive dependence on the initial conditions.

Cell 7.4

```
Table[ParametricPlot[Evaluate[c[x0, y0]], {t, 0, 10},
   DisplayFunction→Identity], {x0, -2, 2}, {y0, -2, 2}];

Show[%, DisplayFunction -> $DisplayFunction,
  PlotRange→{{-30, 30}, {-30, 30}}, AspectRatio→1];
```

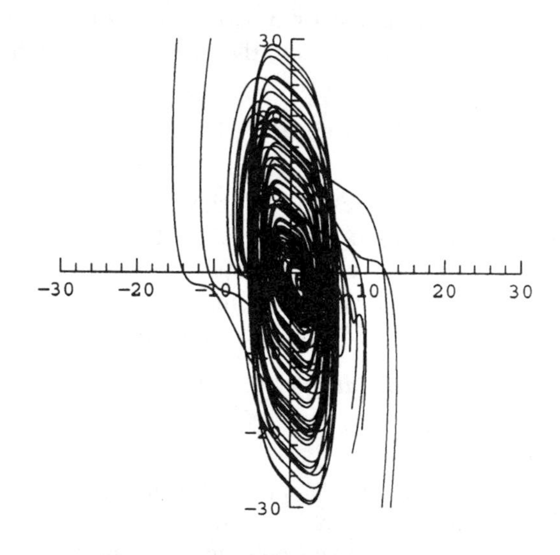

The behavior of a patch of dye in this flow can be followed by computing the positions of the edge of the patch as a function of time. We do so in Cell 7.5 for a patch that is initially circular, following 100 points from the patch edge. Although the patch area remains constant, the dye patch is stretched and folded by the flow in a very complex manner due to the chaotic nature of the characteristics.

Ultimately, the filaments of dye become so thin that they are indistinguishable, and the dye becomes thoroughly mixed through the background fluid.

Cell 7.5

```
patchedge[t1_] := Table[c[Sin[θ], Cos[θ]], {θ, 0, 2 Pi,
  .02 Pi}]/. t→t1;

Table[ListPlot[Evaluate[patchedge[t1]], PlotJoined→True,
   PlotRange→{{-10, 10}, {-10, 10}}, AspectRatio→1],
    {t1, 0, 5, .2}];
```

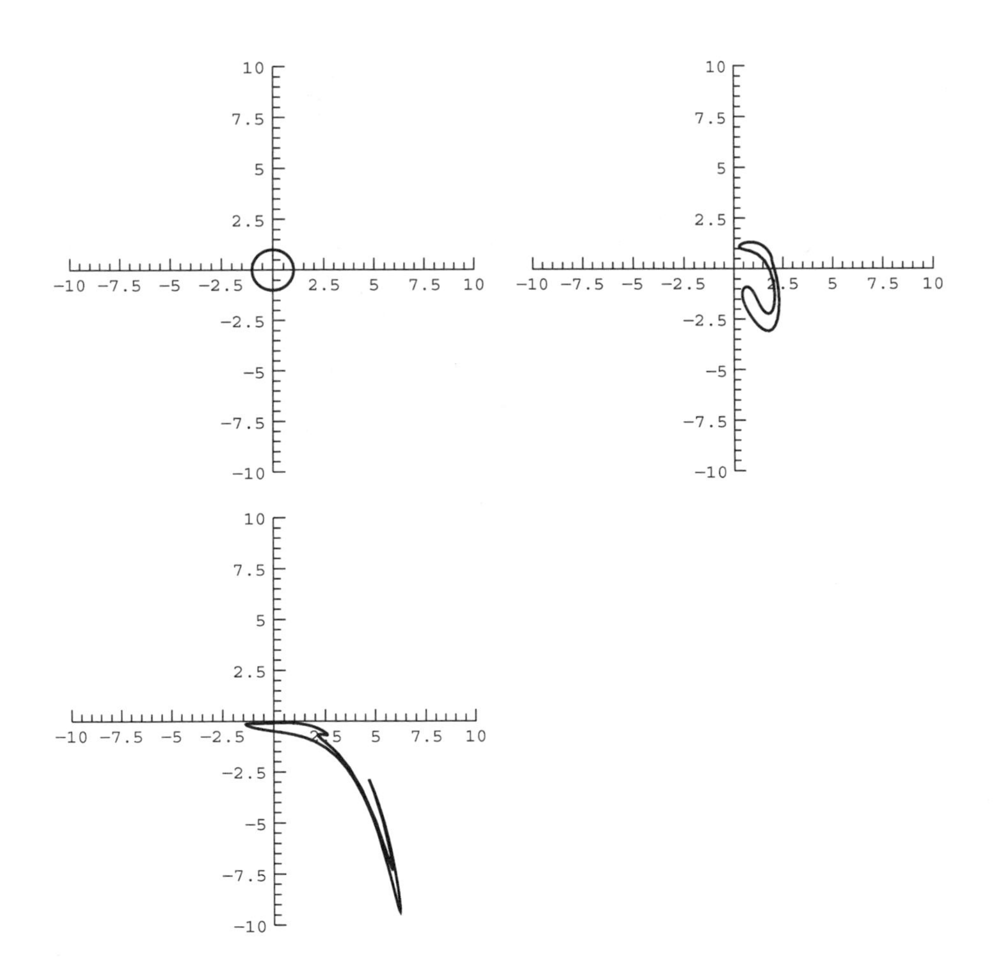

Example 4: The Collisionless Boltzmann Equation Equation (7.1.18) can also be interpreted in a different manner. Rather than thinking of f as the density of dye placed in a fluid, we can instead think of f as the density in phase space (x, p) of a one-dimensional gas of atoms of mass m, where $p = m\,dx/dt$ is the particle

momentum. Now $f = f(x, p, t)$, and Eq. (7.1.18) is written as

$$\frac{\partial f}{\partial t} + \frac{p}{m}\frac{\partial f}{\partial x} + F(x,p,t)\frac{\partial f}{\partial p} = 0. \tag{7.1.21}$$

The physical interpretation of the function $F(x, p, t)$ becomes clear when we write out the characteristic equations:

$$\frac{dx}{dt} = \frac{p}{m},$$

$$\frac{dp}{dt} = F(x,p,t).$$

These are Newton's equations for the motion of particles in a prescribed external force $F(x, p, t)$.

Equation (7.1.21) is called the collisionless Boltzmann (or Vlasov) equation. The number of particles in the element $dx\,dp$ is $f\,dx\,dp$. According to our understanding of convective derivatives, Eq. (7.1.21) is equivalent to the equation $df/dt = 0$ along the characteristics. This means that the particle density in phase space remains constant along the phase-space flow: particles are neither created nor destroyed, but are simply carried through phase space by their dynamics.

Interactions between the particles are neglected in this description of the gas, since the only force acting on the particles is the external force F. Therefore, Eq. (7.1.21) describes the evolution of an *ideal gas* of *noninteracting particles*.

For example, consider the behavior of a gas of particles falling in gravity. If the particles are confined in an evacuated container, the force law could be given by something like $F = -mg$, $x > 0$, and $F = -mg - kx$, $x < 0$, where the term kx describes an elastic collision of the atoms with the bottom of the container at $x = 0$. A few characteristics for this flow in the (x, p) plane are shown in Cell 7.6 for a range of initial conditions, taking $k = 20$, $m = g = 1$.

Cell 7.6

```
Clear[c];

m = g = 1; k = 20;
F[x_] := -m g/; x > 0
F[x_] := -m g - k x/; x≤0
c[x0_, p0_] :=
 c[x0, p0] = {x[t], p[t]}/.NDSolve[{x'[t] ==
   p[t] /m, p '[t] == F[x[t]],
      x[0] == x0, p[0] == p0}, {x, p}, {t, 0, 50},
        MaxSteps→4000][[1]]

Tabel[ParametricPlot[Evaluate[c[x0, p0]], {t, 0, 20},
   DisplayFunction→Identity], {x0, 0, 2}, {p0, -1, 1}];
Show[%, DisplayFunction→$DisplayFunction,
  AxesLabel→{"x", "p"}];
```

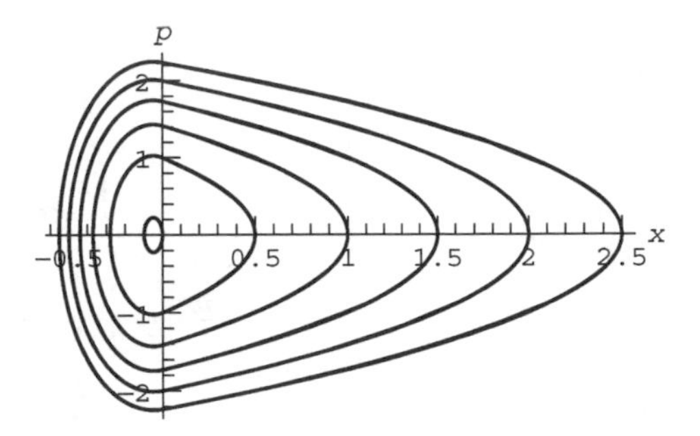

These characteristics show particles falling toward $x = 0$, then rapidly reversing their velocities and returning to their initial heights. There is no damping in the equations of motion, so each particle in the gas continues to bounce off the bottom of the container forever. Unlike the example of mixing dye presented earlier, the characteristics here are not chaotic. Each characteristic curve is parametrized by the constant energy E of particles on the curve. The particle motion is periodic, with a frequency ω_0 that depends on the energy:

$$\omega_0 = \omega_0(E). \tag{7.1.22}$$

Even though the motion is not chaotic, an initial distribution of particles will rapidly filament and mix. This kind of nonchaotic mixing is called *phase mixing*. The phase mixing is due to particles moving around their energy surfaces at different frequencies. The particles get *out of phase* with one another in their periodic motion. An initially localized patch of phase-space density becomes stretched and filamented due to the range of frequencies within the patch.

We can see this phase mixing by following $M = 500$ particles in their motion. We choose the particles with random initial velocities and positions in the range $0-\frac{1}{2}$, and display their subsequent trajectories in Cell 7.7.

A series of filamentary structures form and then are stretched out to such a fine scale that they can no longer be discerned. Although f continues to evolve on finer and finer scales as the filaments are stretched and wrapped up, on larger scales f becomes time-independent. That is, the number of particles in any finite phase-space element $\Delta x \, \Delta p$, namely $f \Delta x \, \Delta p$, eventually becomes independent of time.

Cell 7.7

```
M = 500;

particlepos[t1_] = Table[c[0.5 Random[], 0.5 Random[]],
  {M}] /. t → t1;

Table[ListPlot[Evaluate[particlepos[t1]], PlotRange →
  {{-1, 1}, {-1, 1}},
   AspectRatio → 1, AxesLabel → {"x", "p"}], {t1, 0, 50, .4}];
```

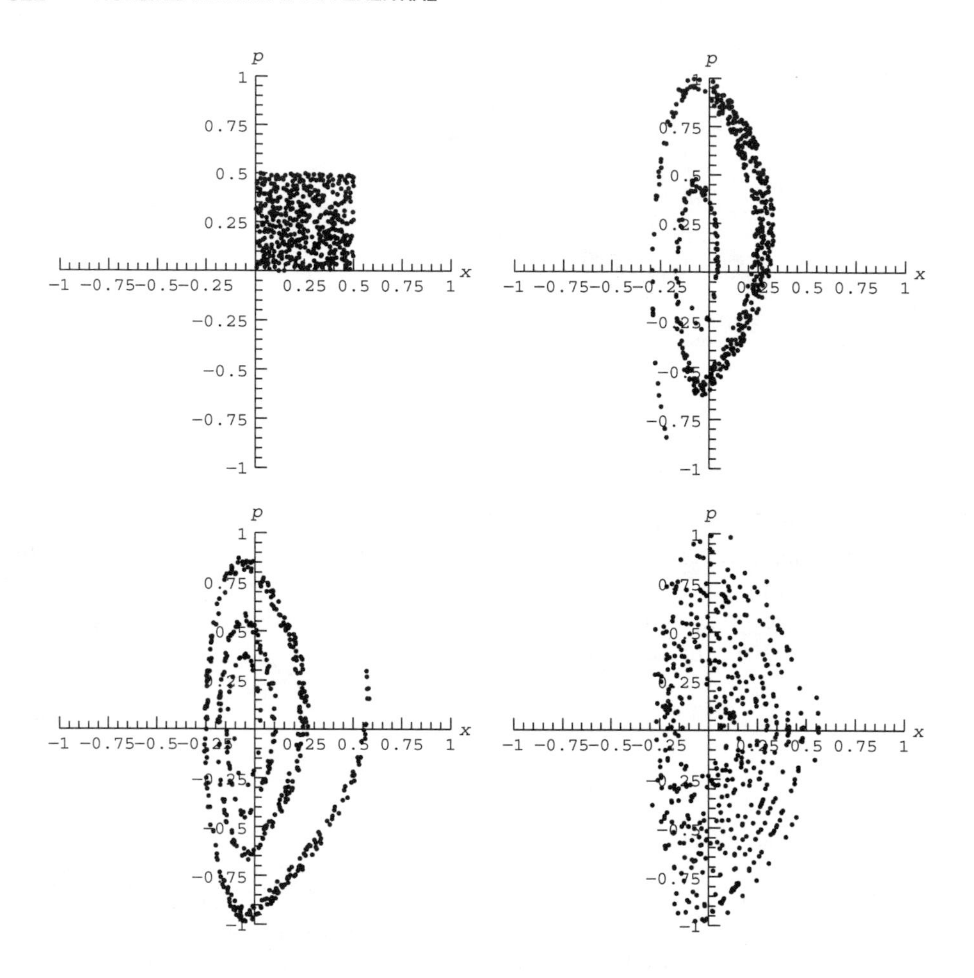

The time-independent solution $f_{\text{eq}}(x, p)$ for f can be found by dropping the partial time derivative in Eq. (7.1.21):

$$\frac{p}{m}\frac{\partial f_{\text{eq}}}{\partial x} - \frac{\partial V}{\partial x}\frac{\partial f_{\text{eq}}}{\partial p} = 0. \tag{7.1.23}$$

Here we have used the fact that the force $F(x)$ in this example is derivable from a potential $V(x)$ via $F = -\partial V/\partial x$. The solution to Eq. (7.1.23) can be found by making a change of variables from (x, p) to (E, p), where $E = p^2/(2m) + V(x)$ is the energy. In these variables, $f_{\text{eq}} = f_{\text{eq}}(E, p)$, and it is left as an exercise in partial derivatives to show that Eq. (7.1.23) becomes

$$-\frac{\partial V}{\partial x}\frac{\partial f_{\text{eq}}}{\partial p}\bigg|_E = 0. \tag{7.1.24}$$

This equation can be integrated, yielding

$$f_{\text{eq}} = f_{\text{eq}}(E). \tag{7.1.25}$$

The equilibrium is a function only of the energy $E = p^2/(2m) + V(x)$. This can be easily understood intuitively: the phase mixing of the particle distribution eventually spreads particles out evenly along each energy surface, so that the particle density becomes constant on each surface, but may vary from surface to surface.

The functional form of $f_{\text{eq}}(E)$ is determined by the initial conditions. For an initial distribution $f_0(x, p)$, energy conservation implies that the number of particles between the energy surfaces E and $E + dE$ remains constant over time. This number can be written either as $f_{\text{eq}}(E)\,dE$ or as $\int dx\,dp\,f_0(x, p)$, where the area integral is over the infinitesimal area between energy surfaces E and $E + dE$ $[E \le H(x, p) \le E + dE]$, and where $H(x, p) = p^2/(2m) + V(x)$ is the Hamiltonian. Therefore, the equilibrium is determined by the equation

$$f_{\text{eq}}(E)\,dE = \int_{E \le H(x,p) \le E + dE} dx\,dp\,f_0(x, p). \qquad (7.1.26)$$

Even though the area integral is an infinitesimal, it is nontrivial to evaluate because the distance between energy surfaces varies from place to place along the surfaces. (The integral can be evaluated using action-angle variables, but we will not involve ourselves with the canonical transformations required to define these variables.) On the other hand, the integral can be performed numerically if we convert dE to a finite quantity ΔE.

We first define the initial particle density. Since there are $M = 500$ particles spread uniformly over a region $0 < x < \frac{1}{2}$ and $0 < p < \frac{1}{2}$, we define the initial distribution:

Cell 7.8

```
f0[x_, p_] := 4 M /; 0 < x < 1/2 && 0 < p < 1/2
f0[x_, p_] := 0 /; Not[0 < x < 1/2 && 0 < p < 1/2]
```

With this definition, $\int dx\,dp\,f_0(x, p) = M$. Next, we define the potential $V(x)$ for this problem, and the Hamiltonian function $H(x, p)$:

Cell 7.9

```
m = g = 1; k = 20;
V[x_] = m g x + k x^2/2 UnitStep[-x];
H[x_, p_] = p^2/(2m) + V[x];
```

Finally, we integrate over the area between two constant-energy surfaces separated by ΔE, using `Boole` notation:

Cell 7.10

```
<<Calculus`;

feq[e_] :=
 NIntegrate[Boole[e < H[x, p] <e + ΔE] f0[x, p],
   {x, 0, 1/2}, {p, 0, 1/2}]/ΔE

ΔE = 0.001;
```

A plot of the equilibrium energy distribution is shown in Cell 7.11. (Using **ListPlot** rather than **Plot** speeds up the evaluation considerably by reducing the number of function evaluations.)

Cell 7.11

```
ListPlot[Table[{e, feq[e]}, {e, -.2, 1, .025}],
  PlotJoined→True, AxesLabel→{"E", "feq(E)"}];
```

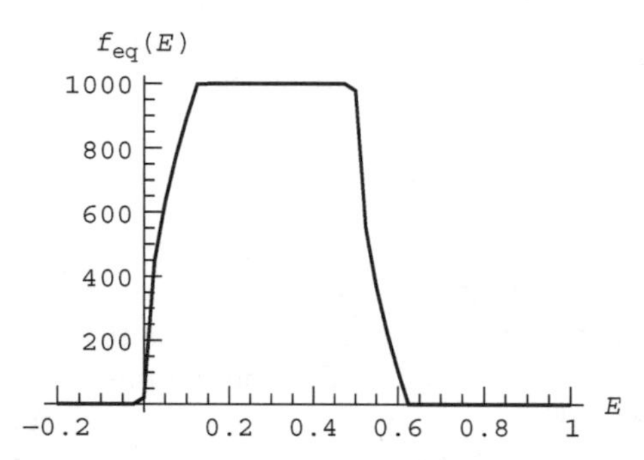

The distribution goes to zero for $E < 0$ and $E \gtrsim 0.6$ due to our choice of initial conditions, with all particles concentrated in this energy range. Also, it can be easily verified that

$$\int_{-\infty}^{\infty} f_{eq}(E)\, dE = M.$$

We must now translate this distribution over energies back to a distribution in the (x, p) plane. Since $f_{eq}(E)\,dE$ is the number of particles between energy surfaces E and $E + dE$, the distribution $f_{eq}(x, p)$ is related to $f_{eq}(E)$ by the same formula as we used in Eq. (7.1.26):

$$f_{eq}(E)\, dE = \int_{E \le H(x,p) \le E+dE} dx\, dp\, f_{eq}(x, p).$$

In fact, since $f_{eq}(x, p)$ is constant along the energy surfaces, we can take it out of the integral and write

$$f_{eq}(E)\, dE = dA(E)\, f_{eq}(x, p),$$

where

$$dA(E) = \int_{E \le H(x,p) \le E+dE} dx\, dp$$

is the area between adjacent energy surfaces. Thus,

$$f_{eq}(x, p) = \frac{f_{eq}(E)}{\frac{dA}{dE}(E)}, \tag{7.1.27}$$

where $E = H(x, p)$. [Using action-angle variables, one can show that $dA/dE = 2\pi/\omega_0(E)$, where $\omega_0(E)$ is the orbit frequency of particles with energy E.]

In order to plot $f_{eq}(x, p)$ we must first work out dA/dE. We do so numerically by first evaluating $A(E)$, the phase space area within energy surface E, and then taking a derivative:

Cell 7.12

```
A[e_] := NIntegrate[Boole[H[x, p] < e],
  {x, -Infinity, Infinity}, {p, -Infinity, Infinity}]
```

In mechanics, this phase-space area is called the *action* for a given energy surface. To take the derivative of this numerical function, it is best to first interpolate the function, and then take a derivative of the interpolating function:

Cell 7.13

```
areadata = Table[{e, A[e]}, {e, 0, 2, .1}];

Ainterp = Interpolation[areadata];

dAdE[e_] := D[Ainterp[e1], e1]/. e1 → e
```

This derivative of the action is plotted in Cell 7.14.

Cell 7.14

```
Plot[dAdE[e], {e, 0, 1}, PlotLabel → "dA/dE",
  AxesLabel → {"E", ""}];
```

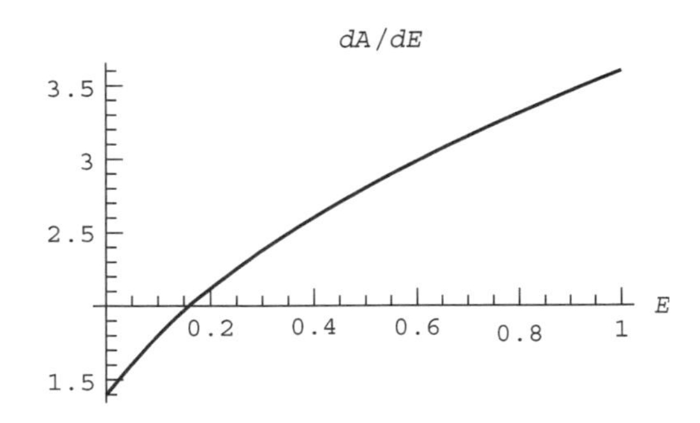

To speed up the evaluation it is best to also create an interpolation function for $f_{eq}(E)$:

Cell 7.15

```
feqinterp =
  Interpolation[Table[{e, feq[e]}, {e, 0, 1, .025}],
   InterpolationOrder → 1];
```

We use `InterpolationOrder` of only 1 because $f_{eq}(E)$ has a discontinuous first derivative. We then define the function $f_{eq}(x, p)$ according to Eq. (7.1.27):

Cell 7.16

```
feq[x_, p_] := feqinterp[H[x, p]]/dAdE[H[x, p]]/;
  0 < H[x, p] < 1;

feq[x_, p_] := 0/; Not[0 < H[x, p] < 1]
```

The use of the conditional statements allows us to extend the definition to energies beyond the range of the interpolation function. We plot the equilibrium in Cell 7.17. Note the hole in the distribution near the origin, which is also clearly visible in the previous animation of the particle positions, Cell 7.7.

Cell 7.17

```
Plot3D[feq[x, p], {x, -1, 1}, {p, -1, 1}, PlotRange→All,
  PlotPoints→40, AxesLabel→{"x", "p", ""},
   PlotLabel -> "feq(x,p)"];
```

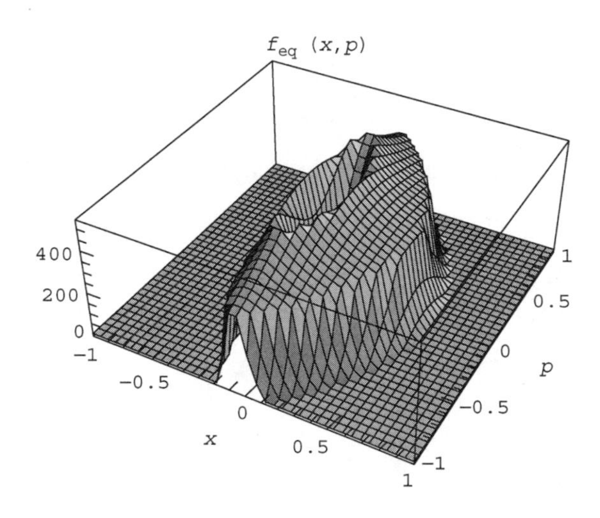

It is interesting that this gas comes to an equilibrium at all, because we have not included interactions between the particles that would be needed to drive this system to a *thermal* equilibrium state. The equilibrium achieved here is a *partial* equilibrium due to phase-mixing alone. Thermal equilibrium corresponds to a particular form for $f_{\text{eq}}(E)$: $f_{\text{eq}}(E) = Ce^{-E/k_B T}$, where T is the temperature, C is a constant, and k_B is Boltzmann's constant. We will have more to say concerning thermal equilibrium in Chapter 8.

Another view of the equilibrium process considers only the particle positions. In this view, we extract the positions, neglecting momenta.

Cell 7.18

```
pos[t1_] := (a = particlepos[t1]; Table[a[[n, 1]], {n, 1, M}])
```

Then we plot the particle positions in Cell 7.19 (only the initial and final states are displayed in the printed version).

Cell 7.19

```
Table[ListPlot[pos[t1], PlotRange→{-1, 1}],
  {t1, 0, 50, .4}];
```

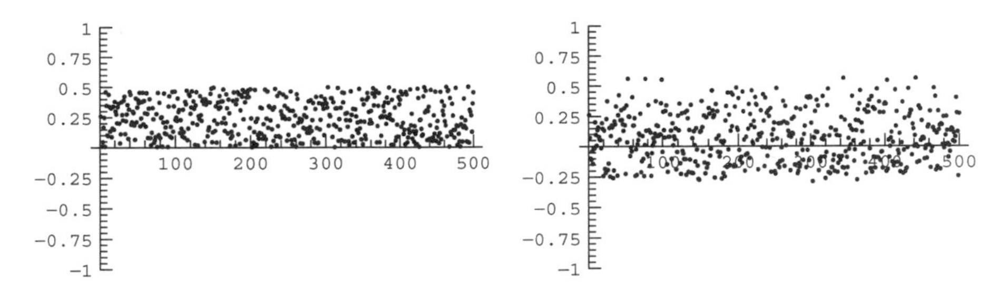

The phase mixing should be quite clear in this animation. After a few bounces, the gas approaches equilibrium. The evolution toward equilibrium can be seen more directly by evaluating the particle density $n(x,t)$. It is the integral over the momenta of f:

$$n(x,t) = \int_{-\infty}^{\infty} f(x,p,t)\, dp. \tag{7.1.28}$$

We can use the data generated above to approximately determine $n(x,t)$ for this evolution. The number of particles in the range $x - x + dx$ equals $n(x,t)\, dx$. Therefore, we can determine $n(x,t)$ by counting particles that fall in small boxes of size Δx. If, at time t, the number of particles in the range $x - x + \Delta x$ equals $H(x,t)$, then

$$n(x,t) = H(x,t)/\Delta x.$$

This binning can be accomplished using the **Histogram** plotting function, available in the graphics add-on packages:

Cell 7.20

```
<<Graphics`
```

Histogram takes as its primary argument a list of data values that it bins into given categories. The **HistogramCategories** option is used to determine the location and size of the bins. Here the bins are all of size 0.1, and run from -1 to 1. We use the **Show** function in order to actually display the histograms, because this allows us more flexibility in choosing the plot range and the axis origin (see Cell 7.21). (Only the first histogram is displayed in the printed version.)

Cell 7.21

```
H[t_] := (plt = Histogram[pos[t],
    HistogramCategories→Table[x, {x, -1, 1, .1}],
     DisplayFunction→Identity];
  Show[plt, PlotRange→{{-1, 1}, {0, 300}},
   DisplayFunction→$DisplayFunction,
  AxesLabel→{"x", "H[x, t]"}, AxesOrigin→{0, 0}]);

hist = Table[H[t], {t, 0, 50, .4}];
```

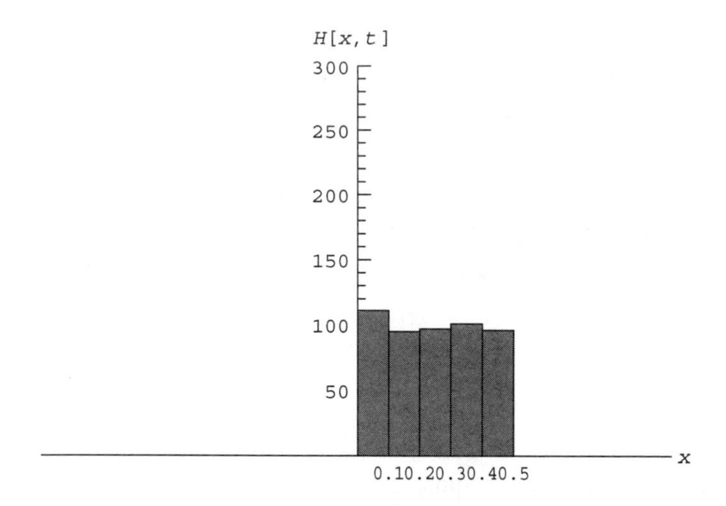

This method of solving the Boltzmann equation, by following a large collection of particles, is called a *particle simulation*. The fluctuations that we see in the density at late times are due to the finite number of particles used in the simulation. We will discuss other particle simulation methods in Sec. 7.3 and in Chapter 8.

We can also determine the expected equilibrium form of the density using Eq. (7.1.26) and (7.1.27):

Cell 7.22

```
neq[x_] := NIntegrate[feq[x, p], {p, -2, 2}, MaxRecursion→25]
```

In Fig. 7.3 we plot the density multiplied by the bin size Δx, so that we can compare it with the histogram function $H(x, t)$ evaluated at the last timestep. The theoretical equilibrium density matches the late-time density found in the particle

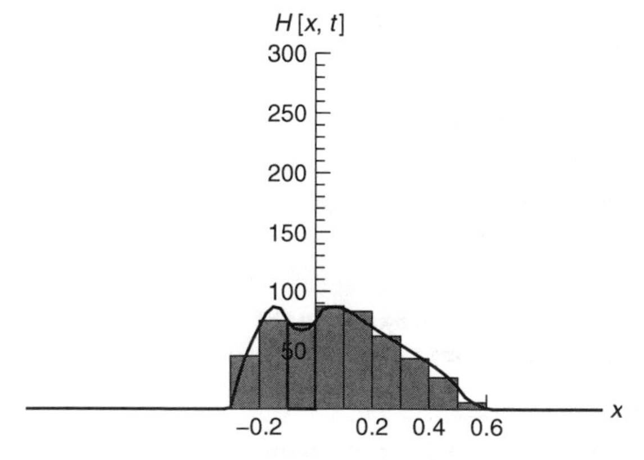

Fig. 7.3 Equilibrium density of particles vs. height x: theory (line) and simulation (histogram).

simulation. Note that this density, with a dip near the origin, is far from what one might expect intuitively in an equilibrium gas. This is because we are used to thinking of thermal equilibrium states. The equilibrium shown above is a partial equilibrium caused by collisionless phase mixing of the initial condition. It is what one would see before collisions between particles have a chance of acting. In many systems, the collisions act so rapidly that partial equilibrium is not observed. However, in other systems, collisions occur rarely so that partial equilibrium is easily seen. For instance, a gas can enter this collisionless regime if the density is sufficiently low so that atom–atom collisions can be ignored over the time range of interest.

7.1.3 Nonlinear Waves

A First-Order Nonlinear Wave Equation We now consider a nonlinear PDE that describes the evolution of shallow-water waves with small but finite amplitude:

$$\frac{\partial z}{\partial t} + c_0\left(1 + \frac{3}{2}\frac{z}{h}\right)\frac{\partial z}{\partial x} = 0, \tag{7.1.29}$$

where $z(x,t)$ is the wave height, $c_0 = \sqrt{gh}$ is the linear wave speed, g is the acceleration of gravity, and h is the equilibrium depth of the water [see Eq. (3.1.79)]. The term $3zc_0/2h$ is the lowest-order nonlinear correction to the wave speed due to finite wave height [see Whitham (1999) for a derivation of this correction]. Simply put, a wave peak increases the water depth, resulting in a local increase in wave speed. Without this term, solutions merely propagate to the right with speed c_0, describing linear shallow water waves [see Eq. (7.1.14)]. Equation (7.1.29) neglects dispersion and viscous effects, and allows propagation in only one direction. Nevertheless, its solutions have much to tell us concerning the behavior of real finite-amplitude water waves.

We will first put the equation in a standard format. This is accomplished by making a Galilean transformation to a moving frame, moving at speed c_0, via the change of variables $\bar{x} = x - c_0 t$. In this frame Eq. (7.1.29) becomes

$$\frac{\partial z}{\partial t} + \frac{3}{2}c_0\frac{z}{h}\frac{\partial z}{\partial \bar{x}} = 0.$$

Next, we scale the wave height z, defining $f(\bar{x},t) = 3c_0 z/(2h)$. Using these scaled variables, and dropping the overbar on x for notational convenience, we arrive at

$$\frac{\partial f}{\partial t} + f\frac{\partial f}{\partial x} = 0. \tag{7.1.30}$$

The simple form of the nonlinearity in Eq. (7.1.30) appears in many other physical contexts, making the equation a paradigmatic nonlinear PDE. It is sometimes referred to as Burgers' equation without diffusion. The version with diffusion is discussed in Sec. 7.2.4.

Solution Using the method of characteristics, we can obtain a complete analytic solution to Eq. (7.1.30). (This is one of a mere handful of nonlinear PDEs where such a complete analytic solution is possible.)

The characteristics for Eq. (7.1.30) are solutions to the equation

$$dx/dt = f(x,t). \tag{7.1.31}$$

Thus, the velocity along the characteristic *depends on the amplitude of the solution along the characteristic*. However, this is not as complicated as it sounds, because Eq. (7.1.30) implies that

$$df/dt = 0 \qquad \text{along the characteristics}. \tag{7.1.32}$$

Thus, $f =$ constant along each characteristic, retaining its initial value:

$$f(x(t,x_0),t) = f_0(x_0), \tag{7.1.33}$$

where $f_0(x) = f(x,0)$. Therefore, we can integrate Eq. (7.1.31) to obtain an equation for the characteristics:

$$x = x_0 + f_0(x_0)t. \tag{7.1.34}$$

Five of these nonlinear characteristics are depicted in Fig. 7.4. They are straight lines in the (x,t) plane, with slope that depends on $f_0(x_0)$.

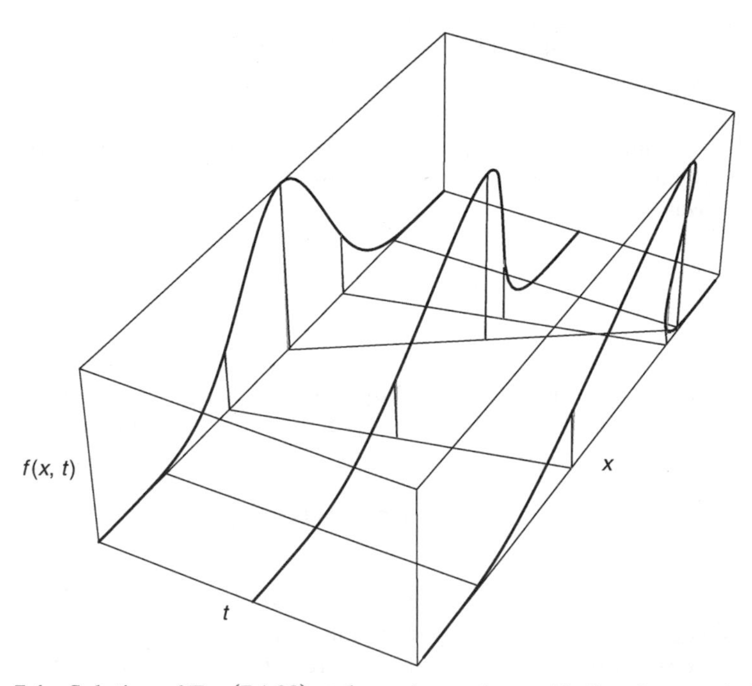

Fig. 7.4 Solution of Eq. (7.1.30) at three times, along with five characteristics.

In principle, one could invert Eq. (7.1.34) to obtain $x_0 = x_0(x,t)$. Then, using this solution in Eq. (7.1.33), one would obtain the solution

$$f(x,t) = f_0(x_0(x,t)).$$

However, as the required inversion usually cannot be accomplished analytically, it is often easier to simply determine the solution for $f(x,t)$ numerically using Eqs. (7.1.33) and (7.1.34). This can be done as follows. First, we choose an initial condition:

Cell 7.23

```
f0[x_] = Exp[-x^2];
```

Next, define a table of values taken from the initial condition, and call it **fval[0]**:

Cell 7.24

```
fval[0] = Table[{x0, f0[x0]}, {x0, -3, 3, .1}];
```

This table is a discrete representation of the continuous initial condition. It can be plotted using a **ListPlot**, or an interpolation could be taken to reproduce $f_0(x)$.

Next, we use Eqs. (7.1.27) and (7.1.28) to follow the evolution of each data point in the table. The positions of the points satisfy Eq. (7.1.28), and the values of f remain fixed in time:

Cell 7.25

```
x[t_, x0_] = x0 + f0[x0] t;
f[t_, x0_] = f0[x0];

fval[t_] := Table[{x[t, x0], f[t, x0]}, {x0, -3, 3, .1}]
```

The table **fval[t]** provides us with a discretized solution to the problem at all times t. In Cell 7.26 we plot this solution. The bottom of the pulse does not move, because the wave amplitude is zero there. (Recall that this pulse is a wave as seen in a frame that moves with the linear wave speed. Linear waves would not evolve at all in this moving frame.) The peak of the pulse moves the fastest because wave speed increases with increasing amplitude. Eventually, the peak overtakes the rest of the pulse. This solution exhibits *wavebreaking*: the solution becomes double-valued at around $t = 1$. This phenomenon is familiar to anyone who has enjoyed watching waves curl and collapse on the beach. Our nonlinear wave equation cannot describe the collapse of the wave, which occurs because the wave crest is no longer supported by the fluid below it. However, it has captured the essential behavior of a breaking wave up to the point where the solution becomes double-valued.

Cell 7.26

```
Table[ListPlot[fval[t], PlotJoined→True,
  PlotRange→{{-3, 4}, {0, 1}},
   PlotLabel -> "t=" <>ToString[t]], {t, 0, 2.4, .2}];
```

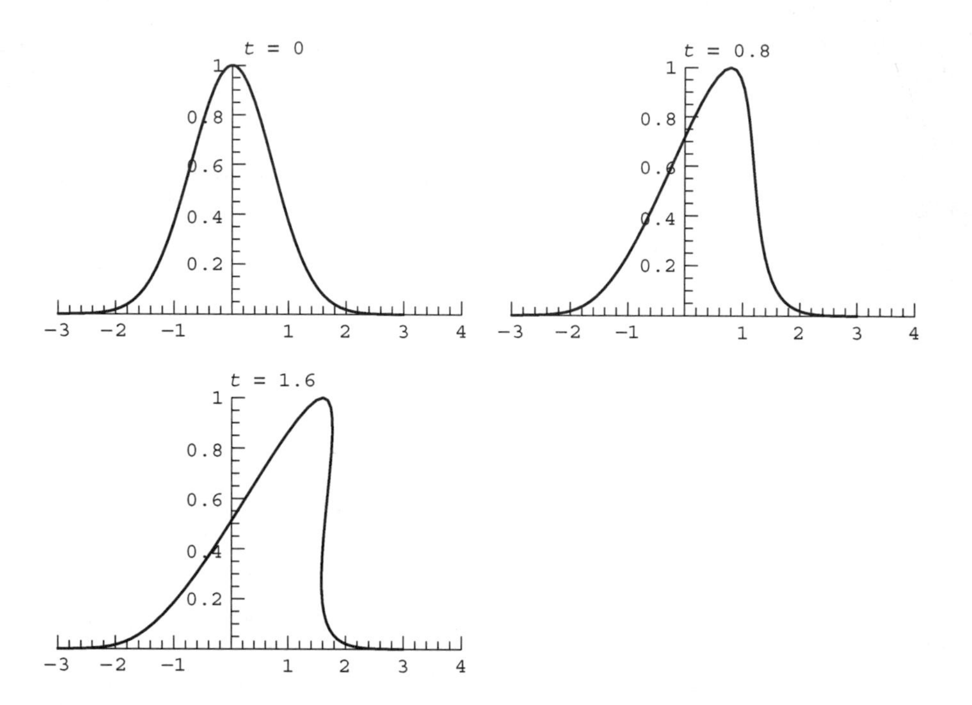

Wavebreaking Time The wavebreaking can also be understood by examining the characteristics. One can see from Eq. (7.1.34) or Fig. 7.4 that there are places where the characteristics cross. At such points values of f that were initially separated in x come to the same position, implying that the solution has become double-valued in x.

Although different characteristics cross at different places and times, it is evident that there is a time t_0 when the crossing of characteristics first occurs. At this instant, the *wavebreaking time*, the solution goes from being a single-valued function of x to being double-valued.

In order to determine the wavebreaking time, let us consider two characteristics that arise from initial conditions x_{01} and $x_{02} = x_{01} + \Delta x$. These characteristics cross when $x(t, x_{01}) = x(t, x_{02})$. Applying Eq. (7.1.34) then yields a crossing time t given by

$$x_{01} + f_0(x_{01})t = x_{02} + f_0(x_{02})t.$$

Solving for t yields

$$t = -\frac{x_{01} - x_{02}}{f(x_{01}) - f(x_{02})} = \frac{\Delta x}{f(x_{01}) - f(x_{01} + \Delta x)}. \tag{7.1.35}$$

Assuming for the moment that Δx is small, we Taylor-expand $f(x_{01} + \Delta x)$ to first order in Δx and so obtain

$$t = -\frac{1}{df_0/dx_{01}}. \tag{7.1.36}$$

To find the wavebreaking time t_0, we find the minimum (positive) value for t:

$$t_0 = \min\left(-\frac{1}{df_0/dx_{01}} \right). \tag{7.1.37}$$

That is, we look for the maximum value of $-df_0/dx_{01}$. This maximum will occur at the inflection point in f_0, where

$$d^2 f_0/dx_{01}^2 = 0 \tag{7.1.38}$$

and

$$d^3 f_0/dx_{01}^3 > 0. \tag{7.1.39}$$

For our Gaussian initial condition, there is only one inflection point with $-df_0/dx_{01} > 0$ and $d^2 f_0/dx_{01}^2 = 0$. It may be found easily using *Mathematica*:

Cell 7.27

```
Factor[D[f0[x0], {x0, 2}]]

2 e^(-x0^2) (-1 + 2 x0^2)
```

Thus, the inflection point happens at $x_{01} = 1/\sqrt{2}$. At this point $-df_0/dx_{01}$ is given by

Cell 7.28

```
-f0'[1/Sqrt[2]]
```

$$\sqrt{\frac{2}{e}}$$

Therefore, the wavebreaking time for our Gaussian initial condition is $t_0 = \sqrt{e/2} = 0.858\ldots$. This agrees with the results of the previous animation.

We found the wavebreaking time by assuming that nearby characteristics were the first to cross, so that a first-order Taylor expansion of Eq. (7.1.35) was allowed. We must now show that the minimum wavebreaking time really does occur for nearby characteristics. To do so, we keep up to third-order terms in Δx in Eq. (7.1.35), and vary with respect to Δx. Equation (7.1.36) is then replaced by

$$t = -\frac{1}{\dfrac{df_0}{dx_{01}} + \dfrac{\Delta x}{2}\dfrac{d^2 f_0}{dx_{01}^2} + \dfrac{\Delta x^2}{6}\dfrac{d^3 f_0}{dx_{01}^3}}. \tag{7.1.40}$$

However, at the inflection point, $d^2f_0/dx_{01}^2 = 0$ and this expression becomes

$$t = \frac{1}{-\dfrac{df_0}{dx_{01}} - \dfrac{\Delta x^2}{6}\dfrac{d^3f_0}{dx_{01}^3}}. \tag{7.1.41}$$

Since both $-df_0/dx_{01}$ and d^3f_0/dx_{01}^3 are greater than zero, the minimum value of t occurs at $\Delta x \to 0$, proving that the minimum wavebreaking time happens for closely separated characteristics, and is given by Eq. (7.1.37).

EXERCISES FOR SEC. 7.1

(1) For the following PDEs,

(a) find the characteristics, and plot them for $x_0 = -1$, 0, and 1 and $0 < t < 2$;

(b) solve the PDEs for $f(x,t)$:

(i) $\dfrac{\partial f}{\partial t} + e^{-x}\dfrac{\partial f}{\partial x} = 0$, $f(x,0) = e^{-x^2}$. Animate the solution for $0 < t < 2$ on $-3 < x < 3$.

(ii) $\dfrac{\partial f}{\partial t} + 6\dfrac{\partial f}{\partial x} = -f$, $f(x,0) = \sin x$. Animate for $0 < t < 2$ on $-2\pi < x < 2\pi$.

(iii) $\dfrac{\partial f}{\partial t} - x\dfrac{\partial f}{\partial x} = \log(x^2)$, $f(x,0) = 0$. Animate for $0 < t < 2$ on $-5 < x < 5$.

(2) Prove Eq. (7.1.24).

(3) A radio wave packet with central frequency ω_0 travels in the $+x$ direction through a nonuniform plasma. The dispersion relation for the waves is $\omega^2 = \alpha^2 x^2 + c^2 k^2$, where c is the speed of light and α is a constant.

(a) Using the equations of geometrical optics, Eqs. (5.2.48)–(5.2.75), show that the local wavenumber satisfies $c^2k^2(x) = \omega_0^2 - \alpha^2 x^2$.

(b) Find the characteristics for the energy equation (5.2.59). In particular, show that $x(t) = (\omega_0/\alpha)\sin[(c\alpha/\omega_0)t + a]$, where a is a constant of integration.

(c) Use the method of characteristics to solve the energy equation. Show that for a given initial energy density $\epsilon_0(x)$, $\epsilon(x,t)$ is given by $\epsilon(x,t) = [\epsilon_0(x_0)\cos(a)]/\cos[(c\alpha/\omega_0)t + a]$, where $a = \sin^{-1}(\alpha x/\omega_0) - c\alpha t/\omega_0$, and where $x_0 = (\omega_0 \sin a)/\alpha$ is the initial condition. Animate this result on $-5 < x < 10$ for $0 < t < 1.5$, assuming that $\epsilon_0(x_0) = e^{-x_0^2}$, $c = \alpha = 1$, and $\omega_0 = 10$.

(4) The dispersion relation for the waves in a time-dependent inhomogeneous 2D system is $\omega = xk_x + tyk_y$. A wave packet initially has wave action $N(x,y,t=0) = e^{-2(x^2+y^2)}$. Solve Eq. (5.3.23) for the wave action $N(x,y,t)$, using the method of characteristics. Animate the result using a series of `Plot3D` commands in the time interval $0 < t < 2$.

(5) In the Galerkin method, it is often rather difficult to find basis functions that satisfy von Neumann boundary conditions in two- or three-dimensional domains. The method of characteristics can be used to determine the basis functions in this case. Take, for example, a two-dimensional domain with a surface S determined by the equation $z(x,y)=0$. We wish to find basis functions v_{ij} that satisfy $\mathbf{n}\cdot\nabla v_{ij}(x,y)|_S=0$, where $\mathbf{n}$ is any normal vector to the surface; for example, $\mathbf{n}=\nabla z(x,y)$. Consider the following first-order PDE:

$$\mathbf{n}\cdot\nabla v_{ij}(x,y)=z(x,y)x^i y^j. \tag{7.1.42}$$

If we solve this PDE inside the domain, then by construction we will have found functions v_{ij} that satisfy $\mathbf{n}\cdot\nabla v_{ij}(x,y)|_S=0$, because $z(x,y)=0$ on S. For instance, consider the elliptical domain specified by $z(x,y)=x^2/a^2+y^2/b^2-1$. A normal vector to this domain is $\mathbf{n}=(1,\ ya^2/xb^2)$. Using this normal vector in Eq. (7.1.42), and applying the method of characteristics, show that

$$v_{ij}(x,y)=x^i y^j\left[(b^2 i+a^2 j)\left(\frac{x^2}{a^2(2b^2+b^2 i+a^2 j)}+\frac{y^2}{b^2(2a^2+b^2 i+a^2 j)}\right)-1\right],\qquad i,j\geq 0.$$

(6) **(a)** Particles in an ideal gas are confined in a harmonic well of the form $V(x)=m\omega_0^2 x^2/2$. Initially, the particles are distributed in phase space according to $f(x,p,t=0)=N\delta(p)/a$, $|x|<a$; $f=0$ otherwise. Find the phase-mixed equilibrium solution $f_{\text{eq}}(x,p)$ for this initial condition.

(b) Simulate this problem for $0<t<10\pi/\omega_0$ for $N=100$ particles. Show from the movie of the particle phase-space orbits that phase mixing does not occur, and explain why, in a few words.

(7) Find the wavebreaking time t_0 for the following initial condition in Eq. (7.1.30): $f(x,0)=A\sin kx$. Animate the solution for $k=A=1$ on $-2\pi<x<2\pi$ for $0<t<2t_0$.

(8) Find the solution to the following nonlinear first-order partial differential equation using the method of characteristics on $-\infty<x<\infty$:

$$\frac{\partial f(x,t)}{\partial t}+f^2(x,t)\frac{\partial f(x,t)}{\partial x}=0 \qquad \left[\text{initial condition } f(x,0)=\sin 2x\right].$$

Find the time to the formation of a singularity in the solution (the wavebreaking time t_0). Make an animation of the evolution of $f(x,t)$ for $0<x<\pi$ and for $0<t<2t_0$.

(9) **(a)** Find the solution to the following nonlinear PDE using the method of characteristics on $-\infty<x<\infty$:

$$\frac{\partial u(x,t)}{\partial t}+u(x,t)\frac{\partial u(x,t)}{\partial x}=-u \qquad \left[\text{initial condition } u(x,0)=a\exp(-2x^2)\right].$$

(b) For this initial condition, with $a=1$, find the time t_0 to the formation of a singularity in the solution.

(c) Make an animation of the evolution of $u(x,t)$ for $-5<x<5$ and for $0<t<2t_0$.

(d) Find the value of a below which no singularity forms, because the solution decays away before wavebreaking can occur. (Answer: $a=0.824631\dots$.)

(10) Consider the following simple model for the flow of traffic on a freeway. The cars are modeled as a fluid of density $f(x,t)$. Cars move to the right with a speed that depends on this density. At low densities, the cars move at the speed limit c. But at higher traffic density, their speed is reduced. In our simple model, we assume that the speed depends on density as $c-f$. According to the Boltzmann equation, the density will then obey

$$\frac{\partial f}{\partial t}+(c-f)\frac{\partial f}{\partial x}=s, \tag{7.1.43}$$

where s is a source of cars (such as an on ramp). Show that for $s=0$ and for any nonuniform initial condition with regions of positive slope (i.e., traffic density increasing in the direction of flow) the slope increases until the solution becomes double-valued, after which it is no longer valid. These regions of increasing slope propagate along the freeway (the phenomenon of sudden traffic slowdowns for no apparent reason, familiar to any freeway driver). Illustrate this by means of an animation, taking $c=1$, $s=0$, and $f(x,0)=0.2+0.5\sin^2 x$. See the exercises in Sec. 7.2 and Whitham (1999) for other freeway models.

7.2 THE KdV EQUATION

7.2.1 Shallow-Water Waves with Dispersion

In the previous discussion of nonlinear shallow-water waves, dispersion was neglected: the wave speed was independent of wavenumber. In this section we will see that adding dispersion has an important effect: it suppresses the wavebreaking. The basic reason is easy to understand: dispersion causes waves to spread, and this acts against the steepening effect of the nonlinearity.

For linear water waves without surface tension, we have previously seen that the dispersion relation is

$$\omega(k)=\sqrt{gk\tanh kh}\,, \tag{7.2.1}$$

where h is the depth [see Exercise (1), Sec. 6.1]. For long wavelengths where $kh\ll 1$, this returns to the shallow-water result, $\omega(k)=c_0k=\sqrt{gh}\,k$. When small nonlinearity is added, the resulting wave equation is Eq. (7.1.29). However, if kh is not infinitesimally small, then we need to keep dispersion in the equation. If we expand $\omega(k)$ in small kh, we obtain a correction to $\omega(k)$:

Cell 7.29

```
Series[Sqrt[g k Tanh[k h]], {k, 0, 3}]
```

$$\sqrt{gh}\ k - \frac{1}{6}\left(h^2 \sqrt{gh}\right) k^3 + O[k]^4$$

Thus, the lowest-order correction to $\omega(k)$ is

$$\omega(k) = c_0 k\left(1 - \tfrac{1}{6}k^2 h^2\right). \tag{7.2.2}$$

This correction can be added to Eq. (7.1.29) by appending a term:

$$\frac{\partial z}{\partial t} + c_0\left(1 + \frac{3}{2}\frac{z}{h}\right)\frac{\partial z}{\partial x} + \frac{1}{6}c_0 h^2 \frac{\partial^3 z}{\partial x^3} = 0. \tag{7.2.3}$$

If we neglect the nonlinear term (assuming small-amplitude disturbances), and Fourier transform this equation in time and space, we obtain

$$\left(-i\omega + ic_0 k + i\tfrac{1}{6}c_0 h^2 k^3\right)\tilde{z}(k, \omega) = 0.$$

This yields the linear dispersion relation of Eq. (7.2.2).

We can put Eq. (7.2.3) into a standard format using the same transformations as we used for Eq. (7.1.30): $\bar{x} = x - c_0 t$, and $f(\bar{x}, t) = 3c_0 z/(2h)$. The result is the *Korteweg–de Vries (KdV) equation*:

$$\frac{\partial f}{\partial t} + f\frac{\partial f}{\partial x} + \alpha\frac{\partial^3 f}{\partial x^3} = 0, \tag{7.2.4}$$

where $\alpha = c_0 h^2/6$, and where we have dropped the overbar on x for convenience.

Because the KdV equation is not first-order, we can no longer apply the method of characteristics. However, analytic solutions of the equation can be found using advanced inverse-scattering methods that are beyond the scope of this book. Here, for general initial conditions, we will content ourselves with a purely numerical solution to the problem.

However, there is one type of analytic solution that can be derived with relative ease: steady solutions, where f is time-independent when observed in a comoving frame. These solutions will be the subject of the next subsection.

7.2.2 Steady Solutions: Cnoidal Waves and Solitons

The KdV equation admits steady solutions of the form $f = f(x - vt)$. In a frame moving with speed v, such solutions are time-independent. These solutions occur because nonlinear steepening of the waves is in balance with dispersive broadening, allowing an equilibrium as viewed in the moving frame.

In order to find the form of these steady solutions, we define the variable $s = x - vt$. Then we assume $f = f(s)$, so that $\partial f/\partial x = df/ds$ and $\partial f/\partial t = -v\,df/ds$. The KdV equation then becomes the following ODE:

$$-v\frac{df}{ds} + f\frac{df}{ds} + \alpha\frac{d^3 f}{ds^3} = 0.$$

Writing $f\,df/ds = \frac{1}{2}(d/ds)f^2$ allows us to integrate this third-order ODE once, yielding the following second-order ODE:

$$-vf + \tfrac{1}{2}f^2 + \alpha\frac{d^2f}{ds^2} = A,$$

where A is a constant of integration. If we now rearrange terms, writing $\alpha\, d^2f/ds^2 = A + vf - \frac{1}{2}f^2$, we can see that this equation is of the same form as Newton's second law, $m\,d^2x/dt^2 = F(x)$, where the mass m replaces α, the position x replaces f, the time t replaces s, and the force $F(x)$ replaces the right-hand side. Equations of this type have a constant of integration: the energy $E = \frac{1}{2}m(dx/dt)^2 + V(x)$, where the potential $V(x)$ is given by $F = -dV/dx$. If we now translate back to our variables, we find a constant "energy" given by

$$E = \frac{1}{2}\alpha\left(\frac{df}{ds}\right)^2 + V(f), \tag{7.2.5}$$

where the "potential" $V(f)$ is

$$V(f) = -Af - \tfrac{1}{2}vf^2 + \tfrac{1}{6}f^3. \tag{7.2.6}$$

This potential is a cubic function of "position" f, and the behavior of f as a function of "time" s can be determined by examining this potential function. The function $V(f)$ can have one of two basic forms, as shown in Cell 7.30.

Cell 7.30

```
<<Graphics`;
V[f_] = -Af - 1/2 vf^2 + 1/6 f^3;
Plot[Evaluate[{(V[f]/.{A→1, v→1}), V[f]/.{A→-1, v→1}}],
  {f, -6, 6}, PlotStyle→{Red, Dashing[{0.05, 0.03}]}, Blue}];
```

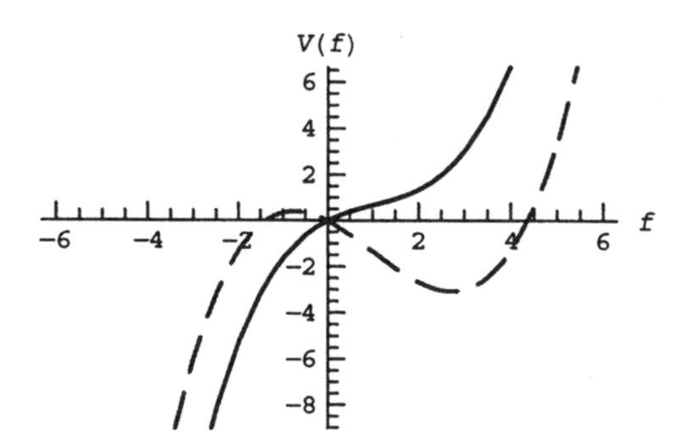

Solutions with $A < -v^2/2$ (the solid curve) proceed monotonically from $-\infty$ to $+\infty$. This can be seen analytically by solving for local extrema:

Cell 7.31

```
Solve[V'[f] == 0, f]
{{f→v - √(2 A + v²)}, {f→v + √(2A+v²)}}
```

Thus, for $A < -v^2/2$ "particles" will slide down the potential hill as "time" s increases, and this implies $f \to -\infty$ for large s. Such unbounded solutions are not of physical interest. On the other hand, for $A > -v^2/2$, $V(f)$ is no longer monotonic (the dashed curve), because there are now two real local extrema, a maximum and a minimum. In this regime, "particles" can get trapped in the potential well, resulting in periodic oscillations of f as a function of "time" s.

The form of these periodic nonlinear wavetrains depends on the value of A and of the "energy" E, and may be found by solving Eq. (7.2.5) for df/ds:

$$\frac{df}{ds} = \sqrt{\frac{2}{\alpha}[E - V(f)]}\,. \tag{7.2.7}$$

This is an ODE that can be reduced to quadratures:

$$\int^f \frac{d\bar{f}}{\sqrt{(2/\alpha)\left[E - V(\bar{f})\right]}} = s + B, \tag{7.2.8}$$

where B is another constant of integration, which we will set to zero in what follows, as it merely corresponds to a change in the origin of s. The integral involves the square root of a cubic polynomial, which makes it rather unpleasant, but it can nevertheless be carried out analytically in terms of special functions called *elliptic functions*:

Cell 7.32

```
Sqrt[α/2] Integrate[1/Sqrt[e - V[f]], f]
```

$$\Bigg(\sqrt{2}\,\sqrt{\alpha}$$

```
EllipticF[ArcSin[√((-f + Root[-6 e-6 A#1-3v#1^2+#1^3 &, 3])/(-Root[-6 e-
  6 A#1-3 v#1^2 +#^3 &, 2] + Root[-6 e-6 A#1-3 v#1^2 +#1^3 &, 3]))],

  Root[-6 e-6 A#1-3 v#1^2 +#1^3 &, 2]-Root[-6 e-6 A#1-3 v#1^2 +#1^3 &, 3]
  ----------------------------------------------------------------------- ]
  Root[-6 e-6 A#1-3 v#1^2 +#1^3 &, 1]-Root[-6 e-6 A#1-3 v#1^2 +#1^3 &, 3]

 (f-Root[- 6 e-6 A#1-3 v#1^2 +#1^3 &, 3])

  Sqrt[ (f-Root[- 6 e-6 A#1 -3 v#1^2 +#1^3 &, 1]) /
        (-Root[- 6 e-6 A#1-3 v#1^2 +#1^3 &, 1] + Root[- 6 e-6 A#1-3 v#1^2 + #1^3 &, 3]) ]

  Sqrt[ (f-Root[- 6 e-6 A#1-3v #1^2 +#1^3 &, 2]) /
        (- Root[- 6 e-6 A #1-3 v #1^2 +#1^3 &, 2] +Root[- 6 e-6 A #1-3 v#1^2 +#1^3 &, 3]) ] ) /

 ( Sqrt[ e + A f - f^3/6 + f^2 v/2 ]

  Sqrt[ (f-Root[- 6 e-6 A#1-3 v #1^2 +#1^3 &, 3]) /
        (Root[- 6 e-6 A #1-3 v #1^2 +#1^3 &, 2]-Root[- 6 e-6 A#1-3 v #1^2 + #1^3 &, 3]) ] )
```

This mess can be simplified somewhat if one notices that the **Root** functions in the above expression are simply placeholders for the roots of the cubic polynomial $E - V(f)$. That is, if we write

$$E - V(f) = -\tfrac{1}{6}(f-a)(f-b)(f-c) \tag{7.2.9}$$

with $a \geq b \geq c$, then we can substitute for these roots and obtain

Cell 7.33

```
FullSimplify[% /.
{Root[-6 e-6 A #1-3 v #1^2 + #1^3 &, 1] → c,
 Root[-6 e-6 A #1-3 v #1^2 + #1^3 &, 2] → b,
 Root[-6 e-6 A #1-3 v #1^2 + #1^3 &, 3] → a,
e → V[f] - 1/6 (f-a) (f-b) (f-c) }, c < b < a ]
```

$$-\frac{2\sqrt{3}\sqrt{(a-b)\,(a-f)}\sqrt{\frac{b-f}{-a+b}}\sqrt{\frac{c-f}{-a+c}}\sqrt{\alpha}\,\text{EllipticF}\left[\text{ArcSin}\left[\sqrt{\frac{a-f}{a-b}}\right],\frac{a-b}{a-c}\right]}{\sqrt{(a-f)\,(-b+f)\,(-c+f)}}$$

The special function **EllipticF[θ,m]** is an incomplete elliptic integral of the first kind, referred to as $F(\theta|m)$ in most textbooks. The definition of this function is

$$F(\theta|m) \equiv \int_0^\theta \frac{d\phi}{\sqrt{1-m\sin^2\phi}}.$$

The next step is most easily done by hand: the square roots involving f all cancel, leaving us with

$$\frac{-2\sqrt{3\alpha}}{\sqrt{a-c}}F\left(\arcsin\sqrt{\frac{a-f}{a-b}}\,\middle|\,m\right) = s, \tag{7.2.10}$$

where

$$m = \frac{a-b}{a-c}. \tag{7.2.11}$$

Since $c \leq b \leq a$, it is worthwhile to note for future reference that $0 \leq m \leq 1$.

Equation (7.2.10) can be inverted to obtain $f(s)$:

$$\frac{a-f(s)}{a-b} = \sin^2\left[F^{-1}\left(\sqrt{\frac{a-c}{12\alpha}}\,s\,\middle|\,m\right)\right].$$

The right-hand side can, in turn, be expressed as another special function, a *Jacobian elliptic function* $\mathrm{sn}(\theta|m)$, defined by $\mathrm{sn}(\theta|m) \equiv \sin F^{-1}(\theta|m)$. In other words, if $u = F(\theta|m)$ then $\mathrm{sn}(u) \equiv \sin\theta$. Thus,

$$f(s) = a - (a-b)\,\mathrm{sn}^2\left(\sqrt{\frac{a-c}{12\alpha}}\,s\middle|m\right). \tag{7.2.12}$$

The function sn is a nonlinear version of the sine function. In *Mathematica* it is referred to as **JacobiSN[θ,m]**. There is a second function, cn, which is analogous to the cosine function, and is defined by the identity $\mathrm{sn}^2 + \mathrm{cn}^2 = 1$. Our periodic solution for f is often called a *cnoidal wave* (because it is often expressed in terms of cn rather than sn). The function sn is plotted in Cell 7.34 for $m = \frac{1}{2}$.

Cell 7.34

```
Plot[JacobiSN[θ, 1/2], {θ, -10, 10},
  PlotLabel→ "sn(θ|m), m=1/2", AxesLabel→{"θ", ""}];
```

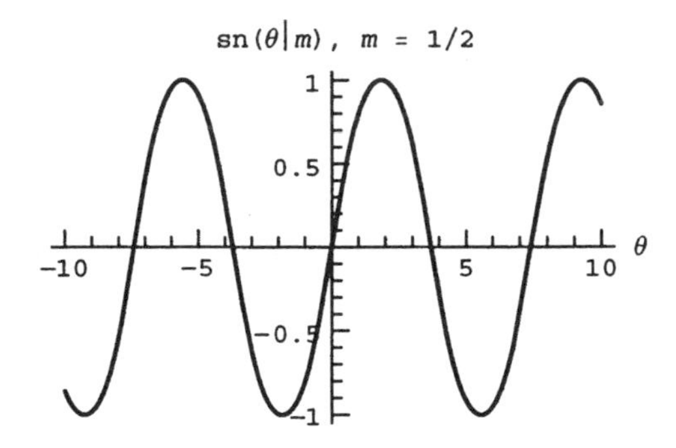

Since the sn function, like the sine function, is zero at the origin and runs from -1 to 1, we see from Eq. (7.2.12) that the maximum and minimum of the wave is a and b respectively. This makes sense, since these are the turning points in the motion of the "particle" in the "potential well" $V(f)$.

However, unlike the sine function, the period $P(m)$ of the sn function is not fixed, but rather varies, depending on m. As m increases away from zero, the period $P(m)$ of the sn function increases. In fact, one can show that

$$P(m) = 4K(m), \tag{7.2.13}$$

where $K(m)$ is another special function, the complete elliptic function of the first kind (called **EllipticK[m]** in *Mathematica*). This function is related to F by $K(m) = F(\pi/2|m)$. The period of sn is plotted in Cell 7.35 vs. m for $0 < m < 1$.

Cell 7.35

```
Plot[4 EllipticK[m], {m, 0, 1},
  PlotLabel -> "Period of sn(θ|m)", AxesLabel→{"m", ""}];
```

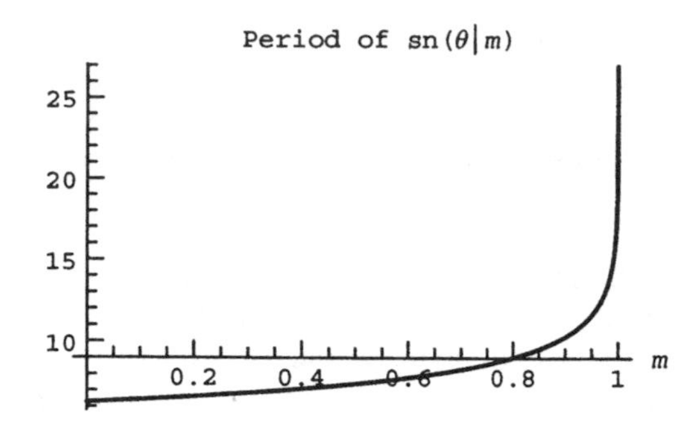

For $m \to 0$, corresponding to $b \to a$, the period of sn is 2π, and in fact, $\text{sn}(\theta, 0) = \sin\theta$:

Cell 7.36

```
JacobiSN[θ, 0]
```

```
Sin[θ]
```

This low-amplitude limit of the cnoidal waves [$m \to 0$ implies that $b \to a$ in Eq. (7.2.12)] corresponds to the limit of linear waves.

Since sn^2 has a period half that of sn, Eq. (7.2.12) implies that the wavelength $\lambda(a,b,c)$ of the cnoidal waves is given by

$$\lambda(a,b,c) = \frac{1}{2}\sqrt{\frac{12\alpha}{a-c}}\,P(m). \tag{7.2.14}$$

In the linear limit $b \to a$ ($m \to 0$) this becomes

$$\lambda(a,a,c) = \sqrt{\frac{12\alpha}{a-c}}\,\pi. \tag{7.2.15}$$

Some examples of cnoidal waves for increasing amplitude are shown in Cell 7.37, keeping the wave height constant and increasing the depth of the troughs ($a = 1$, $c = 0$, and b decreasing away from a toward c).

Cell 7.37

```
f[s_, a_, b_, c_] =
  a-(a-b) JacobiSN[Sqrt[(a-c)/12] s, m]^2/. m→(a-b)/(a-c);
```

```
Table[Plot[f[s, 1, b, 0], {s, -20, 20},
   PlotLabel→"m =" <>ToString[(a-b)/(a-c)/. {a→1, c→0}],
   AxesLabel→{"s/√α", ""}, PlotRange→{-1, 1}],
    {b, .9, 0., -.2}];
```

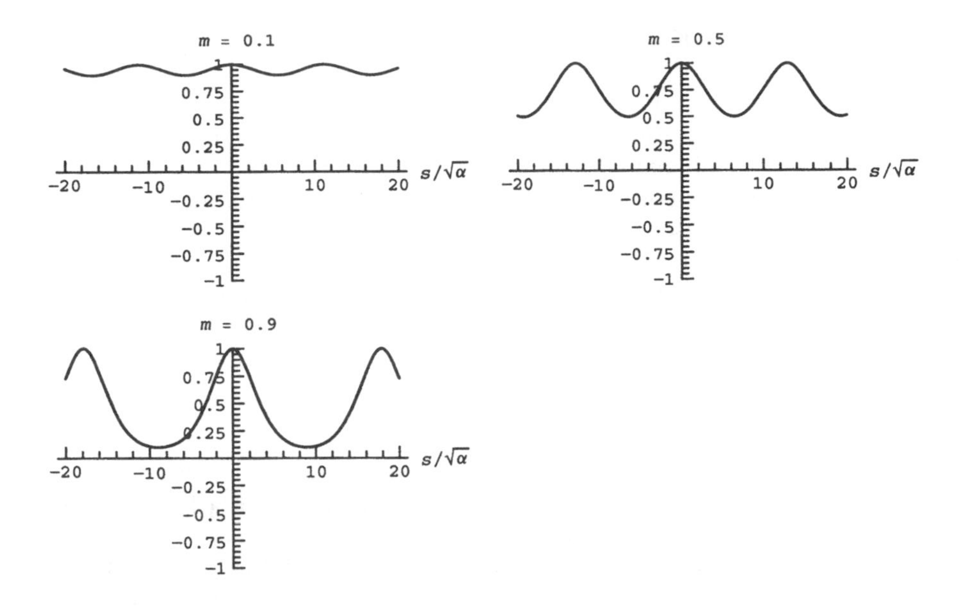

Different values of a and c, $a \geq 0 \geq c$, lead to the same basic behavior, but with different values for the peak height and trough depth. (The reader is encouraged to vary the parameters a and c to verify this.) Note how, for finite wave amplitude, the troughs of the wave broaden and the peaks narrow. This corresponds qualitatively to the peaking of real water waves, shown in Fig. 7.5.

One can see from this photo that real nonlinear water waves are even more sharply peaked than cnoidal waves. This has to do with the approximate form of

Fig. 7.5 Nonlinear water waves, created in a wave tank at the Ship Division of the National Physical Laboratory, England. [Photograph by J. E. Frier, in Benjamin (1967).]

dispersion used in the KdV equation. Equation (7.2.2) is a poor approximation to the exact linear dispersion relation (7.2.1) at large wavenumbers, which contribute to the sharp peak. More accurate equations that keep the exact linear dispersion can be constructed, and show the desired slope discontinuities at the wave peaks. [See, for instance, Whitham (1999).]

The wavelength λ is not all that depends on the wave amplitude. The speed v of the cnoidal wave also depends on a, b, and c. This can be easily seen by substituting Eq. (7.2.6) into (7.2.9). The coefficient of f^2 on the right-hand side of Eq. (7.2.9) is $\frac{1}{6}(a+b+c)$, but the coefficient of f^2 on the left-hand side is $v/2$. Therefore, we find that the wave speed is

$$v = \tfrac{1}{3}(a+b+c). \tag{7.2.16}$$

In the limit of a linear wave oscillating about $f=0$, $b \to a \to 0$ and we have $v \to \frac{1}{3}c$. This is the correct result for the linear wave speed as seen in a frame moving at velocity c_0. In this frame, the linear wave speed is given by $v = \omega/k - c_0$. Using Eq. (7.2.1) for ω yields $v = -c_0 k^2 h^2/6$. This does not look like $v = c/3$. However, remember that the wavelength $\lambda = 2\pi/k$ for these waves is given by Eq. (7.2.15) with $a=0$: $\lambda = \sqrt{12\alpha/(-c)}\,\pi$. Squaring this equation and substituting $c = 3v$ implies that $v = -\alpha k^2$, which may be seen to agree with the linear dispersion relation if we recall that $\alpha = c_0 h^2/6$.

Another important case is the limit $m \to 1$ (i.e. $b \to c$). In this limit, the period $P(m)$ of sn approaches infinity [see Eq. (7.2.13) and Cell 7.35], and as a result so does the period of the cnoidal waves. The reason for this divergence in the period can be understood by examining motion in the potential $V(f)$. The limit $b \to c$ corresponds to an energy E for which there is a turning point just at the top of the potential curve (Fig. 7.6). Particles that start with zero velocity at the top of the hill take an infinitely long time to fall down the slope, which is why the period of the oscillatory motion goes to infinity for this energy.

Since the period of the motion is infinite, the wave peaks becomes infinitely separated, resulting in a single isolated peak. This peak is called a *soltion*, or

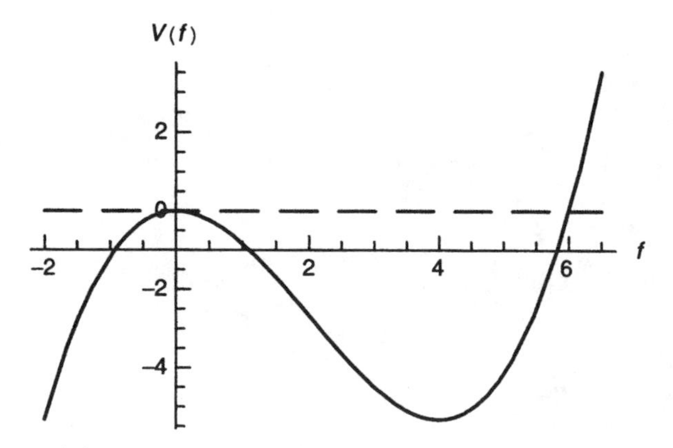

Fig. 7.6 Potential $V(f)$ for $v=2$, $A=0$ (solid), and energy E (dashed) for which a soliton solution exists.

solitary wave. In Cell 7.38, we plot the case of a soliton of unit amplitude ($a = 1$) for which $f \to 0$ at large s (i.e. $c \to b \to 0$).

Cell 7.38

```
Plot[f[s, 1, 0, 0], {s, -20, 20}, PlotLabel→
  "m =" <>ToString[1] <>"(soliton)",
  AxesLabel→{"s/√α", ""}, PlotRange→{-.2, 1}];
```

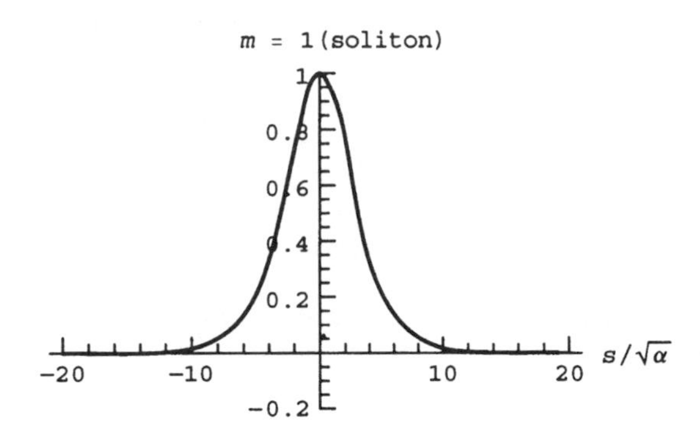

This soliton has a simple analytic form:

Cell 7.39

```
Simplify[f[s/√α, a, 0, 0]]
```

$$a\,\mathrm{Sech}\left[\frac{\sqrt{a}\ s}{2\ \sqrt{3}\ \sqrt{\alpha}}\right]^2$$

Written as an equation, the soliton solution is

$$f(s) = \frac{a}{\cosh^2\left(\sqrt{a/12\alpha}\, s\right)}. \tag{7.2.17}$$

From this expression, one can see that the length of the soliton increases as the amplitude a decreases. Cell 7.38 shows that the length (from end to end) is roughly

$$L = 20\sqrt{\alpha/a}\ .$$

Also, according to Eq. (7.2.16), the speed of the soliton increases as the amplitude increases:

$$v = a/3.$$

(This is the speed as viewed from a frame already moving at speed c_0.) These expressions for the length and velocity are in scaled units. If we go back to real units, the soliton height is $z_0 = 2ha/(3c_0)$, and the soliton speed v_{lab} as seen in the lab frame depends on height z_0 according to

$$v_{\text{lab}} = c_0\left(1 + \frac{z_0}{2h}\right), \tag{7.2.18}$$

and the length is

$$L = \tfrac{20}{3}\sqrt{h^3/z_0}\,. \tag{7.2.19}$$

The earliest recorded observation of a soliton in the wild is found in a famous paper by Scott Russell (1844). There, he describes a soliton that he had witnessed ten years earlier:

> I was observing the motion of a boat which was rapidly drawn along a narrow channel by a pair of horses, when the boat suddenly stopped—not so the mass of water in the channel which it had put in motion; it accumulated around the prow of the vessel in a state of violent agitation, then suddenly leaving it behind, rolled forward with great velocity, assuming the form of a large solitary elevation, a rounded, smooth and well-defined heap of water, which continued its course along the channel apparently without change of form or diminution of speed. I followed it on horseback, and overtook it still rolling at a rate of some eight to nine miles an hour, preserving its original figure some thirty feet long and a foot to a foot and a half in height. Its height gradually diminished, and after a long chase of one or two miles I lost it in the windings of the channel. Such, in the month of August, 1834, was my first chance encounter with that singular and beautiful phenomenon.

This exciting narrative shows that solitons are not mere artifacts of some mathematical idealization. In fact, it turns out that solitons are ubiquitous in nonlinear systems with dispersion. Some examples of other systems exhibiting solitons may be found in the exercises. They may even have useful practical applications: it has been proposed that soliton pulses could be used to transmit information in fiber optics networks. Even the best optical fibers are slightly dispersive, so that pulse trains eventually spread and lose coherence. However, pulse trains consisting of optical solitons (pulses with sufficiently large amplitude so that nonlinear effects in the index of refraction become important) are not affected by dispersive spreading. As of the year 2000, 10-Gbit/s bit rates have been demonstrated in fibers over distances of 10^4–10^5 km (see nearly any issue of *Optics Letters* from around this time). This is still an area of ongoing research, and the interested reader is referred to A. Hasegawa and Y. Kodama (1995).

7.2.3 Time-Dependent Solutions: The Galerkin Method

Introduction We have seen that steady soliton and cnoidal wave solutions to the KdV equation occur because of a balance between nonlinear steepening and dispersive spreading. But how do we know that such a balance is stable? If a steady solution is slightly perturbed, does the result remain close to the steady solution, or diverge from it? If the latter is true, then these steady solutions are not of much more than academic interest. Although Scott Russell's description indicates that solitons are in fact stable, it would be nice to see this fact arise from the mathematics.

There has been considerable analytical work on the stability of steady solutions to the KdV equation. This work is of a highly technical nature and will not be discussed here. Rather, we will use numerical techniques already at our disposal to test the stability of the solutions. We will also study the interaction between solitons, and wavebreaking.

We will use the Galerkin method, solving the KdV equation with periodic boundary conditions on an interval running from $-L$ to L. A set of basis functions for these boundary conditions are

$$v_n(x) = e^{ikx}, \qquad k = n\pi/L, \quad -M \le n \le M. \tag{7.2.20}$$

The solution for $f(x,t)$ is then written in terms of these functions:

$$f(x,t) = \sum_n c_n(t)\, e^{ikx}. \tag{7.2.21}$$

The basis functions are orthogonal with respect to the inner product $(f,g) = \int_{-L}^{L} f^* g\, dx$, so this is the inner product we will use. Substituting this series into the KdV equation, we obtain

$$\sum_n \frac{\partial}{\partial t} c_n(t)\, e^{ikx} + \sum_{\bar{\bar{n}}} \sum_{\bar{n}} c_{\bar{\bar{n}}}(t)\, c_{\bar{n}}(t)\, i\bar{k}\, e^{i(\bar{\bar{k}}+\bar{k})x} - \alpha \sum_n i k^3 c_n(t)\, e^{ikx} = 0. \tag{7.2.22}$$

We now take an inner product with respect to e^{ikx}, and note that $(e^{ikx}, e^{i(\bar{\bar{k}}+\bar{k})x}) = L\delta_{k,\bar{\bar{k}}+\bar{k}}$. Therefore, $\bar{\bar{k}} = k - \bar{k}$, and Eq. (7.2.22) becomes

$$\frac{\partial}{\partial t} c_n(t) + \sum_{\bar{n}} c_{n-\bar{n}}(t)\, c_{\bar{n}}(t)\, i\bar{k} - \alpha i k^3 c_n(t) = 0. \tag{7.2.23}$$

We will solve these coupled nonlinear ODEs for $c_k(t)$ using **NDSolve**. Before we do, however, we note that $f(x,t)$ is a real function, so this implies

$$c_{-n}(t) = c_n(t)^*. \tag{7.2.24}$$

Therefore, we do not need to solve for the coefficients with $n < 0$:

Cell 7.40

```
c[n_][t_] := Conjugate[c[-n][t]] /; n < 0;
```

Also, note that the nonlinear term in Eq. (7.2.23) couples modes with different wavenumbers. Even though the sum over $\bar{n}$ runs only from $-M$ to M, coefficients $c_{n-\bar{n}}(t)$ may fall outside this range. In fact, this is sure to happen eventually, even if $c_n(t)$ is initially zero for $|n| > M$. For example, say that initially only c_{-1} and c_1 are the only nonzero coefficients (i.e., the solution is a sinusoidal wave with wavenumber π/L). Then according to Eq. (7.2.23), there is an initial rate of change for $c_{\pm 2}$

given by

$$\frac{\partial}{\partial t} c_2 + i \frac{\pi}{L} (c_1)^2 = 0,$$

$$\frac{\partial}{\partial t} c_{-2} - i \frac{\pi}{L} (c_{-1})^2 = 0.$$

When $c_{\pm 2}$ has grown to large amplitude, these coefficients couple in turn to modes with $n = \pm 3, \pm 4$, and so on, until a broad range of wavenumbers are excited. This is called a *cascade*. If it is unchecked, eventually waves with arbitrarily high wavenumbers are excited. This is what leads to wavebreaking in the case of Eq. (7.1.30): as the wave attains infinite slope, this corresponds to Fourier modes with infinite wavenumber.

We can hope that the cascade is suppressed by the dispersive term in the KdV equation. However, one can see in Eq. (7.2.23) that this term, proportional to α, causes only a rapid oscillation in the phase of c_n for large n, and does not affect its amplitude. On the other hand, we might imagine that this rapid oscillation does cause phase mixing, which has an equivalent effect to the suppression of the amplitude, so we will simply neglect all modes with $|n| > M$, explicitly setting them equal to zero:

Cell 7.41

```
c[n_][t_] := 0 /; Abs[n] > M;
```

The hope is that these modes will not be needed.

Example 1: Cnoidal Wave For our first test, we will try a cnoidal wave as the initial condition:

Cell 7.42

```
f[s_, a_, b_, c_] =
 a-(a-b) JacobiSN[Sqrt[(a-c)/12] s, m]^2/. m→(a-b)/(a-c);

f0[x_] = f[x, 1, .1, 0];
```

In order to fit this cnoidal wave into periodic boundary conditions, we choose L to be a multiple of $\lambda(a, b, c)$:

Cell 7.43

```
λ[a_, b_, c_] = 2 EllipticK[m] Sqrt[12/(a-c)]/.m→(a-b)/(a-c);

L = 2 λ[1, .1, 0]

35.7231
```

As usual, the initial conditions on the Fourier modes are determined by $f_0(x)$ according to

$$c_n(0) = \frac{\left(e^{ikx}, f_0(x)\right)}{2L}. \tag{7.2.25}$$

We will keep 15 modes in our solution:

Cell 7.44

```
M = 15;

k[n_] := n Pi/L;

ics = Table[
   c[n][0] == NIntegrate[Exp[-I k[n] x] f0[x],
    {x, -L, L}]/(2 L), {n, 0, M}];

NIntegrate: :ncvb :
 NIntegrate failed to converge to prescribed accuracy after
  7 recursive bisections in x near x = 35.72309221520074'.

NIntegrate: :ncvb :
 NIntegrate failed to converge to prescribed accuracy after
  7 recursive bisections in x near x = -35.7231.

NIntegrate: :ncvb :
 NIntegrate failed to converge to prescribed accuracy after
  7 recursive bisections in x near x = 35.72309221520074'.

General: :stop : Further output of
   NIntegrate: :ncvb will be suppressed during this
    calculation.
```

The errors in the integration can be ignored: they arise from the fact that several of the modes have zero initial amplitude. These errors can be avoided by setting these mode amplitudes equal to zero directly without evaluating the integrals.

We now integrate the ODEs forward in time using **NDSolve**, and plot the result over a time range $0 < t < t_{max}$ in steps of Δt, with plot range from f_{min} to f_{max}. We do so using a module, since we will be studying several examples. The ODEs that we will solve are given by Eq. (7.2.23), taking $\alpha = 1$:

Cell 7.45

```
kdVmod[tmax_, Δt_, fmin_, fmax_] := Module[{k, ODEs},

  k[n_] = n Pi/L;

  ODEs = Table[D[c[n][t], t] - I k[n]^3 c[n][t] +
       Sum[I k[nb] c[nb][t] c[n - nb][t], {nb, -M, M}] == 0,
        {n, 0, M}];

  vars = Table[c[n][t], {n, 0, M}];

  sol = NDSolve[Join[ODEs, ics], vars, {t, 0, tmax},
    MaxSteps → 6000];

  f[x_, t_] = Sum[c[n][t] Exp[I k[n] x], {n, -M, M}];
  f[x_, t_] = f[x, t]/. sol[[1]];

  Table[
   Plot[f[x, t], {x, -L, L}, PlotRange → {{-L, L},
    {fmin, fmax}}, PlotLabel → " M = "<>ToString[M]<>",
     t = "<>ToString[t]], {t, 0, tmax, Δt}];]
```

We now evaluate the solution over the time range $0 < t < 200$, as shown in Cell 7.46.

Cell 7.46

```
kdVmod[200, 5, 0, 2]
```

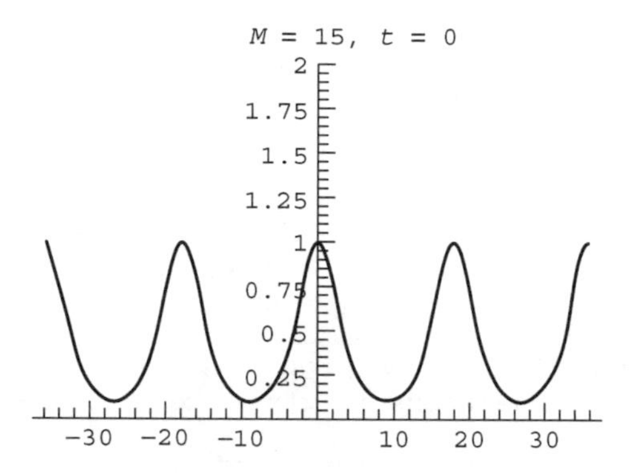

The solution appears to be well behaved and stable over the time scale of interest. It moves to the right with a speed given by Eq. (7.2.16). Other cnoidal waves and solitons can be handled by simply varying the parameters. One can also add small perturbations to these solutions. One finds by this sort of empirical approach that steady solutions to the KdV equation are stable to small perturbations. This is born out by the previously mentioned analytic work on this problem.

Example 2: Interacting Solitons In this example, for initial conditions we take two soliton solutions with different amplitudes:

Cell 7.47

```
L = 60;

f0[x_] = 1/3/Cosh[Sqrt[1/36] (x + 30)]^2 + 1/4/
  Cosh[Sqrt[1/48] (x - 30)]^2;
```

We will again evaluate the solution keeping 15 modes:

Cell 7.48

```
M = 15;

k[n_] := n Pi/L;

ics = Table[
   c[n][0] == NIntegrate[Exp[- I k[n] x] f0[x],
  {x, -L, L}]/(2L), {n, 0, M}];
```

In order to follow the dynamics we must integrate over a longer time, $0 < t < 6000$, because our initial conditions are solitons of rather small amplitude. This implies their velocities are small. The solution is displayed in Cell 7.49. The soliton in back, being of larger amplitude, moves with higher speed and catches up to the soliton in the front. It then feeds some of its amplitude to the soliton in the front, until they exchange amplitudes, and the soliton in the front then moves away from the one in the back. Effectively, the two solitons have exchanged places: they have passed through one another without change. Thanks to the periodic boundary conditions, this interaction then repeats itself ad infinitum.

Cell 7.49

```
kdVmod[6000, 100, 0, .6]
```

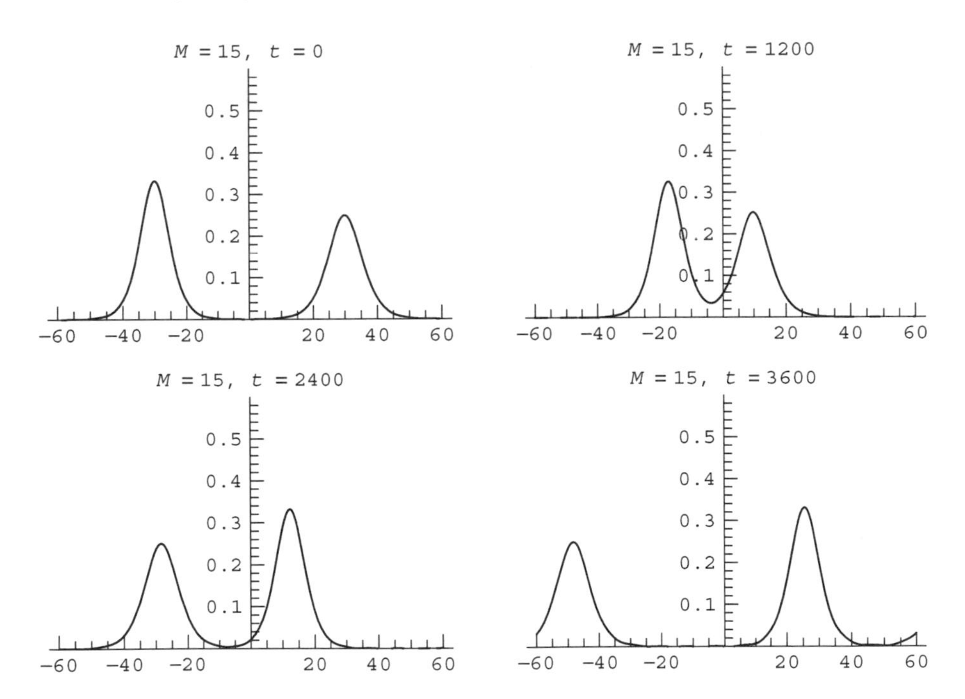

The fact that solitons effectively retain their shape during binary interactions implies that a description of nonlinear processes can often be greatly simplified—one can think of the nonlinear system as a gas of discrete solitons, which act almost like discrete particles, interacting with one another but retaining their identity.

Example 3: Suppression of Wavebreaking In this example, we investigate whether wavebreaking occurs in the KdV equation. Recall that wavebreaking was observed in the solutions to Eq. (7.1.30), where dispersion is neglected. Although we cannot hope to observe similar double-valued solutions using the Galerkin method, which expands in a set of single-valued modes, we might expect to see some initial conditions steepen, and the spectrum of waves should display a cascade to high wavenumbers.

For our initial condition, we take a single finite-amplitude sine wave. We now keep $M = 20$ modes in order to be certain of capturing the high-wavenumber dynamics properly, as shown in Cell 7.50.

Cell 7.50

```
L = 40; M = 20;
ics = Table[c[n][0] == 0, {n, 0, M}];
ics[[2]] = c[1][0] == 1/8;

kdVmod[800, 10, -1, 1]
```

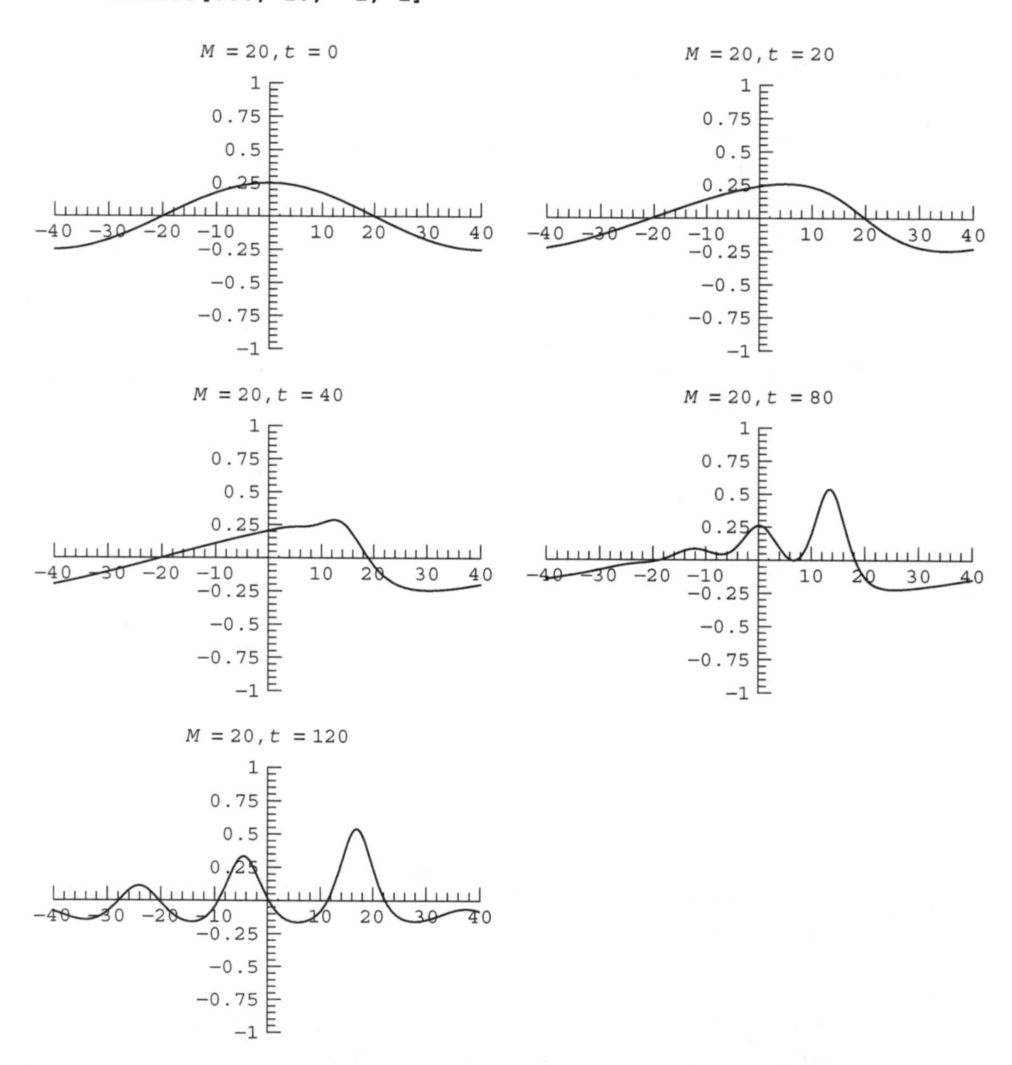

Initially the wave steepens, just as when dispersion is neglected. However, instead of breaking, the wave dissociates into several solitons that proceed to interact with one another in a complex manner. The spectrum of the solution is peaked at low wavenumbers, as can be seen in the plot of $|c_n(t)|$ shown in Cell 7.51. Evidently the cascade to large wavenumbers has been suppressed by the dispersive term in the KdV equation. Modes in the upper half of the spectral range have negligible amplitude. This implies that our cutoff of the spectrum at $n = \pm M$ does not affect the result. This can be verified by taking larger (or somewhat smaller) values of M: the answer does not change.

Cell 7.51

```
spectrum = Abs[vars/.sol[[1]]];
Table[ParametricPlot3D[{k[n], t, spectrum[[n + 1]]},
   {t, 0, 800}, DisplayFunction→Identity], {n, 0, M}];
Show[%, DisplayFunction→$DisplayFunction, BoxRatios→
  {1, 1, 1},
   AxesLabel→{"k", "t", "|cn(t)|   "}, PlotLabel ->
    "wave spectrum\n "];
```

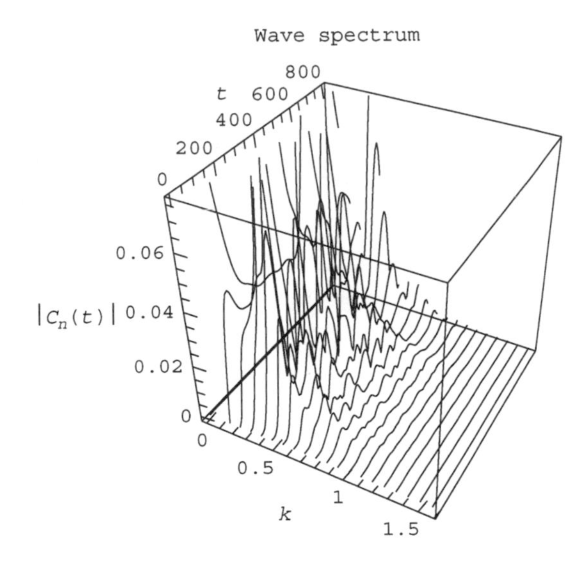

Thus, when dispersion is added in the KdV equation, wavebreaking no longer occurs. Is this due to the relatively small amplitude of the wave chosen in the above example? The answer is no—even for large amplitudes, the wavebreaking is suppressed. However, it is harder to see this numerically because more modes must be kept, making the problem somewhat harder to integrate. In Fig. 7.7 we show an example of a sinusoidal initial condition with eight times the previous amplitude. The solution still breaks up into solitons, and the spectrum (Fig. 7.8) remains peaked in the lower half of the spectral range.

This leaves us in a bit of a quandary. By including dispersion in the KdV equation, which should make the equation a more accurate descriptor of water waves than Eq. (7.1.30), we have actually lost wavebreaking, one of the main qualitative features of such waves.

The solution to this puzzle lies in the approximate form of dispersion in the KdV equation. As already mentioned in connection with the rounded peaks of cnoidal waves, Eq. (7.2.2) is a poor approximation for large wavenumbers, yielding a phase velocity that can actually be negative for large k. The exact dispersion

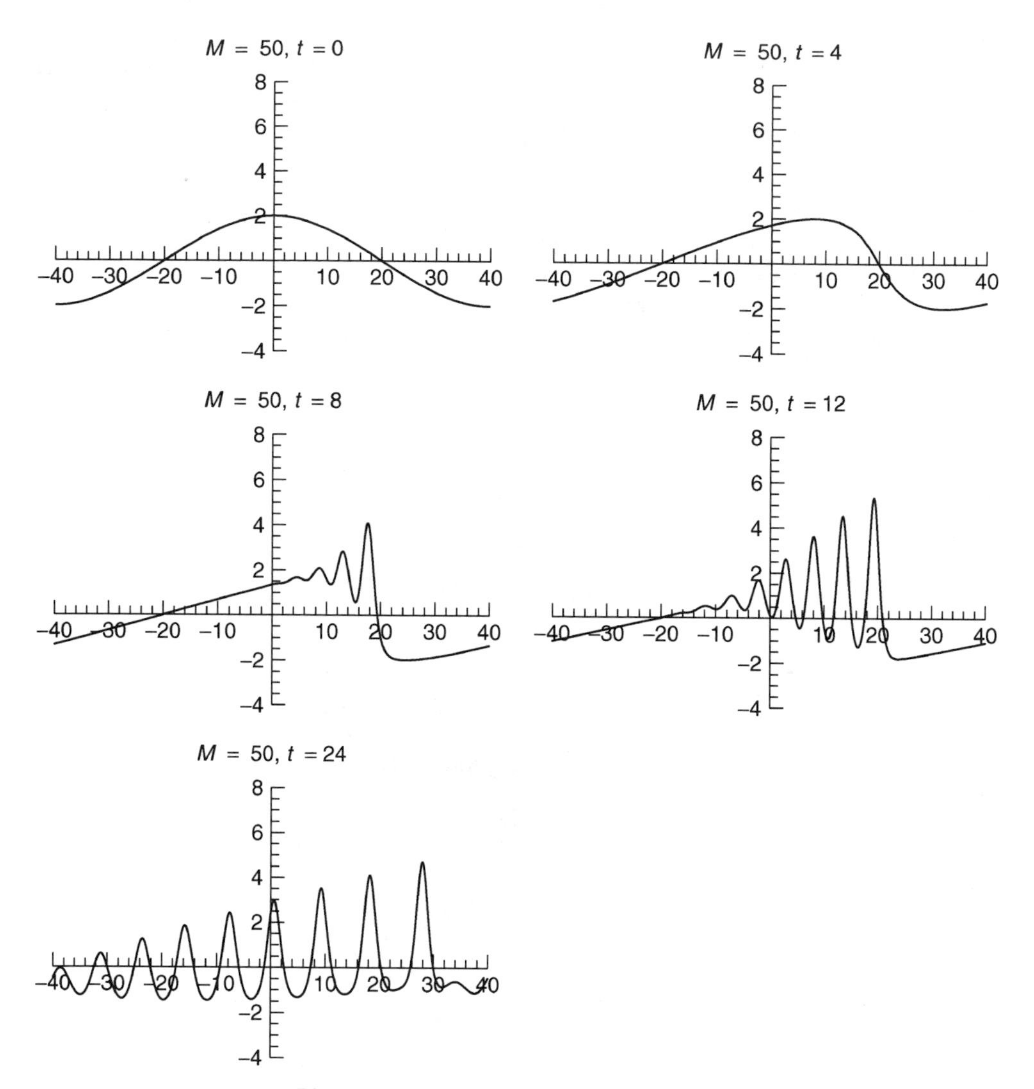

Fig. 7.7 A solution of the KdV equation.

relation (7.2.1) varies much less rapidly at large k, and results in weaker dispersion that allows real water waves to break, provided their amplitude is sufficiently large.

7.2.4 Shock Waves: Burgers' Equation

Steady Solutions If we take the dispersive term $\partial^3 f/\partial x^3$ in the KdV equation, change the sign, and lower the order of the derivative by one, we obtain Burgers' equation,

$$\frac{\partial f}{\partial t} + f\frac{\partial f}{\partial x} - \chi\frac{\partial^2 f}{\partial x^2} = 0. \tag{7.2.26}$$

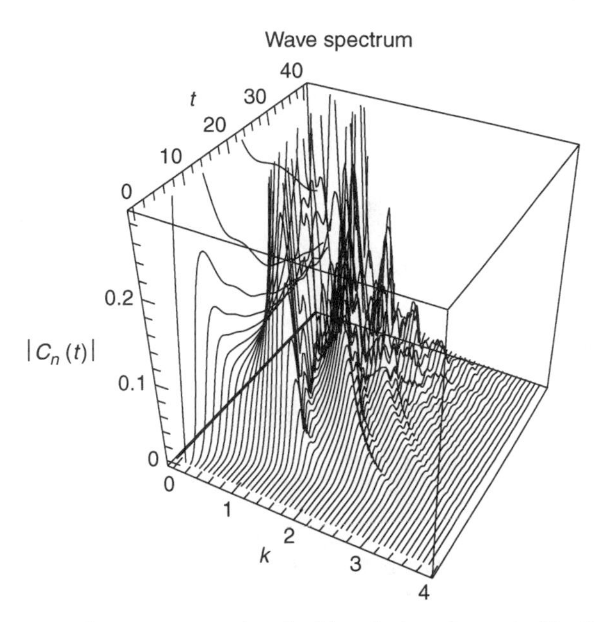

Fig. 7.8 Spectrum associated with solution shown in Fig. 7.7.

This nonlinear equation is a simplified model for the propagation of shock waves. Now f can be thought of as either the perturbed pressure or density of a gas due to a compressional wave, as seen in a frame moving with the linear wave speed. The term proportional to χ describes diffusion of these quantities due to viscosity or thermal conduction, and the nonlinear term qualitatively describes a nonlinear increase in wave speed when the wave pressure is large. The equation also has application to other areas, such as the study of nonlinear patterns in traffic flow (see the exercises).

A steady solution for f can be found using the same techniques as were used for the KdV equation. The solution travels to the right with speed v, takes on one value, $f_{-\infty}$, at $x = -\infty$, and takes on a lower value, f_{∞}, at $x = +\infty$. The form of the solution is

$$f(x,t) = v - \frac{\Delta f}{2} \tanh\left(\frac{\Delta f}{4\chi}(x - vt) \right), \tag{7.2.27}$$

where $\Delta f = f_{-\infty} - f_{\infty}$ is the jump in f across the shock (see the exercises). The speed v of the shock is related to $f_{-\infty}$ and f_{∞} according to

$$v = \tfrac{1}{2}(f_{-\infty} + f_{\infty}). \tag{7.2.28}$$

This relation between shock speed and the amplitude at $\pm\infty$ is called a *Hugoniot relation*.

A typical shock solution is plotted in Cell 7.52 in the comoving frame. In this example, we assume that $f_{\infty} = 0$. According to Eq. (7.2.27) the width of the shock depends on the diffusivity χ and on the jump Δf, scaling roughly as $\chi/\Delta f$. The smaller the value of χ, or the larger the jump, the narrower the shock.

Cell 7.52

```
Plot[1/2 - 1/2 Tanh[x/4], {x, -20, 20},
  AxesLabel→{"(x- v t)Δf/χ", "f(x,t)/Δf"}];
```

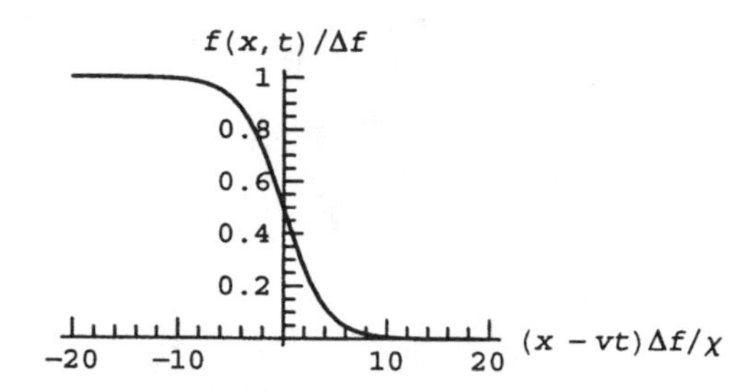

CTCS/Crank-Nicolson Approach For numerical solutions of Burgers' equation, the Galerkin method used in Sec. 7.2.3 for the KdV equation can be easily adapted. This is left to the exercises. Here, we will instead use a grid method, for practice. We solve the problem under periodic boundary conditions on $0 \le x < L$ and take $x = j\,\Delta x$, $j = 0, 1, 2, \ldots, M-1$, $\Delta x = L/M$, and $t = n\,\Delta t$.

The first two terms of Eq. (7.2.26) are of the form of the wave equation, (7.1.30), and hence we expect that the CTCS method will work for these terms. The equation is nonlinear, which hinders von Neumann stability analysis, but if we replace $f\,\partial f/\partial x$ by $c\,\partial f/\partial x$, where c is a constant, then one can show that the CTCS method is stable provided that timestep that satisfies the Courant condition $c\,\Delta t/\Delta x \le 1$ [see Eq. (6.2.44)]. Thus, we might hope that if $\Delta t \le \Delta x/\max|f|$, the CTCS method will be stable. Actually, it turns out that sharp gradients in f can also destabilize the method, but for sufficiently small timestep this method works well.

However, the last term in Eq. (7.2.26) is diffusive, which (as we saw in Sec. 6.2.1) causes instability in the CTCS method. For the diffusion term we therefore use a variant of the Crank–Nicolson scheme, taking the average of the diffusion operator at two timesteps centered around timestep n. The resulting difference equation is given below:

$$\frac{f_j^{n+1} - f_j^{n-1}}{2\,\Delta t} + f_j^n \frac{f_{j+1}^n - f_{j-1}^n}{2\,\Delta x} - \frac{\chi}{2}\left(\frac{f_{j+1}^{n+1} - 2 f_j^{n+1} + f_{j-1}^{n+1}}{\Delta x^2} + \frac{f_{j+1}^{n-1} - 2 f_j^{n-1} + f_{j-1}^{n-1}}{\Delta x^2} \right) = S_j^n. \quad (7.2.29)$$

For future reference, note that we have added a finite-differenced source function $S(x,t)$ to the right-hand side of the equation. Equation (7.2.29) is an implicit method. To solve it, we could apply the Crank–Nicolson algorithm that we used in Sec. 6.2.1, but instead we will write the procedure in a different way, using a form of operator splitting that we haven't seen before. First, we set up the grid and choose values for Δt and χ:

Cell 7.53

```
Clear["Global`*"]
M = 40; L = 4; Δx = L/M; Δt = 0.03; χ = 0.1;
```

Next, in order to simplify notation, we split the operator, taking the CTCS step and keeping that part of the Crank–Nicolson diffusion operator that depends on previous times. We place the result in an intermediate array F:

$$F_j^{n+1} = f_j^{n-1} - 2\,\Delta t f_j^n \frac{f_{j+1}^n - f_{j-1}^n}{2\,\Delta x} + \chi\,\Delta t\, d_2(f, j, n-1) + 2\,\Delta t s_j^n, \quad (7.2.30)$$

where we have introduced a finite-difference operator $d_2(f, j, n) = (f_{j+1}^n - 2f_j^n + f_{j-1}^n)/\Delta x^2$ in order to further simply the notation. Defined as a recursion relation, the left-hand side of Eq. (7.2.30) must be evaluated at time step n rather than $n+1$:

Cell 7.54

```
F[j_, n_] := f[j, n-2] - Δt (f[j, n-1]/Δx) (f[j + 1, n-1]-
  f[j-1, n-1]) +
χΔt d2[f, j, n-2] + 2 Δt S[j, n-1]
```

Cell 7.55

```
d2[f_, j_, n_] := (f[j + 1, n] - 2 f[j, n] + f[j - 1, n]) / Δx^2
```

We then solve coupled equations for f_j^{n+1} according to

$$f_j^{n+1} = F_j^{n+1} + \chi\,\Delta t\, d_2(f, j, n+1). \quad (7.2.31)$$

Equations (7.2.30) and (7.2.31) are equivalent to Eq. (7.2.29). Note that this form of operator splitting differs in concept from that used previously in Eqs. (6.2.31). There, different forms of the *full* operator were used in two substeps of size $\Delta t/2$. Here, the full time step Δt is taken for each substep, but only *pieces* of the operator are used in each.

To solve Eq. (7.2.31) we introduce dummy variables ρ_j^n to replace f_j^n. The reason is that the equations in **Solve** must be in terms of unknown quantities, but the f_j^n's are determined recursively by the procedure and so are not unknown variables. After the equations are solved for ρ_j^n, we use these values to update f_j^n.

The variables used in the equations are

Cell 7.56

```
vars[n_] = Table[ρ[j, n], {j, 0, M-1}];
```

In these variables, Eq. (7.2.31) becomes a list of equations:

Cell 7.57

```
eqns[n_] := Table[ρ[j, n] == F[j, n] + χΔt d2[ρ, j, n],
  {j, 0, M-1}];
```

Now we solve these equations (once) at the nth step:

Cell 7.58

```
varsn[n_] := varsn[n] = vars[n]/. Solve[eqns[n], vars[n]][[1]]
```

Then we substitute the results into f_j^n:

Cell 7.59

```
f[j_, n_] := (f[j, n] = varsn[n][[j + 1]])/; n > 1 &&
  0 ≤ j ≤ M -1
```

Since we will solve for points on the grid from $j=0$ up to $M-1$, the above equations will require values for points $j=-1$ and $j=M$. According to the periodic boundary conditions, these points are equivalent to $j=M-1$ and $j=0$ respectively. Boundary conditions for both f and the dummy variable ρ must be specified:

Cell 7.60

```
ρ[-1, n_] := ρ[M - 1, n];
ρ[M, n_] := ρ[0, n];
f[-1, n_] := f[M - 1, n];
f[M, n_] := f[0, n];
```

Also, the first timestep must be treated differently, since in the CTCS step $n=1$ calls $n=-1$. The simplest approach is to apply FTCS to the equation for this one timestep. This is only first-order accurate in Δt, but that is sufficient to our needs here. A second-order scheme involving the predictor–corrector method is discussed in the exercises. Here we take

Cell 7.61

```
f[j_, 1] := f[j, 1] =
  f[j, 0] - Δt f[j, 0] (f[j + 1, 0] - f[j - 1, 0])/Δx +
  χΔt d2[f, j, 0] + Δt S[j, 0]
```

For the source, we take $S=0$,

Cell 7.62

```
S[j_, n_] = 0;
```

and for the initial condition, we take a simple sine wave, offset from the origin so that it propagates to the right, as shown in Cell 7.63.

Cell 7.63

```
f[j_, 0] := 1 + Sin[j Δx Pi];

Table[ListPlot[Table[{j Δx, f[j, n]}, {j, 0, M}],
  PlotRange→{0, 2},
   PlotLabel -> "t = "<>ToString [n Δt]], {n, 0, 80, 2}];
```

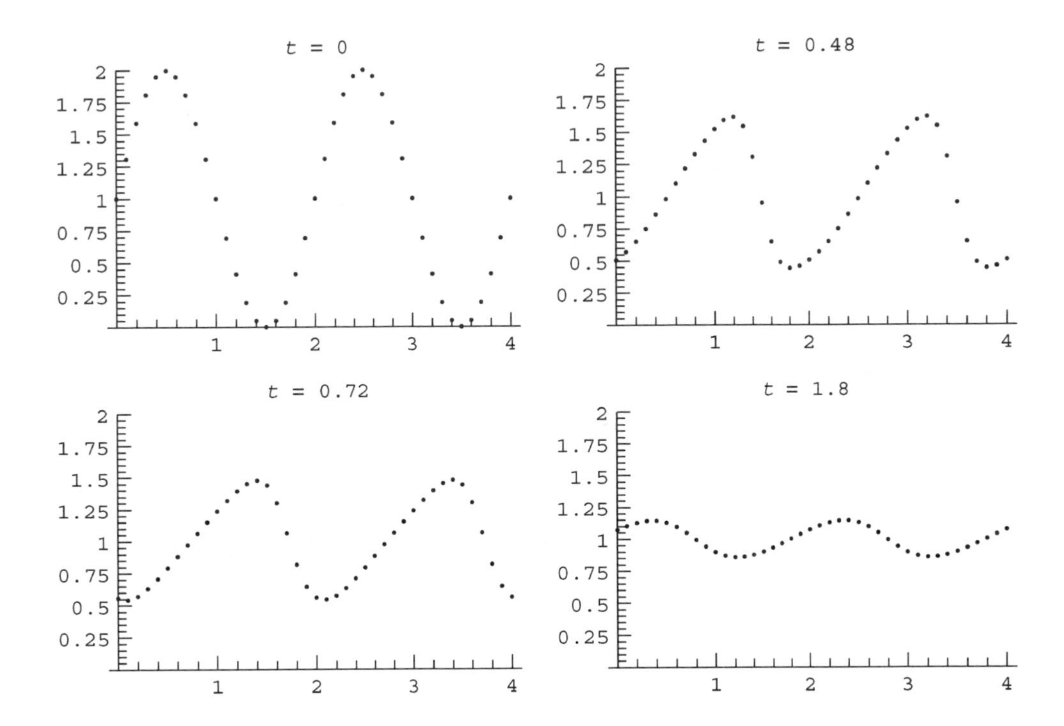

The wave steepens, but also decays due to the finite diffusivity. In order to see a shock wave form, it is necessary to drive the system with a source. This is done by reevaluating the previous cells, but now we take the following source:

Cell 7.64

```
S[j_, n_] := 0 /; j > 2;
S[0, n_] = S[1, n_] = S[2, n_] = 10;
```

along with a zero initial condition.

Cell 7.65

```
f[j_, 0] := 0;

Table[ListPlot[Table[{j Δx, f[j, n]}, {j, 0, M}],
  PlotRange→{0, 5},
   PlotLabel -> "t = "<>ToString[n Δt]], {n, 0, 80, 2}];
```

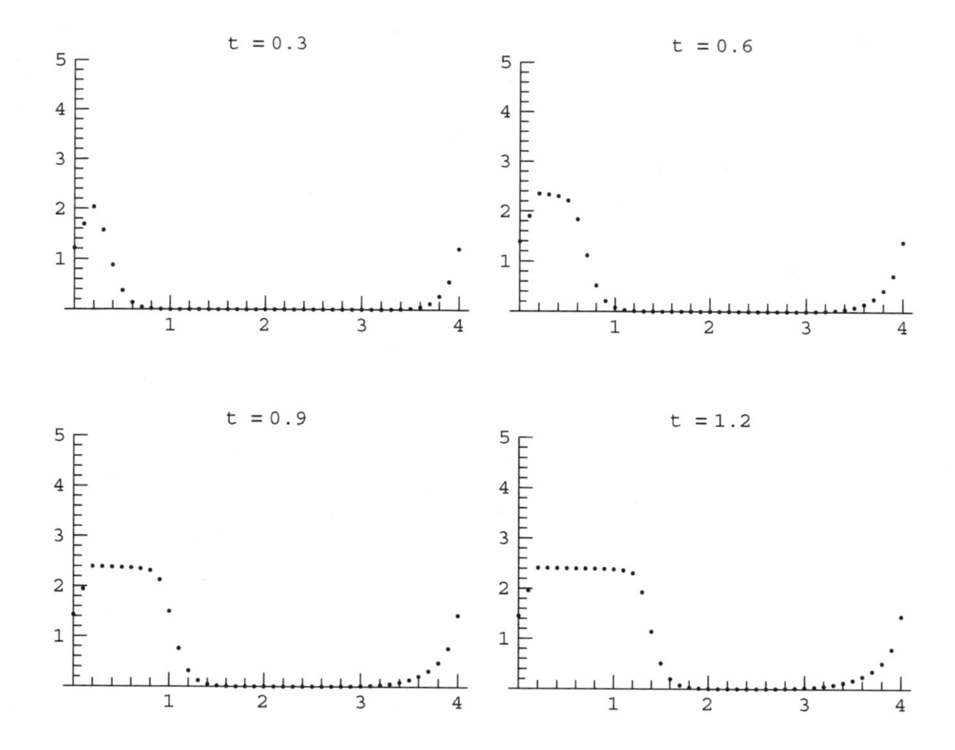

The result is shown in Cell 7.65. The effect of the localized source is to create a shock wave moving to the right. (Think of the source as an explosive release of gas.) The height of the shock wave is roughly 2.5 units, which depends on the strength of the source: if we define $N(t) = \int_0^L f(x,t)\,dx$, then according to Eq. (7.2.26), $dN/dt = \int_0^L S\,dx$. However, for this moving shock front, $dN/dt = vf_- = f_-^2/2$, where f_- is the height of the shock, and in the second stop we have used the Hugoniot relation, Eq. (7.2.28). Thus, for this shock wave the height of the shock depends on the source in the following way:

$$\frac{f_-^2}{2} = \int_0^L S\,dx.$$

Since $\int_0^L S\,dx \simeq 3\,\Delta x \times 10 = 3$ for our source, we find $f_- \simeq \sqrt{6} = 2.45\ldots$, very close to the value observed in the simulation.

Also, between times 0.6 and 2.4 the shock covers a distance of roughly 2 units, making the speed roughly $2/1.8 = 1.1$. This is close to the value of 1.25 expected from the Hugoniot relation, Eq. (7.2.28).

EXERCISES FOR SEC. 7.2

(1) **(a)** According to the description by Scott Russell, the length of the soliton that he observed was $L = 30$ ft, and the height was roughly 1.25 ft. Use these facts to determine the depth of the water in the channel.

(b) Predict the speed of the soliton using the results of part (a). How does your prediction compare to Scott Russell's observation of $v_{\mathrm{lab}} = 8\text{--}9$ miles per hour?

(2) **(a)** Show that the coefficient A in the potential function $V(f)$ is given in terms of the roots a, b, c by $A = -(ab + ac + bc)/6$.

(b) Show that $E = abc/6$.

(3) The Boussinesq equations describe slightly dispersive shallow-water waves propagating both to the left and to the right, as viewed in the lab frame. These are coupled equations for the horizontal velocity of the fluid surface, $u(x,t)$, and the water depth $h(x,t)$, where $h(x,t) = h_0 + z(x,t)$, h_0 is the equilibrium depth, and z is the wave height. The equations are

$$\frac{\partial h}{\partial t} + \frac{\partial}{\partial x}(hu) = 0,$$

$$\frac{\partial u}{\partial t} + u\frac{\partial u}{\partial x} + g\frac{\partial h}{\partial x} + \frac{1}{3}h_0\frac{\partial^3 h}{\partial x\,\partial t^2} = 0.$$

(a) Linearize these equations, assuming that z and u are small quantities, and show that the resulting linear equations describe waves with the following dispersion relation:

$$\omega^2 = \frac{c_0^2 k^2}{1 + k^2 h_0^2/3}.$$

To second order in kh_0 this is equivalent to Eq. (7.2.2).

(b) Defining $\bar{h} = h/h_0$, $\bar{u} = u/c_0$, $\bar{x} = x/h_0$, and $\bar{t} = tc_0/h_0$, the Boussinesq equations become

$$\begin{aligned} &\frac{\partial \bar{h}}{\partial \bar{t}} + \frac{\partial}{\partial \bar{x}}(\bar{h}\bar{u}) = 0, \\ &\frac{\partial \bar{u}}{\partial \bar{t}} + \bar{u}\frac{\partial \bar{u}}{\partial \bar{x}} + \frac{\partial \bar{h}}{\partial \bar{x}} + \frac{1}{3}\frac{\partial^3 \bar{h}}{\partial \bar{x}\,\partial \bar{t}^2} = 0. \end{aligned} \tag{7.2.32}$$

Show that steady solutions to these equations depending only on $s = \bar{x} - v\bar{t}$ are described by the motion of a particle with position $\bar{h}$ and mass $\frac{1}{3}$ moving in a potential

$$V(\bar{h}) = -\frac{B^2}{2\bar{h}} + \frac{\bar{h}^2}{2} - \frac{1}{2}\bar{h}(2A + v^2),$$

and where A and B are constants of integration, with B relating $\bar{u}$ and $\bar{h}$ through $\bar{u} = B/\bar{h} - v$.

(c) Show that a soliton solution exists with $\bar{h} \to 1$ as $s \to \pm\infty$, provided that $A = 1 + (B^2 - v^2)/2$, and find the particle energy E corresponding to this solution. Show that $\bar{u} \to 0$ at $s \to \pm\infty$ only if $B = v$, which implies $A = 1$ for solitons.

(d) Show that the height $\bar{h}$ of this soliton is related to the speed by $\bar{h} = v^2$. Solve the equation of motion numerically on $0 < s < 4$ for the case $B = v = 2$, $A = 1$, and plot the form of this soliton solution for $\bar{h}(s)$ and $\bar{u}(s)$.

(4) The *sine–Gordon equation,*

$$\frac{\partial^2 f}{\partial t^2} - \frac{\partial^2 f}{\partial x^2} + \sin f = 0, \tag{7.2.33}$$

describes, among other things, the motion of magnetic domains in a ferromagnetic material. Here, f is proportional to the magnetization in the material, which changes rapidly as one moves from one magnetic domain to another. The linearized version of this equation is the *Klein–Gordon equation*

$$\frac{\partial^2 f}{\partial t^2} - \frac{\partial^2 f}{\partial x^2} + f = 0,$$

which has applications in several areas, including particle and plasma physics.

(a) Find the dispersion relation for the Klein–Gordon equation, and show that the waves are dispersive, with phase speed $c = \pm\sqrt{1 + 1/k^2}$.

(b) Since the $\sin f$ term in the sine–Gordon equation causes both nonlinearity and dispersion, one expects that the equation has steady solutions, depending only on $s = x - vt$. Show that such solutions do exist provided that $|v| > 1$, and find the functional form of soliton solutions for which $f \to 0$ as $s \to +\infty$ and $f \to 2\pi$ for $s \to -\infty$. Show that the range in s over which f varies from 2π to 0 is proportional to $\sqrt{v^2 - 1}$. Plot $f(s)$.

(5) The nonlinear Schrödinger equation

$$i\frac{\partial \psi}{\partial t} + \frac{\partial^2 \psi}{\partial x^2} + K|\psi|^2\psi = 0 \tag{7.2.34}$$

describes the evolution of a quantum wave function $\psi(x,t)$ trapped in a potential $V(x) = -K|\psi|^2$. For $K > 0$ this particle is attracted to regions of large probability density $|\psi|^2$, i.e., it is attracted to itself. This can lead to nonlinear steepening of the wave function, but this steepening can be balanced by dispersion, allowing steady solutions and solitons. This equation has applications in many areas, including, of course, quantum physics, as well as nonlinear optics (where ψ is the electric field in a high-intensity wave moving through a nonlinear dielectric material). Solve for the steady solutions to Eq. (7.2.34), and find the functional form of solitons for which $\psi \to 0$ at $x = \pm\infty$.

(6) Analyze the steady solutions of Burgers' equation, Eq. (7.2.26). Show that these shock waves have the form given by Eq. (7.2.27).

(7) Models of compressional shock waves in a gas require the solution of the *Navier–Stokes equations,* coupled equations for the pressure $p(x,t)$, velocity

$u(x,t)$ and mass density $\rho(x,t)$ of the gas. The three equations are the continuity equation for mass density,

$$\frac{\partial \rho}{\partial t} + \frac{\partial}{\partial x}(u\rho) = 0; \tag{7.2.35}$$

the momentum equation,

$$\frac{\partial u}{\partial t} + u\frac{\partial}{\partial x}u = -\frac{1}{\rho}\frac{\partial}{\partial x}p + \eta\frac{\partial^2}{\partial x^2}u, \tag{7.2.36}$$

where η is the kinematic viscosity of the gas (units m^2/s); and an equation of state relating pressure to density,

$$p = p_0(\rho/\rho_0)^{\gamma}, \tag{7.2.37}$$

where $\gamma > 1$ is the ratio of specific heats in the gas, and where p_0 and ρ_0 are the equilibrium values of the pressure and mass density respectively. (This model neglects thermal conduction for simplicity.)

(a) Linearize these equations for small deviations from the equilibrium $p = p_0$, $u = 0$, $\rho = \rho_0$, to show that sound waves propagate with a dispersion relation given by

$$\omega^2 = k^2(c_0^2 - i\omega\eta),$$

where $c_0 = \sqrt{\gamma p_0/\rho_0}$ is the speed of sound. Show that these waves damp exponentially due to finite viscosity, at a rate r given by $r = k^2\eta/2$ when $k\eta \ll c_0$.

(b) Analyze steady shock-wave solutions to Eqs. (7.2.35)–(7.2.37), which depend only on the variable $s = x - vt$ (where v is the shock speed). Show, for a shock moving to the right into still air where $\rho = \rho_0$ and $u = 0$ at $x \to \infty$, that the fluid velocity in the shock satisfies

$$\eta\frac{\partial u}{\partial s} = \frac{c_0^2}{\gamma - 1}\left[\left(\frac{\rho}{\rho_0}\right)^{\gamma-1} - 1\right] + \frac{1}{2}u^2 - uv, \tag{7.2.38}$$

where $\rho/\rho_0 = v/(v-u)$.

(c) Use the result of part (b) to derive the following Hugoniot relation:

$$\frac{c_0^2}{\gamma-1}\left[\left(\frac{v}{v-u_-}\right)^{\gamma-1} - 1\right] + \frac{1}{2}u_-^2 - u_-v = 0,$$

where u_- is the velocity of the gas well behind the shock.

(d) Show that the shock always travels with speed $v > c_0$ and that v/c_0 is an increasing function of u_-/u. (Hint: Scale u_- by v in the Hugoniot relation, and note that $[1/(1-x)]^{\gamma-1}$ is an increasing function of x.)

(e) Analytically solve for the shock assuming it is *weak*, so that $v \to c_0$ and $u_- \ll c_0$. Show that in this limit,

$$u_- = \frac{4c_0}{\gamma+1}\left(1 - \frac{c_0}{v}\right) + O\left(\left(1 - \frac{c_0}{v}\right)^2\right)$$

and

$$u(s) = \frac{u_-}{1 + \exp[u_-(1+\gamma)s/2\eta]}.$$

Plot u vs. s, assuming that $\gamma = \frac{5}{3}$, $c_0 = 340$ m/s, $v = 360$ m/s, and $\eta = 1 \times 10^{-2}$ m^2/s. Roughly speaking, what is the width of the shock in meters? (Hint: Expand the Hugoniot relation and Eq. (7.2.38) to *second* order in u_-/v.)

(8) Use the Galerkin method to numerically solve Burgers' equation with constant nonuniform forcing, in periodic boundary conditions, on $0 < x < 20$:

$$\frac{\partial f}{\partial t} + f\frac{\partial f}{\partial x} - \frac{1}{10}\frac{\partial^2 f}{\partial x^2} = \frac{2}{15}e^{-(x-5)^2}, \qquad 0 < x < 20.$$

The forcing function on the right-hand side describes an explosive local increase in temperature, pressure, or density. Take 13 Fourier modes ($M = 12$), and solve for $0 < t < 25$. Note the formation of a shock front propagating to the right.

(9) Use the CTCS method to solve the sine–Gordon equation (7.2.33) in periodic boundary conditions on $0 < x < L$, with $L = 10$. For the initial condition, take $f(x,0) = 3\sin(\pi x/L)$. Integrate the result for $0 < t < 10$.

(10) Use the CTCS method to solve the traffic flow problem (7.1.43), taking $f(x,0) = 0$ (i.e. an initially empty freeway) with $s = 1$ for $0 < x < 1$ and $s = 0$ otherwise (i.e., there is an on ramp at this location, feeding cars onto the freeway at a constant rate). Take $c = 1$ and use periodic boundary conditions on $0 < x < 10$, and integrate in time up to a time as close as possible to the formation of a singularity (the CTCS method will become unstable at some point before this time, depending on your choice of step size).

(11) A diffusive term is often added to the traffic flow problem, in order to model diffusive spreading of the cars in the absence of nonlinearity. The equation now becomes

$$\frac{\partial f}{\partial t} + (c - f)\frac{\partial f}{\partial x} - \chi\frac{\partial^2 f}{\partial x^2} = s.$$

Using the numerical method of your choice, redo the previous problem, taking $\chi = 0.2$. Solve the problem over the time interval $0 < t < 15$. Note the formation of a backward-propagating shock (a traffic jam, propagating back from the on ramp).

(12) Here's another numerical method for solving PDEs: finite-difference the spatial operator on a grid, and then solve the resulting coupled ODEs in time. These ODEs can be interpreted as the equations of motion for a system of coupled masses and springs, which provides us with another way of thinking about the problem. Consider, for instance, the following nonlinear PDE:

$$\frac{\partial^2 \eta}{\partial t^2} = c^2 \frac{\partial^2 \eta}{\partial x^2}\left(1 + \alpha \frac{\partial \eta}{\partial x}\right).$$

Centered-differencing of the spatial operator on a uniform grid of positions $x = j\,\Delta x$ yields

$$\begin{aligned} \frac{d^2\eta_j}{dt^2} &= c^2 \frac{\eta_{j+1} - 2\eta_j + \eta_{j-1}}{\Delta x^2}\left(1 + \alpha \frac{\eta_{j+1} - \eta_{j-1}}{2\,\Delta x}\right) \\ &= \frac{c^2}{\Delta x^2}\left((\eta_{j+1} - \eta_j) - (\eta_j - \eta_{j-1}) + \frac{\alpha}{2\,\Delta x}\left[(\eta_{j+1} - \eta_j)^2 - (\eta_{j-1} - \eta_j)^2\right]\right). \end{aligned} \tag{7.2.39}$$

These equations are the same as the dynamical equations for a series of masses of unit mass, at positions $\eta_j(t)$, connected to nearest neighbors by nonlinear springs with a nonlinear spring force vs. displacement η given by

$$F = -k\eta - \gamma\eta^2,$$

where $k = c^2/\Delta x^2$ and $\gamma = c^2\alpha/(2\,\Delta x^3)$. In a famous paper from the early days of computational physics, Fermi, Pasta, and Ulam (FPU, 1955) studied the behavior of this coupled mass–spring system. The authors were expecting that the nonlinearity of the system, combined with the many degrees of freedom, would result in chaotic dynamics that would lead to a thermal equilibrium state, where energy was shared equally between all degrees of freedom. Surprisingly, however, this is not what the authors observed. In this problem, we will repeat a portion of their work.

(a) Use the molecular dynamics code in Chapter 2 (modified for 1D motion and nearest-neighbor interactions) to solve Eqs. (7.2.39) for $M + 1$ masses, where $M = 15$, with boundary conditions $\eta_0 = \eta_M = 0$ (i.e., the end masses are fixed to walls). For initial conditions take $\eta_j = \sin(\pi j/M)$ (i.e., all energy in the lowest normal mode of the linearized system), and $\dot{\eta}_j = 0$. Taking $k = 1$ and $\gamma = \frac{1}{4}$, integrate the equations for $0 < t < 1000$. Plot $2j + \eta_j(t)$ vs. t on a single graph for $j = 1, \dots, M - 1$. (The $2j$ is used to separate the different curves.)

(b) Perform the following analysis on your results from part (a). FPU expected that the nonlinearity in the force would result in coupling between normal modes, and a cascade of energy to higher and higher Fourier modes, resulting in eventual energy equipartition where, on average, all modes have the same energy. The amplitude of the kth Fourier mode vs. time is

$$a_k(t) = \frac{2}{M} \sum_{j=0}^{M} \eta_j(t) \sin \frac{\pi k j}{M}, \qquad k = 1, 2, \ldots, M-1. \quad (7.2.40)$$

The energy in a normal mode (neglecting the nonlinear terms) is

$$E_k(t) = \tfrac{1}{2}\dot{a}_k^2 + \tfrac{1}{2}\omega^2(k) a_k^2,$$

where $\omega^2(k) = 4\sin^2[\pi k/(2M)]$ is the frequency squared of these normal modes. [See Eq. (4.2.30).] Evaluate these mode energies and plot them versus time to show the spectral distribution of mode energies. Is energy being shared equally among these modes?

(c) What do you think would happen if we used this technique on the KdV equation? Would energy be shared equally among the modes?

REFERENCES

T. B. Benjamin, *Instability of periodic wavetrains in nonlinear dispersive systems*, Proc. R. Soc. London Ser. A **299**, 59 (1967).

J. Binney and S. Tremaine, *Galactic Dynamics* (Princeton University Press, Princeton, NJ, 1987).

C. K. Birdsall and A. B. Langdon, *Plasma Physics via Computer Simulation* (Adam Hilger, Bristol, 1991).

C. F. Driscoll and K. Fine, *Experiments on vortex dynamics in pure electron plasmas*, Phys. Fluids B **2**, 1359 (1990).

E. Fermi, J. Pasta, and S. Ulam, *Studies of Nonlinear Problems*, Los Alamos Report LA-1940 (1955).

K. Fine, W. Flynn, A. Cass, and C. F. Driscoll, *Relaxation of 2D turbulence to vortex crystals*, Phys. Rev. Lett. **75**, 3277 (1995).

J. Glanz, *From a turbulent maelstrom, order*, Science **280**, 519 (1998).

L. Greengard, *Fast algorithms for classical physics*, Science **280**, 519 (1998).

A. Hasegawa and Y. Kodama, *Solitons in Optical Communications* (Oxford University Press, New York, 1995).

J. Scott Russell, *Report on Waves*, Report of the Fourteenth Meeting of the British Association for the Advancement of Science, York, September 1844 (London, 1845), pp. 311–390.

G. B. Whitham, *Linear and Nonlinear Waves* (John Wiley and Sons, New York, 1999).

CHAPTER 8

INTRODUCTION TO RANDOM PROCESSES

8.1 RANDOM WALKS

8.1.1 Introduction

How is heat conducted through a material? We know from experiment that the heat flux obeys Fick's law, $\mathbf{\Gamma} = -\kappa \nabla T$, and as a result the temperature is described by the heat equation,

$$\partial T / \partial t = \chi \nabla^2 T(\mathbf{r}, t). \tag{8.1.1}$$

Why, however, does the heat flux obey Fick's law? What are the underlying microscopic processes that lead to this result?

In a conductor, it is known that heat energy is carried from place to place mostly by the conduction electrons, which are free to move throughout the material. In a hot region of the conductor, the electrons on average move more rapidly than in the surrounding colder regions. As a result, they disperse through the colder electrons and carry their energy with them. The resulting diffusion of heat is intimately connected to the diffusion of the electrons themselves. For this reason, good electrical conductors are also good thermal conductors.

The electrons don't simply move in straight-line trajectories; otherwise they would be described by the collisionless Boltzmann equation, not the heat equation. Rather, the electrons collide with one another and with nearly stationary ions in the material. As a result of these collisions, the electron dynamics is chaotic, and the electrons move randomly. We will see that this random component to the dynamics is essential to the diffusive spreading of the temperature observed in solutions of the heat equation.

These complex collisional processes require many-body quantum theory for their description. However, Fick's law is a very simple result. Furthermore, this law applies to a broad range of systems for which the details of the dynamics differ greatly. For example, thermal conduction in nonconductors is due to the propagation of phonons (sound wave packets) rather than electrons, yet Fick's law still applies. Can we find a simplified model of the electron dynamics that still leads to diffusion of heat, but which avoids the niceties of quantum collision theory?

Many of the ideas discussed in the following sections were developed by Ludwig Boltzmann, the father of statistical physics. Important theory contributions were also made by Poisson, Maxwell, Gibbs, Einstein, von Smoluchowski, and others. However, we will begin with some elementary ideas in probability theory, first put forward by Jacob Bernoulli in the seventeenth century.

8.1.2 The Statistics of Random Walks

The Random Walk Model We will describe thermal diffusion in a conductor with a random walk model. First, we will neglect quantum effects and treat the electrons as a gas of classical point particles. Next, we will assume that the conduction electrons move only in one dimension. Third, we will assume that electron positions are restricted to a grid of values, $x = j\,\Delta x$, $j = \dots, -2, -1, 0, 1, 2, \dots$. Fourth, we will assume that there are only two groups of electrons, *hot* electrons with energy 1 in scaled units, and *cold* electrons with energy zero, which remain stationary and so are unimportant. The hot electrons diffuse, and carry their energy with them, causing thermal diffusion.

We will employ the following simple dynamical model for the hot electrons: an electron suffers a random collision at regular time intervals Δt. Thus, Δt^{-1} *is roughly the electron collision frequency*. After a collision, the electron ends up at one of the two neighboring grid points to its initial location. Thus, Δx *is roughly the mean free path*, or mean distance traveled between collisions. If the electron is at grid point j at time $t_n = n\,\Delta t$, at the next timestep t_{n+1} the electron is at either grid point $j-1$ or $j+1$, with equal probability.

We will show that in the limit that the number of electrons $M \to \infty$ and as $\Delta t \to 0$ and $\Delta x \to 0$, the number density $n(x,t)$ of the hot particles obeys the diffusion equation

$$\frac{\partial n}{\partial t} = D\,\nabla^2 n, \tag{8.1.2}$$

where D is the *particle diffusion coefficient* (unit m^2/s), related to Δx and Δt by $D = \Delta x^2/(2\,\Delta t)$. Hence, in this model the thermal diffusivity is $\chi = D$.

This version of the electron dynamics is called a random walk in one dimension. It is not realistic as it stands, but it contains much of the basic physics behind the chaotic collision processes that are actually responsible for thermal conduction. [Some of the pitfalls associated with the neglect of quantum effects are considered in Exercise (12).] The same model can be applied to any number of other diffusive processes, such as thermal diffusion in nonconductors where phonons take the place of electrons, or the diffusion of atoms through a gas. The model is simple enough to solve analytically using concepts from probability theory. Before we do so, however, it is entertaining and instructive to view the dynamics via a simulation of the random walk.

Simulations *Mathematica*'s intrinsic function **Random** can be used to perform random walk simulations. **Random[]** generates a random number in the range from 0 to 1. **Random[t,range]** generates a random number of type **t** (**t** being either **Integer**, **Real**, or **Complex**) in the range **range**. For instance, **Random[Integer,{i,j}]** yields any integer in the list $\{i, i+1, \ldots, j\}$ with equal probability. this means that, if we evaluate this random number $N \gg 1$ times, the number of times $H(i)$ that integer i occurs is *approximately* the same as for any other integer in the list. More precisely, in the limit as $N \to \infty$, the probability $P(i)$ that i occurs is defined as

$$P(i) = \lim_{N \to \infty} \frac{H(i)}{N}. \tag{8.1.3}$$

This is simply the fraction of times that i occurs. In any random process, this probability is well defined and is independent of N in the limit as $N \to \infty$. Integers i and j are equally probable when $P(i) = P(j)$. However, for finite N, $H(i)$ may differ substantially from the prediction based on the *expected value*, $\langle H \rangle(i) = NP(i)$. One of the objects of this chapter will be to examine how quantities such as $H(i)$ vary from their expected (*average*) values.

In order to make an electron take random steps of size Δx to either the left or right, define the random number function **s[]**:

Cell 8.1

```
s[] := Δx (2 Random[Integer, {0, 1}] -1)
```

This function, which has no arguments, is designed to return either $-\Delta x$ or Δx with equal probability. (Try evaluating it several times to test this. Make a table of 20 values. How many times does Δx occur?)

We will follow the dynamics of M electrons simultaneously. For the kth electron, with position x_k^n at timestep n, the dynamics is given by the equation

$$x_k^n = x_k^{n-1} + s. \tag{8.1.4}$$

This equation can be programmed as a recursion relation in the usual way:

Cell 8.2

```
x[k_, n_] := x[k, n] = x[k, n - 1] + s[]
```

We will follow the dynamics of $M = 200$ electrons for $N_{\text{step}} = 50$ steps. As an initial condition, we will start all electrons at $x = 0$:

Cell 8.3

```
x[k_, 0] = 0;

Δx = 1; M = 200; Nstep = 50;
positions = Table[x[k, n], {k, 1, M}, {n, 0, Nstep}];
```

Each row in the matrix positions is the trajectory of a particular electron, as shown in Cell 8.4. The electron wanders randomly, sometimes returning to the origin, but over time tending to take large excursions from its initial location. This can also be observed by viewing the positions of all the electrons vs. time using an animation.

Cell 8.4

```
ListPlot[positions[[10]], AxesLabel→{"n", "x/Δx"}];
```

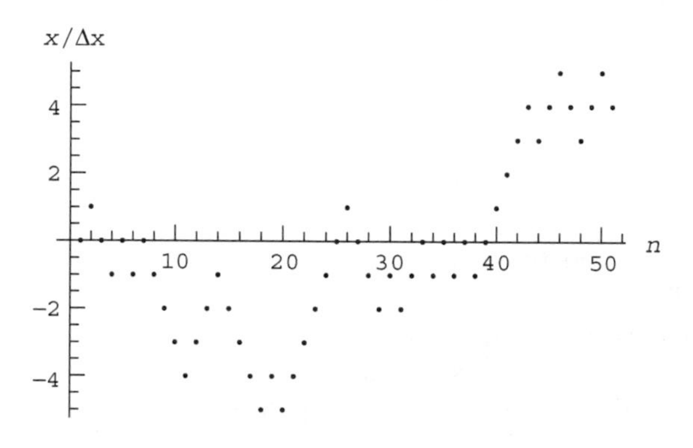

At timestep n the positions of the M electrons are given by the following list:

Cell 8.5

```
pos[n_] := Table[positions[[j, n + 1]], {j, 1, M}]
```

These positions are followed in Cell 8.6 for 50 steps.

Although the dynamics of any one particle is completely unpredictable, as a group the cloud of particles spreads in a quite predictable and reproducible fashion. If the experiment is repeated several times, the cloud will spread in the same manner every time. This remarkable predictability of large numbers of randomly moving particles is actually a simple consequence of the laws of probability. In a large group of particles starting from the origin, we expect in the first step that roughly half the particles will move in one direction, and the other half will move in the other direction. In the next step, half of these two groups will step in each direction again, and so on, producing the observed smooth and even spreading of the distribution. The random walk shown above is one of the simplest examples of how predictable behavior can arise from the chaotic dynamics of large groups of particles. This effect is at the root of several fields of physics, notably statistical mechanics—not to mention other fields of endeavor (such as insurance).

The smooth spreading of the particle distribution can be quantitatively studied by making a histogram of particle positions as a function of timestep. We have used histograms before, in Sec. 7.1, but here we will go into a little more detail. A histogram is a graph that shows the number of particles that fall into given bins. Define the histogram function $H(j, n)$ as the number of particles located at grid position j at timestep n. This function satisfies

$$\sum_j H(j,n) = M, \tag{8.1.5}$$

where the sum runs over all possible grid points. A histogram is simply a plot of H vs. j at any given time.

Cell 8.6

```
Table[ListPlot[pos[n],
   PlotRange→{-30, 30}, AxesLabel→
    {"k", "x/Δx, n ="<>ToString[n]}], {n, 0, 50}];
```

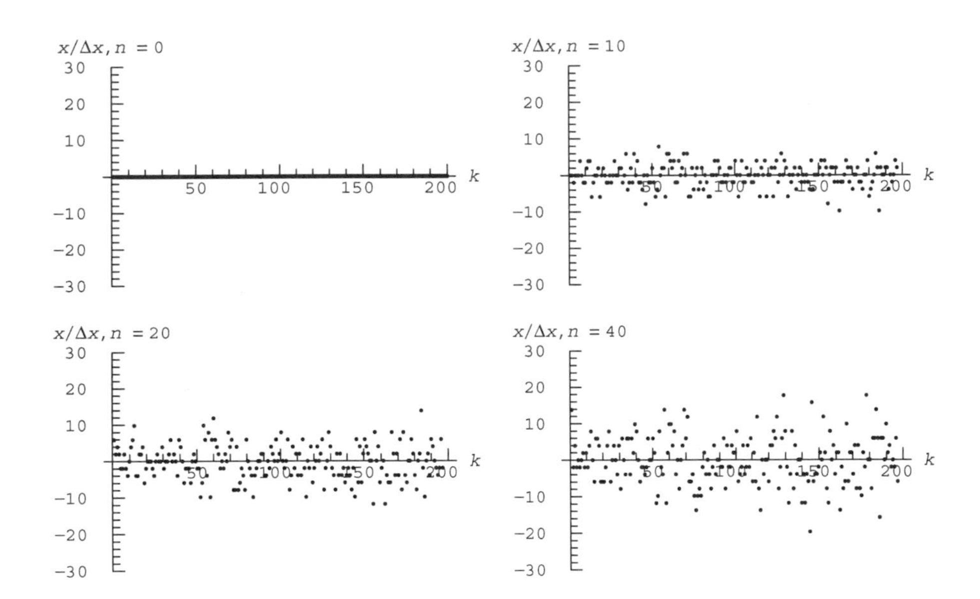

The function $H(j, n)$ can be evaluated directly from a list of positions using the **BinCounts** function, available in the **Statistics** add-on packages. **Bincounts[data,{xmin,xmax,Δx}]** determines the number of elements in the list **data** that fall in bins of size **Δx**, running from **xmin** to **xmax**. For instance,

Cell 8.7

```
<<Statistics`;

H = BinCounts[{0, 1, 3, 1, 5, 5, 4, 3, 3, 2},
  {-1/2, 5 + 1/2, 1}]

{1, 2, 1, 3, 1, 2}
```

There are one element in the range from $-\frac{1}{2}$ to $\frac{1}{2}$, two in the range $\frac{1}{2}$ to $1\frac{1}{2}$, and so on. The function H is simply the numbers associated with each bin. These values can be plotted as a bar chart using the **Histogram** plotting function. In order to manipulate the axes ranges in the nicest way, we first generate a histogram and then use **Show** to set the axis ranges and tick marks, as shown in Cell 8.8.

Cell 8.8

```
<<Graphics`;

histograms = Table[p1 = Histogram[pos[n], DisplayFunction→
 Identity,
     HistogramCategories→50, HistogramRange→{-25, 25},
     PlotLabel -> "H[j, n],n= "<>ToString[n] <> "\n",
       AxesLabel→{"j", ""}];

   Show[p1, PlotRange→{{-25, 25}, {0, 100}},
    DisplayFunction→$DisplayFunction, AxesOrigin→
     {0, 0}], {n, 0, 50, 1}];
```

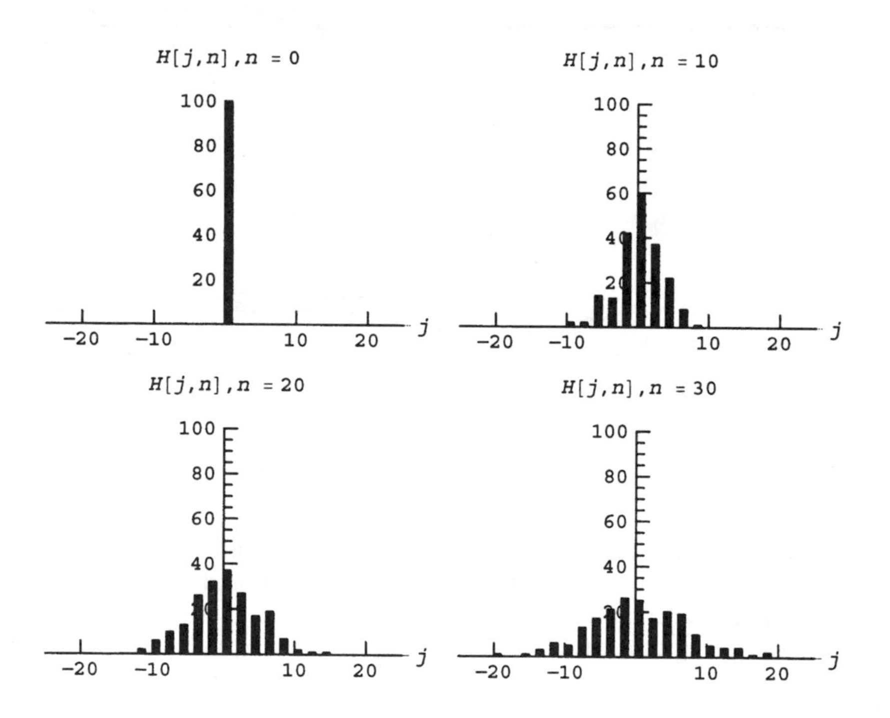

The figure shows that the distribution of particles spreads out smoothly over time in a manner reminiscent of solutions to the heat equation, although there is some noise. The noise is associated with the finite number of particles used in the simulation ($M = 200$). It is only by using many particles that smooth results are obtained. In the limit $M \to \infty$, the fraction of particles at any given position, $H(j, n)/M$, is a well-defined quantity between zero to 1. Following along with the definition of probability given in Eq. (8.1.3), we define the probability $P(j, n)$ of finding particles at a given grid position at timestep n:

$$P(j,n) \equiv \lim_{M\to\infty} \frac{H(j,n)}{M}. \tag{8.1.6}$$

According to Eq. (8.1.5), the probability satisfies

$$\sum_j P(j,n) = 1; \tag{8.1.7}$$

that is, there is unit probability of finding the particles somewhere on the grid.

From Eq. (8.1.6), we can see that the probability $P(j,n)$ is the fraction of times that particles fall at position j at time n, taken out of an infinite number of different trajectories. For instance, at $n=0$ we have $P(0,0)=1$, there is unit probability of finding particles at the origin (and therefore, there is zero probability of finding them anywhere else). At $n=1$, we would expect from our random walk dynamics that $P(1,1)=P(-1,1)=\frac{1}{2}$. How to characterize $P(j,n)$ for $n>1$ will be the subject of the next section.

First, however, we consider another way to examine the spreading of the distribution: by evaluating the average position of the particles vs. time, $\langle x\rangle(n)$, and the mean squared position vs. time, $\langle x^2\rangle(n)$.

The average position is defined as

$$\langle x\rangle(n) = \lim_{M\to\infty} \frac{1}{M} \sum_{k=1}^{M} x_j^n. \tag{8.1.8}$$

This quantity can be obtained from the simulation data using the function **Mean** available in the **Statistics** add-on packages. **Mean** determines the mean value of a list of numbers (thought one can just take the sum and divide by M), as shown in Cell 8.9.

The average position fluctuates around zero, because the particles spread equally in both directions in an unbiased random walk. Fluctuations in this quantity are related to the finite number of particles in the simulation, in this case, $M=200$. For example, if there were only one particle, the average position would fluctuate considerably, but as $M\to\infty$ these fluctuations are averaged out and we would expect that, for $M\to\infty$,

$$\langle x\rangle(n) = 0. \tag{8.1.9}$$

In the next section we will prove this result.

Cell 8.9

```
<<Statistics`;
xav[n_] := Mean[pos[n]];

ListPlot[Table[{n, xav[n]}, {n, 0, 50}],
  AxesLabel → {"n", "<x>/Δx"}];
```

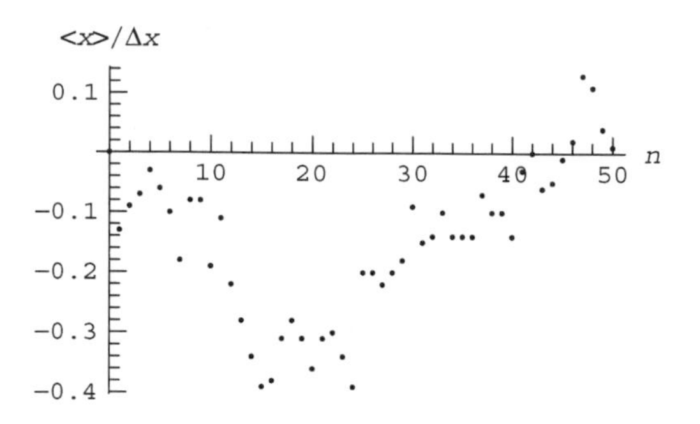

The spreading of the distribution is characterized by $\langle x^2 \rangle(n)$, defined as

$$\langle x^2 \rangle(n) = \lim_{M\to\infty} \frac{1}{M} \sum_{k=1}^{M} \left(x_j^n\right)^2. \tag{8.1.10}$$

This function is the mean squared distance of particles from the origin. The rms (root-mean-square) width of the distribution is the square root of $\langle x^2 \rangle$, and is a measure of the mean distance of particles from the origin. $\langle x^2 \rangle$ can also be evaluated using **Mean**:

Cell 8.10

```
x2av[n_] := Mean[pos[n]^2]
```

(Note: This command squares each element in **pos[n]**, and then takes the mean.) The result is plotted in Fig. 8.1. (Of course, we cannot take the $M \to \infty$ limit in a simulation, so our numerical results have some statistical error. The size of this error will be considered in Sec. 8.3.2.) The dots are values from the simulation, and

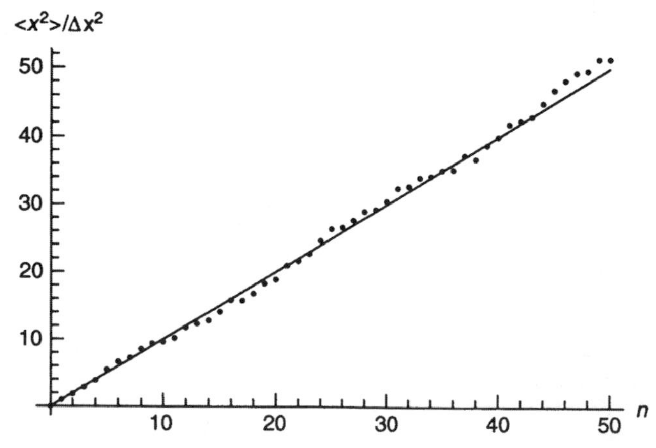

Fig. 8.1 Spreading of particles in a random walk simulation (dots) together with theory (line).

the straight line is a theory for $\langle x^2\rangle(n)$ that we will derive presently [see Eq. (8.1.31)]. It seems that $\langle x^2\rangle$ increases linearly with time, so that the rms width of the particle distribution increases like $\sqrt{t}$. This is a signature of diffusive processes, and was observed previously in solutions of the heat equation [see Eq. (5.1.100)]. In fact, the width $w(t)$ defined by Eq. (6.100) is simply the rms width of the solution.

The Binomial Distribution We can analytically determine the probability distribution $P(j,n)$ for the preceding random walk process by applying elementary ideas from probability theory. To illustrate the random walk, we first construct a *decision tree*. Such a tree is shown in Fig. 8.2 for the first three steps of the walk. Starting at $j=0$, a particle can move either to the left or the the right to $j=\pm 1$, and each location has an equal probability of $\frac{1}{2}$. In the next step, the particles can either return to the origin, or move further away, to ± 2. There are now four possible paths taken by the particles, shown as dashed lines. Each path is equally likely. Only one out of four of the paths leads to $j=-2$ and similarly for $j=+2$. However, two out of four of the paths return to $j=0$. Therefore,

$$P(-2,2)=P(2,2)=\tfrac{1}{4} \quad \text{and} \quad P(0,2)=\tfrac{2}{4}=\tfrac{1}{2}.$$

In fact, one can see that for an unbiased random walk

$$P(j,n)=\frac{\text{number of paths arriving at } j \text{ after } n \text{ steps}}{\text{total number of paths}}. \tag{8.1.11}$$

For instance, for $n=3$ there are now eight possible paths, and

$$P(-3,3)=P(3,3)=\tfrac{1}{8},$$
$$P(-1,3)=P(1,3)=\tfrac{3}{8}.$$

(Exercise: In Fig. 8.2 can you see the three paths that lead to $j=1$ for $n=3$?)

It is easy to generalize these results to consider a random walk with a bias, where the probability of a step to the right equals p and that for a step to the left

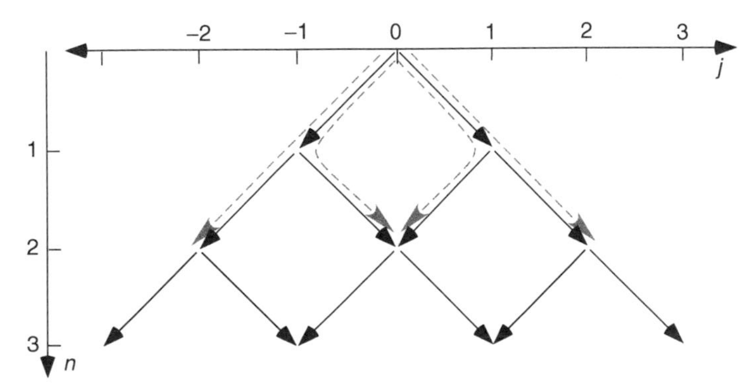

Fig. 8.2 Decision tree for the first three steps of a random walk, starting at $j=0$. Dashed lines: possible paths for $n=2$ steps.

equals $q = 1 - p$. Physically, such bias can occur if there is some reason why particles should move more in one direction than the other, as would be the case if gravity or an electric field acts on the system.

Clearly, after one step $P(-1,1) = q$, $P(1,1) = p$. After two steps, $P(-2,2) = q^2$ (two steps to the left in a row have total probability $q \times q = q^2$), while $P(2,2) = p^2$ for the same reason. Also, the two paths that lead back to the origin each have probability pq, so the total probability $P(0,2)$ is a sum of the probabilities for each separate path, $P(0,2) = 2pq$. These results agree with the unbiased case given above when $p = q = \frac{1}{2}$.

Similar reasoning leads to the following results for $n = 3$:

$$\begin{aligned}
P(-3,3) &= q^3, \\
P(-1,3) &= q^2p + qpq + pq^2 = 3q^2p, \\
P(1,3) &= pqp + qp^2 + ppq = 3p^2q, \\
P(3,3) &= p^3.
\end{aligned}$$

These results also agree with Eq. (8.1.11) for the unbiased case $p = q = \frac{1}{2}$.

We are now prepared to consider an arbitrary number of steps. In order to get to grid point j after n steps, a particle must move n_R steps to the right and n_L steps to the left, with

$$n = n_R + n_L \tag{8.1.12}$$

and

$$j = n_R - n_L. \tag{8.1.13}$$

These two equations imply that the position j can be written in terms of n_R and the total number of steps n:

$$j = 2n_R - n. \tag{8.1.14}$$

The decision tree for this random walk has the following form. Any given path that ends up at j will consist of n_R steps to the right, and $n - n_R$ steps to the left. Each path is simply a sum of steps, as shown below for $n = 15$:

$$j = -1+1-1-1+1+1+1-1-1+1-1-1+1-1-1 = -3.$$

The steps to the left or right in this sum can be in any order and the path will still end up at $j = -3$, provided that there are $n_R = 6$ steps to the right and $n_L = 9$ steps to the left. The total probability of any such path is $p^{n_R} q^{n_L}$. Therefore, *all paths* ending at j after n steps have the *same probability*, $p^{n_R} q^{n_L}$.

Now all we need is the number of different paths that lead to position j. This is simply the number of distinct permutations of the left and right steps making up the path. This combinatoric factor is given by the formula

$$\text{number of paths} = \frac{n!}{n_R! n_L!}.$$

This factor can be understood as follows. If there were n distinct elements in the list of steps making up a path, then there would be $n!$ ways of arranging these elements: n possibilities for the first element, $n-1$ possibilities for the second element, and so on. Now, however, there are subgroups consisting of n_R right steps and n_L left steps within this list for which the elements are all identical. Permutations which correspond to a mere rearrangement of these elements within their own subgroups are not distinct paths. The number of such permutations is $n_R!$ for the right steps and $n_L!$ for the left steps. We divide by these factors in order to be left with only the distinct permutations of the list.

Since the total probability is the probability per path times the number of paths, we arrive at

$$P(j,n) = p^{n_R} q^{n_L} \frac{n!}{n_R! n_L!}, \qquad \text{assuming } n+j \text{ even and } |j| \le n, \quad (8.1.15)$$

where n_R is given in terms of n and j by Eq. (8.1.14), and $n_L = n - n_R$. Note that the equation is valid only for $n+j$ even, because if $n+j$ is odd then n_R is not an integer: see Eq. (8.1.14). This is related to the fact that for n even, only even grid points are populated, and for n odd, only odd grid points are populated. Thus, we also have

$$P(j,n) = 0, \qquad n+j \text{ odd or } |j| > n. \quad (8.1.16)$$

Equations (8.1.15) and (8.1.16) give the famous *binomial distribution*, first derived by Jacob Bernouilli. In Cell 8.11 we compare this theoretical formula with the histograms obtained previously. For simplicity in the graphics, we evaluated $P(j,n)$ using Eq. (8.1.15) only, for real values of j, although only integer values of j with $j+n$ even should have been allowed. Fortunately, Eq. (8.1.15) is a smooth function of j, and we can see that $MP(j,n)$ nicely tracks the histograms from the simulation. [Question: Why do we need the factor of M in $MP(j,n)$ in order to match the histogram?]

Cell 8.11

```
p = q = 1/2;
nR = (n + j)/2;
nL = n - nR;
P[n_, j_] = p^nR q^nL n !/(nR ! nL !);

Table[Plot[M P[n, x], {x, -n - 1, n + 1},
    PlotRange→{0, 200}, DisplayFunction→Identity],
     {n, 0, 50}];

Table[Show[histograms[[n]], %[[n]], PlotRange→
   {{-25, 25}, {0, 200}},
    DisplayFunction→$DisplayFunction], {n, 1, Length[%]}];
```

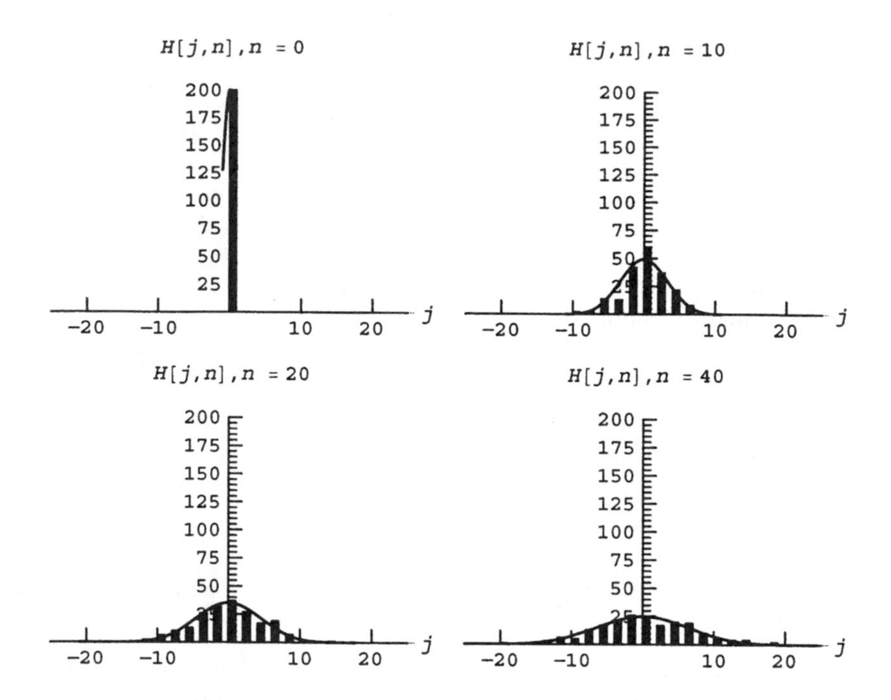

Averages We can use the binomial distribution to understand the behavior of average quantities such as $\langle x\rangle(n)$ and $\langle x^2\rangle(n)$. The average of a general function $f(x)$ of the particle positions can be written as

$$\langle f\rangle(n) = \lim_{M\to\infty} \frac{1}{M} \sum_{k=1}^{M} f(x_k^n), \tag{8.1.17}$$

where the sum is over the particles. This is the method used previously to evaluate $\langle x\rangle$ and $\langle x^2\rangle$; see Eqs. (8.1.18) and (8.1.10). However, there is a better way to evaluate this average, in terms of the probability $P(j,n)$. First, we note that the particles can be grouped onto their grid locations j, with $H(j,n)$ giving the number of particles on each grid point. Then an equivalent way to do the sum over particles in Eq. (8.1.17) is

$$\langle f\rangle(n) = \lim_{M\to\infty} \frac{1}{M} \sum_{j=-n}^{n} H(j,n) f(x_j),$$

where $x_j = j\,\Delta x$ and now the sum runs over all the possible grid positions. Using Eq. (8.1.6) then yields the important result

$$\langle f\rangle(n) = \sum_{j} P(j,n) f(x_j). \tag{8.1.18}$$

We need not sum over all the particle positions to determine $\langle f\rangle$; rather, we can sum over the probabilities of different states of the system. With this formula, we can prove several other useful results. For example, for any two functions $f(x)$ and $g(x)$, and for any constant c, Eq. (8.1.18) immediately implies that

$$\langle f + cg \rangle = \langle f \rangle + c \langle g \rangle. \tag{8.1.19}$$

Thus, taking an average is a linear operation.

We can apply Eq. (8.1.18), along with Eqs. (8.1.15) and (8.1.16), to calculate $\langle x \rangle(n)$ and $\langle x^2 \rangle(n)$. These averages can be calculated by means of the *binomial theorem*. This theorem states that

$$\sum_{n_R=0}^{n} \frac{n!}{n_R!(n-n_R!)} p^{n_R} q^{n-n_R} = (p+q)^n. \tag{8.1.20}$$

However, the argument of the sum can be directly related to $P(j,n)$. According to Eq. (8.1.14),

$$\sum_{n_R=0}^{n} = \sum_{j=-n,-n+2,-n+4,\ldots,n} . \tag{8.1.21}$$

Also, the argument of the sum in Eq. (8.1.20) is just the value of the binomial distribution $P(j,n)$ for $j+n$ even. Therefore, Eq. (8.1.20) can be written as

$$\sum_{j=-n}^{n} P(j,n) = (p+q)^n = 1, \tag{8.1.22}$$

where in the last step we used the fact that $p+q=1$. Of course, Eq. (8.1.22) is merely a restatement of the general property of probability distributions, Eq. (8.1.7). However, we can also use Eq. (8.1.22) to calculate averages.

For $\langle x \rangle(n)$, we note that Eq. (8.1.14) implies

$$\langle x \rangle = \langle j\,\Delta x \rangle = \Delta x \langle 2n_R - n \rangle = \Delta x \,(2\langle n_R \rangle - n), \tag{8.1.23}$$

where in the third and fourth steps we applied Eq. (8.1.19). However, according to Eqs. (8.1.18), (8.1.15), and (8.1.16) the average number of steps to the right after n steps, $\langle n_R \rangle(n)$, can be evaluated as

$$\begin{aligned} \langle n_R \rangle(n) &= \sum_{j=-n,-n+2,-n+4,\ldots,n} n_R P(j,n) \\ &= \sum_{n_R=0}^{n} n_R \frac{n!}{n_R!(n-n_R)!} p^{n_R} q^{n-n_R}, \end{aligned} \tag{8.1.24}$$

where in the second step we have applied Eq. (8.1.21). This sum can be carried out analytically by noticing that $n_R p^{n_R} = p\,\partial(p^{n_R})/\partial p$. Applying this identity to Eq.

(8.1.22) yields

$$\langle n_R\rangle(n) = p\frac{\partial}{\partial p}\sum_{n_R=0}^{n}\frac{n!}{n_R!(n-n_R!)}p^{n_R}q^{n-n_R}$$

$$= p\frac{\partial}{\partial p}(p+q)^n = np(p+q)^{n-1} = np. \tag{8.1.25}$$

In the second step we have employed the binomial theorem, Eq. (8.1.20). Equation (8.1.25) is eminently reasonable: in any one step, the probability of a step to the right is p, so after n steps one expects to have taken an average number of steps np to the right.

Substitution of this result into Eq. (8.1.23) then yields

$$\langle x\rangle = \Delta x\,(2pn - n) = n\,\Delta x\,(p-q). \tag{8.1.26}$$

If $p = q = \frac{1}{2}$, then $\langle x\rangle = 0$ for all time, as we expected; see Eq. (8.1.9). However, for $p \neq q$, there is a net drift of the particle distribution at constant velocity v, $\langle x\rangle = vt$, where the drift velocity v is given by

$$v = (p-q)\frac{\Delta x}{\Delta t}. \tag{8.1.27}$$

We will have more to say concerning this drift later, but first let us turn to the evaluation of $\langle x^2\rangle$. We follow the same procedure as for (x). Using Eqs. (8.1.14) and (8.1.19) we obtain

$$\langle x^2\rangle = \langle j^2\,\Delta x^2\rangle = \Delta x^2\left\langle (2n_R - n)^2\right\rangle = \Delta x^2\left(4\langle n_R^2\rangle - 4\langle n_R\rangle n + n^2\right). \tag{8.1.28}$$

We already know $\langle n_R\rangle = np$, so we only need to determine $\langle n_R^2\rangle$. Using a procedure analogous to Eq. (8.1.24) leads to

$$\begin{aligned}
\langle n_R^2\rangle(n) &= \sum_{n_R=0}^{n} n_R^2\frac{n!}{n_R!(n-n_R!)}p^{n_R}q^{n-n_R}\\
&= p\frac{\partial}{\partial p}p\frac{\partial}{\partial p}\sum_{n_R=0}^{n}\frac{n!}{n_R!(n-n_R!)}p^{n_R}q^{n-n_R}\\
&= p\frac{\partial}{\partial p}p\frac{\partial}{\partial p}(p+q)^n = p\frac{\partial}{\partial p}np(p+q)^{n-1}\\
&= np(p+q)^{n-1} + np^2(n-1)(p+q)^{n-2}\\
&= np + n(n-1)p^2,
\end{aligned}$$

where in the second step we employed the identity $n_R^2 p^{n_R} = p(\partial/\partial p)p(\partial/\partial p)p^{n_R}$. Applying this result to Eq. (8.1.27) yields, after some algebra,

$$\langle x^2\rangle = \langle x\rangle^2 + 4npq\,\Delta x^2, \tag{8.1.29}$$

where $\langle x\rangle$ is given by Eq. (8.1.26).

For the case where $p = q = \frac{1}{2}$, Eqs. (8.1.29) and (8.1.26) yield $\langle x^2 \rangle = n\,\Delta x^2$. If we define a diffusion coefficient D as

$$D = \frac{\Delta x^2}{2\,\Delta t}, \tag{8.1.30}$$

then Eq (8.1.29) becomes $\langle x^2 \rangle = 2Dt$: the squared width of the distribution increases linearly with time, with slope $2D$. This linear increase agrees with the behavior observed in the previous simulation (see Fig. 8.1) and also agrees with the solution to the heat equation starting from a concentrated heat pulse, Eq. (5.1.100).

For $p \neq q$, $\langle x^2 \rangle$ by itself is no longer a good measure of the change in the width of the distribution. A better measure is $\langle \delta x^2 \rangle \equiv \langle (x - \langle x \rangle)^2 \rangle$, the mean squared change in the position from the average position. However, if we expand the expression and use Eq. (8.1.29) we obtain

$$\begin{aligned} \langle \delta x^2 \rangle &= \left\langle (x - \langle x \rangle)^2 \right\rangle = \langle x^2 - 2x\langle x \rangle + \langle x \rangle^2 \rangle \\ &= \langle x^2 \rangle - \langle x \rangle^2 = 2D't, \end{aligned} \tag{8.1.31}$$

where

$$D' = 4pqD. \tag{8.1.32}$$

Thus, for $p \neq q$, in addition to the mean drift of the distribution given by Eq. (8.1.26), there is also a diffusive spreading around the mean with diffusion coefficient D'. The form of D' makes intuitive sense: for instance, if $p = 1$ and $q = 0$, then particles always step to the right, and the entire distribution moves to the right without spreading: $D' = 0$.

The Heat and Fokker–Planck Equations from the Random Walk Model

The Master Equation. So far we have considered the evolution of the probability distribution $P(j,n)$ in a random walk that starts with all particles at $j = 0$ at timestep $n = 0$. Now let's consider a random walk that starts from a general initial distribution. We will derive a dynamical equation for $P(j,n)$, called the *master equation.* In the limit that grid spacing Δx and timestep size Δt approach zero, the master equation becomes the *Fokker–Planck equation.* This PDE is a generalization of the heat equation.

Given the distribution on the grid at timestep n, $P(j,n)$, the master equation predicts the distribution at the next timestep, $P(j, n+1)$. In general,

$$P(j, n+1) = P(j,n) + \Delta P_+(j,n) - \Delta P_-(j,n), \tag{8.1.33}$$

where ΔP_+ is the number of particles arriving at x_j in this step (divided by M), and ΔP_- is the number of particles leaving x_j in the step (also divided by M). However, in one step, the random walk prescribes that all particles that were at x_j more to the left or right, so $\Delta P_-(j,n) = P(j,n)$. Also, particles arrive at x_j only from the adjacent sites to the right and left. The site to the left at x_{j-1} contributes a fraction p of the particles resident there, and the site to the right contributes a fraction q of its particles. Consequently, $\Delta P_+(j,n) = pP(j-1,n) + qP(j+1,n)$. Substituting our results for ΔP_+ and ΔP_- into Eq. (8.1.33) yields the master equation for the random walk,

$$P(j,n+1) = pP(j-1,n) + qP(j+1,n). \tag{8.1.34}$$

This simple difference equation can be used as a recursion relation to solve for P starting from any initial distribution. It can be shown that the solution corresponding to an initial condition with all particles at $j=0$ is the binomial distribution, Eqs. (8.1.15) and (8.1.16) (see the exercises).

The Fokker–Planck Equation. If we assume that the distribution $P(j,n)$ is slowly varying in position and time compared to Δx and Δt (i.e., $\Delta x \to 0$ and $\Delta t \to 0$), we can derive a PDE from the master equation: the Fokker–Planck equation. First, let us define a *probability density* $\rho(x,t)$ according to

$$\rho(x_j, t_n) = P(j,n)/\Delta x. \tag{8.1.35}$$

This probability density is the unknown function whose dynamics are described by the Fokker–Planck equation. In the limit that $\Delta x \to 0$, the probability density can be used to determine the probability $P_{ab}(t)$ that a particle falls in the range $a < x < b$ at time t, according to

$$P_{ab}(t) = \int_a^b dx\, \rho(x,t). \tag{8.1.36}$$

Note for future reference that probability densities are always normalized so that

$$\int_{-\infty}^{\infty} dx\, \rho(x,t) = 1; \tag{8.1.37}$$

i.e., there is unit probability that a particle is somewhere on the x-axis. This equation also follows from Eq. (8.1.7).

If we write the master equation in terms of ρ rather than P, it becomes

$$\rho(x_j, t_{n+1}) = p\rho(x_j - \Delta x, t_n) + q\rho(x_j + \Delta x, t_n). \tag{8.1.38}$$

Dropping the index j and taking the limit $\Delta x \to 0$, a Taylor expansion of the right-hand side to second order in Δx yields

$$\rho(x, t_{n+1}) = p\left(\rho(x,t_n) - \Delta x \frac{\partial}{\partial x}\rho(x,t_n) + \frac{1}{2}\Delta x^2 \frac{\partial^2}{\partial x^2}\rho(x,t_n) \right)$$
$$+ q\left(\rho(x,t_n) + \Delta x \frac{\partial}{\partial x}\rho(x,t_n) + \frac{1}{2}\Delta x^2 \frac{\partial^2}{\partial x^2}\rho(x,t_n) \right).$$

After collecting terms and using the identity $p+q=1$, we obtain

$$\rho(x,t_{n+1}) - \rho(x,t_n) = -\Delta x\,(p-q)\frac{\partial}{\partial x}\rho(x,t_n) + \frac{1}{2}\Delta x^2\frac{\partial^2}{\partial x^2}\rho(x,t_n). \quad (8.1.39)$$

If we now divide the left and right-hand sides by Δt, and take the limit as $\Delta t \to 0$, the left-hand side becomes a forward-differenced time derivative and we obtain the Fokker–Planck equation,

$$\frac{\partial \rho}{\partial t} = -v\frac{\partial \rho}{\partial x} + D\frac{\partial^2 \rho}{\partial x^2}. \quad (8.1.40)$$

Here $v=(p-q)\,\Delta x/\Delta t$ is the drift velocity of the distribution [see Eq. (8.1.26)], and $D=\Delta x^2/2\,\Delta t$ is the diffusion coefficient obtained previously in Eq. (8.1.32) for the case $p=q=\frac{1}{2}$. In fact, when $p=q=\frac{1}{2}$, we have $v=0$ and Eq. (8.1.40) becomes the heat equation for the temperature, Eq. (8.1.1) [or the diffusion equation for the particle density $n=N\rho$, Eq. (8.1.2)]. We have therefore shown that unbiased random walk dynamics does lead to the heat and diffusion equations.

The Fokker–Planck equation can be used to model many physical systems for which diffusion is coupled with a net drift. One example is the application of an electric field to a conductor, causing a current to flow, along with the diffusion caused by collisions. Another example is the behavior of a layer of dust added to a fluid such as water. Gravity causes the dust to settle, but collisions of the dust grains with the molecules in the fluid also cause the dust to diffuse, and create a drag force on the dust grains. For dust grains in gravity, we would expect that the drag force would balance gravity, causing a net drift at the terminal velocity

$$v = -g/\gamma, \quad (8.1.41)$$

where γ is the drag coefficient. For conduction electrons, we would expect v to be related to the applied electric field and the electrical resistivity (see the exercises).

While diffusion of electrons can be observed only indirectly through its effect on thermal diffusion, the random motion of dust in a liquid can be directly observed with a microscope, and was first noted by the Scottish botanist Robert Brown in his famous work on *Brownian motion*. Einstein, along with von Smoluchowski, then applied the statistical theory of random walks to describe this motion (more on this later).

Although the Fokker–Planck equation has been derived from artificial random walk dynamics, it can often also be justified on more fundamental grounds. The probability for steps of different sizes can in many cases be calculated directly from the microscopic collisional dynamics of the particles. For instance, Boltzmann carried out such calculations for a low-density gas of classical particles interacting via a short-ranged potential through two-body collisions. With such calculations, the thermal conductivity, diffusion coefficient, and viscosity of the gas can be directly related to the density, temperature, and form of the interparticle potential. We will not discuss this rather involved subject here, but interested readers can find more material in Lifshitz and Pitaevskii (1981).

In other cases, such as in the Brownian motion of dust grains in a fluid, or the motion of electrons in a conductor, the probability for steps of different sizes cannot be easily determined from first principles calculations. Nevertheless the drift velocity and diffusion coefficient are not arbitrary functions. We have observed that the drift velocity is related to the applied external fields that are causing the drift. Also, we will find that the diffusion and the drift are related to one another by an *Einstein relation*. We will discuss Einstein relations in the exercises and in Sec. 8.2.3.

It may give one pause that when $p \neq q$, the diffusion coefficient D in the Fokker–Planck equation differs from $D' = 2pq\,\Delta x^2/\Delta t$, which describes the diffusion associated with the binomial distribution [see Eq. (8.1.32)]. After all, both the binomial distribution and the Fokker–Planck equation are supposed to describe the same random walk. The resolution of this apparent paradox lies in the assumption inherent in Eq. (8.1.40), that ρ is slowly varying in both space and time compared to Δx and Δt. In the limit that Δx and Δt approach zero, the ratio $\Delta x^2/\Delta t$ must remain finite in order to provide a finite diffusion coefficient. This implies that Δx is of order $\sqrt{\Delta t}$. However, this also implies that the drift velocity v is of order $(p-q)/\sqrt{\Delta t}$. For v to remain finite in the limit as $\Delta t \to 0$, we therefore require $p-q$ to be of order $\sqrt{\Delta t}$, in other words, $p \sim q \sim \frac{1}{2} + O(\sqrt{\Delta t})$. This implies that $D' = D + O(\sqrt{\Delta t})$, so there is no paradox.

Entropy and Irreversibility. In describing the evolution of solutions to the Fokker–Planck and heat equations, it is useful to define a function of the probability density called the *entropy*

$$S(t) = -k_B \int dx\, \rho(x,t) \ln \rho(x,t), \tag{8.1.42}$$

where k_B is Boltzmann's constant. The importance of the entropy stems from the following fact, which we will prove below:

$$dS/dt > 0 \tag{8.1.43}$$

for any solution of the Fokker–Planck (or heat) equation. The proof of Eq. (8.1.43) is as follows. If we take a time derivative of Eq. (8.1.42) we obtain

$$\frac{dS}{dt} = -k_B \int dx\, [\ln \rho(x,t) + 1] \frac{\partial}{\partial t} \rho(x,t).$$

However, $\int dx\, (\partial/\partial t)\rho(x,t) = (d/dt)\int dx\, \rho = 0$ according to Eq. (8.1.37), so we can drop the second term in the square brackets. Then substituting from the Fokker–Planck equation for $\partial\rho/\partial t$ yields

$$\frac{dS}{dt} = k_B \int dx \ln \rho(x,t) \left(v \frac{\partial \rho}{\partial x} - D \frac{\partial^2 \rho}{\partial x^2} \right).$$

Integrating oncc by parts, and assuming that $\rho \to 0$ at infinity, yields

$$\frac{dS}{dt} = -k_B \int dx \frac{1}{\rho}\frac{\partial \rho}{\partial x}\left[v\rho - D\frac{\partial \rho}{\partial x}\right] = k_B \int dx \left[-v\frac{\partial \rho}{\partial x} + \frac{D}{\rho}\left(\frac{\partial \rho}{\partial x}\right)^2\right].$$

However, the first term in the square bracket vanishes upon performing the integration, since ρ vanishes at infinity, and the second term is positive definite because $\rho \geq 0$ and $D > 0$. Therefore,

$$\frac{dS}{dt} = k_B D \int dx \frac{1}{\rho}\left(\frac{\partial \rho}{\partial x}\right)^2 > 0. \tag{8.1.44}$$

Because S must always increase, the Fokker–Planck and heat equations have the property of *irreversibility*: solutions of these equations never repeat themselves over time because this would require that S return to a previous value. Initial conditions never recur. This is as opposed to solutions of other PDEs such as the wave equation or the KdV equation, where we observed that initial conditions often repeat themselves, as in a normal mode of oscillation.

The irreversibility of the heat and Fokker–Planck equations is caused by the random motion of many particles. We have seen that this random motion causes diffusive spreading of the distribution, and this is what is responsible for the increase of the entropy. The broader the distribution becomes, the smaller its mean amplitude must be in order to satisfy the normalization condition (8.1.37), and hence the larger the value of $-\langle \ln \rho \rangle$. Irreversibility implies that once the distribution has spread, it never becomes narrower again. Irreversibility is an important property of many systems that exhibit chaotic dynamics, and is a fact of everyday life. Once the eggs are scrambled they can't be unscrambled, or more precisely, they can't unscramble themselves.

Or can they? Actually, in chaotic systems with many degrees of freedom, it is possible for initial conditions to recur. It is not impossible for a broken glass lying on the floor to suddenly draw kinetic energy from the random motion of molecules in the floor tiles, and fly back up onto the table; it is just (highly) unlikely. This is analogous to the notion that a roomful of monkeys seated in front of typewriters could eventually bang out the works of Shakespeare (or a physics textbook). It *could* happen. (Maybe it has already!)

In fact, Poincaré *proved* that for closed Hamiltonian systems, recurrence back to within an arbitrarily small difference from the initial conditions *must* always occur. It is not only possible for the glass to fly up off the floor, it is *required by the laws of physics*. But don't get out your video cameras yet. The catch is that this recurrence can take a long time to happen. For chaotic systems with many degrees of freedom, the Poincaré recurrence time is typically much longer than the age of the universe.

We can see why this is so from our unbiased random walk simulation. It is possible for all the particles to just happen to end up back at $x = 0$ at the same timestep. If there were only one particle, this recurrence would occur quite often, but for $M \gg 1$ particles, the likelihood is incredibly small, equal to $P(0, n)^M$. For $p = q = \frac{1}{2}$ and only $n = 10$ steps, $P(0, n)$ is already down to $63/256 = 0.236\dots$.

Thus, for only $M = 100$ particles executing 10 unbiased steps, the probability of recurrence is $P(0, n)^M = 0.236^{100} = 1.3 \times 10^{-61}$.

On the other hand, if the dynamics were not random, but were instead a simple oscillation of the particles as in the wave equation, recurrence could be a relatively common event even for a great many particles. It is the chaotic nature of the dynamics, coupled with the many degrees of freedom in the system, that makes recurrence unlikely.

But this leaves us with one final question: if recurrence occurs (with small probability) in a random walk, and the Fokker–Planck equation describes a random walk, then doesn't that contradict Eq. (8.1.44), which states that initial conditions cannot recur? No. The point is that Eq. (8.1.44) describes the evolution of the *probability distribution* ρ for a system of particles. The probability distribution describes the fraction of particles found in given states in the limit as the number of particles M approaches infinity. But for finite M there are fluctuations away from the expected behavior based on ρ. Recurrence is simply a very large fluctuation. We will consider the size of these fluctuations in Sec. 8.3.2, Example 1.

The fact that the Fokker–Planck equation describes the probability distribution, and therefore does not directly exhibit recurrence, was a source of considerable controversy in the early days of the theory of statistical mechanics. This tiny weakness in the probabilistic theory of thermal relaxation caused Boltzmann considerable heartache, and is thought by many to have been a factor contributing to his eventual suicide.

EXERCISES FOR SEC. 8.1

(1) **(a)** Analytically determine the probability of throwing 8 points or less with three dice, each numbered from 1 to 6.

(b) Write a simulation using the *Mathematica* random number generator in order to determine this probability numerically. (Hint: Average results over 1000 realizations to get a good value for the probability.)

(2) **(a)** Write a simulation to numerically determine the probability that two or more students in a 20-student class have the same birthday (assuming a 365-day year, with no leap years). (Hint: Assign a number from 1 to 365 for each day. Use `Sort` to sort the list of 20 birthdays before hunting for identical birthdays; this will speed up the search. The `While` or `If` statements may come in handy. Average results over 1000 realizations of the class.)

(b) Find an analytic expression for this probability, and compare with the numerics. (Hint: Determine the probability that all the birthdays are different, and subtract this from unity.)

(3) **(a)** The game of craps is played with two dice (numbered from 1 to 6). If you roll 2, 3, or 12 on the first try, you lose. If you roll 7 or 11 on the first try, you win. If you roll something else (i.e., $4, 5, 6, 8, 9, 10$), to win you must roll that number again before you roll 7, but if you roll 7 you lose. You can roll as many times as you need to until you either win or lose. Using a decision tree, analytically evaluate the probability that you will win.

(b) Evaluate this probability numerically with a simulation, using 1000 realizations of the game.

(4) **(a)** In a game of blackjack, drawing from a fresh 52-card deck, the dealer gives himself two tens. He deals you a 6 and a 10. Work out the decision tree to analytically determine the probability that you will beat the dealer, scoring 21. You can draw as many cards as you like from this single deck until you win (or lose). Remember, aces can be either 1 or 11 points, and face cards (jack, queen, king) are worth 10 points.

(b) Simulate the game, and evaluate the probability of winning numerically out of 1000 tries.

(5) **(a)** Two particles start random walks from the origin at the same instant, with steps of equal probability to the left and right. Determine the probability that the particles will meet again after n steps. (Consider the relative displacement between the walkers.)

(b) Perform the same calculation for an unbiased 2D random walk on a square grid.

(c) Find an expression involving a sum for the mean number of times, $\langle N \rangle$, that the particles meet after taking n steps, in 1D, 2D, and 3D unbiased walks. By evaluating the sum numerically, show that $\langle N \rangle$ diverges like $\sqrt{n}$ in 1D and like $\ln n$ in 2D as n increases (i.e., over a long time period the particles meet many times), but for 3D, on average, particles only meet about 0.4 times, even if one waits forever (i.e., particles lose one another).

(6) The binomial distribution is the probability $P(j,n)$ that a particle ends up at position j after n steps. However, since $j = 2n_R - n$, $P(j,n)$ can also be thought of as the probability that n_R steps are taken to the right out of a total of n steps. Denoting this probability by $W(n_R, n)$, we have $W(n_R, n) = P(j,n)$, where $j = 2n_R - n$. Then

$$W(n_R, n) = p^{n_R}(1-p)^{n-n_R}\frac{n!}{n_R!(n-n_R)!}. \tag{8.1.45}$$

This distribution is the probability that n_R *events* (i.e. steps to the right) of probability p occur in a total of n trials. It can be simplified in the limit that n is large. In this limit, we can show that W is peaked around the average value $n_R = np$. Since $n \gg 1$ is assumed we then have $1 \ll n_R \ll n$ (assuming $0 < p < 1$).

(a) Take the logarithm of W and use *Stirling's formula*

$$\ln N! \simeq N \ln N - N \qquad \text{for} \quad N \gg 1. \tag{8.1.46}$$

With the aid of *Mathematica*, take the derivative of $\ln W$ with respect to n_R to show that an extremum in $\ln W$ exists at $n_R = np$.

(b) Show that this extremum is a maximum by evaluating the second derivative of $\ln W$ with respect to n_R. In particular, show that $(\ln W)''|_{n_R = np} = -1/(npq)$.

(c) Use the results of (a) and (b) to show that

$$W(n_R, n) \approx C e^{-(n_R - np)^2/(2npq)}, \tag{8.1.47}$$

where C is a constant. This is a *Gaussian* distribution. (Hint: Perform a Taylor expansion of $\ln W$ about its maximum, keeping only up to quadratic terms.)

(d) Stirling's formula as given by Eq. (8.1.46) is too crude to determine the constant C. However, C can be found using the following argument. Since $2npq$ is a large number by assumption, W varies little as $n_R \to n_R + 1$, and can be thought of as a continuous function of n_R. Also, $W(0, n) \approx W(n, n) \approx 0$. Therefore, $\Sigma_{n_R=0}^{n} W(n_R, n) \approx \int_{-\infty}^{\infty} W(n_R, n)\, dn_R = 1$. Use this to show that $C = (2\pi npq)^{-1/2}$.

(e) Plot Eq. (8.1.47) in comparison with Eq. (8.1.45) for the case $p = q = \frac{1}{2}$, $n = 10$.

(7) The distribution W can also be simplified in the limit that the probability p is small and n is large. In this limit, $W(n_R, n)$ is negligible unless $0 \ll n_R \ll n$. In these limits, show that

(a) $(1-p)^{n-n_R} \approx e^{np}$. [Hint: Use $\ln(1-p) \approx -p$.]

(b) Using Stirling's formula show that $n!/(n-n_R)! \approx n^{n_R}$.

(c) Use these results to show that $W(n_R, n)$ becomes

$$W(n_R, n) \simeq \frac{(np)^{n_R}}{n_R!} e^{-np}, \tag{8.1.48}$$

Equation (8.1.48) is called the *Poisson distribution.* Plot W vs. n_R for $n = 20$ and $p = 0.1$.

(d) Show that Eq. (8.1.48) is properly normalized: $\Sigma_{n_R=0}^{\infty} W(n_R, n) = 1$. (The sum can be continued to infinity because W is negligible for $n_R \to n$.)

(8) A penny is tossed 600 times. What is the probability of scoring 372 tails? Evaluate this using (i) Eq. (8.1.45), (ii) Eq. (8.1.47).

(9) **(a)** In radioactive decay, in a small time Δt the probability p that one atom decays in a sample of M atoms is $p = M\Delta t/\tau$, where τ is the lifetime of any one atom. Assume that Δt is chosen sufficiently small so that $p \ll 1$. Then over a time $t \gg \Delta t$ (but $t \ll \tau$), show that the probability that m atoms have decayed, $W(m, t)$, is given by Poisson distribution.

(b) Find the mean number of particles that have decayed in time t, $\langle m \rangle(t)$.

(10) A missile has a probability of 0.3 of hitting its intended target. How many missiles should be fired at the target in order to have a probability of at least 80% that at least one missile will hit the target?

(11) Show by substitution that the binomial distribution satisfies the master equation (8.1.34).

(12) **(a)** Conduction electrons in an electrical conductor have number density n. When an electric field is applied E, there is a current density j (units of amperes per square meter) given by $E = jr$, where r is the resistivity (units of Ω m). Show that the conduction electrons attain a drift velocity v given by $v = -E/enr$. Show that this is equivalent to a collisional drag coefficient $\gamma = e^2 nr/m$.

(b) In copper, the density of conduction electrons is roughly 10^{28} m^{-3}, and the resistivity at room temperature is $r = 1.7 \times 10^{-8}$ Ω m. Find the drag coefficient γ, and find the drift velocity v in a 1-m-long wire across which 100 V is applied.

(c) If we treat electrons as classical particles, then their mean speed is given by the temperature T, according to $\bar{v} = \sqrt{k_B T/m}$. The time between collisions can be estimated as roughly $\Delta t = \gamma^{-1}$ and the electron mean free path is then roughly $\Delta x = \bar{v}/\gamma$. Using these estimates, show that $D \sim k_B T/(2m\gamma)$. We will prove this *Einstein relation* for classical particles (except for the factor of 2) in Sec. 8.2.3. Use this result to estimate the thermal diffusivity of copper at $T = 300$ K. Compare with the actual thermal diffusivity, $\chi = 1.1 \times 10^{-4}$ m^2/s.

(d) It's surprising that this classical estimate works as well as it does, considering the fact that electrons in a conductor are degenerate, with much larger kinetic energy than the classical formula would suggest, given by the *Fermi energy* $\epsilon_F \gg k_B T$. Thus, the mean free path is much larger than the classical result, and one would think the diffusion would be commensurately larger. However, only a small fraction of these degenerate electrons actually participate in the conductivity, that fraction being of order $k_B T/\epsilon_F$. (This is the fraction within energy $k_B T$ of the *Fermi surface*.) Then show that the classical formula of part (c) still works approximately to describe the thermal diffusivity. For more on this topic, see Reif (1965, Sec. 12.4).

(13) *The Langevin equation.* Consider the following model of Brownian motion: a dust particle's *velocity* is randomly changed by a small random amount s_n every timestep $t_n = n\,\Delta t$, due to a randomly fluctuating force. (This force arises from collisions between the dust particle and atoms in the surrounding fluid.) In addition, there is a drag force on the particle's velocity, so that on average $d\langle v\rangle/dt = -\gamma\langle v\rangle$, where γ is a drag coefficient. In differential form the equation for the velocity is then $dv/dt = -\gamma v + f(t)$, where $f(t)$ is a random force with zero mean, $\langle f\rangle = 0$. The random velocity step is therefore $s_n = \int_{t_n}^{t_n+\Delta t} f(t')\,dt'$. Position x evolves according to $dx/dt = v$. Finite-differencing the derivatives leads to the following random walk model for the particle velocity and position:

$$\begin{aligned} v_n &= \alpha v_{n-1} + s_{n-1}, \\ x_n &= x_{n-1} + \Delta t\, v_{n-1}, \end{aligned} \tag{8.1.49}$$

where $\alpha = 1 - \gamma\,\Delta t$. (These equations comprise the finite-differenced *Langevin equations* for position and velocity.)

(a) Write a particle simulation for this problem for 1000 particles, taking $\Delta t = 1$, $\gamma = 0.1$, and $s_n = \pm 1$ with equal probability for either sign. Start all particles at $v_0 = 10$, and $x_0 = 0$, and follow the dynamics for 50 timesteps.

(b) How do the average velocity and position evolve in the simulation?

(c) Find analytic expressions for $\langle v\rangle(t)$ and $\langle x\rangle(t)$, and compare with the simulation. [Hint: It may be simpler for you to solve differential equations rather than difference equations for $\langle x\rangle(t)$ and $\langle v\rangle(t)$.]

(d) Plot the mean squared change in velocity, $\langle \delta v^2\rangle = \langle v^2\rangle - \langle v\rangle^2$, vs. time from the simulation. Does the particle velocity diffuse?

(e) Plot the mean squared position change, and estimate the value of the particle diffusion coefficient D from your simulation data.

(f) Solve the difference equation for v_n analytically, and use this solution to show that

$$\langle \delta v_n^2\rangle = \langle s^2\rangle \frac{1-\alpha^{2n}}{1-\alpha^2}.$$

Compare this result with the simulation result obtained in part (d). (Hint: $\langle s_n s_m\rangle = 0$ for $n \neq m$. Also, note that *Mathematica* can find many sums analytically using the function **Sum**.)

(g) Show in the limit that $n \to \infty$ and $\Delta t \to 0$ but $t_n = n\,\Delta t$ remains finite, that this result becomes

$$\langle \delta v^2\rangle(t_n) = \frac{\langle s^2\rangle}{2\gamma\,\Delta t}(1 - e^{-2\gamma t}), \tag{8.1.50}$$

For short times $t_n \to 0$, show that $\langle \delta v^2\rangle = 2D_V t_n$, where $D_V = \langle s^2\rangle/2\,\Delta t$ is the *velocity diffusion coefficient*. Evaluate D_V, and compare this short-time theory with the simulation result for $\langle \delta v^2\rangle$ by graphing both together. [Hint: use the same method to simplify $(1-\gamma\,\Delta t)^{2n}$ as was used in Exercise (7)(a).]

(h) In the long-time limit $t_n \to \infty$, statistical physics predicts that the particle will come to thermal equilibrium with a mean squared velocity equal to the square of the *thermal velocity*, $\langle v^2\rangle = k_B T/m$, where m is the particle mass and T is the temperature. In this limit, prove the Einstein relation

$$k_B T = m D_V/\gamma. \tag{8.1.51}$$

(We will prove another Einstein relation in Sec. 8.2.3.)

(i) Analytically evaluate $\langle x_n v_n\rangle$, plot the result vs. n for the parameters of the simulation, and compare with the simulation. Explain why this quantity is positive for short times, but vanishes at large times $t_n \gg \gamma^{-1}$.

(j) Evaluate $\langle x_n^2\rangle$, and show that, at large times $t_n \gg \gamma^{-1}$ and for $\Delta t \to 0$, one has $(x_n^2) = 2Dt_n$, where $D = D_V/\gamma^2$ is the spatial diffusion coefficient. Plot the result vs. n for the simulation parameters, and compare with the simulation. (Solution: $\langle x_n^2\rangle = 2D_V\,\Delta t^3[n(1-\alpha^2) - 2\alpha - \alpha^2 + \alpha^{1+n}(2 + 2\alpha - \alpha^{1+n})]/[(1-\alpha)^3(1+\alpha)] + (x_n)^2$.)

(14) **(a)** A gas of N particles is initially confined to the left half of a container of volume V by a partition. The partition is then removed, and the particles distribute themselves randomly throughout the container. What is the probability that, at any instant, all the particles are all found back in the left half of the container? (This is a crude example of recurrence, and is much *more* likely than Poincaré recurrence, where each particle is found with nearly the same position and velocity as it had initially.)

(b) What is the probability $P(N_L)$ that N_L particles are in the left half of the container at any instant (and $N - N_L$ are in the right half)? Show that for N large this probability is given by a Gaussian distribution.

(c) Use this probability to determine the mean squared fluctuation $\langle \delta N_L^2 \rangle$ in the number of particles in the left half away from the mean value $N/2$.

(15) A polymer is a long chain of atoms connected by bonds. Consider the following simplified model of a polymer. Assume that the polymer lies in the x-y plane, and the bonds are all of unit length. Furthermore, assume that the bonds can only be in the $\pm x$ or $\pm y$ directions. Then starting at one end of the chain, at $(x, y) = (0, 0)$, the next atom can be at one of four locations, $(\pm 1, 0)$ or $(0, \pm 1)$, with equal probability. The next atom can be at any of its four possible locations, and so on. Thus we see that the atoms are at the locations corresponding to a two-dimensional random walk. One important question, which can be addressed experimentally, is the mean size of a polymer chain consisting of n bonds. The mean size is typically evaluated as $\langle r^2 \rangle^{1/2}$, where $\langle r^2 \rangle(n) = (1/n)\Sigma_{i=0}^{n} |\mathbf{r}_i|^2$, the sum runs over the positions of all $n + 1$ atoms in the chain, and the average is over all possible configurations of the polymer, all of which are assumed to be equally probable. This is equivalent to an average over all possible random walk processes. Determine $\langle r^2 \rangle(n)$ in this random walk model, and in particular show that $\langle r^2 \rangle \propto n$.

(16) Experiments on polymers show that the size of a polymer with n bonds is larger than suggested by the result given in the previous problem. The reason is that atoms in a polymer take up space, and so the random walk is excluded from regions already occupied by the polymer. This causes the polymer to spread out more than the simple random walk of the previous problem would predict. One way to estimate the effect of the excluded volume on the size of the polymer is to use a *self-avoiding* random walk. This is a random walk where, as before, a step in any of the four directions is equally probable, *except* that the step cannot move to an already occupied position. The possible polymer configurations for $n = 2$ bonds are shown in Fig. 8.3, assuming that the first bond is always to the right (the direction of this bond can be chosen arbitrarily because of the fourfold symmetry of the random walk on a square lattice). A considerable effort over the years has gone into determining the statistics of such self-avoiding walks. Little can be accomplished analytically without great effort, but numerical methods have proven to be very fruitful.

(a) Determine $\langle r^2 \rangle(n)$ analytically for a self-avoiding random walk for $n = 2$ and 3. Average over all possible configurations of the chain, assuming that each configuration is equally likely. Compare to the results of the previous problem. Which result is larger? (Amazingly, Dr. E. Teramoto of Kyoto University evaluated $\langle r^2 \rangle(n)$ analytically for n up to 20 bonds, using pencil and paper [according to Rosenbluth and Rosenbluth (1955)].

(b) Write a simulation for a self-avoiding random walk. Starting at the origin, the first step is always to the right, and the steps thereafter can be in any

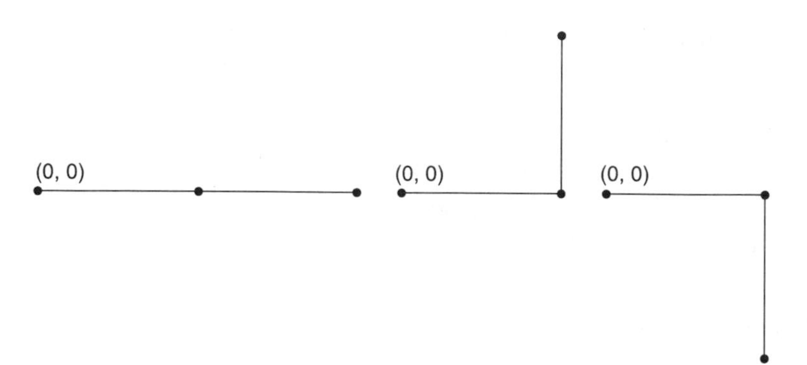

Fig. 8.3 The three possible paths for a self-avoiding random walk consisting of two steps.

direction. In determining $\langle r^2\rangle(n)$, throw away any configuration that crosses itself at timestep n or earlier. (Such configurations can still be counted in the average for timesteps before a crossing in the configuration has occurred.)

There is one addendum to this method that must be made. In order to increase the efficiency of the simulation, do not allow the walk to double back on itself. This implies that at any timestep there are only *three* possible directions for the next step, each with equal probability. There are many different ways to code this, and the method is left up to you. Run your simulation of the polymer for $M = 5000$ trials, up to a length of $n = 50$. The reason that M must be chosen this large is that by the time $n = 50$ only about 100 configurations will be left in the average. Make a log–log plot of $\langle r^2\rangle$ vs. n. It should be nearly a straight line with slope b. Thus, $\langle r^2\rangle \propto n^b$. The exponent b equals one in a simple random walk, but should be larger than one in a self-avoiding random walk. Thus, the polymer grows in size with increasing n more rapidly than expected from a simple random walk. This is the excluded-volume effect. To find the exponent b, for $n > 10$, fit $\log\langle r^2\rangle$ to a form $\log\langle r^2\rangle = a + b\log n$. For large n, it has been predicted that b is roughly 1.5 for a random walk in 2D, and 1.2 in 3D. How close did your simulation result come to the 2D estimate?

8.2 THERMAL EQUILIBRIUM

8.2.1 Random Walks with Arbitrary Steps

Averages In a random walk the steps need not be of fixed size. Let's consider a more general random walk where the step size s is distributed according to a probability density $w(s)$. In one step, the particle position changes from x_n to $x_{n+1} = x_n + s$. As is always the case with probability densities,

$$P_{ab} = \int_a^b w(s)\, ds \tag{8.2.1}$$

is the probability that the step falls in the range $a < s < b$ [see Eq. (8.1.36)].

The average step size is $\langle s\rangle = \int_{-\infty}^{\infty} s w(s)\, ds$. For instance, in our previous random walk, $w(s) = p\,\delta(s - \Delta x) + q\,\delta(s + \Delta x)$, and then $\langle s\rangle = \Delta x\,(p-q)$.

Consider N steps in a random walk process. We will label each step s_n, $n = 1, 2, 3, \dots, N$. The probability dP that each step falls in the range $s_n \to s_n + ds_n$, $n = 1, 2, \dots, N$, is the product of the probabilities for the steps:

$$dP = w(s_1)\, ds_1 \cdots w(s_N)\, ds_N. \tag{8.2.2}$$

This provides an example of a *multivariate* (i.e. multidimensional) probability distribution. The function $\rho(s_1, s_2, \dots, s_N) \equiv w(s_1)w(s_2)\cdots w(s_N)$ on the right-hand side of the equation is the probability density for the N-step process. This multivariate distribution is normalized in the usual way:

$$\int \rho(s_1, \dots, s_N)\, ds_1\, ds_2 \cdots ds_N = 1,$$

and we can calculate average quantities over this distribution just as we did for averages over a single variable. For any function $f(s_1, s_2, \dots, s_N)$, the average is

$$(f) = \int_{-\infty}^{\infty} ds_1 \int_{-\infty}^{\infty} ds_2 \cdots \int_{-\infty}^{\infty} ds_N\, \rho(s_1, \dots, s_N) f(s_1, s_2, \dots, s_N). \tag{8.2.3}$$

For example, if we wish to determine the average position after N steps, $\langle x\rangle(N)$, we can write the position as a sum of the individual steps, $x = s_1 + s_2 + \cdots + s_N$. Taking an average, we obtain

$$\langle x\rangle(N) = \langle s_1 + s_2 + \cdots + s_N\rangle = \langle s_1\rangle + \langle s_2\rangle + \cdots + \langle s_N\rangle,$$

where in the second step we used the linearity property of averages, Eq. (8.1.19). Now, for a random walk, with $\rho = w(s_1)w(s_2)\cdots w(s_N)$, integrals over each term in the sum yield the same result. For instance, for the nth term,

$$\langle s_n\rangle = \int_{-\infty}^{\infty} ds_1\, w(s_1) \int_{-\infty}^{\infty} ds_2\, w(s_2) \cdots \int_{-\infty}^{\infty} ds_N\, w(s_N)\, s_n = \int_{-\infty}^{\infty} ds_n\, w(s_n)\, s_n = \langle s\rangle;$$

that is, the average step size $\langle s\rangle$ is independent of the time step n. Therefore, we find that

$$\langle x\rangle(N) = N\langle s\rangle, \tag{8.2.4}$$

so there is a net drift of the particle distribution with velocity $v = \langle s\rangle/\Delta t$. We can perform a similar analysis for $\langle x^2\rangle(N)$. According to Eq. (8.2.3),

$$\langle x^2\rangle(N) = \left\langle \sum_{i=1}^{N} s_i \sum_{j=1}^{N} s_j \right\rangle = \sum_{i=1}^{N} \langle s_i^2\rangle + \sum_{\substack{i,j=1 \\ j\neq i}}^{N} \langle s_i s_j\rangle,$$

where in the second step we separated out terms in the sums for which $i = j$. Applying our distribution to these averages, we find that $\langle s_i^2\rangle = \langle s^2\rangle$, independent of i, and $\langle s_i s_j \rangle = \langle s_i \rangle \langle s_j \rangle = \langle s \rangle^2$. Thus, we find

$$\langle x^2 \rangle (N) = N \langle s^2 \rangle + N(N-1) \langle s \rangle^2. \tag{8.2.5}$$

The factor $N(N-1)$ is the number of terms in the sum over i and j, $i \neq j$. If we now consider the mean squared change in position away from the average, $\langle \delta x^2 \rangle = \langle x^2 \rangle - \langle x \rangle^2$, [see Eq. (8.1.31)] and use Eqs. (8.2.4) and (8.2.5) to calculate this quantity, we obtain

$$\langle \delta x^2 \rangle = N \left[\langle s^2 \rangle - \langle s \rangle^2 \right]. \tag{8.2.6}$$

The mean squared change in position increases linearly with the number of steps in a random walk. In analogy to Eq. (8.1.31), we then define a diffusion coefficient D for this process, so that $\langle \delta x^2 \rangle = 2Dt$:

$$D = \frac{\langle s^2 \rangle - (\langle s \rangle)^2}{2\,\Delta t} \tag{8.2.7}$$

This generalizes our previous expression $D = 2pq\,\Delta x^2/\Delta t$.

Conditional Probability and the Master Equation An important property of random walks is that the steps are *uncorrelated*. As an example, consider a random walk with fixed size $s = \pm 1$, and define the probability of a step s to be $P(s)$, where $P(1) = p$ and $P(-1) = q$. The probability of a step is not connected to the results of previous steps, and this means that the steps are uncorrelated with previous steps. For such a process, we know that the probability of several consecutive steps is given by the product of the individual probabilities. For instance, the probability $P(s_1, s_2)$ of two consecutive steps with outcomes s_1 and s_2 is

$$P(s_1, s_2) = P(s_1) P(s_2). \tag{8.2.8}$$

We have already uses this result several times; see, for instance, Eq. (8.2.2). However, this is only true for uncorrelated steps. More generally, if the second step depends in some way on the first step, we define a *conditional probability* $P(s_2|s_1)$. This is the probability that the outcome of the second step is s_2, *on condition that the outcome of the first step was* s_1. In other words, if we take these two steps M times where $M \to \infty$, $P(s_2|s_1)$ is the fraction of the outcomes that yield s_2 taken out of those that first yielded s_1. In the form of an equation, $P(s_2|s_1) = N(s_2|s_1)/N(s_1)$, where $N(s_1)$ is the total number of outcomes that yielded s_1, and $N(s_2|s_1)$ is the number of outcomes that yielded s_2 given that s_1 first occurred. However, it is also the case that $P(s_1, s_2) = N(s_2|s_1)/M$, and $P(s_1) = N(s_1)/M$. These results can be combined to yield the following equation:

$$P(s_1, s_2) = P(s_1) P(s_2|s_1). \tag{8.2.9}$$

It is instructive to examine this important result from several perspectives. For an uncorrelated random walk, $P(s_2|s_1) = P(s_2)$ independent of the first step, and Eq. (8.2.9) returns to Eq. (8.2.8). At the other extreme, the case of perfect correlation, the second step is always the same as the first. In this case $P(s_2|s_1) = 1$ if $s_2 = s_1$

and is zero otherwise. Then Eq. (8.2.9) yields $P(s_1, s_2) = P(s_1)$ if $s_2 = s_1$ and $P(s_1, s_2) = 0$ if $s_2 \neq s_1$, as we would have intuitively guessed when s_2 is perfectly correlated to s_1.

An example of random variables in between these two extremes of no correlation and perfect correlation is the set of position x_n of a particle at timesteps n. The position is correlated with its value at the previous timestep. Taking the case of a random walk on a grid (with fixed step size Δx), the conditional probability $P(j|i)$ for a particle to be at grid position j given that it was at position i on the previous step is $P(j|i) = p\delta_{j-1,\,i} + q\delta_{j+1,\,i}$, where δ_{ij} is the Kronecker delta function. Then the probability $P(i, n; j, n+1)$ that a particle will be at position i at timestep n *and* at step j at timestep $n+1$ is, according to Eq. (8.2.9),

$$P(i,n;j,n+1) = P(i,n)\left(p\delta_{j-1,i} + q\delta_{j+1,i}\right). \tag{8.2.10}$$

If we now sum both sides over all locations i, the left side becomes $\Sigma_i P(i, n; j, n+1) = P(j, n+1)$, the probability that a particle is at step j at timestep $n+1$ *for any* previous position. On the right side we obtain $pP(j-1, n) + qP(j+1, n)$. Therefore we return to the master equation for this random walk, Eq. (8.1.34).

For a general uncorrelated random walk with a probability distribution $w(s)$ for each step, the conditional probability that particles are in a range x to $x + dx$ given that they were at position x' at the previous timestep, is $P(x, x + dx|x') = w(x - x')\,dx$. We can use this result to obtain a master equation for a general random walk. Let us assume that the position x' occurred at timestep n. Then the probability for consecutive steps at times t_n and t_{n+1}, $\rho(x', t_n; x, t_{n+1})\,dx'\,dx$, can be obtained from Eq. (8.2.9):

$$\rho(x', t_n; x, t_{n+1})\,dx'\,dx = \rho(x', t_n)\,dx'\,w(x - x')\,dx,$$

where $\rho(x', t_n)\,dx'$ is the probability at timestep n; see Eq. (8.1.35). Integrating both sides over x', the left-hand side becomes $\rho(x, t_{n+1})\,dx$, the probability at the $n+1$st timestep that the particle is in the range $x \to x + dx$, for any previous position. Dividing both sides by dx, we obtain the master equation for the evolution of the probability density ρ in a general random walk,

$$\rho(x, t_{n+1}) = \int dx'\,\rho(x', t_n)\,w(x - x'). \tag{8.2.11}$$

This master equation can be solved directly for any given initial condition using Fourier transform techniques (see the exercises).

Correlation Functions How can we tell if two outcomes of a random process, i and j, are correlated or not? One way is to examine the *correlation* c, defined by

$$c = \frac{\langle ij\rangle - \langle i\rangle\langle j\rangle}{\sqrt{\left(\langle i^2\rangle - \langle i\rangle^2\right)\left(\left(\langle i\rangle^2\right) - \langle j\rangle^2\right)}}. \tag{8.2.12}$$

The average $\langle ij\rangle$ is given by $\langle ij\rangle = \Sigma_{i,\,j} P(i, j)ij$, where the sum is over all possible outcomes for i and j, and $P(i, j)$ is the probability that both i and j occur.

However, for uncorrelated outcomes we can use Eq. (8.2.8), in which case we obtain $\langle ij \rangle = \Sigma_i P(i) i \Sigma_j P(j) j = \langle i \rangle \langle j \rangle$. Therefore, for uncorrelated outcomes, Eq. (8.2.12) yields $c = 0$. For perfectly correlated outcomes where $j = i$, Eq. (8.2.12) implies $c = 1$. For perfect *anticorrelation*, where $j = -i$, it is easy to see that $c = -1$. Generally one can prove that c must be somewhere between -1 and 1, with negative values implying some degree of anticorrelation, and positive values corresponding to positive correlation.

Take, for example, the correlation between positions in a random walk that starts at the origin, with general steps s_n. The correlation between positions at timesteps n and $n+m$ is

$$c(n, n+m) = \frac{\langle x_n x_{n+m} \rangle - \langle x_n \rangle \langle x_{n+m} \rangle}{\sqrt{\langle \delta x_n^2 \rangle \langle \delta x_{n+m}^2 \rangle}}.$$

We already know that $\langle x_n \rangle = n \langle s \rangle$ for any n. However, what is the value of $\langle x_n x_{n+m} \rangle$? This is given by

$$\langle x_n x_{n+m} \rangle = \sum_{i=1}^{n} \sum_{j=1}^{n+m} \langle s_i s_j \rangle = \sum_{i=1}^{n} \langle s_i^2 \rangle + \sum_{i=1}^{n} \langle s_i \rangle \sum_{\substack{j=1 \\ j \neq i}}^{n+m} \langle s_j \rangle$$

$$= n \langle s^2 \rangle + n(n+m-1) \langle s \rangle^2,$$

where in the second step we separated out the terms with $i = j$. Then using Eqs. (8.2.6) and (8.2.4), a little algebra leads to the result $c(n, n+m) = n / \sqrt{n(n+m)}$. This shows that correlations between positions x_n and x_{n+m} are always nonnegative, and slowly fall off as m increases. Over time, the particle loses memory of its previous position at timestep n as it steps randomly away from that position. If we consider the case $n = 0$, we see that $c = 0$ for $m > 0$: the loss of correlation is immediate because the first step is completely uncorrelated with the origin.

Inhomogeneous Random Walks In a real system, eventually the dust reaches the bottom of the fluid container; or the electrons reach the edge of the conductor. How can we allow for the effect of this barrier on the random walk? More generally, external forces applied to the dust grains or the electrons may vary with position. We would then expect that the distribution of the random steps should depend on the position x of the particle: $w = w(x, s)$. (Here we take x to be the position *before the step is taken*.) For example, for the case of a barrier at $x = 0$ with particles confined to $x > 0$, we would require that $w(x, s) = 0$ for $s < -x$, so that no steps below $x = 0$ are allowed.

The master equation for an inhomogeneous random walk follows from the same analysis as for Eq. (8.2.11). Now the conditional probability that particles are in range around x given that they were at previously at x' is $P(x, x + dx | x') = w(x', x - x')\, dx$, and the resulting master equation is

$$\rho(x, t_{n+1}) = \int dx' \, \rho(x', t_n) w(x', x - x'). \tag{8.2.13}$$

A drift velocity and a diffusion coefficient can still be defined for such an inhomogeneous random walk, provided that the steps are small compared to the inhomogeneity scale length. If this is so, then a Fokker–Planck equation can be obtained from Eq. (8.2.13). Defining the step size $s = x - x'$ and changing the integration variable to s, Eq. (8.2.13) becomes

$$\rho(x, t_{n+1}) = \int ds\, \rho(x - s, t_n) w(x - s, s).$$

Assuming that $\rho(x, t_n)$ and $w(x, s)$ are slowly varying in x, we can Taylor-expand these functions to second order in s:

$$\rho(x, t_{n+1}) = \int ds \left\{ \rho(x, t_n) w(x, s) - s \frac{\partial}{\partial x} [\rho(x, t_n) w(x, s)] + \frac{1}{2} s^2 \frac{\partial^2}{\partial x^2} [\rho(x, t_n) w(x, s)] \right\}.$$

Dividing by Δt and performing the integration over s then yields the inhomogeneous Fokker–Planck equation:

$$\frac{\partial \rho}{\partial t} + \frac{\partial}{\partial x} \left(\bar{v}(x) \rho - \frac{\partial}{\partial x} [D(x) \rho] \right) = 0. \tag{8.2.14}$$

The quantity in large parentheses is the particle flux due to diffusion and drift, where

$$D(x) = \frac{\langle s^2 \rangle}{2\, \Delta t} = \frac{\int s^2 w(x, s)\, ds}{2\, \Delta t} \tag{8.2.15}$$

is the diffusion coefficient, and where

$$\bar{v}(x) = \frac{\langle s \rangle (x)}{\Delta t} = \frac{\int s w(x, s)\, ds}{\Delta t} \tag{8.2.16}$$

is related to the drift velocity $v(x)$ of the particles. This relation follows from noting that diffusive flux has the form of Fick's law, $-D\, \partial \rho / \partial x$, and flux due to drift has the form $v\rho$. Then we can write the particle flux appearing in Eq. (8.2.14) as a sum of these two fluxes:

$$\bar{v}(x) \rho - \frac{\partial}{\partial x} [D(x) \rho] = v(x) \rho - D \frac{\partial \rho}{\partial x},$$

from which we obtain the drift velocity v:

$$v(x) = \bar{v}(x) - \frac{\partial D}{\partial x}. \tag{8.2.17}$$

The inhomogeneous Fokker–Planck equation, written in terms of v rather than $\bar{v}$, is

$$\frac{\partial \rho}{\partial t} + \frac{\partial}{\partial x}\left(v(x)\,\rho - D(x)\,\frac{\partial \rho}{\partial x} \right) = 0. \tag{8.2.18}$$

Because this equation is inhomogeneous, it is difficult to solve analytically except in certain special cases. However, the numerical methods discussed in Chapter 6 can be applied—see the exercises. We can also use random walk simulations to obtain approximate numerical solutions to Eq. (8.2.18). This method is considered in the next subsection.

8.2.2 Simulations

In order to create a random walk simulation with steps distributed according to some general distribution $w(s)$, we require a method of choosing points randomly from this probability distribution. Random number generators such as **`Random`** distribute the particles uniformly over a given range. How do we generate a nonuniform distribution? The two most commonly used approaches are the transformation method and the rejection method.

Transformation Method The transformation method works as follows. We can use **`Random`** to generate a uniform distribution $p(x)$ of particles over some range of x. In order to generate a new distribution of particles, $w(x)$, we apply a transformation to the particle positions $y = y(x)$. A range of positions $x \to x + dx$ transforms to a new range $y(x) \to y(x + dx)$. Taylor-expanding, this range may be written as $y(x) \to y(x) + dy$, where $dy = dx\, dy/dx$. The probability distribution for y, $w(y)$, is obtained from the fact that every particle in dx transforms to a particle in dy, so the number of particles in the range dx equals the number of particles in the range dy:

$$w(y)\,|dy| = p(x)\,dx. \tag{8.2.19}$$

In other words,

$$w(y) = p(x)\left|\frac{dx}{dy}\right|. \tag{8.2.20}$$

Thus, to generate some distribution $w(y)$, we can use Eq. (8.2.20) to determine the required transformation $y(x)$, given that $p(x)$ is constant. Then according to Eq. (8.2.20),

$$x(y) = \int_{-\infty}^{y} w(y)\,dy, \tag{8.2.21}$$

where the function $x(y)$ is the inverse of the required transformation $y(x)$, and we have dropped the constant p because this constant enters only as a normalization to $w(y)$.

For example, let's assume that $w(y) = y^2/9$ for $0 < y < 3$, and is zero otherwise. [This distribution has the proper normalization, $\int_{-\infty}^{\infty} w(y)\,dy = 1$.] Then according

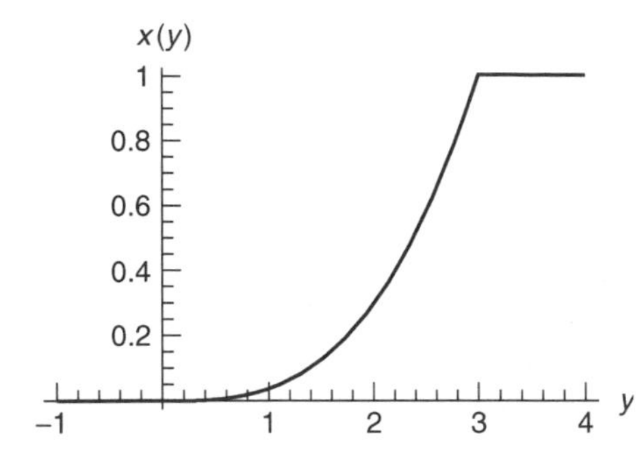

Fig. 8.4 The function $x(y)$ for the choice $w(y) = y^2/9$, $0 < y < 3$.

to Eq. (8.2.21), $x(y) = 0$ for $y < 0$, $x(y) = y^3/27$ for $0 < y < 3$, and $x(y) = 1$ for $y > 3$. This function is plotted in Fig. 8.4. The function $y(x)$ is then $y(x) = 3x^{1/3}$ over the range of interest, $0 < x < 1$. In Cell 8.12, we define a random number r that is distributed according to $w(y)$, and generate the resulting distribution out of 10,000 random numbers. We compare the histogram in bins of size $\Delta x = 0.1$ with the function $10{,}000 w(y) \Delta x$, which is the expected number of particles in a bin. The correct quadratic distribution is properly reproduced over the required range $0 < y < 3$.

Cell 8.12

```
y[x_] = 3 (x)^(1/3);
s[] := y[Random[]];

data = Table[s[], {10000}];
<<Graphics`;
h = Histogram[data, HistogramRange→{-1, 4},
  HistogramCategories→50, DisplayFunction→Identity];
t = Plot[0.1 10000 y^2/9, {y, 0, 3}, DisplayFunction→
  Identity];
Show[h, t, DisplayFunction→$DisplayFunction]
```

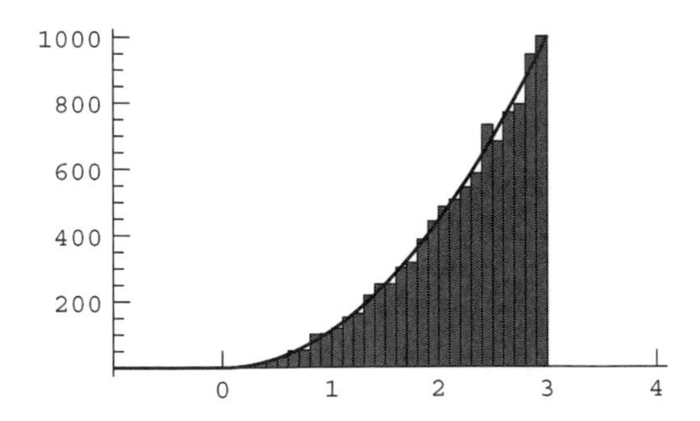

Rejection Method The transformation method is only useful if it is easy to find the function $y(x)$. This is not always the case. Another method can often be applied in such situations: the *rejection* method. In this method, we distribute particles in x according to some distribution $w(x)$ by using *two* uniformly distributed random numbers.

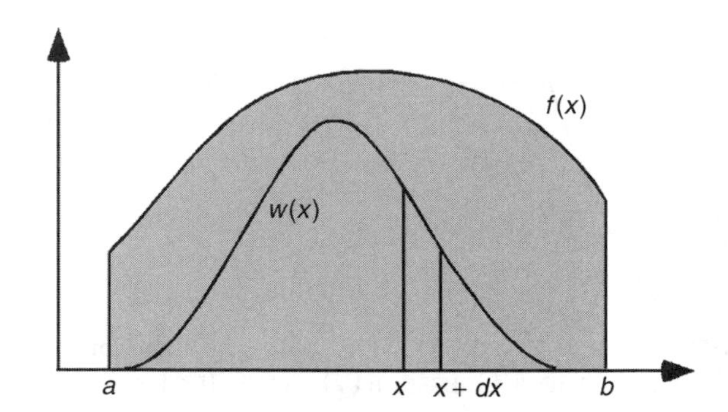

Fig. 8.5 Rejection method. Random points are distributed uniformly in the grey region. Points are rejected if they fall above the curve $w(x)$.

The method is depicted in Fig. 8.5. Around the function $w(x)$ we draw a curve $f(x)$. This curve is arbitrary, but must contain $w(x)$. If $w(x)$ is nonzero only on the range $a < x < b$, then $f(x)$ must also be nonzero over this range, and must be greater than or equal to $w(x)$. Typically $f(x)$ is just chosen to be a constant that is greater than or equal to the maximum of $w(x)$, and that is what we will do here: $f(x) = f_0 \geq \max \omega(x)$ for $a < x < b$.

We will now use our uniform random number generator to distribute particles *uniformly* below the line $f(x) = f_0$. That is, we choose one random number s_x in the range $a < x < b$, and a second random number s_y in the range $0 < y < f_0$:

Cell 8.13

```
Clear["Global`*"];

sx[] := Random[Real, {a, b}];
sy[] := Random[Real, {0, f0}];
```

Now, notice that the number of points (s_x, s_y) that fall below the curve $w(x)$ in the range x to $x + dx$ is simply proportional to the area $dx\,w(x)$. Therefore, if we keep only those points that fall below the curve $w(x)$ and reject the rest, the x-position of the points will be distributed according to $w(x)$: the number of values of x falling in the range x to $x + dx$ is proportional to $w(x)\,dx$. The program for the rejection method is given below:

Cell 8.14

```
s[] := (xx = sx[]; While[sy[] > w[xx], xx = sx[]]; xx)
```

The **While** statement repeatedly executes the test **sy[]>w[xx]** and the command **xx=sx[]** until the test is not true. This allows us to reject points above the curve $w(x)$ until a point falls below the curve, when it is accepted. The x-position of this point is the desired random number.

Below, we show how the method can be applied to determine a random number distributed according to a Gaussian density, $w(x) = e^{-x^2}/\sqrt{\pi}$. In principle, this distribution extends to $\pm\infty$, but large values of x are highly unlikely, so we take $a = -10$ and $b = 10$. We should not take a, b, or f_0 to be overly large, or we waste too much time rejecting almost all of the points. We choose f_0 as the maximum of $w(x)$:

Cell 8.15

```
w[x_] = Exp[-x^2]/Sqrt[Pi];
a = - 10; b = 10;
f0 = 1/Sqrt [Pi];
```

We now plot distribution arising from 5000 points as a histogram in Cell 8.16. It is clearly an excellent match to a Gaussian.

Cell 8.16

```
Table[s[], {5000}]; Histogram[%];
```

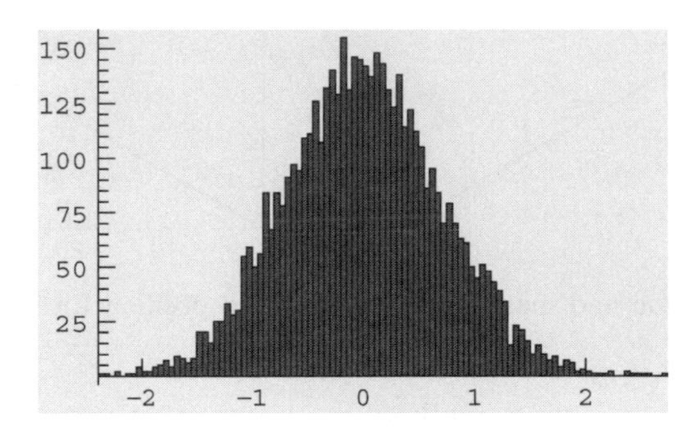

Two Simulations of a Random Walk with Bias

Uniform System. We can now simulate a random walk with a given probability distribution $w(s)$ for the steps. We first take the case of a homogeneous system for which

$$w(s) = -\frac{1}{2a(1-a)}(1-s-a), \qquad -a<s<a, \tag{8.2.22}$$

and $w(s)=0$ otherwise. Thus, the maximum step size is a. This distribution is plotted in Cell 8.17 for the case $a=\frac{1}{4}$. According to Eq. (8.2.4) we expect a uniform drift speed v given by $v=\langle s\rangle/\Delta t$. For this distribution, $\langle s\rangle = -\frac{1}{36}$. Also, the particle distribution should spread with diffusion coefficient D given by Eq. (8.2.7). For this distribution, $\langle s^2\rangle = \frac{1}{48} = 0.021\ldots$, so $D=(\frac{1}{48}-\frac{1}{36}^2)/(2\,\Delta t) = 13/(1296\,\Delta t) = 0.0100\ldots/\Delta t$.

Cell 8.17

```
Clear["Global`*"];
a = 1/4;
w[s_] := 1/(2 a (1 - a)) (1 - s - a) /; -a ≤ s ≤ a;
w[s_] := 0 /; Abs[s] > a;
Plot[w[s], {x, -1, 1}, AxesLabel→{"s", "w(s)"}];
```

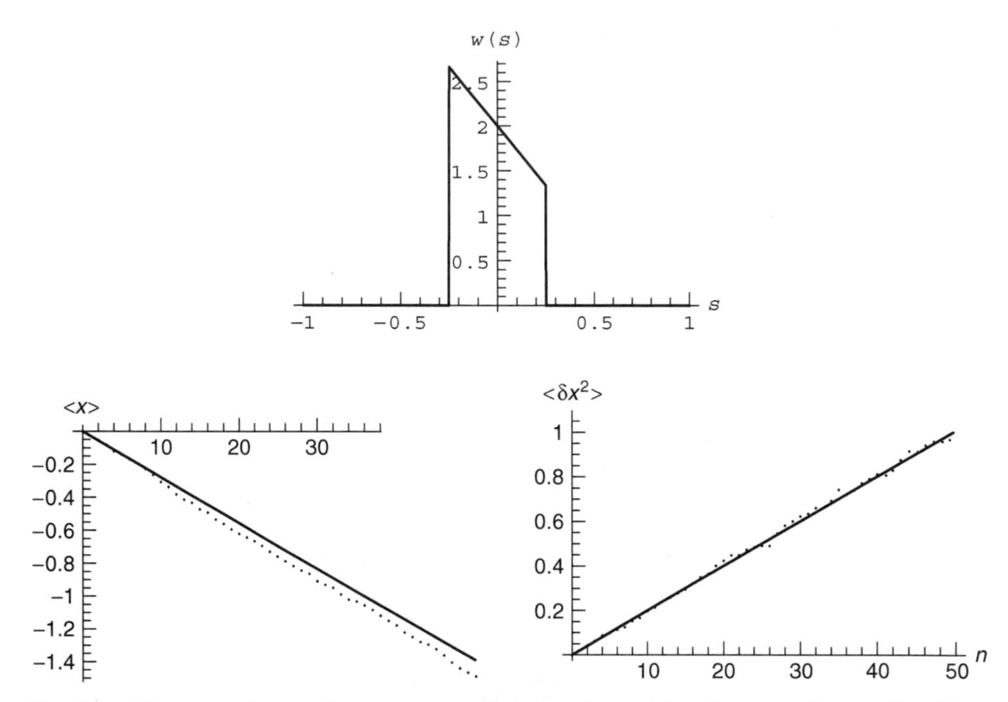

Fig. 8.6 Mean position and mean squared change in position for a random walk with a bias.

We will use the rejection method to create a random variable **s[]** with distribution $w(s)$, and test these predictions for the resulting random walk, starting all particles at $x = 0$. The resulting simulation for $M = 200$ particles is displayed in Cell 8.18. Particles settle toward lower x, and spread as expected. The mean position of the distribution and the mean squared change in position are shown in Fig. 8.6, together with the theory predictions of Eqs. (8.2.4) and (8.2.7). Both quantities closely adhere to their predicted values.

Cell 8.18

```
(* define the random step *)
f0 = 3.;
sx[] := Random[Real, {-a, a}];
sy[] := Random[Real, {0, f0}];
s[] := (ss = sx[]; While[sy[] > w[ss], ss = sx[]]; ss);
 (* the random walk *)
x[k_, n_] := x[k, n] = x[k, n - 1] + s[];
(* initial condition and parameters*)
 x[k_, 0] = 0;
M = 200;
Nstep = 50;
 (* analysis*)
  positions = Table[x[k, n], {k, 1, 200}, {n, 0, Nstep}];
pos[n_] := Table[positions[[j, n + 1]], {j, 1, M}];
Table[
   ListPlot[pos[n], PlotRange→{-6, 2}, AxesLabel→
    {"k", "x"}], {n, 0, 50}];
```

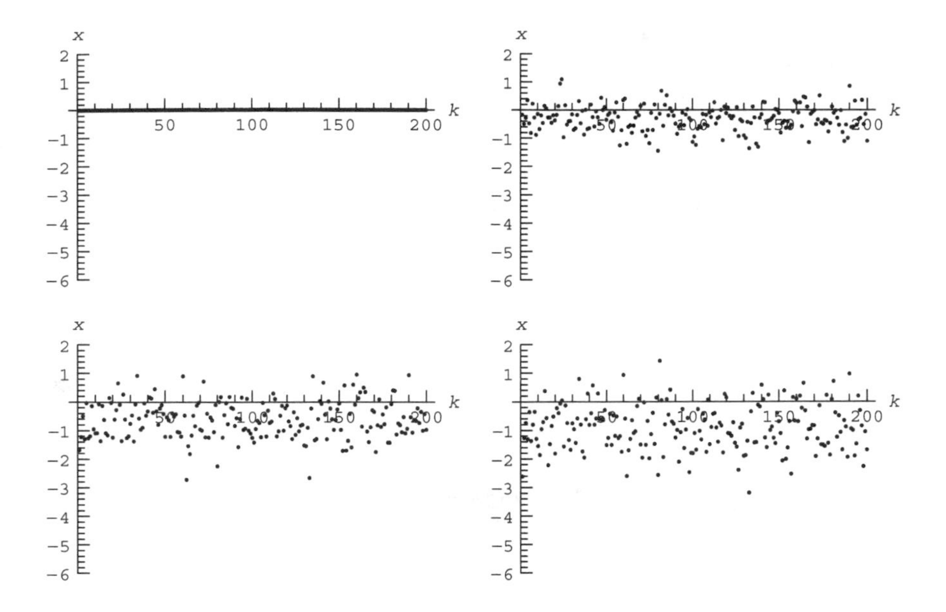

Nonuniform System. Here we simulate a nonuniform system where the particles drift toward smaller x but are confined to the range $x > 0$ (as occurs, for instance, when dust settles through a fluid to the bottom of a container at $x = 0$). We therefore require that no particles step below $x = 0$: $w = 0$ if $s < -x$. The previous random walk can be easily modified to include this constraint by shifting the step distribution to the right, so that

$$w(x, s) = -\frac{1}{2a(1-a)}[1 - s - a + b(x)], \qquad -a + b(x) < s < a + b(x), \tag{8.2.23}$$

where $b(x) = 0$ if $x > a$ and $b(x) = a - x$ if $x < a$. Thus, $w(x, s)$ is independent of x for $x > a$, but for $0 < x < a$ it shifts so that no particle ever steps below $x = 0$. Since the distribution of steps shifts without changing shape, the diffusion coefficient D is unaffected, and for $a = \frac{1}{4}$ remains $D = 13/(1296\,\Delta t)$ for all $x > 0$. The resulting random walk is shown in Cell 8.19. The only change to the previous rejection method is that now the random number s is a function of position x, and we choose the random number s_x from a range that depends on x. We now begin particles at $x = 2$ so that we can see them settle toward the container bottom at $x = 0$.

Cell 8.19

```
Clear["Global`*"];
a = 1/4;
b[x_] := 0/; x > a;
b[x_] := a - x/; x ≤ a;
w[x_, s_] := 1/(2 a (1 - a)) (1 - s - a + b[x])
(* rejection method *)
f0 = 3.;
```

```
sx[x_] := Random[Real, {-a + b[x], a + b[x]}];
sy[] := Random[Real, {0, f0}];
s[x_] := (ss = sx[x]; While[sy[] > w[x, ss],
   ss = sx[x]]; ss);

(* the random walk *)
x[k_, n_] := x[k, n] = x[k, n - 1] + s[x[k, n - 1]];
(* initial condition and parameters*)
 x[k_, 0] = 2;
M = 200;
Nstep = 150;
 (* analysis*)
positions = Table[x[k, n], {k, 1, 200}, {n, 0, Nstep}];
pos[n_] := Table[positions[[j, n + 1]], {j, 1, M}];

Table[
  ListPlot[pos[n], PlotRange→{0, 3}, AxesLabel→{"k", "x"}],
   {n, 0, 100, 2}];
```

The random diffusion causes the particles to dance above the bottom of the container, so that the density distribution attains an equilibrium. This equilibrium can also be seen in the behavior of average quantities, such as $\langle x \rangle$, shown in Cell 8.20.

Cell 8.20

```
ListPlot[Table[{n, xav[n]}, {n, 0, 150}],
  AxesLabel→{"n", "<x>"}, PlotRange→{0, 2}];
```

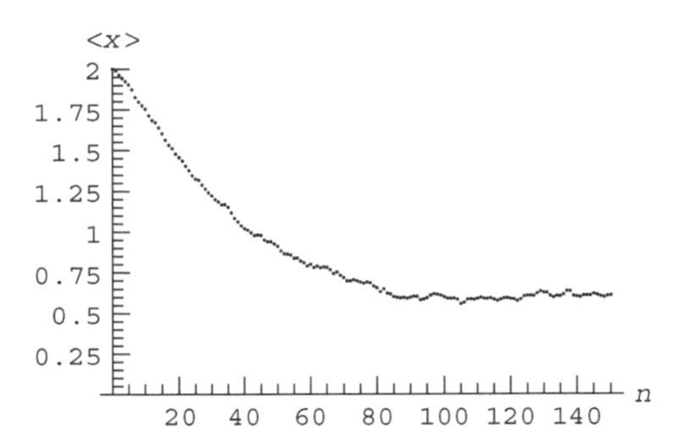

In a real system, this equilibrium is called a *thermal equilibrium state*. To what extent does our random process model true thermal equilibrium? In order to answer this question, we need the proper functional form of the thermal equilibrium distribution.

8.2.3 Thermal Equilibrium

The Boltzmann Distribution The thermal equilibrium distribution of particles in an external potential was first derived by Boltzmann, and is referred to as the Boltzmann distribution $\rho_B(x)$. One way to derive this distribution is as a time-independent solution of the Fokker–Planck equation (8.2.18):

$$0 = \frac{\partial}{\partial x}\left(v(x)\,\rho_B - D(x)\frac{\partial \rho_B}{\partial x}\right).$$

Integrating this equation once yields

$$\Gamma = v(x)\,\rho_B - D(x)\frac{\partial \rho_B}{\partial x},$$

where Γ is a constant of integration. However, the right-hand side is the particle flux (divided by the total number of particles) caused by diffusion and convection at velocity v. Since we assume that $x = 0$ is impermeable, the flux must be zero there, and this implies $\Gamma = 0$, as one would expect in equilibrium. The resulting first-order differential equation for ρ_B has the general solution

$$\rho_B(x) = C\exp\left(\int_0^x \frac{v(x')}{D(x')}\,dx'\right),$$

where C is another constant of integration. [Note that $v < 0$ is assumed, so that $\rho_B(x) \to 0$ exponentially as $x \to \infty$.] The constant C can be obtained from the normalization condition for all probability densities, Eq. (8.1.37). The result is

$$\rho_B(x) = \frac{\exp\left(\int_0^x \frac{v(x')}{D(x')} dx'\right)}{\int_0^\infty dx'' \exp\left(\int_0^{x''} \frac{v(x')}{D(x')} dx'\right)}. \tag{8.2.24}$$

The Boltzmann distribution can also be derived directly using the theory of equilibrium statistical mechanics, without reference to the Fokker–Planck equation. It is the distribution that maximizes the entropy S for particles trapped in a given potential $\phi(x)$, with given mean potential energy $\langle \phi \rangle$ per particle. This form of the Boltzmann distribution is written as

$$\rho_B(x) = \frac{e^{-\phi(x)/k_B T}}{\int dx'' \, e^{-\phi(x'')/k_B T}}, \tag{8.2.25}$$

where T is the temperature of the system, and where k_B is Boltzmann's constant.

Einstein Relations At first glance, Eqs. (8.2.24) and (8.2.25) appear to be unrelated. The fact that they are actually identical distributions leads to a nontrivial result, called an *Einstein relation*. In order for Eqs. (8.2.24) and (8.2.25) to be identical distributions, it must be the case that

$$\frac{v(x)}{D(x)} = -\frac{1}{k_B T} \frac{\partial \phi(x)}{\partial x}. \tag{8.2.26}$$

On the other hand, the drift velocity of a particle should be determined by the balance between the force $-\partial\phi/\partial x$ on the particle and the collisional drag: $0 = -m\gamma v - \partial\phi/\partial x$, where m is the particle mass and γ is the drag coefficient. This implies that

$$v(x) = -\frac{1}{m\gamma} \frac{\partial \phi(x)}{\partial x}, \tag{8.2.27}$$

which is simply the usual expression for terminal velocity. Comparing Eqs. (8.2.26) and (8.2.27) leads to the Einstein relation

$$D = \frac{k_B T}{m\gamma}. \tag{8.2.28}$$

This important and nontrivial relation between seemingly unconnected quantities was first discovered by Einstein in his doctoral research on Brownian motion of dust grains in a fluid. Diffusion and frictional drag are inversely proportional to one another in a thermal-equilibrium system. Note that the coefficients D and γ often cannot be calculated directly from the detailed microscopic collision processes that are responsible for the diffusion and drag. However, the drag rate γ can be easily *measured* in experiments, and then Eq. (8.2.28) provides a prediction for D. This prediction has been verified time and again in experiments on disparate systems.

Irreversibility and the Free Energy Solutions to the inhomogeneous Fokker–Planck equation approach thermal equilibrium monotonically over time, in the following sense. The following function, called the *Helmholtz free energy per particle F*, decreases monotonically:

$$F(t) = \langle \phi \rangle(t) - TS(t) = \int \phi(x)\,\rho(x,t)\,dx + k_B T \int dx\,\rho(x,t)\ln\rho(x,t). \tag{8.2.29}$$

Through an analysis analogous to that used to prove Eq. (8.1.44), one can show that

$$\frac{dF}{dt} = -k_B T \int \frac{dx}{\rho(x,t)D(x)} \left[v(x)\,\rho(x,t) - D(x)\frac{\partial \rho}{\partial x} \right]^2 \le 0 \tag{8.2.30}$$

(see the exercises). When thermal equilibrium is achieved, the particle flux (the square bracket) vanishes and $dF/dt = 0$. Thus, solutions of the inhomogeneous Fokker–Planck equation display irreversibility: initial conditions that are out of thermal equilibrium never recur. The dust never re-forms in a layer back at the top of the fluid. (Of course, this neglects the extremely unlikely possibility of Poincaré recurrence back to the initial state, as discussed previously in Sec. 8.1.2.)

Comparison with Simulations Let's now compare our previous simulation of dust grains settling in a fluid with the predictions of thermal equilibrium. For these simulations, $v = -\frac{1}{36}$ and $D = \frac{13}{1296}$ are independent of x, so Eq. (8.2.24) predicts that the Boltzmann distribution is

$$\rho_B = C e^{-|v| x/D} \tag{8.2.31}$$

with $C = |v|/D$. This exponential distribution is also what one would obtain from Eq. (8.2.25) for dust grains in a gravitational field $\phi(x) = mgx$: the grains pile up against the bottom of the container with an exponential tail in their density due to finite temperature. In Cell 8.21 this thermal distribution is compared with a histogram of the particle positions, obtained from averaging over the last 20 timesteps in the previous simulation of $M = 200$ particles. We take histogram bins of size $\Delta x = 0.1$, so the number of particles in each bin should be $H(x) = 20\,\Delta x M \rho_{eq}(x)$.

Cell 8.21

```
p1 = Histogram[Flatten[Table[pos[n], {n, 131, 150}]],
  HistogramRange→{0, 2},
   HistogramCategories→20, DisplayFunction→Identity];

d = 13/1296; v = 1/36; Δx = 0.1;
```

```
ρB[x_] = v/d Exp[-v x/d];
p2 = Plot[20 Δx M ρB[x], {x, 0, 2}, DisplayFunction→
  Identity];

Show[p1, p2, AxesLabel→{"x", "h(x)"}, DisplayFunction→
  $DisplayFunction];
```

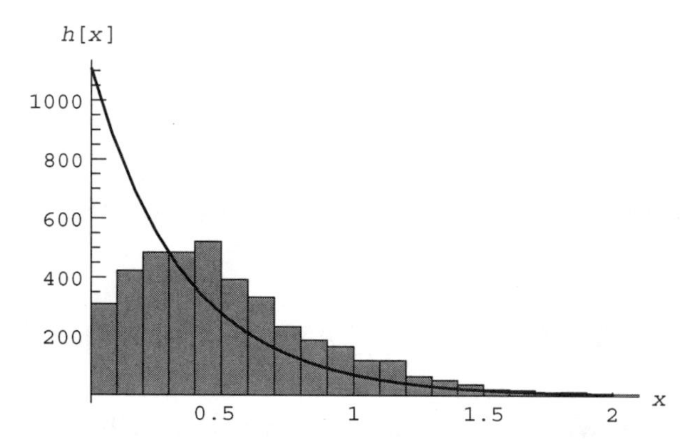

The thermal equilibrium theory does not explain our simulation very well. This is because the Fokker–Planck equation assumes very small steps are being taken, but $w(x,s)$ has finite width $2a = \frac{1}{2}$, which is not much smaller than the width of ρ_B. Our steps are too large to use the Boltzmann distribution.

Actually, much of this error can be removed by being more careful in determining the drift speed $v(x)$. For our choice of $w(x,s)$, the drift speed $v(x)$ is *not*

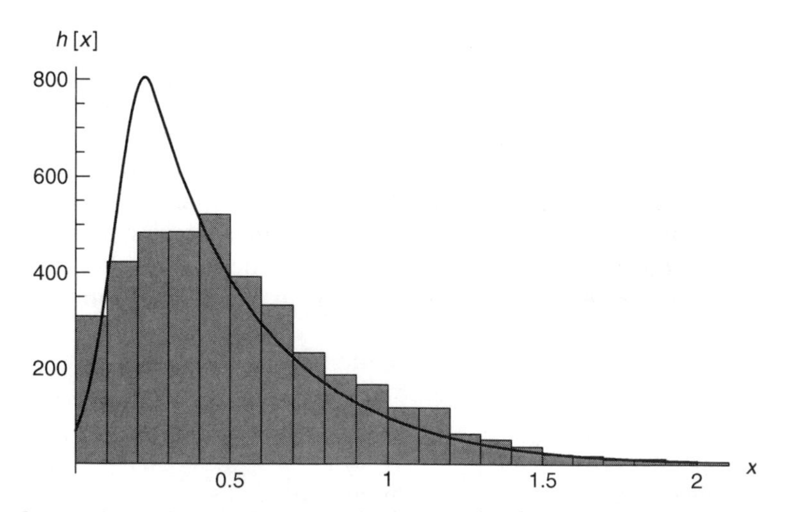

Fig. 8.7 Comparison of the Boltzmann distribution (line) with histograms from a random walk with bias, for $D(x) = \frac{13}{1296}$, $v(x) = \frac{2}{9} - x$, $x < \frac{1}{4}$, and $v(x) = -\frac{1}{36}$, $x \geq \frac{1}{4}$.

constant when $x < a$. Instead, $w(x, s)$ varies with x, and it is left as an exercise to show that

$$v(x) = \tfrac{2}{9} - x, x < a$$

(for $a = \frac{1}{4}$ and $\Delta t = 1$). However, as previously noted, D remains fixed at $D = \frac{13}{1296}$ for all x. The true Boltzmann distribution using this nonuniform drift speed, described by Eq. (8.2.24), is shown in Fig. 8.7 and is a much better fit to the data from the simulation, although there is still considerable error for $x < a$. According to Eq. (8.2.27), there is now an effective repulsive potential for $x < a$ that reduces the probability of particles being in this region.

One can improve the agreement between theory and simulation by reducing the width a of $w(x, s)$, but this increases the time required for the system to come to equilibrium. A better simulation method is discussed in Sec. 8.3.

EXERCISES FOR SEC. 8.2

(1) In a random walk in three dimensions, a particle starts from the origin and takes steps of fixed length l but with equal probability for any direction. Determine the rms distance the particle moves away from the origin in N steps.

(2) An integer is chosen randomly in the range 1–20. What is the probability that it is prime? If the number is odd, what is the probability that it is prime?

(3) In a certain town of 5000 people, 750 have blond hair, 500 have blue eyes, and 420 have both blond hair and blue eyes. A person is selected at random from the town.

- **(a)** If he/she has blond hair, what is the probability that he/she will have blue eyes?
- **(b)** If he/she has blue eyes, what is the probability that he/she does not have blond hair?
- **(c)** What is the probability that he/she has neither blond hair nor blue eyes?

(4) In a certain college, 7% of the men and 3% of the women are taller than 6 ft. Furthermore, 60% of the students are men. If a student is selected at random, and is over 6 ft, what is the probability he/she's a woman?

(5) **(a)** You are somehow transported back to the year 1975, where you find yourself in a rabbit costume, as a contestant on the then-popular TV game show *Let's Make a Deal*. The host Monty Hall approaches you and demands that you choose either door number 1, door number 2, or door number 3. Behind one of these doors there are fabulous prizes; behind the other two there are piles of useless junk. You choose door number 1. Monty, knowing the right door, opens door number 3, revealing that it is one of the no-prize doors. So the right door is either door 1, or door 2. Monty then gives you the opportunity to switch doors to door 2. Explain why you should switch to door number 2, by using conditional probabilities to determine the probability that door 1 is correct versus the probability that door 2 hides the prizes. (Hint: It's easiest to see the answer for the more general case of $N \gg 1$ doors and 1 prize door, where Monty opens all the doors but yours and one other.)

(b) Prove that it is better to switch by creating a simulation of the game with three doors. For 1000 tries, take the case where you don't switch. How many times do you win? Now repeat for the case where you switch doors. Now how many times do you win?

(6) **(a)** Solve the master equation for a general random walk, Eq. (8.2.11), using Fourier transforms, starting all particles at $x = 0$. (Hint: Use the convolution theorem.) Show that the solution is

$$\rho(x,n) = \int \frac{dk}{2\pi} e^{ikx} [\tilde{w}(k)]^n, \tag{8.2.32}$$

where $\tilde{w}(k)$ is the Fourier transform of the probability density $\omega(s)$ for a step of size s.

(b) Prove using Eq. (8.2.32) that $\rho(x, n)$ is properly normalized: $\int_{-\infty}^{\infty} dx\, \rho(x,n) = 1$.

(c) Prove the following identity for this random walk using Eq. (8.2.32):

$$\langle x^m \rangle(n) = i^m \frac{\partial^m}{\partial k^m} [\tilde{w}(k)]^n \bigg|_{k=0}. \tag{8.2.33}$$

(d) For a step distribution $w(s) = a e^{-a|s|}/2$, evaluate $\langle x \rangle(n)$, $\langle x^2 \rangle(n)$, and $\langle x^4 \rangle(n)$ using Eq. (8.2.33), and show that $\rho(x, n) = 2a(8\pi)^{-n/2}(a|x|)^{n-1/2} K_{1/2-n}(a|x|)/(n-1)!$, where $K_n(x)$ is a modified Bessel function. Animate this result for $n = 5$ steps.

(7) **(a)** In Brownian motion, the velocity v_n at timestep n obeys the finite-differenced Langevin equation (8.1.49), $v_n = \alpha v_{n-1} + s_{n-1}$, where $\alpha = 1 - \gamma \Delta t$, γ is the drag coefficient, and s_n is a random step in velocity with zero mean, $(s) = 0$. Solve this equation for v_n, assuming an initial velocity v_0, and prove that

$$\langle v_n v_{n+m} \rangle = v_0^2 \alpha^{2n+m} + \langle s^2 \rangle \alpha^m \frac{1-\alpha^{2n}}{1-\alpha^2}.$$

(b) In the long-time limit $n\,\Delta t \to \infty$, but taking $\Delta t \to 0$, show that $\langle v_n v_{n+m} \rangle \to (D_V/\gamma) e^{-\gamma t_m}$, where $t_m = m\,\Delta t$ and D_V is the velocity diffusion coefficient—see Eq. (8.1.51). Hence, in thermal equilibrium, the velocity correlation function $c(n, n+m)$ falls off exponentially with time, as $e^{-\gamma t_m}$.

(8) Create a random walk simulation using 1000 particles with step distribution $w(s)$ given in Exercise (6)(d) with $a = 1$. Use the rejection method. Follow the distribution for five steps, and compare the resulting histograms with the theory of Exercise (6)(d).

(9) **(a)** Create a random walk simulation with $N = 1000$ particles with steps of fixed size ± 1, but taking $p = 0.4$, $q = 0.6$. Follow the simulation of $N = 50$ steps, starting all particles at $x = 0$.

(b) Evaluate $\langle x\rangle(n)$ and $\langle\delta x^2\rangle(n)$, and compare with the theoretical predictions based on the binomial distribution.

(c) Make an animation of the histogram of positions, and compare it with the binomial distribution itself at each timestep.

(10) Use the transformation method to write a code that distributes a random variable s according to the probability distribution $w(s) = e^{-s}$, $s > 0$, $w(s) = 0$, $s < 0$. Test the code for 1000 tries, and compare the resulting histogram for s with $w(s)$.

(11) In a 2D random walk process, the step probability density $w(x, y) = 2$, $0 < x < 1$ and $0 < y < x$, and is zero otherwise. Particles start from the origin.

(a) Analytically determine $\langle x_n\rangle, \langle y_n\rangle, \langle x_n^2\rangle, \langle y_n^2\rangle$ vs. the timestep n.

(b) Analytically evaluate the following correlation function:

$$c_{xy}(n, m) = \frac{\langle x_n y_{n+m}\rangle - \langle x_n\rangle\langle y_{n+m}\rangle}{\sqrt{\langle \delta x_n^2\rangle\langle \delta y_{n+m}^2\rangle}}.$$

(c) Create a 2D random walk simulation for 1000 particles, and test part (b) over 50 steps by plotting the correlation function vs. m for $n = 0$, 5, and 10, along with the theory.

(12) **(a)** A particle undergoes Langevin dynamics with *zero* random forcing, taking steps according to $v_n = \alpha v_{n-1}$, where $\alpha = 1 - \gamma\,\Delta t$. Show that a distribution of velocities $\rho(v, n)$ evolves in time according to the equation $\rho(v, n) = \rho(v/\alpha, n-1)/\alpha$. (Hint: Use the transformation method.)

(b) Solve this equation numerically and animate the result over 20 steps for the case $\alpha = 0.9$, $\rho(v, 0) = e^{-v^2}/\sqrt{\pi}$.

(13) **(a)** Show that the master equation for the velocity distribution $\rho(V, n)$ of a particle of mass m undergoing Langevin dynamics, $v_n = \alpha v_{n-1} + s_{n-1}$, is

$$\rho(v, n) = \frac{1}{\alpha}\rho(v/\alpha, n-1) + \int w(v - v')\,\rho(v', n-1)\,dv',$$

where $w(s)$ is the distribution of the steps.

(b) In the limit as Δt and the step size approach zero, show that this master equation becomes the Fokker–Planck equation for velocity diffusion,

$$\frac{\partial\rho}{\partial t} = \frac{\partial}{\partial v} D_V\left(\frac{mv}{k_B T}\rho + \frac{\partial\rho}{\partial v}\right), \qquad (8.2.34)$$

where $D_V = \langle s^2\rangle/(2\,\Delta t)$ is the velocity diffusion coefficient. [Hint: Recall the Einstein relation (8.1.51).]

(c) Show that the thermal equilibrium form for the velocity distribution is a *Maxwellian* distribution, $\rho = C e^{-mv^2/2k_B T}$, and find C.

(14) **(a)** Assuming that D_V is constant, apply the method of characteristics to Eq. (8.2.34) to show that the equation can be written as

$$\left.\frac{\partial \rho}{\partial t}\right|_{v_0} = \gamma\rho + D_V\, e^{2\gamma t}\frac{\partial^2 \rho}{\partial v_0^2},$$

where $v_0 = v e^{\gamma t}$ is the initial velocity, and $\gamma = mD_V/k_B T$.

(b) Solve this equation by Fourier transforming in velocity v and solving the resulting first-order ODE in time, with initial condition $\rho(v,0) = \rho_0(v)$. Show that the solution is

$$\rho(v,t) = \int \frac{dk}{2\pi}\tilde{\rho}_0(k)\exp\left(ikve^{\gamma t} + \gamma t - \frac{Dk^2}{2\gamma}(e^{2\gamma t} - 1)\right).$$

(c) Perform the Fourier transformation for the case of an initial condition $\rho_0(v) = \delta(v - v_0)$ to show that

$$\rho(v,t) = \frac{1}{\sqrt{2\pi\langle \delta v^2\rangle(t)}} e^{-[v - \langle v\rangle(t)]^2/2\langle \delta v^2(t)\rangle}.$$

where $\langle v\rangle(t) = v_0\, e^{-\gamma t}$ is the mean velocity, and $\langle \delta v^2\rangle(t)$ is given by Eq. (8.1.50). Animate the solution for $v_0 = 2$ and $D_V = \gamma = 1$, over the time $0 < t < 3$.

(15) Prove Eq. (8.2.30), which implies that nonequilibrium solutions to the inhomogeneous Fokker–Planck equation exhibit irreversibility.

(16) For particles diffusing in a potential $\phi(x)$, the Fokker–Planck equation (8.2.18) is

$$\frac{\partial \rho}{\partial t} = \frac{\partial}{\partial x} D(x)\left(\frac{1}{k_B T}\frac{\partial \phi}{\partial x}\rho + \frac{\partial \rho}{\partial x}\right), \tag{8.2.35}$$

where we have substituted for $v(x)$ using Eq. (8.2.26). Consider the case $D = k_B T = 1$ and $\phi = x^2$ (i.e., particles trapped in a harmonic potential). Using any numerical method you choose taken from Chapter 6 or 7, solve this problem for $\rho(x,t)$ with initial condition $\rho(x,0) = 1$, $0 < x < 1$, and $\rho(x,0) = 0$ otherwise. For boundary conditions take $\rho = 0$ at $x = \pm 5$. Animate the solution for $0 < t < 2$, and show that the solution approaches the Boltzmann distribution $\rho_B = e^{-x^2}/\sqrt{\pi}$.

(17) **(a)** Solve the Fokker–Planck equation *analytically* for dust particles of mass m at temperature T falling in gravity g to the bottom of a container at $x = 0$. Assume that the diffusion coefficient D = constant, and take as the initial condition $\rho(x) = \delta(x - x_0)$. Boundary conditions are that the flux is zero at the bottom of the container. (Solution:

$$\rho(x,t) = \kappa e^{-\kappa x} + \frac{2}{\pi}\int_0^\infty dk\, e^{-D(k^2+\kappa^2/4)t+\kappa(x-x_0)/2} \times \frac{(2k\cos kx_0 - \kappa \sin kx_0)(2k\cos kx - \kappa\sin kx)}{4k^2+\kappa^2},$$

where $\kappa = mg/k_B T$.) (Hint: Solve as an eigenmode expansion, assuming that ρ vanishes at $x = L$, and then take the limit as $L \to \infty$.)

(b) For $D = \kappa = x_0 = 1$, evaluate the integral numerically and create an animation of the solution over the time range $0 < t < 0.3$.

(18) *Simulation project: The Eden model.* Random process models have diffused into many fields of science. Consider the following model for the spread of tumors, rumors, disease, urban sprawl, or any number of other uncontrolled growth processes: the Eden model, named after the biologist M. Eden. A small cluster of tumor cells infect cells on the perimeter of the tumor, and these infect other adjacent cells, and so on. At each step in the process, one of the perimeter cells is chosen randomly to become infected, after which it is added to the cluster and a new perimeter is calculated, from which a new infected cell will be chosen in the next step. The resulting tumor growth is surprisingly realistic (and rather horrifying to watch as an animation). This model is one of a large group of random processes, variously referred to as *kinetic growth* or *percolation* models. We will perform the Eden model on a 2D square lattice. Initially, only one cell at the origin is in the tumor cluster:

Cell 8.22

```
cluster[0] = {{0, 0}},
```

The perimeter points of the cluster are held in the list **perim**:

Cell 8.23

```
perim[0] = {{0, 1}, {1, 0}, {-1, 0}, {0, -1}};
```

From this perimeter list, a newly infected cell is chosen at random. We define a function **newsite[n]**, which is the position of this cell, chosen from the previous perimeter list, **perim[n-1]**:

Cell 8.24

```
newsite[n_] :=
newsite[n] = perim[n - 1][[Random[Integer,
  {1, Length[perim[n - 1]]}]]]
```

The new cluster is the union of the old cluster with the new site:

Cell 8.25

```
cluster[n_] := cluster[n] = Append[cluster[n - 1],
  newsite[n]]
```

Finally, we must calculate the new perimeter, by first removing the new site from the old perimeter, and then adding the group of nearest neighbors to the new site onto the perimeter (making sure to avoid those nearest neighbors already in the cluster):

Cell 8.26

```
perim[n_] := perim[n] = (perim1 = Complement[perim[n-1],
  {newsite[n]}]
   (*removes newsite from the perimeter *);
   nn = Table[newsite[n] + {{0, 1}, {1, 0}, {-1, 0},
  {0, -1}} [[m]], {m, 1, 4}];
   nn = Complement[nn, cluster[n]];
   (*nn is the nearest neighbors to the new site,
   excluding those in the cluster *)
  Union[perim1, nn] (* add nearest neighbors to the
    perimeter list *))
```

(a) Create an animation of the cluster growth for up to $n = 2000$, showing only every 20th step. (Use a **ListPlot** to display the positions.)

(b) The edge of the tumor cluster is highly corrugated. The length L of the perimeter is simply the number of cells in the list **perim**. Show using a log–log plot that $L \propto n^b$, and find a value of b from your simulation. What value of b would you expect if the tumor had a smooth edge?

(c) The Eden model can be made more realistic by assigning a probability of immunity p to each member of the perimeter. When a perimeter cell is chosen, use the rejection method to determine whether to infect the cell: Evaluate a random number r with $0 < r < 1$; if $r < p$, the cell is not allowed to be infected in any future step: set it aside in a new list **immunes**, which can never enter the tumor cluster, and choose another perimeter cell until one is found that can be infected.

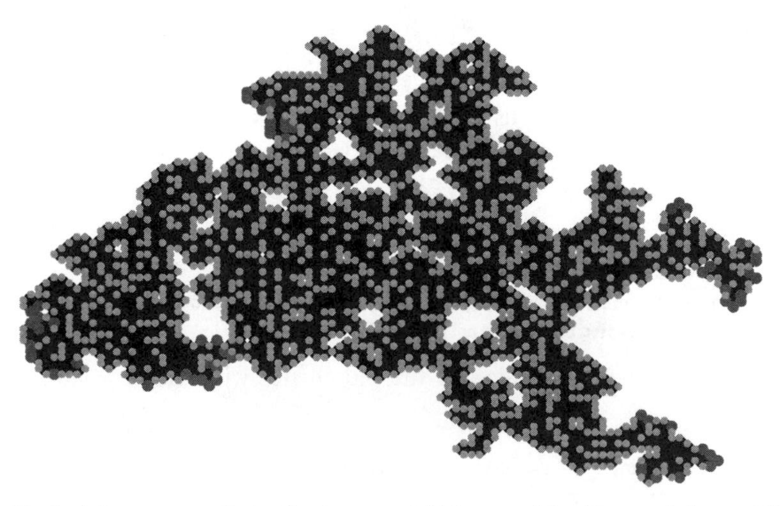

Fig. 8.8 Typical tumor growth in the improved Eden model with $p = 0.4$, $n = 2000$ steps. Light grey cells are immune; dark grey cells are on the perimeter where more growth can occur.

(i) Reevaluate the improved Eden model taking $p = 0.5$ for all cells and $n = 200$. Create an animation as in part (a), and reevaluate the exponent b for the length of the perimeter (include the immunes in the perimeter). (See Fig. 8.8.) Note: If p is chosen too large, then the tumor has a good probability of not growing, as it can be surrounded by immune cells. Did the tumor grow without bound or stop?

(ii) Repeat for $p = 0.2$ and $n = 2000$.

REFERENCES

M. Eden, "A two-dimensional growth process," in *Proceedings of Fourth Berkeley Symposium of Mathematics, Statistics, and Probability*, volume 4, pp. 223–239 (University of California Press, Berkeley, 1960).

E. M. Lifshitz and L. P. Pitaevskii, *Physical Kinetics* (Pergamon Press, Oxford, 1981).

N. Metropolis, A. W. Rosenbluth, M. N. Rosenbluth, A. H. Teller, and E. Teller, *Equation of state calculations by fast computing machines*, J. Chem. Phys. **21**, 1087 (1953).

M. Newman and G. Barkema, *Monte Carlo Methods in Statistical Physics* (Clarendon Press, Oxford, 2001).

M. Plischke and B. Bergersen, *Equilibrium Statistical Physics* (Prentice Hall, Englewood Cliffs, NJ, 1989).

F. Reif, *Fundamentals of Statistical and Thermal Physics* (McGraw-Hill, New York, 1965).

M. N. Rosenbluth and A. W. Rosenbluth, *Monte Carlo calculation of the average extension of molecular chains*, J, Chem. Phys. **23**, 356 (1955).

APPENDIX

FINITE-DIFFERENCED DERIVATIVES

Say we know the value of some smooth function $y(x)$ only at a sequence of evenly spaced grid points $x = j\,\Delta x,\ j = 0,1,2,3,\ldots$. How do we determine an approximate numerical value for the nth derivative of this function, $y^{(n)}(x)$, evaluated at one of the grid points $x = x_i$? What will be the error in this derivative compared to the true value?

We will refer to this finite-differenced form of the derivative as $y_{\mathrm{FD}}^{(n)}(x)$, in order to distinguish it from the exact derivative $y^{(n)}(x)$ of the function. We will write the finite-differenced derivative, evaluated at the grid position x_i, as a linear combination of the values $y(x_j)$ at M consecutive grid points, starting at $j = l$:

$$y_{\mathrm{FD}}^{(n)}(x_i) = \sum_{j=l}^{l+M-1} \frac{a_j y(x_i + j\,\Delta x)}{\Delta x^n} + O(\Delta x^p), \tag{A.1}$$

where the a_j's are constants that remain to be determined, the order of the error p also must be determined, and l is arbitrary. Typically for best accuracy l and M are chosen so that $x_{i+l} \le x_i \le x_{i+l+M-1}$, but this is not required by the mathematics. We have anticipated that each term in the sum will be of order $1/\Delta x^n$ and have divided this term out, so that the a_j's are of order unity.

In order to find the a_j's, we will Taylor-expand $y(x_i + j\,\Delta x)$ up to order $M-1$:

$$y(x_i + j\,\Delta x) = \sum_{k=0}^{M-1} \frac{1}{k!}(j\,\Delta x)^k y^{(k)}(x_i) + O(\Delta x^M). \tag{A.2}$$

Substituting this expression into Eq. (A.1) yields

$$y_{\mathrm{FD}}^{(n)}(x_i) = \sum_{k=0}^{M-1} \frac{1}{k!}\Delta x^{k-n}\, y^{(k)}(x_i) \sum_{j=l}^{l+M-1} a_j j^k + O(\Delta x^{M-n}). \tag{A.3}$$

In order for this expression to be valid for any choice of the function $y(x)$, we require that only terms on the right-hand side that are proportional to $y^{(n)}(x_i)$ survive the summations. Since, for general $y(x)$, the values of $y^{(k)}(x_i)$ are independent variables, we therefore require that

$$\sum_{j=l}^{l+M-1} a_j j^k = 0 \qquad \text{for} \quad k = 0, 1, 2, \ldots, M-1, \quad k \neq n, \tag{A.4}$$

and for $k = n$,

$$\frac{1}{n!} \sum_{j=l}^{l+M-1} a_j j^n = 1 \tag{A.5}$$

When these equations are used in Eq. (A.3), that equation becomes

$$y_{\mathrm{FD}}^{(n)}(x_i) = y^{(n)}(x_i) + O(\Delta x^{M-n}). \tag{A.6}$$

Equations (A.4) and (A.5) provide us with M linear equations in the M unknowns a_j, $j = l, \ldots, l+M-1$. Their solution provides us with a finite-differenced form for the derivative, Eq. (A.1). Furthermore, the order of the error in the finite-differenced derivative scales as $p = M - n$. Therefore, to reduce the error to at most $O(\Delta x)$, an nth derivative requires at least $M = n+1$ points to be used in Eq. (A.1). Furthermore, the larger the value of M, the more accurate the approximation.

For example, consider the first derivative of $y(x_i)$. In order to find an $O(\Delta x)$ form for this derivative, we require two points, $M = 2$. If we choose these points as $y(x_i)$ and $y(k_{i+1})$, then Eq. (A.4) becomes

$$\sum_{j=0}^{1} a_j j^0 = a_0 + a_1 = 0,$$

and Eq. (A.5) is

$$\frac{1}{1!} \sum_{j=0}^{1} a_j j^1 = a_1 = 1.$$

Therefore, we obtain $a_1 = -a_0 = 1$, and Eq. (A.1) becomes the standard form for a forward-differenced first derivative,

$$y_{\mathrm{FD}}^{(n)}(x_i) = \frac{y(x_{i+1}) - y(x_i)}{\Delta x} + O(\Delta x).$$

However, for derivatives where n is even, such as y'', we can do a bit better if we use a *centered-difference* form of the derivative. In such a form, l is chosen so that the number of grid points in Eq. (A.1) below x_i equals the number of points above x_i, that is, $l = -(M-1)/2$ with M odd.

The error estimate in Eq. (A.3) assumes that the $O(\Delta x^{M-n})$ term in the power series expansion has a nonvanishing coefficient. However, if we choose a centered-difference form for the derivative, one can show that this term actually vanishes thanks to the symmetry of the sum, and the error is actually $O(\Delta x^{M-n+1})$. Thus, for a centered-difference form of the second derivative, taking $M=3$ terms involving $y(x_{i-1})$, $y(x_i)$, and $y(x_{i+1})$ results in an error not of order Δx, but rather of order Δx^2:

$$y''_{\rm FD}(x_i) = \frac{y(x_{i+1}) - 2y(x_i) + y(x_{i-1})}{\Delta x^2} + O(\Delta x^2).$$

Centered-difference forms for odd derivatives also exist, for which one again takes $l = -(M-1)/2$ with M odd. For these derivatives, one can show that $a_0 = 0$ using symmetry, and as a result these centered-difference odd derivatives are independent of $y(x_i)$. For instance, one finds that the centered-difference form of y' taking $M=3$ terms is

$$y'_{\rm FD}(x_i) = \frac{y(x_{i+1}) + 0y(x_i) - y(x_{i-1})}{2\,\Delta x} + O(\Delta x^2).$$

Of course, it is possible to use *Mathematica* to solve Eqs. (A.4) and (A.5) for derivatives of any order, keeping any number of grid points. Below, we provide a module that does so. It evaluates $y^{(n)}_{\rm FD}(0)$ keeping grid points from $j=l$ to $j=l+M-1$:

Cell A.1

```
Clear["Global`*"]
FD[n_, M_, l_] := Module[{p, eqns1, eqns2, eqn3, coeff, a},
p[j_, k_] = j^k;
p[0, 0] = 1;
eqns1 = Table[Sum[a[j] p[j, k], {j, l, l + M - 1}] == 0,
  {k, 0, n - 1}];
eqns2 = Table[Sum[a[j] p[j, k], {j, l, l + M - 1}] == 0,
  {k, n + 1, M - 1}];
eqn3 = {Sum[a[j] p[j, n], {j, l, l + M - 1}] / n ! == 1};
coeff = Solve[Join[eqns1, eqns2, eqn3], Table[a[j],
  {j, l, l + M - 1}]];
Together[Expand[Sum[a[j] y[j Δx], {j, l, l + M - 1}]/
  Δx^n] /. coeff[[1]]]]
```

For instance, a second-order-accurate backward-difference version of the first derivative, $y'_{\rm FD}(0)$, is given by

Cell A.2

```
FD[1, 3, -2]
```

$$\frac{3\,y[0] + y[-2\,\Delta x] - 4\,y[-\Delta x]}{2\,\Delta x}$$

This can be checked by Taylor-expanding the result in powers of Δx:

Cell A.3

```
Series[%, {Δx, 0, 1}]
```

$$y'[0] + O[\Delta x]^2$$

As a second example, the second-order-accurate centered-difference form of the fourth derivative is

Cell A.4

```
FD[4, 5, -2]
```

$$\frac{6\ y[0] + y[-2\,\Delta x] - 4\ y[-\Delta x] - 4\ y[\Delta x] + y[2\,\Delta x]}{\Delta x^4}$$

Again, the form and the error can be checked by Taylor-expanding the result in powers of Δx:

Cell A.5

```
Series[%, {Δx, 0, 1}]
```

$$y^{(4)}[0] + O[\Delta x]^2$$

INDEX

Note: Items that appear only in the electronic version are indexed according to the section numbers in which they occur.

THEOREMS FROM VECTOR CALCULUS

In the following theorems, V is a volume with volume element $\boldsymbol{d}^3r$, S is the surface of this volume, and $\hat{\boldsymbol{n}}$ is a unit vector normal to this surface, pointing out of the volume.

$$\int_V \nabla\cdot\boldsymbol{A}\,\boldsymbol{d}^3r = \int_S \boldsymbol{A}\cdot\hat{\boldsymbol{n}}\,\boldsymbol{d}^2r \quad \text{(divergence theorem)}$$

$$\int_V (f\nabla^2 g + \nabla f\cdot\nabla g)\,\boldsymbol{d}^3r = \int_S f\hat{\boldsymbol{n}}\cdot\nabla g\,\boldsymbol{d}^2r \quad \text{(Green's first identity)}$$

$$\int_V (f\nabla^2 g - g\nabla^2 f)\,\boldsymbol{d}^3r = \int_S (f\nabla g - g\nabla f)\cdot\hat{\boldsymbol{n}}\,\boldsymbol{d}^2r \quad \text{(Green's theorem)}$$

EXPLICIT FORMS FOR VECTOR DIFFERENTIAL OPERATIONS

Cartesian coordinates (x, y, z):

$$\nabla\psi = \hat{\mathbf{x}}\frac{\partial\psi}{\partial x} + \hat{\mathbf{y}}\frac{\partial\psi}{\partial y} + \hat{\mathbf{z}}\frac{\partial\psi}{\partial t}$$

$$\nabla^2\psi = \frac{\partial^2\psi}{\partial x^2} + \frac{\partial^2\psi}{\partial y^2} + \frac{\partial^2\psi}{\partial z^2}$$

$$\nabla\cdot\boldsymbol{A} = \frac{\partial A_x}{\partial x} + \frac{\partial A_y}{\partial y} + \frac{\partial A_z}{\partial z}$$

$$\nabla\times\boldsymbol{A} = \hat{\mathbf{x}}\left(\frac{\partial A_z}{\partial y} - \frac{\partial A_y}{\partial z}\right) + \hat{\mathbf{y}}\left(\frac{\partial A_x}{\partial z} - \frac{\partial A_z}{\partial x}\right) + \hat{\mathbf{z}}\left(\frac{\partial A_y}{\partial x} - \frac{\partial A_x}{\partial y}\right)$$

Cylindrical coordinates (r, θ, z):

$$\nabla\psi = \hat{\mathbf{r}}\frac{\partial\psi}{\partial r} + \hat{\boldsymbol{\theta}}\frac{1}{r}\frac{\partial\psi}{\partial\theta} + \hat{\mathbf{z}}\frac{\partial\psi}{\partial z}$$

$$\nabla^2\psi = \frac{1}{r}\frac{\partial}{\partial r}\left(r\frac{\partial\psi}{\partial r}\right) + \frac{1}{r^2}\frac{\partial^2\psi}{\partial\theta^2} + \frac{\partial^2\psi}{\partial z^2}$$

$$\nabla\cdot\boldsymbol{A} = \frac{1}{r}\frac{\partial}{\partial r}(rA_r) + \frac{1}{r}\frac{\partial A_\theta}{\partial\theta} + \frac{\partial A_z}{\partial z}$$

$$\nabla\times\boldsymbol{A} = \hat{\mathbf{r}}\left(\frac{1}{r}\frac{\partial A_z}{\partial\theta} - \frac{\partial A_\theta}{\partial z}\right) + \hat{\boldsymbol{\theta}}\left(\frac{\partial A_r}{\partial z} - \frac{\partial A_z}{\partial r}\right) + \hat{\mathbf{z}}\left(\frac{1}{r}\frac{\partial}{\partial r}(rA_\theta) - \frac{1}{r}\frac{\partial A_r}{\partial\theta}\right)$$

Spherical coordinates (r, θ, ϕ):

$$\nabla\psi = \hat{\mathbf{r}}\frac{\partial\psi}{\partial r} + \hat{\boldsymbol{\theta}}\frac{1}{r}\frac{\partial\psi}{\partial\theta} + \hat{\boldsymbol{\phi}}\frac{1}{r\sin\theta}\frac{\partial\psi}{\partial\phi}$$

$$\nabla^2\psi = \frac{1}{r^2}\frac{\partial}{\partial r}\left(r^2\frac{\partial\psi}{\partial r}\right) + \frac{1}{r^2\sin\theta}\frac{\partial}{\partial\theta}\left(\sin\theta\frac{\partial\psi}{\partial\theta}\right) + \frac{1}{r^2\sin^2\theta}\frac{\partial^2\psi}{\partial\phi^2}$$

$$\nabla\cdot\boldsymbol{A} = \frac{1}{r^2}\frac{\partial}{\partial r}(r^2A_r) + \frac{1}{r^2\sin\theta}\frac{\partial}{\partial\theta}(\sin\theta A_\theta) + \frac{1}{r\sin\theta}\frac{\partial A_\phi}{\partial\phi}$$

$$\nabla\times\boldsymbol{A} = \hat{\mathbf{r}}\frac{1}{r\sin\theta}\left(\frac{\partial}{\partial\theta}(\sin\theta A_\phi) - \frac{\partial A_\theta}{\partial\phi}\right) + \hat{\boldsymbol{\theta}}\left(\frac{1}{r\sin\theta}\frac{\partial A_r}{\partial\phi} - \frac{1}{r}\frac{\partial}{\partial r}(rA_\phi)\right) + \hat{\boldsymbol{\phi}}\frac{1}{r}\left(\frac{\partial}{\partial r}(rA_\theta) - \frac{\partial A_r}{\partial\theta}\right)$$